FAUNA BRITANNICA

FAUNA BRITANNICA

FOREWORD BY
HRH THE PRINCE OF WALES

STEFAN BUCZACKI

hamlyn

To Clinton Keeling, who taught me much of what I know about animals and inspired me to discover the rest.

First published in Great Britain in 2002 by
Hamlyn, a division of Octopus Publishing Group Ltd
2–4 Heron Quays, London E14 4JP
This paperback edition published 2005

ISBN 0 600 61392 5

EAN 9780600613923

A CIP catalogue record for this book is available from the British Library

Printed and bound in China

10 9 8 7 6 5 4 3 2

Contents

ST. JAMES'S PALACE

A dispassionate observer of life on Earth at the beginning of the 21st Century would be forced to conclude that our intensely demanding way of life is making it increasingly difficult, and in some cases impossible, for other species to co-exist with us. 'Fauna Britannica' is a timely reminder of the richness and complexity of the past and present interactions between British people and the wide variety of wild mammals, birds, reptiles, amphibians, fish and insects that also inhabit these islands.

Relations between man and wildlife are in a constant state of flux. As otters begin to make a welcome return to the now cleaner rivers of England, so the water vole is being driven to the verge of extinction by the depredations of introduced mink. Conservation efforts enable a few more people to hear the calls of corncrake and bittern once again, yet intensive agriculture means fewer can hear the song of the skylark and corn bunting, or the evocative sound of the grey partridge. These are not isolated examples. Every generation has had an impact on the environment and wildlife, but none has ever before had the technological capacity to make such rapid and sweeping changes as our own. The damage caused by such technological capability has been intense and far-reaching throughout the latter half of the 20th Century. The great challenge now, it seems to me, is how to re-create lost habitats so that what is left of our more vulnerable fauna has a better chance of survival in the future.

When the environmental impacts of decisions and developments are assessed, any consideration of wildlife takes place under its scientific guise of 'biodiversity'. This is not a bad place to start. No-one would want to see decisions taken without a full knowledge of the underlying science. But 'sound science' is a narrow and utilitarian ethic, which needs to be kept in proportion. Any full and proper consideration of the role of wildlife in our lives must also take account of the historical, cultural and aesthetic factors which cannot be quantified, but nevertheless add richness and meaning to our lives.

Treating what would once have been called natural history as a narrowly scientific subject allows the views of so-called 'ordinary people' to be disregarded. I hope that 'Fauna Britannica', with its wonderfully diverse and eclectic approach, will encourage millions of gardeners, bird-watchers, anglers, walkers and others to have the courage of their convictions. Anyone who simply loves the British countryside and its wildlife for its own sake should not be afraid to say so.

Charles

Introduction

This book describes and celebrates the wild animal life of Britain, especially its relationship with the human population as it has been expressed down the centuries through folklore, custom, tradition, language and literature.

Since our ancestors lived in caves, man has coexisted alongside the country's wild animals and each has served the other for good or ill. My account encompasses the practices and beliefs, founded and unfounded, logical and mysterious, that reflect this association. Sometimes it is a tale of much beauty and charm, sometimes of great cruelty and apparently illogical persecution. The story tells of hunting and being hunted, of eating and being clothed, of fear and fondness, use and abuse, all balanced, especially in recent times, by much caring, concern and conservation. My starting point is the folklore and folkloric traditions associated with British animals but, although we might imagine that we know what folklore is and what it encompasses, the reality is rather different and there simply is no general agreement on its real meaning.

Folklore is 'the beliefs, legends and customs current among the common people' according to one dictionary. However, it is, rather differently, 'the unwritten literature of a people as expressed in folk tales, proverbs, riddles, songs etc.' according to another. A more comprehensive attempt is given by yet a third: 'a department of the study of antiquities or archaeology, embracing everything relating to ancient observances and customs, to the notions, beliefs, traditions, superstitions, and prejudices of the common people – the science which treats of the survivals of archaic beliefs and customs in modern ages'. This source even attributes the first use of the term 'folklore' to W.J. Thomas in 1846, and it is closest to my brief here, where I am using 'folklore' in a very general sense to embrace all of these things, and more. Nonetheless, by definition, there can be no folklore without folk; it is a manifestation of human response. I am conscious of the notion that, strictly, it is an unwritten literature and therefore handed down only by word of mouth, and it is certainly possible that some of the beliefs and practices that I include might date back to a time when all lore was handed on in this way and all literature was unwritten, a time before recorded history. The belief in the 'Pent Cuckoo', that by entrapping a Cuckoo, man could capture eternal spring (see Cuckoo, page 300), might be an example of this and may have its origins in the Neolithic period. The mysterious association of woodpeckers with oak trees and rain (see Green Woodpecker, page 315) might be another.

However, I had to draw a line somewhere, to find a way of separating out, on the one hand, beliefs which might be based on something factual and, on the other, those that are the stuff of genuine legend, in which the link with real life, if

The magnificent Red Deer *(Cervus elaphus)* is the largest British land mammal.

A lone Marsh Harrier *(Circus aeruginosus)* flies low over misty marsh land.

it ever existed, has long been lost. In my compass of creatures therefore, I have excluded mythical beasts (although I have been unable to avoid at least passing references to dragons and unicorns) and have limited myself to describing those real animals that have coexisted in Britain with more or less literate people, those animals that have lived here since the beginning of properly recorded history, although not all of them are present today. Extinct animals such as the Aurochs *(Bos primigenius)*, Wild Horse *(Equus ferus)* and Irish Elk *(Megaloceros giganteus)* were once real enough, certainly lived alongside man in Britain and, in their day, no doubt many traditions and beliefs grew up around them. However, apart from self-evidently hunting and eating them, man's only other known response was to scratch images of some of them on bones or antlers. They are 'out' therefore, along with all other creatures, extinct or extant, known or unknown, that disappeared from Britain in pre-Roman times. The Brown Bear *(Ursus arctos)*, however, which probably survived until as recently as the eighth century, and has left some mark of its existence in place names and customs, is 'in', along with the European Beaver *(Castor fiber)*, the Wolf *(Canis lupus)*, the Black-veined White butterfly *(Aporia crataegi)*, the Burbot *(Lota lota)*, the Pool Frog *(Rana lessonae)* and others. These animals may be extinct in Britain now, although they all live on elsewhere, but their former existence in this country is well, and in some cases voluminously, recorded.

Much old belief and folklore was born of an age of intellectual ignorance. It pre-dates natural history as the science we know today. A good deal of it, like the belief that bats will entangle themselves in your hair, can only be a manifestation of the fear of the unknown. However, I have not simply written an account of old, misguided and outdated fantasies, of obscure literature and outmoded habits. Folklore, tradition, custom and practice blend imperceptibly, one into the other. Thus, the pigeons that foul our city streets, the wasps that we feel obliged to swat as we pick our plums, the slugs that we trap to protect our lettuces and the farmed trout that we buy from the supermarket shelf all have as important a part to play in our inter-relationship with our fauna as the 'eye of newt' and 'wool of bat' that were added to the witch's brew in *Macbeth*, the Robin that signalled death if it entered a house, or the spider that (reputedly) so inspired Robert the Bruce.

I feel fortunate to have grown up in the English countryside and to have spent most of my life living there. I have personally seen most of the creatures that I describe in this book and have been touched by many of their associated traditions. Nonetheless, it would be a grave mistake to believe that animals exist only in the countryside. The bird population of our cities is enormous; witness the countless urban gardens with bird tables. The Fox is arguably more of an urban than a rural creature today (I saw one recently flattened into the tarmac outside the Victoria and Albert Museum in central London) and both town- and city-dwellers are certainly not immune from the attentions of flies, woodlice or mice. With 90 per cent of the population of Britain now living in urban communities, it is there that the bulk of modern animal customs and traditions originate.

The conservation of and caring for our fauna have become a minor growth industry over the past 25 or 30 years, but it is not necessarily straightforward. When I was growing up after the Second World War, the Royal Society for the Protection of Birds had a relative handful of members; their numbers only passed 10,000 in 1960 but there are over a million today. Fifty years ago, Otter-hunting was still legal and egg-collecting was the stock in trade of every rural boy. Few will regret their passing but, conversely, I can recall still the pleasure, pride and satisfaction of nursing back to health a Daubenton's Bat with an injured wing that I found in the roof of our house, something that would now be illegal under the Wildlife and Countryside Act 1981. This highlights the dilemma facing anyone attempting to legislate for the well-being of wild animals. The Act does its best to allow injured individuals, even animals that are listed for protection in its Schedule 5, to be tended and then released, but bats are specifically excluded from this, unless they are taken within 'the living area of a dwelling house'; this does not include the roof space where they roost. Nothing is perfect.

In collecting folklore and customs, I have relied on two principal sources: published literature, old and new, and the personal knowledge and reminiscences of numerous correspondents. Contributions have come from every county in England (with Somerset top of the list), and from many in Wales, Scotland and Ireland. There have also been valuable additions from the Isle of Man and the Channel Islands. Most folklore, and most contributions, relate to birds, followed by mammals, insects, amphibians, reptiles, fish and other groups. By and large, I am sure this is simply a reflection of the frequency with which people encounter these animals, how big and obvious they are and/or the effects that they have on us, rather than any deep, psychological reason. Almost everyone in this country sees at least one bird every day of their lives, but they may see a wild mammal no more than once every week or two, an amphibian a few times a year, and a reptile or a freshwater fish very rarely.

Wild animals have inspired British writers and embellished and graced our national literature and I quote liberally from Shakespeare, John Clare and other authors. Two widespread usages of animal names and animal images with historical interest that I also refer to frequently are in heraldry and as inn names, while place names can reveal much of interest and value about the history of our fauna and its relative importance at different times. The multitude of place names that have some reference to the Wolf, for instance, are ample testimony to its importance and widespread occurrence in former times. I have made no attempt, however, to enter into the thorough analysis that has taken place under the aegis of the English Place Name Society, but have made reference to the names of larger settlements that are relevant. Because of my own linguistic limitations, I have made little inroads into the animal significance of place names of Celtic origin.

It is one thing to record and chronicle folklore and custom, but quite another to understand and explain it. What reasoning can lie behind the rhyme, found all over England: 'If you want to live and thrive, let the spider run alive'? Or what logic can attach to the belief that seeing two Magpies will bring better fortune than seeing one? In some instances, people have attempted an analysis; in some places I have offered my own, but, like many an author before me, I remain receptive to suggestions.

Above all, I hope that this book will stimulate further interest in our British wild fauna, which collectively has never been under greater threat than it is today, mainly through loss of habitats. I hope it may open some people's eyes and create a greater awareness of the importance of appropriate conservation measures. However, man in Britain today cannot live in isolation from the rest of the animal population. We have to coexist and, as long as we do so, the interaction – the interface between us – will be the stimulus for new customs, practices both good and bad, traditions, language and literature. I urge that it is chronicled and I ask readers to write to me, not only with instances of historic and regional traditions and folklore that I may have missed, but also of modern and evolving customs, practices and uses that should be recorded for the interest of future generations.

Arrangement of this book

Classification and taxonomy

I have listed the main groups of creatures in what appears to be the most widely accepted sequence and, although there can never be universal acceptance of one system over another, I have generally followed that published by BIOSIS (www.biosis.org), with some reference to older schemes, such as those in Borradaile et al. (1961). By and large, these sequences attempt to rank creatures in increasing order of complexity, an arrangement that, rather commonly, seems also to follow an evolutionary trend. Within groups, I have similarly followed convention in the arrangement of smaller taxonomic units.

Inevitably, much tradition, custom and folklore relating to the smaller and more primitive creatures applies to groups rather than individual species. There is not much distinction between our perceptions of one type of sponge over another, for instance, and so, for most invertebrates, I have described groups of creatures rather than single species; although there are odd exceptions, such as butterflies, oysters, ladybirds (although even here, it is the combined total of around 40 species rather than individuals that are referred to) and woodlice. Depending on the creatures concerned, the 'group' may be a Family, an Order, a Class or even a Phylum. I have excluded the numerous groups of invertebrates that may be zoologically very interesting or important but are seldom if ever seen by the public, mainly because they are very small and/or live in the sea.

Vertebrates are generally larger than invertebrates and are more obvious, seen more frequently, and have accumulated a much greater association of folklore and custom. I have therefore described almost all wild species of British vertebrate individually, in accepted family sequence, making no distinction between those that are indigenous (native) and those that are introduced. I have excluded most accidental or

With a life span of up to 40 years, the Common Toad *(Bufo bufo)* is one of the longest lived British wild animals.

The Chalkhill Blue *(Polyommatus coridon)* has a delicate, lace-like fringe on its wings.

rare vagrant birds and have included only those sea fish that either live in coastal waters or are routinely caught from deeper waters and are therefore brought to our attention. Within vertebrate groups, the arrangements of species I have used were included in the following more popular and easily accessible publications: Maitland and Campbell (1994) for fish, Beebee and Griffiths (2000) for amphibians and reptiles, Mead (2000) for birds and Macdonald and Barrett (1993) for mammals.

All the animals in this book, from the most primitive (the sponges) to the most advanced, (the mammals) are multicellular. This is not to deny the existence of vast quantities of unicellular creatures. In terms of numbers of individuals, and probably in numbers of species, they exceed the multicellular types several times over. However, because of their nature, most people never see them and, historically in this country, no-one did see them until Robert Hooke, with his compound microscope, observed 'animalcules' in the second half of the seventeenth century. Today, these unicellular organisms tend to be placed in a kingdom distinct from animals, variously given a name such as Protozoa, Protista or Protoctista. Animals or not, there is clearly no folklore, custom and tradition attaching to their like.

Names of animals

For each creature, I have given the currently accepted scientific name (which is sometimes erroneously called 'the Latin name', although many are based on Greek or other languages and adapted to Latin grammatical form). I have also given the currently accepted common vernacular (English) name. Much of the interest in the history of our fauna, however, attaches to alternative common names, most of regional origin, and I have listed those that are known to have been used within the past hundred years or so. There was an inevitable temptation to list only those known to have been used very recently. I resisted this, partly because of a wish to help preserve and record a valuable linguistic heritage and because the names often have links with other aspects of the animal's folklore, but also because many names that have widely been thought extinct have, surprisingly, been referred to in correspondence to me as having been used in the writers' lifetimes and so it is clearly dangerous to be dogmatic. Moreover, some names that are seldom if ever now used in speech (such as 'coney' for the Rabbit, or 'yuttick' for the Whinchat) are perpetuated in the names of inns or in other ways. Names that are self-evidently obsolete have been excluded but, where they may be of interest or are valuable, for instance, in giving a clue to the origin of introduced animals, I have referred to them in the text. I have made only token inclusions of Scots, Gaelic, Irish, Welsh and Cornish names, partly because of space limitations but more importantly because, again, I am not competent to judge inferences from their nature and meaning. The Cornish language is particularly rich in the names of animals, especially, and not surprisingly, fish and other sea creatures. Useful summaries of common animal names in Gaelic and Cornish can be found in Couch (1864), Forbes (1905), Nance (1963) and Rodd (1880), among other sources.

In outlining the history of the common names of animals, I have not attempted any deep linguistic analysis but have followed the *Oxford English Dictionary* practice and referred to the precursors of our modern language as 'Middle English' (*c.*1150–*c.*1450) and 'Old English' (pre-c.1150; sometimes popularly called 'Anglo-Saxon'). Among other language names, 'Old Teutonic' especially is referred to from time to time; this is the hypothetical prehistoric Germanic language from which English and other western European languages evolved. When referring to name origins in Latin or Greek, I have not attempted any consideration of which branches of these classical languages were involved; an interesting and relevant discussion of the classical origins of bird names, which also has resonances in the names of other animals, can be found in Lockwood (1984). Many British wild creatures have acquired several common names, most of them of local, regional origin, although the number of names seems to bear little relation to the size or importance of the animal. The Nightingale and Kingfisher, for instance, have one name each, while the woodlouse has over 200. Where there are several regional and other common names for particular species, I have listed them in alphabetical order.

Names of counties

I have used the names for the counties of England, Wales and Scotland that pre-date the 1974 revisions, largely because the wealth of folklore from before that time relates to them but also because most of my corresponding contributors have preferred to use them.

Invert

ebrates (Invertebrata)

Invertebrates are animals without a backbone (see Vertebrates, page 115). Numerically, they are an enormous group, far outnumbering vertebrates in both individuals and species, but their structure and physiology mean that they have never been able to rival vertebrates in dominating the earth, although in certain conditions and habitats, they can be extremely successful. Many have an external skeleton conferring some degree of rigidity, while others are little more than masses of soft, pliable tissue.

They are generally divided into eight or ten main groups or phyla, all of which have representatives in or around Britain, the arthropods being much the most significant. Although some are little known, seldom seen and obscure, others have long had major influence on the lives of humans.

SPONGES

(Porifera)

Who first decided that a sponge was an ideal object with which to wash themselves we shall never know. It certainly occurred a long time ago and, in classical times, a sponge was not only a conventional item of washing equipment but, mounted on a stick, was also the precursor of toilet paper.

Sponges are biologically the simplest of all multicellular animals. They do not move but remain attached to rocks and other objects, mainly in the sea, although there are also some freshwater species. They draw water into their colonial bodies and filter out minute food particles. Not all have the familiar 'bath-sponge' form, and many are relatively two-dimensional objects, flattened against rocks. Many sponges are brightly coloured and, in their way, attractive, but they are difficult to identify.

The name 'sponge' is Old English, originally from Greek, but it was not applied specifically to this group of organisms until the sixteenth century. Over the years, it has taken on a range of related meanings, such as 'living parasitically on others' or simply 'absorbing large volumes of liquid', as in one who drinks to excess.

CNIDARIANS

(Cnidaria [Coelenterata])

The cnidarians are an extremely large and extremely diverse group of creatures that have a very simple body structure. They are radially symmetrical, with a single body cavity (and therefore no sophisticated internal organs) and a single body opening, the mouth, which is surrounded by tentacles. Most have two phases in their life cycle: a free-swimming phase and a sedentary one.

These Jewel Anemones *(Corynactis viridis)* are related to the true anemones, but occur in large groups not singly.

The Beadlet Anemone *(Actinia equina)* is the most familiar of the sea anemones.

They include some of the most familiar of sea creatures, although most are more numerous in warmer waters, and have featured little in British lore. Some are colonial, others solitary, and many are beautiful. The stony corals (Scleractinia) are almost exclusively warm-water creatures that play an important part in shaping the geomorphology of many tropical islands.

In British seas, small numbers of sea pens (Pennatulacea), shaped like an old-fashioned quill pen, and the descriptively named sea fans (Gorgonacea) may be found, together with soft corals (Alcyonacea) and other groups, but the most significant are the sea anemones (Actiniaria) and the jellyfish (Scyphozoa).

Sea anemones

(Actiniaria)

All sea anemones are solitary and are usually attached to rocks or other matter by a large basal sucker. They are among the most familiar and beautiful creatures in rock pools, waving their tentacles until disturbed by a paddler's foot, when they withdraw rapidly to become a lump of coloured jelly. The nineteenth-century Victorian naturalist Philip Henry Gosse could, nonetheless, find beauty in anything and thought them then to be 'glossy and plump, like some ripe pulpy fruit, tempting the eye and mouth'.[1]

Most species are carnivorous and use their stinging tentacles, which can sometimes be extremely numerous, to trap fish and other passing creatures. Among the commonest and most familiar is the bright red Beadlet Anemone *(Actinia equina)*.

Despite their familiarity, one of the puzzles about sea anemones lies in their reproduction: some reproduce sexually; others are parthenogenetic. In some species, both sexes disgorge already well-developed juveniles, while other species simply divide by binary fission.

In reality, they bear no more resemblance to anemones than to any other type of garden flower and, to confuse matters further, some species have acquired multiple floral associations: the Daisy Anemone *(Cereus pedunculatus)* and the Dahlia Anemone *(Urticina felina)* for instance.

In Yorkshire, sea anemones were once called 'herring-shine' because of a mysterious belief that, over the course of time, they turn into Herrings. I have been quite unable to discover or even guess how such an idea could have come about.

Jellyfish

(Scyphozoa)

OTHER COMMON NAMES: *blovers; blubber; blue-slutter; cruden; donal blue; follieshat; fyke; galls; loch-liver; morge; nettle; paps; roother; scadder; scalder; schnap; scoudre; sculder; sealch's bubble; stinging blood-sucker; whale-blubs; whale's food.*

Jellyfish always seem to strike fear into bathers because of the painful stings that some species can give from their tentacles. In reality, most of the seriously stinging species are oceanic and come close to the shore only when carried there by storms and strong currents. Nonetheless, it is wise to avoid contact with any swimming jellyfish. They are all fascinating, and many are compellingly beautiful creatures as they pulsate their way through the water, catching fish and other creatures, which they paralyse with their stinging cells. A beached jellyfish, however, is a sad and sorry sight.

They have intrigued people for centuries and, almost alone among this group of animals, have acquired a large number of local names. The name 'jellyfish' (or 'jelly-fish') is as recent as the nineteenth century; the Middle English 'medusa', from Greek, was an earlier name.

The most infamous and potentially most dangerous 'jellyfish' in British waters is in reality a member of a quite distinct group of organisms called the Siphonophora. This is the Portuguese Man-of-War *(Physalia physalia*; Family Physalidae), a colonial animal that consists of a collection of three different types of 'polyp' (feeding, stinging and reproductive). These are suspended beneath a floating bladder with a sail, by which means the siphonophores are carried across the oceans. Strong southwesterly winds may bring them close to British beaches. The stinging polyps may be many metres long, and their sting is extremely dangerous and potentially fatal to humans.

The Common or Moon Jellyfish *(Aurelia aurita)* has four pale violet circles or crescents around its centre.

SEA GOOSEBERRIES AND COMB JELLIES

(Ctenophora)

These fragile-looking, transparent, gelatinous creatures are extremely beautiful when seen swimming in large numbers, especially at night when they display an eerie phosphorescence. They are seen in their full glory most commonly in open water but are sometimes found in rock pools or in a deflated state along the tide-mark. They feed on plankton, which they trap on long, sticky tentacles.

ECHINODERMS

(Echinodermata)

Echinoderms are unusual among invertebrate groups in being wholly marine. There are five main groups, of which two – the starfish (Asteroidea) and sea urchins (Echinoidea) – are very widespread in British waters. The sea cucumbers (Holothurioidea), which really do look like the short, bristly outdoor varieties of cucumber (one species is even called the Sea Gherkin), and the plant-like sea lilies (Crinoidea) occur around Britain but are much more common in warm seas. Many types of echinoderm are extremely familiar as fossils, especially in the chalk of the Cretaceous.

> There have been innumerable cases of fossils being used in folk medicine ... but the medicinal use of fossil sea-urchins in the Chalk of Kent is perhaps less well known. John Woodward published the statement that *Echini marinae* dug out of the chalk pits at Purfleet, Greenhithe and Northfleet contain specially pure fine chalk, and that the diggers drive a trade with the seamen who pay good prices for these 'Chalk-eggs'. The fine clean chalk within these fossil shells was, he said, 'one of the finest remedies for subduing acrid humours of the stomach'. He wrote further: 'Those who frequent the sea, and are not apt to vomit at their first setting, fall frequently into loosenesses, which are sometimes long, troublesome and dangerous. In these they find Chalk so good a remedy that the experienced seaman will not venture on board without.'[2]

Echinoderms have a chalky 'skeleton' of external plates that, in the sea urchins, forms a complete, more or less spherical covering; in the starfish, there is flexible tissue between the plates. They are unusual creatures in having a pentameral (five-part) symmetry; this is most obvious in the starfish and the brittle

'Sea urchins' are so called because of their resemblance to Hedgehogs. This is an Edible Sea Urchin *(Echinus esculentus)*.

The Common Starfish *(Asterias rubens)* typically has five arms, and can regenerate damaged or missing limbs.

stars (Ophiuroidea), which have five arms, although the fact that starfish can regenerate lost or partially lost arms means that individuals with unusual numbers of arms are sometimes seen. One family of sea urchins, the heart urchins (Family Spatangidae) are heart-shaped, with a bilateral symmetry imposed on the overall pentameral symmetry.

Echinoderms also have tube-like 'feet'. In starfish these are concealed on the underside and enable the animal to glide over the sea floor. In sea urchins, the feet are generally fewer and take the form of 'spines'. Starfish are predatory creatures, feeding on other starfish and marine worms, or on shellfish, which they prise apart with aggressive vigour. Sea urchins browse on algae, while sea cucumbers are usually filter-feeders.

> Here and there on the East Anglia coast can still be seen, in old fishermen's house, the sea urchin shells which were known as 'fairy' or 'fairy loaves'. It was thought a good thing to keep these in the house or on the outside windowsill, for they insured their owners against want and protected them from storms. Fairy loaves, though could bring luck to anyone who found and kept one.[3]

> There's a story here [Jersey] of a fisherman ripping one leg off a huge spiny starfish in order to make the sign of a cross to ward off a witch that was demanding a proportion of every fisherman's catch as he went past a certain rock.[4]

> Sea-urchins thrusting themselves into the mud, or striving to cover their bodies with sand, foreshows a storm; for the windy exhalations disturb the lowest waters first, in the bottome of the sea, which makes the other fishes rise and trust in their swimming; and the Urchin unapt for that, and fearing to be hurried away with the tumultuous waves, gets near the shore, and there stayes itself by creeping into the earth.[5]

The name 'starfish' is sixteenth century and self-explanatory, and the origin of the name 'urchin' for the sea urchin and its relationship with the seventeenth-century name 'sea hedgehog' are discussed elsewhere (see Hedgehog, page 385). Among other common regional names for sea urchins are: canniber, cauniber, conan-mara, cragan, crogan, echini, garbhan, gibneach, ivegar, ivigar, piper, sea-cracken, sea-egg and sea hog.

The shells of some species of sea urchin, minus their spines, are rather common along the tide-mark, and the yellow shells of *Echinocardium cordatum* are common enough to have acquired the name 'sea potatoes'. Despite a rather widely held belief that both starfish and sea urchins can sting, in many parts of Britain some sea urchins have long been eaten, more or less as a luxury food (called 'sea-eggs'), although the only edible parts are the star-shaped ovaries. The most frequently eaten species is the large, red and purple Edible Sea Urchin *(Echinus esculentus)*. The 'shells' are commonly used as household ornaments or as playthings.

PLATYHELMINTHS

(Platyhelminthes)

There are three main groups of Platyhelminthes: the flatworms or planarians (Turbellaria), the flukes (Trematoda) and the tapeworms (Cestoda). The turbellarians are almost all free-living but the other two groups are entirely parasitic. Although the name 'flatworm' tends to be restricted to the free-living turbellarians, all platyhelminths are flattened in shape. Perhaps the most familiar to the British public in recent years have been several free-living Antipodean species: the New Zealand Flatworms *(Arthurdendyus triangulata, A. albidus* and *A. australis)* and the Australian Flatworm *(Austroplana sanguinea)*.

Platyhelminths have some strikingly unusual features. Almost all are hermaphrodite, and they have only one body opening, which serves as both mouth and anus. The turbellarians especially are able to exist without food for periods of many months, during which time they become smaller and smaller, with a gradual absorption of their internal organs. They come as close as any living animal to disappearing into themselves, although, if food becomes available again, the organs regenerate and the creature returns to its original size.

Perhaps the best known of the flukes is the Liver Fluke *(Fasciola hepatica)* because of its remarkable life history. Its story begins in the bile duct of a sheep, where the eggs are laid. These are carried in the bile to the intestine and ultimately pass to the exterior in the sheep's faeces. Under precise and appropriate conditions of temperature and other factors, each egg hatches to form a larva, which seeks out a particular species of snail, *Lymnaea truncatula*, and penetrates its body. There, it passes through a series of further stages, eventually entering the snail's liver and feeding upon it. The snail, unsurprisingly, dies and the fluke then develops into yet another stage, which escapes from its dead host and attaches itself to grass, until it is eaten by a passing sheep. A further journey takes it from the sheep's intestine into the liver, where it causes the most serious damage, and ultimately to the bile duct once more. Given appropriate conditions, Liver Flukes may live their extraordinary existence for up to 11 years. There are many other notorious species of parasitic platyhelminth, especially in the tropics, where genera such as the trematode *Schistosoma*, the cause of bilharzia, bring about great suffering in humans.

The most complete parasites are the cestodes, which have dispensed with both mouth and gut and absorb food from their hosts through the skin. Many are known familiarly as tapeworms and may attain barely credible lengths; *Taenia saginata* for instance, which affects man and cattle, can reach 25 metres (82 feet). Although rarely important in humans in Britain, species such as the Dog Tapeworm *(Dipylidium caninum)*, which affect domestic animals, are familiar enough and both farm stock and household pets require routine preventative or curative treatments.

NEMATODES
(Nematoda)

OTHER COMMON NAMES: *eelworms; hookworms; lungworms; roundworms; threadworms; whipworms.*

Although most nematodes are microscopic and apparently featureless, even when seen under a microscope, they have become familiar to people in two main ways. While some species are free-living in soil, the sea or fresh water, other species are parasitic, some in plants, some in invertebrates and some in vertebrates, including humans and domestic animals. In some species, the larva is parasitic and the adult free-living; in others, the situation is reversed and, in several cases, more than one host is involved, a situation that has sometimes hindered the development of effective control measures.

The roundworm *Ascaris lumbricoides*, which parasitizes both man and pigs, is one of the largest and may be 30 centimetres (12 inches) long. More recently, some species have become familiar to gardeners as a means for the biological control of certain garden pests. *Phasmarhabditis hermaphroditica* is sold as dried cultures for the control of slugs, while *Heterorhabditis megidis* and *Steinernema carpocapsae* are available for use against Vine Weevils (see page 112) and certain other soil-inhabiting insect pests. However, contrary to common belief, they achieve this by by introducing disease-causing bacteria into the body of the victim rather than by direct parasitism. Free-living roundworms occur in prodigious quantities in soil and are probably the most numerous of all soil animals.

Among the most notorious parasites are the hookworms *(Ancylostoma* and related genera), which have a crown-like device at one end, by which they attach themselves to their host's intestine. The eggs pass to the exterior in the host's faeces, hatch in the soil, and the larvae subsequently reinfect the host through the skin.

With few exceptions, parasitic species are far more important in the tropics and, although some have been serious in Britain in the past, modern chemical treatments mean that they are relatively insignificant in modern Western societies.

HAIR WORMS
(Nematomorpha)

OTHER COMMON NAMES: *Gordian worms; horse-hair worms.*

Although their biology is imperfectly understood, hair worms are fairly familiar to country-dwellers because they sometimes forsake the ponds and ditches in which they live and are found after heavy summer rain among nearby vegetation. The old belief that they arise by the spontaneous animation of horse-hairs is then perhaps understandable.

They are very slender, hair-like, up to 20 centimetres (8 inches) long, and otherwise apparently featureless. Some species are probably parasitic in other invertebrates, rather in the manner of some nematodes.

The name 'Gordian worm' is a reference to the fact that the creatures twist themselves into knots and to the legend of the knot tied by Gordius, king of Gordium, and the story that whoever loosened it would rule all of Asia. Alexander the Great sliced through the knot with his sword.

RIBBON WORMS
(Nemertea [Nemertinea])

Few people have seen one of these smooth, slimy, highly elongated marine creatures, which live on the bottom of the sea but are sometimes found along the lower shore and in pools, voraciously hunting almost anything that comes their way and is capable of being swallowed. Ribbon worms have a sharply pointed proboscis with a poisonous tip. They are impressive enough creatures but have achieved their greatest fame through being among the longest animals in existence. A specimen of the Bootlace Worm *(Lineus longissimus)*, found dead on the shore at St Andrews, Fife, in 1864, was more than 55 metres (180 feet) long.

SEGMENTED WORMS
(Annelida)

Think of an earthworm and you will conjure the image of a typical annelid; the fact that their bodies are segmented is readily evident from the multitude of external, tyre-like rings. The annelids are a large group of animals that are found in marine, freshwater and terrestrial environments; the terrestrial species especially are both very familiar and very numerous. These are the archetypal worms and, today, when the term 'worms' is used colloquially, it is almost invariably in reference to a member of this group.

The name 'worm' is Old English and, over the centuries, has been widely used for any superficially similar creature, ranging from caterpillars to snakes. Even today, it is correctly used for nematodes (see above), ribbon worms (see above), platyhelminths (see page 21) and some other groups, too. It is related to the Latin word *vermis*, which finds modern expression in 'vermiform', meaning 'worm-like'. There is an interesting sixteenth-century verb 'to worm', meaning 'to catch worms', but it is now generally limited to clinical or veterinary treatments for parasitic platyhelminths or nematodes. An alternative meaning of the verb 'to worm' is 'to wriggle or inveigle one's way into something', such as another's affections.

There are three main groups of segmented worms: the Oligochaeta, the Hirudinea and the Polychaeta. Most are hermaphrodite, each individual functioning as both male and female but requiring cross-fertilization.

Oligochaetes
(Oligochaeta)

This is a large group of very diverse worms that all have very few bristles ('Oligochaeta' means just that). It includes the sludge worms (Family Tubificidae), most notable of which are the tiny, bright, red aquatic bloodworms or tubifex worms *(Tubifex costatus)*, which live in vast quantities in mud and are familiar to aquarists as live food for ornamental fish; the pot worms (Family Enchytraeidae), which are often found among the roots of ornamental plants, although causing no harm; and, of course, the earthworms themselves, of which there are several families.

There are 25 British species of earthworm and the most important family is the Lumbricidae, which contains the genera *Lumbricus*, familiar from school biology lessons, and *Allolobophora*, the most important for producing casts of soil on lawns. The largest British earthworm is *Lumbricus terrestris*,

Lumbricus terrestris **is the largest of the British earthworms and plays an invaluable part in maintaining soil fertility.**

which may, when extended, reach 35 centimetres (14 inches) in length, although this is small when compared with the giant earthworm of South Africa, which may exceed 6 metres (20 feet). Particularly large earthworms, of whatever species, are commonly called 'lob-worms', while a familiar and related type is the fisherman's red-coloured Brandling *(Eisenia foetida)* of manure and compost heaps.

> Now these be most of them particularly good for particular fishes: but for the trout, the dew-worm (which some also call the lob-worm) and the brandling are the chief; and especially the first for a great trout, and the latter for a less. There be also of lob-worms some called squirrel-tails (a worm that has a red head, a streak down the back, and a broad tail) which are noted to be the best, because they are the toughest and most lively, and live longest in the water: for you are to know that a dead worm is but a dead bait.[6]

The importance of earthworms in maintaining soil fertility is hard to overvalue. They ingest soil containing organic matter and then, having removed the organic component, expel the mineral particles. In the course of this activity, they mix the soil, providing access for air and assisting drainage. Charles Darwin, among others, calculated that up to 100 tonnes of soil per hectare are moved by the earthworm population in the course of a single year. In recent years, parts of Britain have suffered from infestations of several species of introduced southern hemisphere platyhelminths (see Platyhelminths, page 21), which feed on earthworms and have seriously depleted the worm population in Northern Ireland and other areas.

Earthworms have long fascinated people but, curiously, such numerous creatures seem to have acquired few regional names. Apart from being fed to birds and fish, they have also acquired relatively few uses, although they are immensely important as a natural food for a great number of other creatures, including many types of bird and mammal. Almost inevitably, people have eaten earthworms, occasionally out of necessity and commonly out of bravado (it is a favourite playground pursuit for schoolboys, who often find they can earn small sums in return for the 'dare'); and they have certainly found medical applications.

> Boiled worms were a remedy for stomach disorders. A stranger who claimed to be a seventh son of a seventh son was tested by an earthworm put in the palm of his hand. If the man was telling the truth, it was believed the worm would die at once.[7]

> In one case, known to the writer, a woman, in the hope of getting rid of the pain [of toothache], carried in her mouth an earthworm in the early morning into a neighbouring parish, a distance of about two miles, and there spat it out. On her return she reported the pain gone.[8]

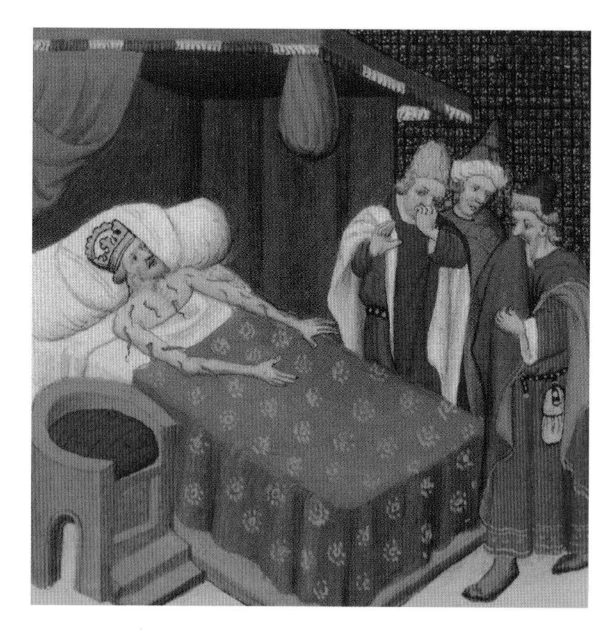

Physicians watch over the Roman emperor Galerius who has Medicinal Leeches *(Hirudo medicinalis)* attached to his torso.

It was once very widely believed that toothache was caused by an earthworm in the tooth. More recently, a bizarre pursuit known as 'worm-charming' has gained a following. By agitating the soil with garden forks and other implements, earthworms can be 'charmed' to the surface, and competitors in worm-charming contests have managed to persuade 500 or more of the creatures to appear within the allotted time of 30 minutes.

Leeches

(Hirudinea)

Leeches are annelids with relatively short bodies and few segments, some of which are modified to form sucking organs at each end; they have an evil reputation as parasitic bloodsuckers. There are four British families and, although there are a few marine leeches, the natural home of most is in freshwater or very damp habitats, such as under stones. Leeches feed on the blood of other creatures. Many, as befits their habitat, are fish parasites; others attack other invertebrates, some of which they swallow whole. A very few feed on mammalian blood and, of these, there are two in Britain. One species, the Horse Leech *(Haemopis sanguisanga)* is often thought to suck the blood of horses but, in reality, it feeds on invertebrates and the name 'horse' is used, as in other animal and plant names, simply to mean 'large' or 'coarse'.

The most significant mammalian parasite is the Medicinal Leech *(Hirudo medicinalis)*, which also feeds on fish and amphibians. It has been used medicinally since ancient times because of the long-held but totally false belief that bleeding a patient assists their recovery. At one time, Medicinal Leeches were especially common in ponds close to old monastery sites, where such activities took place and where collecting them provided an income for the humble.

> He told, that to these waters he had come
> To gather leeches, being old and poor:
> Employment hazardous and wearisome!
> And he had many hardships to endure:
> From pond to pond he roamed, from moor to moor;
> Housing, with God's good help, by choice or chance;
> And in his way he gained an honest maintenance.
> (William Wordsworth, 'Resolution and Independence')

Today, the Medicinal Leech is a rare and protected animal. There are probably no more than 20 isolated populations remaining in Britain: in Argyll, Islay, Cumbria, Anglesey, Norfolk, Kent and Dorset, with the largest at Lydd in Kent. Unexpectedly, it is again being used medically, not for bleeding but because it secretes an anticoagulant that prevents wounds and surgical scars from healing too quickly, thus encouraging the formation of new tissue. To satisfy this renewed demand, a company in south Wales now farms leeches and exports them widely. Leeches have also found other applications.

> Leeches too, make good barometers. I have two leeches in a long bottle, which generally will indicate what sort of weather is coming in the next twenty-four hours.[9]

> Previous to slight rain or snow it creeps to the top of the bottle, but soon sinks; but; if the rain or wind is likely to be of long duration; the leech remains a longer time at the surface. If thunder approaches, the leech starts about in an agitated and convulsive manner.[10]

> Leeches were made into a broth (by boiling the creatures in water and straining off the liquid for drinking) which had a good reputation as a cure for tuberculosis and other wasting diseases and was also given for persistently bad coughs ... .[11]

The name 'leech' is Old English and there is an old, little-used verb 'to leech', meaning 'to cure or heal'. 'Leech' also came to mean one skilled in medicine, or 'leech craft'. There is a modern surname 'Leech', which probably has this origin.

> And straightway sent, with careful diligence,
> To fetch a leach the which had great insight
> In that disease.
> (Edmund Spenser, 'The Fairie Queene')

The use of the word to describe a person who stays close to another in the expectation of obtaining something is eighteenth century.

Bristle worms

(Polychaeta)

The bristle worms are the largest and most varied of all the annelid groups and are almost exclusively marine. They are distinctive worms, each body segment bearing a mass of bristles on a small extension. These extensions commonly function as legs or paddles and enable the animals to crawl very effectively and also to swim. Many species are relatively large, fantastically shaped and brightly coloured.

Among the most familiar groups are the ragworms (Family Nereidae), which are common in marine mud and widely dug by fishermen as bait. The largest species of British annelid worm, the King Rag *(Nereis virens)*, which may exceptionally

The Peacock Worm *(Sabella pavonina)*, a type of bristle worm, extends a crown of tentacles to catch food particles in the water.

reach almost 100 centimetres (39 inches) in length, belongs here. Other distinctive types are the rapidly swimming paddle worms (Family Phyllodocidae); the short, scaly scaleworms (Family Aphroditidae), which include the bizarre Sea Mouse *(Aphrodita aculeata)*, broad and covered with long, fur-like bristles; the earthworm-like lug-worms (Family Arenicolidae), which are also an important fishing bait and a significant food for many wading birds; the fan worms (Family Sabellidae), with a fan-like array of feathery tentacles; and the several families of tube worms, which secrete a calcareous tube and attach themselves to rocks and often to other living things, such as seaweeds, molluscs or crustaceans.

SEA MATS AND SEA MOSSES

(Bryozoa)

Bryozoans are aquatic colonial animals, common in both the sea and fresh water, and usually fixed to rocks, aquatic plants or other objects. Many are encrusted with mineral matter and assume the form of masses of fine, branched tubes; others are simply gelatinous lumps. They feed by pushing out a small crown of tentacles from the tubes and filtering particles from the water. Many take on a most attractive overall form, like some strange underwater plants, but, although they are commonly seen on the sea shore or elsewhere, they are seldom recognized for what they are. They are usually thought to be something else, such as coral or seaweed.

Perhaps the commonest is the Sea Mat *(Membranipora mebranacea)*, which is almost invariably present as flat, white, crusty patches on the flat blades of the large, brown oarweeds (*Laminaria* species). A species that itself looks rather like a seaweed is *Alcyonidium diaphanum*, which is familiar enough along the tide-mark to have acquired the name Sea Chervil.

MOLLUSCS

(Mollusca)

The molluscs constitute an enormous group of predominantly marine animals that embraces many very evidently closely related creatures but also some that superficially are quite distinct. All have soft, fleshy bodies that are not divided into segments (unlike those of insects, crustaceans and arachnids) but comprise a head, a large muscular foot and a central part, the visceral mass, covered with a structure called a mantle, which has a skirt-like frill around its periphery. The mantle secretes the hard, protective calcareous shell that most molluscs possess (many are collectively called 'shellfish'), although, in some, such as the slugs, sea hares and squids, the shell is reduced or absent.

> Fool: Canst tell how an oyster makes his shell?
> Lear: No.
> Fool: Nor I neither; but I can tell why a snail has a house.
> Lear: Why?
> Fool: Why, to put's head in ...
> (*King Lear*, Act 1, Scene 5)

The facts that most molluscs are fairly large and totally unlike any other groups of animals and that many are very common and sometimes highly edible mean that they have always attracted attention. They have been used since ancient times as food, as decoration and, in some cases, more unexpectedly, as a source of dyes. Because the shells of dead molluscs are washed up on the sea shore, many species that habitually live in fairly deep water may be as familiar as those that inhabit the coastal shallows. The following account covers only a small selection of those that have had a particular impact on human life and cuisine in Britain.

Chitons

(Polyplacophora)

OTHER COMMON NAMES: *coat-of-mail shells.*

Chitons are marine shellfish with a single dorsal shell composed of eight, arching, overlapping plates, superficially and, with a little imagination, resembling part of a suit of armour. Rather like a Pill Woodlouse (see Woodlice, page 49), they roll into a ball when disturbed.

The name 'chiton' is from Greek, meaning 'a coat of mail', although it was also used in classical times to mean a long woollen tunic. Chitons browse on algae in shallow water, and one or more species are common on most British coasts.

Gastropods

(Gastropoda)

This is the largest of all groups of molluscs, accounting for perhaps as many as 80 per cent of all living species. Gastropods have a single, often more or less spiral shell, from which emerges a conspicuous head and a large, substantial foot. These are often sealed 'indoors' by a horny disc (the operculum) when the animal withdraws into its shell.

The name 'Gastropoda' means 'stomach-footed' and snails are archetypal gastropods. There are many subdivisions but most of the familiar types of gastropod molluscs belong to individual families and their English names tend to be derived from a superficial resemblance to some everyday object.

The shells of some gastropods are remarkably beautiful and, inevitably, they have become collectable objects. In many parts of the world, several species are now seriously threatened because of the activities of divers, who collect and kill them for the value of their shells. Like those of many bivalve molluscs (see Pelecypods, page 37), some gastropod shells are beautifully endowed with the iridescent secretion called 'mother-of-pearl'.

MARINE SNAILS AND LIMPETS (Prosobranchia)

Ear shells (Haliotidae)

OTHER COMMON NAMES: *ormers.*

The ear shells have flattened, shallow, oval shells, with a distinctive curving line of small holes and brilliant mother-of-pearl iridescence within. They are found in shallow water, clinging tenaciously to rocks. Only one species occurs in British waters, the Green Ormer *(Haliotis tuberculata)*. This reaches no further north than the Channel Islands, where its collection is now strictly controlled.

> We have an abalone type mollusc here called the ormer. Found only in the Channel Islands and the adjacent coast of France, its name comes from the Guernsey French meaning ear of the sea. Ormers live in shallow water and can only be gathered on extreme low spring tides to which they have given their name – ormering tides. Ormers are still gathered and people still cook that good old Jersey dish, ormer stew – some love it, others hate it. It takes hours and hours to cook otherwise it's like eating pieces-of-rubber stew.[12]

> In Guernsey, ear-shells are used by farmers to frighten away small birds from the standing corn, two or three of these shells being strung together and suspended by a string from the end of a long stick, so as to make a clattering noise when moved by the wind.[13]

Limpets (Patellidae and related families)

OTHER COMMON NAMES: *bonnet-shell; brennygen (Cornwall); croggan (Cornwall); flitter (Isle of Man); lampit (Scotland); lempot (Cornwall).*

There are three main families of marine limpets, the Fissurellidae, Acmaeidae and Patellidae, all of which have a neat, low, more or less symmetrically conical shell and a legendary ability to cling to rocks. 'Limpet' has become a by-word for tenacity and the Second World War 'limpet mine' was aptly named.

The slit limpets and keyhole limpets (Family Fissurellidae) have a tiny slit or keyhole-shaped opening in the apex, while

Common Limpets *(Pabella vulgata)* cling with tenacity to a seaside rock.

the tortoiseshell limpets (Family Acmaeidae) have beautifully patterned shells remarkably like tortoiseshell in coloration.

The various species of true limpets (Family Patellidae) have more or less plain shells and sometimes occur in vast numbers between the tide-marks. Their ability to cling tightly is a water-conservation mechanism, which ensures that they do not dry out at low tide.

Less familiar than the limpets of the sea shore are the freshwater limpets (Family Ancylidae), which are much smaller but occur in clean rivers and lakes, where they can be extremely numerous.

> At distance viewed, it seems to lie
> On its rough bed so carelessly,
> That 'twould an infant hand obey,
> Stretched forth to seize it in his play.
> But let that infant's hand draw near,
> It shrinks with quick instinctive fear,
> And clings as close, as though the stone
> It rests upon, and it, were one;
> And should the strongest arm endeavour
> The Limpet from its rock to sever,
> 'Tis seen its loved support to clasp,
> With such tenacity of grasp,
> We wonder that such strength should dwell
> In such a small and simple shell.
> (William Wordsworth)[14]

Limpets may appear to be permanently glued to rocks, but, in reality, they wander short distances under water to browse on algae before returning to the same spot; on soft rocks and over a long time (limpets can live for up to 16 years) this becomes worn to form a small depression. People have long been attracted to limpets, and they have been collected as food for centuries; a sharp lateral knock with a knife or similar implement will remove them.

A limpet mine is handed to a frogman in preparation for an Allied raid in Bordeaux, 1942.

In Scotland, the juice is used, mixed with oatmeal, and they also once had a valued medicinal role:

> The Limpet being parboil'd with a very little quantity of Water, the Broth is drank [sic] to increase Milk in Nurses, and likewise when the Milk proves astringent to the Infants. ... I had an Account of a poor Woman, who was a Native of the Isle of Jura, and by the Troubles in King Charles the First's Reign was almost reduc'd to a starving Condition; so that she lost her Milk quite, by which her Infant had nothing proper for its Sustenance; upon this she boil'd some of the tender Fat of the Limpets, and gave it to her Infant, to whom it became so agreeable, that it had no other Food for several Months together; and yet there was not a Child in Jura, or any of the adjacent Isles, wholesomer than this poor Infant, which was expos'd to so great a Strait.[15]

Limpets clearly once had symbolic importance as well because masses of limpet shells, sometimes strung together, apparently as necklaces, have been found in chambered tombs in the Channel Islands.

The name 'limpet' is Old English, from Latin, and may stem from the same root as 'lamprey' (see Lampreys, page 118). There are a number of places in Cornwall said to be named after limpets; Scrawston (from the Cornish name 'croggan') is one, while Porthe Brenegan, near Portreath (after the Cornish 'brennygen') means 'limpet cove'.

Top shells (Trochidae)

These are more or less spherical or pyramidal little shells, characteristically and prettily mottled on the outside and with a silvery mother-of-pearl lining that commonly shows through as the outer layers are worn away. This is almost invariably apparent on dead shells, and the common Grey Top Shell *(Gibbula cineraria)* is known as 'silver Willie' in Orkney. Many species occur commonly on rocks on all British coasts, where they browse on algae.

Winkles (Littorinidae)

The winkle-picker shoe with its sharply pointed toe was the painful height of fashion in the 1960s. The connotation was familiar even to people who had never seen a real winkle-picker, let alone the winkle it is used to pick. However, all the references are testimony to the importance of a group of shells that have been laboriously collected, and even more laboriously eaten, for centuries. There are many species of

Winkle-picker shoes were essential wear for the fashion-conscious in the 1960s.

winkle, all superficially similar in their small spirals, wide apertures and rather thick texture, and all grazing on seaweeds.

The Common Periwinkle (corvin [northern England]; gwean [Cornwall]; horse-winkle [Ireland]; sea-snail; whelk [Northern Ireland]) *(Littorina littorea)* sometimes occurs in vast numbers and is often the most numerous of all coastal gastropods. It is dark brown or almost black, although other species are attractively bright yellow.

Winkles are collected and boiled, and then the contents are extracted with a winkle-pinker; this may be purpose-made but a stout pin will serve just as well. Eaten in the traditional manner, with brown bread and butter, they make a pleasant, if unexciting feast, but it is important to remove the hard, horny operculum first, and to remember not to drench them in vinegar.

The name 'winkle' is a late sixteenth-century contraction from the early sixteenth-century 'periwinkle', which seems to originate in Old English but is of unknown meaning and appears to have a different basis from the plants *(Vinca* species) of the same name. There are many place names incorporating 'winkle' or similar words, but few if any have any relationship with molluscs, referring instead to personal names, meadows, military camps and other origins. The use of the verb 'to winkle', meaning 'to extract with difficulty' is more recent.

Cowries (Trividae)

OTHER COMMON NAMES: *cowry; gowry; groat; grottie-buckie (Orkney); nun; stick-farthing.*

Cowries resemble tiny, ridged eggs, flattened on one side, where there is a slit-like aperture. Elsewhere in the world, larger species have been used as currency, but there seems be no tradition of this in Britain. The Orkney name 'grottie-buckie' (or 'groatie-buckie') might be imagined to have some connection with the old coin, the groat, but is most commonly thought to be related to the locality of John O'Groats. 'Cowrie' (or 'cowry') sounds Gaelic but in reality came into English in the seventeenth century from Hindi and Urdu; other species are abundant in the Indian Ocean.

Murexes (Muricidae)

The murexes have elongated shells with marked ribs and ridges and, in some species, spiked protrusions. They are aggressive, carnivorous animals, often preying on other molluscs. In parts of the world, some species are extensively collected for food, and one species, the Mediterranean *Bolinus brandis*, was valued as the source of the important dye Tyrian purple.

Among the British species are the Dog Whelk (cattie-buckie; dog-winkle; horse-winkle [Ireland]; purple) *(Nucella lapillus)* and the sting winkles, which are important predators of other mollusc species. The American Sting Winkle *(Urosalpinx cinerea)* was introduced, with oysters, to southeast England where it is now a common predator of oysters. Extract of Dog Whelk, on exposure to light, was once used to produce a purple dye which was widely used, especially in Ireland, to dye linen.

The European Cowrie *(Trivia monacha)* typically has three brownish-purple spots on its shell.

Stalls selling cockles and mussels, whelks and winkles were once a familiar sight in London.

Whelks (Buccinidae)

OTHER COMMON NAMES: *conch; cuckoo shell; goggle (Ireland, south Wales); googawn; sting winkle; whelk-tingle.*

Whelks contribute to the tide-mark debris in two ways. Their empty shells are common (and are among the largest gastropod shells to be found on many British beaches), but their sponge-like egg masses, which roll along the beach in the wind, are less usually identified correctly. The commonest are those of the Common Whelk *(Buccinum undatum)* and the Red Whelk (buckie) *(Neptunea antiqua)* and they are familiarly known as 'sea wash balls'.

Whelks have tall, ridged shells with a flared aperture and feed carnivorously, although often on dead rather than living marine animals. If this were more widely known, they might be less popular as food. As it is, Common Whelks are collected in large quantities and have a quiet, subtle taste but are distinctly rubbery in texture. They are still to be found for sale on whelk stalls in London and at specialist sea-food restaurants, but tend now to be thought of as rather modest fare, although they once formed a significant part of the menu for grand feasts. At the enthroning of William Warham (*c.*1450–1532) as Archbishop of Canterbury in 1504:

> ... 8000 whelks were supplied at five shillings a thousand and were served as an accompaniment to sturgeon.[16]

As with other shellfish, different coastal ports vie with each other for the finest; Wells-next-the-Sea, Norfolk, considers its whelks to be peerless.

Sometimes whelks are dredged from the sea but sometimes carrion is buried under a basket covered in stones to trap them. Whelks are still used extensively as a fishing bait. The Red Whelk is a very large species, occasionally up to 20 centimetres (8 inches) in length, and was once common around the northern coasts of Britain. Like its exotic relative, the conch, it was blown to produce a note for calling both cattle and farm-workers. It was also used as food, and in Shetland:

> ... the cottagers ... made use of the shell as an elegant lamp by suspending it horizontally, filling it with oil, and allowing a wick to lie in the canal, and the lighted portion protruding. It owes its name antiqua to the fact that it is plentiful as fossils in the Crag ... .[17]

The name 'whelk' is Old English but its origin is unknown. So-called 'dog whelks', which are inferior or worthless for eating, belong to other, related families. Whelks are the second most popular shellfish in heraldry, after scallops; the Shelly family of Lincolnshire, appropriately, bear 'argent a chevron gules, between three whelks sable'.[18]

SEA SLUGS AND SEA HARES (Opisthobranchia)

'Sea slug' is the collective name given to members of the numerous families of usually fairly large, fleshy gastropods that have an internal and/or an external shell or no shells at all. In some families there are species displaying all possible options. Some are free-swimming, others merely crawl like their terrestrial counterparts.

They are compelling animals and many have fantastically ornamented bodies with amazing tentacles and protuberances; some, especially in warmer waters, are astonishingly beautiful in their rich coloration. Many are herbivorous, feeding on seaweeds, but some eat sponges, sea squirts and other marine animals.

The members of one family of free-swimming species are known as 'sea hares', presumably because someone once thought that their long tentacles, elongated shape and rapid movement gave them some resemblance to a terrestrial hare. Having watched both sea and land hares at close quarters, I find this somewhat fanciful. Sea hares are curious animals, swimming by means of wing-like lobes and releasing clouds of coloured fluid into the sea, rather like squids (see Octopuses, Squids and Cuttlefish, page 45). One rare British species, *Aplysia fasciata*, can reach 40 centimetres (16 inches) in length.

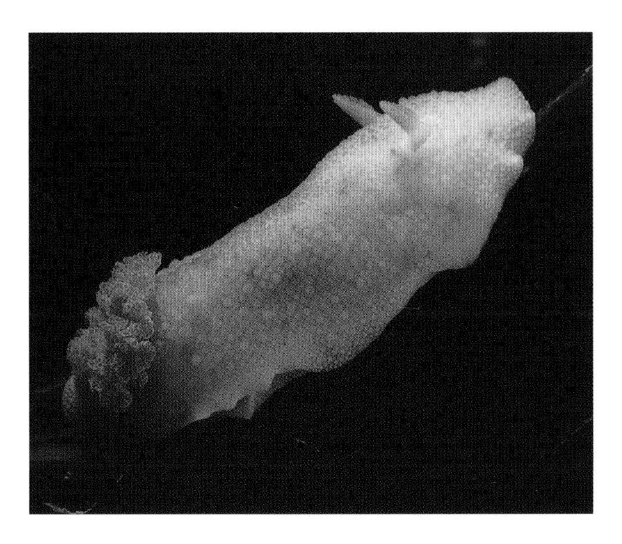

The Sea Lemon *(Archidoris pseudoargus)* feeds on sponges and and lays coiled white egg ribbons.

It must have been another fevered imagination that produced the name 'Sea Lemon', for the very common, large, warty, yellowish sea slug, *Archidoris pseudoargus*.

FRESHWATER AND TERRESTRIAL SNAILS AND SLUGS (Pulmonata)

Some gastropods have evolved for a life largely away from the marine environment. Many have reduced or internal shells, or, while having a substantial shell, are unable to withdraw into it fully and lack the closing horny operculum that typifies such superficially similar creatures as winkles. These gastropods are conveniently divided into the freshwater snails and the terrestrial snails and slugs.

FRESHWATER SNAILS

It is always important, or so aquarists and, to some extent, pond-keepers are told, to have at least some water snails in order to 'keep the water clean'. The theory is that they feed on algae, fish droppings and other debris but, in my experience, their impact is minimal. Nonetheless, in the natural environment, the freshwater snails are an important group of gastropod molluscs.

There are several families, several genera and many species, although the most familiar are those in the genera *Lymnaea*, the pond snails with the more typical snail-shaped shell, and *Planorbis*, the ramshorn snails, which, as their name suggests, have shells with flattened spirals.

A number of snails are the hosts for stages in the life cycle of parasitic worms, and one species, *Lymnaea truncatula*, is particularly important as the intermediate host of the Liver Fluke *(Fasciola hepatica)*, which affects sheep and cattle (see Platyhelminths, page 21).

TERRESTRIAL SNAILS AND SLUGS

OTHER COMMON NAMES: *Mainly snail names but sometimes also applied to slugs: bulhorn; biljinks; bull-jig; bulorn (West Country); clare (East Yorkshire – a snail shell); cogger (striped); conger; conker; dod/dodden/dodman (Norfolk); dodman/hodmedod (various spellings; Cambridgeshire; Hertfordshire; Norfolk; Suffolk); drutheen (white); grey; guggle; hornywink; jack-sna; jan-jake (Cornwall); jin-jorn; john jago (Cornwall); lobury one; malorn; oddie/oddy/oddy-doddy (Hertfordshire; Oxfordshire); snayle; snele; sniggle (East Yorkshire); snyle.*

There are approximately a hundred species of terrestrial snails and slugs in Britain, of which about a dozen are familiar because they occur in gardens. Consequently, they have attracted a considerable amount of interest, lore and tradition – and a fair bit

Aquarists and pond-keepers will be familiar with the ramshorn snails (genus *Planorbis*), which have shells with flattened spirals.

of cursing. Although they seem to form neat and discrete groups – snails with shells and slugs without shells – in reality the boundaries between them, as with other shelled and shell-less gastropods, are blurred. There are some slugs with small shells, and some snails almost without shells.

In practice, snails and slugs are very closely related. All are herbivorous, although, as far as freshness of vegetation is concerned, their preferences vary; all reproduce by laying eggs, and their relationship to creatures that live in water is betrayed by the fact that they must have cool, damp habitats if they are not to dry out.

One particularly large species, the Roman Snail (apple snail; edible snail; vine snail) *(Helix pomatia)*, which occurs in south-east England and has a shell up to 5 centimetres (2 inches) in diameter, has attracted a great deal of interest. Its localized distribution has led to the belief that it is not native but was introduced by the Romans for culinary purposes; it is the species most widely 'farmed' for the table. It certainly occurs in some areas where Roman settlements were significant but, conversely, it is absent from others. This distribution seems to be related more to areas of chalk than to areas of habitation, and hence the link with the Romans is a fantasy. Its common alternative name of 'apple snail' is also a fantasy because the scientific name *pomatia* comes not from the Latin *pomum*, 'an apple', but from the Greek *poma*, 'a pot lid'.

> I was born eighty odd years ago. My mother was a real country girl, having been born and brought up on a small farm in what was then rural Essex. She always referred to snails as hodmedods and ... there is an old name for a snail quoted by Dickens in 'David Copperfield', he has Mr Peggotty saying 'I'm a regular Dodman, I am', by which he meant he was a snail. The name hodmedod or hodman for the snail can also refer to the ancient pre-historic surveyor or dod-man because he carries on his head the dodman's implements, the two sighting staves.[19]

> Most land snails have shells that have a clockwise coil, (dextral). Very infrequently an animal with a mirror image of itself (sinistral) turns up in the wild, at a rate of about 1 specimen in 30,000. If this happens, as the reproductive organs are on the right side of the head and mating snails adopt a head to head position, it is impossible for the luckless freak to mate unless it finds another with a similar deformity, which is unlikely. During surveys of *Cepaea hortensis*, in separate colonies in Somerset, two specimens were found and bred from. The resulting crosses were dextrally coiled; in other words the shell spiralled clockwise. The next generation produced a few sinistral individuals. Further sinistrals were found and I repeated the experiment and it is the descendants of these that are now living wild in the ivy in the garden.[20]

Today, snail-farming, using *Helix aspersa*, is being promoted in Britain as a cottage industry.

> Ninety per cent of culinary snails are collected from the wild so you can never guarantee supply. They hibernate for about half the year and hide away so you can't find them. The only way to guarantee a year-round supply of live snails – what most restaurants now demand – is to farm them indoors. Europe currently eats 30,000 tons of snails a year but not more than 100 tons are farmed. So the market is wide open. We're importing 15 million snails – that's 150 tons – a year into Britain when we should be growing our own. Wild snails will get even more scarce – there are now close seasons in France and a total ban on commercial collection in Switzerland. If you can put the car in the drive and farm snails in the garage, why not go for it?[21]

The commonest way of eating snails is grilled or baked with garlic butter and served, these days, in a small dish with a dozen holes. Why snails should almost always be served in dozens rather than tens has long mystified me, especially as the dish is so characteristically French. Perhaps it is a prerequisite for mollusc-eating (oysters, too, are counted similarly). Nonetheless, there are other ways of eating snails, although they are seldom seen today. Snail salad and snail sauce have been widely recommended, as has snail soup:

> Use only the large edible snails in grey shells, and select those of which the shell is closed up. They are first dropped into boiling water, slightly salted, and boiled for an hour, then extracted from their shells by means of a fork. The black skin that will be found on top must be cut away, as well as the horns and the ring that, starting from this black skin, encircles the snail. They are now laid on a dish and strewn with a handful of salt, whereby the slime is extracted. They next require washing in three or four waters, and must finally be squeezed quite dry. To about fifty of the snails thus carefully cleansed are added two or three quarts of good stock. Boil until they are soft, when they should be taken out and two-thirds of them chopped small. Fry these lightly in butter, and add to them so much of the stock as will be required for soup. Boil them up once more, with the addition of a little mace as well as the whole snails remaining; stir in the yolks of two or three eggs, and serve with toast cut into dice.[22]

Until at least the 1970s, the Garden Snail *(Helix aspersa)* was collected extensively in the northeast of England, boiled and offered for sale in public houses and working men's clubs. This is an area well known for its extremely enthusiastic gardeners:

> ... one man, I am told, collects them in his garden and puts them on the top plate of his greenhouse coke heater. When the snails are roasted he extracts them from their shells and eats them immediately.[23]

Snails move with legendary slowness and this has led to the sport of snail-racing. Snail races have been held at various times in many parts of the country, often on an *ad hoc* basis as part of a village fête. At Congham, in Norfolk, it has been more enduring:

> The trophy is a silver tankard stuffed with lettuce leaves. The race over 13 inches is held on a covered table top from a starting circle to an outer finishing ring. Although this is an annual event, the world record of 2 minutes 20 seconds, set up in Burnaby in Canada in 1992, remains unbeaten.[24]

The Roman Snail *(Helix pomatia)* is notable for its size and is the species most widely farmed for the table.

Snail-racing prompted an interesting comment from one distinguished author, P.G. Wodehouse, in his book *Over Seventy – An Autobiography with Digressions*:

> The news in my daily paper that in certain parts of Wales the latest craze is snail-racing has turned my attention to these gastropods after months and months during which I do not suppose I have given more than a passing thought to them. As a writer I have always rather kept off snails, feeling that they lacked sustained dramatic interest. With a snail nothing much ever happens, and, of course, there is no sex angle. An informant on whom I can rely says they are 'sexless or at least ambivalent'. This means, broadly speaking, that there are no boy snails and no girl snails, so that if you want to write a novel with a strong snail interest, you are dished from the start. Obviously the snail-meets-snail, snail-loses-snail, snail-gets-snail formula will not help you, and this discourages writers from the outset. But this snail-racing should mean a change for the better and give authors more of a chance. The way it works, I understand, is that each entrant pays a small fee and the owner of the first snail to pass the judge's box takes the lot. The runners have their owner's colours painted on their shells and 'are attracted to the winning-post by a pile of wet ivy leaves', with a delirious crowd, no doubt, shouting 'Come on, Steve' or words to that effect.[25]

Snails in general, and *Helix aspersa* in particular, have been used extensively in traditional medicine. It was widely believed that the broth obtained from the snail's mucilage was soothing to a dry throat, and the glass-blowers of Newcastle upon Tyne 'held an annual feast of snails to strengthen their lungs for the coming year'.[26] The snails were usually boiled, shelled, baked with butter, garlic, pepper, salt and parsley, and served with brown bread and butter.

Snails prepared in various ways were also used against such diverse ails as ague, jaundice, hangovers, corns, poor eyesight and, very extensively, to cure warts, as Flora Thompson, in *Candleford Green*, describes:

> A few innocent charms and superstitious practices were all that remained of magic. Warts were still charmed away by binding a large black slug upon the wart for a night and a day. Then the sufferer would go by night to

These children are following the progress of their snails in the World Snail Race Championship held at Brighton on 13 September 1969.

> the nearest cross-roads, and, by flinging the slug over the left shoulder, hope to get rid of the wart.[27]

One widely used alternative involved impaling the slug on a spike or thorn after it had been applied to the wart; as the slug shrivelled, so did the wart. Roy Palmer tells of a practice at Pillerton, in Warwickshire, in which a sick child was given a piece of bread and butter every morning while the dew was on the grass, with a second piece being put out for the snails. The theory was that the illness passed to them.[28]

> That snail remedies were taken seriously in bygone days is evidenced by the entry in the London Gazette of March 23rd. 1739 which tells that Mrs. Joanna Stevens received from the Government then in power £5,000 for revealing the secret of her famous cure against 'stones'. The cure consisted chiefly of snails, egg-shells, soap, honey and herbs.[29]

The behaviours of snails and slugs have also been taken as tokens. If a snail extruded its horns, this was a clear sign that rain was to be expected, for good or ill. In Scotland, they said:

> Snailie, snailie, shoot out your horn,
> And tell us if it will be a bonnie day the morn.[30]

In Somerset:

> Snail, snail, put out your horn,
> We want some rain to grow our corn.
> Out, horn, out.[31]

And in Dorset:

> When black snails cross your path
> Black clouds much moisture hath ... .[32]

In Cornwall, the implication was more sinister:

> Bulorn, Bulorn, put out your long horn, your father and mother is dead;
> Your sister and brother is to the back-door, a-begging of barley bread![33]

In my copy of *Songs for the Nursery* (1825), there is something rather pointlessly aggressive:

> Snail, snail, come out of your hole,
> Or else I'll beat you black as a coal
> Snail, snail, put out your horns,
> Here comes a thief to pull down your walls.[34]

It is strange that animals as familiar as snails and slugs should have had to wait until comparatively recently for an accepted common name, although regional names and a number of now-obsolete names were evidently used in earlier times. The name 'snail' is Old English but, until the eighteenth century, it was used fairly indiscriminately for any shell-bearing gastropod. In the sixteenth century, it came to mean 'a slow or indolent person' and from this usage rather than its current one have come such expressions as 'snail-like', 'snail's pace' and so forth. 'Slug' is Middle English and originally meant someone slow or slothful; there was an appealing old verb 'to slug', meaning 'to be lazy or slow'. (The more familiar, mainly American verb 'to slug', meaning 'to strike', which appears most familiarly in expressions like 'slug it out', is mid-nineteenth century.) 'Slug' gradually came to mean any slow-moving animal, and even a slow-moving sailing ship, but it was not until 1704 that it was first used in its modern context. It clearly took time to gain wide acceptance and, 70 years later, Gilbert White, in his *Naturalist's Calendar*, was still writing enigmatically:

> ... the house-snail begins to appear: the naked black snail comes forth much sooner. Slugs, which are covered with slime ... are moving all winter in mild weather.[35]

The various other uses of 'slug' ('a bullet', 'a dose of strong drink' and so forth) are relatively recent but it is not clear whether they have the same etymology.

> In September 1972 I was asked to settle an argument. An English schoolboy had found a *Limax flavus* in the garden, and was calling it a slug, but a lass from Lanark thought it must be a snail, because, to her, a slug was 'wee and white'. I explained that I was quite happy to call the creature a slug, and a snail too: in England it would generally be called a yellow slug, but in Scotland it was likely to be called a snail. I first met this Scottish usage a few years ago. A retired lady doctor, in conversation after dinner in a Scottish hotel, was astonished to hear that I was looking at slugs, because she said that there would only be snails in the district. She remembered having had a black snail to dissect in an examination in her youth at Edinburgh, and she had never heard anyone call it a slug. Unfortunately, she could not describe just what she would call a slug. I was surprised that she was as ignorant of the English use of the term as I had been of her Scottish use. Last summer, at Cowdenbeath, a policeman assured me that the big black chaps were always called snails, and the little white fellows who ate his tomatoes were slugs. His wife, who had been in England more, was uncertain about the distinction.[36]

Slugs and snails have featured widely in children's literature and rhyme. After all, what are little boys made of but 'Slugs and snails and puppy-dogs' tails'? Young girls (but not apparently young boys) looking for an indication of whom they were to marry need look no further than the nearest mollusc. If a snail was placed in the ashes of the hearth, it would trace out the initial of their intended:

> I seiz'd the vermin, home I quickly sped,
> And on the hearth the milk-white embers spread.
> Slow crawl'd the snail, and if I right can spell,
> In the soft ashes marked a curious L;
> Oh, may the wondrous omen lucky prove!
> For L is found in Lubberkin and Love.
> (John Gay, 'The Shepherd's Week')

As recently as the early twentieth century, the practice was alive in Brecon as a form of Hallowe'en divination:

> We caught snails and put them under a cover or box lid to prevent their crawling away. By the morning they were supposed to have traced out in their slime the initials of our future husbands. But my snail always crawled up on the underside of the lid and there it remained.[37]

Slugs and snails are unquestionably genuine garden and commercial horticultural pests. Many cause serious damage to foliage (hostas are notoriously and inexplicably prone to attack), although the subterranean keeled slugs *(Milax* species) are especially damaging to potatoes. Slugs and snails can be controlled reasonably effectively by the use of pellets or baits, most of which contain the anaesthetic poison metaldehyde. Understandably, however, gardeners have now become increasingly loath to use wide-spectrum poisonous pest controls, not least because of a not altogether well-founded fear that the many creatures (like Hedgehogs, see page 385) that feed on slugs will come to grief. Sprinkling slugs with salt has been the traditional, although fairly inhumane, method of control. Hand-picking is both possible and practicable, and a visit to a garden after dark with a torch will generally reveal the astonishing richness of the local snail and slug population.

The Great Black Slug *(Arion ater)* shows a wide colour range, even orange, but it is certainly large (up to 20 centimetres/8 inches).

To many people's amazement, such nocturnal expeditions also reveal not only the variety of slugs and snails, but also the numbers that exist and the size of some individuals. The so-called 'Great Black Slug' *(Arion ater)*, which is not a serious pest, exhibits a wide colour range, sometimes being a rich orange rather than black, and may reach 20 centimetres (8 inches) in length. Recently a biological control method, employing a culture of nematodes (see Nematodes, page 22) that introduce bacteria into the creatures' bodies with fatal results, has become available for garden use.

However, it is not only in horticulture that snails have created problems. They are probably unique in this country as postal pests:

> A letter reached us last week in a sorry state. Frayed and torn in one corner, it looked as if it had been chewed. The explanation lay in a regrets note from the postmaster where it had been posted. It said simply – 'Eaten by snails in the post-box.'[38]
>
> Royal Mail workers in Devon and Cornwall have solved the problem of snails ... eating the letters posted in rural letter boxes by placing two strips of bristles across the front of wall-set letter boxes. The brushes, which are rather like those found in a carpet sweeper, prevent the snails from entering the boxes without harming them. A Royal Mail spokesman, Adrian Booth, said that the snails are attracted by the glue on the envelopes and the saliva used to stick them down.[39]

Paradoxically, the slowness of snails has also more recently spawned a relevant neologism. In an age of electronic, almost instant communication, the traditional postal service has become known, disparagingly, as 'snail-mail'.

The slime trails that both slugs and snails produce betray their activities and also serve conveniently to distinguish the damage that they cause to plants from the superficially similar

Various preparations of the Garden Snail *(Helix aspersa)* have been used to cure a diversity of ailments.

effects of caterpillars and other pests. In turn, however, slugs and snails form a major part of the diet of a great many other animals, including several mammals, reptiles and amphibians. Many birds also feed on them voraciously and the so-called 'thrush's anvil', a stone surrounded by the remains of smashed snail shells, is a hall-mark of bird activity.

> The snail crawls out with his house on his back,
> You many know whence he comes by his slimy track.
> Creep, creep, how slowly he goes,
> And you'd do the same if you carried your house.
> The snail crawls out with his house on his back
> But a blackbird is watching him on his track.
> Tack, tack! on the roof of his house,
> He gobbles him up as a cat does a mouse.
> (Sabine Baring-Gould, 'The Snail')

Nonetheless, the fact that a snail's shell is often likened to a house has, on occasion, been its salvation. The Romany people will not kill a snail because, in carrying its home with it throughout its life, it has something in common with their own lifestyle.

The word 'snail' appears rather commonly in place names but, while the names of fields and other minor places, such as Snail Down in Wiltshire, Snail House Meadow and Snailbeach in Shropshire, and comparable places, may have a genuine snail connection, most do not.

Pelecypods

(Pelecypoda [Bivalvia])

The pelecypod molluscs, sometimes known as lamellibranchs, differ from the gastropods in having two shells, or valves, that are hinged together. Consequently, they are popularly called 'bivalves'. In some, the two shells are identical, while, in others, they are markedly different. Although a few species occur in fresh water, most are marine and their shells, which usually become separated after death, form the bulk of the shell debris found on the sea shore. Unlike gastropods, pelecypods tend to remain relatively stationary, usually attached to rocks, opening the two shells to sift food from the water as currents bring it within reach; some move to a small extent as adults and a few others do so much more freely. Pelecypods include many of the most familiar of shellfish and many of the most familiar edible species, including cockles, mussels and oysters.

The shells of the numerous species found in the bounteous sea-shore debris are representative of many groups, usually descriptively named from the colour or shape of their shells. These include the nut shells (Family Nuculidae), ark shells (Family Arcidae), dog cockles (Family Glycymeridae), saddle oysters (Family Anomiidae), fan mussels (Family Pinnidae), file shells (Family Limidae), hatchet shells (Family Lucinidae), coin shells (Family Leptonidae), Venus shells (Family Veneridae) and many others.

Some pelecypods are edible and some tend collectively to be sold as 'clams', although this is a rather imprecise term (sixteenth century; 'a shell that clamps') and the British species are different from those that make up such familiar North American dishes as clam chowder. A few shellfish families, however, outweigh all others in their impact on human life.

Mussels (Mytilidae and related families)

The true mussels (Family Mytilidae) comprise about 40 species but one surpasses all others in importance. The familiar Common Mussel *(Mytilus edulis)*, with its pointed, deep blue-purple shells, has been a significant food since ancient times. The mussels are cut or pulled from rocks, piers and other structures, to which they firmly attach themselves with a fibrous structure called a 'byssus'. (This remarkably ancient word, in its present form, is seventeenth century from Latin, but then from Greek for 'flax', or 'linen', and originally perhaps from Egyptian.)

The Common Mussel is a very variable species, the range in size, overall shape, colour and degree of ribbing on the shells being considerable. (It should not be confused with the large Green-lipped Mussel, *Perna canaliculus*, which does not occur in Britain but is imported from Australia and New Zealand, and which has become a popular food item. Because of its glycosaminoglycan content, it is also promoted medicinally as an anti-inflammatory agent and for stimulating the growth of connective tissues).

The name 'mussel' is Old English, from the same Latin root as *mus*, 'a mouse'; the word 'muscle' has the same origin because it was once thought that some muscles look like mice.

Although mussels are still dredged from the wild, many are also farmed: the UK production of farmed mussels in 1998 was nearly 11,000 tonnes. Farming involves dredging juvenile mussels and transporting them to sheltered coastal waters where they are grown on to marketable size. The Scottish method is to suspend the mussels on ropes hung from rafts or long lines.

Apart from their culinary appeal, mussels have also long been recognized for their role in protecting sea defences. As long ago as the mid-nineteenth century, *The Times* was reporting that:

> So valuable are mussels towards the protection of our shores from the inundations of the sea on some parts of our coasts, that it becomes necessary to prevent their being gathered in some places. An action for trespass was brought some time ago for the purpose of establishing the right of the Lord of the Manor to prevent the inhabitants of Heacham from taking mussels from the seashore.
>
> Mussel-beds on some parts of our coasts have been jealously watched because of their power to hold the mud and sand, and prevent the foreshore shifting away with the tides and currents. The story of Bideford Bridge [Devon] is founded upon the knowledge of the binding power of the byssus. Most books mentioning Mussels repeat the legend that Mussels instead of mortar bind the stones of the bridge together, and that, therefore, the town authorities forbid the removal of the molluscs that throng its piers. It is true that a prohibition of this sort exists, but it is due to a conviction that the Mussels keep off the wear and tear of the tide.[40]

Mussels were once picked by hand, as shown in this painting *Gathering Mussels* by John Maclauchlan Milne (1920).

Low tide on this Cornish shoreline reveals extensive beds of mussels *(Mytilus edulis)*, firmly attached to the rocks.

The commonest and generally most popular way of eating mussels is steamed (not boiled), with wine and a little onion and parsley or other herb flavouring. Because of their popularity in France, they almost invariably appear on British menus as *Moules*, or *Moules marinière*. Despite their appeal as food, many people are, or become, allergic to mussels, as to other shellfish. Mussels can also sometimes induce other, extremely unpleasant symptoms, which seem to arise either through eating a dead mussel (when fresh, the shells should be tightly closed; any that remain open should be discarded) or eating mussels from stagnant water, such as harbours; these mussels may be physiologically different and perhaps accumulate toxins. The general rule not to eat mussels or oysters when there is an 'r' in the month is a sound one because, even if they are not harmful at such times, they may be less flavoursome and also they may be breeding.

A related species, the Horse Mussel *(Modiolus modiolus)*, can be distinguished by its generally larger size (most creatures prefixed with 'horse' are abnormally large or coarse) and by the protruding 'beaks', a slight distance from the extreme pointed ends of the shells. It, too, is edible and is caught commercially. They are known as 'clabbydoos' in Scotland and are, I am told, to be found for sale as specialities in fish shops in Glasgow.

Closely related to the marine mussels are several freshwater species, of which three have achieved special fame. The Zebra Mussel (*Dreissena polymorpha*; Family Dreissenidae) is thought to have reached Britain attached to ships from the Baltic; it was discovered in London Docks in the nineteenth century (see Tufted Duck, page 240) and has now spread widely throughout many British freshwater habitats. The Pearl Mussel (*Margaritifera margaritifera*; Family Margaritiferidae) buries itself in the bed of clean, fast-flowing rivers, generally in acidic areas and has been known since Roman times for its fine-quality pearls.

> The pearl mussel, a biodiversity species, will be studied in a partnership project involving Environment Agency Wales, the National Museum and Galleries of Wales and the Countryside Council for Wales ... they were historically fished for black pearls but are now a protected species under the Wildlife and Countryside Act. In recent years, it has become apparent that existing colonies have stopped breeding, probably because of eutrophication and increased sedimentation in rivers. Although the mussel can live for up to 100 years, it is important to study any new colonies to identify if they are breeding

> and how they can be protected so that their future existence is assured. It has long been rumoured that they were present in the Tywi catchment ... earlier this year an Agency conservation officer found a live washed up specimen ... .[41]

The Swan Mussel *(Anodonta cygnea)* and its relative the Duck Mussel *(Anodonta anatina)*, of the Family Anodontidae, similarly bury themselves in the river bed but they have little to do with freshwater birds, as their names might suggest, and rather more to do with freshwater fish. The female of the small, introduced carp-type fish, the Bitterling (see page 147), has a modified ovipositor through which it lays its eggs into the mussel's shell. Here they are fertilized by the male fish and here they remain until they hatch some three weeks later.

Freshwater mussels have been eaten but, it seems, out of necessity rather than choice. The Pearl Mussel has been described as 'extremely bad, unsavoury and unwholesome',[42] but the Swan Mussel was certainly eaten in Ireland, by peasants and during the famine of the mid-1840s, especially in the poorest regions such as Leitrim, as reported by the surgeon and antiquary William Wilde (1815–76), father of the poet and playwright Oscar, and medical commissioner for the Irish Census in 1841.

In common with some other molluscs, the importance of mussels is underlined by their association with a number of places around the coast, of which Musselburgh, east of Edinburgh, is the best known and largest, although, on a smaller scale, there is Mussel Point near Zennor in Cornwall, Mussel Bank on the Isle of Sheppey in Kent, and Musselwick Sands in Pembrokeshire.

Scallops (Pectinidae)

OTHER COMMON NAMES: *clams (Ireland); frills; queens; squins (Dorset).*

Sadly, scallops are probably most familiar in the form of the ash-trays often found in more modest seaside boarding houses; the beautiful, fan-shaped shells of some species are up to 15 centimetres (6 inches) in diameter and sufficient to accommodate the output of the most addicted smoker.

Several species are caught commercially for food and stocks have been depleted in consequence. The largest and finest, the Great Scallop (king scallop) *(Pecten maximus)* and the Queen Scallop *(Aequipecten (Clamys) opercularis)*, have been particularly affected. Most are caught by dredging but some are also collected by divers. Some are then farmed by being grown on in mesh bags, or 'lanterns', the Great Scallop taking about five years but the Queen Scallop only two years to reach marketable size. However, only about 200 tonnes are produced each year.

Unlike most pelecypods, scallops are free-swimming, propelling themselves through the water by clapping their shells together; this can come as a great surprise to anyone seeing it for the first time. By ejecting a torrent of sea water very rapidly, they can also show a surprising turn of speed when attacked. They are sometimes called 'sea butterflies', or 'butterflies of the ocean', because they appear to fly through the water. Great Scallops have one round and one flat shell while the shells of their relatives, the Queen Scallops, are both rounded.

All scallops, like this Great Scallop *(Pecten maximus)*, swim freely through the water by a form of jet propulsion.

They are excellent fare and, like mussels, oysters and cockles, their shells are found in vast numbers in ancient kitchen middens. The less complicated the recipe, the better they taste. Merely poached is the best, but the most famous scallop recipe is the French *Coquilles St Jacques*, in which they are fried with onion, mixed with wine, parsley and breadcrumbs, and ultimately browned under a grill.

The name 'scallop' is Middle English, from the Old French *escalope*, and by the seventeenth century the name came to be applied to anything with a similar shape, hence the adjective 'scalloped'. There is a long, extensive and complex tradition attached to scallop shells throughout the ancient and historical world. Aphrodite emerged from a scallop shell, and many forms of ancient domestic utensil were shaped like scallop shells. Scallops have long been important heraldic symbols in England; 5 per cent of all English coats of arms include scallops and, among relatively recent arms, those of Sir Anthony Eden and Sir Winston Churchill both bore them.

> It is likely that their appeal was mainly derived from the distinctive shape of the two forms used. Each coat of arms had to identify one person, and the use of distinctive devices was essential. Also another reason of some importance was the punning or canting association

with the names of the users of arms containing shells. The two forms of shells regularly appearing on canting arms were scallops (described as 'escallops') and whelks, other forms being only rarely used. Scallops and whelks were used as a pun on the Shell element in a surname ... Not all users of scallops had puns in mind, and the device was also used by people with the James surname, this shell being the attributed emblem of St. James, the Apostle [and hence Coquilles St Jacques]. Other families have claimed that the scallop in their arms shows that an ancestor took part in a pilgrimage to Santiago de Compostela, in Spain.[43]

The shrine at Santiago de Compostela is dedicated to St James the Great and pilgrims would make two holes in scallop shells in order to attach them to their clothing. These two holes commonly appear as two eyes in heraldic representations of the shells. In some parts of the country, scallop shells, or at least silver or wooden representations of them, are used to dispense water in the baptismal service and, although this is an ancient practice, there is no obvious reason why scallop shells should have been chosen for this purpose. For parishes at any distance from the coast, they must have been difficult to acquire. In modern times, much the most familiar and widely seen scallop shell is probably seldom recognized as such. The Shell oil company uses a scallop shell as its symbol, a tradition dating back to 1904, when the rather indeterminate shell used by the founders of the company became more formalized. When Shell celebrated its diamond jubilee in 1957, a beautiful book, dedicated to the scallop and with many eminent contributors, was produced as a memento.[44]

The scallop shell is traditionally associated with St James the Great, depicted here wearing a scallop-shell brooch.

Oysters (Ostreidae)

There is no more famous, fabled and celebrated mollusc in the world than the oyster. Oysters have inspired artists and poets, song-writers and novelists and have most recently been eulogized by Hector Bolitho in his book *The Glorious Oyster.*[45] They have been cherished and eaten since ancient times and, although today they are perceived (and charged for) as a luxury food, the vast quantities of oyster shells to be found in Roman and other archaeological excavations indicates that they were once an everyday, if not a staple, commodity.

Like snails, oysters are almost invariably served in batches of twelve (see Terrestrial Snails and Slugs, page 31), but why eating great quantities of them should be a matter for celebration is a mystery. It happens with elvers (see European Eel, page 150) and it certainly happens with oysters. Many famous individuals have contributed to the statistics: the German statesman Bismarck is said to have devoured them in huge numbers, but, for an Englishman, there can be no more legendary an oyster-eater than the England cricket captain, A.P.F. (Percy) Chapman of Cambridge University and Kent. He played one of the great innings of between-the-wars cricket when, in 1927, he went in to bat with Kent on 70 for 5 and hit the Lancashire bowling for 260 in 3 hours. Even more significantly, he once ate 208 oysters at a sitting 'before witnesses'.

Like some other shellfish, oysters have a reputation for inducing strange and unpleasant symptoms in the unwary. As with mussels, an inherent allergy to oysters is rather common and an induced one not rare, while eating oysters with too great a quantity of alcohol may also bring about unfortunate gastric responses. Nonetheless, the two great traditional accompaniments of oysters are both alcoholic: champagne and, in Ireland, Guinness.

The combination of oysters and Guinness provides the stock in trade of the Galway Oyster Festival, which began in 1954 and is perhaps the largest and grandest of many such events that take place around the coasts of the British Isles. It features, among many other diversions, the Irish oyster-opening championships. Opening an oyster has always required some skill. Because the oyster is alive, the two shells are held tightly together and can be prised apart only by a combination of manual dexterity, a specially designed knife and a knowledge of where the creature's muscles are situated.

Although today, most gastronomes eat their oysters alive (ideally with a raspberry vinegar dressing, certainly not drenched in chilli sauce), most possibilities seem to have been explored at different times. Hector Bolitho cites a feast at Merchant Tailors Hall on 20 January 1622 when the bill of fare included:

> ... roast mutton with oysters for the first course,
> followed by boyled oysters, broiled oysters, oyster pie
> and pickled oysters.[46]

Today, perhaps the most popular alternative to raw oysters is the strangely named 'Angels on Horseback', in which each oyster is wrapped in a rasher of bacon, baked and served on toast.

Why oysters should have such a special role in human gastronomy is not particularly obvious. Undoubtedly, to the addicted, they are delicious and have a distinctive flavour, but the fact that they are also usually eaten alive is distasteful to many people. Oysters are also reputedly aphrodisiac, although the general modern view is that any effect is more likely to be the result of the ambience and the alcohol that accompany their consumption. Nonetheless, the animal's own sexual conduct is somewhat plastic, in common with that of certain other molluscs, and this may influence human perception. Some oyster species remain sexually constant, some are born bisexual but then orientate one way or the other as they mature, while individuals of the British native oyster, *Ostrea edulis*, are usually born male but change sex several times during their lives; and they do always spawn at a full moon.

Although there are four European oyster species, the Common Oyster (flat oyster) *(Ostrea edulis)* outweighs all others in importance. It is a very variable species and specimens from different localities differ greatly in the appearance of the shell and also in taste. Among the more celebrated localities are Whitstable in Kent, Colchester (which also has an oyster festival of some renown in October) and Pyefleet in Essex, and Strangford Loch in Northern Ireland; aficionados can instantly place any particular specimen from its look and flavour. The Romans clearly appreciated British oysters because their identifiable remains have been found not only in Britain but in Rome itself (how did they arrive fresh?) It is said that those collected from Richborough in Kent were the most highly prized and were known as 'rutupians'. The expression 'close as a Kentish oyster' means 'totally correct, as tightly sealed as a Kentish oyster'.

Whitstable, in Kent, is famous for its oysters *(Ostrea edulis)*. These oystermen are at work on the oysterbeds in nearby Seasalter.

The Colchester Oyster Season, here opened by Mayor Mrs C.B. Alderton, features gin and gingerbread, as well as oysters.

Although oysters were invariably dredged from the sea bed in the past, for today's market, farming is a further option. This is based on the Pacific Oyster (rock oyster) *(Crassostrea gigas)*, which was introduced to Britain in the 1970s and is now bred in hatcheries and then generally placed in plastic mesh bags attached to submerged trestles. Over 1,000 tonnes are now produced in this way annually. Most native oysters are still dredged, although some may then be relaid on sheltered inshore beds for 'fattening', a misleading term because the food reserve that they accumulate is not fat but glycogen. Although native oyster stocks once became seriously depleted, partly through over-dredging but also because of disease, careful conservation has enabled them to recover and such famous localities as Whitstable are now producing them again. Because native oysters spawn in the summer, they are not sold in Britain from May to August (when there is no 'r' in the month), whereas the Pacific Oyster, which rarely spawns at the temperatures found in British waters, can be sold all year round and is generally farmed in the traditional native oyster localities; hence such menu designations as 'Strangford Loch Rock Oysters'.

There is an old expression 'Who eats oysters on St James's Day will never want', which is slightly odd because St James's Day is 25 July and therefore in the middle of the 'inedible' season. The logic seems to be that anyone who, for whatever reason, was in a position to obtainable an unseasonable luxury must of necessity have been extremely wealthy.

There is more to an oyster than its culinary merits, however; there is mother-of-pearl and the pearl itself. Mother-of-pearl is the innermost of the three layers of material that make up the shell and consists of layers of aragonite or calcite (calcium carbonate) bound together by an organic substance called 'conchiolin'. It is produced by a secretion from the entire body surface and increases in thickness as the oyster ages. It is not unique to oysters, being found in many other molluscs. If a small particle of sand or other impurity becomes trapped within the shells, it too will become coated with the same nacreous substance and, in time, this will also increase in thickness; in due course, a pearl is formed. The production of pearls is not peculiar to oysters (see Mussels, page 37) and, when the whole process occurs naturally, the pearl is called a 'natural pearl'. However, it is now a common and widespread practice (and the basis of a major industry, especially in Japan) artificially to introduce a small particle of mother-of-pearl into living oysters, mainly warm-water species of *Pinctata*, to produce what is known as a 'cultured pearl'. The more perfectly spherical and the larger the pearl, the greater is its value. The finest pearls are white, sometimes with a hint of pink or cream, but other colours exist and black pearls are especially prized because of their rarity. An 'artificial pearl', by contrast, is produced by machine and has nothing whatever to do with oysters.

> ... rich honesty dwells like a miser, sir, in a poor-house;
> as your pearl in your foul oyster.
> (*As You Like It*, Act 5, Scene 4)

The name 'oyster' is Middle English, originally from Latin and Greek, and it has become attached to other things, such as the Oystercatcher (see page 269), which does not in fact catch oysters, and the Oyster Mushroom *(Pleurotus ostreatus)*, which is roughly similar in shape and colour. Although it does not seem to have been applied to a fisherman's surname, many places, mainly small ones, have acquired an 'oyster' tag; Oysterfleet on Canvey Island in Essex, Oyster Ness in Yorkshire and Oyster Island in Sligo Bay, County Sligo, are among the best known. It has also given rise to the odd expression 'the world is your oyster', which originally meant that the world was the place from which to extract a profit, just as the pearl could be extracted from the oyster.

> Why then, the world's mine oyster,
> Which I with sword will open.
> (*Merry Wives of Windsor*, Act 2, Scene 2)

Today, it tends to have the more general meaning of 'having an infinite range of opportunities'. I have been unable to trace the even odder 'and when did you see an oyster walk upstairs', although it is applied to people telling incredible tales of their past exploits.

Cockles (Cardidae)

'Cockles and mussels, alive, alive O' were most famously sold by Molly Malone in Dublin's fair city, but they were sold by countless other girls and women in many other places too. Cockles are distinctive molluscs that burrow into sand and gravel and have two almost identical, markedly ribbed shells.

Several species are dredged or collected at low tides ('cockling') and eaten but one, the Common Cockle *(Cerastoderma edulis)*, outweighs all others in importance and is sometimes present in truly vast numbers on many British coasts. Some cockle-collectors use rakes but others watch for the spurt of water that betrays the cockles' whereabouts, while others simply feel for the shells below the surface of the sand with their toes. As with many other shellfish, the East Coast traditionally supplied large quantities for the London market. Cockling is still popular near Heacham in Norfolk and other places around the Wash, while Leigh-on-Sea in Essex, at the mouth of the Thames, has long claimed to produce the finest, although there are rivals, such as Stiffkey in Norfolk. The cockles were generally boiled on the spot, as they were landed, the cockle-boilers traditionally being housed in small shacks on the shore.

A house decorated with sea shells, including cockles, at Polperro, Cornwall.

The shells also were put to good use. At coastal resorts, there are several 'shell houses' whose walls are covered with cockle shells, while in King's Lynn, in Norfolk, a path from the London Road to King's Lynn Town football field was paved with shells and known in consequence as 'Cockleshell Walk'. Cockle Sand, at Exmouth in Devon, Cockleshell Hard on the Isle of Grain in Kent and Cockleridge in Devon are among other places that owe their names to the one-time prevalence locally of cockles.

The name 'cockleshell' achieved its own immortality through C.E. Lucas Phillips' 1956 book *Cockleshell Heroes*, which later appeared as a film.[47] This told of the true and daring wartime raid by marine commandos on Bordeaux harbour, using tiny boats shaped like cockles. In truth, it was a double mollusc event because the commandos attacked the German shipping with limpet mines. The familiar expression 'to warm the cockles of the heart' comes from a supposed resemblance between the chambers of the heart and the shape of the cockle shell.

Razor shells (Solenidae)
There is no mistaking the razor shells because their elongated shells, like an old-fashioned cut-throat razor, are quite distinctive. They are also almost impossible to catch by hand because they bury themselves in sand with astonishing rapidity. Some species are edible, and the Common Razor Shell *(Ensis ensis)* has been seriously depleted by overfishing in many areas. Like other shellfish, when sold for eating they are sometimes called 'clams'. A North American species known as the Jack Knife Clam *(Ensis directus)* appeared in the North Sea in 1979 and has since spread very widely and aggressively.

Razor shells (Solenidae), which normally live well below the surface of the sand, are sometimes washed up in huge numbers.

Piddocks and ship-worms (Pholadidae and Teredinidae)
The piddocks (Family Pholadidae) are elongated shells that have achieved their fame and significance, not through any edibility, but from their ability to bore into materials as hard as wood or even limestone, producing characteristic circular holes. They do this by using a cutting action of the sharp ridges on the shells rather than some chemical dissolution, as might be imagined.

The ship-worms (Family Teredinidae) are remarkably modified wood-boring creatures that cause enormous damage to the hulls of wooden ships. The commonest and most important species, the Great Shipworm *(Teredo nivalis)*, occurs throughout much of the world, having travelled the oceans in ships and in driftwood. They have short, widely gaping shells and line their bore holes with a calcareous deposit, producing something much larger that does indeed look like a worm with a shell.

Octopuses, Squids and Cuttlefish

(Cephalopoda)

As 'Gastropoda' means 'stomach-footed', so 'Cephalopoda' means 'head-footed'. The cephalopods are superficially the most distinct of molluscs, and the most modified, and include the largest and probably the most intelligent of all invertebrates. Three of the main groups – the octopuses, squids and cuttlefish – can be found in British seas, but they are not nearly as numerous as they are in warmer waters, such as those of the Mediterranean. Consequently, these curious and bizarre creatures have featured much less in the traditions of Britain than in those of classical and other cultures. Nonetheless, octopuses especially feature in many children's stories and are very familiar as decorative embellishments on buckets and spades and other sea-side paraphernalia, even though the creatures themselves are rarely found on the shore or stranded in rock pools.

None of the British species has an external shell, unlike other types of cephalopod, such as the *Nautilus* of warm seas and also the extinct ammonites. The British species simply have a chalky internal shell, commonly called a 'cuttlebone', or 'cuttlefish bone', which is commonly found both along the tide-mark and in Budgerigar cages. Who first decided that these obscure objects would provide exactly what a Budgerigar needs to keep its beak in trim does not seem to be recorded. Cuttlebone has also long been used as a polishing material: 'Their bones were gathered during the occupation [Jersey] and ground up to make a form of toothpaste.'[48]

The octopuses, squids and cuttlefish have large, well-developed heads, a mouth with hard, beak-like jaws, and large and complex eyes. Superficially, the most obvious difference between the groups is that the octopuses (Family Octopodidae) have eight tentacular 'legs' or 'arms', equipped with suckers, while the cuttlefish and squids (Families Sepiidae, Sepiolidae and Loliginidae) have an additional pair of long, retractile tentacles, giving a total of ten. Squids and cuttlefish are therefore sometimes referred to collectively as 'decapods'. They are all good swimmers and move by jet propulsion, forcing water through a short funnel. They produce a dark-coloured, ink-like substance, known as sepia, as a defence mechanism. This has been used in the past both as ink and a dye, and cuttlefish are sometimes called 'ink-fish'.

Octopuses and cuttlefish feed mainly on crabs and other crustaceans, although squids tend to take a wider range of prey. They, in turn, are eaten by many types of fish and other sea creatures. The fabled abyssal Giant Squid *(Architeuthis dux)*, of which the largest found was over 16 metres (52½ feet) in length, is the favoured prey of the Sperm Whale (see page 441). The largest squid ever found in British waters was one over 7 metres (23 feet) long that was washed ashore in Shetland in 1959.

If you should see an octopus
And stop him for a chat,
I warn you that an octopus
Will never raise his hat.
He'll boast of his accomplishments
In kingdoms underseas,
And then he'll bore you with complaints
Of water on the knees.
(Peter Wesley Smith, 'The Octopus')[49]

Many species are caught commercially as food, the squids *Loligo forbesii* and *Loligo vulgaris* and the Common Octopus *(Octopus vulgaris)* being especially important. However, because they have not really been part of traditional British cuisine, squid tend to appear on British menus under the southern European name *calamari*. The young animals are eaten whole, but large octopuses and squid are sold as pieces of tentacle. When young and deep fried, they can be both crisp and delicious; when older, they are rubbery and their edibility really depends on the quality of the accompanying sauce.

The name 'octopus' is as recent as the eighteenth century and comes straight from the Greek word for 'eight-footed'. The Latin plural 'octopi' is often erroneously used; in reality 'octopuses' and 'octopodes' are both acceptable. 'Squid' is seventeenth-century but its origin is obscure; 'sepia' was an earlier name. 'Cuttle' is Old English but, before that, its origin is also obscure.

The Curled Octopus *(Eledone cirrhosa)* is widely distributed but commonest in the north.

Men with brooms remove barnacles (Cirripedia) from the propellors of the *Queen Elizabeth* during her refit in August 1946.

CRUSTACEANS

(Arthropoda: Crustacea)

The Crustacea comprise a huge group of creatures, familiar to the general public through the relatively small number of species that are important as food and are to be found on every fishmonger's slab. Most, such as crabs, lobsters and shrimps, are marine, although a significant minority live in fresh water and a very few, such as woodlice, live on land.

A typical crustacean is a highly jointed, armour-plated animal but the seemingly complex structure is made up of three main body parts, a head, thorax and abdomen, rather like an insect. There are varying numbers of legs and other body appendages and the external skeleton is commonly very thick and very hard. Crustaceans lay eggs, which hatch into larval forms that pass through a series of increasingly complex stages before becoming adults.

The lifestyles and appearance of crustaceans are bewilderingly diverse. Most feed on minute plant and animal life, which they filter from the water, but there are many very tiny parasitic species as well as the more familiar large ones.

There is no everyday English name for the group and they tend collectively to be referred to as 'crustaceans', simply by expansion from the scientific name.

Barnacles

(Cirripedia)

Barnacles are perhaps least like typical crustaceans and were widely believed to be shellfish (molluscs) until their true relationships and life cycles were unravelled in the early nineteenth century. Although they have free-swimming larvae, the adults are sedentary, living attached to rocks or other submerged objects. Unlike most other crustaceans, barnacles are hermaphrodite. The razor-like covering on rocks exposed at high tide, which makes walking over them in bare feet such a painful experience, is formed by barnacles.

The image conjured by the word 'barnacle' has led to the term 'barnacled' being used for anything comparably rough, and to such names as 'Barnacle Bill' being popularly applied to crusty old fishermen or sailors. Some species grow on the hulls of ships and must be scraped away periodically because they impede the ships' progress through the water. One distinct group, the goose barnacles (Family Lepadidae) has given rise to a quite extraordinary belief: they were once thought to develop into birds (see Barnacle Goose, page 231).

The name 'barnacle' is Middle English and was originally applied to the Barnacle Goose; it was not used for crustaceans until the late sixteenth century.

Copepods

(Copepoda)

Copepods are tiny (mostly less than 2 millimetres/1⁄16 inch long) and mainly have elongated, narrow bodies. Many are planktonic, others live along the sea shore, and some are highly specialized and parasitize fish and other sea creatures, including other crustaceans.

Almost all pass unremarked, with one notable exception, *Lepeophtheirus salmonis*. These parasitic copepods are the first thing that salmon-anglers look for when they pull a fish from the river. Known to anglers as 'sea lice', they survive for only a short time in fresh water and their presence on a salmon's body is certain evidence that the fish is 'fresh-run' (see Atlantic Salmon, page 130). A related species, the Freshwater Fish Louse *(Argulus foliaceus)*, is a small, free-swimming, indiscriminate parasite of freshwater fish, to which it can cause considerable harm.

Malacostracans

(Malacostraca)

This huge subdivision of the crustaceans embraces many tiny, strange, superficially shrimp-like creatures, such as cumaceans, opossum shrimps, skeleton shrimps, sandhoppers (which leap in all directions when piles of seaweed are disturbed) and freshwater shrimps, or gammarids (which occur in clean rivers and lakes almost everywhere). Few of these attract the interest of anyone other than specialists (the familiar and edible true shrimps are Decapods, see below), but the remaining groups of malacostracans certainly do.

Isopods

(Isopoda)

Isopods have a flattened body, two pairs of antennae, prominent eyes, an abdomen with six segments and a thorax with seven segments, each of which bears one pair of legs. Almost all of them only have scientific names, but one, a small, rather dull brown marine species, has acquired an utterly charming English name, possibly derived from the word 'grub'. This is the Gribble *(Limnoria lignorum)*, which burrows into soft, wet wood and has achieved both a name and reputation for itself because it actively eats wooden piers.

Representatives of other isopod groups can be found on the sea shore and in the tidal zones; some are fantastic-looking animals that would be truly nightmarish if they were longer than a few millimetres. However, no-one who has looked closely at the animals of the sea shore can have missed the sea slaters, superficially woodlouse-like animals with freshwater equivalents in the water slaters, or hog-lice. Nonetheless the most familiar of all is a group of animals that astonishingly, have acquired far more names than any other British animal: the truly terrestrial isopods, the woodlice themselves.

WOODLICE

OTHER COMMON NAMES: *armadillo; armourhead; bakers; ballyspider; beer barrel; bibble bug (Staffordshire); billy button (Dorset); bues; button crawler; carbender; carpenter; carpenter pig; carpenter's bug; carpenter's flea; cheese bug; cheeselip; cheeselipp worm; cheeseybob (Surrey); cheesy bug; cheesy log (Oxford); cherwood; chesbug; chestlock; chestworm; chewbug; chicky-pig; chiselbob (Devon – Yealmpton); chiselhead (Birmingham); chitty bob; chizzleball; chizzler; chooky pig Dorset); chucky pig (Dorset); chuggy pig; chuggyslice; church pig; cobber; coffin cutter (Ireland); crawley baker; crawley pig; cricket ball; criller; crumple ball; cudworm (Shropshire); curley button; doodlybug; earywig; fairies pigs (Lancashire); fat sow; footballer; freds; gammer louse; gammer-zow; God's pigs (Devon); gramma-sow (Wales); grammer-zow; grampian; grampus wood-bug; grampy gravy; granfather Dick; granfer (Wiltshire, Dorset); granfer greeks (Devon – Modbury); granfer grey; granfer grig; granfer jig; granny grig; granny picker; grannysmith; greyback; grumasil; guinea pig; hardback; hardyback; hobthrust; hobthrust louse; hoglouse; horace (Devon – Wembury); humbug; Johnny bug; Johnny carpenter; jovial live; killimbobs; kirklouse; kitchen ball; leatherback; little armadillo (Oxford); loafer; lobster louse; lobstrous louse; lockchester; loop; loose pig; mackintosh; monkey pea (Kent); monkey pease; monk's pease; odimadod; old sow; parson's pig (Gloucestershire); pea-bug (Kent); penny pea bug; penny sow; pennypig; pennymouse (Ireland); Peter grandad; pig; pig-louse; piggy; piggy-wiggy; pig's buttons; pill bug; pill worm; pollybaker; pollydishwasher; reudan (Hebrides); rolinto ball; Saint Anthony's buttons; Saint Anthony's pig; scabby sow bug; shell back; show laces; show maker; sink lice; slater; snot; soda-pig; sour-bug; sow; sow bug; sow lice; sow peg; sow pig; sow-beetle; spaceman; steelback; Susie pig; tailor; tank; thrush louse; thurse louse; thurslows; tiddyboar; tiddyhog; tiggy-hog; tiler's lice; tillerlouse; toebiter (East London); trilobite; tylus; wedderclock; William button; wood bug; wood flea; woodly bug; wood pig; woozy pug; yekersterse.*

This list is not exhaustive and is based on those assembled by Walter Collinge, H.G. Hurrell and Gillian Moore.[50] I can vouch for a considerable number of names still being in use; 'slater', for instance, is very widely used, especially in the north of England and Scotland. The other names can mostly be grouped as: pig names, grandfather/grandmother names, names associated in some way with wood, names alluding to the ability to roll into a ball, names that suggest an ability to bite, names that suggest smallness, and names that refer to the armoured appearance of the body. Some of these associations are obviously inaccurate.

Woodlice certainly occur very commonly in and among old and rotting wood, and masses of them may be uncovered when bark is pulled away from a dead tree trunk. They are typically woodland animals but wood is not their only food. The

40 or so species recorded in Britain eat a wide variety of plant material, and about five or six species can cause significant damage to young seedlings and other soft fleshy plant parts in gardens. By breaking down plant material, they play an important and beneficial role in the recycling process. I am not alone in believing that woodlice populations have increased significantly over the past 30 years or so but have found no accurate data to support this view.

There is no obvious reason why the names 'pig', or 'hog', should be linked with woodlice, unless because of some perceived ugliness or perhaps because they are scavengers. Suggestions that they bite people are erroneous, and 'grandfather', or 'granny', can surely only be terms of endearment. The name 'woodlouse' (or 'wood-louse') is early seventeenth century. 'Slater' is also seventeenth century, but later, and has always tended to be more common in Scotland and the north of England. 'Cudworm' stems from an old practice of feeding woodlice to cattle, supposedly to improve the chewing of the cud; perhaps someone imagined them 'scuttling around inside the rumen and encouraging the cow to regurgitate its food',[51] although it is more likely to be allied to another of the old woodlouse names, the fascinating 'pill bug'. This name is used for those species that can roll themselves into a ball. The Pill Woodlouse *(Armadillidium vulgare)* is the commonest and best known, and superficially resembles the Pill Millipede *(Glomeris marginata)*. Because of their similarity to medicinal pills, both were believed to have comparable value and people were prescribed woodlice to swallow for the treatment of a wide range of ills.

> There is a farm at Oake, near Milverton, Somerset, where the pill woodlice *Armadillidium vulgare*, are particularly large. The other interesting thing about this population is that about five per cent of them, instead of being the usual slate grey, are bright orange in colour. In the nineteen sixties, the wrinkled and ancient farmer who lived there told me that his own parents had considered these orange ones lucky because they were a cure-all for various medical problems, ranging from sore throats, whooping cough and colic, in humans, if swallowed whole and unopened. I collected a few specimens in the early nineteen eighties even though the red Devonian walls of the farm buildings had been dismantled and my little culture still breeds true to orange colour. I haven't tried them as a medicine though.[52]

The association of woodlice with damp places is not mere chance nor a matter of individual preference. They have an unusual requirement for dampness because they effectively breath through their legs. Each of the first five pairs of legs is composed of two fragile flaps, which act as gills and enable the animal to absorb oxygen from the air, provided it is damp. 'Woodlice don't just dry up in a dry atmosphere; they suffocate as well.'[53] There is a Hertfordshire expression 'As much use as a dehydrated woodlouse'.[54] They are also repelled by light. Woodlice secrete foul-tasting fluids that serve to protect them from some potential predators, but certainly not all, and they form an important food for insects, some types of spider, toads and shrews, among others.

Woodlice have an endearing method of rearing their young. The female develops a 'false floor' to her body – a brood pouch – into which she lays her eggs. These hatch after a few weeks into pale, miniature versions of their parents and soon leave the pouch to begin an independent existence.

Decapods
(Decapoda)

'Decapoda' means 'ten legs' and most members of this very important group of crustaceans do have ten genuine walking legs. However, the first pair are commonly modified to form claws and there are also antennae and other body appendages, so the overall impression is often of a creature 'having bits sticking out all over the place'. The Decapoda includes the prawns and shrimps, the lobsters and crayfish, and the crabs.

PRAWNS AND SHRIMPS
(Caridea [Natantia])

Prawns are fairly immediately recognizable, elongated animals with two pairs of long antennae, a prominent, often toothed projection (the rostrum) between the stalked eyes, and a downward-curving abdomen culminating in a tail-fan (telson). Most are marine, although many occur in the brackish water of estuaries and a few live entirely in fresh water, although the handful of British marine species also found in fresh water are able only to survive there but not to breed. The burrowing prawns (Family Callianassidae) remain out of sight in the mud of the sea bed, but most prawns are free-swimming.

There are many species, varying considerably in size; some of the larger forms are very like small lobsters and several have been caught for centuries as a very important food resource. Like most other crustaceans, female prawns retain their eggs in a brood pouch; rather disconcertingly, this is often found still attached to prawns that are served for eating

The name 'prawn' is Middle English but its origin is unknown. The Family Crangonidae comprise mainly small species with a short rostrum and a figured, often spiny shell, or carapace. They include the edible types commonly called 'shrimps', especially the important commercial species, the Brown Shrimp *(Crangon crangon)*, but members of other prawn families are also sometimes called 'shrimps'. There is no hard-and-fast rule governing the terminology; although the culinary shrimp is generally smaller than a culinary prawn, this rule does not hold zoologically, and another important commercial species, the Pink Shrimp (*Pandalus montagui*; Family Pandalidae) is in reality one of the bigger prawns. Shrimps tend to stay fairly close to the sea bed and are traditionally caught by means of a net that first agitates the sand or mud and then entraps the creatures as they swim off, an operation known as 'shrimping'.

The name 'shrimp' is Middle English and has come to be applied to many things that are small and feeble; it may originate in the old German verb *schrimpen*, 'to shrink up'.

> Alas! This is a child, a silly dwarf!
> It cannot be this weak and writhled shrimp
> Should strike such terror to his enemies.
> (*Henry VI Part 1*, Act 2, Scene 3)

Although some species of prawn and shrimp are naturally pink, many people are surprised to discover that, as with lobsters and some crabs, the bright pink colour familiar from the fishmonger's slab develops only after cooking. Both shrimps and prawns have given rise to a characteristic cuisine. Potted Shrimp, based on the Brown Shrimp *(Crangon crangon)*, is a traditional English dish and, like game and roast beef, is part of a long culinary tradition. Shelled shrimps are boiled, mixed with melted butter and flavoured with nutmeg, cayenne and mace.

The prawn has always been best grilled or fried but has, unfortunately, became devalued as that bland icon of the 1970s' restaurant trade, the Prawn Cocktail. Today, a large proportion of the prawns sold for eating, such as the King Prawn and the Tiger Prawn, are imported exotic species. However, many people share my view that the finest of all crustaceans, of whatever kind, are those caught in cold water. Ironically, the Dublin Bay Prawn (also known as 'scampi', or 'langoustine') is a lobster (see Lobsters, below).

LOBSTERS AND CRAYFISH
(Reptantia: Astacura)

Lobsters (Nephropidae)

Lobsters are rather like giant prawns, with narrow, elongated bodies, well-developed walking legs, a very thick exoskeleton and greatly enlarged claws. The name 'lobster' was once the Latin *locusta*, 'a locust' (an insect that is also armoured and jointed), but, at some stage, the 'c' became 'p', hence the Old English word *lopustre*, from which the modern name derives. By the seventeenth century 'lobster' was being used to describe a man with a red face, but it also came to mean, none too flatteringly, a British soldier. This is said to derive from the jointed armour, known as 'lobster-tail', that was worn by Parliamentarian troops in the English Civil War.

The Common Lobster *(Homarus gammarus)* occurs all around the British coast and is an impressive animal, to be handled with care; a nip from a large one is memorable. It

In the English Civil War, Parliamentarian troops wore jointed armour, called 'lobster-tail'.

lives in holes in rocks in fairly deep water, emerging to scavenge for food – and to be caught in the traditional baited traps called 'lobster pots' ('The Lobster Pot' must be one of the commonest names for a sea-side inn). Lobsters have been an important and valuable food since ancient times, although I find the high price differential between them and crabs baffling because crabs are immeasurably tastier. As with almost all other creatures of the sea, numbers have been seriously depleted by overcatching, but management of lobster fisheries is difficult because, curiously, the lobster's life cycle is still imperfectly known. Young lobsters are rarely found and they are thought to spend the first few years of their lives burrowing in sea-bed sediments. Like some other crustaceans, lobsters are long-lived; 20 years is not unusual, by which time they may be one metre (3¼ feet) in length and weigh 5 kilograms (11 pounds).

The Norway Lobster (Dublin Bay prawn; langoustine; nephrops; scampi) *(Nephrops norvegicus)* is a much smaller species, seldom more than 20 centimetres (8 inches) long, but hugely important commercially and seriously depleted through overcatching. These lobsters live in large numbers in networks of extensive, branched sea-bed tunnels, but the special association with Dublin Bay is not obvious.

There has been much argument and passion expended on the least cruel way to kill a lobster or crab. Slow boiling is patently inhumane, but killing with a knife can also be a slow death unless done expertly. Plunging it into fiercely boiling water is thought, on balance, to kill it quickest.

Spiny lobsters (Palinuridae)
Spiny lobsters are short, stout lobsters with a very spiny carapace. Unlike the Common Lobster, they do not burrow but some species are similarly prized for eating, most notably the Crawfish (langouste) *(Palinurus elephas)*, which occurs on the western coasts of the British Isles. 'Crawfish' is a nineteenth-century American name, a variant of 'crayfish'.

Squat lobsters (Galathidae)
Squat lobsters are well named. They are short, stout lobsters, but with a ridged rather than a spiny carapace. The rostrum is characteristically short and spiny and the legs are bristly. They look relatively innocuous compared with a Common Lobster but anyone who has handled one carelessly will tell you otherwise. The nip is extraordinarily fierce and tenacious.

Several species occur around the British coast and are caught commercially. Most are red-brown or red but one species, *Galathea strigosa*, is remarkably attractive when washed clear of gravel, being deep red, with longitudinal blue stripes.

Hermit crabs (Paguridae)
Hermit crabs, despite their name, are more closely related to lobsters and, once seen out of their shells, their elongated, most uncrab-like bodies reveal why. However, it is because their bodies lack a hard and protective exoskeleton that they have adopted their curious lifestyle. They tuck their abdomen into an empty gastropod shell, changing to a larger one as they grow. Each species of hermit crab tends to use the shells of a particular type of mollusc, which they then drag around with them, apparently laboriously.

Crayfish (Astacidae)
Crayfish are freshwater relatives of the lobsters. There is one native British species, the White-clawed Crayfish *(Austropotamobius pallipes)*, which occurs in clean rivers throughout the British Isles. Like its marine cousins, it has been caught and

The Squat Lobster *(Galathea strigosa)* has longitudinal blue stripes and is occasionally found under stones on the lower shore.

eaten for centuries by humans, as well as by Otters and other animals. A Warwickshire speciality was Crayfish and Bacon Savoury in which the two meats were cut into tiny pieces, fried together and poured over hot, buttered toast. Crayfish are naturally green or brown but become strikingly red when boiled.

Over the past hundred or so years, its numbers have been seriously depleted through fungal disease, and a number of North American species, immune from the disease, have been introduced and farmed.

The name 'crayfish', from which 'crawfish' is derived, is Middle English but with a confused etymology. It is derived from the Old French *crevice* and originally from the Old High German word for a crab. 'Crevice' became 'crayfish', although it has nothing, zoologically or linguistically, to do with fish.

CRABS
(Reptantia: Brachyura)

OTHER COMMON NAMES: *partan (Scotland).*
Apart from the hermit crabs (see Hermit crabs, page 51), and the two species of porcelain crab (Family Porcellanidae), which have almost circular carapaces and are considered to be related to the lobsters, all true crabs belong in this group. They live mainly on the sea bed, some hiding in holes, others burying themselves, often with great rapidity, in sand or mud. There are about 70 species of native and introduced crabs in British waters and all are marine. (Freshwater and also more or less terrestrial species, known as 'land crabs', occur in other parts of the world.) They fall into a number of distinct families but relatively few are seen by the casual observer and even fewer are important commercially.

The Common Spider Crab *(Maja squinado)* is Britain's largest species, its carapace up to 20 centimetres (8 inches) long.

They all have a flattened, sharp-edged carapace, which is fused to the exoskeleton of the underside. There is no tail-fan and the antennae are almost always short. They are a fairly distinctive group and most people know a crab when they see one. They are perhaps most famous for walking sideways, if not backwards; 'crabwise' has become a well-used word.

> ... for you yourself, sir, shall grow old as I am, if, like a crab, you could go backward.
> (*Hamlet*, Act 2, Scene 2)

The name 'crab' is Old English and may originate in an old German word meaning 'to crawl'. The scientific generic name of the common British edible crab is *Cancer* (also a sign of the zodiac) and this word occurred in Middle English, coming directly from the Latin name for a crab. It was also used for a creeping infection, such as a gangrene and, more recently, has been restricted to the diseases that we now call cancer.

Shore crabs and swimming crabs (Portunidae and related families)

Among the commonest crabs likely to be seen close inshore are those in the Family Portunidae, which includes the Green Shore Crab *(Carcinus maenas)* that must have been caught by any child searching rock pools with a bucket and net. Other common species in the same family are the fierce and very fast-moving Velvet Swimming Crab *(Necora puber)* and the reddish Harbour Crab *(Liocarcinus depurator)*, with its purple-blue-tipped hind legs.

On rocky shores, the brown Hairy Crab (*Pilumnus hirtellus*; Family Xanthidae), with its extremely hairy legs, is common, while encrustations of mussels or other bivalve molluscs may well have colonies of the little soft-shelled Pea Crabs (*Pinnotheres* species; Family Pinnotheridae) living within them.

Spider crabs (Majidae)

The most impressive crabs are the spider crabs, with their very long, slender legs and claws. Unlike real spiders, they tend to move slowly and deliberately, and although small ones can be found in rock pools and slightly larger ones dead along the tide-mark, much larger ones will be found in deep water.

The Common Spider Crab *(Maja squinado)*, which is caught commercially, is the largest British species and its pear-shaped spiny carapace may be 20 centimetres (8 inches) long. In comparison, the related Japanese species *Macrocheira kaempferi* is the largest of all crustaceans and specimens measuring over 5 metres (16½ feet) across the claws have been reported.

Edible crabs (Cancridae)

Much the most important British crab is the Edible Crab *(Cancer pagurus)*, with its distinctive black-tipped claws. It can be a large, heavy and fearsome creature. One of my most vivid

The 'pie-crust' edging to the carapace of the Edible Crab *(Cancer pagurus)* is perhaps an indication that it is good to eat!

memories is of exploring the mud-laden fossil forest just offshore at Sutton-on-Sea in Lincolnshire, which was exposed only at abnormally low tides. Among the old tree stumps, my friend and I, then aged about eight, found crabs that had to be seen to be believed. Although the maximum carapace width for this species is said to be 16 centimetres (6¼ inches), I am sure that several of the ones we found were much larger.

Enormous numbers of Edible Crabs are caught off most British coasts and many are exported to continental Europe. Few finer flavours come from the sea than cold cooked crab; if served hot, I find it loses all subtlety. However, if you must have hot crab, Isle of Skye Partan Pie is probably the way to prepare it. The meat is removed and mixed with pepper, salt, nutmeg and breadcrumbs. Butter, vinegar and mustard are heated together and beaten with the crab-meat mixture. The whole is put back in the shell and browned under a grill. A popular East Anglian dish is Cromer Crab. Crabs from the nearby Norfolk coast can be particularly weighty, but are generally too small to obtain high prices at London markets. As a result, they tend to be sold locally from roadside stalls, or even from people's front rooms. Baked Cromer Crab is traditionally prepared with breadcrumbs and grated cheese. A cooked crab, with the flesh removed and served cold, is always said to be 'dressed'. Many people baulk at dressing a crab but it is really a matter of logical method and patience (although a zoological background does help, as well as adding to the interest). The finest meat is in the claws but everything else is edible, except for the stomach and the gills (or 'dead-man's fingers').

Crabs have fascinated people since time immemorial, as evidenced by their appearance on coats of arms, the names of inns in coastal towns and villages (although strangely, less often than lobsters) and the number of uses of the word 'crab' in other contexts. Curiously, despite their desirability as food, crabs have always been thought of as inferior creatures. Whenever the adjective 'crab' is used (as in 'crab-apple'), it denotes a poor relation and the description 'crabby' simply means to be bad-tempered and disagreeable. One might expect some little fishing port somewhere to have been named after crabs but this does not seem to have happened. 'Crabwall' is in Cheshire, a long way from the sea, although it might mean 'a stream where crayfish were found'. The name 'crab' is from the Old English *crabba* and there is no Old English word for crayfish.

SPIDERS, INSECTS AND OTHER ARTHROPODS

(Arthropoda: Arachnida, Insecta and other groups)

Sea spiders

(Pycnogonida)

Among the red seaweeds or tufts of bryozoans on the sea shore, you may come across what appears to be a sea spider, with long, tangled legs. The name, however, is misleading. The pycnogonids are in reality primitive creatures and, although they have eight legs, they are quite unrelated to real spiders, nor are they apparently related to anything else. They feed on bryozoans, sea anemones and similar creatures.

Spiders and Scorpions

(Arachnida)

Arachnids are arthropods that almost invariably have four pairs of legs (a very few have fewer). They have no wings, no antennae and a body divided into, at most, two parts (unlike the three parts of insects). Only four groups of arachnids occur in Britain: the spiders, harvestmen, ticks and mites, and pseudoscorpions.

The most important and best known of the other groups, the true scorpions, are found in parts of continental Europe but have not existed in Britain as native animals since prehistoric times. Notwithstanding, the fearsome-looking but harmless *Euscorpius flavicaudis* may sometimes be found in dockyard buildings.

There are far fewer arachnids than insects, but they have made up for this in the emotions that they evoke. Spiders have graced the English language with the word 'arachnophobia', which was even used as the title and subject of a horror film (although the condition, because it is evoked only by spiders, would more accurately be termed 'aranophobia').

Spiders

(Araneae)

OTHER COMMON NAMES: *aftercrop; aitercap; arain; aran; arand; aranee; arran; arrand; arrant; arrian; arrin; attercap; attercob; attercop; attercrop; capper; cop; eathercrop; eddercop; eddercrop; eddicrop; eddircrop; eddycrop; erayne; ethercap; ettercap; ettircap; hatter; irain; meratoo; mooratoog; mooratow (moorland spider); nattercrop; nedikrop; ottercop; shepherd-spinner (long-legged spider); willie-buck (Shetland); wyver.*

Few people view spiders with equanimity. Like any expert, a committed arachnologist will find his subject utterly fascinating but, for the majority of the population, arachnids are objects of loathing and even of fear bordering on naked terror. Most people dislike rats and snakes but neither group has spawned an everyday word to describe the emotion in the same way as 'arachnophobia'. Yet this fear defies most logical explanation. Some years ago, I developed a theory based on the capacity of the human brain to analyse movement. We can see and discern the motion of up to six independent legs, which takes account of all vertebrates and insects. When the number of legs becomes really large, as in chilopods and diplopods, we see these creatures not as individual moving objects but merely as a wave motion, so that creates no problem either. However, spiders have eight legs moving independently and, the bigger they are, the more evident this becomes. Our brains cannot work this out, we do not know where each leg will go next, and so the fear of the unknown takes over.

Spiders are carnivorous but they do not spread disease and they commonly feed on and control other creatures (such as flies) that people find unpleasant. While a few exotic species have an extremely poisonous bite, most are quite harmless. With the exception of the Water Spider *(Argyroneta aquatica)*, few British spiders will even give a painful nip.

> An improbable, but true, report of spider bites at a sewage farm in Birmingham, England, in 1974, involved thousands of tiny 'money spiders'. It had never been imagined that such spiders, little more than one-tenth of an inch (2 mm) in length, could cause the slightest annoyance to man. But when maintaining the filter beds in the month of July, a number of workmen were bitten by swarms of these spiders, which dropped down their necks or crawled up their arms, causing inflammation and swelling. Subsequent investigation of the filter beds found astonishingly high numbers of the tiny arachnids living among the clinker – about 10,000 per cubic metre – vastly more than their normal density in a natural habitat.[55]

The identification and classification of spiders is based on a wide variety of often microscopic characteristics, such as the structure of the mouthparts and genitalia and the number of eyes (most spiders have either six or eight). Apart from a small number of common and distinctive species, the identification of the 600 or so British spiders is a task for the expert and, because of this, most spider lore and beliefs relate to the creatures generally rather than to individual species. Many animals, large and small, feed on spiders but people have generally chosen not to do so, other than for dubiously effective medicinal purposes (see page 59).

Spiders reproduce by laying eggs that hatch into miniatures of the adults called 'spiderlings'. They do not undergo a metamorphosis. There are spider species adapted to live in most environments but those few that have acquired a fondness for a life in our homes (often as an alternative to caves) far outweigh the remainder in terms of the attention that they attract. Most species of spider spin silk, which they use for many purposes but

A field strewn with dew-bedecked cobwebs is a wonderful sight and a far cry from the dark, webbed dungeons of horror films.

most obviously and famously to construct webs for snaring their prey (not all spiders species trap their prey; some are hunters). Different types of spider produce characteristically different web patterns and many spiders can be identified, at least to genus level, on web appearance alone; group names such as 'orb-webs' and 'funnel-webs' refer to the appearance of the web.

The web was first called a 'cobweb' in Middle English ('cob-waaf' is a West Country variant still used) and the name is closely allied to old terms for the spider itself. 'Cop' was a Middle English name for the spider, perhaps meaning 'a top' and an allusion to the rounded, top-like body shape that many possess. There is, however, an Old English word, *attercop*, originally meaning 'a poison cup', 'poison top' or 'poison head' and this came to be used for the spider in the seventeenth century. The modern word 'spider' is from the Old English *spithra*, meaning 'a spinner' but it has since been applied to other things that do not spin but have long legs, such as spider crabs and spider monkeys. A 'spider man' is someone who climbs, spider-like, to erect scaffolding or the metal framework of a building.

Cobwebs, like spiders, have long been fascinating and few will deny that they inspire many emotions. Black and dust-laden, covering some dark old room, they are sinister and chilling indeed and an artificial cobweb-making machine is part of the stock in trade of the horror-film maker. However, those same silk threads, bedecked with dew and festooning fields and hedgerows on a summer's morning, can be as uplifting a sight as any in the British landscape.

> On September 21st, 1741, being then on a visit, and intent on field-diversions, I rose before daybreak: when I came into the enclosures, I found the stubbles and clover-grounds matted all over with thick coat of cobweb, in the meshes of which a copious and heavy dew hung so plentifully that the whole face of the country seemed, as it were, covered with two or three setting-nets drawn one over another. When the dogs attempted to hunt, their eyes were so blinded and hoodwinked that they could not proceed, but were obliged to lie down and scrape the incumbrances from their faces with their forefeet, so that, finding my sport interrupted, I returned home, musing in my mind on the oddness of the occurrence ... .[56]

Finding an abundance of cobwebs in the morning is widely thought to indicate fine weather and seeing orb-web spiders sitting at the entrance to their creations is particularly favourable. Oddly, as learned a soul as Robert Hooke (1635–1703) suggested to the Royal Society in 1664 that, having looked at cobwebs, 'it is not unlikely that those great white clouds which appear all the summertime may be of the same substance'.[57]

Cobwebs have found widespread use medicinally, or at least as first-aid, and a mass of thick cobwebs applied to a wound is fairly effective at staunching the flow of blood. Unfortunately, it is not always possible to obtain sufficient cobwebs at short notice, but the evidence for their value is widespread.

> I was born in a rural part of Berkshire in 1913. At about eight years old I sliced a small piece of flesh from the top of a finger with a sliver of broken glass. My Aunt who dealt with our cuts and bruises immediately replaced it, went to our outdoor shed and returned with a dirty cobweb which she placed over the wound and bound up tightly. It healed well and quickly.[58]

> I shall desire you of more acquaintance, good Master Cobweb: if I cut my finger, I shall make bold with you.
> (*A Midsummer Night's Dream*, Act 3, Scene 1)

> He sweeps no cobwebs here, but sells 'em for cut fingers.
> (Ben Jonson, *The Staple of News*)

> Spider's webs were used after de-horning cattle. They were placed over the cut to stem the flow of blood. I have personally seen this used and it does work. It was also used when fighting cocks were de-combed prior to a contest.[59]

In the late nineteenth century, it was claimed that an effective treatment for fever had been isolated from cobwebs.

Few British spiders have unambiguous English names and any name, if it does exist, usually refers to more than one species. Even among the 600 or so British species, there is a huge range in size, from the many tiny types, collectively known as 'money spiders', some of which are barely 1 millimetre (1/16 inch) across, to the two largest species: the Great Raft Spider *(Dolomedes plantarius)* and the Cardinal Spider *(Tegeneria parietina)*, which can exceed 13 centimetres (5 inches) across the legs.

The Great Raft Spider is huge, rare and localized, while the Cardinal Spider, although also uncommon, sometimes occurs in old buildings. If you check its size with a ruler, you may understand how it gained its name: from giving apoplexy to Cardinal Wolsey (*c.*1476–1530) when they both shared a residence at Hampton Court Palace. I am told that, in the flickering, shadowy candlelight of the old rooms at Hampton Court, a Cardinal Spider scurrying across the floor at night can easily be mistaken for a mouse. In Leicester, where Wolsey died and is buried, it is said that to kill a spider will result in the ghost of the Cardinal rising from his grave.

We have not all had quite such an unnerving experience, but surely everyone, at some time or another, has found a spider in the bath.

The Raft Spider *(Dolomedes fimbriatus)* lives in or near damp places, where it feeds on bluebottles, damselflies and even small fish.

I have fought a grizzly bear,
Tracked a cobra to its lair,
Killed a crocodile who dared to cross my path;
But the thing I really dread
When I've just got out of bed
Is to find that there's a spider in the bath ...
(Michael Flanders and Donald Swann, 'The Spider')

Some people may have wondered how the spider came to be in the bath. Contrary to common belief, the answer is not up the waste pipe; most modern wastes have a water trap that provides an impassable barrier. The answer is simply that spiders, often rather big ones and commonly *Tegeneria domestica*, a close relative of the Cardinal Spider, wander around our homes at night and may fall into the bath, from whence they are unable to escape because of the slippery sides. Nonetheless, I am sure that many people share my memory of my first school rhyme and the immortal lines:

Incy Wincy Spider climbed up the spout.
Down came the rain and washed poor Incy out.
Out came the sun, and dried up all the rain,
So Incy Wincy Spider climbed the spout again.[60]

I have never discovered who named the spider Incy Wincy, nor what the name means, but when reciting the lines we usually made a tickling, groping action on a school chum, crudely imitating the actions of Incy.

A slightly similar rhyme involving spiders and tickling comes from Orkney, where Kirsty Kringlik was a long-legged species of hill spider. Schoolchildren would place it in the palm of their hands, asking it not to tickle them and reciting a list of possible foods to be served at supper. If the spider left behind a drop of water in the hand before it was given its freedom, this was a sure sign of a fine meal to come.

Kirsty, Kirsty Kringlik
Gae me nieve a tinglik
What shall yeh
For supper ha'e?
Deer, sheer, bret an' smeer
Minch-meat sma' or nane ave?
Kirsty Kringlik rin awa![61]

Although 'Incy Wincy Spider' is well enough known as a children's spider rhyme, it cannot hold a candle to 'Little Miss Muffet', arguably the most familiar nursery rhyme of them all and one which is still recited by children today, although both the children and their teachers may have little idea of who Miss Muffet was, what she sat on, or even exactly what she was eating.

***Little Miss Muffet* was painted by Sir John Everett Millais in 1884.**

Little Miss Muffet
Sat on a tuffet,
Eating her curds and whey;
There came a big spider,
Who sat down beside her
And frightened Miss Muffet away.[62]

The rhyme first appeared in 1805, although Bill Bristowe claims to have identified Miss Muffet with Patience Muffet, the daughter of the entomologist Dr Thomas Muffet (1553–1604), also known as Moffett, or Moufet, and author of the *Theater of Insects*.[63] Muffet was evidently very fond of spiders and used them to treat his daughter's ailments, although another version of the story tells of Muffet himself being attacked by wasps in Epping Forest while on a picnic and changing the facts to fit his poetry.

The logic of the connection with the Muffet family in general seems sound enough, although Iona and Peter Opie have pointed out that the time difference of over 200 years must be significant and that the rhyme does appear in other versions: 'Little Mary Ester sat upon a tester' (1812) and 'Little Miss Mopsey sat in the shopsey' (1842). It also has a noticeable similarity to 'Little Polly Flinders', 'Little Poll Parrot', 'Little Tommy Tacket', 'Little General Monk' and 'Little Jack Horner'.[64] 'Tuffet' is more easily explained; this is a sixteenth-century word for 'a low seat or mound, derived from tuft'. 'Curds' is coagulated milk; 'whey' is the watery liquid left behind. There is a famous picture entitled *Little Miss Muffet*, which was painted by Sir John Everett Millais in 1884.

It was widely thought, although as far as I can gather with no real evidence, that spiders found certain types of wood repellent and that they and their cobwebs would never be found in buildings made from these woods. At Winchester Cathedral, there was an old belief, which has some connection with St Patrick and his banishing of vermin from Ireland, that spiders would not make webs on the chapel or cloister roofs because they were made from Irish oak. Chestnut wood is said to be particularly distasteful to spiders and the roofs of New College and Christ Church in Oxford, and of Turner's Court in Tewkesbury, are said to be always free from cobwebs. (This is certainly untrue as far as New College is concerned, as I can vouch.) At Goodrich Castle in Herefordshire, built in 1828 but now demolished and a house that Wordsworth called 'an impertinent structure', it was said that Irish wood was chosen especially to keep the place free from cobwebs; by a similar token, Irish soil was brought for the cellar floors to keep them free from toads.

> O what a tangled web we weave,
> When first we practise to deceive!
> (Sir Walter Scott, 'Marmion')

Sir Walter Scott is an appropriate source for a comment on spiders because the most famous spider in British history must be the Scottish one that featured in the legend of the Scottish King Robert the Bruce (Robert de Bruce I, 1274–1329). Following his defeat by Edward I at Methven in 1306, Bruce wandered the western highlands and islands, finding shelter in a cave, where he watched a spider try seven times before it succeeded in attaching its web to the ceiling. Inspired by its perseverance, he decided to try yet again, after his own setbacks. He left his island retreat in 1307, landed at Carrick and surprised the English garrison. From then on, he gradually gained the upper hand, finally routing the English at Bannockburn in 1314. The locality of the cave and the identity of the spider are mysteries. Most accounts place the cave on the island of Rathlin, but the King's Cave on Jura is another possibility, and there are plenty of cave-inhabiting spider species in contention. The great spider expert Bill Bristowe described an experience in one cave on Arran, also reputed to be that of Bruce:

> I watched with interest the faltering footsteps of *Zygiella x-notata* on the roof of the cave mouth. This species builds her orb-webs across the windows of millions of Englishmen's windows and if this were to remind us of the discomfiture she caused our ancestors many years ago, we may console ourselves by reflecting that she now obliges us by trapping numerous flies intent on entering our homes.[65]

In Thomas Bewick's engraving (*c.*1815), Robert the Bruce is inspired by the humble spider that shares his hiding place.

The Robert the Bruce story first appeared, or at least gained a wide audience, when Sir Walter Scott wrote of it in *The Tales of a Grandfather* (1827–9).[66] However, it could all be a big mistake, and recent research at the Scottish Record Office and a rediscovered letter put the tale in a different light.

> Scott's version was thought to be the first and definitive written account of a folk tale which mysteriously does not appear in John Barbour's 14th century 'The Brus', the earliest work on the king. Scott ... may have pinched the tale from the writer Hume of Godscroft's history of the Douglas family written a full 200 years earlier, and which has been passed down to the present Duke of Hamilton. Some historians have always known that Hume claimed the incident actually happened to Sir James Douglas – Good Sir James to the Scots, Black Douglas to the English ... in this version, it is Sir James who tells Bruce that 'I spied a spider clymbing by his webb to the height of an trie and at 12 several times I perceived his web broke, and the spider fel to the ground. But the 13 tyme he attempted and clambe up the tree'.[67]

For many years it was thought that all spiders were, to some degree, poisonous. The most celebrated reference to spiders in a courtroom was on exactly this theme. It occurred during the investigation into the murder in the Tower, by poisoning, of Sir Thomas Overbury (1581–1613) at the hands of agents working for the Countess of Essex. One of the witnesses claimed that the Countess had asked him to obtain the strongest possible poison and that, accordingly, he had brought her seven large spiders. A more recent and charming courtroom reference to spiders was at Norwich in 1948, when His Honour Judge David Carey Evans (1899–1982) ruled that the presence of spiders on the bedroom ceiling provided no excuse for a woman who walked out of a hotel without paying her bill.[68]

Shakespeare appeared uncertain whether spiders were poisonous:

> There may be in the cup
> A spider steept, and one may drink, depart,
> And yet partake no venom; for his knowledge
> Is not infected: but if one present
> Th'abhorr'd ingredient to his eye, make known
> How he hath drunk, he cracks his gorge, his sides,
> With violent hefts.
> (*The Winter's Tale*, Act 2, Scene 1)

But he very evidently didn't like them:

> Weaving spiders, come not here;
> Hence, you long-legg'd spinners, hence!
> (*A Midsummer Night's Dream*, Act 2, Scene 2)

Their supposed poisonous nature has not prevented spiders from falling into the armoury of folk medicine. They are especially thought to be have been beneficial in warding off the effects of ague, or malaria (see Mosquitoes, page 90), the treatment being applied in various ways. The old folk tradition of hanging them around the neck was particularly popular and was referred to by Elias Ashmole in his *Diary*:

> I took early this morning a good dose of elixir, and hung three spiders round my neck, and they drove my ague away.[69]

Rolling them in butter or treacle before swallowing them was recommended in some areas and was also believed to be effective in the treatment of rheumatism.

> As a teenager in the Second World War, I was told that the English garden spider was contributing to the war effort. The story went as follows. The English garden spider spun its web from a cluster of four spinnerets equally placed circularly. Furthermore, the thread is spun straight, whereas other (foreign) spiders put a helix or twist on the thread. This 'spin-straight' characteristic allowed British optical instrument makers to split the thread, lengthways, into four pieces. The split-down thread was then used for graticules on such devices as submarine periscopes and telescopic gunsights.[70]

An intriguing paradox is that, while spiders are so widely disliked and feared, it is equally widely believed that to kill one will bring bad luck. 'If you want to live and thrive, let the spider run alive' is a very old rhyme, found all over England. It is usually said to originate with an ancient Christian belief that a spider spun a web either across Christ's manger or to conceal the Holy Family in a cave and so protect them from danger, although protection by spiders is a custom in non-Christian religions too. Another old North Country English rhyme on a similar theme is:

> Kill a spider and bad luck yours will be,
> Until of flies you've swatted fifty three.[71]

Frank Gibson told of an experience in Yorkshire where it was believed that to kill a spider would result in rain.[72]

The tiny money spiders are conferred with special virtues: not only is it bad luck to kill them, but they will bring good

The poet Sir Thomas Overbury (1581–1613) was allegedly poisoned in the Tower of London by the venom of seven spiders.

luck and good fortune. A money spider running over your clothes is believed to have come to spin you some new ones, or at least to ensure that you have the money to obtain some.

> Money spiders were to be spinned three times around your head to bring you riches.[73]

> As a child in the 20s I always remember my mother saying 'If you wish to live and thrive, let the spider run alive'. She would never kill one and a 'money-spinner' spider had to waved round the head three times for luck – not that it seemed to have much effect. Her family came from Suffolk so it could be a Suffolk tradition.[74]

In many areas of England it was deemed fortunate for a bride or groom to see a spider on their way to church.

> 'Will you walk into my parlour?' said the spider to the fly;
> 'Tis the prettiest little parlour that ever you did spy.
> The way into my parlour is up a winding stair,
> And I have many curious things to show when you are there.'
> 'Oh no, no,' said the little fly, 'to ask me is in vain,
> For who goes up your winding stair can ne'er come down again'.
> (Mary Howitt, 'The Spider and the Fly')

The obligatory association between spiders and flies is real enough and, as children, we used to pose the riddle: 'Why did the fly fly? Because the spider spied 'er'. There is not a great deal of biological truth in this explanation. Spiders are an ancient group, the earliest fossils dating from the famous 300-million-year-old beds of Devonian chert at Rhynie in Aberdeenshire. There were almost no flying insects at that time and, if insects did develop the powers of flight in response to being chased by spiders, they took an inordinately long time to do it; flies certainly did not evolve until 100 million years later in the Jurassic.

Mites and Ticks
(Acarina)

The acarines are among the smallest of all arthropods, so small that the word 'mite' has entered the language as meaning anything extremely tiny. The creatures themselves are often barely visible without a magnifying lens so it is surprising to find that the word is as old as Middle English, when such lenses were hardly commonplace. Originally it meant any minute creature and was used indiscriminately for many small insects. 'Tick' is possibly Old English, certainly Middle English, but its origin is unknown. Most species are more or less globular and some have fewer than the regulation arachnid complement of eight legs. The main difference between the two groups is that mites are variously scavenging, carnivorous, parasitic or vegetarian while ticks are all blood-sucking.

Mites impinge on our lives in numerous ways. Gardeners will be familiar with such species as the Red Spider Mites (*Tetranychus* and *Panonychus* species), which feed on plants and can cause serious damage; affected plants have a bronze appearance and are sometimes festooned with webbing. Gall mites (Family Eriophyidae) are responsible for abnormal swellings on plants, such as the familiar big-bud symptoms on blackcurrants caused by *Cecidophyopsis ribis*. Control of mites in crops is difficult because they are not affected by most insecticides. Growers often rely on maintaining a moist atmosphere around the plants, which plant-attacking mites find unappealing, although there are other mite species that live in water. Gardeners and growers are turning to biological control of mites, sometimes using non-parasitic mite species as the control agent. The Glasshouse Red Spider Mite (*Tetranychus urticae*), for instance, is controlled with the South American predatory mite *Phytoseiulus persimilis*. Other species make themselves apparent by causing skin rashes on people who handle plant matter or walk through dry grass in summer.

The house-dust mites (Family Pyroglyphididae) have become familiar because they are among the commonest causes of household-dust allergies. The Flour Mite (*Acarus siro*) was once among the most serious pests of foodstuffs. The Cheese Mite (*Tyrophagus casei*) produces tiny holes in cheese, which appears to be covered with a grey mould, formed from the mites and their cast skins; it caused much damage before the present strict food hygiene regulations came into force. The only significant human pest among mites in Britain today is the appropriately named Itch Mite (*Sarcoptes scabei*), which lives in the skin and causes an eczema-like inflammation.

The Red Spider Mite (*Tetranychus* and *Panonychus* species) is a common pest on the leaves of house and garden plants.

The false scorpions (Pseudoscorpiones) are tiny and seldom seen. They hold out their pincers as they walk, like the true scorpions.

Mites reproduce by laying eggs which hatch into larvae. They then pass through further immature stages before finally metamorphosing into adults. Many species are parthenogenetic, males being rare or absent.

The blood-sucking ticks swell to many times their normal size when gorged with their host's blood and some may carry diseases; relapsing fever in man, for instance, is caused by an organism transmitted by the tick *Ornithodorus moubata*. However, in Britain today, most people encounter ticks only when the household cat or dog picks them up after walking through grass in summer. The traditional remedy is still the best: smear the ticks with petroleum jelly so that they are starved of air and die, and then pull them out.

Harvestmen
(Opiliones)

OTHER COMMON NAMES: *shepherd spiders.*
Harvestmen are commonly confused with those other extremely long-legged creatures that find their way into our homes, the Crane-flies (see page 90). Close examination, however, will soon reveal that they have no wings and eight legs, not six. They differ from spiders in having a globular, undivided body. Also unlike spiders, they spin no webs and cannot produce venom, although they are carnivorous, feeding on a wide range of small creatures, alive or dead.

There are about 20 British species, all of which are completely harmless to humans and cause no damage to our products. Nonetheless, their long legs inevitably make them unpopular. The name 'harvestmen' betrays the fact that they are particularly common in late summer; they have no special association with the harvest.

Pseudoscorpions
(Pseudoscorpiones)

OTHER COMMON NAMES: *false scorpions.*
I have always found pseudoscorpions utterly fascinating, partly because very few other people do and partly because they look rather like real scorpions, although are very much smaller. There are about 25 British species and they can be found, if you look carefully enough, among leaf litter and compost, although some are common in birds' nests. They feed on mites, springtails and other comparably small arthropods, using their claw-like pedipalps that both hold the prey and secrete venom.

Insects
(Insecta)

Insects are arthropods that, at some time in their lives, have six legs. There is little more that can be generally stated about arguably the most successful group of animals. If numbers are any criterion, insects would win every time. They comprise about 80 per cent of all animal species and there are over 20,000 known kinds in Britain, with certainly a great many more yet to be identified or discovered.

In terms of numbers of individuals, insects also defy superlatives; for instance, there may be 50,000 Honey Bees in a single colony. However, because many are small, inconspicuous and apparently innocuous, insect beliefs and traditions do not match those of any of the vertebrates. Nonetheless, those species that have made themselves known or noticed, either for the good or ill of mankind, have certainly attracted some extraordinary lore.

WINGLESS INSECTS
(Apterygota)

Most insects are wingless in one or more phases of their life cycles and a good number, such as lice and fleas, are always apparently wingless, although close examination will reveal that wings are present but vestigial. The Apterygota, by contrast, have never had wings and, in most respects, are thought to be evolutionarily primitive. The young do not undergo metamorphosis but resemble their parents. Many are very small and secretive and, consequently, are usually little known by the general population, the main exceptions being the bristletails and springtails.

Bristletails
(Thysanura)

The bristletails include some of the largest species of wingless insects. Most houses have at some time provided a refuge for the Silver-fish *(Lepisma saccharina)*, which is most commonly seen scurrying into a crevice in a kitchen cupboard or emerging at night to feed on fragments of carbohydrate. In fireplaces and other warm spots, its relative the Firebrat *(Thermobia domestica)* may be seen, although it seems less fond of central heating than it was of old-fashioned hearths, where cooking was done on the spit.

Springtails
(Collembola)

Springtails are tiny, more or less cylindrical, wingless insects that are most often seen on the surface of the compost of pot plants, or in greenhouses where they feed on plant leaves and in large numbers can cause damage. Their name refers to their ability to catapult themselves into the air by means of a spring-like appendage at the rear.

WINGED INSECTS
(Pterygota)

Mayflies
(Ephemeroptera)

Mayflies (Ephemeroptera) cannot fold their wings over their bodies and, like this *Chloeon dipterum*, have elongated tails.

OTHER COMMON NAMES: *dayflies; upwing flies.*
The name 'Ephemeroptera' says it all because these are the ephemeral insects *par excellence*. The adults of many species live for less than 24 hours, and May is typically the month when they appear, although some species can be found throughout much of the rest of the year. They are graceful but feeble creatures, much beloved by birds, bats and fish – and by trout-anglers, who go to extraordinary and skilful lengths to imitate them with artificial creations of feathers and silk. They have one or two pairs of beautiful and fragile wings, which they are unable to fold over their bodies (hence the name 'upwing flies') and two or three, characteristically elongated tails. The immature wingless nymphs live in water, the clearer and cleaner the better, where they feed on algae and other plant life. The nymphs typically live for a year; some species live longer and others not as long.

Mayflies are so important to trout-anglers that a complete alternative vocabulary has built up around them. Different species and different stages of the life cycle have acquired their own names over the past couple of hundred years as entomology and fly fishing have taken parallel courses. The 'nymph' is a 'nymph' to both entomologist and angler. The creature that emerges from the final nymph stage is entomologically a 'sub-imago' but a 'dun' to an angler (they tend to be dull brown in colour). The mature insects, which appear after the sub-imago has moulted (mayflies are unique among insects in moulting in the winged state), is entomologically an 'imagine' but a 'spinner' to an angler. After a brief but hectic life in which there is time only to mate in the air and then drop eggs into the water, but not to feed, the imagine/spinner becomes, for the angler, a sad, floating, 'spent gnat' – and food for something bigger.

Therefore, while each species has its scientific name, it has two or more angling names. To confuse matters further, in many cases, individual life-cycle stages may be imitated by more than one artificial fly. Commonly, different species in the same genus are given the same name by anglers and imitated by the same artificial fly; in addition, certain artificial flies with the same name can be tied in various different ways or patterns. For example, *Ephemerella* species (Family Ephemerellidae) are called 'blue-winged olives', so their dun is a 'blue-winged olive dun' and their spinner is a 'blue-winged olive spinner'. Similarly, *Rhithrogena* species (Family Heptageniidae) are called 'March browns', with a 'March brown dun' and a 'March brown spinner'. Both of the large mayflies *Ephemera danica* and *Ephemera vulgata* (Family Ephemeridae) are called 'greendrake', and so their dun is called a 'greendrake dun', but the spinners are always 'spent gnats'. Among other well-known mayflies, the tiny *Caenis* species (Family Caenidae) are known in all their stages as 'the angler's curse'. If the existence of a mayfly called a March brown is not confusing enough, however, there are also 'late March browns', 'August duns' and even 'summer mayflies'.

Dragonflies and Damselflies
(Odonata)

OTHER COMMON NAMES: *adderbolt (East Anglia); balance fly; boult; bullstang (Cumberland, Yorkshire); devil's darning needle; devil's needle; devil's riding horse; eather; feeder spear; fleeing eather (Northumberland); hawk; heather-bill; hobby-horse; horse-adder; horse-long-cripple; horse-stanger (Worcestershire); horse-stinger; hoss stinger (Wiltshire, Dorset); horse-teng (Yorkshire); peacock (Lancashire); snake doctor; snake's stang (Isle of Wight); stinger; tang; tanging nadder; yedward.*

Although there are barely 40 British species, the dragonflies include some of the largest and most spectacular of native insects and, significantly, they are mainly day-flying. They are hard to ignore and, after butterflies, have probably attracted more interest among the general public than any other insect group. They are readily identified, with their two pairs of rigidly outstretched wings, enormous compound eyes, often brilliantly iridescent colours and generally powerful flight that includes an ability to hover.

The immature nymphs live in water for up to five years, aggressively feeding on other aquatic life as large as small fish and fish fry. In turn, they provide food for some fish and ducks. However, unlike some aquatic insects with active immature stages, such as mayflies, the adults also hunt and catch other insects, mainly flies, hawking alongside and over water, often with clearly defined territories. The adults become sexually mature, mate, produce eggs and usually live for up to about ten weeks.

Dragonflies have a long and distinguished history, whose understanding and appreciation is helped by the fact that, being so large and robust, they have formed excellent fossils. They were certainly well developed by the later Carboniferous period, over 200 million years ago, and truly enormous species, with wing-spans of over 60 centimetres (24 inches), flew among the plants of the coal swamps. They may have been the largest insects that have ever existed.

There are two main groups: the true dragonflies (which used to be called 'horse-stingers') and the smaller, more slender damsel flies (which used to be called 'devil's darning needles'). The name 'dragonfly' (or 'dragon fly', or 'dragon-fly'), derived from their aggressive appearance, has been in general use since the early seventeenth century. 'Damselfly' (and, once, 'lady fly') has been used for a similar time and is derived from the French *demoiselle*, indicative of a gentler, more delicate body. Most other common names allude in some way to the fear and misunderstanding attached to the creatures.

> In 1995, when talking to a robust, elderly farmer from the Somerset Levels, he told me that he always felt frightened if a big dragonfly flew into the cab of his tractor; he said he would either kill the insect or stop the tractor so that it could fly off. Then, in August 1995, at Westhay Reserve, Somerset, I saw a boy aged about eight years old walking towards a perched male Southern Hawker *(Aeshna cyanea)*. Suddenly the boy was shouted at by his father and told to come away because 'it will sting you if you go near'.[75]

The striking bright blue colour of the Emperor Dragonfly *(Anax imperator)* makes it one of our most spectacular native insects.

The Southern Hawker *(Aeshna cyanea)*, one of the commonest dragonflies, is often seen hawking along hedgerows and in woodland glades.

They do not sting horses, nor anything else, although there was a widespread belief that they were guardians of fish and would sting all 'improper persons' who threatened them. The name 'snake's stang' implied something as fearsome as an Adder. A rhyme from the Isle of Wight was said to have been sung by children when a dragonfly appeared:

> Snake's-stang, Snake's-stang, vlee all about the brooks,
> And sting all the bad bwoys that vor the fishes looks;
> But let all the good bwoys ketch all the fish they can,
> And car'em away hooam to fry 'em in the pan.
> Bren butter they sholl yet at supper wi' their fish
> But all the little bad bwoys shall onny lick the dish.[76]

The association with Adders is found time and again. In some areas the presence of a dragonfly was thought to indicate that an Adder was nearby (improbable as the creatures frequent different habitats), while the elongated shape may also have been considered snake-like. The name 'yedward' commemorates King Edward I, 'the Hammer of the Scots', but I have been unable to determine whether this was in praise or condemnation. However, it is another, more recent king who is commemorated in the only species of dragonfly to have acquired a genuine English name: *Calopteryx virgo* is sometimes called 'King George': 'Presumably a flattering but delightfully unsuitable allusion to George III.'[77]

Other dragonflies have been given peculiar, contrived names that were concocted from Latin with the best of intentions, largely by Cynthia Longfield in the 1930s. These include the Common Ischnura for *Ischnura elegans* and the Variable Coenagrion for *Coenagrion pulchellum*. However, the common names of dragonflies are frankly a mess because the author of the current standard work on British dragonflies, Cyril Hammond, discarded most of Longfield's names and invented his own. The Common Ischnura thus became the Blue-tailed Damselfly and the Variable Coenagrion became The Variable Damselfly. Even 'King George' was replaced in favour of 'The Beautiful Demoiselle'.[78]

Dragonflies have featured extensively in the literature, art and lore of the Far East, where many kinds are common. However, in that part of the world, it is red species that attract attention; when colours are referred to in English literature, it is blue that seems to take the eye.

> Deep in the sun-searched growths the dragonfly
> Hangs like a blue thread loosened from the sky ...
> (Dante Gabriel Rossetti, *The House of Life*, 'The Choice')

> ... a dragonfly
> Shot by me like a flash of purple fire.
> (Alfred, Lord Tennyson, 'Lover's Tale')

> When o'er the vale of Balbec winging
> Slowly, she sees a child at play,
> Among the sunny wild flowers singing,
> As rosy and as wild as they;
> Chasing with eager hands and eyes
> The beautiful blue damsel flies,
> That fluttered round the jasmine stems,
> Like winged flowers or flying gems.
> (Thomas Moore, 'Paradise and the Peri')

Most species of dragonfly have declined significantly in Britain over the past century, very largely due to the loss of wetland habitats by drainage. In the third quarter of the twentieth century, three species are thought to have become extinct in Britain, although, conversely, at least one new breeding species has been recorded and sometimes vagrant species make their presence known in the most improbable of ways. The seventh British record of the rare vagrant North African dragonfly *Hemianax ephippiger* was made in the New Forest, Hampshire, in 1984, when one was hit by a car and became stuck behind the windscreen wipers.

Stoneflies
(Plecoptera)

Rather like mayflies, stoneflies are creatures familiar only to entomologists and anglers. The nymphs are aquatic, somewhat like earwigs in appearance, and most live for up to a year before crawling onto land to mature into adults superficially similar to stumpy mayflies. The adults of some species have two long, mayfly-like tails and, like mayflies, are unadventurous fliers.

> When they fly, which is neither often nor far, they exhibit a characteristic lack of skill, travelling mostly in a straight line until they land or collide with something.[79]

Stoneflies take their name from the nymphs' habit of clinging tenaciously to stones on the stream bed, where they provide food for many types of fish. Anglers call them 'creepers' and sometimes use them as bait, while the adults are imitated, like mayflies, by artificial flies.

Crickets and Grasshoppers
(Orthoptera [Saltatoria])

The orthopterans are generally large, very obvious, sometimes noisy, sometimes brightly coloured and apparently armour-plated insects with big heads and big legs. They are not easily missed, because of their appearance and their sound, and they have inspired and sometimes alarmed people for centuries. Sadly, the British orthopteran fauna is poor compared with that of warmer places; when you sit in a field at midsummer, watching the apparently spring-loaded grasshoppers flit from plant to plant, the trilling of the crickets may seem impressive but it is is a poor echo of the choruses that other countries enjoy. About three families of crickets and two families of grasshoppers, perhaps 30 species in total, are all that Britain can muster, and many are mainly southern in distribution.

They are among the few insects regularly kept as pets and, in parts of Europe, they are still commonly seen in small cages, singing their wings, if not their hearts, out. The characteristic sound is produced by rubbing one part of the body over another; grasshoppers tend to rub a wing on a leg and crickets a wing on another wing. Grasshoppers have short antennae while crickets have long ones.

CRICKETS

OTHER COMMON NAMES: *bruck (field cricket); charker; cheiper; crackel; grig (Cornwall); hama; knid (West Country).*

The 'Cricket on the Hearth' that inspired Charles Dickens's story of the same name was a House Cricket (*Acheta domesticus*; Family Gryllidae). This is an introduced Asian and African insect that has become well established in bakeries, kitchens and other warm places, including old-style hearths, where it feeds on whatever domestic scraps may come its way. There is a fairly credible belief that they were brought to Britain by the Crusaders. (A similar claim made for the Mute Swan, see page 216, is almost certainly untrue.) Although crickets are still found in warm commercial premises, they are uncommon in

Crusaders may have introduced the House Cricket *(Acheta domestica)*, which originates from Asia and Africa, to Britain.

modern homes but to have one chirping in the hearth was once widely thought to bring good luck and to bring riches to the household. As Dickens's heroine, Mrs Peerybingle, said:

> It's sure to bring us good fortune, John! It always has been so. To have a Cricket on the hearth is the luckiest thing in the world.[80]

Conversely, if the crickets left, it was regarded as a very bad sign and, if they suddenly reappeared in a house where they had been previously seen, the possibility of an imminent death was to be considered. Enid Porter told of a Cambridgeshire man, as recently as 1960, saying:

> We used to listen to the crickets every evening sixty years ago, and then one evening we didn't hear them. My father said, 'That means George has passed away'. George was my uncle; he was living in Leicestershire at the time and we knew he was very ill. Sure enough, a day or so later a letter came from my aunt telling us of Uncle's death.[81]

Gilbert White called them 'the housewife's barometer', foretelling not only good or ill luck but also the coming of rain. Today, most people simply shudder at what is now a fairly unusual sight.

> Where glowing embers through the room
> Teach light to counterfeit a gloom,
> Far from all resort of mirth,
> Save the cricket on the hearth.
> (John Milton, 'L'Allegro')

The outdoor crickets are close relatives of the House Cricket and are similarly rather unattractive, scurrying, dark brown creatures. However, the fact that they are rarely seen, only heard, has given them a special place in people's affection. They entranced John Clare:

> Sweet little minstrel of the sunny summer,
> Housed in the pleasant swells that front the sun.
> Neighbour to many a happy yearly comer,
> For joys glad tidings when the winters done.
> How doth thy music through the silk grass run
> That cloaths the pleasant banks with herbage new.
> A chittering sound of healthy happiness
> That bids the passer bye be happy too.
> Who hearing thee feels full of pleasant moods,
> Picturing the cheerfulness that summers dress,
> Brings to the eye with all her leaves and grass,
> In freshness beautified and summers sounds
> Brings to the ear in one continued flood
> The luxury of joy that knows no bounds.
> I often pause to seek thee when I pass
> Thy cottage in the sweet refreshing hue

Of sunny flowers and rich luxuriant grass.
But thou wert ever hidden from the view,
Brooding and piping o'er thy rural song
In all the happiness of solitude.
Busy intruders do thy music wrong
And scare thy gladness dumb – where they intrude.
I've seen thy dwelling by the scythe laid bare,
And thee in russet garb from bent to bent,
Moping without a song in silence there.
Till grass should bring anew thy home content,
And leave thee to thyself to sing and wear
The summer through without another care.
(John Clare, 'Field Cricket')

The largest of the outdoor species, the Field Cricket (*Gryllus campestris*; Family Gryllidae) is one of only a dozen or so British insects to have specific legal protection under the Wildlife and Countryside Act 1981. Another is the Mole Cricket (fen cricket) (*Gryllotalpa gryllotalpa*; Family Gryllotalpidae), a large, burrowing subterranean creature, virtually confined to the New Forest. Yet in other, warmer parts of the world, it is a very serious crop pest and, when I have told foreign colleagues that Britain affords it special protection, I have been greeted with incredulity. These two protected species are the only British crickets or grasshoppers that hibernate.

The little bush crickets (Family Tettigoniidae) are much prettier creatures than their big, brown relatives and, apart from their enormously long antennae, which are far longer than their bodies, they are often referred to as 'grasshoppers', which they are popularly thought to be.

The name 'cricket' is Middle English, from an Old French name for the cicada and originally imitative of its sound. 'Cricket', as in the game, has no connection and nor do such places as Cricket St Thomas and Cricket Malherbie in Somerset, where 'cricket' is a diminutive derived from the Old English *crūc*, meaning 'a hill'.

GRASSHOPPERS

OTHER COMMON NAMES: *grigam; griggan; grypsope; skipjack.*

Six or seven grass-hoppers pulverized, with as many grains of pepper, are good against the Cholic, being taken inwardly. The ashes of them taken in a proper vehicle, waste the stone, provoke urine, and expel the gravel. Grass-hoppers reduced to powder, and given to the quantity of one scruple, in white wine, are good to cure a diarrhoea, or looseness.[82]

Thus wrote the predictably outrageous John Keogh.

Grasshoppers (Family Acrididae) have been called 'grasshoppers' since the Middle Ages and the name has since been applied to almost anything that moves around quickly outdoors; there are countless 'Grasshoppers' sports teams. The insects, however, not only hop from one grass stalk to another, they also eat them, although fortunately not in such quantity, nor in such numbers, as their warm-climate relatives, the locusts. Their sound and the misleading impression it can give inspired Edmund Burke to use them as a convenient euphemism in his *A Letter to a Member of the National Assembly*:

Because half a dozen grasshoppers under a fern make the field ring with their importunate chink ... pray do not imagine that those who make the noise are the only inhabitants of the field; that, of course, they are many in number; or that, after all, they are other than the little, shrivelled, meagre, hopping, though loud and troublesome insects of the hour.[83]

An insect of dry grassland, the Common Field Grasshopper *(Chorthippus brunneus)* is one of many similar species.

Earwigs
(Dermaptera)

OTHER COMMON NAMES: *alliwig; arrawig; arrawiggle; arrywiggle; arrywinkle; arwgyll; arywiggle; battle-twig; clepshires; clipshears; coach/ coch-bell/ bill; cock-tail; codge-bell; cody-bell; coffin-cutter; devil's coachman; earvrig; earwag; earwiggle; earwike; earwrig (Somerset); earywig; ermit; errewig; erriwig; erriwiggle; firkin/forkin-robin; forkit; forkit-tail; forky-tail; furkin; furkin-robin; gailick; gallacher; gavelock; gelick; gelloch; gewlick; golach; golack; gollach; goulock; gowlick; gowlock; harrywig; horn-golach/ gollooch; narrow-wriggle; pincher-wig; reox; scotch-bell; touch-bell or spale; twitch-bell; twinge (northern England); yarwig; yerriwig; yerwig.*

The Common Earwig *(Forficula auricularia)*, here seen on a daisy, can be found high on trees and shrubs during summer.

Few things are more surprising to young naturalists than the first time they see an earwig fly, although earwigs seldom do so, perhaps because of the difficulty of unfolding their wings, which are pleated, concertina-fashion, about 40 times. They are somewhat similar to rove beetles (see Rove beetles, page 105), with elongated but shining brown bodies, and are among the most familiar of British insects. The multitude of colloquial names is a testimony to this, although, in reality, there are only two common species and a couple more rare ones in Britain. The commonest is *Forficula auricularia* (Family Forficulidae), whose scientific name, *auricularia*, reflects the common association of earwigs with ears, although why this should be is an almost complete mystery.

There is no doubt that the connection is an old one, and certainly dates back to the Old English name *éarwicga*; the name 'worm' and the verb 'to wiggle' or 'to wriggle' come from the same root. Certainly, if you lie down in a grassy field, there is a chance that an earwig may enter your ear (I know someone with personal experience of this), but I would have thought an ant or a woodlouse would be the more likely invader. However, there is a Cornish story of a man who was driven to distraction by an odd sensation inside his head. Eventually, when he could tolerate it no more, he went to his butcher to request that his ear be chopped off. As he laid his head on the block, out crawled an earwig and the pain disappeared. In Yorkshire, it was thought that an earwig, if it entered your ear, would eat its way into your brain, with inevitably fatal results.

There is no denying that the earwig's association with ears is reflected in the French name *perce-oreille* ('ear-piercer') and the German *Ohrwurm* ('ear worm'), although there is no link in most other European languages. One explanation is that the earwig's wings, when unfolded, are more or less ear-shaped, which belies the fact that hardly anyone ever sees an earwig's wings. I prefer the explanation that, in the centuries when we slept on beds of straw rather than on sprung mattresses, earwigs were much more likely to crawl into the ear. Indeed, an old means of trapping earwigs entails using upturned plant pots (hollow hooves were used in times past) that were filled with straw and placed on top of canes or sticks among crop plants. Being nocturnal, the insects retreat there during the daytime, when they can be collected and disposed of.

There was once a somewhat more unexpected anatomical association in the north of England, where the earwig was known as a 'twich-ballock', or:

> ... as if you would call it Scrotomordium ... for where ever it findes a rivled pleated skin, it will cause very great pain, either by biting with the mouth, or by winding about it with its forked tail.[84]

Most people today, however, are more likely to find earwigs nibbling their dahlias and chrysanthemums than their ears or scrotums. They cause annoyance and modest damage by taking pieces from the flowers, but straw traps work well.

Earwigs are among the few insects outside the colonial species, such as the ants and bees, that exhibit parental care, and the female, who can be recognized by her straight, not curved rear pincers, tends her eggs throughout the winter in the soil.

Cockroaches
(Dictyoptera)

Cockroaches are neither attractive nor appealing. I have a guide to hotels in one less well-developed country that grades accommodation not by stars but by numbers of cockroaches: a five-cockroach establishment is one to avoid. Cockroaches are omnivorous, often rather large, superficially beetle-like creatures, with shiny brown bodies, long legs and an ability to move very fast. They were once grouped with crickets and grasshoppers but lack the abnormally long jumping legs. They also tend to fly fairly infrequently, although most have perfectly good wings. If you look closely enough, you will note that, unlike beetles, their wing cases (elytra) overlap.

Because they are attracted to food scraps and domestic waste, cockroaches are associated with dirty, unhealthy places. Nonetheless, as anyone who has kept them will know, they are very clean creatures and make rather interesting pets, although *en masse* they have an unpleasant and very characteristic smell.

The three small cockroach species native to Britain are restricted to the south; the most familiar species are all introduced from warmer climates.

The commonest, the Common Cockroach (black beetle; black worm; Oriental cockroach) *(Blatta orientalis)*, is the most beetle-like and reached Britain in the sixteenth century, almost certainly by ship although perhaps, despite its name, not from the Far East. North Africa seems the more likely source, although its true origin is unknown, but it now occurs widely in houses and other buildings in all except the northern and western parts of Britain. Unlike the other common domestic species in Britain, it is flightless.

The German Cockroach (croton bug; shiner; steamfly) *(Blattella germanica)* is a smaller, pale yellow-brown insect that can fly but seems reluctant to do so. Its name is also misleading. It certainly did not originate in Germany and probably followed the Common Cockroach from North Africa some time later.

The largest species in Britain, the American Cockroach *(Periplaneta americana)*, is also probably North African. This shiny creature, nearly 4 centimetres (1½ inches) long, is familiar through being the standard insect in school biology dissection classes. It is common on ships but less frequent in houses. It has been suggested that this species was picked up by Sir Francis Drake's ship *Golden Hind*, which returned to England with a cargo of treasure and invertebrates in 1580. In 1578, Drake had captured a Spanish galleon called the *San Felipe*, loaded with spices, and found the ship over-run with 'a wonderful company of winged Moths, but somewhat bigger than ours and of a more swarthy color'.[85]

The American Cockroach *(Periplaneta americana)* may have been brought to Britain on Sir Francis Drake's *Golden Hind.*

Until the seventeenth century, cockroaches, like many another insect, were called 'moths'. The Common Cockroach was the stinking moth that 'doth not only annoy those that stand near it, but offends all the place thereabouts with its filthy favour'.[86] The name 'cockroach' dates from 1624; which explains why it never appeared in any works of Shakespeare. It originated in the Spanish *cucarachas*, literally 'an objectionable little caterpillar', and is said to have entered English in its present form through the adventurer Captain John Smith (long-time friend of Pocahontas), who, having misheard the Spanish pronunciation, wrote of it as:

> ... a certaine India Bug, called by the Spaniards a
> cacarootch, the which creeping into Chests they eat
> and defile with their ill scented dung.[87]

Closely related to cockroaches are perhaps the most magnificent and fascinating of all insects, the aggressively carnivorous praying mantids. A few species of these warm climate insects reach southern Europe but sadly, not Britain.

Termites
(Isoptera)

These notorious timber-eating insects, sometimes erroneously described as ants, are restricted to warm climates, especially the tropics. However, some notoriety in the press and on television was achieved by the colonies of *Reticulitermes* that invaded and became established in two properties at Saunton in North Devon in 1994. Their origin was uncertain but it was speculated that they had arrived with plants from the Canary Islands. The official attitude was that there was no need for concern and the colonies would be exterminated, but much local anxiety was expressed about their possible spread to other comparably mild areas, especially with the effects of climate change.

Lice
(Mallophaga, Anoplura and Psocoptera)

Apart perhaps from 'rat', few animal names today can be so guaranteed to produce a shudder as 'louse', or its plural 'lice'. These animals are the archetypal parasites: minute, wingless, scarcely visible, but taking their living from the flesh and blood of other creatures, including people. Perhaps this reaction arises because lice infestations of humans are now less common in Britain (although still far more numerous than many imagine) and are associated with unhygienic living conditions. Perhaps it is because an older generation can still remember men returning from the trenches of the First World War, where encounters with lice were an everyday experience. Or perhaps it is because of an ingrained knowledge that, historically, lice have been important carriers of human disease and misery (see Black Rat, page 479). There is no doubt that being described by someone as 'a louse' is not easily dismissed. The name itself is Old English of Teutonic origin; in modern usage, the derived adjective 'lousy' has acquired a much more diluted meaning than its old, literal connotation.

> Your majesty hear now, saving your majesty's
> manhood, what an arrant, rascally, beggarly, lousy
> knave it is ...
> (*Henry V*, Act 4, Scene 8)

There are two main groups of animal-infesting lice: the biting lice, or bird lice (Mallophaga) and the sucking lice (Anoplura). A third group, the book lice (Psocoptera), live naturally on bark, algae, fungi and in similar places, but can cause problems in libraries and on wallpaper, where they eat, not the paper itself, but the mould growing on it.

Of the two main groups, the biting lice occur mainly on birds, where they chew feathers and scrape the skin with their jaws. Some species are important pests of poultry, a few affect hoofed mammals, but none attacks people.

The sucking lice are all blood-sucking parasites of mammals and two species attack humans. The Human Louse *(Pediculus humanus)* exists as two 'geographical' races, the Head Louse and the Body Louse, effectively separated by the human neck. They tend to keep to their own devices, and to their own body parts. The Head Louse is a recurring problem, especially in children, and, contrary to popular parental perception, does not single out individuals with the dirtiest hair. It has found a new lease of life in recent years because some populations are resistant to previously effective insecticides, such as malathion and permethrin. The tell-tale eggs of the louse, which health visitors look for attached to hair, are called 'nits', an Old English word that can be traced to the Greek for 'dust'. The Body Louse is now much rarer than the Head Louse in Western society and is really only a problem in dirty and deprived conditions.

> To find a louse on one's linen, is a sign of sickness. To find two, indicates a severe illness. If three lice are so found within a month, it is a 'token to prepare'.[88]

The Human Louse is a vector of typhus and, in earlier centuries, of bubonic plague, although the relative importance of this louse, the Black Rat and the rat louse in bringing about the great plague epidemics of the Middle Ages and later are matters for conjecture (see Black Rat, page 479).

The second human-infesting louse, the Crab Louse (pubic louse) *(Phthirus pubis)*, so called from its crab-like appearance, is confined to the lower parts of the body and carries no disease. Nonetheless, it meets with a great deal of social opprobrium because it is commonly transferred from one individual to another by sexual contact.

True bugs
(Hemiptera)

To many people, the word 'bug' is synonymous with insect or, indeed, with any small, imprecisely described, creeping thing; or even just something small and undesirable, such as a hidden microphone. The true bugs are not that much more precisely defined. They comprise two large groups of rather diverse insects, united by little more than the possession of mouthparts modified to feed by sucking, either on the juices of plants or, in some cases, the juices of animals. They include many familiar and superficially disparate creatures, such as aphids, pond skaters, shield bugs and cicadas. Although many have had a considerable impact on human endeavour, especially if that endeavour has involved the growing of crops, few have made a lasting impact on British traditions and literature.

The name 'bug' is early seventeenth century, although of unknown origin, and it is therefore slightly more recent than 'insect'. The many other words beginning with 'bug' appear to be unrelated.

The usually bright green Common Green Shield Bug *(Palomena prasina)* becomes bronze in autumn prior to hibernation.

HETEROPTERAN BUGS
(Heteroptera)

The first group of bugs, the Heteroptera, includes many species that superficially resemble beetles.

The shield bugs (Families Scutellaridae, Pentatomidae and others) have a remarkably shield-like appearance and are often brightly coloured. There are many plant-feeding species and some are common in gardens. Among them, the charmingly named Parent Bug *(Elasmucha grisea)* is unusual in being a caring species: once the female has laid her legs upon leaves, she sits over them until they hatch.

Many other groups of heteropteran bugs are named from their general appearance or behaviour: lace bugs (Family Tingidae) have finely patterned wings; spurge bugs (Family Stenocephalidae) live on spurges; ground bugs (Family Lygaeidae) live on the ground (hardly a unique situation), capsid bugs (Family Miridae) are a very large group of plant-attacking pests (the Apple Capsid, *Plesiocoris rugicollis*, causes the familiar bumps on apple fruits); and no special qualification is needed to appreciate the lifestyle of the assassin bugs (Family Reduviidae), which feed principally on other arthropods and certainly make their presence felt if you pick one up.

The bed bugs (Family Cimicidae) have played a more intimate role in human history. They suck not plants but animals: some attack birds and are found in bird's nests; some attack other invertebrates; others attack mammals, and some of them tend to do so when their hosts are asleep. This group of insects may have originated in caves because they tend to attack species that are, or have been, cave-dwellers, such as bats and humans.

> A bug and a flea
> Went out to sea
> Upon a reel of cotton;
> The flea was drowned
> But the bug was found
> Biting a lady's bottom.[89]

> Bugs pulverized are very good. Success given to provoke Urine, to expel the Birth, and after Birth. Seven of them swallowed before the Fit approacheth, cure the Quartan Ague, as Dioscorides testifies. The stinking Smell which proceedeth from them, cures the Suffocation of the Matrix. Four being taken three Mornings in Wine, cure the Collic as Gesner affirms, Being pulverized, and mixt with the Oil of Roses, and Honey, they cure the Pains of the Ears, being applied to them. Being pulverized, and taken in any fit Vehicle, they cure Fevers, and the spitting of Blood.[90]

The smell of the Bed Bug (wall louse; mahogany flat; crimson rambler) *(Cimex lectularius)* is not easily forgotten but clearly it is a creature that our ancestors, who slept on straw, bracken or heather, came to know and tolerate. Once established, Bed Bugs take some moving; 'as snug as a bug in a rug' is no idle description. Their geographical origin is not known and they were not mentioned in Britain until 1503. However, barely 200 years later, they were common enough for a man named John Southall, who ran a bug-exterminating business in London, to publish, with much observation, *A Treatise of Buggs.*[91] As recently as 1939, 4 million people (half the population) of Greater London were thought to be affected by Bed Bugs, but modern hygiene and insecticides have improved matters beyond measure. The bugs do not appear to transmit disease but their nocturnal blood-sucking bite certainly causes a painful red swelling the following morning. In the days when, like many young people, I travelled the world with modest means and stayed in modest places, 'bug hostels' soon became known on the grapevine and were, if possible, avoided.

The Common Pond Skater *(Gerris lacustris)* has water-repellent hairs that enable it to skim over the surface of still waters.

A considerable number of bugs live near, in, or on water; and you can scarcely look into a still pool of any sort without seeing one or more of them. Children (and others) have long been fascinated by those that live at the interface between water and air. The Pond Skater (*Gerris lacustris*; Family Gerridae) appears to row itself across the surface film using its extra-long middle pair of legs, the hind pair acting like a trailing rudder. The Water Scorpion *(Nepa cinerea)* and the Water Stick Insect *(Ranatra linearis)*, both of the Family Nepidae, have elongated breathing tubes (siphons) to enable them to lurk at or close to the bottom of shallow water yet retain contact with the air. Most fascinating of all, however, are the backswimmers and the water boatmen, wherry-men or boat-flies (all in the Family Notonectidae and others). The backswimmers swim upside down and aggressively devour smaller insects, fish and tadpoles. Intriguingly, their inverted lifestyle is a response to light and, if they are artificially lit from below, they will swim the right way up. Water boatmen always swim the right way up and are unusual among water bugs in feeding on plant not animal matter.

HOMOPTERAN BUGS
(Homoptera)

The second large group of bugs, the Homoptera, is an even more diverse collection. All are plant-feeders and many are very serious crop pests.

They notably include the New Forest Cicada (*Cicadetta montana*; Family Cicadidae), the only British member of this important and noisy family and one of Britain's most splendid and rare insects. This insect lives in an immature state below ground for many years and survives as an adult for only a few weeks. Discovered in 1812, it was later found at a handful of sites, but was not seen after 1961 and was believed to be extinct. Then, in 1992, it was rediscovered. It appears to be associated with woods containing Small-leaved Lime *(Tilia cordata)* and, like all cicadas, hardly ever seems to stop its constant trilling. The sound is produced by two resonating cavities on either side of the abdomen, a noise-producing system quite unlike that possessed by any other insect.

The froghoppers (Family Aphrophoridae) are among the most familiar of insects, although they are rarely seen. However, the presence of the immature insects is betrayed by the 'Cuckoo spit', which they produce to conceal and protect themselves. The froth is produced, rather like cappucino coffee, by forcing air into a liquid, except that the liquid in this case is produced from the anus of the froghopper. The name 'froghopper' is doubly appropriate because these insects not only leap, frog-like, when disturbed but also have large, protruding eyes like a frog. The name 'cuckoo-spit' alludes to the belief that the insect appears at the same time as the Cuckoo (see page 300).

Leafhoppers (Family Cicadellidae) are another generally very common group of hopping insects. Their activities cause a characteristic mottling of leaf surfaces, which is more frequently seen than the insects themselves. Psyllids, or suckers (Family Psyllidae), like many other plant bugs, are common causes of the appearance of sooty mould on garden plants, species such as the Bay Sucker *(Trioza alacris)* sucking the sap of plants and secreting sticky honeydew on which dark-coloured moulds grow. Their close relatives, the whiteflies (Family Aleyrodidae),

Nymphs of the froghoppers (Aphrophoridae) live in a mass of froth, suspended from vegetation and known as 'Cuckoo spit'.

are among the most serious of greenhouse pests and important carriers of plant viruses in the tropics.

The aphids (Family Aphididae and others), however, are the most numerous of all. The name 'aphid', or 'aphis', is eighteenth century and of unknown origin, but aphids are also commonly called (depending on the species) 'plant lice', 'greenfly' or 'blackfly' and are arguably the most important plant pests in temperate climates, causing damage directly and by virus transmission. Many aphids have complex life cycles and require different host plants at different times. Their fecundity is legendary; it has been calculated that a single aphid could produce about 10 million tonnes of progeny during 100 days of summer breeding, a feat assisted by many species being parthenogenic and not needing the inconvenience of a mating. This theoretical maximum is, of course, never achieved because many natural factors, such as birds, other insects and diseases, limit the populations.

Among other relatively obvious and related groups are the adelgids (Family Adelgidae), which infest conifers and secrete a wax-like white wool for protection; mealy bugs (Family Pseudococcidae), which are similarly woolly, superficially like tiny woodlice and appear on many indoor plants; and scale insects (various families in the Coccoidea), which are more like miniature limpets than insects and are mobile only in their early stages.

Nonetheless, however important many of these creatures are as crop pests, few have had a greater individual impact on human activity and endeavour, and the enjoyment of life, than the Vine Phylloxera (*Daktulosphaira vitifoliae*; Family Phylloxeridae). This small, sap-sucking creature virtually wiped out commercial vineyards in many parts of western Europe after being accidentally introduced from its native North America in *c.*1863. It was never properly controlled, but its effects were avoided and the renaissance of the wine industry was facilitated by grafting European varieties of grapevine onto North American rootstocks that were naturally resistant to the pest. Fortunately, this insect has never become established in Britain, although it has sometimes been introduced accidentally.

Thrips
(Thysanoptera)

A friend of mine refers to thrips as as 'picture-frame' insects. They are the tiny black creatures that miraculously find their way behind the glass of framed pictures and that cannot then be removed without taking the whole thing apart. They are more widely known as 'thunder-flies', because masses of them become apparent, in people's hair as well as in their pictures, after warm humid weather in summer.

Many species are important plant pests: they scrape away the surface tissues of plants in order to gain access to the sap, bringing about a characteristic mottling of the leaf surface in the process.

'Thrips' is a late eighteenth-century word, coming, through Latin, from the Greek word for 'wood-worm', to which they are not remotely related. As the *Oxford English Dictionary* points out, it is a word that is 'often erroneously taken as plural, with the false singular thrip'. Even many entomologists make this mistake.

Neuropterans
(Neuroptera)

The name 'Neuroptera' means 'net-winged' and this group of insects is unusual in possessing a net-like pattern of veining on the wings. Many members of the three subgroups, the alder flies, snake flies and lacewing flies, are very beautiful and graceful creatures, and some, as their names suggest, have elongated, snake-like bodies and delicate wings. Nonetheless, their appearance belies an aggressive, carnivorous nature. They feed on smaller insects and other small creatures, the lacewings often being described as among the 'gardeners' friends' because they are important predators of aphids. Cultures of one common species, *Chrysoperla (Chrysopa) carnea*, may be purchased as a biological control agent for use in greenhouses.

The large, green-winged lacewings (Family Chrysopidae), which I always think look rather like giant aphids, are the most familiar members of the group because they often come into houses on summer evenings. They are sometimes called 'golden-eyes' or 'stink-flies', and if you hold one in your hand, you will soon discover how the latter name originated. Curiously, this production of an evil odour is confined to the adult because the larvae have one particularly interesting and peculiar feature: their digestive tract ends in the hind intestine and there is no connection to the anus. As a result they accumulate waste matter in their bodies and cannot dispose of it until they become adults.

The fragility of the green lacewings, such as this *Chrysopa carnea*, conceals an aggressive, carnivorous nature.

Scorpion flies
(Mecoptera)

Scorpion flies are a small group of small insects that look superficially like mosquitoes. There are only four British species, and all root around in the bottom of hedgerows searching for carrion. They are so named because, in the male, the last three body segments are lengthened and the final segment is enlarged, so that when they are curled over the body, the fly bears a slight resemblance to a scorpion.

They are worthy of more than a passing glance, however, because they are among the most ancient of all insect types. They are now believed to be the ancestors of butterflies and moths, caddis flies, all two-winged flies and fleas.

Butterflies and Moths
(Lepidoptera)

The lepidopterist with happy cries
Devotes his days to hunting butterflies.
The leopard, through some feline mental twist
Would rather hunt a lepidopterist.
That's why I never adopted lepidoptery;
I do not wish to live in jeopardoptery.
(Ogden Nash, 'The Lepidopterist')

Let no-one tell you that beauty does not matter. By most people's reckoning, butterflies, and to some extent moths, which together comprise the Lepidoptera, are the most beautiful and obvious of insects. They have therefore attracted an attention and interest far out of proportion to their numerical or zoological importance. On the list of species protected under the Wildlife and Countryside Act 1981, there are four butterflies and five moths, but only four other insects. There are societies for the conservation of butterflies but they pay rather little attention to the larger moths, and even less to the minor ones; and almost no-one seems in the least interested in conserving even the rarest of aphids, lice or fleas.

Both butterflies and moths have four large and obvious wings that are covered with minute scales ('Lepidoptera' means 'scale-winged'), possess a long, sucking tube (the proboscis) with which they drink liquids, lay eggs and undergo a metamorphosis from a larval form called a caterpillar through a pupal stage (usually called a chrysalis in butterflies but a pupa in moths), which has no external legs. Caterpillars, by contrast, seem to have many legs because the six legs that they have in common with other insects are supplemented by a number of stump-like prolegs. Many other groups of insects, including bees and flies, also have a larval form totally different from the adult. This has the very positive advantage that mature and immature creatures do not compete for the same food resource and any given environment can therefore support a great many individuals. The mature insects are generally much shorter lived than their larvae.

I'm a little butterfly
Born in a bower,
Christened in a teapot,
Died in half an hour.[92]

All caterpillars possess modified salivary glands that secrete a liquid that soon hardens to form a very fine thread. This is put to various purposes. Some species, called 'tent caterpillars' (of various families), construct cobweb-like tents from the secretions for protection; sometimes many hundreds of larvae are concealed beneath these 'tents' and entire shrubs or trees are shrouded with the webbing. The process reaches its ultimate development, however, in the exotic Silk Moth *(Bombyx mori)*, of which the cocoon that protects the pupa is fashioned from a single strand of silk almost a kilometre (1,100 yards) long.

The little caterpillar creeps
Awhile before in silk it sleeps.
It sleeps awhile before it flies,
And flies awhile before it dies,
And that's the end of 3 good tries.
(David McCord, 'Sing Cocoon')

The caterpillars of the Large and Small White *(Pieris brassicae* and *P. rapae)* are serious garden pests, especially on brassicas.

Local children help a lepidopterist catch butterflies in the cabbage fields of a Kent farm in 1940.

The differences between butterflies and moths are not clear cut and there are some insects that appear to straddle the boundary. However, most British species can be distinguished on the premise that the antennae of butterflies are long, with clubbed ends, while those of moths are straight. Most butterflies can fold their wings when at rest, are active in the daytime (to see one at night was once thought to be an omen of death) and are brightly coloured. But there are exceptions. The caterpillars of many species in both groups are important crop pests, although the Large White *(Pieris brassicae)* (see page 78) is the only really important pest butterfly in Britain.

There are about 70 species of butterfly in Britain, including some migrants and some vagrants, and over 2,000 moths. The majority of the moths are very small insects, known collectively as 'microlepidoptera', or 'micros'; and most entomologists try to avoid having to identify them. The subdivision based on size is nonetheless a crude one that cuts across their biological relationships.

For many centuries, all butterflies and moths, like so many other insects, were often referred to simply as 'flies' (see Flies, page 89). The name 'butterfly' is in reality as ancient as Old English and the general conclusion today is that it originally meant 'butter-coloured fly' and was probably applied to the most obviously butter-coloured species in Britain, the Brimstone *(Gonepteryx rhamni)* (see page 78). To call someone 'a butterfly' is to suggest that they lead a flippant, light-hearted if not futile existence; *Butterflies* was a popular television comedy series of the 1980s about a family with just such a lifestyle.

'Moth' is also Old English, which implies some early recognition of a difference between the two groups. However, moths were generally thought of as night-time butterflies and the name 'moth' was also used for other creatures, including cockroaches (see page 68). It may originate in the same Teutonic root as 'midge'. 'Caterpillar' is fifteenth century, possibly from Old French meaning 'a hairy foraging cat'; it is strange that so familiar a creature has not acquired more names, although they are called 'nanny-vipers' or 'snortywinks' in Cornwall. 'Chrysalis' is seventeenth century, from Latin and Greek meaning 'gold coloured', although most chrysalises (or chrysalides) are more green than gold. (As I have indicated below in reference to 'aurelian', I do not find this very convincing.)

BUTTERFLIES

I am sure that people have collected butterflies throughout recorded history: Bill Bristowe found a depiction of people chasing butterflies with nets (for no very obvious reason) on a mid-fourteenth-century Flemish manuscript in the Bodleian Library. By the seventeenth century collecting was being carried out reasonably systematically, although, curiously, the specimens

Lulworth Cove, Dorset, is the home of the Lulworth Skipper *(Thymelicus acteon)*.

were almost never labelled. By the eighteenth century, it existed as a more or less organized pursuit, the collectors calling themselves 'aurelians'. This name is always said to be derived from the Latin word for gold, because some butterfly chrysalises are gold coloured, but frankly I find the notion that the most significant thing about butterflies is that some have, largely imaginary, gold colouring on the chrysalis scarcely credible. Nonetheless, the name achieved immortality through a beautifully illustrated, if quaint, book called *The Aurelian*, which was published in 1766 by Moses Harris, one of the early collectors.[93]

Because butterflies have not been considered worthy of serious note until relatively recently, most of them, like moths and indeed many other insects, do not have ancient names and those in use today, with some modification, have mostly evolved over the past 200 or so years. Many originated with the 'founding father' of British butterfly study, the entomologist, botanist and apothecary James Petiver (1663–1718), who was a friend of John Ray and who, in 1717, published the first book devoted solely to British butterflies, *Papilionum Brittaniae Icones*.[94] Many of the names are passable descriptions of their appearance or colour (Marbled White); some also include an indication of habitat (Heath Fritillary); some recall the place where the butterfly was first found (Lulworth Skipper); some are simply expressions of wonderment (Red Admiral/Admirable); others commemorate people who may have played some part in the capture of early specimens. Also, because there are fewer species of butterflies than of moths, it has not been necessary merely to name them after the plant on which they occur or to invent any truly fanciful names (see Moths, page 86).

Nine families of butterflies are represented in Britain although some only barely. The individuals range in size from the huge 47-millimetre (2-inch) wing-span of the Monarch (milkweed) *(Danaus plexippus)* to the barely 10-millimetre (½-inch) wing-span of the Small Blue *(Cupido minimus)*. Many have declined significantly in recent times, primarily because of habitat loss, and five species have become extinct in Britain during the years since serious butterfly study began.

In Britain as well as many other parts of the world, a butterfly was once thought to be the soul of the departed and, in some areas of rural England, they were believed, rather like Kittiwakes (see page 288), to be dead children who had died unbaptized and were unable to enter Heaven.

Skippers (Hesperiidae)

The skippers are small, rather moth-like butterflies with a quick, darting flight (they skip from flower to flower). Unlike most butterflies, they rest with their wings more or less flattened over the body rather than raised vertically, which is also somewhat moth-like. They were once known, for no very evident reason, as 'hogs'. There are eight resident species, only one of which, the Dingy Skipper *(Erynnis tages)*, occurs in Ireland.

The Chequered Skipper *(Carterocephalus palaemon)*, with its very dark brown, yellow-spotted wings, was last seen in England, in Rutland, in 1976 and is now confined to moist, grassy localities in the west of Scotland.

The Small Skipper *(Thymelicus sylvestris)* is still a common russet-brown insect of places with tall, undisturbed grasses throughout much of England and Wales except the far north.

The Essex Skipper *(Thymelicus lineola)* is found throughout the southeastern half of England in sunny, grassy places. It is very similar to the Small Skipper, with which it remained confused until the end of the nineteenth century, making it the last resident British butterfly species to be recognized.

The Lulworth Skipper *(Thymelicus acteon)* is a tiny brown insect and one of the most locally distributed of all British butterflies. It is confined to grasslands around the village of Lulworth on the Dorset coast, where it was first discovered in the summer of 1832 by the outstanding lepidopterist J.C. Dale, the squire of the Dorset village of Glanvilles Wootton (sometimes called Wootton Glanville). These skippers can occur in very large numbers at Lulworth: descending the cliffs to Durdle Door cove some years ago, I found myself surrounded by them and it was hard to believe that, just a few miles away, my chances of seeing one would be reduced to nought, so local are they.

The Silver-spotted Skipper *(Hesperis comma)* is now a rare butterfly of chalk grassland in the south of England, where its food plant, the Sheep's Fescue *(Festuca ovina)*, occurs. It can be distinguished by silver-white spots on the underside.

The Large Skipper *(Ochlodes venata)*, the largest British species in the family, is still relatively common and occurs throughout most of England and Wales in sheltered spots where tall grasses grow.

The Dingy Skipper *(Erynnis tages)* is also still relatively frequent in open, sunny places, where it is found in association with its vetch and trefoil food plants It is found throughout England and Wales, with scattered colonies in Scotland and Ireland, although it has declined significantly.

The Grizzled Skipper *(Pyrgus malvae)*, despite its scientific name, feeds not on mallow (*Malva* species) but on members of the rose family. It is a distinctive butterfly, with a black-and-white checked pattern, and still occurs in a wide range of habitats throughout England, although it has retreated southwards.

Swallowtails (Papilionidae)

By general reckoning the Swallowtail *(Papilio machaon)* is the most beautiful of British butterflies. It is the largest breeding species and the only member of a very important but mainly tropical family, mostly characterized by the two tail-like projections of the hindwings. In Britain, the Swallowtail is confined to small areas of the Norfolk fens, where the caterpillars feed on Milk Parsley *(Peucedanum palustre)*. Elsewhere in

The Swallowtail *(Papilio machaon)* is the largest breeding species in Britain and confined to the moist habitats of the Norfolk Broads.

Europe, it is a butterfly of open pastures and feeds on other umbellifers of the plant family Apiaceae.

It seems always to have been a creature of wet places in Britain and the earliest authentic specimen is said to be one caught on marsh land around St James's Palace in the seventeenth century. The specimen was given to James Petiver and, in recognition of the source of the insect, Petiver called it 'The Royal William' in honour of King William III.

Whites and Yellows (Pieridae)

Six species of this large family of mainly white or yellow butterflies are resident in Britain, with four more arriving each year in greater or lesser numbers from continental Europe.

The most striking of the white species, the Black-veined White *(Aporia crataegi)*, after suffering a reduction in range over many years, finally disappeared from its last locality in Kent in the early part of the twentieth century. The reasons for its disappearance are not fully understood and attempts to reintroduce it have failed.

Another species that seems likely to follow it is a fragile, delicate-looking insect with a very slim body called the Wood White *(Leptidea sinapis)*. This has declined dramatically in England and Wales (while strangely expanding in Ireland), mainly as a result of the shading of the woodland rides that contain the vetches and related plants which it prefers.

In addition to the whites, there are three predominantly yellow species in this family. The Clouded Yellow *(Colias crocea)* and its paler counterpart, the Pale Clouded Yellow *(Colias hyale)*, migrate to Britain in summer, the Pale generally in few numbers along the south coasts. A third species, Berger's Clouded Yellow *(Colias alfacariensis)*, once known as the New Clouded Yellow, or the New Pale Clouded Yellow, occasionally reaches Britain and was first distinguished from the other two species in 1947. This proved to be a propitious year for the genus. Clouded Yellows occasionally occur in vast numbers but, for 1947, C.B. Williams recorded a mass migration over the English Channel that was estimated to involve over 100,000 butterflies on a front of about 80 kilometres (50 miles).[95]

It is generally thought to be the Clouded Yellow that most fits the description of 'gilded butterfly':

> ... has such a confirm'd countenance. I saw him run after a gilded butterfly; and when he caught it, he let it go again; and after it again; and over and over he comes, and up again; catcht it again: or whether his fall enrag'd him, or how 'twas, he did so set his teeth, and tear it: O! I warrant, how he mammockt it!
> (*Coriolanus*, Act 1, Scene 3)

The most striking and widespread of the yellows, however, is the Brimstone *(Gonepteryx rhamni)*, in which the male has the beautiful sulphur-yellow coloration that was the original inspiration for the name 'butterfly'. It has generally been known as 'Brimstone' since the early nineteenth century. Inexplicably, this species, whose caterpillars feed inoffensively on Buckthorn *(Rhamnus catharticus)* and Alder Buckthorn *(Frangula alnus)*, once had a bad reputation among gardeners. To see it in the spring among the daffodils was thought a sign of bad weather. It occurs throughout much of England and Wales and parts of Ireland and has begun to spread northwards in recent years.

The commonest white, and one of the commonest British butterflies, is the Large White *(Pieris brassicae)*, which in some years can cause devastating damage on brassica crops. In Derbyshire, it was invariably called the Cabbage White, as it still is in other areas, although, strangely, this does not seem to be one of its officially recognized names.[96] I can remember years during the 1950s when our garden vegetable plot was alive with the foul-smelling, black, green and yellow caterpillars. Fortunately, many natural factors operate to control them, most notably parasitism by ichneumon flies (see Ichneumon and Braconid flies, page 97). In Gloucestershire, it used to be said that, if the first butterfly seen in summer was a white one, observers would eat fine, white bread all the year and be prosperous, but if the first butterfly seen was brown, they would eat only basic brown bread. It is curious how our perceptions of the relative merits of white and brown bread, and perhaps our perceptions of the Large White as well, have changed over the years.

It is certainly true that Large Whites have not had such a warm welcome everywhere. There was once a rather widespread custom of naming Large Whites after groups of people who were locally unpopular. In Lincolnshire they were called 'Frenchmen' and, in strongly Protestant parts of the north of England, they were known as 'Papishes' and were hunted to destruction by gangs of youths on Oak Apple Day (29 May). According to one tradition, if the first Large White seen could be caught before it had laid its eggs, the crops would be free from caterpillars all season. A more general view was that, if you were to crush the first butterfly seen each year, of whatever hue, this would enable you to crush your enemies and have good fortune. Sadly, this absurd belief has been employed in more recent time by sports teams who believe that, by disposing of the first butterfly, they will similarly dispose of their opponents.

Two very common smaller species, the Small White *(Pieris rapae)* and the Green-veined White *(Pieris napi)*, are almost equally widespread but of much less importance as pests, not least because they lay their eggs singly rather than in large groups.

The Bath White *(Pontia daplidice)* takes its name from 'the circumstance that a young lady of that city executed a piece of needlework in which an example of the butterfly was represented'.[97] Who the lady was and why she did such a thing does not seem to have been recorded. The Bath White was previously known as Vernon's Half-mourner, an early specimen having been captured, probably in 1702, by the entomologist William

The Clouded Yellow *(Colias crocea)* is one of a large family (Pieridae) of basically white or yellow butterflies.

Vernon. Presumably it was Vernon who alluded to the partial blackening of the white wings. Today, it is a very rare vagrant.

The last of this related group of widespread whites has always been my personal favourite among British butterflies, possibly because of the contrast that its vivid orange wing splashes offer to the run-of-the mill, fluttering white species of early spring. The Orange-tip *(Anthocharis cardamines)* appears in my garden when the Hedge Mustard (Jack-by-the-hedge) *(Alliaria petiolata)* is in flower and, I find, is the most authentic sign that spring has arrived. It was once known as 'the lady of the woods', not entirely appropriately because it is an insect of hedgerows rather than woods and because the female of the species lacks the distinctive orange coloration.

Coppers, Hairstreaks and Blues *(Lycaenidae)*

The Lycaenidae is a large family of mainly small butterflies, several of which are difficult to identify unless examined very closely; this is not easy to do in the field because they flit swiftly from plant to plant. The males of most species are generally blue, brown or a copper-orange, while the females are much more sombre in colour. The family falls naturally into three groups: the so-called 'hairstreaks', which tend to feed on trees and shrubs; the 'coppers', which feed on sorrels; and the 'blues', which feed mainly on legumes. Among the blues are some of the commonest and rarest species in Britain, and one with quite the oddest of all butterfly life histories, the Large Blue (see below).

The curious name 'hairstreak' is early nineteenth century and alludes to the thin, hair-like line or pattern of tiny dots in the undersides of the wings.

The Green Hairstreak *(Callophrys rubi)* is actually brown above but has a green flush on the underside. I have found it most often on gorse (*Ulex* species) but it also occurs on other shrubs in scattered localities throughout the British Isles.

The Brown Hairstreak *(Thecla betulae)* is very dark brown, with orange wing blotches, and is not as common on birch (*Betulus* species) as its scientific name suggests, being most usually found among Blackthorn *(Prunus spinosa)*. It is now mainly confined to parts of south and western England and to the west of Ireland. Its decline has been attributed to the flaying of hedgerows, which destroys the eggs.

The Purple Hairstreak *(Neozephyrus quercus)* is a species of tall oak trees and is found throughout much of England, with some evidence of an expansion of its range in some areas.

The White-letter Hairstreak *(Satyrium w-album)* has a hairstreak line so wavy that it takes the form of the letter 'w'. As a species, its decline matches that of its food plant, the elm (*Ulmus* species), although it is now regaining some territory throughout England and eastern Wales.

The Black Hairstreak *(Satyrium pruni)* is the rarity of the group, being confined to parts of the southeast English Midlands, despite the fact that its food plant, the Blackthorn *(Prunus spinosa)*, is common and widespread enough. It is black only in so far as having marginal black spots on the undersides; it is not as black as the White-letter Hairstreak, which the older collectors called the Dark Hairstreak.

The Small Copper *(Lycaena phlaeas)* was one of the most familiar garden butterflies during my childhood. They always seemed especially interested in the tagetes and African marigold flowers, although there was a large area of docks and sorrels close to the river that undoubtedly was the real attraction. Today, I rarely see one in a garden, although it is still a common and widespread species throughout Britain. It is appropriately named, being not only coppery in colour but also possessing the shining reflectivity of genuine burnished copper. It has a larger relative that, sadly, no-one alive today has ever seen in Britain.

The Large Copper *(Lycaena dispar)* was a butterfly of the fens, where it fed on the Great Water Dock *(Rumex hydrolapathum)*. It became extinct as the fens were drained in the mid-nineteenth century, although the collectors of the time may have been at least partly responsible for hounding it out of existence. The native, and now lost, British race was a magnificent insect. Although the rather smaller Dutch race has been introduced to Britain several times, most notably to Woodwalton Fen in Cambridgeshire, all introductions have ultimately failed.

The Small Blue (Little Blue) *(Cupido minimus)* is a very dark, dusky blue and is the smallest British butterfly. Although it occurs in widely scattered grassland areas throughout Britain, it is declining almost everywhere as its habitat is lost.

The Silver-studded Blue *(Plebeius argus)* is a heathland species, mainly of southern England, but it too is declining. The male is blue but the female is brown; both have silvery spots on the hindwings.

The Brown Argus *(Aricia agestis)* is alone among British blues in having both sexes brown and orange (it was once called the Brown Blue). It is still relatively common on chalk grassland in the southeastern half of England, with isolated colonies elsewhere. Its relative, the Northern Brown Argus *(Aricia artaxerxes)*, is its northern and Scottish counterpart.

The Common Blue *(Polyommatus icarus)* is still the commonest and most widespread of the blues. It is found in sunny, grassy places throughout most of the British Isles wherever vetches, trefoils and related plants occur.

> On Tean [Isles of Scilly], the Common Blue butterflies have a different pattern of spots and colouring with silvery-blue scales, when on other islands, the normal appearance is usual. This butterfly rarely strays far from its place of hatching so the mutations are highly localised.[98]

E.B. Ford thought this Tean form of the butterfly seemed to be 'a stage in the evolution of an independent sub-species'.[99]

The Chalkhill Blue *(Polyommatus coridon)* is the largest British resident species and is a lovely silvery blue overall, with a pale,

The Chalkhill Blue *(Polyommatus coridon)* is found typically in the chalk grasslands of southern England.

lace-like fringe to the wings. It is appropriately named because its habitat is the chalk grasslands of southern England.

The Adonis Blue *(Polyommatus bellargus)* is still fairly common and also occurs on chalk grasslands in the south of England where its food plant, the Horseshoe Vetch *(Hippocrepis comosa)*, grows.

The Holly Blue *(Celastrina argiolus)* feeds largely on Holly *(Ilex aquifolium)* and is common throughout most of England and Wales and parts of Ireland. It is the commonest blue butterfly in gardens and the earliest to appear in the spring.

The Long-tailed Blue *(Lampides boeticus)* and the exceedingly rare Short-tailed Blue (Bloxworth Blue) *(Cupido argiades)* are southern European species that very occasionally reach Britain. A pair of Short-tailed Blues were taken on Bloxworth Heath in Dorset in 1885, hence the alternative name. I can still recall the thrill of finding a Long-tailed Blue in Warwickshire in the 1970s, although it subsequently turned out to be an escaped specimen.

The Mazarine Blue *(Polyommatus semiargus)* is one of the butterflies of Britain's entomological past. It was referred to in some of the butterfly books of the early eighteenth century and, in some years, seems to have been reasonably frequent in Dorset and other parts of southern England. It mysteriously and finally disappeared in *c.*1904, although migrant insects still turn up.

The Large Blue *(Maculinea arion)* is the biggest of the blues, and the strangest. It declined in Britain from the early nineteenth century onwards, eventually becoming extinct in 1979, but Swedish insects have been reintroduced on a number of occasions since 1983 and the colonies are surviving. Although a number of other insects associate with ants (see Ants, page 97), including several members of the Lycaenidae, the Large Blue is in a special category. It was not until the early years of the twentieth century that its life history was fully elucidated, by two outstanding British lepidopterists, F.W. Frowhawk (1861–1946) and E.B. Purefoy (1868–1960). They showed that the caterpillars feed on Wild Thyme *(Thymus serpyllum)*, and also on their weaker siblings, until they are chanced upon by a particular type of wandering red ant, usually a species of *Myrmica*. The ant caresses the caterpillar, thus stimulating it to produce a honey-like secretion, and then picks it up in its jaws and bears it away to its underground nest. There the caterpillar remains from late summer until the following spring, feeding on young ant larvae with a winter pause for hibernation. Finally it pupates and metamorphoses into the adult butterfly, which finds its way to the soil surface and the outside world.

Riodinids (Riodinidae)

Closely related to the Lycaenidae, but with more or less functionless forelegs, the Duke of Burgundy (Duke of Burgundy Fritillary) *(Hamearis lucina)* is a mainly rare butterfly of southern and central England, although it does occur sparsely in the north, where its food plants, the Cowslip *(Primula veris)* and

These Small Tortoiseshells *(Aglais urticae)* are often to be found on herbaceous species of *Sedum*.

Primrose *(Primula vulgaris)*, occur. It displays the brown-and-white chequered pattern typical of the true fritillaries. The relationship with the eponymous Duke of Burgundy is obscure.

Nymphalids, Fritillaries and Browns (Nymphalidae)

As in the Lycaenidae, there are three rather distinct groups within this large and important family: the so-called 'typical nymphalids', the 'fritillaries' and the 'browns'. Many of the nymphalids and browns are still common and familiar and not under threat, but many of the fritillaries are in serious decline.

The White Admiral *(Limenitis camilla)* is no great rarity in central southern parts of England, where mature woodland and its food plant, the Honeysuckle *(Lonicera periclymenum)*, occur, and it is now spreading again rather effectively. When I was a child in Derbyshire, it was always a butterfly of faraway places and one of the delights of holidays in the New Forest, but now I see it in Warwickshire every year. I always think it an especially marvellous sight when seen in a woodland ride, as it is in the Forest, because I consider it the only British butterfly that truly glides.

Another woodland butterfly whose sighting has always been a special triumph is the Purple Emperor *(Apatura iris)*, once called the Emperor of Morocco. The purple-brown iridescence of its wings is lost to view as it flits high among the oaks and poplars, although its food plants are willows (*Salix* species). It is thought to be expanding slightly in the woodlands of central southern England that are its stronghold.

Plant a *Buddleja davidii* or a clump of Michaelmas daisies in your garden, wait for their flowers to open, choose a sunny day and, within a short space of time, you could be watching a significant proportion of the most brightly coloured members of the Nymphalidae. They are among the most familiar of garden butterflies and include the Red Admiral *(Vanessa atalanta)*; Painted Lady *(Cynthia cardui)*; Small Tortoiseshell *(Aglais urticae)*; Peacock *(Inachis io)* with its peacock-feather eyemarkings; and the ragged Comma *(Polygonia c-album)*, with a shape like a fallen leaf and a 'punctuation' mark on its underside. The only one whose population has changed significantly in my lifetime is the Comma, which was in serious decline for much of the twentieth century but is once again a very common species and spreading northwards by the year. The caterpillars of all of them feed primarily on nettles and every wildlife garden should have a clump of Stinging Nettles *(Urtica dioica)* somewhere for this purpose. The Peacock, Small Tortoiseshell and Comma are three of the butterflies most frequently found hibernating in sheds and outhouses. The Painted Lady arrives as a migrant from southern Europe, sometimes in very large numbers, and it is thought that some may survive the British winter.

The Red Admiral, formerly the Red Admirable, appears to arrive mainly as a migrant, although a few may hibernate in Britain. As this is the only British species with such intense red coloration, it was clearly the 'red butterfly' that was hunted in the north of England and Scottish borders in the belief that it was a witch. Its 'black with scarlet' fans were wonderfully evoked by the long-neglected Essex poet Coventry Patmore (1823–96), who recalled the legend of Cupid and the butterfly-winged Psyche:

> She brought me 'Wiltshire Butterflies',
> The Prize-book; then we paced the lawn,
> Close-cut, and with geranium-plots,
> A rival glow of green and red,
> They counted sixty apricots
> On one small tree; the gold fish fed;
> And watched where, black with scarlet fans,
> Proud Psyche stood and flash'd like flame,
> Showing and shutting splendid fans,
> And in the prize we found its name.
> (Coventry Patmore, *The Angel in the House*, 'The Morning Call')

Just as the Large White was called 'Frenchman' in Lincolnshire, so:

> ... in my youth in Barnsley, any showy Vanessid species was called a Frenchie – most often applied to the Small Tortoiseshell and Red Admiral.[100]

The name seems to be fairly widespread, at least in northern England, and was certainly used in the Chesterfield area.[101] The

Red Admiral was known as 'King George' in Cornwall, for no obvious reason, and with no obvious connection to the similarly named dragonfly (see Dragonflies and Damselflies, page 63).

This branch of the family Nymphalidae embraces other species that epitomize the lost and elusive treasures of British butterfly history. The Large Tortoiseshell *(Nymphalis polychloros)* was almost certainly lost as a resident species *c.*1980, although a few immigrants are still seen in some years.

Most evocative of all is a very rare migrant, the Camberwell Beauty *(Nymphalis antiopa)*, a butterfly that has acquired almost legendary status since the first British specimens were taken in the middle of August 1748 in Cool Harbour Lane, near the then rural village of Camberwell in Surrey. In subsequent years, others were caught there, attracted it is thought by the willows that were then so numerous, and at least one of these eighteenth-century specimens survives. The Camberwell Beauty has never bred in Britain, but migrant insects, mainly from Scandinavia, are seen in most years and occasionally arrive in large numbers. Cool Harbour Lane has now become Coldharbour Lane and a busy road (the A2217) from Brixton to Camberwell, but although I had been told that its most famous resident was depicted on the street sign, on a recent visit I could find no hint of the memorial. The fritillaries belong to the family Nymphalidae and are probably the most difficult group of British butterflies to identify: all nine species are broadly similar, red-brown or yellow-brown in colour, with black, often chequered markings on the upper surfaces, and some with a pronounced silver flush on the undersides.

The name 'fritillary' is seventeenth century, from the Latin word *fritillus*, meaning 'dice-box'. It was first used for the flowers of the genus *Fritillaria*, among which the lovely and rare Snake's Head Fritillary *(Fritillaria meleagris)* displays the same chequered patterning on its petals. Two centuries later, it was adopted for butterflies. Some of the common names reflect the silver coloration, for instance the Small Pearl-bordered Fritillary *(Boloria selene)*, the Pearl-bordered Fritillary *(Boloria euphrosyne)* and the Silver-washed Fritillary *(Argynnis paphia)*. The Pearl-bordered Fritillary was referred to by early entomologists as the April Fritillary, a name that, as E.B. Ford pointed out, is inappropriate because this insect does not usually appear until the end of the first week of May. However, Ford perceptively added:

> Ray wrote before the change in the calendar which took place in 1752. Eleven days were omitted from that year, so that in the early eighteenth century the Pearl-bordered Fritillary would normally be seen in the woods at the end of April.[102]

The White Admiral *(Limenitis camilla)* is now a fairly common sight in central southern England.

The Small Pearl-bordered Fritillary was then called the May Fritillary, a name that would still be reasonable, although, as with other animal names that relate to calendar months, it may require reassessment as climate change becomes ever more apparent.

The Dark Green Fritillary *(Argynnis aglaja)* and the High Brown Fritillary *(Argynnis adippe)* also have distinctive colours, or so some would have you believe, while the Marsh Fritillary *(Euphydryas aurinia)* (once Dandridge's Black Fritillary, after Joseph Dandridge, a butterfly-collector otherwise largely forgotten) and the very scarce and local Heath Fritillary *(Melitaea athalia)* are supposedly distinguished by their habitat. The Marsh Fritillary feeds on Devil's-bit Scabious *(Succisa pratensis)*. In the nineteenth century, there was a report of an infestation of caterpillars in County Fermanagh so serious that the local populace had to barricade their homes with blocks of peat and questions were asked in The House. Presumably, there must have been an equivalent infestation of Devil's-bit Scabious to provide the wherewithal for them to flourish.

Most of these fritillaries tend to occur in scattered localities where their food plants grow, which is predominantly in the western parts of Britain. But what of the Glanville Fritillary *(Melitaea cinxia)* and that rare visitor to Britain, the Queen of Spain Fritillary *(Issoria lathonia)*? The Glanville Fritillary was once called the Dullidge Fritillary, presumably because it was found at Dulwich. Logically, therefore, the later name might seem to commemorate that classical site of British butterfly-collecting, the village of Glanvilles Wootton, which lies south of Sherborne in Dorset. (This was the home of J.C. Dale, who made many important collections of British insects in the early years of the nineteenth century.) In fact, the butterfly was named after Eleanor Glanville (*c.*1654–1709), who captured the first British specimens in Lincolnshire in the latter part of the seventeenth century. This lady was clearly a remarkable woman and butterfly-collector, but her will was disputed on the grounds that, by collecting butterflies, she could not have been of sound mind. She was even mentioned in a House of Commons debate on environmental conservation in 1992 and was often erroneously referred to as 'Lady Glanville' (a common appellation in the seventeenth and eighteenth centuries), a mistake traceable to Moses Harris and *The Aurelian*. Her butterfly is a rare insect in Britain today, being confined almost entirely to the southern coast of the Isle of Wight.

The Queen of Spain Fritillary, which is often confused with the Dark Green Fritillary, is a rare visitor to the southern and eastern coasts of England. There seems to be no record of which Queen of Spain so generously donated her name to an insect, nor why.

The group of butterflies known as the 'browns' rely mainly on grasses as food plants. Until recently, they were distinct enough to have been placed in a separate family, the Satyridae. With one or two notable exceptions, no group of insects has been more aptly named: these are brown butterflies; some are pale brown, some darker; some have orange markings and most have spots, at least on the forewings; some are big, some small; but all are certainly more brown than anything else.

The only serious exception among British butterflies is the Marbled White *(Melanargia galathea)*, in which the brown is slightly outshone by the white, although the butterfly as whole certainly has a pronounced marbled appearance. This insect is on the increase, spreading eastward and northward from its former stronghold in central southern England, where its food grasses occur. It is a butterfly that, along with the Chalkhill Blue, I always associate with the South Downs' localities that I know well.

> I recently had a most interesting experience when I pulled into a lay-by for an emergency 'comfort stop'. As I stood knee-high in the grass, discreetly obscured behind my car, I saw three of my favourite butterflies, just out of 'range' on the same thistle: a Marbled White, a Speckled Wood and a Common Blue.[103]

The Speckled Wood *(Pararge aegeria)*, with its white-spotted forewings and eyed hindwings, is appropriately named. It can be seen flitting among the dappled light of a woodland glade,

The Small Pearl-bordered Fritillary *(Boloria selene)* used to be called the May Fritillary.

although I increasingly find it with the nymphalids on my garden flowers, and there are no significant woods nearby. It has a curious distribution: common throughout much of England, it is absent from the north, and common throughout Ireland but also occurs in parts of northern Scotland.

Unfortunately, the butterfly that I grew up to know simply as the Wall *(Lasiommata megera)*, so called because of its habit of resting on walls in the sunshine, and known for a time as the Wall Brown, can no longer be included on the list of common browns. For obscure reasons, its range throughout many of the central areas of England and Ireland has declined. It is particularly pretty for a brown, with a chequered pattern that almost equals that of the fritillaries.

The ringlet butterflies have greater or lesser arcs of small, haloed spots on both wings. The basic ringlet, called simply the Ringlet *(Aphantopus hyperantus)*, is arguably the dullest and most uninteresting of any British butterfly in appearance. It is generally found along hedgerows and the edges of woodland almost anywhere except the far north of Scotland, and in a swathe to the west of the Pennines, but is barely noticeable. The Mountain Ringlet *(Erebia epiphron)*, its much rarer northern and upland counterpart, occurrs in a few isolated localities in the English Lake District and the western Scottish Highlands. The genus *Erebia* is essentially northern and Arctic. Its other British representative, the Scotch Argus *(Erebia aethiops)*, is also confined to the northern half of Britain, although it is not exclusively Scottish. An additional species, the Arran Brown *(Erebia ligea)*, is a different matter. This is a fairly common European butterfly but, in 1804, Sir Patrick Walker caught specimens on the moors behind Brodick Castle on the Isle of Arran. It was caught there only once, and, arguably, only once on the Scottish mainland, although the specimens are authentic.

The Grayling *(Hipparchia semele)*, which borrowed the name of a fish (see Grayling, page 139), is, appropriately enough, usually common near the sea. It is found around most British coasts where its grass food plants grow, but it is an undistinguished-looking insect, often overlooked because it is well camouflaged against bark and dark rocks. The fish is far prettier.

The Gatekeeper *(Pyronia tithonus)*, a pretty, orange-brown butterfly with dull brown wing edges, is appropriately, and typically, a hedgerow species. Common throughout most of England and Wales, except the far north, it also found around the southern Irish coasts.

The Meadow Brown *(Maniola jurtina)* must vie with the Large White and the Red Admiral as the commonest British butterfly. It has the added advantage of being one of the few species that seem content to fly in dull, overcast weather and often 'flocks' in large numbers.

> The Meadow Brown is a clumsy flier and seldom travels far and on Tean [Isles of Scilly], there appear to be three distinct types with different distribution of spots on the underwing, the populations only a few yards from each other.[104]

Noted for its extensive migrations, the Monarch *(Danaus plexippus)* is a vagrant from North America.

The paler and smaller Small Heath *(Coenonympha pamphilus)* is almost as common, but its larger relative, the Large Heath *(Coenonympha tullia)*, is a northern butterfly, occurring on moorlands in association with Hare's-tail Cottongrass *(Eriophorum vaginatum)*. It is therefore found at high altitudes in northern England and Wales, as well as Scotland and Ireland.

Monarchs (Danaidae)

The Monarch (Milkweed) *(Danaus plexippus)* is the only representative of this family to be found in Britain. It is at, once, the largest and one of the rarest British butterflies and arrives here as a vagrant from its native North America. To spot one is a triumph for any British lepidopterist, although at least one or two are seen most years; occasionally a hundred or more occur, almost invariably and understandably towards the west of the country. It is a rich orange-brown butterfly with black vein markings and tiny white marginal spots.

The Puss Moth *(Cerura vinula)* is sometimes called a 'miller' because it appears to be dusted with flour.

MOTHS

Relatively few of the 2,000 or more British species of moth are known or recognized as anything more specific than a moth by the casual observer. Even such large and distinctive species as the hawk moths are seldom identified with any accuracy. Therefore, even more than with butterflies, much of what is said and believed about moths is very general. Even those species whose caterpillars are significant as pests usually pass unremarked or unrecognized as adults. How many farmers or gardeners know that the cutworm caterpillars that sever the stems of their vegetables are the larvae of distinctive and colourful noctuid moths such as the Large Yellow Underwing *(Noctua pronuba)*?

While there are, at most, seven families of British butterflies, there are many dozens of moth families and, unlike butterflies, there are many individual species, especially of the smaller moths, that have no English name. However, the countless names that do exist are a curious and fascinating assembly. Most have arisen through the enterprise of eighteenth- and nineteenth-century collectors, and one can only be impressed by their ingenuity and inventiveness. Many names, such as Alder *(Acronicta alni)* and Sycamore *(Acronicta aceris)*, simply refer to the plant on which the caterpillar feeds. Some are often rather clever and ingenious attempts to describe the insect's coloration or patterning. These include Magpie *(Abraxas grossulariata)*; Triangle *(Heterogenea asella)*; Chinese Character *(Cilix glaucata)*, which has a mark like a Chinese character, although its overall appearance suggests that it would be better named 'The Bird Dropping'; Heart and Dart *(Agrotis exclamationis)*; Muslin Moth *(Diaphora mendica)*; Marbled Green *(Cryphia muralis)*; Beautiful Yellow Underwing *(Anarta myrtilliv)*, which must be one of few other animals with such a subjective adjective as part of their name; and Silver Y *(Autographa gamma)*, so often misconstrued by well-meaning typesetters as 'Silvery Moth'. The idea of a descriptive name is fine in theory but can be rather unimaginative in practice: Black-veined Moth *(Siona lineata)* and Pale Shoulder *(Acontia lucida)* could apply to a great many species.

Some moth names describe the habitat, although in a fairly vague way, such as the Common Heath *(Ematurga atomaria)*, which could apply to countless other moths. Some names describe the time of year when the moths are most likely to be seen, such as March Moth *(Alsophila aescularia)* and Winter Moth *(Operophtera brumata)*. A few names describe neither appearance nor behaviour but smell. You only need a single encounter with the massive tree-boring larvae of *Cossus cossus*, which take three or more years to mature, to understand why the species is called the Goat Moth. Some names, as with butterflies, relate to a geographical location, generally the place where the creature was first found but which may later turn out to be relatively untypical, insignificant, or simply no longer applicable. The Jersey Tiger *(Euplagia quadripunctaria)*, for example, does occur on Jersey but you would have some difficulty in finding a Kentish Glory *(Endromis versicolora)* in Kent.

The name 'looper' exemplifies another moth-collector's wile: describing the behaviour of the caterpillars. Looper caterpillars walk by stretching forwards and then arching their back while they bring their hind parts to catch up. Scientifically their family is called the Geometridae, or geometers, because they measure their way along.

Another ploy is to put the word 'scarce' in front of the name of an existing species to indicate a very similar but possibly rarer species, such as Blackneck *(Lygephila pastinum)* and Scarce Blackneck *(Lygephila craccae)*; Burnished Brass *(Diachrysia chrysitis)* and Scarce Burnished Brass *(Diachrysia chryson)*. Sometimes, collectors found a creature so entrancing that they were all but lost for words; after capturing a Clifden Nonpareil *(Catocala fraxini)* what else was there to do? However, just occasionally, even among such linguistic inventiveness, there is a real gem. *Hydriomena ruberata* is a slightly reddish coloured moth that is occasionally found flitting along hedges and sallow bushes; I like to think of an early entomologist, brandishing his net above his head as it soared out of reach and, thinking about his experience when he returned home, simply calling it the Ruddy Highflyer.

It always seems strange that creatures so addicted to a nocturnal lifestyle as moths should so hanker for the brightness of artificial light. By and large, night-loving animals shy away from light, yet the expression 'like moths to a lamp' has become a euphemism for irresistible attraction.

> How, like a moth, the simple maid
> Still plays about the flame!
> (John Gay, *The Beggar's Opera*)

The serious entomologist catches night-flying species by using a light-trap, a device with a bright light that lures the insects into a chamber from which they cannot escape. However, the genuine scientific student, unlike the mere collector, will release them in the morning after checking and recording the catch.

Puss moths (Notonitidae)

Understandably, the moths that have attracted the greatest attention have been the day-flying species, the large species, and also the white species, whatever their size. The Puss Moth *(Cerura vinula)* has attracted attention on more than one count. As with several species, its name is always given in full: it is the Puss Moth, not just Puss or the Puss. It is a fairly large, stout, hairy moth (like a pussycat), pale coloured, almost white, but with a most extraordinary caterpillar. In its normal posture, feeding on the leaves of sallow (*Salix* species), the caterpillar simply appears green. However, when threatened, it withdraws its head, displays its last pair of prolegs, which are modified to form elongated horns, and discharges an irritating fluid from a special thoracic gland.

In several parts of the country, The Puss Moth is called a 'miller', or 'millard', because the adults appear to be dusted with flour, like a miller. There is an old Dorset rhyme that accompanied a macabre children's ritual in which the moth was put on trial:

> Millery, millery, dousty poll!
> How many zacks hast thee a-stole?
> Vour an' twenty an' a peck.
> Hang the miller up by's neck.[105]

An old name for large white moths in general was 'owl', or 'owlet' (a Cornish variant is 'maggieowler'). In some districts, this name was, in truth, more likely to apply to an insect than a bird. Just like butterflies, moths too were commonly believed to be the souls of the departed and the fact that many fly at night served simply to enhance this belief. In Cornwall, they were thought to be fairies and were called 'pisgies' (pixies).

Tiger moths (Arctiidae)

The beautifully coloured and patterned tiger moths are unusual in that their caterpillars too have special names. They are particularly large and hairy and, in consequence, are known as 'woolly bears'. The seventeenth-century naturalist Edward

Found in fens and along riverbanks, the Scarlet Tiger Moth *(Callimorpha dominula)* feeds on shrubs and herbaceous plants.

Topsell called them 'Palmer worms', not after a man named Palmer but because a palmer was 'a wandering monk who had roving habits and a rather ragged appearance late in the season'.[106] In Shropshire, for no very obvious reason, they are called 'Tommy-tailors' and it was widely thought lucky if one should creep up on you. In truth, it could prove unlucky if you picked it up because the hairs on many species of caterpillar can cause an irritant rash and so they are best handled with gloves. The hairs are said to be 'urticating', a name derived from *Urtica*, the Stinging Nettle, whose hairs have similar effects.

Hawk moths (Sphingidae)

Although nothing in any British naturalist's experience can match seeing some of the giant tropical butterflies and moths, little in British entomology rivals the thrill of seeing one of the hawk moths for the first time. There are probably nine resident British species and some visitors from the European mainland, but they include not only some of the largest of British moths but also some of the most beautiful. Most are declining in numbers in Britain, largely through habitat loss. However, there is still a good chance of seeing the Narrow-bordered Bee Hawk Moth *(Hemaris tityus)*, which so closely resembles a bumble bee, and the Humming-bird Hawk Moth *(Macroglossum stellatarum)*, which hovers, like a humming-bird, in front of summer flowers, into which it darts its long proboscis, just as the bird does its tongue. The commonest species in my experience is the Elephant Hawk Moth *(Deilephila elpenor)*, named I am sure, not because it is so large, but because its truly huge caterpillar, with its two eye-spots, looks almost like an elephant's trunk as it browses among garden fuchsias.

Among this family of wonderful insects, the Death's Head Hawk Moth *(Acherontia atropos)* is truly outstanding. This huge black-and-yellow insect is the largest British moth and takes its name from the remarkably skull-like marking on the top of its thorax. Its vast larvae feed mainly on potatoes and, in the West Country, where they are found most frequently, they were called 'tatur-dogs'. Its rarity means that it is never a pest, although its size has prompted a country belief that the caterpillars eventually turn into moles. The species did not acquire its present name until 1778, although the skull-and-crossbones flag of the pirates who roamed the high seas at that time was probably less familiar than it has become today, through the medium of films and books. A later, fanciful belief was that it first appeared in England in 1649, the year of the execution of Charles I, but it fairly evidently arrived rather earlier, and less romantically, with some of the first potato imports.

The day-flying Narrow-bordered Bee Hawk Moth *(Hemaris tityus)* is so called because it resembles a bumble bee when at rest.

The Death's Head Hawk Moth *(Acherontia atropos)* takes its name from the skull-like pattern on its thorax.

The Death's Head Hawk Moth is a night-flying moth and, when disturbed during the daytime, sometimes produces a squeak. This is as unexpected as it is alarming and is produced by forcing air through the proboscis. An old name for the moth was 'bee tiger', which stems from its extraordinary habit of entering bee hives and nests to feed on the honey. However, it sometimes never leaves, for the simple reason that this creature, so associated with mortality, is stung to death.

Clothes moths (Tineidae)

Among the commonest and most pestilential of moths is one that few people see. The Common Clothes Moth *(Tineola bisselliella)* and its relatives are insignificant, pale brown species whose caterpillars have a fondness for the chemical keratin, an ingredient of hair, feathers and many other fibrous materials. They have found it convenient to turn their attention from raw to processed materials, and large sums of money are still spent annually on mothballs containing naphthalene or similar substances in order to prevent them damaging fabrics.

Their activities have led to the expression 'moth-eaten' entering everyday currency. Curiously, however, the use of the word 'mothballed', to describe equipment, especially ships, that have been put into storage really has more to do with cocoons than mothballs. The vessels are sprayed with a protective plastic material that makes them look as if they festooned with silk.

Caddis flies
(Trichoptera)

OTHER COMMON NAMES: *caddis worm (larva); case worm (larva); cor bait (larva); sedge flies (Ireland); stick worm (larva); straw worm (larva); rails (Ireland).*

Caddis flies are small insects with two pairs of wings. Many species are superficially like small moths, but with hairy, not scaly, wings. They are typically found near water and almost all have aquatic larvae.

These larvae are very familiar to any student of freshwater life, and to anglers, who use them, or artificial versions of them, as bait. The caddis-fly larva is so distinctive because many species build small, protective cases for themselves from tiny pieces of twig and other woody matter, tiny fragments of gravel, or even the discarded shells of minute molluscs. The materials used, and the overall appearance of the case, is characteristic of individual species. As the larva grows, it adds pieces to the case, which thus 'grows' as well.

The larvae are omnivorous or carnivorous although adult caddis flies have very weak mouthparts and seldom feed on anything. The name 'caddis' is seventeenth century but its origin is obscure.

Flies
(Diptera)

OTHER COMMON NAMES: *flies; two-winged flies.*

Unlike most other insects, the true flies have only two wings, the hind pair being reduced to vestigial stumps. The Diptera is a huge order, with over 5,000 British species, although they are subdivided into a number of reasonably distinct groups. Some are among the most familiar of all insects.

The name 'fly' is commonly used in a general sense to mean any flying insect. It originates in Old English from Old Teutonic and has become familiarly applied to a great many other things that move through the air or simply travel quickly. Throughout the nineteenth century 'a fly' was 'a fast carriage' (by contraction from 'fly-by-night'), originally pulled by men. Later it came to mean 'a one-horse hackney-carriage'; usually given the unexpected plural 'flys'.

By convention, the names of the groups of true flies are hyphenated (house-fly, fruit-fly, soldier-fly and so forth), whereas the names of other insects, referred to less accurately as flies, have no hyphen (for example, dragonfly, whitefly, greenfly). Many species of fly are important crop pests and carriers of disease of man and animals, both adult and larval forms being implicated in different species. They are also an essential food resource for vast numbers of other creatures. The larvae of some groups have distinct names, although the word 'maggot', a Middle English name, is used generally for many of them; it seems unrelated to 'maggot' in the sense of 'Magpie' (see page 358). The following account includes only those true flies that are sufficiently large, common or important to have achieved some sort of recognition by the general population.

Over the centuries, man has attempted to control flies for as long as he has coexisted with them. However, it is only recently, with the improvement of hygiene in Western society, that there has been any major decline in the importance of flies in relation to human health. In combating the relatively few that are still a nuisance, the fly-swatter and fly-paper have given way to the aerosol insecticide.

Because they are scarcely endearing, flies have made little contribution to the fine arts. A modern generation might most readily think of William Golding's 1954 novel, *Lord of the Flies*, but even that featured the less than appealing image of a rotting dead pig: 'the pile of guts was a black blob of flies that buzzed like a saw.'[107] The 'Lord of the Flies', or 'Beelzebub' in Arabic, is one of the manifestations of the devil.

Crane-flies (Tipulidae)

OTHER COMMON NAMES: *crane-fly; daddy-long-legs; father-long-legs; friar; harry/ horry-long-legs; leather-jacket (larva); tom-tailor (Cornwall).*

Leatherjacket larvae are the curse of the gardener and green-keeper because they feed on grass roots, causing yellow patches in the turf. However, few people associate these subterranean pests with the large, leggy, fluttering creatures that enter our homes in late summer. One species of crane-fly, *Tipula maxima*, has a wing-span of 65 millimetres (2¾ inches), making it the largest of all British flies.

The name 'crane-fly' derives from the insect's resemblance to a crane (the long-legged bird, not, as is often assumed, the lifting machinery). 'Daddy-long-legs' is early nineteenth century, the suffix 'daddy' serving the same familiarizing function as personal names such as 'Jack' and 'Jill'.

Mosquitoes (Culicidae)

Almost any biting insect is likely to be called a mosquito, but the true mosquitoes are in a rather special league because the females have an elongated proboscis with which they pierce mammalian skin and suck blood. In the process, they can transmit several very important and debilitating diseases, including malaria, yellow fever and elephantiasis. It is not always appreciated that true mosquitoes exist in Britain, but there are about 30 species, of which *Culex pipiens* is the commonest. It is frequently found in our homes, its presence betrayed, like that of most mosquitoes, by the high-pitched buzz that it emits. It was this noise that led to the famous wooden De Haviland fighter-bomber of the Second World

One of the most famous aeroplanes of the Second World War, was the 'Mosquito' fighter-bomber, here being constructed in May 1943.

War being called the 'Mosquito'. *Culex pipiens* is harmless but *en masse* its appearance can be dramatic:

> In the year 1736 the Gnats, *Culex pipiens*, were so numerous in England, that, as it is recorded, vast columns of them were seen to rise in the air from the steeple of the cathedral at Salisbury, which, at a little distance resembling columns of smoke, occasioned many people to think the edifice was on fire.[108]

There is no yellow fever or elephantiasis in this country, but there are five species of British mosquito capable of transmitting the *Plasmodium* malarial parasites. The most efficient, *Anopheles atroparvus*, tends to breed in brackish water along river estuaries and, although malaria is no longer indigenous to Britain, this was not always the case. In early centuries, these estuaries were unpleasant and foul-smelling places; the name malaria itself is from the Italian *mala aria*, meaning 'bad air'. In England, the disease was called 'ague'. 'Mosquito' is sixteenth century, from Spanish and Portuguese, meaning 'a small fly'.

There are detailed descriptions of coastal areas of Kent and Essex in the sixteenth and seventeenth centuries that paint an altogether grim picture. At Iwade in Kent, it was said that:

> ... in summer dry weather, the stench of the mud in the ponds and ditches ... contributes so much to its unwholesomeness, that almost everyone is terrified to live in it ... .[109]

> gentlemen who ... go so far for it, often return with an Essex ague on their backs, which they find a heavier load than the fowls they have shot.[110]

By the late eighteenth century, children under the age of five in the so-called 'marsh parishes' suffered a mortality rate of almost 10 per cent, double the rate in settlements away from the marshes. For a variety of reasons, most importantly the draining of the marshes, malaria gradually declined and the last indigenous case in England was in the 1950s. In 1975 the World Health Organization declared Europe free of malaria, but climate change may yet show that nothing can be certain.

Gnats and midges

(Chironomidae and Ceratopogonidae)

OTHER COMMON NAMES: *buver; stut (West Country).*

> ... Then in a wailful choir the small gnats mourn
> Among the river sallows, borne aloft
> Or sinking as the light wind lives or dies ...
> (John Keats, 'To Autumn')

All small, dark-coloured flies that tend to occur in large numbers are popularly called 'gnats' or 'midges', and most belong to one of two families: the Chironomidae or the Ceratopogonidae. In angling circles, many artificial flies imitate chironomid midges and are familiar through such names as 'blae and blacks' in Scotland and 'buzzers', 'racehorses', 'black-flies' or 'duck-flies' in Ireland.

However, it is the members of the Ceratopogonidae that are the most significant because, being blood-sucking species, they cause irritation and annoyance out of all proportion to their size. This is especially the case in northern parts of Britain in summer, when they seem to descend on people with deliberate malice. In reality, they suck, not bite, but that is small compensation for the victim and enormous sums of money are spend each year by walkers, campers and anglers on fly-repellent preparations. Most irritation is caused by one species only, *Culicoides obsoletus*, a creature that seems anything but obsolete, as its scientific name suggests, when you are on the receiving end of its attentions. It occurs in vast numbers in the west of Scotland.

The larval forms of both families of midge live in water and, in the Chironomidae, are better known than the adults. These are small, red and wriggling and are commonly known as 'blood worms' (not to be confused with the *Tubifex* worms, which are similarly named (see Oligochaetes, page 23).

Both 'gnat' and 'midge' are Old English words and from 'midge' has come several other words that are applied to small creatures, such as 'midget'. Generally, 'gnat' is used for the larger species, but the names have no precise meaning, and mosquitoes are also sometimes called gnats.

> Countrey people suppose them, and that not improbably, to be procreated of some corrupt moisture of the earth. These small Summer Gnats are most frequent in the moneth of May, and seem to be nourished with a watery vapour, for their intestine or ventricle is very small, white and welnigh invisible, full of a white frothy thin moisture, and of little or no tenacity; sometimes they fly farther off from the water, and gather themselves in great companies about houses, as men passe over bridges they swarm about their heads, they love places that are without wind, the shun what they can a turbulent air, for by the troublesomenesse of the air they are dispersed hither and thither. Those kind of Gnats are properly called in English Midges.[111]

Bibionids (Bibionidae)

These are among the commonest of the small, black, hairy flies that are especially abundant in grassland from early spring onwards. The old name 'fever-fly' for *Dilophus febrilis* is ill-founded because it has nothing to do with fevers. 'St Mark's fly' for *Bibio marci* is more accurate because the fly often makes its first appearance around St Mark's Day (25 April), although why this particular insect should be one of the few flies with a genuine folk name is obscure.

Fungus-gnats (Mycetophilidae)
Any collector of wild mushrooms will be familiar with the larvae of fungus-gnats because it is their maggots that so frustratingly find the choicest edible specimens first. The adults are more familiar to the house-plant owner: these are the small, dark insects that, after hatching from pupae among the roots, flutter above the compost when it is allowed to become too wet and acidic.

Gall-midges (Cecidomyiidae)
Gall-midges are responsible for many of the puzzling abnormalities that appear on plants and cause largely unwarranted concern to gardeners. They are the commonest causes of plant galls, the swellings in which their larvae live, although most galls are merely disfiguring.

Horse-flies (Tabanidae)
OTHER COMMON NAMES: *breese; burrel-fly; cleg; clinger; doctor; gad-fly; gleg; goad-bee; goad-fly; stowt.*
Horse-flies are big insects with bulging eyes. The females are blood-sucking and, as their name implies, they feed most commonly on horses and other animals. That said, any insect capable of penetrating the skin of a horse can penetrate human skin rather more easily, and an attack by a horse-fly is not to be dismissed. The commonest species is the Common Cleg *(Haematopota pluvialis)* and, unlike most 'biting' flies, it gives absolutely no warning of its presence. It does not buzz, it does not hum; it simply appears and starts to suck your blood before you are aware of it.

The name 'cleg' seems appropriately unpleasant; it was originally Old Norse and came to England in the fifteenth century through Scots and dialect use. 'Gad-fly' is sixteenth century and can be traced back to an Old Norse word *gaddr*, meaning 'a spike'. The fact that the names are so old indicates that these creatures have plagued people for many centuries.

> They suck out bloud with such force and in so great abundance, that a friend of mine whom I dare believe told me, that his horse being tyed to a tree, was by reason of the multitude of them, killed in less than six hours, they had drawn out so much blood that the spirits failing he fell down dead.[112]

Robber-flies (Asilidae)
Robber-flies are large, strong, fearsome-looking insects with an ominous name and, in several species, black-and-yellow warning coloration. Nonetheless, it is other insects which they lie in wait for and rob; they are quite harmless to humans.

Hover-flies (Syrphidae)
Syrphids, or hover-flies, have now become as well known as House-flies, Blow-flies and a handful of other groups of flies.

The larvae of hover-flies, such as this *Syrphus ribesii* taking nectar from an *Aubrieta* flower, are significant predators of aphids.

This is because the larvae of many species are extremely significant natural predators of aphids and it has happened since gardeners began to use non-chemical pest controls and the organic growing of crops became more widespread. Nonetheless, it is still sometimes difficult to persuade people not to swat an insect that, in the adult state, can be mistaken for a wasp. The name 'hover-fly' is self-explanatory: the flies hover above flowers before landing.

One species with non-predatory larvae is superficially less endearing. The Drone-fly *(Eristalis tenax)* has an aquatic larva with an elongated, telescopic breathing tube. It is sometimes found in garden ponds and water butts, feeding on decomposing organic matter, where it is known as a 'rat-tailed maggot'. The adult fly, called a 'Drone-fly' because of its remarkable resemblance to a Honey Bee, was responsible for the ancient belief that bees arose spontaneously from putrefaction, a view that was not disproved until the seventeenth century (see Bees, page 99).

Bot-flies (Gasterophilidae)
Despite their large size, bot-flies are relatively little known to the general public but are very familiar to horse-owners because they are troublesome pests. The large adult fly lays its eggs on the leg of a horse, which licks off the resulting larvae; these then attach themselves to the lining of the animal's stomach. A severe attack can seriously weaken the horse and lead to infections.

'Bot', or 'bott', of unknown origin, is a sixteenth-century word for a maggot.

Piophilids (Piophilidae)

My grandfather was of the view that 'the more maggots, the better the cheese' and, even more picturesquely, that a really good cheese would walk out to meet you. However, I do not recall our cheese being noticeably mobile and am inclined to think that it was all talk on his part. Nonetheless, modern food hygiene regulations have all but done for the maggoty cheese and so the larvae of the Cheese-skipper *(Piophila casei)* are less familiar on cheese and bacon today than they were to previous generations.

> There is a small long shining fly in these parts very troublesome to the housewife, by getting into the chimneys, and laying its eggs in the bacon while it is drying; these eggs produce maggots called jumpers, which, harbouring in the gammons and best parts of the hogs, eat down to the bone, and make great waste ... it is to be seen in the summer in farm-kitchens on the bacon-racks and about the mantelpieces, and on the ceilings.[113]

Psilids (Psilidae)

Among the most troublesome of pests for generations of gardeners has been the Carrot-fly *(Psila rosae)*, an insect superficially like a small house-fly, although generally unnoticed in its adult state. Its larvae, however, are all too apparent in the roots of carrots and, over the years, all manner of wiles have been used to deter it from laying its eggs close to the plants. Apart from using noxious insecticides, which taint the carrots, much the most ingenious solution has been to erect low barriers around the plants, using the barrage-balloon principle and capitalizing on the fact that the female flies seldom rise far above the ground. Today, the development of carrot varieties that are resistant to the pest has been a major advance.

Vinegar-flies (Drosophilidae)

OTHER COMMON NAMES: *fruit-flies.*

The larvae of these small, relatively insignificant flies feed on rotting fruit and other fermenting products. They would be relatively unknown, other than to people whose livelihood depends on such things, were it not for one species that has entered the everyday currency of language. The Vinegar-fly *(Drosophila melanogaster)* has been lured away from vinegar to become the most important of all creatures for genetic study. It has a very short life cycle, very large chromosomes, which can readily be seen and counted under a microscope, and a ready propensity to mutate; these and other features have made *Drosophila* arguably the most studied animal in the world.

Warble-flies (Oestridae)

Warble-flies have for centuries been among the farmer's nightmares. All members of this family are internal parasites of mammals and, like bot-flies (see page 92), lay their eggs on animals' legs.

Two species are major pests of cattle in Britain, *Hypoderma bovis* and *Hypoderma lineatum*. Instead of entering the gut, they migrate through the body until they reach the skin beneath the back, where they create a small breathing pore to the exterior. The tissues around the larvae become inflamed and irritant, the hides are damaged and the animals weakened. The inflamed area constitutes the 'warble' and in due course, the larvae fall to the ground and pupate. A concerted campaign, largely with organophosphorus insecticides, has come close to eradicating warble-flies from Britain. A related species, the Sheep Nostril Fly *(Oestrus ovis)* has a similar life cycle but is confined to the nasal passages.

The name 'warble' is sixteenth century and preceded the name of the insect that causes it.

Blow-flies (Calliphoridae)

OTHER COMMON NAMES: *flesh-flies; Miss Margaret (Cornwall); old maid.*

> God in His Wisdom
> Made the Fly
> And then forgot
> To tell us why.
> (Ogden Nash, 'The Fly')

These must be among the most familiar of all flies and, because they include many species that enter our homes, they have also entered our lives, language and beliefs. They are large, they buzz, they are often strikingly coloured and they are fairly universally considered unpleasant. Among the best known are the Bluebottle *(Calliphora vomitoria)*, Greenbottle (*Lucilia* species), Flesh-fly *(Sarcophaga carnaria)* and Cluster-fly *(Pollenia rudis)*, which is an earthworm parasite and sometimes accumulates in huge numbers as it hibernates in buildings (thus giving a dramatic answer to the old question 'Where do flies go in winter?')

Many species feed on carrion and other waste material and therefore carry disease. Meat is said to be 'blown' when it has blow-fly eggs or maggots on it. The use of the word 'blown' in this way is of unknown origin. but 'blow-fly' is as recent as the nineteenth century. when it tended to replace the older 'flesh-fly'. The expression 'fly-blown' is often used today to mean anything dirty and contaminated, with the corollary 'there are no flies on ...'.

Nonetheless, blow-flies do have their uses. The familiar anglers' maggots are blow-fly larvae, commonly dyed with bright colours in order to be more attractive to fish. Few planning applications attract more interest, concern and hostility than those concerning the establishment of maggot farms for raising larvae. Perhaps to soften the impact on themselves and the public at large, anglers tend to use the sixteenth-century

This scanning electron micrograph of a Bluebottle fly (*Calliphora vomitoria*) shows the compound eyes that occupy most of the head.

and otherwise obsolete word 'gentle' (gentil) for their bait, rather than 'maggot'. I still occasionally hear the old word 'jumper' used similarly.

Blow-fly larvae are flesh-eating and:

> ... it is interesting to recall that this very condition was put to good use during the First World War. The larvae of some Blow-fly species will feed on wounds, but will restrict their attentions to putrefying tissues and will not attack healthy ones. Furthermore, the bactericidal properties of the maggots encouraged the growth of healthy tissues. This discovery was made by Baron Larrey, one of Napoleon's surgeons, but it seems not have been applied medically to any extent until the First World War. During that war, maggots were reared under sterile conditions and applied to the wounds of soldiers in the field, especially in cases on osteomyelitis (bone infection). The method was often successful, but it sometimes failed because, in certain cases, the maggots did not confine their attentions to the diseased tissues, but attacked healthy tissues as well. This happened because the medics did not distinguish between different species of Blow-fly, and sometimes they inadvertently included species which habitually attack healthy tissues. This was not understood at the time, and the method of treatment fell out of favour. Recently, there has been a renewed interest in this method, especially in the United States ... .[114]

Adult blow-flies also find an unexpected use in plant-breeding laboratories where they are used for the controlled pollination of caged flowering plants as they are much more manageable and cheaper than bees.

House-flies (Muscidae)

The members of the Muscidae are similar to the blow-flies and their relatives but are generally smaller and less hairy. There are many British species, and many are common in houses, but *the* House-fly *(Musca domestica)* is possibly the commonest and most widely distributed animal in the world. Unlike blow-flies, it lays its eggs in decomposing material and manure; this is the fly whose maggots will be found in uncollected rubbish and unemptied dustbins. It is thought to be responsible for a large number of stomach upsets and other gastrointestinal disorders by transmitting harmful bacteria to our food.

> A classic British study early this century showed that the decline of horse-drawn vehicles correlated with a decline in the death rate caused by summer diarrhoea, and it was concluded that the decrease in the amount of horse manure on the roads reduced numbers of fly-breeding sites resulting in fewer flies and hence less diarrhoea. The fly species in this case was the House-fly, *Musca domestica.*[115]

This family includes other flies that share a fondness for manure, such as the Stable-fly *(Stomoxys calcitrans)*, which resembles the House-fly but bites, and the Yellow Dung-fly *(Scatophaga stercoraria)*. Incredible as it seems, and despite this unwholesome behaviour, House-flies have been put to medicinal use:

> A fly chewed and swallowed doth vehemently provoke vomiting ... I have heard of a certain man that was wont to take three or four flies into his body, which gave him a very good stool.[116]

Possibly there is more than a grain of truth in the old rhyme after all:

> There was an old woman who swallowed a fly;
> Perhaps she'll die.[117]

Other species in the same family that are familiar, at least to gardeners and growers, are the Cabbage Root Fly *(Delia radicum)* and the Onion Fly *(Delia antiqua)*, which are among several important plant pests.

Louse-flies (Hippoboscidae)

These flies are all blood-sucking parasites of birds and mammals, and include a number of species that are perhaps better known by name than by sight. Deer are attacked by the Deer-fly *(Lipoptena cervi)*, sheep by the Sheep Ked *(Melophagus ovinus)* – also written as 'kade', a sixteenth-century word of unknown origin – and New Forest ponies by the Forest-fly *(Hippobosca equina)*, but it is fairly unusual for any of them to bite people. Nonetheless, the Forest-fly is a rather fearsome-looking insect that can cling tenaciously to hair with its long, curved claws.

At one time (and perhaps still) small boys would gain much amusement from capturing a Forest-fly in a match-box and later releasing it into their sisters' hair.

Fleas
(Siphonaptera)

> So, naturalists observe, a flea
> Hath smaller fleas that on him prey;
> And these have smaller fleas to bite 'em,
> And so proceed *ad infinitum.*
> Thus every poet, in his kind,
> Is bit by him that comes behind.
> (Jonathan Swift, 'On Poetry; A Rhapsody')

This often misquoted verse is not a bad stab at natural history for the eighteenth century, although, in reality, fleas do not themselves have fleas. Considering how few species of flea there are (perhaps 60 in Britain), they have certainly made their mark on mankind in more ways than one.

Fleas are tiny, wingless, and, uniquely among insects, flattened laterally (to enable them to wend their way through hair or fur); they are also parasitic, mainly on mammals, but a few on birds, sucking their host's blood. They are irritating and they cause you to scratch. Unlike lice, however, they are much less closely tied to their hosts and wander off from time to time. Only the adults are parasitic, the legless, blind larvae living in debris, and they are not usually host-specific. The Human Flea *(Pulex irritans)* is far from exclusive to humans and occurs on other, hole-inhabiting mammals, such as Foxes, and adopted our ancestors when they lived in caves. The fact that cat fleas and dog fleas also bite people is testimony to their relatively catholic tastes. Also unlike lice, fleas can transmit diseases and the possible role of fleas and Black Rats in the historical outbreaks of plague is discussed elsewhere (see Black Rat, page 479). In Europe, the Rabbit Flea *(Spilopsyllus cuniculi)* is the main vector of the Myxoma virus, the cause of myxomatosis (see Rabbit, page 490), while the extremely common Cat Flea *(Ctenocephalides felis)* is important not only because it affects both humans and dogs as well, but also for its ability to transmit the Dog Tapeworm *(Dipylidium caninum,* page 22*)*.

The Human Flea is now far less important in Western society, but as recently as the nineteenth and early twentieth centuries such beliefs as the need to take special care on St David's Day (1 March), were widely followed. This was thought to be the day when fleas returned to people's homes but, paradoxically, different parts of the country had different beliefs. In some places, it was thought important to keep the house securely closed to prevent the fleas from entering. In others, all doors and windows were flung wide open to rid the house of them and everything was then cleaned from top to bottom. In Arundel, Sussex, the residents stood on the local bridge and shook themselves in order to be rid of infestation for the coming year. Aromatic plants, especially the appropriately named Common Fleabane *(Pulicaria dysenterica)*, were brought into the house.

Much the best-known attribute of fleas is their ability to jump enormous distances, a feat that facilitates their transfer from one host to another. The record leaps for *Pulex irritans* appear to be 300 millimetres (about 12 inches) laterally (a long jump) and 197 millimetres (about 8 inches) vertically (a high jump); it is often said that if a man, in proportion to his weight, could jump as high as a flea then St Paul's Cathedral would be well within his scope. It is this ability to leap that is believed, at least in part, to have been the means for the spread of plague and is the reason why close, crowded and none-too-clean conditions represent ideal opportunities for the insects to infest new hosts. In the twentieth century it was not for nothing that modest and poorly managed cinemas became known as 'flea-pits'; nor was the name 'flea-market' ill-founded.

However, much the oddest incorporation of fleas into human society has to have been the flea circus. These circuses seem to have originated in the nineteenth century and were an essential entertainment in travelling fairs. The fact that the principal performers were all but invisible (*Pulex irritans* is only about 1.5 millimetres/1⁄20 inch long) was no deterrent. If you could barely see the fleas, at least you could see what they did and, by harnessing the insects to pull coaches and perform similar feats, their strength and prowess were amply demonstrated. The few modern flea circuses sometimes cheat by using video cameras to project images of the show onto giant screens.

An odd phrase that has become part of our everyday language is to send someone away 'with a flea in his ear'. There is clearly something special about human ears that attracts insect life (see Earwig, page 67) and the association between ears and fleas dates back to the fifteenth century in English, although there are older versions in French. It is suggested that the ears in question are not those of humans but of dogs, which display particular discomfort when so afflicted.

Hymenopterans
(Hymenoptera)

This is an enormous group of insects that includes the bees, wasps and ants. Although they range widely in size, they can be recognized relatively easily by their two pairs of wings (although many species also have wingless forms) and, in most cases, their very narrow 'waist'. So characteristic is this feature that 'wasp-waisted' has passed into the language of women's fashion.

The group has a number of fascinating and intriguing characteristics and, not surprisingly, has had a considerable impact on lives and customs in Britain. Bees are hugely important in the pollination of flowers, as well as in the production of honey and wax, and, although it is inaccurate to describe them as truly domesticated, they have come under human direction and influence more than any other insect.

Many hymenopterans are social, living in large colonies with a remarkably complex collective structure and function. A considerable number of species also either bite or sting, with painful results. Apart from the sawflies, the larvae are legless and all species undergo metamorphosis from egg, through larva and pupa, to adult.

SAWFLIES
(Symphyta)

Sawflies take their name from the saw-like ovipositor and are the only hymenopterans not to have a waist. They include species that, while not especially common, probably look more alarming than almost any other British insect. Sawfly larvae bear a remarkable resemblance to butterfly or moth caterpillars but have more prolegs.

The most familiar, however, are known not by their appearance, but by their actions. Many are important pests of both farm and garden crops; the larvae of the Gooseberry Sawfly *(Nematus ribesii)*, for example, will defoliate gooseberry or currant bushes in gardens almost literally overnight in the spring. Among other common effects of sawfly larvae attack are the stripping of the surface of rose leaves by the Rose Slug Sawfly *(Endelomyia aethiops)*, the tiny red galls on willow leaves caused by the Willow Bean-gall Sawfly *(Pontania proxima)*, the ribbon scarring of apple fruit caused by the Apple Sawfly *(Hoplocampa testudinea)* and the serious defoliation of pine trees caused by *Neodiprion sertifer* and other species.

The really terrifying-looking insects, however, are the so-called 'wood wasps', or 'horntails' (Family Siricidae), the

The huge ovipositor of the Wood Wasp, or Horntail *(Uroceras gigas)*, is quite harmless, although it is often mistaken for a sting.

commonest of which, *Uroceras gigas*, is a huge black-and-yellow insect with an enormous ovipositor. Understandably, but quite mistakenly, this is often thought to be a prodigious sting but it is so large and so long merely because the insects use it to bore into trees. An obscure species of wood wasp *(Xyphidria camellus)* that attacks alder trees (*Alnus* species) achieved enormous but unexpected fame in the 1950s when it became the subject of an award-winning macro-photographic film, *The Alder Woodwasp and its Insect Enemies*, by the pioneering British natural-history film-makers Gerald Thompson and Eric Skinner, which was shown around the world.

ICHNEUMON AND BRACONID FLIES
(Ichneumonoidea)

The effects of ichneumon flies are seen more frequently than the insects themselves. They are parasitic, principally on other insects, and play a vital role in the regulation of pest species. Ichneumon flies (Family Ichneumonidae) are superficially rather like flying ants, although the females commonly have an elongated ovipositor, often mistakenly thought to be a sting. It is with this organ that they lay their eggs in the living bodies of other insects which are, in due course, eaten by the ichneumon larvae.

The name 'ichneumon fly' is seventeenth century but its origin is in the Greek word for 'a tracker', or 'follower'. In the sixteenth century, it was first applied to the Egyptian Mongoose *(Herpestes ichneumon)*, which was highly regarded by the Romans for its supposed habit of eating crocodile eggs.

GALL WASPS
(Cynipoidea)

I have often said to gardeners that, if your plants are going to be attacked by a pest, at least be thankful if the symptoms are attractive, and this is the case with many of the gall wasps. The insects themselves are minute and lay their eggs in plant tissue. The resulting larvae cause the host to develop an abnormal swelling, or gall, which can sometimes be so attractive that it is mistaken for an ornamental fruit or flower.

Perhaps the commonest gall wasps (Family Cynipidae) are the four species of *Neuroterus* that cause masses of small, brightly coloured spangle galls that dot the underside of oak leaves. In more recent years, the 'knopper' galls on distorted acorns, caused by Acorn Cup Gall Wasp *(Andricus quercuscalicis)*, have attracted much attention but perhaps the most striking of all are 'Robin's pincushions'. These remarkably flower-like, 'hairy' red galls on roses are caused by the Bedeguar Gall Wasp *(Diplolepis rosae)*, and, I believe, are named after the mythical Robin Goodfellow, or Puck, not the bird (see Robin, page 330).

The flower-like 'Robin's pincushions', found on rose bushes, are caused by the Bedeguar Gall Wasp *(Diplolepis rosae)*.

ANTS
(Formicoidea)

OTHER COMMON NAMES: *emmet; emmut; meryan/meryon/muryan/myryan (Cornwall); pismire; yammet.*

All ants (Family Formicidae) live in organized social colonies consisting of a female (the queen), some males, and the workers, which make up the vast majority and are wingless, sexually undeveloped females. There are about 36 species of ant in Britain and, in some, certain of the workers develop slightly differently and serve as 'soldiers', protecting the colony.

Ants become most obvious when they enter houses in search of sweet food materials; when they undermine garden plants with their tunnelling activities; when they bite (the general perception is that red ants bite, black ants do not and, among common species, this generally holds true); and when, at certain times of the year, vast numbers of them take to the air. Flying ants are fully developed winged males and females, and individual nests seem to produce mostly one sex or the other. Because the development and emergence of the flying ants is controlled by weather, all nests in a district erupt simultaneously as the insects take off on mating flights before subsequently establishing new colonies.

The Wood Ant *(Formica rufa)* builds a large mound-like nest of leaves and other debris, which may contain many queens.

Different species of ant exhibit different levels and types of social organization and many parallels have been drawn with human society. Thus, some species are nomadic, some keep 'slaves', while others have been likened to farmers in cultivating fungi and even herding aphids and other creatures for their own ends.

> They have Officers of all sorts, as Purveyours for Corn, Gleaners, Storers, Yeomen of the Larder, Householders, Carpenters, Masons, Arch-workers, Pioners; for such is the vertue and skill of every one, that each Ant knowes what is needfull to be done, and willingly doth his best to help the Common-wealth.[118]

The most impressive British species is the Wood Ant *(Formica rufa)*, whose enormous mounded nests of twigs and other materials are especially familiar in the New Forest. Ants themselves are eaten by a large number of creatures, including other insects and many birds, most conspicuously the Green Woodpecker (see page 315).

The name 'ant' originates in the Old English *émete*, which survives in the colloquial 'emmet'. However, despite their carefully organized and efficient social structure, there is no obvious reason why they should be thought prudent, although Milton wrote of them as the 'parsimonious emmet'. 'Pismire', which also survives in some areas, comes from the verb 'to piss', which is derived from old French (ant-hills smell of urine), and 'mire', another Old English word for ant, which originates in the Greek (and from whence comes 'myrmecology', the study of ants). Shakespeare used both 'ant' and 'pismire' in *Henry IV, Part 1.*

Curiously, there appear to be no major settlements named after ants in any of their linguistic guises, although places such as Pismire Hill, near Sheffield, testify to a thriving insect population in the past.[119] The Latin generic name *Formica* may ring bells, however, because it is the trade name of a type of laminated plastic sheet, which is chemically related to formic acid, the astringent chemical that is produced by ants and that causes the painful but fairly short-lived sensation after a bite. 'Formicary' and 'formicarium' are nineteenth-century names for an ant-hill but I am particularly saddened that we seem to have lost the seventeenth-century verb 'to formicate', meaning 'to crawl around like an ant'.

There was a Cornish belief that ants were fairies in the final stages of their earthly existence, gradually, generation by generation, becoming ever smaller. According to a related story, unbaptized children become piskies (see Puss moths, page 87) when they die, and finally, through further transformations, become meryons (ants). Special interest has attached to ants' eggs, which have been used for various medicinal purposes; in Scotland, ants' eggs mixed with onion juice were a treatment for deafness.

WASPS AND HORNETS
(Vespoidea)

OTHER COMMON NAMES: *apple-drane/drone (West Country); cercole (hornet); haeps; hornicle (hornet); hyrnet (hornet); waesp; weaps; whamp; wype.*

Wasps must be the most misunderstood and unjustifiably maligned of common British insects. They can be undeniably troublesome in late summer, when the workers have finished looking after the young and are attracted by ripening fruit – and are pursued with rolled-up newspapers in consequence. This is understandable but they should not be destroyed unless their nests are very close to habitations where they might come in contact with very young or very old people, or people who are allergic to their stings. It is not generally realized that, whereas bees feed their young on pollen and nectar, wasps feed their young on animal matter, including a great many insect pests. As a result, their role as biological control agents is largely unappreciated. One family of wasps has a more dubious benefit, however, in feeding solely on other valuable creatures; they are known, with good reason, as the spider-hunting wasps (Family Pompilidae).

Most of the best-known British wasps, including the largest, the Hornet (*Vespa crabro*; Family Vespidae), are social and live in colonies; 'to stir up a hornet's nest' is a familiar enough phrase and means 'to create trouble'. Yet, in reality, most British wasps are solitary species and some make truly beautiful miniature nests for their offspring.

Anyone with an old house whose bricks are held together by lime mortar will be familiar with mason wasps (*Odynerus* species; Family Eumenidae), which tunnel into the walls and make small cavities in which the tiny, chambered nests are built. Into each chamber, the female wasp places caterpillars to feed her brood and then lays a single egg before flying away. Rather touchingly, the adult wasps never see their offspring and the wasp larvae have no contact with their siblings.

The large mated females of the social wasps (Family Vespidae) are seen in spring, when they emerge from hibernation sites, in roof spaces and similar sheltered spots, to seek a suitable hole in which to build their nests. Unlike bees, wasps are unable to make wax for their nests but, instead, build them of papery material; like real paper, this is made from chewed-up wood fragments. Having built a small nest, the queen lays eggs and raises a few workers. These then take over the task, enlarging the structure until, at the height of the season, it may contain over 20,000 insects.

The characteristic black-and-yellow warning colours of the wasps have been mimicked by many other insects as a form of protection and, as far as wasps themselves are concerned, it is remarkably effective; very few birds will touch them. However, if the warning fails, wasps have something rather more robust in reserve. Most people are aware that wasps, unlike bees, can sting more than once because they are able to withdraw the modified ovipositor that forms the stinging organ from their target and use it again. The venom in a wasp sting is chemically very complex, but the most significant component is an allergy-provoking protein called Antigen 5. In Britain the majority of allergic reactions to insect stings are provoked by wasps. A single sting may be sufficient to sensitize someone, although most people are allergic to either wasp venom or bee venom, but not both. Wasp stings are obviously painful but, generally, that is all and the best treatment is a dose of oral antihistamine. There is no danger of long-term or fatal injury except in people who have very serious allergies or who have been subjected to multiple stings. In Britain, in an average year, there are only four deaths from anaphylactic shock following wasp or bee stings, usually stings to the face, scalp or neck. This is a minute number in relation to the number of insects, the number of people and the number of stings. There is some evidence, however, that certain people secrete a pheromone that is attractive to the insects, thus making them them more likely to be stung.

The Hornet *(Vespa crabro)* is the largest British wasp and nests in hollow trees, chimneys and wall cavities.

The name 'wasp' is from Old English (both *waeps* and *waesp*) and was originally Old Teutonic, possibly related to 'weave', in reference to the way that the nests are constructed. 'Hornet' is also Old English and perhaps has some connection with 'horn'. 'Waspish', to describe irritating, aggressive, tetchy behaviour, is sixteenth century.

> If I be waspish, best beware my sting.
> (*The Taming of the Shrew*, Act 2, Scene 1)

Wasps have never been popular creatures, which must explain their rarity in heraldry, and as inn names, and the fact that no-one will admit to being named after them. Nor are there are any places of significance named in their honour; a likely candidate in Warwickshire is Wasperton but that seems to originate in the Old English word for 'swamp'.

BEES
(Apoidea)

For most people, there are only two kinds of bees, the bumble bees (mostly *Bombus* species) and the Honey Bee *(Apis mellifera)*, although, in reality, there about 250 species in Britain; a tenth of these are social, the remainder being solitary. The main difference between bees and wasps is that both adult and immature bees feed on nectar and pollen, whereas wasps feed on other creatures. It is in collecting nectar and pollen that bees perform such a crucial role in the pollination of flowers, the two familiar groups of bumble bees and honey bees, with their extremely long tongues, being especially important.

Bee stings are usually less significant than wasp stings and the active principle is different: the major allergy-producing component of bee venom is an enzyme called phospholipase. Bees are able to sting only once because the sting and venom are usually pulled out of the insect's body during the stinging process and the insect dies within a short time. If you are stung, it is important to flick the sting away from the skin rather than squeezing it, which may liberate more venom into the wound. People who are likely to be stung by bees, perhaps in the course of their work or hobby, can be treated by venom desensitization immunotherapy. This can be very effective, although it does carry the risk of inducing an allergic reaction.

As with wasps, relatively few other creatures will feed on bees. Nonetheless, among birds, that now almost vanished British species, the Red-backed Shrike *(Lanius colloria)*, together with the Green Woodpecker *(Picus viridis)* and the Honey Buzzard *(Pernis apivorus)*, which has a special fondness for the larvae, will take them with relish. Many other birds have the colloquial name 'bee-bird' but, generally, this indicates only that bees are components of a varied insect diet.

This worker Bumble Bee *(Bombus terrestris)* is collecting pollen, which it will take back to the colony's underground nest.

The word 'bee' is Old English and may have an older origin that meant 'buzzing', or even 'quivering' as in fear, but it is odd that this hugely important creature has acquired so few other names down the centuries.

Leaf-cutting and burrowing bees
(Megachilidae and Andrenidae)

Among the common but less familiar bees are the leaf-cutting bees (*Megachile* species). These creatures are responsible for removing neat, semicircular pieces of leaf, which they carry away to construct their nests. Every rose-grower must have noticed this damage, even if he or she failed to appreciate the cause. Burrowing bees (*Andrena* species) may make a nuisance of themselves by excavating holes in lawns. However, it is the bumble and honey bees that really attract attention.

Bumble bees (Bombidae)

The hairy bumble bees are all social bees, mostly species of *Bombus*, and some are very imposing insects. A big queen bumble bee, emerging from hibernation in spring to begin the search for a nesting site (old mouse holes are very popular), might seem alarming as it bumps and buzzes its way around a window pane. In reality, the sting of a bumble bee bears no relation to the creature's size and is relatively insignificant.

The name 'bumble bee' is sixteenth century, from a Middle English word meaning 'to boom' or, in its more picturesque older form, 'to bum'. 'Humble bee', which is less used today, is slightly older and perhaps alludes to the humming sound. The delightful and still fairly widely used regional name 'dumbledore' (with variants such as 'doombledore', 'dor', 'dory', 'drane', 'drumbee', 'drumble' and 'drumble-do') is eighteenth century and is sometimes also used for the Cockchafer (see Chafers and scarab beetles, page 106).

Like many birds and other creatures, bees have declined with the change in appearance of the countryside and loss of once widespread habitats.

> The short-haired bumble bee is the 154th species to become extinct in Britain this century, said the World Wide Fund for Nature yesterday. *Bombus subterraneus* is, or was, one of 21 species of bumble bee in Britain ... it was one of the long-tongued bumble bees with a proboscis that could take nectar from the flowers of traditional hay meadows. The ninety five percent decline of ancient meadows and pastures since 1945 is thought to be the main reason for its decline ...[120]

Honey bees (Apidae)

The Honey Bee *(Apis mellifera)* is probably the best-known insect in the world. All honey bees in Britain today are derived from stock that was introduced artificially from elsewhere in Europe for honey production in hives. The Honey Bee is so

No substitute has yet been found for honey and so the age-old craft of bee-keeping is still an important commercial activity.

inseparable from the artificial nesting site of the hive that it is sometimes called the 'hive bee'.

Modern technology has produced many things but never an artificial substitute for honey, and bee-keeping is still an important commercial activity. Like most intensively 'farmed' creatures, honey bees are subject to a number of pests and diseases, of which much the most serious has been the mite *Varroa jacobsoni*. This has transferred its attention from its natural host, the Asian Honey Bee, and was first found in England in 1992. It is a blood-sucking creature that attacks larvae, pupae and adults, and it is now probably present throughout most of the British Isles, threatening the very existence of the bee-keeping industry. It is now combated by a range of chemical acaricides.

As with other social insects, there are three castes: the fertile female (queen), the male (drone) and the worker. However, compared with ants and wasps, and even other types of bee, the devotional roles of all three are exaggerated in the Honey Bee. The workers especially are possessed of legendary energy ('as busy as a bee'), while the queen is quite unable to secrete wax to build nest cells or to collect pollen. So devoted is she to her procreative role that the expression 'queen bee' has become used for a woman who is very demonstrably matriarchal. A queen bee cannot therefore start a new colony on her own; she must either take over an existing hive or take a few workers with her.

The production of queen bees is controlled entirely by the way that the larvae are fed by the workers. All bee larvae are fed on a substance popularly called 'royal jelly', a protein-rich material that the workers produce in their salivary glands, at least for the first three days. Most larvae are then fed on nectar and pollen and develop into workers. The larvae in a few cells of the hive continue to be fed on royal jelly and consequently develop into queens.

The first new queen to emerge usually stings the other potential queens to death, before flying away with the drone bees, mating with one (who afterwards dies) on a 'nuptial' flight. She then usually returns to her own hive, although not necessarily directly, or in a 'bee-line'. If the existing queen is old and tired, she may be killed by the workers in favour of the new one, or she may simply fly off with some of the workers in the familiar swarm to found a new colony. In the autumn, the drones, which have no role other than to fertilize the queens, die or are ejected from the hive. Many of the workers also die and the reduced colony of some workers and the queen passes the winter living on the pollen and honey that it has stored.

There's a whisper down the field where the year has shot her yield,
And the ricks stand grey to the sun,
Singing: – 'Over then, come over, for the bee has quit the clover,
And your English summer's done'.
(Rudyard Kipling, 'The Long Trail')

The belief that bees are produced in the carcasses of dead animals dates back to classical times, especially to Virgil, and you need not look very far to see the modern manifestation of it: the Tate and Lyle Golden Syrup tin bearing a picture of bees emerging from a dead lion with the slogan 'Out of the strong shall come forth sweetness'. The notion is generally thought to have originated with a superficially bee-like fly, such as the Drone-fly (see Robber-flies, page 92), which breeds in dead animals.

Beside, who doth not see, in daily practice,
Art can beget bees, hornets, beetles, wasps,
Out of the carcases and dung of creatures;
Yea, scorpions, of a herbe, being rightly placed ...
(Ben Jonson, 'The Alchemist')

The bee hive, the place where bees are kept, is, of course, as old as bee-keeping itself, although its form has changed over the centuries. The word comes from the Old English *hyf*, and perhaps from the Latin word for a cask; older bee hives tended to be cask-shaped and to be made of wicker or similar material. Oddly, they were then usually called a 'skep', an Old Norse word for a basket, which survives today in the word 'skip', meaning 'a receptacle for rubbish'. 'Hive' has become used to mean any place where people are very busy – 'a hive of activity' – although it seems a shame that the verb 'to hive', meaning 'to gather together', has only really survived in the expression 'to hive off'.

Because of their role in providing human society with an invaluable source of sweetening, and also because of the importance of beeswax in the manufacture of church candles (especially the Paschal candle, symbolic of the risen Christ), bees have long had a religious significance. They have been endowed with numerous religious and quasi-religious attributes and, in parts of Britain, they were believed to hum the Hundredth Psalm on Christmas Eve. At least one Parish has proved it:

At Muchelney Church, near Langport in Somerset, the nave interior roof is painted with angels. It also has a very old barrel organ. Sometime during one summer in the1800s, the bees that were owned by Mrs White (farmer on opposite side of the road) swarmed and could not be found. On Christmas Eve, the sexton who was setting the barrel organ to play the 100th Psalm for the church service, heard a big hum as if the angels were singing. Somehow the bees had got into the roof.[121]

Because of this special significance, killing a Honey Bee has always met with special opprobrium. Bees are thought to be sensitive, becoming distressed when there is strife in a household, as well as being lovers of chastity and bringers of good luck. There is a very old belief that it is unlucky to sell or buy bees and, in the north of England, it was held that bees should never be owned by one person but by partners from different households.

However, perhaps strongest of all is the very widespread view that the bees must be allowed to share in the joy and sadness of the family. When the head of a household dies, the news must always be told first to the bees. This is done in slightly different ways in different parts of the country, although it is not a practice peculiar to England and may originate with the ancient Egyptians.

The old method of notifying an owner's death was for his eldest son, or his widow, to strike the hives three times with the iron door-key and say 'The master is dead'. If this was not done, the bees would die, or fly away, and fresh misfortune would follow. The hives had then to be put into mourning by tying pieces of black crêpe upon them. In some districts, they were turned round, or moved to a new place, at the moment when the corpse was being carried out of the house. At the funeral feast, sugar, or biscuits soaked in wine or samples from the dishes served to the mourners, were taken to the bees.[122]

The following tale, concerning bees, was told to me by my mother. Her father was a nurseryman of repute, taking prizes for his pot plants and growing, assisted by the family, a wide variety of fruits, vegetables and flowers to supply the family shop in Wiltshire. Her

The Tate and Lyle Golden Syrup tin depicts bees emerging from a dead lion.

> mother sadly died one day during the summer of 1900. The next morning, her father went to the hives and found that all the bees were dead. He solemnly declared that this has happened because he had not gone to tell them that his wife had died.[123]

As with countless other creatures, a bee entering the house has been regarded as foretelling some event, although not necessarily a bad one.

> A stranger is coming if a bumble bee enters the house.[124]

> My grandmother, who was born in 1897 in Dunchurch, Warwickshire and lived there for most of her life, believed that a bumble bee entering the house foretold the arrival of a visitor. As there was an almost constant stream of people calling in, she was invariably correct. When my grandfather died and his stocks of honey bees passed to me, my grandmother told the bees, so that they would know for whom to work.[125]

While much folk belief is as mysterious as it is obscure, I always find the imagined association between the swarming of bees and death to be among the most understandable. Such massed and vibrant life leaving the colony would inevitably attract superstition and, if a swarm landed on a dead tree, or even on dead wood, it was widely taken as a death warning. In Cornwall, anyone finding a swarm and throwing a handkerchief over it could claim it as his own.

Until very recently, there was a widespread practice of 'tanging', or 'ringing', a swarm of bees; this involved banging kitchen utensils together in order to make a loud noise and thus persuade the bees not to travel far. It was said that an owner could pursue swarming bees onto another's land without being accused of trespass, provided that he kept up the clamour. There is a saying found in many parts of the country:

> A swarm of bees in July is not worth a butterfly.[126]

A Sussex variant runs:

> A swarm of bees in May is worth a load of hay;
> A swarm in June is worth a silver spoon;
> A swarm in July is not worth a butterfly.[127]

While in Wales, it was said that:

> A swarm of bees in May betokens a good crop of hay ... A July swarm of bees means a needy winter.[128]

Uniquely among insects, 'bee' and associated words crop up everywhere in personal and place names, some suggesting the keeping of bees, others simply the collecting of their produce, but all are testimony to their very special importance in human endeavours: Beckett in Berkshire ('bee cot'), Beoley in Worcestershire ('bee wood') and, most evocatively, Bickerstaffe and Bickershaw in Lancashire ('the copse and the landing place of the bee-keepers'); Honeyborne in Gloucestershire ('stream where honey could be collected'), Honiley in Warwickshire ('a wood were honey was obtained'), Honington in Warwickshire ('homestead where honey was produced'), Hunnington in Worcestershire ('a tun where honey was produced'). Nonetheless, not all names like these can be assumed to have bee associations because, although *hunig* was Old English for 'honey', Hūna was an Anglo-Saxon personal name.

In this 1837 engraving by G.S. Tregear, the British hierarchy is compared with the caste system in a beehive.

It is easy today, when honey is little more than an item to spread on toast, to forget how important it was in the days before cane sugar and food preservatives were available. Honey was used to preserve natural materials as diverse as meat and leather, and among its other more unexpected applications was the technique of honey-gilding; this was widely used to decorate early porcelain, powdered gold being mixed with honey before being applied like paint. Also fairly unfamiliar today is mead, an alcoholic drink made by fermenting a mixture of honey and water. This beverage is of very ancient origin (the word itself probably comes from Sanskrit) and was widely drunk in England, especially before the introduction of beer in the sixteenth century. It is an acquired taste for the modern palate; most people find it sickly.

The odd expression 'a bee in your bonnet', meaning 'to have a strongly held or irrational view about some matter' is of unknown origin, but the belief that bees might inhabit bonnets is old:

> Ah! Woe is me, woe, woe is me,
> Alack and well a-day!
> For pitty, sir, find out that Bee
> Which bore my love away.
> I'le seek him in your bonnet brave,
> I'le seek him in your eyes;
> Nay, now I think th'ave made his grave,
> I'th' bed of strawberries ...
> (Robert Herrick, 'The Mad Maid's Song')

There is an old Scottish saying 'his head is full of bees', which was applied to drunkards.

Beetles
(Coleoptera)

Among a remarkably successful and numerous group of animals, the beetles are the most successful and numerous of all. They are the largest of all the insect groups, with more than 4,000 species in Britain alone, although relatively few are seen by the casual observer because so many are small and live well-concealed lives among vegetation at soil level.

Despite the huge range of species, and the variety of their diets and habitats, most beetles are readily recognizable as such because their basic form is fairly constant. Almost all can fly and have two pairs of wings, the hind pair generally delicate but well protected by the tough front pair (elytra), which meet neatly down the middle of the back when folded. Most have strong, biting jaws. After hatching, all species undergo a larval stage, which is usually six-legged, with a clearly defined head, and a pupa stage. Unlike butterflies, moths and flies, the larvae usually eat the same type of food as the adults.

The vividly coloured Green Tiger Beetle *(Cicindela campestris)* is common in Britain.

'Beetle' is a word with origins in the Old English *bitan*, meaning 'to bite'. Surprisingly for such common and numerous insects, the name has found relatively few other applications. 'Beetle-browed', meaning 'overhanging', or more specifically 'with bushy eyebrows', might be related, but the hammering implement called a 'beetle' originates in the Old English *beatan*, 'to beat'. In recent times, of course, the word 'beetle' has become inseparable from the Beatles of popular music fame. A 'beetle drive' is a strange game, often associated with whist, in which the contestants endeavour to draw the outline form of a beetle.

Tiger beetles (Cicindelidae)
The often brightly coloured tiger beetles are among the commonest large species and are found in sunny places among grass and other vegetation.

Ground beetles (Carabidae)
Relatives of the tiger beetles, the generally black ground beetles are among those creatures that dash for cover whenever large logs or stones are turned over. It is largely these beetles that have given rise to the beliefs attached to black beetles generally: that one running inside a house is bad luck and that, if it crawls over someone's foot, death may follow. Nonetheless, killing a black beetle is also ill-advised, bringing at worst, death and hardship and, at best, heavy rain.

Water beetles (Gyrinidae and related families)
The appropriately named whirligig beetles (Family Gyrinidae) are familiar on the surface of ponds as they spin around, apparently aimlessly, in company with water boatmen (see Heteropteran bugs, page 71) and other surface-dwelling creatures. The Great Water Beetle (*Dytiscus marginalis*; Family Dytiscidae) is one of the largest British species. A diving beetle almost 4 centimetres (1½ inches) long, it is an extremely aggressive carnivore, well able to take on creatures such as newts and fish that are many times larger. Living in the same habitat but even larger is the Great Silver Beetle (*Hydrophilus piceus*; Family Hydrophilidae) the second largest British beetle, but vegetarian and therefore a popular creature for an aquarium; collectors have brought about a significant decline in its population in recent years.

Burying beetles (Silphidae)
The burying beetles perform an important although unsavoury task in disposing of dead animal carcasses, which they bury by digging away the soil from beneath before laying their eggs in the grave.

When disturbed, the Devil's Coach Horse beetle *(Staphylinus olens)* adopts a threatening posture and emits an offensive liquid.

Rove beetles (Staphylinidae)

The best-known of the rove beetles is often mistaken for an earwig and has earned itself a sinister common name. The Devil's Coach Horse (cock-tail) *(Staphylinus olens)* is a black, elongated creature which turns up its tail in an almost scorpion-like attitude of aggression. It has been disliked and held in awe in many places, and has been regarded as a symbolic representation of corruption. If anyone had dealings with the Devil and took money in consequence, a Devil's Coach Horse would appear in the hand. It has given rise to a particular story in Ireland, where it is called the 'Dar Daol':

> The day before Our Lord's betrayal He came to a field where the people were sowing corn. He blessed the work, and as a result the crop grew up miraculously, so that when the Jews searching for Our Lord the next day came to the spot they found a field of wheat. They inquired if the Saviour had gone that way, and were told He had passed when the corn was being sowed. 'That is too long ago,' they said, and turned back. Then the Evil One, taking the form of a Dar Daol, put up his head and said, 'Yesterday, yesterday,' and set His enemies on His track. Wherefore the Dar Daol should be killed whensoever met. But there is only one safe way of doing it, they say in Co. Wicklow, for if you kill it with your thumb, as is done in the south of Ireland, or crush it with your boot, a stone, or a stick, the slightest blow from the thing used for its destruction occasions a mortal injury to either men or animals. Therefore the Dar Daol should be lifted with a shovel and burnt in the fire, and no harm will result.[129]

Stag beetles (Lucanidae)

There are only three British species in the Family Lucanidae but one of them is the biggest British beetle, one of the biggest British insects, and a truly wonderful and somewhat fearsome-looking creature: the Stag Beetle *(Lucanus cervus)*. The male, with its huge, antler-like jaws, can easily be 5 centimetres (2 inches) long – the largest ever found was 8.7 centimetres/3½ inches) long – but it is often smaller, as is the less impressively endowed female. Today, it is only at all common in the New Forest and, despite its appearance, it is relatively harmless, feeding on rotting wood. Although now among the more cherished of British animals, it was persecuted in the past. In the New Forest, it was once called 'the devil's imp' and thought to damage farmers' crops, and it was pelted with stones in consequence. Its close relative, the Lesser Stag

Beetle *(Dorcus parallelopipedus)* is much more common and more widespread; I can almost guarantee to find at least one in my Warwickshire garden every year.

Dor beetles (Geotropidae)

The dor beetles play an important role in recycling organic matter in the environment because they feed on and bury dung, different species preferring the dung of different mammals. The word 'dor' is Old English but its meaning is unknown; it is used in connection with a number of other beetles, including cockchafers, and is also used for some unrelated insects, such as bumble bees and hornets.

Chafers and scarab beetles (Scarabidae)

Many other types of dung beetle occur in the Family Scarabidae, which includes a huge number of species as well as some huge beetles. Many forms, collectively called 'scarabs', have Rhinoceros-like horns and among them is the tropical African Goliath Beetle *(Goliathus goliathus)*, which is up to 12 centimetres (4¾ inches) long and the heaviest insect in the world. Chafers belong to the same family and, in Britain, the largest is also the commonest. The Common Cockchafer (maybug) *(Melolontha melolontha)* is frequently found, somewhat dazed, on windowsills on humid summer evenings, after being attracted by the light and crashing into the glass. Unlike scarabs, the chafers are plant-eaters and some, both as adults and larvae, are serious crop pests, especially damaging to plant roots. Birds often tear up grass and turf searching for them and cockchafer larvae are sometimes known as 'rookworms' because of their appeal to Rooks and other crows. There are numerous historical accounts of cockchafers appearing in plague proportions in some seasons.

'Chafer' is a Middle English word, perhaps meaning 'to gnaw' and possibly related to the word 'chaff' for the husks of grain. 'Cockchafer' is early eighteenth century but the prefix 'cock' may not necessarily be related to the bird; it could simply imply maleness, or, as in 'cock sparrow', just be a term of familiarity. Among related and common garden pests is the strikingly iridescent green Rose Chafer *(Cetonia aurata)*.

Click beetles (Elateridae)

OTHER COMMON NAMES: *blacksmith; skip-jack; snap; spring beetle; watch.*

Among the smaller beetles, the click beetles are both common and familiar from their habit of catapulting themselves into the air to land on their legs after they have fallen on their backs (as they are prone to do when in the hands of small boys). Gardeners and farmers are more familiar with their elongated, pale brown larvae, which are generally called 'wireworms' and cause considerable damage to grass, cereals and root crops. They are responsible for most of the tiny circular holes found in potato tubers.

Glow-worms (Lampyridae)

OTHER COMMON NAMES: *glass worm; shine worm.*

Were it not for one remarkable feature, hardly anyone would have heard of a species of beetle known as *Lampyris noctiluca* that belongs to the Family Lampyridae. The adults are insignificant, elongated, pale brown beetles that rarely feed and the larvae dine usefully on snails and slugs, but it is their chemistry that has brought them enduring fame. The insects, especially the females, contain a chemical called luciferin, which is oxidized in the presence of water and oxygen to produce a pale blue-green light; the insects in consequence are called Glow-worms and a large colony can be an impressive sight at night. However, the purpose of the light is not to impress humans, but does serve very effectively to attract males to the females.

These beetles have been called Glow-Worms since Middle English was spoken:

> Glories, like glow-worms, afar off shine bright,
> But looked to near, have neither heat nor light.
> (John Webster, *The Duchess of Malfi*)

The sight can be so impressive that in Tewkesbury, Gloucestershire, the Glow-worms have become a tourist attraction.

> 'They're like stars on a dark night aren't they? When I first saw them, I was absolutely smitten, mesmerised' says Jan Lucas who, every Friday night from June to September, leads visitors on a Glow-Worm Spotting Tour ...The best sighting in one evening so far, says Jan, was last year: '204 in one night. Magical'.[130]

> ... a son and I went one July evening intending to watch an old badger's sett to see if it was still occupied. We stayed until it was really dark and were delighted to find on our way home the path on the woodland edge was lit by the greenish lights of Glow Worms – dozens of them. At Bearwood Church [Berkshire], (about four miles away), the 1849 Journal of the Reverend Robert Wilmott includes this entry:
>
> I have been turning glow-worms to a use this evening, which no naturalist ever thought of – reading the Psalms by their cool green radiance ... moving six of them from verse to verse. The experiment was perfectly successful.[131]

The insects glow mostly at dusk but also on and off throughout the night:

> The glow-worm shows the matin to be near,
> And 'gins to pale his uneffectual fire ...
> (*Hamlet*, Act 1, Scene 5)

The Stag Beetle *(Lucanus cervus)* derives its name from the antler-like jaws of the male, used to fight off rivals during the breeding season.

Dermestid beetles (Dermestidae)
The dermestids are small, dull, scavenging beetles, but it is what they scavenge that has brought them into unfortunate contact with humans. Although holes in stored clothes and other materials are usually attributed to 'moth', much of the damage is in reality caused by the Carpet Beetle *(Attagenus pellio)*. This creature has a distinctive white spot on its black elytra and has larvae with a tuft of hairs at one end, commonly known as 'woolly bears'. The Larder Beetle (bacon beetle) *(Dermestes lardarius)* and Hide Beetle *(Dermestes maculatus)* can be damaging, not only to bacon in larders, but more significantly to stored animal hides.

Their close relatives, the museum beetles (*Anthrenus* species), cause similar damage but, as their name suggests, can also be very troublesome in museums, where they can destroy taxidermal specimens and, indeed, collections of dried insects.

Wood-boring beetles (Lyctidae and Anobiidae)
The two families Lyctidae and Anobiidae include some of the most troublesome of timber-destroying pests. The powder post beetles *(Lyctus fuscus* and its relatives) are slim, brown insects whose larvae reduce both living and dead wood to a very fine dust (hence the name). Their presence is betrayed by small exit holes in the wood, made by the adult insects when they emerge. The Common Woodworm (furniture beetle) *(Anobium punctatum)* feeds similarly on dead wood and is the commonest pest of old furniture. This is the creature whose holes are sometimes faked by those seeking to 'age' reproduction antique furniture.

Much the most damaging of these creatures, however, is the Death Watch Beetle *(Xestobium rufovillosum)*, which readily takes on entire wooden houses and causes very significant structural problems. It has earned its name from the ticking sound made as it repeatedly bangs its head against the wood, apparently seeking to attract a mate. Heard in the dead of night in an old and unfamiliar house, it can make the hairs on the back of your neck stand on end. At such times, the old belief that it was a watch counting down the time to a death does not seem so outlandish.

> You that lie with wasted lungs
> Waiting for your summons ...
> In the night, O the night!
> O the death watch beating!
> (Alfred, Lord Tennyson, 'Forlorn')

These and other timber-boring beetles are most easily identified, even when the insects cannot be seen, by the size, shape and pattern of the exit holes and the nature of the discharged sawdust. The obligatory chemical treatment of timbers for the eradication of infestations, or for the protection of new wooden buildings, has had an incidental and serious impact in an unexpected way in reducing the potential roosting sites for bats. Modern legislation requires less toxic products to be used.

A Cardinal Beetle *(Pyrochroa coccinea)* raises its elytra prior to flight. This family of beetles is found on flowers and old trees.

Cardinal beetles (Pyrochroideae)
The Cardinal Beetle *(Pyrochroa coccinea)* is named for its vivid red colour; unlike the Cardinal Spider (see Spiders, page 54), which has a more direct association with men of the cloth.

Oil beetles (Meloidae)
The Meloidae is a family of species with a curious lifestyle, all being parasites of bees and grasshoppers. The larvae are inadvertently picked up from flowers and carried to the hosts' nests.

Flour beetles (Tenebrionidae)
The Family Tenebrionidae is a large group of nocturnal, dark-coloured ground beetles, but one, *Tenebrio molitor*, has become particularly familiar because it feeds on flour and cereals. Its larvae are known as 'mealworms' and anyone who has kept aviary birds or lizards is likely to have bought them as food for their pets. The adults can still sometimes evade hygiene procedures:

> Three million Twix bars have been destroyed after common flour beetles were found at Mars's factory in Slough, Berkshire. Production was halted for three days when staff saw 'black specks' in biscuits.[132]

Ladybirds (Coccinellidae)
OTHER COMMON NAMES: *bishie barni-bee (Norfolk; Suffolk); bishop barnabee; bishop is burning; bishop that burneth; bishy bishy barnabee (Norfolk; Suffolk); burnie-bee; bushy bandy (Suffolk); clock-o'clay; cow lady; cushcow; cushcow lady (Yorkshire); farmer's friend (Warwickshire); God Almighty's cow; God's little cow; goldie bird; King Galowa (Scotland); lady beetle; ladybug; lady clock; ladyclock (Scotland); lady-couch (Scotland); ladycow; lady fly; lady lanners; Mary gold; Mary-gold; sodger (Northumberland).*

Among a group of mainly scavenging beetle families, the Family Coccinellidae, known as ladybirds, is predatory. Although this small family contains only just over 40 British species, it has achieved a disproportionate amount of attention, recognition, lore and tradition. There is probably a greater folk literature attached to ladybirds than to all other British insects combined.

Ladybirds are well known as being small, round, spotted and fairly brightly coloured, but the range is nonetheless considerable and there are exceptions to the general form. The species with most spots is *Subcoccinella vigintiquattuorpunctata*, which, like other spotted ladybirds, tends to be referred to in shorthand as *Subcoccinella 24-punctata.* Other species are comparably called *5-punctata*, *2-punctata*, *7-punctata* and so on, although the designations can be misleading because, in some, the numbers of spots are variable. Indeed, there are ladybirds that have no spots

The principal food of most British ladybird species, both larvae and adults, is aphids, which is why they feature high on the list of 'gardener's friends' and why people are told to encourage them into their gardens (although with only the vaguest indication of how this might be done – the presence of aphids is usually good enough).

There is no entirely satisfactory explanation for why ladybirds should have eclipsed all other insects in terms of popularity, especially with children. In Britain, the Ladybird children's book imprint began in the First World War when Wills and Hepworth, a firm of jobbing printers in Loughborough, Leicestershire, began to produce 'pure and healthy literature for children', registering the logo in 1915. It has been an enduring brand name ever since and, some years later, the Woolworth company began to use the name for its range of children's clothes.

However, the appeal of ladybirds has been global and A.W. Exell has recorded well over 300 names for them in over 50 languages.[133] The English name 'ladybird' (sometimes 'lady-bird') was first recorded for the insect in 1592 as a dedication to the Virgin Mary. One explanation is that it was applied originally to the commonest species, the Seven-spot Ladybird *(Coccinella 7-punctata)*, which represented the Virgin's red cloak with seven spots for her seven joys and seven sorrows. A very large number of the names used for ladybirds in other languages relate to the Virgin Mary, many others to God and yet others to various saints, although names with a female connotation certainly predominate. Even in non-Christian countries, ladybird names are

Ladybirds (Coccinellidae) are popular with gardeners because they feed mainly on aphids.

often of religious significance. Other names and beliefs relate to ladybirds being fortune-tellers, bringers of good luck and especially of success in love. Michael Majerus, who has made the most intensive of all studies of ladybirds suggests that this is because they are highly promiscuous:

> Two-spot ladybirds *(Adalia 2-punctata)* mate on average 20 times, and have strong powers of endurance; a successful Two-spot mating rarely takes less than an hour and a pair may stay together; *in copula*, for over nine hours ... .[134]

A Norfolk and Suffolk name is 'bishie barnie-bee'. This is usually said to be a reference either to Bishop Barnabas or to Saint Barnabas, the first-century apostle, and an improbable link between this religious association and success in love comes in a local rhyme:

Bishie, Bishie Barnabee
Tell me when my wedding be
If it be tomorrow day
Take your wings and fly away!
Fly to east, fly to west
Fly to him that I love best.[135]

There is no logical connection between Saint Barnabas and ladybirds and, although it is said that Bishop Barnabas was burned at the stake ('bishop is burning' is another local name), this is all rather confusing. Saint Barnabas and Bishop Barnabas appear to be the same person, Saint Barnabas reputedly being the first Bishop of Milan, but there is no evidence that he was burned. Moreover, 'burnie-bee' might also refer to the insect's burnished appearance:

Back o'er thy shoulders throw thy ruby shards,
With many a coal-black freckle deck'd;
My watchful eye thy loitering saunter guards,
My ready hand thy footsteps shall protect.
So shall the fairy train, by glowworm light,
With rainbow tints thy folding pennons fret,
Thy scaly brest is deep azure dight
Thy burnish'd armour deck with glossier jet.
(Robert Southey, 'The Burnie-bee')

Michael Majerus has collected over 200 poems and rhymes featuring ladybirds but none is better known in England than the one beginning:

Ladybird, ladybird, fly away home,
Your house is on fire and your children are gone.[136]

There are many variations on the next couplet but most are on the theme of:

All except one and that's little Anne
And she has crept under the warming pan.[137]

In Derbyshire, we always sang:

Fly away east, fly away west,
Show me where lives the one I love best.

Once again, this associates the insect with foretelling romance. Surprisingly, however, Michael Majerus thinks the most likely explanation is that it relates to hop-picking in Kent. Many London women went down to Kent every year to earn money during the hop harvest, taking their children with them. The rhyme was supposedly devised to entertain the children and alluded to the fact that the hops would be burned at the end of the season, although surely its widespread occurrence in places where there are no hops must argue against this:

> I originate from East Kent and went hop picking many times but I cannot say I ever heard the rhyme with that association ... I seem to recall 'your house is on fire and your children at home' cropping up in rhymes on mainland Europe.[138]

Children invariably balance a ladybird on one finger while reciting the rhyme, whatever its meaning, but it is thought bad luck to brush it away. It must fly off of its own accord, propelled by no more than a puff of breath. In Northumberland, however, the ritual was different. There:

> ... children pick it up as soon as they see it and throw it high in the air, saying:
> 'Reed, reed sodger, fly away,
> And make the morn a sunny day'.[139]

Dark-coloured ladybirds are thought especially lucky and many districts have beliefs related to the number of spots: foretelling the success of the harvest, the months in the year, and so on. There is a widespread view that ladybirds should not be killed and, in East Anglia, if this should happen by accident, the creature must be carefully buried, the grave stamped on three times and the 'house on fire' rhyme recited.

Handling ladybirds will often result in your fingers being tainted by the characteristically strong-smelling yellow liquid that they exude from the leg joints. This has in some places bizarrely been thought a cure for toothache. But is it true that ladybirds will bite you? Like other beetles, they have respectable jaws and can give a nip that injects a tiny amount of enzyme, giving rise to a painful lump for a short while.

Ladybirds hibernate over winter as adults and gardeners sometimes discover vast numbers of them among vegetation, in cracks in fence posts, or similar places. When they begin to emerge in March or April, I can guarantee that someone will tell me that their numbers, whether large or small, are an indication that there will be a comparable number of aphids in the

The wood-devouring Longhorn Beetle (Family Cerambycidae) is characterized by long antennae.

coming season. Certainly, huge numbers are seen in some summers, but this is generally a reflection of warm weather and good breeding conditions; both of which will be equally beneficial to aphids.

> During July and August 1976, at least 400 miles of tideline on the south and east coasts of England consisted of little but solid ladybirds. If all were 7-spots, a conservative estimate gives a figure of 23,654,400,000 beetles in the tide-line at any one time ... the number of ladybirds ... was more than double the number of humans that ... have ever lived on earth. And this figure is for a single day, and takes no account of the ladybirds which stayed on dry land, nor those drowned at sea and were never washed up.[140]

Numerous residents of Worthing and other south coast towns sought treatment for bites and irritations.

Individual ladybird species are generally associated with particular types of vegetation and habitat (there are even species favouring water and wetlands), but at least some occur throughout the British Isles, with an overall tendency for there to be a greater number of both species and individuals towards the southeast of England. Despite their extreme familiarity, however, there appear to be no places named after ladybirds and I know of no-one with 'Ladybird' as a personal name, although it was most famously adopted by the American Claudia Alta Taylor, who then became better known as Lady Bird Johnson, First Lady of the United States.

Longhorn beetles (Cerambycidae)

The Family Cerambycidae embraces a group of wood-eating beetles that are characterized by long antennae and are consequently known as Longhorn Beetles. Among them is a rather impressive insect that has achieved notoriety for living in houses, although it is not usually a serious pest in Britain. The larvae of the House Longhorn *(Hylotrupes bajulus)* can, in favourable conditions, seriously affect the stability of timber-framed buildings.

Leaf beetles (Chrysomelidae)

The leaf beetles form a very large group of leaf-eating beetles. Most are small, many are very brightly coloured in metallic hues and many are important crop pests. A few stand out as of especial importance or interest. The adults of the tortoise beetles (*Cassida* species) are remarkable-looking creatures in which the body is completely covered by the elytra and thoracic plates, hence their common name.

The Lily Beetle *(Lilioceris lilii)* is a bright red creature with larvae that look like bird droppings. It has become very familiar to gardeners, especially in the southeast of England, devouring lilies and seemingly on an unstoppable march across the country.

Flea beetles (*Phyllotreta* species) are responsible for the almost lace-curtain-like holing of the leaves of young vegetables. They gain their name from their habit of jumping vigorously when the foliage is disturbed.

The Bloody-nosed Beetle *(Timarcha tenebricosa)* is a startling big black insect with a remarkable defence mechanism. When attacked, it discharges blood from its mouth and other orifices.

The Rainbow Leaf Beetle *(Chrysolina cerealis)* is that great rarity, a beetle that is legally protected. It is also evidence yet again that a British animal, and especially a British insect, must be big and/or brightly coloured to merit legal protection.

Probably the most notorious chrysomelid beetle in Britain is more likely to be seen on a poster in a police station than eating leaves. The Colorado Potato Beetle *(Leptinotarsa decemlineata)*, is somewhat like a yellow-and-black striped ladybird and was accidentally introduced from its native North America to France in 1921. It is one of the alien pests that, in Britain, are notifiable: its discovery must, by law, be reported to the police or the Department for Food, Environment & Rural Affairs. Remarkably, as a result of this policy, Britain has remained free of self-sustaining colonies. I have found it in quantity at the nearest point to Britain, on crops close behind the cliffs at Cap Gris Nez, but it rarely seems to cross the Channel.

Weevils (Curculionidae)

The weevils are almost all plant-feeders. They are an enormous group, with over 500 British species and 40,000 species worldwide. All have a very characteristic appearance, with a greatly elongated head that forms a beak-like projection (or rostrum) with the jaws at the tip.

The weevils have clearly been recognized as distinct for a very long time because the name 'weevil' is Old English, meaning 'to move briskly' and probably from the same Old Teutonic root as the verb 'to wave'.

Among the numerous important crop pests, the Vine Weevil *(Otiorhynchus sulcatus)* is probably most familiar, especially to gardeners. Its larvae feed on bulbs, corms and plant roots generally, while the adults nibble elongated notches in the leaf margins of evergreen shrubs. It has become much more common and widespread since horticulturists began growing plants in peat-based composts, which seem to provide the conditions it likes for egg-laying. In recent years, the use of species of nematodes for biological control (see Nematodes, page 22) has met with some success.

Bark beetles (Scolytidae)
Closely related to the weevils, but lacking the elongated snout, are the bark beetles. These feed beneath the bark of living and dead trees, their larvae forming rather attractive patterns of tunnelling that becoming visible when the bark is pulled away. Different species cause different patterns.

Two species in particular have almost literally changed the face of much of the British landscape. *Scolytus scolytus* and *Scolytus multistriatus* feed on elms, inadvertently carrying the spores of the fungi that cause Dutch elm disease as they fly from diseased to healthy trees. The fungi grow within the bark tunnels and other members of the same family, known as 'ambrosia beetles', feed on the fungal growth.

Millipedes
(Diplopoda)

Millipedes are slender creatures, superficially similar to centipedes but with more legs. They have about 60 body segments, each with two pairs of legs. They move in a deliberate, slow, snake-like manner and, because they have so many legs, the leg actions appear like a wave-motion; there are too many of them for the process to be described as running.

There are about 50 British species and they are mainly vegetarian, although they will also scavenge dead animal matter. They eat large quantities of fungi and, in gardens, can cause damage to young seedlings and to root crops, such as potatoes. The pink-coloured Spotted Millipede *(Blaniulus guttulatus)* is particularly troublesome, although the commonest species is the harmless Black Millipede *(Tachypodiulus niger)*.

Millipedes lay eggs and generally offer them some parental protection until they hatch into immature six-legged creatures resembling insect larvae. The larvae then undergo a series of moults, acquiring more segments and more legs at each stage.

Spotted Millipedes *(Blaniulus guttulatus)*, seen here coiled and uncoiled, live mainly in cultivated soil.

The hatching of a Millipede brings curious things to light:
The embryo in with its shell is curled up snug and tight
Enclosed inside an inner skin with a thorn upon its neck,
Whose task it is to pierce the shell, as chicks their
prisons peck.
What is this extra covering that thus comes into view?
An heirloom from antiquity here blended with the new?
Another 'Nauplius-coat' around another embryo,
The same that Peracarids on their cradled babes bestow?
(Walter Garstang, ' The Millipede's Egg-tooth')

One distinct group, the Glomerida, comprises the shorter, stouter pill millipedes, which are commonly mistaken for woodlice and similarly acquired the odd epithet 'pill' (see Woodlice, page 48). The name 'millipede' (often spelled 'millepede') is, like 'centipede', early seventeenth century.

Centipedes
(Chilopoda)

OTHER COMMON NAMES: *Jenny-hun'r-legs; lad of the knives (Scotland); Martin of the knives (Scotland); Meggy-monny-legs (northern England); red fox (Scotland); thrush-lice (northern England).*
Almost everyone can recognize a centipede, or at least the two commonest species, and they probably assume that they have a hundred legs. This is not the case, any more than millipedes have a thousand legs. 'Centipede' is an early seventeenth-century word and does literally mean 'hundred legs', but the number is seldom correct, either in terms of individual legs or pairs of legs. Centipedes have one pair of legs to each segment of their elongated bodies; but the number of segments varies, even within

individual species. The number of legs in British centipedes ranges from 30 to 202; there can, of course, never be an odd number, although there are often odd numbers of segments.

There are about 45 British species, among the commonest being the slender golden-yellow species *Necrophloeophagus longicornis*, which is frequently found by gardeners when digging, and the extremely numerous shorter, stouter, rich glossy brown *Lithobius forficatus*, which is often seen scurrying for cover when stones are lifted.

A centipede was happy quite,
Until a toad in fun
Said, 'Pray which leg moves after which?'
This raised her doubts to such a pitch,
She fell exhausted in the ditch,
Not knowing how to run.[141]

Centipedes are carnivorous and are equipped with poisonous claws on the head. The bite of tropical and other exotic species can be painful, even dangerous to humans, but all British species are harmless. Nonetheless, they once posed a different sort of threat: in the Hebrides, bare-footed children were expected to wash their feet in the burn before coming indoors to bed and, if they failed to do this, they were threatened with a centipede.

Symphylids
(Symphyla)

Symphylids are like miniature centipedes, white and up to 1 centimetre (½ inch) long, with elongated antennae. Most people will never knowingly seeing one, unless they are gardeners or keep house plants, in which case they may well come across them attacking the roots and causing general debilitation of their plants. Symphylids are often mistaken for springtails (see page 62), but they have four times as many legs and do not jump.

ACORN WORMS AND RELATED SPECIES
(Hemichordata)

Hemichordates are an improbable group intermediate between the true chordates, which have a spinal chord (and, in most cases – the vertebrates – a backbone), and the true invertebrates, which have no spinal chord. All are marine and they include most importantly the acorn worms (Enteropneusta), which burrow in sand and mud. These should not be confused with such abundant and common mud-burrowing worms as ragworms and their relatives, which are true invertebrates (see Bristle worms, page 251).

The graptolites, a huge and important extinct group, which reached their zenith in the Silurian sea of about 400 million years ago and are familiar to every fossil-collector, also belong here.

SEA SQUIRTS
(Urochordata)

The urochordates are also part way to the true chordates and the most important group among them are the sea squirts (Ascidiacea). The adult forms are more or less tubular, soft-bodied creatures that live permanently attached to underwater rocks or other bodies, sometimes singly, sometimes in colonies. Many are brightly coloured and some of the solitary species are up to 15 centimetres (6 inches) in length.

They feed by filtering particles from sea water, which they drawn into their body through an opening known as a 'siphon'. They then forcibly expel the water through another siphon, which is why they are known as 'sea squirts'.

Adult sea squirts live singly or in colonies, like this cluster of Light-bulb Sea Squirts *(Clavelina lepadiformis)*.

Vert

ebrates (Vertebrata)

Vertebrates are animals with an internal skeletal backbone, also called a vertebral or spinal column, which is composed either of bone or cartilage. The individual units of the backbone are known as vertebrae. Although the Vertebrata probably embrace only about three per cent of the world's animal species, they are immensely important, not least because their skeletal structure and associated body form and function confer the ability to grow much larger than most invertebrates.

The five phyla (groups) of vertebrates – mammals, birds, reptiles, amphibians and fish – are all represented among the British fauna and all have had a highly significant impact on the best known of their number, man himself.

A shoal of Atlantic Salmon *(Salmo salar)* circles around a diver off the coast of Scotland.

Fish

Fish are cold-blooded aquatic vertebrates, breathing by gills, possessing jaws and usually having fins and a scaly skin. Most reproduce by laying eggs; these are fertilized externally and usually hatch into immature forms called 'fry', which gradually metamorphose into miniature versions of the adults. They provide mankind (and many other mammals, as well as birds and reptiles) with an immensely important source of food but, unlike other human food animals, they have only very recently been controlled or farmed; hence, the use of fish for food has had a very direct impact on natural wild populations. In many instances, mankind has learned about this impact the hard way and only within the memories of people alive today has the value of fish conservation been properly appreciated. Because of their great importance to man, they have featured significantly in lore and tradition, although, because fish are seen alive much less frequently than terrestrial vertebrates, the lore is very often that of fishing and its associated activities rather than of the animals themselves.

Jawless fish

(Agnatha)

Although invariably grouped with fish, these creatures are evolutionarily primitive, eel-like aquatic vertebrates which lack a bony skeleton but have discrete nodules of cartilage to provide bodily strength. The body is scale-less and slimy, and they have no jaws (Agnatha means 'lacking jaws') and only very simple gill openings. A marine group known as the hagfishes (Family Myxinidae) occurs in British waters but are seldom seen because they live in or close to sea-bed mud in fairly deep water; these have attracted no tradition or folklore.

Lampreys

(Petromyzonidae)

OTHER COMMON NAMES: *argoseen; argus-eyes; barling; bayrn (Isle of Man); blind lamprey; brennie (Cornwall); cunning; fyke; geyes; horse-eel; king-fish; lamper; lamper-eel; lampern (and numerous similar names); lumper-eel; lumping-eel.*

The scientific name Petromyzonidae means 'stone-suckers'. 'Lamprey' has a similar connotation, being a Middle English word originally from the Latin *lampere* 'to lick' and *petra*, 'stone'. This is just what you may see them doing: apparently licking a stone to which they are attached by the circular, disc-like structure that surrounds the mouth. In reality, they are using the disc as a temporary anchor while they move through a fast-flowing current.

However, this is only a subsidiary use for the disc because the adults of most species of these extraordinary, eel-like creatures lock themselves onto fish, on which they feed by rasping away the skin and dining on the flesh beneath. In parts of the world, and most notably in the North American Great Lakes, they are extremely serious fish parasites. In Henry Williamson's *Salar the Salmon*, the eponymous hero was attacked by 'Petromyzon the lamprey' while living in the sea.[1] Because of this less than endearing custom, lampreys have earned themselves some notoriety and there is still a widely held belief that they will similarly attach themselves to bathers and suck out their blood; this is patently untrue.

There are three species of lamprey in Britain and two are catadromous, that is, they spawn in fresh water, the eggs hatching into small, blind larvae, which live for some years in the silt of the river bed before metamorphosing into adults.

Sea Lamprey *(Petromyzon marinus)*

OTHER COMMON NAMES: *lamprey eel; marine lamprey.*

The largest of the three species, the Sea Lamprey, can reach 1 metre (3¼ feet) in length and, after metamorphosis, migrates to the sea to feed and live its adult life, returning to rivers to breed. However, there is a non-migratory population that lives and feeds permanently in Loch Lomond.

River Lamprey *(Lampetra fluviatilis)*

OTHER COMMON NAMES: *juneba; lamper eel; lampern; nine-eyed eel; nine eyes; seven eyes; stone grig.*

The River Lamprey is only about one-third of the size of the Sea Lamprey. It does not have a multitude of eyes; the common names refer to the circular gill openings on the sides of the head. It too migrates to the sea in most areas, although, in the past, vast numbers were intercepted by netting or basket traps in estuaries and river mouths.

The minor industry of catching River Lampreys on the River Severn no longer exists, although Brian Waters told of old George Cooke, from Purton in Gloucestershire, who spent a lifetime there as 'a private fisherman':

> ... George Cooke had the fondness of a Plantagenet king for a lamprey. And like those royal autocrats, not only did he demand every lamprey caught at Purton for his own table, but like Henry III he fixed the price he paid for his fish; he never paid more than half a crown ... a lamprey cannot be skinned, and George Cooke's lampreys were prepared by scaling the fish and cutting away the gills. The body was stuffed and baked in a covered tin, gently for two hours. The flavour of the fish on the table is not unlike that of a dish of elvers.[2]

They are still an important and highly valued food in other parts of Europe, although Jews were forbidden to eat them because they have no scales and were thought unclean.

Curiously, a creature as biologically obscure and relatively insignificant as the River Lamprey has been implicated in the deaths of two English kings: King John and King Henry I. The supposed experience of Henry I even led to the expression 'a surfeit of lampreys' for comparable disastrous happenings.

> He ate voraciously of a lamprey, which he was accustomed to delight in more than anything else and paid not attention to his physicians when they forbade it him. But when his weakness had overcome his natural strength, King Henry yielded to his fate.[3]

Whatever happened, it occurred on 1 December 1135 in France and all contemporary accounts point very firmly to the lampreys that the king ate after a day's hunting. Clifford Brewer raises the possibility that they were infected with salmonella or some similar organism and that Henry, at the then very great age of 67, was physically weak. However, he concludes that, just as with King John, whose contact with lampreys was more tenuous, an ulcer followed by peritonitis was perhaps the more likely cause of death.[4]

Historically, in Britain and currently elsewhere in Europe, the preferred method of eating lampreys has been hot, grilled or smoked after they have been heavily dosed with salt to remove the clinging mucus. Nonetheless, there has long been a tradition in Gloucester of presenting to the sovereign on accession, and every year at Christmas and certain other occasions, a lamprey pie made from those caught locally in the River Severn. It is said that King John fined the men of Gloucester 40 marks (£26.67) because 'they did not pay him sufficient respect in the matter of his lampreys'.[5] In the Royal Archives at Windsor there is a letter from the then Mayor of Gloucester, dated June 1900 and addressed to Queen Victoria:

> In accordance with an Ancient custom, recently revived and continued by my predecessors in Office with your Majesty's gracious approval, I have the honour to forward a Royal Lamprey Pie as a birthday offering: and I trust your Majesty will be graciously pleased to accept it in token of the continued loyalty and affection of the Citizens of Gloucester.[6]

The 'revival' referred to in the letter was for Queen Victoria's Diamond Jubilee in 1897 and a note attached to the letter states that 'they often send one', but there are no other references in the archive, apart from a note in 1917 saying that the annual gift of a lamprey pie from Gloucester had been suspended during the war; there is no indication of it having restarted later, although a pie was baked by the Home Economics Department of Gloucestershire College of Education for Queen Elizabeth II's Silver Jubilee in 1977,[7] and is revived on other special occasions such as the Millennium celebrations although the local council presently seems somewhat vague over its responsibilities and the Lord Chamberlain's Office can trace no information about the custom.[8]

Brook Lamprey *(Lampetra planeri)*
OTHER COMMON NAMES: *mud lamprey; pride.*
The third British lamprey species is the one I knew well as a boy because, every spring in our local stream, masses of spawning animals appeared like knots of writhing silver cords. The Brook Lamprey is barely 15 centimetres (6 inches) long when mature and, I always think, a rather pathetic creature because, after metamorphosis, it neither migrates nor feeds but simply spawns and dies. It is easily the commonest species and occurs in streams throughout Britain. However, like the other two species of lamprey, it tends to be absent from the far north of Scotland.

Cartilaginous fish
(Chondrichthyes)

These distinctive fish have a cartilaginous rather than a bony skeleton, a jaw placed beneath the head, exposed gill openings and no real rays to the fins. They include some remarkably impressive, famous and infamous creatures, among which the sharks, skates and rays are especially important.

Sharks and Dogfish
(Selachimorpha)

Popular paperback fiction, films such as *Jaws* and some television channels that appear to show programmes about nothing else, have helped to build and reinforce a view that every shark is a fearsome and dangerous creature and that, to all intents and purposes, there is only one significant species, the Great White Shark *(Carcharodon carcharias)*, which lurks permanently off Bondi Beach. The reality is different. Almost all sharks are marine but the fact that there are around 24 species of them and their close relatives in British waters, and that some are enormous, therefore frequently comes not only as a huge surprise but can also cast a shadow over the family seaside holiday.

In practice, I can find no record of anyone ever being seriously attacked by a shark off the British coast because the species that occur here are relatively benign. However, the Blue Shark (*Prionace glauca*; Family Carcharhinidae) and Common Hammerhead (*Sphyrna zygaena*; Family Sphyrnidae) have on occasion attacked bathers in other parts of the world. Sea anglers from the Cornish port of Looe and elsewhere have certainly caught both of them, along with other

Large sharks like this impressive Blue Shark *(Prionace glauca)* are frequently caught off the coast of Cornwall.

The eggs of some dogfish are contained in a horny case called a 'mermaid's purse', which has tendrils that twine around seaweed.

carnivorous sharks, including the Thresher (*Alopias vulpinus*; Family Alopiidae). This shark has a huge, asymmetric tail, which fishermen would sometimes nail to the wheel-house in the hope that it would steer them towards fish.

The Mako (*Isurus oxyrinchus*; Family Isuridae) and the very similar Porbeagle (mackerel shark) *(Lamna nasus)* in particular provide popular sporting quarry. Specimens of 3 metres (10 feet) or more in length and well over 200 kilograms (440 pounds) in weight have been taken off the British Isles, but they are caught from boats, usually well out in the Western Approaches, not off a local beach. Although carnivorous, these sharks eat the flesh of fish, not humans. Ironically, the Porbeagle especially is itself highly prized as a food fish and, when you see shark on a restaurant menu, there is a good chance that it is this species.

One of the most numerous of the medium-sized sharks and a smaller relative of the Blue Shark is the Tope *(Galeorhinus galeus)*. Up to 2 metres (6½ feet) in length and, with a British record approaching 40 kilograms (88 pounds), it is a popular sporting fish; in Britain it is the species that is used most frequently for that Oriental speciality, shark's fin soup.

Sadly, perhaps the most famous British shark of recent years was not truly British. One of the more controversial modern works of art was Damien Hirst's *The Physical Impossibility of Death in the Mind of Someone Living*, which to many members of the public was simply a pickled shark. It seems, however, that no British shark was sufficiently fierce for the artist ('he wanted something that can kill and eat you') and so a Great White Shark was specially imported from Australia.[9]

People diving, or even snorkelling, off British coasts may well encounter some of the smaller relatives of the sharks fairly close inshore, although there is no cause for undue alarm. The Dogfish, the Rough Hound and the Nurse Hound or Bull Huss (Family Scyliorhinidae) and their relatives, the Smooth Hounds (Family Triakidae), are seldom much more than 1 metre (3¼ feet) in length and harmless, feeding on small fish and bottom-living creatures. However, a related species, the Picked Dogfish (spurdog; common spiny dogfish) (*Squalus acanthias*; Family Squalidae), has a spine at the front of each dorsal fin and these can cause an unpleasant sting.

These animals have been known as 'dogfish' since the fifteenth century, for no very obvious reason unless 'dog' is being used in the sense of something inferior or worthless. If so, this is a pity because some of them have fine flesh.

The Lesser Spotted Dogfish (rough hound) *(Scyliorhinus canicula)* is commonly sold as 'rock salmon', 'rock fish', 'rock eel', 'huss', 'flake' or 'rigg', these names being attempts to disguise its true origin and really unnecessary because it is seriously underappreciated fare. However, many people's first experience of a dogfish was neither in the sea nor in a restaurant but in a school biology laboratory, where, reeking of formaldehyde like Damien Hirst's Great White, a preserved specimen was once an obligatory dissection animal.

'Porbeagle', is almost certainly not another dog-inspired name but, like 'Tope', originates in dialect Cornish. The name 'shark' itself is sixteenth century and of unknown origin, although the fact that it is now applied to people who are cheats and charlatans is further evidence of the low esteem in which all these fish have long been held. Their voracious appetite was obviously familiar to Shakespeare, whose witches' brew included:

> ... maw and gulf
> Of the ravin'd salt-sea shark ...
> (*Macbeth*, Act 4, Scene 1)

True sharks and some dogfish give birth to live young but other dogfish lay eggs and the way that they do this has given rise to much charming fantasy. The eggs are laid in characteristic pouch-like, horny egg cases, attached by curly tendrils to seaweed or other suitable anchorage points, where they remain for about nine months while the embryos develop. The empty egg cases are very frequently found on the shore-line and have fascinated children for centuries; long ago someone christened them 'mermaid's purses'. Those seen most commonly are from the Lesser Spotted Dogfish (rough hound) *(Scyliorhinus canicula)* or the Large Spotted Dogfish (nurse hound; bull huss) *(Scyliorhinus stellaris)*, although similar ones, but with straight tendrils, are produced by some rays (see page 122).

There is, however, another important species of British shark that is quite the most dramatic and that has given rise to any number of tales and myths. The Basking Shark (*Cetorhinus maximus*; Family Cetorhinidae) is, after the Whale Shark *(Rhiniodon typus)*, the second largest fish in the world. In British waters, specimens of 10 metres (33 feet) in length are known, but elsewhere they can be even larger. During the summer months, large and hugely impressive groups of them can be seen off the west coast of Scotland, doing what Basking Sharks do best: basking. They laze just below the surface with the mouth open and the top of the dorsal fin and tail protruding above the water. They are feeding on plankton, filtering vast quantities from the water rather like baleen whales (see Dolphins, Whales and Porpoises, page 435); their teeth are minute and the animals are completely harmless.

The way in which they appear in the water, with parts showing above the surface, has given rise to countless sea-serpent stories and, when carcasses have been washed ashore in a partly decomposed state on some remote beach, the imaginations of local people have run riot. In Shetland, they were known as *brigdi* and were feared by fishermen, not so much because they might be man-eating but because their sheer size meant that fishing boats could be inadvertently smashed like matchwood. There is still a considerable mystery about the origin of British Basking Sharks and it is thought that their appearance off British coasts is the result of large numbers of them migrating north.

Perhaps the largest ever found in British waters was a rotting corpse washed ashore on Stronsay in east Orkney in 1808 and entered legend as a giant sea snake. It seems really to have existed and its story was told by Frank Buckland (reality and a story by Frank Buckland do not always equate) but the idea that it was 17 metres (56 feet) long suggests that the tale was seriously stretched in the telling.[10] Better authenticated was one just over 11 metres (36 feet) long and weighing 8 tonnes that was washed ashore at Brighton in 1806.

Apart from being eaten, sharks and dogfish have been put to other uses. Their skins are very rough and were once widely used for a range of purposes: to provide a grip on sword handles, for polishing wood and other materials, and even to form a striking patch for old-fashioned matches. Dogfish skin was

The Basking Shark *(Cetorhinus maximus)* cruises through the sea with its mouth agape, sifting plankton from the water.

commonly known as 'rubskin'; that from sharks and rays tended to be called 'shagreen', a word curiously derived not from the obvious 'shag' (meaning 'matted') and 'green' (the colour), but from the French *chagrin*, and perhaps originally from the Turkish word for 'rump'. There was also a minor industry of hunting sharks in British waters for their liver oil. The Blue Shark and Basking Shark especially were taken and they were caught, rather like whales, with harpoons.

The most recent, and last, serious commercial shark-hunting for oil in British waters was that by Gavin Maxwell in the years after the Second World War. He bought the Hebridean island of Soay off the east coast of Skye (not to be confused with Soay in the St Kilda group, home of the Soay Sheep; see page 459) at the end of the war and saw shark-fishing as an opportunity to bring some prosperity to the island and to himself. It was a risky and poorly funded enterprise that struggled towards any sort of financial reality. The participants learned their craft as they went along.

> I was able to fire into him at almost point-blank range ... and I could feel the decks below me shudder with the recoil as the harpoon went squarely home ... It was a matter of seconds before the heavy thump of the rope snapping off short of the iron ring to which it was tied – a three-inch manila rope with a breaking strain of about seven tons ... .[11]

Although by 1949 Maxwell was still being offered £130 a ton for shark liver oil, Isle of Soay Shark Fisheries Ltd failed through poor equipment, the unpredictability of the shoals and because of a lack of capital and business acumen. The story was told by Maxwell himself in *Harpoon at a Venture* and by his harpooner, Tex Geddes, in *Hebridean Sharker*.[12]

Skates and Rays
(Rajiformes)

Skates and Rays are close relatives of the sharks and also have coarse, abrasive skin, but they are anatomically flattened in the vertical plane (unlike flatfish, which simply lie on their sides; see page 168), which gives them the appearance of having wings. This impression is enhanced when they swim by what appears to be a flying motion through the water. Most are sea-bed fish, feeding on crustaceans and other bottom-living animals, and they may alarm divers when they suddenly rise from the sea floor in a flurry of sand.

There are no real differences between the two groups and most are in the same family, the Rajidae, but the name 'skate' is generally applied to those that have an elongated snout. The name is often said to refer to the manner in which they appear to skate along the sea bed, but the two meanings have different origins. 'Skate' is Middle English, from an Old Norse word. 'Ray' is also Middle English but from French and Latin.

The Marbled Electric Ray *(Torpedo marmorata)* is one of two species of electric rays that occur in the waters around Britain.

There are many species and some are highly prized as food fish; skate 'wings' are now among the more common items on the supermarket fish counter and at one time were generally obtained from the Common Skate (grey skate; blue skate) *(Raja batis)*, which is one of the larger species and can reach 2 metres (6½ feet) or more in length. By this time, it is a little big for a frying pan (this is definitely the best way to cook it). When the fish is first caught, the flesh is extremely tough and so it is usually hung for a few days first. Today, the Common Skate is also suffering from overfishing.

> The common skate is believed to be extinct in the Irish Sea and will disappear from around the British coast unless measures are taken to save it, the World Wide Fund for Nature said ... Conservationists say fish sold as skate in the shops is mostly from other species of ray ... The WWF wants 'fishing-free zones' to allow species to recover but ... the National Federation of Fishermen's Organisations called the proposal; 'simplistic and naive'.[13]

According to a Scottish tradition, the pleasure of skate-eating has a valuable bonus. 'Skate bree' is the liquor in which skate has been boiled and it is said:

> I'll catch the white fish
> To please my lassie's ee,
> But the bonny black-backit fish
> Has aye the sweetest bree.[14]

I wonder how many girls today could be so readily and simply charmed by a bowl of fish broth.

Common species in the same family, which all provide angling sport, are the pale brown-coloured Blonde Ray *(Raja brachyura)*; several beautifully patterned species, such as the Spotted Ray *(Raja montagui)* and the Painted Ray *(Raja undulata)*; the Cuckoo Ray *(Raja naevus)*; and the Shagreen Ray (Fuller's ray) *(Raja fullonica)*, another important food fish.

The sting rays have a rounder outline and one or two spines at the base of the tail, attached to a venom gland. Anyone standing on a spine will experience the most intense and temporarily paralysing pain, although it must be reassuring to know that the effects are rarely fatal. The Common Sting Ray *(Dasyatis pastinaca)* is the species most usually found. These are big fish, reaching 2 metres (6½ feet) or more in length, although why the scientific name should allude to parsnips I have no idea. In Classical Greek legend, Circe gave her son, Telegonus, a spear tipped with a sting-ray's barb with which he killed his father Ulysses.

However, the greatest general interest and notoriety attaches to a peculiar attribute that is possessed by a handful of species and that they share with some eels but relatively few other creatures: this is the ability to generate electricity and give an electric shock. The feature is best known in the electric rays (*Torpedo* species; Family Torpedinidae) but is also present in the Thornback Ray *(Raja clavata)*. Two species of electric ray sometimes occur off Britain: the Marbled Electric Ray *(Torpedo marmorata)* and the Dark Electric Ray *(Torpedo nobiliana)*. The electricity is generated in two organs on either side of the head and serves to stun their prey; it can also give an unpleasant surprise to divers.

Other bottom-living relatives of the skates and rays are the monkfish, of which the commonest is the Angel Shark (*Squatina squatina*; Family Squatinidae). However, these large, heavy fish are not the same as the Monkfish that is found on restaurant menus, which is a quite different and very ugly creature that most people know as the Angler Fish (see page 171).

Bony fish

(Actinopterygii or Osteichthyes)

These are the 'true' fish with a bony skeleton and the vast majority of species belong in this group. Most have scales and fins with supporting rays. They occur in both fresh and salt water, although most families tend to be adapted for life in one or the other. Some species of fish migrate from salt water to fresh water to breed and are described as being 'anadromous'. A few, described as 'catadromous', do the reverse and their migratory habits have long fascinated mankind. It is not surprising that two of the most notable examples, the eel and the salmon, have attracted an abnormally large share of lore and tradition.

Sturgeons

(Acipenseridae)

Common Sturgeon *(Acipenser sturio)*

OTHER COMMON NAMES: *Baltic sturgeon.*

A few years ago, I came across one of those little nuggets that always add special interest to the researcher's life. Among the archives of a Perthshire family, I found an old photograph album containing a picture of a monstrous creature lying on the shores of the River Tay, where it had been hauled by an old family retainer in the 1870s. It served as a reminder of just what an enormous fish the sturgeon is, and what impossible and unexpected delights are now effectively denied us.

There are about 25 species of sturgeon and they are considered to be among the most primitive of true bony fish. They lack scales but have a characteristic row of bony plates, or scutes, embedded in the skin. Internally, much of their skeleton is cartilage rather than real bone. They are green-brown or black above, paler beneath and feed on small invertebrates, fish and other bottom-living creatures. They not only grow to a great size, but also live to a great age. The Common Sturgeon probably lives to about 40 years but other species reach over 100. It can attain 3.5 metres (11½ feet) in length and weigh in excess of 300 kilograms (660 pounds).

Most sturgeons are marine and anadromous, spawning in fresh water, and in times past they were occasional visitors to a number of British rivers. Although they appear never to have bred in Britain, the existence of the odd account of small individuals being seen occasionally suggests that breeding may have taken place. Sturgeon were seen or caught principally in the Severn, Thames, Avon, Ouse, Trent and several Scottish rivers, including the Tay, but today they are found only occasionally off-shore. Trent Bridge in Nottingham was probably the furthest inland that any sturgeon ever reached. The most recent record of a sturgeon in a British river seems to have been one foul-hooked in the Tywi (Towy) in July 1993, which later beached itself. It was 2.79 metres (9 feet) long and weighed 196 kilograms (431 pounds).

That sturgeons were once reasonably common is evidenced by their discovery in the remains of what were presumably ancient feasts (sturgeon bones have been found in mediaeval excavations at Westminster Abbey) and that they were 'by tradition' royal fish and had to be offered to the sovereign. I have traced this custom to a statute of King Edward II, which dates from *c.*1324 and which, in translation, states:

> Also the King shall have ... whales and sturgeons taken
> in the sea or elsewhere within the realm except in
> certain places privileged by the King.[15]

This royal prerogative is thought to have originated with the Kings of Denmark and Dukes of Normandy, who exercised similar rights. Among a few exceptions was a privilege granted

A 180-kg (400-lb) Common Sturgeon *(Acipenser sturio)*, caught in the English Channel, arrives for sale at an Oxford Street shop in 1947.

to the Lord Mayor of London to claim sturgeons caught in the Thames above Tower Bridge. In 1953, it was being argued that 'elsewhere within the realm' could reasonably mean within the then 3-mile (4.8-kilometre) limit of territorial waters. Now that the territorial limit is an EU-designated region of 200 miles (320 kilometres) (see Herring, below), a test case of the ownership of sturgeons might be a legal minefield. Nonetheless, under Edward II's statute, sturgeons, whales, porpoises and dolphins were considered 'Fishes Royal' and could be claimed on behalf of the Crown and should be reported to the local Receiver of Wreck.

Today, the sovereign no longer claims sturgeons when they are caught offshore but, if a sturgeon is offered to the Queen, Her Majesty accepts it as a gift (in other words, no payment is made). There have been a number of offers during the reign of Queen Elizabeth II, beginning before her Coronation with one brought into Grimsby on 18 January 1953 by the skipper of the steam trawler *Merryn*; the Queen was graciously pleased to accept it. In 1971 the then Lord Chancellor commented that the continued existence of the prerogative right to royal fish was of no benefit to Her Majesty and occasionally caused inconvenience to other people. The most recent occasion when a sturgeon was offered to the Queen was in February 1992. It had been caught in Portland Harbour, weighed between 13.6 and 15.9 kilograms (30 and 35 pounds) and measured 1.37 metres (4½ feet).

Because it had not been killed, Her Majesty agreed that it should remain at the Sea Life Centre in Weymouth.[16] For centuries, the appeal of the sturgeon has been its flesh, which is generally smoked, and its roe, which is prized the world over as caviar. The main caviar-producing species from the Caspian Sea are the Beluga Sturgeon *(Huso huso)*, Sevruga Sturgeon *(Acipenser stellatus)* and Oscietra Sturgeon *(Acipenser gueldenstaedtii)*, none of which have ever occurred in British waters, although the British species certainly produces edible roe. Because sturgeon are among the scale-less fish, for a long time Jews were forbidden to eat them.

All sturgeons are now seriously endangered but some success has been achieved in farming them in order to ensure the continuity of the valuable caviar trade. A caviar substitute often seen in Britain is 'lumpfish caviar', which has nothing to do with sturgeons but is the dyed roe of the Atlantic Lumpsucker (sea hen) *(Cyclopterus lumpus)*, while 'red caviar', which is also commonly seen, is derived from salmon. Immature Sterlets *(Acipenser ruthenus)* are sometimes offered for sale at pet stores for garden ponds but, if they survive, they will eventually outgrow whatever home they are offered.

'Sturgeon' is a Middle English word but has its origins in Anglo-French and other European languages

Herrings
(Clupeidae)

The Herring is said to be 'the king of the sea' and, in the Isle of Man, and other places too, people will tell you why. Long ago, the fish decided to meet and elect a king to tell them what was right and what was wrong. It was a special occasion and they all wanted to look their best. But the Flounder spent so long making himself look attractive that, by the time he arrived, the election was over and the Herring had been chosen. The Flounder was disgusted, curled his mouth on one side and said, 'A simple fish like the Herring, king of the sea!' – and its mouth has been on one side ever since. It is perhaps on account of the importance of the Herring that the Deemsters (the two Justices of the Isle of Man), in their oath, swear to execute the laws of the Isle 'as indifferently as the herring's back-bone doth lie in the midst of the fish'.(17)

Charming legend or not, the tale does emphasize the importance with which the Herring has long been perceived, its only rival for the most important of all northern food fish being the Cod. In addition to the Herring itself, the Clupeidae also contains several smaller but also very important fish and it has been calculated that almost one-third by weight of all fish caught by humans belong to this family. Most are marine, although there are two British species, both called 'shads', that enter fresh water. Collectively, they have accumulated much charming lore and tradition.

One popular edible fish that belongs here, but which is never described is the whitebait. This is simply because it is not one species but several, usually the fry of Herrings and Sprats, and it is best eaten, coated in flour and fried in oil, during the summer months. These fish are a traditional London dish and whitebait festivals were once famously held at the Old Ship and Trafalgar taverns in Greenwich. They were especially associated with an annual feast, held until 1894, to celebrate the successful completion of flood-prevention works in Essex. Prime Minister William Pitt the Younger attended one year. Today, there is still a whitebait festival every September in Southend, where the season's catch is blessed and celebrated. A tale is told of a man called Richard Cannon, from Blackwall in London, who persuaded the then Lord Mayor of London that it was a distinct species and, on this pretext, was given permission to sell them. I have met people who still believe this.

Herring *(Clupea harengus)*

The Herring is a beautiful fish when fresh, deep green or blue on its back and silver below, and varying in length up to around 40 centimetres (15 inches). There is an old tradition of catching Herring in British waters. Felix, Bishop of East Anglia (d.647?) founded a monastery at Soham, near Ely, and

> ... a godly man placed in it to pray for the health and
> success of the fishermen that came to fish at
> Yarmouth in the herring season.(18)

Historically, however, as far as large-scale fishing is concerned, most Herring were landed from the Baltic, where a smaller and earlier maturing race of the fish occurs. This industry laid the foundations for the legendary strength and wealth of the Hanseatic League. Mysteriously, the Baltic stocks declined suddenly in *c.*1420 and the supremacy of the Hanseatic League declined with them. Tradition says that the Herring migrated to the North Sea. What is undeniable is that the North Sea fishery then took over, organized primarily by the Dutch.

Coincidentally and fortuitously, a Dutchman named Beukels invented a method of pickling Herrings in brine rather than simply packing them in dry salt and the fish immediately came to be a year-round food resource. On 12 February 1429 such pickled Herrings were responsible for an improbable neologism. It was the height of the Hundred Years War and a force under the distinguished soldier and veteran of Agincourt, Sir John Fastolf (*c.*1378–1459) was taking provisions, largely in the form of pickled Herrings, to the English troops besieging Orleans. They were attacked by a much larger French unit but swiftly drew up a defensive position, forming an *ad hoc* wall with the barrels of Herrings, from behind which they shot at the French, eventually gaining the better of their attackers. Shooting arrows from the protection of such a makeshift defensive wall was a battlefield novelty and, taking its name from the Old French *barrique* and the Spanish *barrica*, meaning 'barrel', the construction became known as a 'barricade'. The encounter itself has forever after been called the 'Battle of the Herrings'.

In the early nineteenth century, a Scottish Herring fleet began seriously to develop and the British Herring industry grew, reaching its peak immediately before the First World War, when around 350 million tonnes (356 million tons) were landed. Peterhead, once a deep-sea whaling port (see Dolphins, Whales and Porpoises, page 435) found a new lease of life, along with Fraserburgh, Banff, Wick and other towns in the northeast of Scotland. A thriving industry grew up in Shetland too and the huge fleet of Scottish boats with their drift nets sailed south to catch fish in the early autumn off Yorkshire and then, in late autumn, off the coast of East Anglia. The ports of Grimsby, Great Yarmouth, Lowestoft and others were packed with hundreds of brown-sailed drifters landing their catch. A vast industry, embracing merchants, packers, coopers and many others, grew up around this sleek, silvery and unbelievably numerous fish. Around the beginning of the twentieth century steam drifters took over and the picturesque sailing boats vanished for ever.

Today, Herrings are caught by different boats and in different ways. Large, diesel-powered vessels, guided by sonar, use trawls and purse-seine nets to scoop up huge quantities of them, sometimes as their primary catch, sometimes as a by-catch of young fish taken with Sprats and other small species. However, not only has the fishing changed, so has the fish. Herring are now present in very much lower numbers, an event with consequences not simply for fishermen but also for the numerous other species that depended on it as food. Today, Herring-fishing, like Cod-fishing (see page 157) has become hugely complicated, both technically and legally, as the European Union attempts to weld the different priorities of the various governments into mutually agreed strategies and quotas that will conserve dwindling stocks. An understanding that there are three principal stocks of North Sea Herrings, spawning at different times and in different places but coming together as a mixture of young and mature fish, has rendered an already complex situation even more difficult to manage. The North Sea Herring fishery is a classic example of man failing to realize that a natural resource can be exhaustible until the damage is almost irreparable.

> The herring are not in the tides as they were of old;
> My sorrow! for many a creak gave the creel in the cart
> That carried the take to Sligo town to be sold,
> When I was a boy with never a crack in my heart.
> (William Butler Yeats, 'The Meditation of the Old Fisherman')

It is all a far cry from the time when the fishermen of Brighton and Hastings saw the great shoals on 5 November and said that they were coming inshore to see the bonfires; or when the Scottish fishermen of Hamilton would pluck a Wren and see which way its feathers fell to determine whether the Herring-fishing would be successful. In many places, it was thought that, if the first fish taken from the water each season was female, the omens were good; if it was a male, the portents were not so rosy. The unpredictable nature of Herring shoals in particular places and at particular times has always readily lent credence to these fishermen's beliefs. The first fish was sometimes nailed to the mast as a sign for others to follow; if it was dropped overboard, this was bad luck indeed. On the Isle of Man they said 'No Herring, no wedding' because, if the fishing failed, the young men would have no money to marry.

Also on the Isle of Man, it was said that you should never turn over a Herring on your plate but should remove the bone so the lower side could be eaten. To turn it over would be like overturning the boat that caught it should that boat be at sea at that moment. There is also a widespread belief, especially in fishing districts, that a Herring should be eaten from the tail to the head; to eat it from head to tail would turn the shoals of fish away from the shore.

I still come across this belief today in connection with one of my personal culinary passions: the kipper. These peculiarly British wonders are whole Herrings that have been split, gutted and smoked, but the head and tail are never removed and dye is never added. There have been smoked fish for centuries but, in their modern incarnation, kippers were invented in Northumberland, traditionally in Seahouses, by a man called John Woodger in 1843, although other towns will tell you differently. The most popular method of smoking is in the hot smoke from a fire of oak chips. Different localities – Northumberland, East Anglia, the Isle of Man, Loch Fyne and other places in the west of Scotland – each claim to produce the best.

The name 'kipper' for a smoked fish is mid-eighteenth century but the origin is not known. The same word is sometimes used for a male salmon during the spawning season and this meaning seems to be from the Old English *cypera*; perhaps both have something to do with another Old English word, *coper*, relating to the colour.

Kippers should never be confused with another British wonder, the bloater. I can still remember the satisfaction (and the smell) when, as boy, I was given the task of opening the wooden boxes of bloaters that were sent periodically as gifts by a family friend in Great Yarmouth. A 'bloater' (a nineteenth-century word meaning simply 'a bloated herring') is a Herring that has been soaked whole, ungutted, in brine before being cold-smoked. Bloaters, therefore, are still raw and so should be eaten within a day or so. They are best grilled. Much less often seen are buckling, which are gutted, headless and hot-smoked and should be eaten soon, preferably cold with horseradish sauce.

The success of Herrings as food fish stems largely from their versatility. Apart from being eaten raw, cooked fresh, pickled in brine (today most often in the form of roll-mops), hot-smoked or cold-smoked, they have excellent roes, both hard (female) or soft (male), and are also delicious soused (baked in vinegar with spices). A Yorkshire speciality is one of the various fish dishes known as Solomon Gundy, in which the Herrings are boiled and taken from the bone without removing the head and tail; the flesh is shredded and mixed with Anchovies, onion, apple and grated lemon peel. The mixture is placed back over the bones in the shape of the fish, and garnished with capers, mushrooms and pickled oysters.

Wick Harbour, near the northern tip of Scotland, was once the base for a huge Herring fleet that worked the North Sea.

Nonetheless, perhaps the most marvellous of all Herring recipes, sometimes used for other fish such as Pilchards, is star-gazy pie – the West Country wonder in which the fish heads are turned heavenwards. This is an old Cornish version.

6 Herrings
3 or 4 eggs
tarragon vinegar
pepper and salt
parsley
breadcrumbs
pastry

Clean and bone the fish but do not cut off the heads. Place in a pie-dish lined with fat and breadcrumbs with the heads facing inwards. Pour the beaten eggs in tarragon vinegar over them. Put a pastry over the dish with potatoes (as with Cornish pasties), leaving a hole in the middle for the heads to stick out. Bake for 1 hour in a moderate oven. Place parsley in the mouths of the fish before serving.[19]

Predictably, John Keogh had other ideas for Herrings:

> The roe of a herring is diuretic. The whole herring well salted and applied to the soles of the feet draws humours from the head ... .[20]

A Hallowe'en custom in the Isle of Man says that a girl:

> ... should eat salt herring, bones and all, without drinking or speaking; she must then retire to bed backwards; in her dreams she will see her future spouse coming to bring her a drink.[21]

At South Queensferry on the Firth of Forth, a Herring-associated ritual, which once also took place in other Scottish towns, including Buckie and Fraserburgh, still survives. Here, the Burry Man procession takes place on the second Friday in August, and the Burry Man, wearing a white flannel costume covered totally with burrs (the fruits of the Burdock, *Arctium minus*), is led slowly around the town. The origin of the ritual is not known but, at Buckie, the man at least had an easier progression, being paraded in a barrow; he served 'as a charm to raise the herring'. It has been suggested that, at Queensferry, the intention is to express gratitude for the previous

A shoal of blue-green and silver Herring *(Clupea harengus)* is a spectacular sight as it wheels and dives through the water.

fishing season rather than anticipation of the next. Why the burrs? Possibly because, in sticking to the clothing, they represent the Herrings, which, trapped by their gills, appeared to be sticking to an old-style drift net. A more obvious Herring connection appears in the Festival of the Fishing Queen, which used to take place in many Scottish towns and is still alive and flourishing at Eyemouth in Berwickshire.

> The Queen wears a mantle of silver herring net, and some of her pages, kilted or wearing naval uniforms, carry replicas of the lifeboat and of the fishing trawlers on which the life of the port depended. Her maids and ladies-in-waiting, like those of similar queens in most other parts of the Western world, wear beautiful long dresses and are bedecked with flowers. The Queen and her retinue come from fishing families, and are chosen from the Girls' High School. They arrive by sea with an escort of decorated vessels, and walk in procession through the town, led by three boys in traditional fishermen's ganseys and breeks.[22]

A Herring that has ejected or shot off its spawn is considered worthless and became known as a 'shotten herring':

> ... if manhood, good manhood, be not forgot upon the face of the earth, then am I a shotten herring.
> (*Henry IV Part 1*, Act 2, Scene 4)

But what about the most famous of all Herrings, the 'red herring'? In the literal sense, red herrings were an early form of kipper, very heavily salted and red in colour; and now rarely seen, although I am told that West Indian food stores may sell them. In its more familiar meaning, of something that diverts attention, it alludes to the practice of 'drawing a red herring across the path', most usually to divert hounds from the scent of a Fox. The familiarity of Herrings is also evidenced in such expressions as 'herring-bone' to describe a pattern of bricks or other structures. Most fish have their bones in the same pattern but Herring's bones are the kind that people see most often.

The name 'Herring' itself is Old English but nothing is known before that, although it has affinities in several other European languages. The Norwegian name 'sild' is used for immature herrings, generally canned. Herrings occur sometimes in heraldry, most famously and appropriately in the arms of Great Yarmouth, which combines the front halves of three lions with the back halves of three Herrings. 'Herring' also exists as a surname, sometimes with one 'r', sometimes two. The most famous of the human Herrings must be Albert, the hero of Benjamin Britten's opera *Albert Herring*, but sadly such wonderful places as Herringfleet and Herringswell in Suffolk, Herrington in Durham and Herringby in Norfolk can lay no claim to fishy inspiration, all originating from Old English personal names.

Sprat *(Clupea sprattus)*

Sprats are small and consequently rather maligned; after all you 'throw a Sprat to catch a Mackerel'. The name is originally Old English and has often been used to refer to someone small or worthy of contempt. They are superficially like small Herrings, although they never exceed 15 centimetres (6 inches) in length and have a saw-like row of scales down the front keel. They are important food fish but fresh ones are rarely seen for sale because most of the catch is canned. There are small-scale Sprat-catching enterprises in a number of British fishing ports and a large industry in Norway, where they are caught in the fjords, canned and called 'brisling' or 'skippers'.

Pilchard/Sardine *(Sardina pilchardus)*

Sardines are young Pilchards and are around the same size as Sprats when fully grown. They are very important food fish, although they are a more southerly species and are caught only by British boats from the south and west coasts of Britain. For many years, there was an important Pilchard-fishing industry in some of the Cornish ports, such as Polperro, although the catch was largely exported to Italy.

The word 'fish' was synonymous with 'Pilchard' in west Cornwall, where they were said to be 'Food, money and light, all in one night,'[23] but the fishery has declined today. The origin of the name 'Pilchard' is mysterious. 'Sardine' finds echoes in French, Italian and Latin; and, of course, in the island of Sardinia, from whence many of them come. A related small fish, the Anchovy *(Engraulis encrasicholus)*, is very familiar to the British palate. It is almost always preserved rather than eaten fresh but is really too southerly a species to have made much impact in British waters.

Allis Shad *(Alosa alosa)*

OTHER COMMON NAMES: *ale wife; alley; chad; damon herring; keinak; king of the herring; May fish; shad.*

See Twaite Shad below.

Twaite Shad *(Alosa fallax)*

OTHER COMMON NAMES: *bony horseman; chad; goureen; herring shad; killarney shad; May queen; queen of the herring; shad.*

The two species of shad resemble the Herring superficially, but they are deeper bodied and more or less anadromous fish that enter estuaries and the lower reaches of rivers to spawn, often in May; hence the name 'May fish'. They are both very similar but the Twaite Shad usually has a row of dark dots along the flanks. The fact that both have acquired a number of colloquial names is evidence of their familiarity and, although they have become much less common almost everywhere because of pollution, there is some optimism that they may return.

> The Allis Shad ... was found by fisheries inspectors carrying out a survey of fish arriving on the flood tide at Allington Weir on the Medway. With the Allis Shad were two Twaite Shad, a species thought to occur only in some of the cleanest rivers such as the Wye, Usk, Severn and Tywi [Towy] and in the Solway Firth ... Dave Scranny, a senior fisheries inspector said ... their return is good evidence that the drive to clean up our rivers is working.[24]

'Shad' is an Old English word; 'Allis' is also said to be an Old English name for the fish. 'Twaite' is seventeenth century but of unknown origin. The name 'goureen' is given to the non-migratory population of slightly smaller Twaite Shad that has existed for many thousands of years in Lough Leane in Killarney.

Opinions vary on the eating qualities of shad. In some areas, they are thought unworthy of attention (hence 'bony horseman'), although, in some European countries, they are highly prized but tend to be cooked for a long time 'to reduce the bones to a manageable pulp'.[25]

Sardines *(Sardina pilchardus)* are caught off the south and west coasts of Britain.

Salmon and Trout
(Salmonidae)

Perhaps the best-known, most celebrated, most valuable and most cherished of fish, the trout and salmon are northern hemisphere species whose sporting fame has spread around the world. They are streamlined, unarguably attractive fish, characterized by the presence of a small fleshy fin, called an 'adipose fin', between the dorsal fin and the tail. All spawn in fresh water, although some live there permanently, while others are anadromous. In some species, both migratory and land-locked races occur. All are carnivorous. There are three native and two introduced species in Britain, one of which, the Rainbow Trout (see page 137), is now very important commercially.

Atlantic Salmon *(Salmo salar)*
At around six o'clock on the evening of Saturday, 7 October 1922, a young woman named Georgina Ballantine, who had spent the day fishing with her father, cast her spinning Dace bait from their boat across the Boat Pool on the Glendelvine water of the River Tay, close to their home in Perthshire. Within minutes she had entered angling folklore and had set in train quite the oddest of the many extraordinary facts, beliefs and traditions that attach to a species that, for anglers the world over, is the king of fish, the supreme prize. Georgina Ballantine's bait was taken by a salmon that, when she landed it in darkness just over two hours later, turned the scales at 29 kilograms (64 pounds). It was, and will almost certainly forever remain, the largest salmon caught on rod and line in Britain. Two years later, almost to the day, another woman, 'Tiny' Morison, landed a 27.7-kilogram (61-pound) fish on a fly in the Lower Shaw Pool of the River Deveron in Aberdeenshire. This is unlikely to be surpassed as the heaviest salmon ever caught on a fly in Britain. Since then, other women in other places have caught big salmon, and big bags of big salmon, in quantities far out of proportion to the number of women salmon-anglers. It has been suggested, perfectly seriously, that an irresistible female hormone present on bait that women have handled can be detected by the fish in the water. This seems as good an explanation as any.

The Atlantic Salmon is the only native British salmon, and all that I say here relates to this fish. Globally, however, there are several other species that, in their own lands, attract comparable interest from anglers; one or more of these species may yet feature prominently in British waters. The great salmon of North America are Pacific species of the genus *Oncorhynchus* and among them are the Coho Salmon *(Oncorhynchus kisutch)*, which has escaped from fish farms in France, and the Pink Salmon (humpback salmon) *(Oncorhynchus gorbuscha)*, which has been introduced to some Russian rivers. Both have already been caught in the sea off Britain.

The reasons for the salmon's pre-eminence as an angling quarry are not hard to see. It is a beautiful fish to behold, streamlined, silvery in overall appearance but darker on the back, although there is considerable variation in relative stockiness, degree of spotting and other features. Some fish look remarkably like large anadromous Brown Trout (Sea Trout) and only microscopic examination of the scales will confirm identification. The male fish may be rather vividly coloured, with brown or reddish flanks, and they are, in consequence, sometimes called 'kippers' or 'tartan fish'. Not only is the salmon beautiful, it makes wonderful eating (and this is still true today, when our palates have become numbed by the ready availability of farmed fish). It also has the potential to grow very large, is never easy to catch, is especially challenging to take on a fly and has a splendidly romantic anadromous lifestyle.

This lifestyle takes it from the sea to spawn in rivers, which it ascends in spectacular fashion, leaping obstacles in its path (the name 'salmon' is Middle English but probably originates in the Latin *salire*, 'to leap'), only to die exhausted a few weeks later. The salmon is a sea fish and, commercially, it is in the sea and estuaries that large numbers are caught by netting. Unlike the catadromous life of the eel (see page 150), which took centuries to unravel, the life history of the salmon was fairly well understood when Izaak Walton wrote:

> Much of this has been observed by tying a ribband, or some known tape or thread, in the tail of some young Salmons, which have been taken in weirs as they have swimmed towards the salt water, and then by taking a part of them again with the known mark at the same place at their return from the sea ... .[26]

Partly because the salmon's life history is complex, and also because the different stages are visually distinct, the salmon has acquired a variety of names. In times past, some stages have even been thought to be distinct species; although, oddly, the species itself seems to have acquired no alternative names. The young fish that emerge from the eggs, which are laid in fresh water, are called 'alevins' but soon become 'fry', a Middle English word which was used generally for all very young fish of whatever species but which, in times past, was applied to other young things and originated in an Old Norse name for 'seed'. The fry develop to become small, trout-like fish known as 'parr', probably an old Scottish word, and bear large bands or blotches called 'parr spots' along the flanks. These fish were long believed to be quite distinct. As the time for the salmon to migrate from river to sea approaches, generally after two years, the spots on the parr become masked by a silver substance and the back of the fish takes on a green-brown appearance; the fish is then known as a 'smolt', another old Scottish name. The change is today called 'smolting' or sometimes 'smoltification'. There is an old saying:

Atlantic Salmon *(Salmo salar)*, returning from the sea, leap over rocks and rapids to reach their spawning grounds at the river's head.

Salmon Fishing at Eildon Hill **by Charles Landseer (1799–1879) reflects the traditional popularity of the sport.**

The first spate of May
Takes the smolts away.[27]

In Izaak Walton's day, the young fish was known, always I think rather more picturesquely, as a 'samlet'. Once in the sea, salmon feed voraciously on Sand Eels and Herring, crustaceans and other fish and marine creatures. They grow rapidly until, a year later, they become mature fish known as 'grilse', perhaps a Middle English word; many of these return to rivers from summer onwards to spawn in the autumn. The entry of the fish into fresh water is called a 'run'. Some, however, remain in the sea for another winter, returning only in the spring of their fourth year, when they are called 'spring salmon'. Because they are generally larger than grilse, and because they are in the river for a longer time, spring salmon have always been a prime quarry for anglers and it was, inevitably, another woman angler, Doreen Davey, who, on 13 March 1923, landed the largest spring salmon ever caught in Britain, from the Winforton water of the River Wye in Herefordshire. It weighed 27 kilograms (59 pounds). Any salmon that do not enter rivers until the summer, or even the early autumn, of their fourth year are sometimes called 'summer fish' or 'autumn fish' and they can be very big indeed. When salmon are caught, anglers will examine them for 'sea lice', small marine copepods (see Copepods, page 48) that can live for only a few days in fresh water. The presence of sea lice indicates that the fish is fresh-run and vigorous rather than being diseased and unwholesome.

Quite astonishingly, salmon usually return to the river in which they were born and, if they survive the many obstacles, including anglers, on their way upstream, the female fish (usually called 'a hen'; the male is a 'cock') selects a clean, fast-flowing part of the river in which to spawn. She uses her tail to open a hole, known as a 'redd', in the gravelly bottom, where she deposits her spawn. This is immediately fertilized by an accompanying male, which, characteristically in the mating season, develops an upturned hook to the lower jaw known as a 'kype'. After depositing the spawn, the fish, in Izaak Walton's words:

... most cunningly cover it over with the gravel and
stones and then leave it to the Creator's protection.[28]

After spawning, the fish are known as 'kelts' (also perhaps originally Middle English) and most are by then 'spent'. They are debilitated, pathetic-looking, often highly emaciated creatures that slowly make their way to deeper water. A few female kelts return to the sea and may spawn again, but many, and certainly most of the males, die, often overtaken by fungal infection while exhausted. This was the fate of the most literary of salmon, Salar, the hero of Henry Williamson's 1935 novel *Salar the Salmon*, which is arguably the only work that has converted a biology lesson into a fictional narrative:

> ... a kelt with fungus on its head and tail and flank, lying on its side in water not deep enough to cover it. Salar had got so far with the last of his strength, and had died in the darkness.[29]

Many other creatures besides humans find salmon appealing food. In the sea, they are eaten by whales, sharks and other large fish; in rivers, young fish especially are eaten by Otters, American Mink and several kinds of birds, including herons.

One of the enduring mysteries of salmon and salmon-fishing is that, as adult fish, the salmon almost never feed in fresh water. The reason why they take an angler's bait is quite elusive; it must simply be that they find it annoying. For this reason, salmon flies, unlike trout flies, are not simulations of any real creature but are fanciful and spectacularly splendid creations of feathers and wire. The presence of salmon in a river is today taken as a sign that the river is clean and the fact that they have returned to a number of hitherto polluted British rivers is welcome.

Nonetheless, modern technology has hampered the salmon in other ways, most especially by the construction of dams for the harnessing of hydro-electric power, and these have, in turn, led to the construction of salmon 'ladders' and passes to enable the migrating fish to by-pass large, man-made obstacles on their way upstream. These devices often allow salmon to be seen under water and have themselves become tourist attractions.

The salmon, as befits one of the kings of fish, has always had special symbolism. For the Welsh soldier Fluellen, it was salmon that provided the link between his Monmouth king, Henry V, and Alexander the Great:

> I tell you, captain, if you look in the maps of the 'orld,
> I warrant you sall find, in the comparisons between
> Macedon and Monmouth, that the situations, look you,
> is both alike. There is a river in Macedon; and there is
> also moreover a river at Monmouth: it is called Wye at
> Monmouth; but it is out of my prains what is the
> name of the other river; but 'tis all one, 'tis alike as my
> fingers is to my fingers, and there is salmons in both.
> (*Henry V*, Act 4, Scene 7)

There are still salmon in the Wye at Monmouth. It was always one of Britain's great salmon rivers, although its fish population is now much depleted. I caught my first salmon in the Wye but I have not caught many more there since. In 1999, the official returns reported just 567 fish being caught on rod and line in the entire river.[30]

For many a poet, the salmon and its life have provided inspiration over the centuries:

> And when the salmon seeks a fresher stream to find,
> Which hither from the sea comes yearly by his kind;
> As he tow'rds season grows, and stems the wat'ry tract
> Where Tivy falling down makes a high cataract,
> Forced by the rising rocks that there her course oppose,
> As though within her bounds they meant her to inclose;
> Here, when the labouring fish does at the foot arrive,
> And finds that by his strength he does but vainly strive,
> His tail takes in his mouth, and, bending, like a bow,
> That's to full compass drawn, aloft himself doth throw;
> Then springing at his height, as doth a little wand
> That, bended end to end, and started from man's hand,
> Far oft itself doth cast; so does the salmon vault:
> And if at first he fail, his second summersault
> He instantly essays; and from his nimble ring,
> Still yerking, never leaves until himself he fling
> Above the opposing stream ...
> (Michael Drayton, *Poly-Olbion*, 1622)

Fish ladders, such as this one on the River Lui in Scotland, help migrating salmon to negotiate man-made obstacles.

The salmon on the Glasgow coat of arms recalls the legend of St Kentigern and the queen who lost her ring.

For the consumer, the salmon has lately become what the chicken was to the immediate post-war generation: a luxury food suddenly made readily available through improved technology. While trout have been easily farmed for many years, the migratory habits of the salmon posed much more of a challenge but, over the past 30 years, a complex technology has been developed to mimic the fish's natural life cycle, culminating in the use of submerged cages in the sea, to which the smolts are transferred. This has resulted not only in a wider availability of the fish for the consumer but has also placed less of a strain on wild stocks. Salmon-farming is now a major Scottish industry, providing employment for over 2,000 people and second in size only to that of Norway. However, there is another side to this success because the farmed fish may be of different genetic origin from those in the wild and inevitably some have escaped and hybridized with wild fish. The genetic purity of the wild Atlantic Salmon is now causing considerable concern, not least because their homing instincts may have become confused.

Despite the success of farming, wild salmon are still caught in some quantity and, at Norham in Northumberland, the Vicar still conducts a service to bless the opening of the netting season at the mouth of the Tweed. At midnight on 14 February, from a coracle moored at Pedwell Beach, fishermen from the surrounding area repeat an old Pedwell prayer:

Good Lord, lead us,
Good Lord, speed us,
From all perils protect us,
From the darkness us direct;
Finest nights to land our fish,
Sound and big to fill our wish.
God keep our nets from snag and break,
For every man a goodly take,
Lord grant us.[31]

However, no means of catching salmon has matched the ingenuity of the basket traps, or 'kipes', used on the lower reaches of the River Severn. These conical structures, which changed little from prehistoric times to the twentieth century, were made from hazel staves about 2 metres (7 feet) long and were said to resemble a crinoline designed for a giantess. The kipe was fixed so that its gaping flared end faced upriver; to its narrow waist downstream was fitted the butt (another basket of closely woven withy), to which, in turn, was fitted the forewheel (yet another, smaller basket, with its end stopped with seaweed or turf). A weir across the river might have consisted of up to 80 kipes anchored to stakes and 'this type of fish trap, surely the most ingenious ever devised by man' would catch not only salmon but anything from a shrimp to a sturgeon.[32]

Despite its universal availability, smoked salmon still retains its image of a luxury item, but the fresh fish – poached, grilled or, as restaurants now tend to call it, seared – features on almost every menu. Gravlax, or gravadlax, is now also familiar on menus and is often confused with smoked salmon, although it is quite different. A Swedish form of pickled salmon, gravlax has obscured an old English tradition of pickling salmon. A Northumberland recipe for salmon pickled in vinegar with mace, cloves and pepper, rather like pickled Herrings, is quite delicious.

Salmon, usually depicted in mid-leap, appear on many an inn sign, and also on many a coat of arms. Those of Glasgow are perhaps the most famous and recall the legend of St Kentigern, who was approached by a messenger from a queen who had lost a ring, a present from her husband, which she had given to a lover. In reality, the king had obtained the ring, thrown it into the sea and challenged his wife to find it in three days or remain in prison. With considerable disregard for the woman's infidelity, the saint told her messenger to fish in the Clyde. When he did so, the messenger caught a salmon which, most fortuitously, bore the ring in its mouth, and so the queen was freed.

'Salmon' is fairly common as a surname and there are plenty of Salmon inns (although the famous Salmon Tail in Stratford-upon-Avon is named for the shape of the adjoining street plan rather than the fish). Sadly, however, there is no town or village of note that commemorates this most wonderful creature. The most promising candidate, Salmonby in Lincolnshire, is named after a man called Salmund.

Brown Trout *(Salmo trutta)*

OTHER COMMON NAMES: *Freshwater trout – breck; brook trout; brownie; bull trout; burn trout; ferox; gillaroo; lake trout; slob trout; sonaghen; yellow trout. Anadromous trout (Sea Trout) – black tail; bull trout; covichie; finnock; fordwich trout; grey trout; herling; lammasman; mort; peal; phinnock; round tail; salmon trout; scurf; sewen; sprod; truff; white trout; whiting; whitling; yellowfin.*

> The Trout is a fish highly valued both in this and foreign nations; he may be justly said, as the old poet said of wine, and we English say of venison, to be a generous fish: a fish that is so like the buck, that he also has his seasons; for it is observed, that he comes in and goes out of season with the stag and buck ... .[33]

While the salmon may be grand and noble, catching a wild Brown Trout on a dry fly from the rippling clear water of an English chalk stream is still a good many anglers' notion of Utopia. I share with quite a few others the view that the Brown Trout is the most beautiful of fish. It has the salmon's streamlined shape, square-ended tail fin and general features but almost invariably has a patterning of large, rich red, brown or black haloed spots on its back and flanks. All of the colours intensify in the male trout at spawning time but it is nonetheless a very variable fish; only relatively recently has the anadromous form, the Sea Trout, been considered merely a race and not a distinct species. Certainly, the Sea Trout looks more like a salmon in its overall colouring, which is silver with small black spots.

Brown Trout are much smaller than salmon and are to be found in streams, rivers and lakes throughout the British Isles; the clearer the water; the more likely they are to occur. They vary greatly in size: while trout in a small stream may seldom exceed 250 grams (9 ounces) in weight (but they always feel bigger on the end of a line), those in rivers will grow larger and lake trout can be enormous. Indeed, so great is the disparity that there are separate rod-caught records for river and lake fish. Correspondingly, their food also varies. Small stream

The Brown Trout *(Salmo trutta)* has beautiful dark spots, as seen in this nineteenth-century painting by A. Roland Knight.

trout live largely on small crustaceans and insects (including the flies that anglers seek to mimic), whereas the large trout, or 'ferox trout', which inhabit big, deep lakes, eat other fish almost exclusively. In many Scottish lochs, Brown Trout feed very largely on their own close relative, the Arctic Charr *(Salvelinus alpinus)*, and many big trout are cannibals.

Sea Trout are generally largely than Brown Trout and, like salmon, feed little in fresh water, if at all. This is despite the fact that some of them stay in fresh water for a considerable time. Therefore, Sea Trout flies, like salmon flies, are fanciful things designed to attract the fish's attention. Traditionally, Sea Trout were almost always fished for at night, an exciting if potentially hazardous occupation when wading thigh-deep in the tumbling waters of a Highland river.

The life cycle of the Sea Trout is essentially the same as that of the Atlantic Salmon, although it recovers much better after spawning and many kelts return to spawn again several times. Unlike salmon, Sea Trout females frequently fail to spawn but retain the eggs within their bodies; they are then known as 'baggots'. Some Sea Trout never move fully into the open sea but remain in coastal waters and in sea lochs, where they are known, unflatteringly, as 'slob trout'. Many young Sea Trout from coastal waters live in large shoals that swim into rivers in late summer; these fish are known by such local Scottish names as 'herling', 'whitling' and 'finnock'.

As with salmon, the return of Sea Trout to waters from where they have long been absent is taken as a good indication of the increased purity of the water. Similarly, the survival of Brown Trout in rivers and lakes stocked with farm-raised fish indicates clean, pollution-free conditions. Big fish are now being caught again in rivers such as the Thames from which they were absent for many years. I have a cased specimen of a Brown Trout, which I bought some years ago and which includes a label 'Caught 29th May, 1890. River Thames. Weight 8 lb'; this means it must have been one of the last of the big Thames fish before the decline began.

Where the environment is unsuitable for Brown Trout to spawn naturally, populations are maintained for angling purposes by stocking with farm-raised fish. However, very few Brown Trout are now farmed commercially for the table; the trout sold in supermarkets today are Rainbow Trout (see below). Many other creatures take trout as food; most fish-eating birds, including herons and Ospreys, some mammals, especially Otters, American Mink and, in coastal waters, seals, and Pike are important predators.

Apart from its undeniable beauty, strength and grace, it is trout-fishing and the skill attached to it that have given the trout a special place in lore and literature. Although trout can be caught by most angling techniques, the artificial fly, and especially the dry fly, is arguably the supreme example of the angler's art. In the finest trout-fishing waters, such as the chalk streams of southern England, catching trout by any other means is both prohibited and very seriously frowned on. Unlike the salmon fly, the trout fly attempts to mimic the natural prey of the fish. The manufacture, from feathers and similar materials, of something that will fool a wily fish in water as clear as crystal requires not only great skill but also a considerable knowledge of entomology (see Mayflies, page 62). Dry-fly fishing was devised around the middle of the nineteenth century; the fly is cast to a rising fish and floats on the surface of the water. Wet-fly fishing is an older technique in which the submerged fly is not necessarily cast to any specific fish but placed where trout are likely to be lying. In nymph fishing, a technique barely a century old, the lure imitates the larva of an aquatic insect and is cast to a rising fish but is slightly submerged.

This Daddy-long-legs fly, fashioned from feathers and wire, is intended to imitate the natural prey of the trout.

Another skilful trout-catching technique, often spoken of but seldom used effectively, is tickling. This is not easy, especially where the water is clear and, as Samuel Butler noted in his book *On a Hypocritical Nonconformist*:

... trouts are tickled best in muddy water.[34]

This entails crawling, worm-like, along a low river bank to where a fish is lying close to the edge. You should lower one hand gently into the water some way behind the fish and gradually bring it ever closer to its underside. With the deftest of finger work, you can then touch and caress the fish. If you are sufficiently gentle, the fish will not swim away but will be lulled into a sensual stupor, at which point you can grab it with your hand and whisk it out of the water and onto the river bank. Trout-tickling is a very clever and very old technique that requires huge patience and consummate skill. It is, of course, illegal.

Trout have often been associated with holy wells. There is no obvious reason for this because no fish is less likely to be found in a well than a trout, although, to be fair, many such holy 'wells' are in reality springs that soon become small streams. From many parts of Britain come ancient tales and legends of trout that embody the soul of whatever saint gave his or her name to the place. Any trout found in such places must never be taken or eaten.

> At Peterchurch in Herefordshire there is a tablet in the church showing a trout with a golden chain around its neck, of which many tales are told. Some merely say that a fisherman caught it nearby in the Golden Well; others, that some poor monks caught it in the River Dore when they were almost starving; others, that St Peter himself baptised converts in the Golden Well, blessed it, and threw into it a trout with a gold hair round its neck, which was to live there for ever.[35]

In Northumberland, there was a curious use for the local trout. A child suffering from whooping cough would have the head of a live trout (presumably a small one) thrust into its mouth. The trout's breath would cure the child and, although this seems bizarre, it is surely no more odd that having a live frog placed in the mouth for the same purpose (see Common Frog, page 174).

Because they require large expanses of clean water, trout do not take readily to domestication but one that evidently did so is commemorated on a gravestone in the garden of Fish Cottage at Blockley in Worcestershire:

> In memory of the old fish. Under the soil the old fish do lie. 20 years he lived and then did die. He was so tame you understand he would come and eat out of our hand. Died April the 20th 1855. Aged 20 years.[36]

'Trout' is a name of old English origin, traceable to Latin and perhaps a Greek word meaning 'to gnaw'. However, I have never understood why the name of a fish of such surpassing beauty should be used to denote 'an aggressive, disagreeable and unattractive woman', 'the old trout' of many a novel and play. Like 'salmon', 'trout' generally has no plural in modern usage. The Sea Trout was for many years called the 'salmon trout' and it is an appropriate name that is still heard from time to time. 'Sewin' has been used since the sixteenth century, especially in Wales.

The most celebrated of the rather few places named after trout must be Troutbeck ('trout stream') in Cumberland, which is forever associated with John Peel (see Fox, page 404). The most celebrated Trout inn is at Godstow near Oxford (see Perch, page 162).

The Rainbow Trout *(Oncorhynchus mykiss)* has a pinkish iridescence on its flanks and numerous small black spots.

Rainbow Trout *(Oncorhynchus mykiss)*

The Rainbow Trout was first introduced to England from the Pacific Coast region of North America in 1884, when eggs were sent to the National Fish Culture Association in Buckinghamshire. Subsequently, further consignments were made and, in the succeeding years, fish were introduced to many rivers and lakes throughout the British Isles for angling. For many anglers, they have the undeniable appeal of being easier to catch than Brown Trout.

They are also very easily farmed and trout farms, now to be seen everywhere in Britain, rely on the Rainbow Trout for their commercial viability. The trout are produced in huge numbers to satisfy the supermarket and restaurant trade, where they are served in every possible way, but probably best when simply grilled or pan-fried. Regrettably, the majority of people have only ever eaten farmed Rainbow Trout, which is but a poor shadow of the wild Brown Trout in terms of flavour and texture.

Farmed Rainbow Trout are also used to restock angling waters because, despite their almost universal distribution in Britain, self-sustaining, spawning populations have developed in only a very few places. However, catching Rainbow Trout is not angling as I understand it. The fish used for stocking angling 'resorts' are often already large and overfed and no real angler can take any pride or satisfaction in hauling one of these artificially inflated creatures from the water.

Compared with a normal Brown Trout, a Rainbow Trout is generally deeper bodied. It is not dissimilar to some ferox lake trout but strikingly different in colour, being imbued with a pinkish iridescence along the flanks and many small, black spots. It is a pretty rather than a beautiful fish. In North America, there is a Sea Trout counterpart of the Rainbow Trout in the form of the anadromous Steelhead Trout, but this has not been introduced to Britain. The taxonomy of these fish is complex, however, and the Kamloops Trout, another well-known American fish, is now thought to be the same species, while others among this very attractive group, including the most beautiful, red-chinned Cuthroat Trout *(Oncorhynchus clarki)*, are thought to be distinct but closely related species.

Arctic Charr *(Salvelinus alpinus)*
OTHER COMMON NAMES: *alpine char; char; charr; cuddy; red-bellied trout; red waimb.*
The name 'Charr' now tends to be used in preference to 'char', because it is a closer approximation to the original name, which was possibly from Gaelic and meant 'red' or 'blood'.

Charr are little known outside the restricted areas of the British Isles where they occur. They are species of deep old lakes in northern England, Scotland, Wales and Ireland, where they are frequently referred to as relics of the Ice Age. All British charr are landlocked, or at least non-anadromous, but elsewhere in their natural range, which is throughout the north of the northern hemisphere, there are anadromous populations, just as there are of Brown Trout.

They are well known to anglers, who traditionally catch them by deep trolling. Because of the depth at which they normally live, Charr are generally seen only when caught, although they do tend to come to the surface at night. They are relatively easy to catch but offer little sport, allowing themselves simply to be hauled to the surface. They are good to eat and, in Lake Windermere, a small commercial fishery grew up to support a luxury local product known as 'potted char'.

Arctic Charr are smaller than trout – a fish of 3 kilograms (6½ pounds) is a very big one – but are very beautiful, being a green-grey overall, with variously coloured pale spots, although they are variable in colour. Formerly, the differently coloured fish from different lakes were considered to be separate species and such names as 'Willoughby's charr', 'Shetland charr', 'Gray's charr' and 'Scharf's charr', among others, will be found in older writings.

The male of the Brook Charr *(Salvelinus fontinalis)*, introduced to Britain in 1868, has a rich red underside in the breeding season.

Brook Charr *(Salvelinus fontinalis)*
OTHER COMMON NAMES: *American brook trout; aurora trout; fontinalis; mud trout; speckled trout.*
This very pretty charr has uniquely characteristic wavy lines down the flanks and the male has rich orange-red colours on the underside. It was introduced to Britain from eastern North America in 1868, well before the Rainbow Trout. It has also been more successful than the Rainbow Trout in forming self-sustaining populations and is unusual among its family in thriving in poor and muddy waters, as the name 'mud trout' suggests. Brook Charr are now present in a number of lakes and rivers throughout Britain and, although the species can be anadromous, there is no evidence that it is in British rivers.

Whitefish
(Coregonidae)

The whitefish are a small family of generally rather rare, little understood and locally distributed fish. They are related to the salmon and trout, with which they share an adipose fin. Otherwise they are rather like Herrings, being uniformly silver (hence the name 'whitefish'), and having no other colours or patterns and very large scales. They are either freshwater or anadromous, and there are presently three British species, almost entirely confined to particular lakes, where local traditions have built up around them. One additional estuarine species, now extinct in Britain, may yet return if the rivers continue to become less polluted.

Houting *(Coregonus oxyrinchus)*
OTHER COMMON NAMES: *hautin; sea whitefish.*
The Houting is the one that got away, the anadromous whitefish that used to enter the estuaries of some of the eastern rivers, such as the Medway. However, it has probably never spawned in Britain and certainly has not been seen for some years. Nonetheless, if shads can come back to the Medway (see Twaite Shad, page 129), then Houting might too. Once seen, it is unlikely to be mistaken for anything else because of its peculiarly elongated, nose-like snout.

The name 'Houting' is nineteenth century but I have been unable to trace its origin.

Powan *(Coregonus lavaretus)*
OTHER COMMON NAMES: *freshwater herring; guiniad; gwyniad; schelly; skelly.*
The Powan occurs in only six British lakes: Llyn Tegid (Lake Bala) in Wales, Loch Eck and Loch Lomond in Scotland, and Haweswater, Red Tarn and Ullswater in the English Lake District. The fish of each lake were formerly believed to be different species and were each given their own name.

It was once fished for (on a commercial basis during the two world wars when sea fish were scarce) and was considered very good eating, with a slightly oily flesh. However, it is now a protected species and fishing for it is illegal.

Nonetheless, Powan still form important prey for other lake inhabitants. The large ferox trout of Loch Lomond used to be called 'powan-eaters' and a recent survey revealed that one third of all Powan in the loch had been attacked by River Lampreys.[37] Powan themselves feed mainly on crustaceans.

'Powan', the Scottish name for the fish, now tends to be the preferred name. It is derived from the Irish name 'pollan' (see Pollan, below) which may perhaps come from the Irish word 'poll', meaning 'a freshwater lake'.

Vendace *(Coregonus albula)*

OTHER COMMON NAMES: *cisco; Cumberland vendace; Lochmaben vendace.*

The Vendace has only ever been found in four British lakes and was once fairly numerous in at least one of them. The English and Scottish populations were formerly thought to be distinct subspecies. In England, it is found in Bassenthwaite and Derwentwater in the Lake District, where it is called the 'Cumberland vendace'. In Scotland, until about 20 years ago, it was found in the Mill Loch and other small lakes around Lochmaben in Dumfries and Galloway, hence the name 'Lochmaben vendace'.

There is no obvious reason for the disappearance of the Lochmaben fish, and it was clearly rather plentiful at one time. There is a local tradition that Mary, Queen of Scots brought it there from Europe in 1565, but this seems rather improbable; she was far too busy with Lord Darnley and affairs of the heart at that time. Certainly, in the nineteenth century, Vendace-fishing clubs with large numbers of members were formed at Lochmaben and considerable numbers of fish were caught, being considered excellent eating.

The name 'Vendace' seems to have originated in the French name for the superficially slightly similar but quite unrelated fish, the Dace (see page 148).

Pollan *(Coregonus autumnalis)*

OTHER COMMON NAMES: *Arctic cisco; cunn; Lough Erne pollan; Lough Neagh pollan; omul; Shannon pollan.*

Like other whitefish, the Pollan was once thought to be several different fish, one in each of the lakes that it inhabited. At one time there was even uncertainty about whether it was really distinct from the Powan and Vendace.

It is an Irish speciality, occurring, as its regional names suggest, in Lough Neagh and Lough Erne, and in some of the lakes in the Shannon river system. It is still numerous there and fish are caught commercially by seine nets and long lines and sold in Irish, and occasionally English, markets.

The Grayling *(Thymallus thymallus)* is grey-green in colour and has a large, sail-like dorsal fin.

Graylings
(Thymallidae)

Grayling *(Thymallus thymallus)*

A good many fish have been called the 'king of ...' or the 'queen of ...', but I was brought up to believe that the Grayling really is 'queen of the stream'. I do not think any fisherman ever forgets the pleasure and satisfaction of catching his first Grayling. They are best and most satisfyingly taken on a dry fly and are popular with fly fishermen for being in season when trout are not; flies constitute the major part of their natural food. An angling society devoted to them and their conservation, the Grayling Society, was founded in 1978.

Grayling are related to trout and salmon and, like them and the whitefish, have an adipose fin. However, the most individual and magnificent feature is the splendid, almost sail-like dorsal fin. The fish are not large – the British record is just over 1.5 kilograms (3¼ pounds) – but they are a beautiful grey-green, with a pattern of small black spots on the flanks, a pattern that seems to be unique to each fish. The name relates to the grey colour and is fifteenth century; the Grayling butterfly *Hipparchia semele* (see Nymphalids, Fritillaries and Browns, page 82) took on the name much later.

Grayling often occur in small shoals and are fish of clean rivers (and a few lakes) throughout England, Wales and the southern half of Scotland, where they were deliberately introduced in the nineteenth century; they are absent from Ireland. Although the Grayling is clearly native, there was long a belief that it had been introduced by monks as a food fish. Grayling were also believed to feed on gold, a modern explanation being the yellowish grit found in the stomach, which is derived from the caddis-fly larvae on which the fish feeds.

The fish make beautiful eating and are perhaps most delicious when served cold. The flesh has a natural fragrance;

The Pike *(Esox lucius)* is a voracious predator and lies in wait for its prey among aquatic vegetation.

the scientific name *thymallus* is derived from *Thymus* (thyme) and old writers called the Grayling 'the flower of fish' in consequence. Izaak Walton recounted an old and additional use for the fish:

> ... the fat of an umber or grayling, being set, with a little honey, a day or two in the sun, in a little glass, is very excellent against redness, or swarthiness, or anything that breeds in the eyes.[38]

Walton's birthplace, Stafford, has long had a special association with Grayling; its seal shows the fish in the River Sow with the castle on the bank and alludes to a charter granted to the borough by King John. Nonetheless, to my mind, the most beautiful place in England to see Grayling is just below the old bridge on the River Wye at Bakewell in Derbyshire.

Smelts
(Osmeridae)

Smelt *(Osmerus eperlanus)*

OTHER COMMON NAMES: *sparling; sperling.*

The name 'Smelt' originates in Old English but seems to have nothing to do with the verb 'to smell'. This is both surprising and a pity because this is a powerfully aromatic little fish with a scent that reminds most people of that of cucumber. The Scottish name 'sparling' derives from an Old French word for 'small fish'. It is yet another silvery relative of the trout and salmon, with an adipose fin, and is superficially very like a small whitefish.

There was once a freshwater population of Smelt of mysterious origin at Rostherne Mere, the largest lake in Cheshire and now a well-known wildlife sanctuary, but the fish disappeared some 80 years ago. The Romans are believed to have cultivated Smelt in freshwater lakes but it seems unlikely that the Cheshire fish are that ancient a survival.

Today, the Smelt is solely an anadromous species, living for most of the year in coastal waters and estuaries but ascending further upriver to spawn. Although small – usually little more than 20 centimetres (8 inches) in length – it has a reputation as a fearsome creature, feeding voraciously on crustaceans and other, smaller, fish. They are still caught commercially in some English and Scottish river estuaries and may sometimes be seen in fish markets. They are really excellent if they can be obtained very fresh when the cucumber aroma remains, rolled in flour, fried in butter and served with parsley.

Pike
(Esocidae)

Pike *(Esox lucius)*

OTHER COMMON NAMES: *gad; gedd; jack (young); luce; pickerel (young).*

> ... A pond I fished, fifty yards across,
> Whose lilies and muscular tench
> Had outlasted every visible stone
> Of the monastery that planted them
> Stilled legendary depth:

> It was as deep as England. It held
> Pike too immense to stir, so immense and old
> That past nightfall I dared not cast ...
> (Ted Hughes, 'Pike')

No fish in Britain, and probably no freshwater fish in Europe, conjures up such an aura of raw ferocity and naked sharp-toothed aggression as *Esox lucius*. With some justification, it has received a plethora of names: 'freshwater shark', 'water wolf', 'king of the lake', and 'lord of the stream'. There is only one species in Britain, but one is enough.

The Pike is a sleek, streamlined fish with an elongated, almost duck-like snout, amply lined with backward-pointing teeth. It has an undeniable beauty, being a dark olive-green overall, with characteristic banding patterns on the back which serve as camouflage among water plants. The patterning is sufficiently distinct to enable individual fish to be identified. The underside is strikingly paler and almost yellow. Pike vary widely in colour, however, and many big fish seen from above seem almost black.

Pike occur throughout Britain but are rare in the Scottish Highlands. They are not native to Ireland but were introduced there, probably in the fifteenth century. Today, some of the largest Pike and finest Pike-fishing are found in Ireland.

To generations of fishermen and fish-keepers, the Pike has been something special. In trout and salmon waters, its presence is detested and almost any measures are taken to be rid of it. However, on large, still lakes, where ledgered Herrings lie at the end of taut, stout lines, or by deep, dark mill pools, where a shiny Devon minnow is tossed and slowly returned to lure a monster from the depths, fishermen wait with an anticipation like no other. Pike-fishermen, like the Pike itself, are a remarkable breed.

Pike are lonely, solitary fish that hang in the current among the aquatic vegetation, from whence they dart to devour whatever passes. Their immense acceleration is effected by the triple propulsion of the tail, dorsal and anal fins, which are mounted close together. Izaak Walton described them thus:

> The pike is also observed to be a solitary, melancholy, and a bold fish: melancholy because he always swims or rests himself alone, and never swims in shoals or with company, as roach and dace and most other fish do: and bold, because he fears not a shadow, or to see or be seen of anybody, as the trout and chub and all other fish do.[39]

Just what prey a big Pike will eat has itself been the stuff of tale and legend. When tiny, Pike begin with small things, eating zooplankton and insects, but soon progress to fish fry, then the adults of small species like Minnows and Dace. For bigger Pike, Roach, trout and salmon do very well. For really big Pike, there are few limits and their appetite has been summarized: 'Adult Pike will try to eat any animal, alive or dead, that is not too big to swallow'.[40] Mammals, birds, amphibians, and other Pike if the opportunity arises, are all taken. Dogs? Small, swimming dogs can probably be taken by big Pike; men's legs and a hand trailing over the side of a boat have been savaged too.

But just how big can a big Pike be? Here fact merges imperceptibly into folklore and fanciful fiction. The fisherman describing, with outstretched arms, 'the big one that got away' will almost always be telling of a Pike-fishing exploit and countless fishing pubs have a decent mounted specimen above the fireplace (sometimes the weight inscribed on the case is accurate). Pike of 10 kilograms (22 pounds) are moderately common and fish of 15 kilograms (33 pounds) are not exactly rare. I grew up believing that the record British Pike was one of 24 kilograms (53 pounds), caught by a Mr John Garvin in Lough Conn, County Mayo, on 18 July 1920, but that has now been struck from the record books. A big Pike was certainly recovered when Whittlesey Mere, south of Peterborough, in Cambridgeshire, was drained in 1851. This was said to be around 24 kilograms (53 pounds) and 1.3 metres (4¼ feet) in length. However, the all-time, reasonably authentic British monster seems to be a fish obtained in 1774 from Loch Ken in Dumfries and Galloway (which still advertises itself as Scotland's most prolific Pike water), which reputedly weighed almost 33 kilograms (73 pounds). I am told that its skull survives and could well have come from a fish of this size. A photograph of a big Pike, taken around 1853, is said to be the earliest known photograph of any living fish.

The name 'Pike' is a contraction of the fourteenth-century 'pikefish', from the Old English word for a point, and is an allusion to the shape of the jaw. The weapon called a 'pike' (with its long handle, the 'pikestaff') has the same origin, as does the northern word for a 'steep hill' (as in Scafell Pike). A 'turnpike' road was one with a pikestaff placed across it as a barrier. A small pike (or more commonly now, a small species of North American Pike) is a 'pickerel', also from the fourteenth century. Unfortunately, that delicious crumpet, the 'pikelet', is unrelated and has its origins in Welsh.

This seventeenth-century woodcut shows a pikeman clasping a pike as he lights a cannon.

The Common Carp *(Cyprinus carpio)* is amenable to being kept in ponds as a food fish, although its flesh has a rather muddy flavour.

Carp
(Cyprinidae)

The carp family is the largest of all fish families, with some 2,000 species and the most important family of freshwater fish in Europe. Yet carp are generally taken for granted. With a few exceptions, they have attracted no lore and tradition, contributed little to literature and seldom make exciting eating.

Nonetheless, they are the friends of the fisherman, especially the match fisherman, who measures success not in the size of individual fish, and most certainly not in the beauty of the trophy, but in sheer overall weight. Because many of the carp family are shoaling fish, once a fisherman attracts a shoal to his allotted stretch of river, or 'swim', he stands a good chance of landing a fine haul.

The species vary in size and include some of the lightest and some of the heaviest British freshwater species. No carp have teeth, some have barbels, they generally have scales, sometimes very large ones, and the fins are mostly soft-spined. Few carp are beautiful to look at and many are essentially silvery and rather Herring-like to the casual observer.

Common Carp *(Cyprinus carpio)*
OTHER COMMON NAMES: *king carp; koi; leather carp; mirror carp.*

> Hops and Turkies, Carps and Beer
> Came into England all in a year.

So runs one of the best known of English historical aphorisms; and one that is entirely wrong. Izaak Walton attributed the saying to Sir Richard Baker (1568–1645). In reality, hops (and beer) came to England from the Low Countries in *c.*1520, turkeys from the United States in *c.*1530, and available evidence suggests that the Common Carp arrived shortly before 1500. In 1486, Jan van Wynkyn (Wynkyn de Worde) published the second edition of *The Boke of St Albans*, written by Dame Juliana Berners, the prioress of Sopwell nunnery in Hertfordshire. In it, she wrote of the carp as 'a deyntous fisshe, but there ben but fewe in Englonde, and therefore I wryte the lesse of hym'.[41] The name 'carp', however, pre-dates Dame Juliana; it is Middle English from Latin and has no connection with the verb 'to carp'.

Whoever introduced the carp brought it from central Asia with the intention of keeping it in ponds as a food fish. The Romans first brought the carp to western Europe for exactly the same reason, mimicking (although almost certainly unwittingly) the domestication of the fish in China many centuries earlier. Carp are very amenable to being kept in this way, feeding readily, growing quickly and making reasonable, if unexciting, eating. They were probably much more acceptable in the days when to live only a short distance from the sea meant being denied fresh fish of any sort. My father used to tell of carp being served as part of the traditional Christmas and New Year fare in Poland.

The Common Carp also has the invaluable attribute of surviving for long periods out of water.

> ... a friend's father ... told me that as a young man, when food was scarce and there were no refrigerators, he would catch a carp, roll it in a wet towel and keep it alive and fresh for a considerable time, in the top drawer of his mother's chest of drawers. The image of this fascinated me and so, after some time I managed to catch a carp, wrap it up in a wet cloth and cycle many miles home. It was very much alive on my arrival but my mother would not let the experiment go any further by keeping it in with my pants and vests. It was released into the garden pond, where it may still be living.[42]

A small or even medium-sized Common Carp is a rather attractive if unspectacular fish. Olive brown on the back, paler on the sides and slightly golden beneath, it bears some resemblance to a large, bland-looking Goldfish. A big Common Carp however – and the record fish is over 19 kilograms (42 pounds) – is a bloated monster with a lugubrious face, rather like an overweight and listless human. Carp are noisy creatures too, especially at night (people fish for carp when decent folk are abed), when the gross slurping sound as they suck air and food from the surface of the water can be heard a great distance away.

They are fish of big, slow-moving rivers and most especially of warm, deep, weed-ridden ponds. They occur throughout the southern half of Britain. Scotland is too cold for them, certainly for continued breeding. They eat a wide range of food, including insect larvae, crustaceans and molluscs, which they suck up in mud from the bottom of the pond with their extensible, tube-like mouth. In small ponds, the water can be almost permanently cloudy because of the activities of mud-sucking carp.

The Common Carp is a variable fish. The usual form seen in Britain is called a King Carp but two common variants are the Mirror Carp, which has a single line of large scales along the flank, and the Leather Carp, which is scale-less. Keeping the mirror-like scale of a Mirror Carp for a year was once said to assure good fortune. The Koi, a domesticated and sometimes unbelievably expensive version that originated in the Far East, is now kept extensively in garden ponds for its numerous, sometimes remarkable, colour variations.

Crucian Carp *(Carassius carassius)*

OTHER COMMON NAMES: *crowger; gibel carp; Prussian carp.*

This is the only true carp native to Britain. A smaller fish than the Common Carp and superficially like a large, deep-bodied, olive-green Goldfish, it reaches 20–30 centimetres (8–12 inches) in length. It is a fish of small, weedy ponds and, like the Common Carp, has been used extensively in the past as a food fish, especially in eastern and central Europe. Like the Common Carp, it will survive journeys out of water and used to be packed in wet grass for transport.

The word 'Crucian' is derived from the Greek name for a large black Nile perch.

Goldfish *(Carassius auratus)*

The typical Goldfish, offered as a prize in a plastic bag at fair-ground shooting arcades is the simplest domesticated form of a creature on which breeders have worked much mischief. In the wild, the Goldfish is an olive-green colour, not dissimilar to a Crucian Carp, but such fanciful variants as veiltails, comets, shubunkins and lionheads bear little resemblance to any natural creation. It was originally an Asian and eastern European species but it was in the Far East that it first attracted attention, perhaps as early AD 700, when Brown Bears still lived in England (see page 411) and 300 years before Wolves were feasting on the corpses at Brunanburh (see Wolf, page 402). It was subsequently bred beyond recognition as an ornamental fish and found its way into much Oriental literature and tradition.

Just when Goldfish reached England is uncertain but a remark in Pepys's diary has been thought to refer to them:

> Then home and to see my Lady Pen, where my wife and I were shown a fine rarity: of fishes kept in a glass of water, that will live so for ever; and finely marked they are being foreign.[43]

This seems very little on which to base an identification and the red-and-blue-banded Paradise Fish (*Macropodus opercularis*) is a more likely candidate.[44] Nonetheless, being a Goldfish 'in a glass of water' soon became a euphemism for being on public view. As Saki wrote in *Reginald*:

***The Goldfish Bowl*, by Walter Frederick Osbourne. Goldfish *(Carassius auratus)* have traditionally been kept in bowls.**

I might have been a gold-fish in a glass bowl for all
the privacy I got.[45]

Goldfish won English literary distinction in 1742 when Thomas Gray, a few years before he wrote his 'Elegy Written in a Country Churchyard', visited his old Etonian friend Horace Walpole at Strawberry Hill and learned of the death of his cat Selima.

'Twas on a lofty vase's side,
Where China's gayest art had dyed
The azure flowers, that blow;
Demurest of the tabby kind,
The pensive Selima reclined,
Gazed on the lake below ...
The hapless nymph with wonder saw:
A whisker first and then a claw,
With many an ardent wish,
She stretched in vain to reach the prize.
What female heart can gold despise?
What cat's averse to fish? ...
Eight times emerging from the flood
She mewed to every watery god,
Some speedy aid to send.
No dolphin came, no nereid stirred:
Nor cruel Tom nor Susan heard.
A favourite has no friend!
From hence, ye beauties, undeceived,
Know, one false step is ne'er retrieved,
And be with caution bold.
Not all that tempts your wandering eyes
And heedless hearts is lawful prize;
Nor all that glisters, gold.
(Thomas Gray, 'Ode on the Death of a Favourite Cat
Drowned in a Tub of Gold Fishes')

Throughout the eighteenth century, Goldfish became increasingly familiar in England and inevitably many escaped into rivers or were deliberately liberated. Feral fish are now seen fairly frequently and are sometimes caught by anglers, generally in the same sort of places as Crucian Carp and mostly of the Crucian Carp coloration, although sometimes fairly large golden fish are found too. Many ornamental lakes contain large, old and certainly golden individuals. People have eaten Goldfish and, having done so myself, I can understand why they have never been of much importance as a food fish.

Barbel *(Barbus barbus)*

Many fishes have barbels around the mouth to help them sense the presence of food in the river bed but only the Barbel is named after them, perhaps appropriately as it has two pairs of rather long ones. It is a big, rather handsome fish of large, fairly slow-moving but clean rivers and was once very much a fish of southeast England; the Kennet and the Hampshire Avon have always been among its strongholds. More recently, however, it has been introduced to the Severn and elsewhere for its sporting appeal, although it does not occur in the north of England, Scotland or Ireland.

A big Barbel is one over about 5 kilograms (11 pounds) and it is a species that always appears attractive even when large, unlike the Common Carp and Tench, for instance, which simply look gross. The flanks are a beautiful bronze and the back is darker green-brown. It is a difficult fish to catch and I always follow the accepted practice of 'long-trotting', with the float drifting down stream a considerable distance from the rod, although many good fish are caught by ledgering.

No-one in England seems to have been very interested in eating Barbel. However, it is caught commercially in parts of Europe, although the flesh and especially the roe are said to contain toxins during the breeding season.

Gudgeon *(Gobio gobio)*

The Gudgeon is like a miniature Barbel, although it has only one pair of barbels and there is always some darker spotting over the back. It is a neat, pretty little fish and once upon a time occurred, like the Barbel, only in the southeast of England. Unfortunately, its use as live-bait by Pike-anglers means that it is now found throughout much of the British Isles. Gudgeon is eaten by people in France but not in England.

The name 'Gudgeon' comes through Middle English and French from the same Latin root as Goby (see page 163), and *Gobio* remains as the scientific name of the fish. Someone must have thought that these little fish had a voracious, or at least omnivorous, appetite because, in the sixteenth century, 'a gudgeon' came to mean 'a gullible person', 'someone who would swallow anything'. The 'gudgeon pin', a metal pivot on which a wheel turns, seems to have the same linguistic origin, although I can see no obvious connection with the fish.

Tench *(Tinca tinca)*

I can remember the excitement in angling circles in the late 1960s when the then British record rod-caught Tench was eclipsed by a monstrous specimen of grotesque proportions. It was later found to be suffering from a pathological condition rather common in these fish, in which the body becomes distended with retained fluid, and its record status was removed. Today the biggest British Tench on record is around 5.5 kilograms (just over 12 pounds); landing it must have been a considerable feat because the fish has a strength out of all proportion to its size.

Tench are classic bottom-feeding species. They are fish of deep, slow rivers and ponds, even stagnant ponds, with muddy bottoms, where they feed on a variety of animal and plant matter, seeming to favour crustaceans. Tench are found throughout much of the British Isles, except in upland areas or places where the water is very cold.

Although seldom seen, they are nonetheless handsome fish, being an unusual dark olive-green in colour, with a lovely orange eye. There is a rather striking golden form, sometimes called a 'Schlei', which is popular in ponds and aquaria. Their most characteristic feature, however, is their remarkable sliminess. The slime was once thought to have a curative effect on the wounds of other fish, which would be healed by touching it. Tench were widely known as 'doctor fish' for this reason and, at one time, it was even thought that the slime would benefit humans in the same way. The belief is entirely fallacious.

> The whole fish divided into two parts, applied to the wrists, and soles of the feet, mitigates the heat of fevers: Living tenches applied to the navel, and region of the liver, until they die, are said to cure the jaundice, for they attract the yellow colour: The skin burnt to ashes, is with success given to women afflicted with the fluor albus; the gall poured into the ears, cures hardness of hearing: There is a stone found in the head of the tench, which is good to cure the collic, stone, and epilepsy.[46]

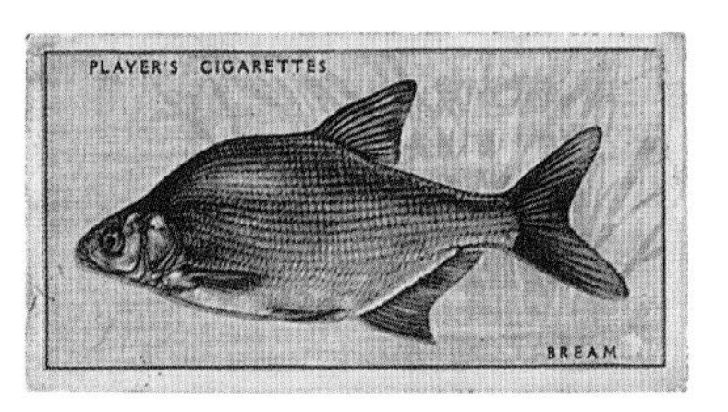

1950s' cigarette cards feature Barbel *(Barbus barbus)*, Gudgeon *(Gobio gobio)*, Tench *(Tinca tinca)* and Bream *(Abramis brama)*.

Tench were also popular fish for ponds, including the stew ponds that were so much a feature of the monasteries, because their firm white flesh makes very good eating.

'Tench' is a Middle English name, coming through Old French from Latin.

Silver Bream *(Blicca bjoerkna)*

OTHER COMMON NAMES: *bream flat (East Anglia); tinplate; white bream.*

'Bream' is a rather common fish name, Middle English in origin and perhaps from a Teutonic word meaning 'to glitter'. Today, it tends to have no plural but is used for a number of silver-sided fish, generally of rather flattened vertical profile.

The Silver Bream is the smaller of the two freshwater breams in Britain, seldom more than about 20–25 centimetres (8–10 inches) in length, lacking barbels and with a rather long dorsal fin. Like many other members of the family, it prefers warm, weedy, shallow lakes, rivers and canals. It is of little sporting significance and is of no culinary interest.

Common Bream *(Abramis brama)*

OTHER COMMON NAMES: *bellows bream; bronze bream; carp bream.*

The match fisherman would be in a poor way without the Common Bream. Weighing up to 6 or 7 kilograms (about 13 to 15 pounds), it does not take many to boost a day's catch. It is a very deep-bodied fish, almost plate-like in profile (hence, presumably, the name 'bellows bream'), occurring in shoals in large, slow-moving rivers. It is also rather handsome, its overall silvery appearance touched with bronze.

Like most members of the carp family, it was originally confined to southeast England, but has been dispersed widely. The slow-moving fenland dykes, in which so much match fishing takes place, now contain large numbers. Most are returned to the water after being weighed, although the fish is eaten in some areas and much thought of. Izaak Walton commented:

> But though some do not, yet the French esteem this fish highly, and to that end have this proverb, 'He that hath Breams in his pond is able to bid his friend welcome'. And it is noted, that the best part of a Bream is his belly and head.

Ever the keen observer, Walton continued:

> Some say, that Breams and Roaches will mix their eggs and melt together, and so there is in many places a bastard breed of Breams ... .[47]

Minnows *(Phoxinus phoxinus)* swim in shoals for safety and to increase the area of search for food.

And so there is. Many members of the carp family hybridize (and so make the affirmation of angling records a difficult matter) but none more so than the Common Bream. Common Bream–Roach hybrids are the commonest of all.

Bleak *(Alburnus alburnus)*

OTHER COMMON NAMES: *blay.*

> There is also a bleak, or fresh-water sprat, a fish that is ever in motion, and therefore called by some the river swallow; for just as you shall observe the swallow to be most evenings in summer ever in motion, making short and quick turns when he flies to catch flies in the air (by which he lives) so does the bleak at the top of the water.[48]

'Bleak' does not mean 'pallid' or 'featureless'. It is a different word, dating from the fifteenth century and possibly originally from Old Norse meaning 'white'. 'Silver' might be even more appropriate because this is a glistening little fish that is found in often large shoals in large, clean, slow-moving rivers and lakes throughout much of England. As the shoals dive and dart below the surface, characteristic flashes can be glimpsed by any observer on the bank. They are also seen by other fish and Bleak form an important food for predatory species such as Pike. Although there is a British record for the fish, it is mainly caught by anglers to use as bait for larger quarry.

In places where it occurs in very large quantities, it has been netted for eating and, even more widely, as a source of fish-meal fertilizer. In France and other places, its iridescent scales were used as a source of a 'paint', known as *essence d'orient*, that was used to coat artificial pearls. As far as I can discover, this has never been done in Britain, although, in the early nineteenth century, large numbers of Bleak were caught in the Thames and their scales exported to France for this purpose. An indication of the abundance of Bleak at that time can be gauged from the fact that 17,000–18,000 fish were required to produce 450 grams (about 16 ounces) of material.

Minnow *(Phoxinus phoxinus)*

OTHER COMMON NAMES: *baggie; baggit; banny; jack barrel; jack sharp; meaker; mennet; mennon; mennot; menon; minim; minnin; peer; penk; pink; shadbrid.*

> Fishes come bite!
> Fishes come bite!
> I have fished all day;
> I will fish all night.
> I sit in the rain on my lily-leaf boat,
> But never a minnow will bob my float.
> Fishes come bite![49]

Minnows bob my float often enough; although not as often as when I was boy. Then, there was no question that they were the commonest of the fish in our local river, with big shoals twisting and turning in all except the fastest-running water. For us, and for countless other children, it was the Minnow rather than the stickleback that was the archetypal 'tiddler'.

The Minnow is a neat, stout little fish, streamlined but blunt-nosed, with a lovely olive-green banding on the back and flanks and, in the breeding season, the most astonishing colours in the male. Then, the lower fins and underside turn vivid red, the green intensifies and the throat becomes dark, almost black; there can be few more striking creatures in British fresh water.

Minnows feed mainly on algae and insect larvae but will take an angler's bait, often hanging on it by sheer tenacity rather than by being hooked. They can also be caught in quantity if the need arises, for instance, when I required reptile food (see Grass Snake, page 194). A wine bottle, with a hole made in the concave base, a gauze placed over the neck and a few scraps of bread placed inside, would catch sufficient Minnows to keep the most hungry of Grass Snakes happy. Many larger fish and many water birds feed on Minnows and, in parts of Europe, sad to relate, they are among the small fish that have been netted in quantity for turning into fish meal.

They are said to be good eating, and were recommended by Izaak Walton, but after trying every culinary variation over the years, I cannot be persuaded. The variety of regional names is evidence of their familiarity and they occur throughout most of Britain. Most of the names are obscure but, as with Minnow itself, which is Middle English, I would like to see a link proved between them and the Latin *minimum* meaning 'small'.

Bitterling *(Rhodeus sericeus)*
The Bitterling is more likely to be seen in an aquarium than a river because it is a fascinating but non-native species, kept for its unique and extraordinary breeding behaviour, which entails the fish inserting a tube into the shell of a living freshwater mussel (see Mussels, page 37) in order to lay its eggs. The young fish are initially protected by the mussel while, in return, the Bitterling helps to disperse the larvae of the shellfish.

Bitterling are silver fish, around 7–8 centimetres (2¾–3¼ inches) in length, with a pronounced iridescent stripe on the flanks. They are native elsewhere in Europe but have been introduced, for no very obvious reason, to rivers in Cheshire and Lancashire, and also to the River Cam.

'Bitterling' is a recent name from German; the flesh is said to taste bitter.

Rudd *(Scardinus erythrophthalmus)*
OTHER COMMON NAMES: *finscale; red eye; roach (Ireland); shallow.*
The Rudd and the Roach (see below) are the twins in the coarse fisherman's vocabulary: two very similar and superficially Herring-like shoaling fish. The Rudd is the deeper bodied and is most easily and accurately distinguished by its dorsal fin, which is further back than its pectoral fins. In the Roach, the dorsal and pectoral fins are aligned, although confusion may arise because hybrids intermediate between the two fish sometimes occur.

Like most of the carp family, the Rudd was originally confined to southeast England but has become widely distributed, although it is still predominately a southern species and barely reaches Scotland.

They are not big fish – a weight of 1.5 kilograms (3¼ pounds) is exceptional – but they certainly occur in very large shoals, which makes them popular with anglers. Rudd are mainly lake fish, although they also are found in large, slow-moving rivers.

The name has nothing to do with rudders; it is seventeenth century and refers to the red coloration of the lower fins. A golden form is popular as an aquarium fish.

Roach *(Rutilus rutilus)*
I can never understand why the Roach and the Rudd are placed in different genera because they are so similar and are clearly sufficiently closely related to hybridize. Both have a similar distribution, although the Roach is more common in Scotland and is now found widely in Ireland as well. Like Rudd, it is an important food for Pike and fish-eating birds.

I would guess that coarse-fish anglers catch more Roach than any other species. I must have landed twice as many

The Roach *(Rutilus rutilus)* is very common in Britain and is a source of food for Pike *(Esox lucius)* and fish-eating birds.

Roach as all other fish combined. The current British record Roach weighed just over 1.8 kilograms (about 4 pounds). However, for very many years, the record for this commonest of fish was held by the Reverend W. Penney, who caught a fish of 2.1 kilograms (3 pounds 10 ounces) in Lambeth Reservoir in 1938. Roach are typically found and caught in reservoirs and large lakes, although they certainly also occur in many larger rivers. I caught all my biggest fish in the lower reaches of the Derbyshire Derwent.

Although both Rudd and Roach are eaten in parts of Europe, they have never been popular in Britain, being thought tasteless and bony. The standard British description of coarse fish, 'cotton wool stuffed with pins', can be applied as fairly to these fish as to any other. The roe is reckoned the only part worth eating. Izaak Walton, among many other fishing writers, has referred to the Roach as the 'water sheep', partly on account of its shoaling habit but also because of its perceived stupidity. The name 'Roach' is Middle English but of obscure origin; it seems to have nothing to do with 'cockroach'.

The Golden Orfe *(Leuciscus idus)* is a familiar sight in garden ponds.

Chub *(Leuciscus cephalus)*

OTHER COMMON NAMES: *bottling; chavender; chevin; knob; lob; loggerhead; poll; pollard; skelly.*

The Chub is the big cousin of the Rudd and the Roach, more elongated, thicker set and reaching about twice the size. It occurs throughout most of England and Wales but is largely absent from Scotland and not present in Ireland. Unlike its relatives, it is not usually found in large shoals and, unlike most British members of the carp family, it is much more of a river fish than a lake fish.

Chub are highly omnivorous feeders, tending to eat almost anything, animal or vegetable, that comes their way; they are not bottom-feeders like many carp. They are also extremely shy and challenging fish for the angler. As with the Barbel, 'long-trotting' is a trusted technique.

The name 'Chub' is late Middle English and, although it is often said to allude to the fish's chubbiness, this seems improbable. The old names 'chavender' and 'chevin' are similarly obscure but have been linked to the French word *chef*, meaning 'head'.

Orfe *(Leuciscus idus)*

OTHER COMMON NAMES: *ide.*

Today, the Orfe is best known in its golden form, which, as a pond and aquarium fish, is almost as popular as the Goldfish. It is closely related to the Chub and of similar shape: sleek and elongated when of aquarium size but later becoming more rotund.

It is not a native fish and, in its golden form, was first introduced to Britain in the 1870s, to the ponds at Woburn Abbey. Both golden fish and the normal silver fish are now fairly well established in a large number of ponds and lakes, especially in the southern half of England.

'Orfe' is a seventeenth-century word, derived originally and mysteriously from the Greek name of a species of sea perch. 'Ide' is a nineteenth-century invention from the scientific name.

Dace *(Leuciscus leuciscus)*

OTHER COMMON NAMES: *dare; dart; graining (northern England).*

The Dace, closely related to the Orfe and the Chub, is the baby of the genus; a 500-gram (1-pound) fish is considered a monster. It is a sleek, silvery, fast-swimming fish of swift, clear rivers throughout most of England, Wales and southern Scotland and is typically seen breaking the water surface as it catches floating insects.

The name 'dart' alludes to its habit; 'Dace' is from a Middle English word meaning the same thing. The name 'graining' is a northern England speciality and arose when the northern species was thought to be different.

Sadly, like the Bleak and other small fish, anglers only take an interest in Dace as bait; it was a Dace that lured Miss Ballantine's record salmon (see Atlantic Salmon, page 130). It was as live-bait that Dace were introduced to Ireland, a tin of them being washed away in a spate on the River Blackwater in County Cork in 1889. The Roach was accidentally introduced in the same way at the same time – but in a different tin.

Izaak Walton went to great lengths to give a recipe for cooking Dace but no-one else seems to have bothered. As Anchovy sauce was recommended as an accompaniment, the conclusion must be that it is pretty tasteless.

Chinese Grass Carp *(Ctenopharygodon idella)*
OTHER COMMON NAMES: *grass carp; white amur.*
This large, vegetarian carp is the latest addition to British angling interests and it has been introduced to, and caught in, a number of ponds and lakes in southern England.

Originally from the River Amur, which forms part of the northern boundary of Mongolia, it became a popular pond fish in China and other parts of the world for two quite different reasons. It provides good and substantial eating – in its native habitat, it will reach 100 kilograms (220 pounds) – but it also maintains very effective control of aquatic vegetation (in captivity, it is commonly fed on grass clippings). It looks superficially like a large, rather scaly, pale-coloured Goldfish.

Loaches
(Cobitidae)

Spined Loach *(Cobitis taenia)*
OTHER COMMON NAMES: *groundling.*
See Stone Loach below.

Stone Loach *(Noemacheilus barbatulus)*
OTHER COMMON NAMES: *beardie; colley; streamer (Derbyshire).*
Catching a loach among the Minnows, sticklebacks and Bullheads is always a special treat for anyone interested in freshwater life because these fast, elusive and largely nocturnal little fish are fascinating when closely observed. All are freshwater, bottom-living species, about 10 centimetres (4 inches) in length, with a number of tiny, fleshy barbels around the mouth and a very pretty lined or spotted back.

The Spined Loach *(Cobitis taenia)* is attractively patterned with spiny fins and a number of fleshy barbels around its mouth.

The Stone Loach is the commoner and much more widely distributed of the two British species, occurring throughout the British Isles, apart from the far north of Scotland, and generally I find in clean, rather fast-flowing streams. By contrast, the Spined Loach is a localized fish of slow-moving rivers and canals, especially in the East Midlands of England. It is absent from Wales, Scotland and Ireland.

Loach are sometimes said to make good eating; I have tried them and was not impressed. The same cannot be said for trout, however, which seem to find them especially appealing, to the extent that Stone Loach have gained a wider distribution through fisherman using them as live-bait.

The name 'loach' is Middle English from French and was at one time used in reference to anyone who was slightly dim-witted. The two-pronged spines that give the Spined Loach its name are minute and are situated just beneath the eyes.

Catfish
(Siluridae)

Danube Catfish *(Silurus glanis)*
OTHER COMMON NAMES: *European catfish; sheatfish; wels.*
I know of no other fish, certainly no other British fish, that, while never beautiful, can look so interesting when seen as an illustration in a book but can be so unbelievably and almost repulsively ugly in the flesh. A big Danube Catfish, and some are very big, is a gruesome brute. Like other catfish (and hence the name), it has long barbels ('whiskers') around its huge, wide-gaping mouth to help it feed on the muddy bottom of large, slow-moving rivers, such as the Danube. It is a solitary, sinister, almost nightmarish creature, lurking in underwater holes and emerging at night to feed on fish, shellfish, birds, and mammals as big as dogs. There are reports from Poland, Russia and Hungary of very large catfish eating people. Certainly human remains have been found in their stomachs, although the general view is that these were individuals who had drowned first and then been eaten whole.

This is not as incredible as it might seem, given that a big Danube Catfish can be enormous – one caught in the River Dnieper in the Ukraine was 5 metres (16¼ feet) in length – and can weigh over 320 kilograms (704 pounds). In Russia, large catfish are caught by trawling, although, where trawlers are unable to operate, lines baited with baby pigs are used. They are the largest European freshwater fish, although no British specimens found so far exceed about 30 kilograms (66 pounds). I stress 'so far' because the Danube Catfish is a relative newcomer to Britain.

It was introduced on a number of occasions in the nineteenth century, mainly into waters in the southern English Midlands and in part through the activities of a misguided group of individuals known as the Acclimatisation Society.

This was formed in 1860 for the purpose of liberating exotic creatures in Britain and Frank Buckland was a leading light. Over the past 30 years, however, the Danube Catfish has both naturally and deliberately been spread further and has been used to stock a number of large lakes because of its sporting appeal. In parts of Europe, it is farmed for its flesh, for its slimy skin, which is used to produce a form of leather, and for its bones and swim bladder, which yield a glue. The flesh is certainly edible but not very exciting, as I can vouch.

Eels
(Anguillidae)

European Eel *(Anguilla anguilla)*

OTHER COMMON NAMES: *astan; broad-nosed eel; bulldog; eel; elver (young); eve; fansen; fausen; frog-mouth; glut; gorb eel; grig; silver eel; snig (Lancashire); yellow eel.*

Today, the European Eel is the classic example of a catadromous fish but with a life history so complex and incredible that it has only been more or less fully unravelled over the last hundred or so years. In its appearance and its characteristics, however, the eel has been known and familiar for as long as human beings have looked into water. With its silvery colour and serpentine shape, it is unlikely to be mistaken for anything else; although lampreys (see page 118) are superficially similar, they have suckers instead of jaws and gill pores on either side of the head. Out of water, and eels do sometimes come out of water, a small eel might at first sight be mistaken for a Slow Worm, but the presence of fins would soon betray its true nature. Eels longer than about 50 centimetres (20 inches) will be female, although few are more than about 90 centimetres (35 inches) in length. In the sea, the Conger Eel (see page 153) is similar but often longer and larger overall.

The name 'eel' is Old English but appears in many other old European languages too. 'Elver' for the young is seventeenth century, a variant of 'eel-fare' from the previous century, which meant 'the passage of the young fish up a river'.

When living in fresh water, European Eels are called 'yellow eels', although they are usually only yellowish beneath, tending to be brown above. They feed on crustaceans, small fish and other aquatic creatures and also scavenge on dead matter. 'Fresh water' is something of a misnomer for the eels' habitats because, as well as rivers and lakes, they are likely to be found in some fairly unsavoury spots, such as ditches, sewers and other places in which organic food can be found and that are not seriously polluted (although, in practice, eels are probably more tolerant of pollution than almost any other native fish). Despite their small eyes, they have good vision and a very acute sense of smell, which are advantages for an animal that feeds mainly at night in dark places.

Eels live in fresh water for many years; although between 10 and 20 years is the average, some have been found that were over 50 years of age. There comes a point, however, in response to an unknown stimulus connected with their body fat content, when their colour changes and they become known, appropriately, as 'silver eels'. They then migrate downstream to the sea and all European Eels eventually find their way into and across the Atlantic Ocean, to finish in the Sargasso Sea between the Bahamas and Bermuda. They are the only European fish to migrate from one continent to another. They travel at around 12 kilometres (7½ miles) per day and apparently navigate by a combination of star position, sensing of the earth's magnetic field and smell.

It has been speculated that this bizarre journey originated at a time in evolutionary history when the European and American continents were closer together. Once the eels have reached their destination, evidently they spawn, although no-one has ever seen it happen. Having spawned, they die. The baby eels are tiny, flattened creatures, very unlike their parents, and take a year to make the return journey across the Atlantic. They undergo some metamorphosis on the way, becoming 'glass eels', and finally arrive in British coastal waters as 'elvers'. After a period of readjustment, the elvers make their way upriver into fresh water in spring, sometimes emerging from the water to negotiate obstacles, such as waterfalls, by wriggling through wet vegetation.

Eels are very important prey for many creatures, including Pike and other large fish, herons, Red-breasted Mergansers, Goosanders, Otters and American Mink. They have long been a significant human food too; they have a high fat content and are considered highly nutritious. Consequently, very large numbers of migrating silver eels never reach the Sargasso Sea, or even the Atlantic Ocean, because they are intercepted by traps set in river estuaries. In ancient times, the eel spear, a vicious barbed weapon like Neptune's trident, was a familiar implement with which the highly skilled caught their meal; more recently, it has been a slightly different weapon that was shovelled into the mud of the river bottom to pinion the eels between its flattened tines.

Eels are excellent smoked, pickled, grilled or otherwise cooked fresh. Historically, the quantities of eels eaten were clearly both important and prodigious. Many Domesday Book entries refer to places sited on rivers yielding 1,000 or 2,000 eels each year, sometimes recording quantities as 'sticks of eels', each stick being 25; the entry for Petersham, in Surrey, stated 1,000 eels and 1,000 lampreys.

A mid-seventeenth century recipe for eel pie was given in *Mistress King's Receipt Book*:

> To bake Eels in a pie. After you have drawn your eels, chop them into pieces three or four inches long, and put them in a pan: season them with Pepper,

Tubby Isaacs's stall in Aldgate is probably the most famous of all the jellied-eel stalls in London's East End.

> Salt, Ginger, great Raisins and Onions small chopped, and cover them with Stock. Then boil them gently until the flesh will easily come from the bones. Put the flesh into a pie dish with a small piece of lemon, a good lump of butter, and enough of your stock to cover them: then put on your cover of pastry and bake your pie very hot, about three quarters of an hour.[50]

Even better are spatch-cocked eels, an Oxfordshire speciality, in which the eels are boned, cut into short pieces, dipped in egg yolks and breadcrumbs and then fried until crisp in butter with plenty of herbs. The most celebrated and excellent of all ways of eating eels, however, are those forever associated with London: eel pie and jellied eels.

> Cry to it, nuncle, as the cockney did to the eels when she put 'em i' the paste alive ...
> (*King Lear*, Act 2, Scene 4)

Jellied eels are a speciality of London's East End, where street stalls, the most famous being Tubby Isaacs's stall in Aldgate, sell them in china bowls sprinkled with hot chilli vinegar. Eel pie and mash was, and to some degree still is, served in traditional London pie houses, although today you are more likely to have meat in the pie with the eels as a side order.

The elvers reaching the rivers are just as vulnerable as the adults going in the opposite direction and they are also caught in large numbers, at one time for the production of elver cakes. Izaak Walton knew of it:

> I have seen in the beginning of July, in a river not far from Canterbury, some parts of it covered over with young Eels about the thickness of a straw; and these Eels did lie on top of that water, as thick as motes are said to be in the sun: and I have heard the like of other rivers, as namely, in the Severn, where they are called yelvers; and in a pond or mere near unto Staffordshire, where about a set time in summer, such small Eels abound so much, that many of the poorer sort of people that inhabit near to it, take such Eels out of this mere with sieves or sheets, and make a kind of Eel-cake out of them, and eat it like as bread.[51]

The Island Hotel on Eel-pie Island in the Thames near Twickenham in the 1880s.

Elver-eating competitions were held until as recently as 1994 at Frampton-on-Severn, with the prize going to the person able to eat the largest quantity in the shortest time.

> In the Fens north of Ely eel traps and lines were prepared ready for St Valentine's day, for it was believed that if the sun shone on February 14th the eels would be on the run before the week ended.
>
> The importance of eels in the early economy of the Fens was very great. Rents and debts were paid and land was sold in 'sticks' of ... eels and the Domesday recorders carefully noted the value of the eels in Fenland parishes ... For the privilege of quarrying stone at Barnack, near Peterborough, for the building of their monastery, the Ely monks paid 4,000 eels a year. This gave rise to the saying that the Cathedral was built with eels.[52]

Ely is certainly one of the few places apparently named after eels, originally being the Old English *el-ge*, 'eel district', although this was later changed to *eleg*, 'eel island'. According to the Venerable Bede, it was named for the great number of eels caught locally. 'Eel-pie Island' is an alternative name for Twickenham Eyot in the Thames, where parties of picnickers used to gather in order to dine on the aforesaid delicacy. Rather surprisingly, eels frequently appear as heraldic symbols and, less surprisingly, as inn names in places where they were once an important food fish.

Eels also had a wide range of medicinal uses. A treatment for warts involved rubbing them with eel's blood. A garter made of eel skin was very widely thought to be beneficial in curing cramp, or indeed preventing it, and swimmers would wear an eel-skin garter around the leg. W.H. Barrett gave the details of how the garters were prepared in Cambridgeshire.

> Only eels caught in the spring provided suitable skins. After the heads and tails had been cut off the skins were removed in one piece and hung up to dry in the sun until they were quite stiff. Then the tow end of each were tied and the skins well greased with fat and rubbed with a round piece of wood until they were pliable again, when the ends were untied and the skins re-stiffened by the insertion of a 'stuffing' of finely chopped thyme and lavender leaves. The skins were next inserted in linen bags which were buried in the peat for the rest of the summer between layers of freshly gathered marsh mint, as the water mint (*Mentha*

> *aquatica*) was locally known. This gave a mottled appearance to the garters. In the autumn the skins were dug up and a final polish was given to them, after the removal of the thyme, lavender and mint, with a smooth stone. The garters, called yorks, were tied just above the knees, men knotting them on the right, women on the left side. Old women declared that in addition to their use as a cure and preventive of rheumatism the yorks stopped mice from running above the garters when the wearers of them were working in the harvest fields.[53]

Before the eel's life history was elucidated, spontaneous generation was frequently offered as an explanation. That aside, the most widely held belief concerning their procreation was that they sprang from horse-hairs placed in water; as recently as the mid-nineteenth century, there were people who not only believed this, but believed they had proved it. Izaak Walton, no mean observer, was as baffled as everyone else:

> Most men differ about their breeding: some say they breed by generation as other fish do, and others that they breed, as some worms do, of mud.[54]

Ely Cathedral lies in the heart of the Fen country and the town is so called because of the large number of eels caught locally.

Eel-anglers are a breed apart, happy to spend their leisure hours in some of the less than endearing places that eels inhabit. However, most anglers detest eels because of their slimy nature ('slippery as an eel' is a by-word for elusiveness) and the enormous difficulty of removing them from a hook as they twist and contort themselves. The simplest and easiest solution is to drop the eel onto sheets of newspaper, which absorb the slime and tend to subdue the eel, rendering it easier to handle. On Orkney, when an eel was caught in a burn, it was traditional to hold it to see whether it would twist itself into a knot. It was then addressed with a variation on the well-known counting rhyme:

> Eeny, meeny, minny mo,
> Cast a knot apin yer tail
> An' then we'll let you go.[55]

Congers
(Congridae)

Conger Eel *(Conger conger)*

OTHER COMMON NAMES: *bumble-bunner; connor, cullach; eve; evil-eel; evil-eye; heevil; hornel; horner; hunter; kinger; kornel; lance; launce-eel; sleekie (Shetland).*

The Conger Eel is a ferocious, powerful, grey-brown fish of rocky shores and shipwrecks, the goal of sea anglers and a fine fish for the table. It has been known and recognized for centuries. The name is fourteenth century, from French and originally Greek, meaning 'sea eel'. Like the European Eel, the Conger Eel migrates to a spawning ground but it is wholly marine and moves from British shores southwards to an area of the Atlantic between the Azores and Gibraltar.

Conger Eels reach 2 metres (6½ feet) in length and are equipped with a large mouth and extremely good teeth. Consequently, a steel trace and large hooks are necessary parts of the angler's equipment. Divers may see them in their characteristic day-time posture, with only the head protruding from some hole or fissure, from which they emerge to feed at night on crustaceans, other fish and cephalopods.

Although in many places, congers have always been recognized as excellent eating, the Scottish tradition was different.

> An eel in the catch is much disliked, on the ground that they believe Old Nick himself to be in it. Indeed, they seemed to dislike an eel anywhere, until they found they could get money from the Sassenach for it. A Scotchwoman, unless her ideas have become Anglicised, will not 'dirty' her frying-pan with an eel, much less eat one. Some fine conger-eels are caught at Reay, but they have all to be sent to the English markets: they are not eaten locally. The fishermen here think it is because they are so like serpents.[56]

The Conger Eel *(Conger conger)*, with its formidable jaws, lives in crevices or holes during the day, emerging at night to feed.

Curiously, therefore, although 500 were on the menu when Margaret, the daughter of King Henry III, married King Alexander III of Scotland in 1251, it is possible that the bride and her party did not partake.[57] By contrast, in the Channel Islands, vast numbers were caught and, in the Middle Ages, they formed a major export industry. Until very recently 'fishermen used to sell congers from door to door in barrows, a tradition now lost with the diminishing popularity of recipes like conger and marigold soup'.[58]

Sticklebacks
(Gasterosteidae)

Three-spined Stickleback *(Gasterosteus aculeatus)*

OTHER COMMON NAMES: *baggie minnow; bandie; barstickle; bulgranade; doctor; jack sharp; pinkeen; prickleback; redthroat; sharplin (Scotland); sprickleback; stickle; stickling; struttle (Northamptonshire); tiddlebat.*

For generations of schoolchildren, the Three-spined Stickleback is the tiddler, the fish that they first catch and one that is pivotal to their introduction to freshwater natural history. It is an aggressive little fish that feeds on tiny crustaceans and other small aquatic life and, I always think, darts rather than swims. It is also arguably the most widely distributed of all the freshwater species in Britain, and surely no British fish has a name that is more apt or descriptive, as it has been since the Middle Ages.

The spines that give all sticklebacks their name are modified rays of the dorsal fin and, although this species usually has three, it varies in the size and number of spines and, indeed, in overall size and scaliness. There are no true scales but a variable number of bony plates, or scutes. It is largely because of this variability that the Three-spined Stickleback has become a popular laboratory animal for genetic studies.

In estuaries, the fish tend to be considerably larger; sometimes up to 10 centimetres (4 inches) in length, compared with the norm, which is around 6 centimetres (2¼ inches). There are also wholly marine Three-spined Sticklebacks, but they are not common.

Overall, they are brown-olive in colour with some silver but, in the breeding season, the males become quite staggeringly coloured: brilliant red on the throat, silver-yellow above, and with a vivid blue eye. They would then not appear out of place in a tropical river or a coral reef.

Male sticklebacks construct a small nest on the river bed from sand grains and plant fragments, held together by a glue that the fish itself secretes. The female is then cajoled into entering the nest, where she lays her eggs, which the male fans with his fins to supply oxygen. At first, the male fish attends to the newly hatched fry but, like fathers the world over, his interest in babies soon wanes.

In parts of the world where they occur in large numbers, Three-spined Sticklebacks have been collected for making into fertilizer or fish meal, or for extracting their oil, although the quantities of fish required for this are prodigious. In Britain, this has not really been successful, although, in the eighteenth century, the fish were said to be so numerous at Spalding, in Lincolnshire, that a man employed by a farmer there earned 4s. (20p) a day for a considerable time by selling them at a halfpenny a bushel as manure.

Nine-spined Stickleback *(Pungitius pungitius)*

OTHER COMMON NAME: *ten-spined stickleback.*

This is the smallest British freshwater fish, barely 5 centimetres (2 inches) long. It has nine or sometimes ten spines, shorter than those of the Three-spined Stickleback. It is not as widespread and tends to be found most often in very shallow, very weedy waters. Although it occurs in estuaries, it is not found in the sea but, overall, its behaviour is very similar and the male builds a comparable nest. Also like the Three-spined Stickleback, this fish been collected for oil production and forms an important food for larger fish and some birds.

Fifteen-spined Stickleback *(Spinachia spinachia)*

The largest British stickleback, this is mainly marine, although it does enter estuaries. It is a very slender, elongated fish, up to 20 centimetres (8 inches) in length and with 14 to 16 (but usually 15) small spines.

Cod and related fish
(Gadidae)

After the herrings (Family Clupeidae), this is the most important family of food fish, especially in the northern hemisphere. They are not pretty fish and have few unifying features, although many have barbels around the mouth and several have three dorsal fins. They include some of the most familiar fish on the fishmonger's slab and in the fish-and-chip shop but most species have declined dramatically in recent years through overfishing.

Much the most important species are the Cod, Haddock, Hake, Pollack, Saithe, Ling and Whiting (see below). Among other, and generally smaller, species that are sometimes seen for sale, or are caught by sea anglers, are the Blue Whiting (poutassou; Couch's whiting) (*Gadus poutassou*), Bib (pout whiting) (*Trisopterus luscus*), Poor Cod (*Trisopterus minutus*), Norway Pout (*Trisopterus esmarkii*), the various bearded Rocklings (*Gaidropsarus*, *Rhinonemus* and *Ciliata* species) – with their young called 'mackerel midges', the Forkbeards (*Phycis* species) and the Tusk (*Brosme brosme*). Many of these names reflect the nature and number of the barbels.

All are marine, but there is one, elusive, extraordinary freshwater species that sadly seems now to be extinct in Britain: the Burbot (see page 160). They all tend to be dark grey-green, brown or grey-blue on the back and silvery on the flanks and beneath, and they usually feed on other fish, molluscs or crustaceans.

Whiting *(Merlangius merlangus)*

OTHER COMMON NAMES: *baivee; darg; fitan; fithin; fitin; kellat; stuffin (fry).*

The Whiting is one of the few fish of this family that has no barbel and it is relatively small, seldom more than about 60 centimetres (24 inches) in length. It is possibly more of a shallow-water fish than the Cod, although the old belief that it truly lives in the shallows is belied by the quantities caught by deep-water trawling.

It is a valuable although under-rated food fish that has suffered through being thought suitable fare only for the sick and elderly. Small split Whiting, strung up through the eye-balls and repeatedly rewetted with sea water as they air-dry, become 'speldings'. These are an eighteenth-century speciality of the north of Scotland, especially popular then, as they tend to be now, around the Dornoch and Moray Firths. They are traditionally grilled and served with oatcakes. The name 'Whiting', alluding to the colour, is Middle English.

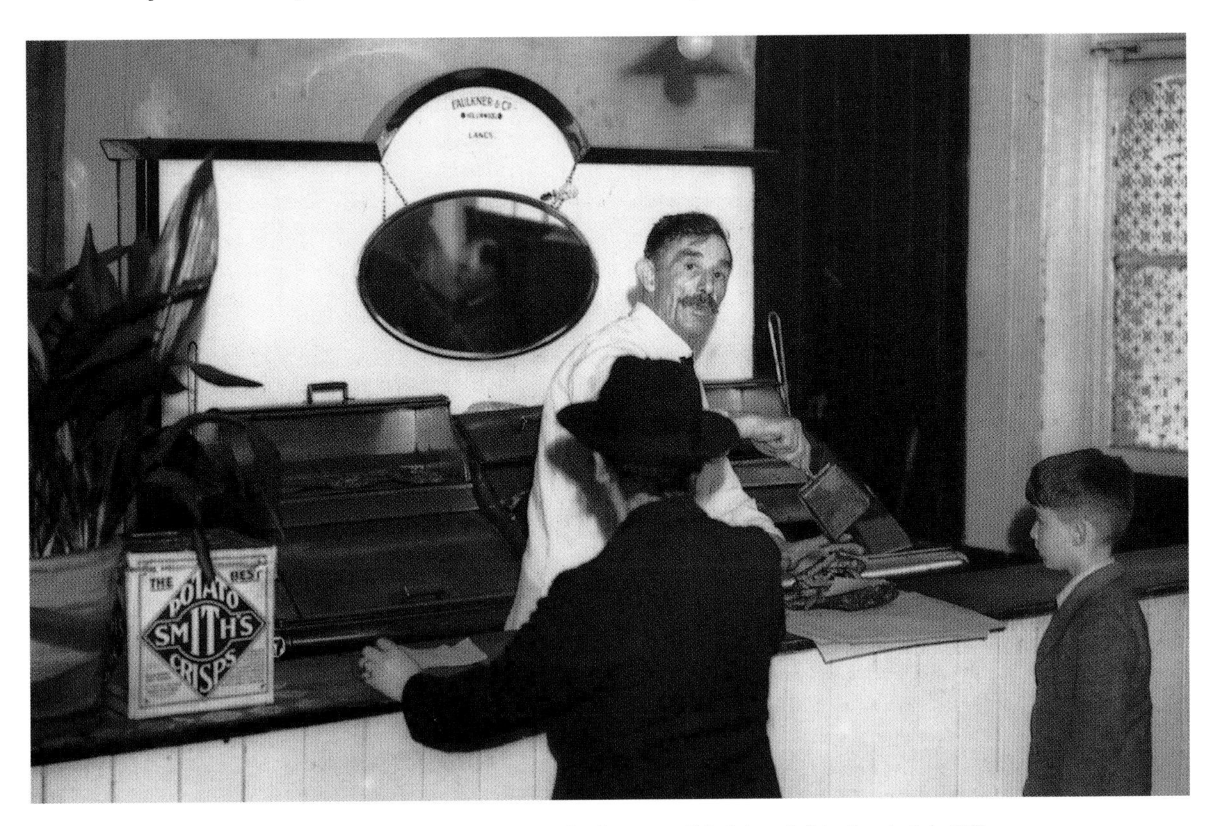

The Mayor Elect of Salisbury, Alfred Courtney, seen through the doorway of his fish-and-chip shop in July 1939.

Pollack *(Pollachius pollachus)*

OTHER COMMON NAMES: *lythe; podley; podlock; pollock.*

The Pollack is superficially similar to the Whiting and also lacks barbels, but it grows much larger, up to about 1.2 metres (4 feet) in length.

It is not a significant commercial food fish, being rather dry, although it can be substituted for Cod in recipes involving sauces and juices. It is very popular with anglers.

The name is seventeenth century but its origin is unknown.

Saithe *(Pollachius virens)*

OTHER COMMON NAMES: *coalfish; coley; cuddy.*

Saithe resemble Pollack but are much more important as food fish and really are excellent, and very much cheaper than Cod or Haddock. When sold for eating, they tend to be called 'coley' which, like 'coalfish', alludes to the dusky, smoky colour of the skin. 'Saithe' is an Old Norse word for the fry of the Cod and, in many places, that is exactly what they were believed to be.

In Scotland and Ireland, young Saithe are especially popular and widely recommended for sufferers from indigestion. In the days when Hebridean spinners held wooden bobbins between their teeth, they would commonly swallow splinters and the resulting discomfort was said to be cured by eating boiled cuddies, complete with the bones, which sounds a very risky business to me.

Cod *(Gadus morhua)*

OTHER COMMON NAMES: *barfus/barvas; codling (young); tomlin (young).*

Once tasted, never forgotten, although I refer not to the fish itself, but to the Cod liver oil with which Britain's post-war generation grew up. Together with concentrated orange juice, it was provided by courtesy of the Ministry of Food. It came in a bottle shaped like those that were once used for poison and adorned with a pale blue label. It was evil-tasting stuff, but is still the world's largest-selling food supplement and Britain's biggest producer currently uses about 800 tonnes (813 tons) of oil every year for its fatty acid content, which is evidently valuable for digestion and heart disease. The current shortage of Cod has led people to investigate alternatives to Cod liver oil and krill seems an option.

The Cod is the largest member of its family; fish approaching 2 metres (6½ feet) in length and weighing over 90 kilograms (198 pounds) have been reported. It is characterized by a single long barbel and eats more or less whatever is available in the waters where it occurs. Other fish, cephalopods, molluscs and crustaceans are all taken, while the Cod itself is eaten by relatively few predators. It is a fecund fish, the females commonly producing in excess of 10 million eggs, and it is adapted for survival in extremely cold waters. Cod are found typically where warm and cold sea currents meet; the point where the Gulf Stream meets cold Arctic water off the British Isles is ideal, as are the Newfoundland Grand Banks, where the same Gulf Stream brushes the cold Labrador Current.

The importance of the Cod as a food fish cannot be denied or doubted, and because of this, and because it occurs not within sovereign states but in more or less international areas of the sea, Cod-fishing has always had major political overtones. In *Cod* by Mark Kurlansky it is described as 'the fish that changed the world'.[59] Certainly, although Britain has had serious differences in centuries past with the Dutch over the North Sea Herring catch, the Cod is the only fish over which this country has taken up arms in recent history. The so-called 'Cod Wars' with Iceland stemmed from the pursuit of national interests, economic pressures and, ultimately, to the decline of the Cod populations due to overfishing.

However, the Cod Wars and the importance of Cod-fishing are nothing new. Wooden pew ends from the chapel of St Nicholas at King's Lynn in Norfolk, which were carved as long ago as the early fifteenth century, show Cod-fishing boats. Throughout the Middle Ages, Cod was a valuable resource and

THE TIMES

The Derby: Wollow's record daunting for his rivals, page 6

Six-month agreement signed in Oslo ends the cod war

200 Syrian tanks thrust into Lebanon

Daughter of Kipling kept MSS secret

Britain calls it common sense, but Iceland says 'We won'

Mr Alastair Burnet gets letter bomb

Soaring fish prices forecast with thousands out of work

More rain likely

Heads' boycott by-passed by Tameside council

Militant builders plan fresh blow to pay restraint despite Murray appeal

IRA men face death penalty for Eire murders

Doubts on Rolls-Royce aero deal with Russia

Pensions court action

Cats shot in rabies scare

Torture report

Dewar's

Smooth and soda.

***The Times* of 2 June 1976 reports the signing of the agreement that signalled the end of the last Cod War with Iceland.**

drying or salting was the means of preserving it for transportation. Salted fish was valuable because, in summer, the North Sea and Icelandic fish could not be dried quickly enough before they went bad. Britain was relatively poorly placed to produce salt fish because the supply of home-produced salt was limited; therefore, while some Cod, called 'green fish', were salted in an undried state, to make the most of the limited salt supplies, some were dried and lightly salted. In addition, some were pickled in brine.

The port of Bristol was well placed to capitalize on the trade in dried Cod, being positioned between the source of the fish in Iceland and the markets in the Mediterranean. In 1475, however, Bristol fell foul of the Hanseatic League merchants, once all-powerful in the Herring trade (see Herring, page 125), who cut off the supply of Icelandic fish as part of their intention to dominate the Cod market. In 1532, an Englishman called John the Broad was killed at Grindavík in Iceland as the dispute with the Hanseatic merchants intensified. On this occasion, the British did not go to war but instead turned their fishing attention westwards to Newfoundland and the Grand Banks.

Throughout the following centuries, Cod loomed large in the economies of all the great northern powers and played its part in many of the conflicts. In the nineteenth century, with the discovery of new fishing grounds south of the Dogger Bank in the North Sea, the east coast fishing towns of Hull and Grimsby became major ports. Old sailing boats with their long lines gave way at first to small boats, called 'smacks', which dragged a net over the beam and along the sea bed, until, in the last two decades of the nineteenth century, steam-powered trawlers took over.

Dried and salt Cod had always taken second place to fresh fish as far as the British palate was concerned and, after various experiments with transporting Cod fresh in cold water, the apparent salvation of the industry came with the twentieth-century inventions of the American Clarence Birdseye and the ability to freeze fish at sea. Automated filleting and the fish finger followed. Radar-guided trawlers worked around the clock to satisfy factory ships and the endless market demand. Catches were huge and the supply apparently limitless. Something had to give, and it was the Cod.

Catches in Icelandic waters dropped dramatically in the 1950s and, in 1958, to conserve its Cod stocks and its own position, Iceland extended its territorial waters from 4 to 12 miles (6.4 to 19.2 kilometres). Britain sent in the navy. She was supported by much of western Europe but eventually the Icelanders more or less prevailed and the extended limit was recognized in 1961. A decade later, Iceland extended its limit to 50 miles (80 kilometres), although, this time, Britain, supported by Germany, held back on the use of force. However, the Icelandic Coastguard, having learned that they could not outgun the Royal Navy if the matter did come to blows, took the simple but effective expedient of cutting the trawl wires.

Four fishwives gather for a chat in the fishing port of Hull in this cover image from the *Picture Post* (3 February 1951).

The conflict was short but unpleasant, and again Britain backed down. Three years later, the situation was resumed when Iceland again extended its fishing limit – to 200 miles (320 kilometres) The Germans opted to catch other types of fish instead – the Icelanders persuaded them to eat Redfish (ocean perch; Norway haddock) (*Sebastes viviparus*), which the British tend to buy only when it is disguised in fish fingers and fish cakes – and the whole affair was overtaken by the decision of the European Economic Community itself to impose a European 200-mile (320-kilometre) zone.

In recent years, Cod have seldom been out of the headlines as government fishery ministers of the European Union have locked horns over fishing quotas and over who can catch what and where, with the overall objective of conserving stocks. In 2000, Britain reluctantly accepted cuts of 50 per cent in Cod catches but, by the winter of 2000–2001, the Cod stocks of the North Sea were in a truly parlous state.

> Fishermen demanded up to £100 million in compensation from the Government yesterday to save them from ruin as Brussels imposed an emergency ban on catches of cod from 40,000 square miles of the North Sea ... It is forecast that about 120 vessels will be driven out of business – a fifth of the main Scottish

> white-fish fleet ... the 12-week ban ... will apply to cod spawning areas off Scotland, Norway, Denmark and Holland. The EU Commission said 'We are facing an extremely serious situation. The cod stock is in danger of irreversible collapse' ... .[60]
>
> Matthew Davis, Director of the World Wide Fund for Nature's Oceans Recovery Plan, applauded the ban.[61] The debate continued however and following a further EU agreement in January 2005, Fisheries Minister Ben Bradshaw apologised to Cornish fishermen for a three-month winter ban on boats under 10 metres, brought in to help stocks of cod recover. Owners of small Cornish boats said their livelihoods were under threat.

As Mark Kurlansky concluded his book:

> It is harder to kill off fish than mammals. But after 1000 years of hunting the Atlantic cod, we know that it can be done.[62]

The name 'Cod' is Middle English but is of unknown origin, although in many places and many languages, it has had an extraordinary sexual connotation. In addition to being the name of a fish, 'cod' is also Middle English for 'sack, bag or scrotum', hence the 'codpiece' purse worn by the gentry of the sixteenth century to contain their valuables and to give the impression of generous sexual endowment. The collecting end of a trawl net is still called the 'cod end'. 'Cod' is slang for 'a prank'; perhaps a reference to the illusion produced by the codpiece purse. In the Caribbean, 'salt cod', or 'salt fish', are slang terms for various sexual functions or activities.

It is presumptuous to say that Cod is wonderful food but there are probably more ways of cooking Cod than any other fish. Despite the centuries-old tradition of preserving it, and the fact that, in some European languages, there is no word for the fresh fish, only the salt, there can be no denying that Cod taken straight from the sea to the kitchen is like nothing else, and totally unlike frozen Cod.

In French the Cod is *cabillaud*, which finds its way to Shetland as *cabbilow* and into Scots as *cabbie-claw*, which is also the name of a dish that requires young, fresh fish, salted, lightly dried and simmered with horseradish and parsley.

North Sea Cod *(Gadus morhua)* populations are in decline and catches like these, at Grimsby in 1907, are unlikely to be seen again.

The Haddock *(Melanogrammus aeglefinus)* is easily recognized by the black 'thumbprint' on its flanks.

Haddock *(Melanogrammus aeglefinus)*

The name 'Haddock' is Middle English but its meaning unknown. The Haddock is usually smaller than the Cod, taking second place to it as a commercial fish, and is readily distinguished by the black 'thumbprint' mark on the flanks. This mark has endowed it with a religious significance like that attached to the similarly marked John Dory (see page 160) and the Tilapia (mainly *Oreochromis* species; Family Cichlidae). The marks are sometimes said to be the thumb- and fingerprints left by St Peter as he took the Tribute out of its mouth. Alternatively, it is Christ's fingerprint, left when He was fishing in St Peter's boat. In Ireland, the bone of the fish that lay beneath this mark was thought to have special curative and other powers. It had to be carried in the pocket and 'always works good to the owner; but it must not be exhibited, and it should never be lent, or touched except by the owner'.[63]

In the north of England and Scotland, Haddock is still preferred to Cod for eating and is often offered as an alternative in fish-and-chip shops. There are many Scottish recipes for 'the haddie' and the finest and best-known smoked Haddock is the Finnan Haddie. (This is also called 'findhorn', 'findon' or 'finnon', names that perhaps originated among the fishermen's wives of Findon, Kincardineshire, south of Aberdeen.) To produce it, Haddock is dried on the sea shore and then smoked over fires of seaweed, green wood, turf and peat to produce the characteristic and wonderful flavour. According to one version of its origin, it was discovered by accident when a fire broke out in one of the fish-curing houses and the proprietor (the maister) sampled the remains from the ashes. The most perfect article, with subtle flavour, is pale-coloured, but artificial dye is now commonly used to give the rich golden colour that many people seem to want. Although the smoked fish is usually poached in milk, there is also a delicious old Moray Firth dish called Cullen Skink, in which it is stewed with onion and mashed potato.

Just as splendid as Finnan Haddock, but much less known outside Scotland, is the Arbroath Smokie, also from the east coast. Smokies are small Haddock that have been cleaned but not split open, tied by their tails in pairs and smoked over a fire of birch or oak chips. They only need heating through because they are already cooked and so soon dry out; steam them and serve them with butter. Another uniquely Scottish dish is Crappit Heids; an intriguing economy when times are hard, it ignores the bulk of the fish and makes use of the 'off-cuts'. Haddock livers are chopped and mixed with oatmeal, salt, pepper and milk. They are then stuffed into Haddock heads and the whole boiled for half an hour.

Hake *(Merluccius merluccius)*

The Hake is another valuable commercial species, but much less numerous than the Cod and Haddock and with a more delicate flavour. It is rather more elongated than either of them, lacks a barbel and has very obvious protruding teeth.

A considerable amount of Hake finds its way into fish-and-chip shops, although much is also exported to Mediterranean countries where it is popular.

Like 'Haddock', 'Hake' is another Middle English word of unknown origin.

Ling *(Molva molva)*

An elongated fish, up to 1 metre (3¼ feet) in length, the Ling has one barbel on the chin and two small pseudo-barbels close to the nostrils.

Sadly much of the commercial catch of Ling is dried and salted before being smoked and sold on as salt or smoked cod.

The name 'Ling' is Middle English, perhaps meaning 'long'. It seems to have a different origin from the plant name often applied to the Heather (*Calluna vulgaris*).

The elongated body of this Ling *(Molva molva)* is barely distinguishable from the rocks behind it.

Burbot *(Lota lota)*

OTHER COMMON NAMES: *barbolt; coney fish; eel pout; freshwater cod; ling; rabbit fish.*

This is the great lost fish of English rivers and the only freshwater member of the Cod family. It is superficially like the Ling, an elongated creature with one barbel on the chin and two pseudo-barbels around the nostrils. The name is fifteenth century, from French and elsewhere in Europe.

Burbot can reach a considerable size but, in the days when they were caught in the deep, slow-moving rivers of the east of England, they were generally about 50 centimetres (20 inches) in length. The Trent was a good Burbot river, as were the Ouse and the Cam. At one time Burbot were so numerous that there are reports of them being netted and sold regularly in local markets; they were even said to have been used as pig food. The last one seen in England was in the River Ouse in 1973.

Pipefishes
(Syngnathidae)

This small family of remarkable fish would certainly have made a greater impact on British folklore had they been larger and/or more common. They have an external skeleton of bony plates that gives them a very curious, jointed appearance.

The pipefishes (*Nerophis* and *Syngnathus* species) are extremely elongated, with a slender snout, and are often mistaken for pieces of the seaweed or sea grass among which they live, swimming feebly and appearing simply to sway in the water. Several species occur in coastal and brackish waters around Britain. The largest, the Great Pipefish (*Syngnathus acus*), occasionally reaches 45 or 50 centimetres (17½–20 inches) in length.

Even odder, but rarer in British waters, is the Sea Horse (*Hippocampus ramulosus*), which is barely 15 centimetres (6 inches) long. As it swims vertically, its prehensile tail grasping water plants for support, it seems the obvious inspiration for the knight on the chess board. Apart from its oddly endearing appearance, the Sea Horse's greatest claim to fame is that, like the pipefish, the male incubates the eggs, which it carries in a brood pouch on the underside of its body.

John Dory
(Zeidae)

John Dory *(Zeus faber)*

OTHER COMMON NAMES: *St Peter; St Peter's fish.*

The John Dory is not a pretty fish and, like the Angler Fish (Monkfish) (see page 171), was for long neglected by the fish-buying public on account of its ugliness. Now, however, it is appreciated for the excellent eating that it offers, although I doubt whether many people who have bought and eaten it have ever seen the fish in its entirety.

With its melancholy drooping mouth and yellow colouring, the John Dory *(Zeus faber)* is a very unattractive specimen.

It is a solitary species that feeds on other, smaller fish in fairly shallow waters. It reaches about 40 centimetres (15 inches) in length and is flattened in the vertical plane, with a huge, long-spined dorsal fin, a downward-sloping, miserable-looking mouth and a large black spot on its side.

It is this spot that has given it the name 'St Peter's Fish', being the supposed thumbprint that St Peter left when he lifted it from the Sea of Galilee. A similar belief attaches to the Haddock (see page 159) and the Tilapia. The fact that neither John Dory nor Haddock occurs anywhere near the Sea of Galilee is of course irrelevant and, in truth, the warm-water Tilapia has a much greater claim.

Who was John Dory? Some extraordinarily ugly individual who gave his name to a look-alike? Unfortunately, no such person existed. The original name of the fish was the Middle English word 'dory', from the French verb *dorer*, meaning 'to gild', and a reference to the fish's glistening yellowish hue. 'John' was added as a familiar term in the seventeenth century, just as other personal names, such as 'Jack', were sometimes tacked onto animal names. However, why a fish with such Christian symbolism should have a scientific name that commemorates the Greek supreme pagan deity, Zeus, eludes me.

Sand smelts
(Atherinidae)

Sand Smelt *(Atherina presbyter)*
The Sand Smelt is one of a small family of superficially Sardine-like fish that form large shoals in shallow water. They are closely related to the grey mullets and, although not themselves a commercially important food fish, they are highly significant as one of the principal foods of the Herring.

Grey mullets
(Mugilidae)

'Mullet' is an old fish name, originally Middle English and applied first to the red mullets (see page 163) and then later to the grey mullets, which are unrelated. It is sometimes used for both perch and bass and seems to have been applied to a large number of species on the grounds of having a spiny dorsal fin. Grey mullets will often be seen swimming in shoals in brackish water, in harbours, close to the harbour wall or near to sewage outfalls, where they feed on algae.

Despite their habitat, they are often seen for sale in fishmongers, where the difference from the much tastier red mullets is not always stated.

There are several very similar British species in the genera *Chelon* and *Liza*, the commonest being the Thick-lipped Grey Mullet (*Chelon labrosus*), which is up to 60 centimetres (24 inches) in length and is the one seen most frequently on the fishmonger's slab. Opinions on its culinary merits are sharply divided; some cooks consider it so inferior to red mullet that they will not touch it, finding it oily in taste and muddy when bought. I incline to the view that it 'makes up in quantity what it lacks in quality'.[64]

The related species, the Thin-lipped Grey Mullet (*Liza ramada*) and the Golden Grey Mullet (*Liza aurata*), are sometimes, with great difficulty, caught by anglers. Although large numbers may often be seen, they are notoriously tricky to take.

Bass
(Serranidae)

The bass are a family of carnivorous, mainly crustacean- and fish-eating species related to the warm-water groupers. They are all superficially similar, with a spined front dorsal fin; bass tend to have two dorsal fins, while the groupers and their relatives, the combers, have only one.

Like 'perch' and 'mullet', 'bass' is another name applied fairly indiscriminately to fish with dorsal spines. The word is Middle English, a corruption of the older words *barse* or *base*, with links to other European languages and alluding to bristles. 'Barse' is still a local name for the freshwater Perch (see page 162).

Sea Bass *(Dicentrarchus labrax)*
OTHER COMMON NAMES: *king of the mullet; salmon bass; school bass; sea perch; white mullet; yawn.*
The Sea Bass is the main species in British waters and reaches about 1 metre (3¼ feet) in length. Silvery overall, it often occurs fairly close to the shore, where it provides very good sport for anglers. Although mainly marine, Sea Bass enter estuaries and sometimes extend so far upstream as to live entirely in fresh water.

Over the past 20 years, the Sea Bass has become one of the new culinary species in Britain and is now almost as familiar as the Herring and the Cod. It is still fairly expensive, although, since the development of farming techniques employing large, submerged cages offshore, the price of at least some of it has fallen. Within Europe, Greece, Spain and other Mediterranean countries, farming has been so successful that the market for farmed Sea Bass and Sea Bream (see below) is now said to be almost saturated.

Sea Bream
(Sparidae)

Sea Bream are mainly warm-water relatives of the bass and have become a familiar culinary species because they are now very successfully farmed. They are generally oval, vertically flattened, silvery fish with large scales and a single dorsal fin with spines towards the front. They have pronounced and prominent teeth in the lower jaw for scraping small creatures from rocks.

The two species seen most commonly in British waters and in British shops are the Black Sea Bream (old wife; stone bream) (*Spondyliosoma cantharus*) and the Red Sea Bream (*Pagellus bogaraveo*). Both need scaling before cooking. The Red Sea Bream is the better of the two and is most readily distinguished from the Black Sea Bream by the black spot behind the gills.

The Sea Bass *(Dicentrarchus labrax)* is found close inshore where it provides good sport and good eating for beach fishermen.

The Red Sea Bream is usually called *la dorade* in France, which leads to confusion with the Dolphin Fish (see True Dolphins, page 438), which is sometimes called the 'dorado' in Britain. Strictly speaking, *la dorade* should be applied only to the non-British Gilthead Bream (*Sparus auratus*), because *dorée* in French means 'gilded' (see also John Dory, page 160). *Dorado* means 'gilded' in Spanish; hence 'El Dorado'.

The less than flattering name 'old wife' alludes to a supposed facial similarity to a disagreeable woman, although not all old wives look disagreeable. The name 'bream' is discussed elsewhere (see Silver Bream, page 145).

Sunfish
(Centrarchidae)

Older books on British fish would have no space for this family. Sometimes called Sunfish (although quite unrelated to the enormous warm-water marine species *Mola mola*, called the Sunfish in Europe), they originate in the lakes and rivers of North America but at least two species have been introduced to Britain, where they may to grow in importance. They are thought excellent for both sport and eating and are aggressively carnivorous.

Largemouth Bass *(Micropterus salmoides)*
OTHER COMMON NAMES: *black bass; green bass.*
This a typical bass/perch-like fish has a spiny dorsal fin and reaches about 40 centimetres (15 inches) in Britain. It has become established in at least two lakes in southern England.

Pumpkinseed *(Lepomis gibbosus)*
OTHER COMMON NAMES: *common sunfish; sun bass; yellow sunfish.*
This fish is smaller, much like a native Perch, although with only indefinite vertical bands and without the separate rear dorsal fin. It is popular with aquarists in Britain and seems to have become established at a few localities in the south of England.

Perch
(Percidae)

The perch family is a small group of distinctive freshwater species distributed across the northern hemisphere. They are aggressively carnivorous and do not readily cohabit with anything smaller.

Perch *(Perca fluviatilis)*
OTHER COMMON NAMES: *bace; barse (northern England); bass.*

> The Pearch is a very good, and a very bold-biting fish; he is one of the fishes of prey, that, like the Pike and the Trout, he carries his teeth in his mouth … .[65]

And where else would he carry them? Izaak Walton thus introduces one of the most highly thought of sporting fish, always a

The Perch *(Perca fluviatilis)* inhabits fairly slow-moving waters and is readily identifiable by the broad black vertical bands.

beautiful animal and always a difficult fish to catch, especially when big; a 2-kilogram (4½-pound) Perch is both huge and rare.

Perch are fish of still or slow-moving waters and, taken from the water, are remarkably handsome. They have a spiny front dorsal fin, between five and eight dark vertical stripes, and a grey-olive-green body, paler and often almost golden beneath but varying greatly in colour overall. When the fish is excited, the lower fins become a vivid orange-red:

> the bright-ey'd perch with fins of Tyrian dye … .
> (Alexander Pope, *Windsor Forest*)

Taxidermists always seem keen, sometimes too keen, to reach for their paintpots when attempting to re-create these colours.

The Perch occurs throughout most of the British Isles, except the far north of Scotland, although it has been artificially introduced over much of the country and is historically native only to southeastern England.

Like many freshwater fish, Perch are very good to eat, although time-consuming to prepare; they must be scaled (and the spines avoided) and it is advisable to fillet them because they have so many bones. They have been caught on a commercial basis and a small fishery was established on Lake Windermere during the Second World War, when they were canned and sold as 'perchines'.

The name 'Perch' is Middle English, from the French *perche*, and so is the old measure, as in 'rod, pole or perch'. However, there the similarity ends because the fish name is originally Greek and the measurement Latin.

There are several inns called The Perch and, like the fish itself, the signs are often very beautiful. One of the most famous Perch inns, and a favourite haunt of Oxford students, lies at the opposite end of Port Meadow to an inn called The Trout.

Ruffe *(Gymnocephalus cernua)*

OTHER COMMON NAMES: *pope; ruff; Tommy parsee (Yorkshire).*

Superficially like a small Perch, for which it may be mistaken, the Ruffe lacks the vertical stripes and its dorsal fins are joined. In Britain, it is mainly a fish of southeast England but it has been introduced to the English Midlands and has spread to parts of the Severn. It now occurs in large numbers in Loch Lomond and other Scottish lochs, apparently having arrived as live-bait for Pike. Like the Perch, the Ruffe is good to eat but, being small, is even more trouble to prepare.

The name 'Ruffe' (or 'ruff') is Middle English and seems to have nothing to do with the bird (see Ruff, page 277), probably originating as 'rough', an allusion to the spines, either on the back or on the gill covers. The alternative name 'pope' has long been a great puzzle to me and I have never come across a satisfactory explanation. Its first recorded use was in 1653, when any name that was evenly remotely connected with Rome could only have been a term of opprobrium. Whatever the reason for inflicting this name upon such an unassuming little fish, it certainly led to unpleasantness because anyone with anti-papal sentiments vented them upon the poor creature. It became a common pastime to stick corks onto the dorsal fin spines and then liberate the fish to its fate.

> Frank Buckland has described how on public holidays the inhabitants of large towns in the East of England and the Midlands would congregate in vast numbers to enjoy the 'sport' of 'corking the pope,' the participants watching with devotional fervour the condemned fish bobbing about on the surface of the water, doomed to a lingering death.[66]

Pike-Perch *(Stizostedion lucioperca)*

OTHER COMMON NAMES: *zander.*

This fish is now generally called the Pikeperch, an unattractive linguistic bastardy; Pike-Perch is preferable, although I am not sure whether the German name 'zander' is any more appealing. The derivation is obvious; this fish does superficially resemble a hybrid between the Pike and the Perch, although it is unrelated to the Pike and is, in effect, an elongated Perch. It is much more a German than a British fish, being native to central and eastern Europe, but has gradually been introduced elsewhere.

Pike-Perch first came to Britain in 1878, when fish from Schleswig-Holstein were delivered to the 9th Duke of Bedford's lakes at Woburn. A further introduction was made by the Duke's grandson, the 11th Duke of Bedford, in 1910. In 1947, and again in 1950, his son, the 12th Duke, gave fish to the Leighton Buzzard Angling Club, which liberated some into nearby river systems. In 1959–60, more fish were netted from the lakes at Woburn and some were released into the River Ouse system. From here, aided by further official and unofficial releases, they have spread widely, especially into eastern England, and seem likely to cover much of the country in the years to come because they are clearly popular with a certain breed of angler. This has really been a wholly uncalled-for and unnecessary introduction of an aggressive carnivore that feeds on smaller coarse fish, such as Roach and Bream. It is therefore unpopular with match fishermen and is likely to appeal only to specimen anglers, although why they should find the Pike inadequate for this purpose is beyond me. The British record is a fish approaching 8 kilograms (17½ pounds).

Gobies
(Gobiidae)

Any small fish darting for cover in a rock pool as your shadow approaches is likely to be a goby. These fish tend to have relatively large brown heads and narrowly tapering bodies. If you can manage to catch one and examine it at close quarters in a collecting jar, you will see the characteristically united pelvic fins.

This is a large family of small, frustratingly similar fish that are difficult to identify. The largest you are likely to find in British waters is the Giant Goby (*Gobius cobitis*), which can reach almost 30 centimetres (12 inches); the smallest species is barely 5 centimetres (2 inches) long. Although most are marine, the little Common Goby (*Pomatoschistus microps*) frequently enters brackish and fresh water.

Gobies are an important food source for many sea birds and other species of fish. Some species, especially in the Mediterranean, are fished commercially, although there is no tradition of this in Britain. The name 'goby' is eighteenth century but comes from the same Middle English, French and ultimately Latin source as 'Gudgeon' (see Gudgeon, page 144).

Red mullets and related species
(Mullidae)

The red mullets are members of a large family of mainly tropical fish that are characterized by a pair of long barbels, a steeply sloping forehead and an excellent flavour. They are far superior to the grey mullets (see page 161) and their culinary splendour was well known to the Romans, who, in somewhat macabre fashion, used to place one in a bowl at their banquets so that guests could watch the colour changes as it died. Arundel Mullets are a Sussex speciality and boiled gently with chopped onions, wine, herbs, lemon juice, nutmeg and Anchovies.

Specialists argue over whether there is one species, *Mullus surmuletus*, or two. The second, *Mullus barbatus*, supposedly has an even steeper forehead and lives in deeper water.

Although called red mullets, these fish tend to be really red only when in deep water. Those caught close to the surface are more of a brown-yellow, although, seen on the fishmonger's slab, a certain redness persists. Izaak Walton, certainly no great sea fisherman, recognized its culinary merits when he compared the red mullet to trout. In his admiration, he turned to Joshua Sylvester's translation of one of his great favourites, the once much-admired but long-forgotten Huguenot poet Guillaume de Saluste, seigneur du Bartas:

> But for chaste love the Mullet hath no peer;
> For if the fisher hath surprised her pheer,
> As mad with wo, to shore she followeth,
> Prest to consort him both in life and death.[67]

Red mullets are sometimes known as the 'sea woodcock' because of the great desirability of the liver, which was considered a particular delicacy by the Romans.

Wrasse
(Labridae)

Wrasse (the word is seventeenth century from Cornish and appears to have no plural) are among the most brightly coloured of all British sea fish and, in the breeding season, the colour and patterning of the males may become even more intense.

They are peculiar fish in other respects: although most are bottom-feeding and carnivorous, some species feed on fungal and animal parasites on the skin of other fish, 'holding court' at cleaning stations to which the client fish come for attention; some engage in elaborate courtship displays and build nests; and some change their sex.

Among many species, all commonest in shallow water, there are few more than about 30 centimetres (12 inches) in length. Among the best known in British waters are the Cuckoo Wrasse (*Labrus mixtus*), with brilliantly blue-and-yellow males, and, commonest of all, the Ballan Wrasse (*Labrus bergylta*), which usually displays an overall pattern of white spots.

You may occasionally see wrasse offered for sale but once you have recovered from the excitement of the colours (which, in any event, fade on death), there is not much more to be said.

Mackerels
(Scombridae)

Mackerel *(Scomber scombrus)*

OTHER COMMON NAMES: *Atlantic mackerel; breal/ breel.*

No fish makes better eating than the Mackerel when it is totally, absolutely fresh from the sea but is such a disappointment after a matter of hours. Mackerel are fewer now than formerly, but, unlike those other great food fish, the Herring and Cod, have suffered nothing like the same catastrophic decline in numbers through overfishing. Nonetheless, they have never achieved the recognition they deserve, mainly

This cloud formation, known as 'Mackerel sky', because it resembles a Mackerel's back, may presage a warm front.

because they must be fresh and do not travel well. After a relatively short time, the flesh becomes inedible because of a build-up of toxins from bacterial decay.

> ... you may buy land now as cheap as stinking mackerel.
> (*Henry IV Part 1*, Act 2, Scene 4)

In various places, and before the advent of modern refrigerated vehicles, it has at times been illegal to sell Mackerel on a Sunday because this was a day when the fisherman did not go out and so any Mackerel could not possibly be fresh. In Fife, Mackerel have long been appreciated but it is said that the Highlander will not touch them because of a belief that they turn to maggots in the stomach. In some places, Mackerel were shunned because they were thought to feed on the bodies of people drowned at sea. Today, Mackerel tend to be eaten smoked with horseradish sauce, as an appetizer, but they are also delicious grilled and served with a sharp sauce.

Mackerel are streamlined, strong-swimming fish with a sharp narrowing of the body in front of the tail. They are bright silver beneath and iridescent green and blue above, and from the characteristic irregular stripes across the back comes the name of the beautiful cloud formation known as a 'Mackerel sky'.

Seldom more than about 50 centimetres (20 inches) in length, Mackerel are found in large shoals, generally close to the surface and feeding on other, smaller fish and crustaceans. They are the easiest of all sea fish to catch and many a young angler's first success has come from hooking one. No real bait is needed, a piece of coloured wool attached to a hook will suffice, and when a shoal is located, it is possible to pull fish from the water one after the other. They are caught commercially with multi-hooked lines trailed from boats. Occasionally, a related but more southerly fish, the Spanish Mackerel (chub mackerel) (*Scomber colias*), is offered for sale in Britain but it will have travelled too far to be worth buying. If eaten in southern Europe, where it has been caught, it is equally good, unlike the unrelated Horse Mackerel, or 'scad' (*Trachurus trachurus*; Family Carangidae).

The big shoals of Mackerel move inshore from late spring onwards. The villagers of Abbotsbury in Dorset, home of the largest collection of Mute Swans in Britain (see page 216), sadly are no longer visited by the Mackerel that once inspired a charming ritual, much like that found at fishing villages the world over. Garlands of flowers were blessed in the Church of St Nicholas on Old May Day (13 May) and then tossed into the sea from the Mackerel boats to ensure a good catch. The colourful scene was captured by the artist Philip Morris in 1893 and exhibited at the Royal Academy.

Today, the tradition survives as Garland Day, when the local children make two garlands from willow frames, one decorated with wild flowers and one with garden flowers. These are taken on a door-to-door collection and then placed on the war memorial and on the grave of a villager who used to take part in the old ceremony. In the 1950s, an overzealous village policeman attempted to put a stop to the collection, on the grounds that it was unlicensed begging, but commonsense prevailed and the Chief Constable apologized.[68]

Until the end of the nineteenth century, on the same date, a similar ritual was observed in Brighton and the fishermen met in the town for a celebration known as 'Bread and Cheese and Beer Day'. The Brighton men would stand bare-headed and pray as the first net was shot and when the tenth net was cast with its barrel floats, the skipper recited:

> Watch, barrel, watch, mackerel for to catch!
> White may they be, like blossom on the tree,
> God send thousands, one, two, three;
> Some by the head, some by the tail,
> God send mackerel and never fail;
> Some by the nose, some by the fin,
> God send as many as we can lift in![69]

Further along the coast, at Rye, the fishermen ritually spat in the mouth of the first Mackerel landed.

In the Dorset village of Abbotsbury, flower garlands were thrown into the sea from boats to mark the opening of the Mackerel season.

Tunny
(Thunnidae)

One entry on my old list of British Rod-caught Fish Records was for the Tunny, the record being a fish of 387 kilograms (851 pounds), caught off Whitby by Captain Mitchell Henry in 1933. I found this puzzling because I had never heard of anyone else catching Tunny in the North Sea and I also wondered whether there was any connection between Tunny and Tuna, which seemed to occur not in the North Sea but only in tins.

The reality is that, while tunnies are among the most important food and sporting fish in the world, most of them are inhabitants of warm seas; the family includes such well-known species as the albacores and bonitos. And, yes, the name 'tuna' is applied to several of them and they are canned and sold widely. The name 'Tunny' is fifteenth century but of obscure origin; 'tuna' is a Spanish-American derivative.

Tunny are big fish, up to 3 metres (10 feet) in length, and superficially like giant Mackerel, to which they are related. Captain Mitchell Henry was one of the British big-game fisherman who caught Tunny (blue-fin tuna) (*Thunnus thynnus*) in the North Sea between the late 1920s and the early 1950s, after which they disappeared, along with the Herring on which they fed.

The Gurnard's Head Hotel in Zennor, Cornwall, is named after this prehistoric-looking fish.

Tunny migrate in large shoals and the North Sea fish are thought to come from the Mediterranean in summer, although others certainly cross the Atlantic. This may explain why a new generation of Captain Mitchell Henrys is becoming excited. In 2000, big fish were caught again, not in the North Sea but by trawlers off the coast of Donegal, and sea anglers who visited the same area caught fish in excess of 200 kilograms (440 pounds). There is, however, one snag for the modern angler: because of overfishing by Japanese fishermen elsewhere in the world – the flesh can command £2 per gram (£56 per ounce) in Japan – the fish is now officially an endangered species.

Sand eels
(Ammodytidae)

No-one is likely to become very excited about sand eels, which are small, elongated, superficially eel-like fish. However, to many larger and grander creatures, such as sea birds or other fish, they represent an enormous food resource.

They spend much of their time buried in the sand of the sea bed, feeding on small worms and crustaceans, and emerging to swim in enormous shoals. They range in size from the Greater Sand Eel *(Hyperoplus lanceolatus)*, at about 30 centimetres (12 inches) long, down to the Sand Lance *(Gymnammodytes cicerellus)*, which is barely 15 centimetres (6 inches) long.

They have been eaten in England, and still are in France, but are seldom seen for sale other than as fishing bait.

Blennies
(Blennidae)

Any small, darting fish in a rock pool that is not a goby (see page 163), because its pectoral fins are not united, will probably be a blenny. In this family, the pectoral fins are reduced to form little spines just behind the gills. Some species of this large group also have tiny, horn-like protuberances on the head, which are helpful in identifying what are often confusingly similar fish. There are also rather similar species in other related families.

Some blennies are barely 5 centimetres (2 inches) long but the Butterfly Blenny *(Blennius ocellaris)*, with its enormous dorsal fin, like a butterfly wing, can reach 20 centimetres (8 inches).

The name 'blenny' is eighteenth century, from a Greek root meaning 'slimy'; they have no scales and are covered in mucus. They are sometimes called 'bulcards' or similar names.

Gurnards
(Triglidae)

The gurnards are prehistoric-looking fish which might not have looked out of place in the Devonian seas of 370 million years ago. This impression is derived from their disproportionately

Although harmless, the Long-spined Sea Scorpion *(Taurulus bubalis)* is related to the venomous Mediterranean scorpion fish.

large armour-plated heads and their peculiar spine-like pectoral fins, with which they probe the sea bed for crustaceans and other food in an action that looks remarkably like walking.

The three commonest species are the Red Gurnard *(Aspitriglia cuculus)*, about 30 centimetres (12 inches) in length, and the Grey Gurnard *(Eutriglia gurnardus)* and Tub Gurnard (yellow gurnard) *(Trigla lucerna)*, which are about twice as long.

Gurnards are commonly found in deep water and are fished for commercially. I find them rather tasteless and bony, although all right in a soup. They were more highly regarded on the south coast of England. Hastings gurnards were dipped in flour then had melted dripping poured over them; they were then grilled, being turned twice, and served hot with melted butter. In Cranbrook, they did things differently:

> Gurnards were caught early in the morning in Hastings and brought to Cranbrook where the people made puddings with them tied in a cloth with suet pastry, and then boiled.[70]

The Red Gurnard is occasionally sold as red mullet, which it resembles only in colour; look for the armour plating and lack of barbels.

Quite the oddest feature of the gurnards is the fact that they grunt, which can come as a considerable surprise to anyone handling a live specimen for the first time. The sound is produced in the air bladder and has probably given rise to the name. 'Gurnard', sometimes 'gurnet' ('I am a soused gurnet'; *Henry IV Part I*, Act 4, Scene 2), is a Middle English word, probably from the French *gronder*, 'to growl'.

Cottids
(Cottidae)

These fairly small, not very beautiful, bottom-living fish are squat and flattened, with a disproportionately large head and a very large mouth. The skin is slimy, mottled and covered with masses of tiny spines. Most are marine but there is one very common freshwater species. There is no general name for them in Britain, although they are called sculpins in North America.

In the sea, the commonest species are the Father Lasher (short-spined sea scorpion; bull rout) *(Myoxocephalus scorpius)*, the Long-spined Sea Scorpion *(Taurulus bubalis)* and the Norway Bullhead *(Taurulus lilljeborgi)*, all of which can reach 30 centimetres (12 inches) or more in length. They are related to the warm-water scorpion fish (Family Scorpaenidae), which are probably the most dangerous animals in the Mediterranean, producing extremely powerful venom in glands at the bases of the fin spines. The Cottidae are not venomous but, nonetheless, do have very sharp spines.

Bullhead *(Cottus gobio)*

OTHER COMMON NAMES: *bully-knob (Derbyshire); chabot; culle; miller's thumb; tom cull; tommy logge.*

Living in Derbyshire, I never knew this fish as anything other than the 'bully-knob', although it is a name I have heard nowhere else. The more widely accepted 'bullhead' is self-explanatory. 'Miller's thumb' alludes to their resemblance to the rough, swollen and gnarled appearance of a miller's thumb, which was acquired from constantly rubbing the flour to test its coarseness as it fell from the grindstone. 'Cull' is a fifteenth-century dialect word.

The Bullhead is a little fish that darts from stone to stone and, because of its extremely large mouth, is annoyingly able to take a big bait being ledgered for something grander. Although a mottled brown overall, its colour intensity alters markedly and rapidly in response to the surroundings and excitement. It is widespread throughout England and Wales, having been spread from its original home in the southeast of England, but is almost entirely absent from Scotland and completely absent from Ireland.

Bullheads feed on a wide range of small aquatic animals and are themselves eaten by trout ; they are also taken by the Kingfisher and herons, although many a bird has suffered a painful end as a result of not swallowing them sufficiently carefully and having the spiny fins stick in the throat.

The largest Bullhead I have ever seen was 13 centimetres (5 inches) long and, although I thought this abnormal (which was why I measured it), I am told that they can grow even bigger. It is, however, the species with the smallest official record angling weight: the British record fish was only 28 grams (1 ounce).

Flatfish
(Bothidae)

Many kinds of fish are more or less flat but some, like the rays and skates, are extremely flat and their flatness is achieved by being apparently squashed in the vertical plane. Members of the Bothidae, Pleuronectidae and Soleidae families achieve their flatness differently. They begin life looking much like any other young fish but undergo a transformation that can, at best, be described as bizarre. When they are about 2 centimetres (¾ inch) long, one of the eyes begins to migrate up the side of the body and over the top until it lies next to the other eye. Simultaneously the dorsal fin grows forwards onto the head, and the fish then lies on its side with both eyes facing upwards and spends the rest of its life flopping around on the bottom of the sea. In the Bothidae and Soleidae, it is usually the right eye that migrates and the fish lie on their right sides; in the Pleuronectidae, it is usually the left eye that moves and the fish lie on their left sides.

Some fish are rounded in outline, while others are characteristically elongated. All are carnivorous, feeding on crustaceans and other fish, and all three families include some extremely important food fish. Because of this, and because of their extraordinary development, they have for long attracted attention.

The Turbot *(Scophthalmus maximus)* lies in wait for its prey on the sea bed, flicking sand over its body to conceal its outline.

There are several species in the Bothidae, including the Topknots (*Zegopterus* and *Phyrnorhombus* species) and the Scaldfish *(Arnoglossus laterna)*, but there are only three important commercial species, which are described below.

Turbot *(Scophthalmus maximus)*

The Turbot is a rounded fish with very thick flesh and no scales but it has odd 'blisters' over its spotted, more or less brown skin. It is among the largest flatfish, sometimes reaching 1 metre (3¼ feet) in length. It is caught commercially in relatively deep water but never in sufficient quantities to satisfy demand, although some are now being farmed on an increasingly successful basis.

The name 'Turbot' is Middle English, from French and Latin, and means 'top-shaped'. The fish has given its name to that most splendid of all cooking utensils: the large, diamond-shaped and expensive fish kettle called a *turbotière.*

Brill *(Scophthalmus rhombus)*

OTHER COMMON NAMES: *logga-lay; lug-a-leaf.*

Closely related to the Turbot but slightly smaller, more oval in shape, and thinner-fleshed, the Brill is usually much cheaper to buy and, although often thought inferior, makes perfectly acceptable eating.

'Brill' is an old name of unknown origin. It certainly has no connection with the modern slang abbreviation for 'brilliant'.

Megrim *(Lepidorhombus whiffiagonis)*

OTHER COMMON NAMES: *sail-fluke; whiff.*

The Megrim is the smallest of the three commercial fish in the Family Bothidae, oval in shape and with large, dark blotches.

It is not often seen for sale and is not usually highly regarded as a food fish. However, while admitting that it is not as good as the best sole or Turbot, I find it far more flavoursome than Plaice. I am told that the French called it *salope*, which means 'slut' and seems very unfair.

'Megrim' is an old name for migraine but I do not think the two are related. 'Whiff' might have something to do with smell but there is also an old verb 'to whiffle' meaning 'to catch fish from a boat'; the coincidence must be too great to be ignored. 'Sail-fluke' seems to derive from a belief that the Megrim used its tail as a sail.

Dabs and related species
(Pleuronectidae)

Dab *(Limanda limanda)*
This very common and very small flatfish, usually little more than 20 centimetres (8 inches) in length, is quite excellent to eat, and I advise grilling it with Mediterranean vegetables such as fennel. The name is fifteenth-century Anglo-French of unknown meaning.

Dabs are commonly found in brackish water close to shore and are frequently trodden on by bathers and paddlers. This resulted in them achieving their greatest immortality in Noël Coward's charming lines, written in 1962:

What ho, Mrs Brisket,
Why not take a plunge and risk it?
The water's warm
There ain't no crabs
And you'll have a lot of fun among the shrimps
and dabs ...
(Noël Coward, 'The Girl Who Came to Supper')

A Suffolk cure for whooping cough required placing a small flatfish, for which the Dab served admirably, onto the bare chest of the sufferer and keeping it there until it died.

Flounder *(Platichthys flesus)*
OTHER COMMON NAMES: *butt; fluke.*
The Flounder is the only catadromous British flatfish and commonly moves into rivers and freshwater lakes at considerable distance from the sea; sometimes the young fish remain in fresh water for a year or more. It is also fairly unusual among flatfish in often swimming considerable distances off the bottom to feed. Seldom much more than 25 centimetres (10 inches) in length, the Flounder is oval in shape, with hard warts at the base of the fins. It is one of the flatfish that is quite commonly anatomically reversed: although most have eyes on the right, some have eyes on the left.

The old Manx tale of the Flounder obtaining its 'sneer' on

The Flounder *(Platichthys flesus)* is found in muddy or sandy bays, in estuaries and even in fresh water all around the British Isles.

discovering that the Herring had been elected king of the fish has already been mentioned (see Herring, page 125). In other places, there are different but just as fanciful explanations for its facial appearance, further evidence that it has long been a familiar species.

Flounders are most commonly seen for sale in places close to where they were caught, because they lose flavour when kept and transported.

The name 'Flounder' would seem logically to have originated from its habit of floundering on the bottom of the sea, but this may not be so. The name is Middle English, apparently from Old Norse, while the verb 'to flounder' is originally Dutch.

Plaice *(Pleuronectes platessa)*

The Plaice is one of the most attractive and unmistakable of flatfish, with its oval shape and dark brown skin, patterned with a number of striking orange spots. Usually up to 60 centimetres (24 inches) in length, Plaice in some areas can be considerably larger; the thickness of the flesh also varies.

They are very important commercially, are highly regarded as culinary fish and are probably the commonest flatfish to be seen for sale. Certainly packs of frozen Plaice fillets must outnumber most other types of frozen fish put together but, by then, the flesh is bland and tasteless. At their best when fresh and mild, Plaice can be delicious and are rather different in taste from most other flatfish. Perhaps this is because they feed almost exclusively on shellfish, which they crush with very powerful teeth.

'Plaice' is a Middle English word, probably originally from Greek, meaning 'broad' or 'flat'.

Lemon Sole *(Microstomus kitt)*

OTHER COMMON NAMES: *lemon dab; Torbay sole.*

Sole are commonly considered the most desirable of all flatfish but this species, while very good, does not match the 'real' sole, the Dover Sole (see below), which is in a different family. However, the Lemon Sole has a similar elongated, oval shape, a similar tough skin and is a dull yellow-brown (although hardly lemon yellow).

Unlike many flatfish, Lemon Sole feed almost exclusively on bottom-living polychaete worms, which they suck from their burrows. They seek their prey by characteristically raising the head and front part of the body from the sea bed.

The name 'sole' is Middle English and originates in the Latin word *solea*, a 'shoe' or 'sandal'. It therefore has exactly the same derivation as the sole of a foot, because this is how it is shaped.

A related and somewhat similar fish, seen increasingly frequently in fish catches, is the Witch (pole dab) *(Glyptocephalus cynoglossus)* perhaps named because it was thought ugly or unearthly in appearance. Before legal accountability for such things, Witch were rather commonly passed off as soles, under

The Plaice *(Pleuronectes platessa)* is dark brown, with orange spots. It feeds on shellfish, which it crushes with its powerful jaw.

such names as 'white sole' or 'Yarmouth sole'. Nonetheless, although they have long been regarded as having less commercial value than soles, the knowledgeable have always considered them quite excellent.

Halibut *(Hippoglossus hippoglossus)*
OTHER COMMON NAMES: *holibut.*
The Halibut is a giant among flatfish, commonly 2.5 metres (8¼ feet) and sometimes almost 4 metres (13 feet) long. Because of its size, you will see parts of it for sale and are very unlikely to have an entire fish served on your plate. Halibut steaks are best poached with something none too spicy so that their delicate flavour is preserved. Fish sold as 'chicken halibut' are simply young individuals, but Greenland Halibut (little halibut) (*Reinhardtius hippoglossoides*), is a quite different and inferior species. As befits its bulk, Halibut eat other fish of considerable size and will swim well off the bottom to catch them.

The origin of the name is Middle English. It is clearly from the old word 'butt', which was applied to several different flatfish, and then the adjective 'holy' was attached because it was eaten on Holy days. This seems to have been a rather widespread practice. In Swedish for instance, the Halibut is *helgeflundra*, 'the holy flounder', whereas the Turbot is called *botta.* The scientific name means 'horse's tongue'; perhaps someone thought the fish looked like one, but it must have been a very big horse.

The charmingly named Long Rough Dab (rough dab) (*Hippoglossoides platessoides*) is a related but very much smaller fish that is not caught commercially.

Soles
(Soleidae)

Dover Sole *(Solea solea)*
OTHER COMMON NAMES: *sole.*
This is the true sole, superficially similar to the Lemon Sole but with its eyes on the opposite side and grey-brown in colour rather than yellowish. Many connoisseurs regard it as the supreme flatfish for the table; it is certainly one of the most expensive. There are dark blotches on the surface but the colour of the Dover Sole varies even more widely than that of most flatfish, and experienced fishermen or fishmongers can often tell where particular individuals have been caught. It is a shallow-water species and, like most of the flatfish, the populations have suffered severely from overfishing.

There are several other species in the same family, including the Eyed Sole (*Solea ocellata*), which has several large round spots; the tiny Solenette (*Solea lutea*), barely 12 centimetres (4¾ inches) long; and the Sand Sole (French sole) (*Pegusa lascaris*). The latter especially is a wonderful culinary fish, greatly overlooked but seldom offered for sale. The Thickback Sole (*Solea variegata*), which has characteristic dark bands across the body, is occasionally seen and is a fine culinary fish.

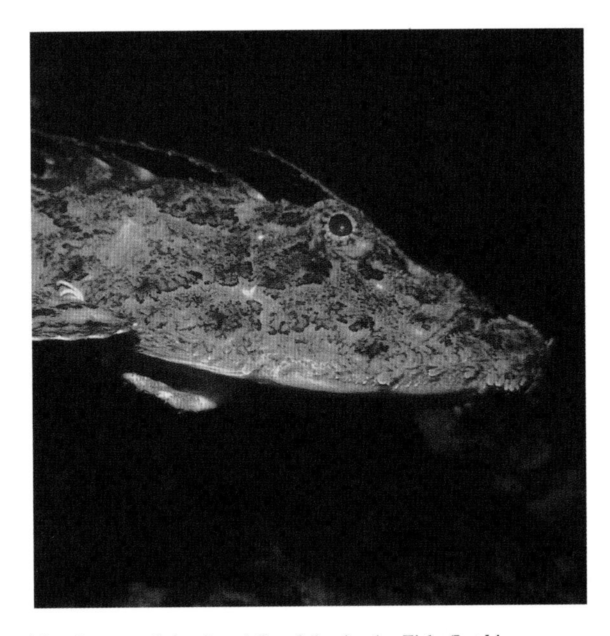

The first ray of the dorsal fin of the Angler Fish *(Lophias piscatorius)* forms a 'lure' that dangles over its carvernous mouth.

Angler fish
(Lophiidae)

Angler Fish *(Lophias piscatorius)*
OTHER COMMON NAMES: *frogfish; monkfish; sea devil.*
This is a remarkably ugly creature with an extraordinary way of feeding. It is flattened from top to bottom, so that its huge, gaping, frog-like mouth is directed upwards, and the first ray of what would be the dorsal fin is isolated and modified to form a structure resembling an angler's fishing line; this dangles, twitching, over the cavernous mouth, attracting other fish which are then engulfed. This adaptation is fairly common in abyssal fish but unusual in one from shallower water, although the term 'shallow' is relative; while Angler Fish may be caught close to the shore in autumn, they move to much deeper water to spawn.

It is a big fish, up to 2 metres (6½ feet) long, and was long considered as merely a biological curiosity. Even relatively recent books on sea fish will tell you that it is 'of no commercial value'. Then someone came up with the idea of cutting off the head and removing the rather flabby skin from the tail part. This left a firm, more or less white lump of perfectly edible flesh and, by calling this 'Monkfish' and ensuring that no-one ever saw the head end, it proved possible to persuade people to buy it. It has since become the *sine qua non* of the modern restaurant trade.

The Natterjack *(Bufo calamita)* is found mainly in coastal areas of East Anglia, Lancashire, Cumbria and southwest Scotland.

Amphibians (Amphibia)

Amphibians are cold-blooded vertebrates with a moist skin that lacks scales. Most live on land for much of the year but require water in which to breed. The majority, including all British species, lay eggs with a characteristic jelly-like protective covering and these hatch into distinctive larval forms, called tadpoles, which slowly metamorphose into the adult state.

'Amphibian' is a seventeenth-century word meaning simply 'living in two places' or 'equally at home in water and on land'. 'Tadpole' is a Late Middle English word from *tadde*, meaning 'a toad', and *poll*, 'a round head'. Amphibians are conveniently divided into two large groups: the frogs and toads, and the newts and salamanders.

The study of reptiles and amphibians collectively is called herpetology, which is a nineteenth-century word from the Greek *herpeton*, meaning 'a creeping thing', although, fortunately, the lamentable twentieth-century invention 'herptile' as a joint term for the two groups seems not to have caught on.

Frogs and Toads
(Anura)

Frogs and toads collectively form the tail-less amphibians (Order Anura). Globally, there are around 4,000 species but only three are native to Britain, although a fourth, the Pool Frog, became extinct very recently. Two aliens are fairly well naturalized, and several others less so, although the interrelationships and genetics of some of the introduced frog species are utterly bizarre.

In addition to the well-established introductions, there are, or have been at different times, small, more or less self-sustaining populations in southern England of other frogs and toads, most notably three European species: the pretty green and noisy European Tree Frog *(Hyla arborea)*, the Yellow-bellied Toad *(Bombina variegata)*, and the Midwife Toad *(Alytes obstetricans)*, which lives in small numbers in the vicinity of Bedford and probably elsewhere. Another intruder is the African Clawed Toad, or Platanna *(Xenopus laevis)*, which was once used in human pregnancy tests and was known as the Clawed Frog when I kept them 40 years ago (not for pregnancy testing).

More recently, a note of caution has been sounded by English Nature concerning the presence in Britain of a huge American species of bullfrog, *Rana catesbeiana*, which can reach 20 centimetres (8 inches) in length and will eat practically anything that will fit into its mouth. This could have a major ecological impact if it was allowed to become established. Although the occasional individual has turned up as an escape or deliberate liberation since they were first imported as pets in the 1970s (the trade is now banned), in 1999 tadpoles were found in an isolated pond in East Sussex; clear evidence of the presence of more than one adult animal. The pond was fenced, drained and every tadpole and baby frog found was humanely destroyed but there are reports of the distinctive call ('like a cello being tuned') being heard elsewhere. There is genuine concern that, if they manage to breed in a system of interconnecting waterways, there may be little to stop them spreading.

The hindlegs of these tadpoles of the Common Frog *(Rana temporaria)* are just beginning to appear.

Scientifically, there is no real distinction between the creatures known in English as frogs and toads, although the name 'frog' tends to be used for those that hop, have smooth, wet skins and spend a considerable amount of time in water. 'Toads' tend to crawl or walk, have drier, warty skins and are usually more terrestrial in their habits, although there are exceptions.

Perhaps because there are so few species in Britain, or perhaps because they are very different from any other members of the native fauna and two at least have long been very common, they have attracted a considerable amount of attention and have accumulated much lore and tradition.

Frogs
(Ranidae)

Common Frog *(Rana temporaria)*

OTHER COMMON NAMES: *Adults – brown frog; charlie (young animals); fresh/fresher/fresk/frosh (northern England); frosk (young animals); glouton; grass frog; gwelsken (Cornwall); laid-lick (young animals); march-bird; packer-poo; paddle-doo; paddock; paddow; pade; padger-paw, paget-poo (Cornwall); pan (Somerset); poddock; puddock; quilken; squilky; tommynoddy; wilkin/wilken/wilky (Cornwall). Frog spawn – junder; paddow-redd; ride; rodd; rode;, rood; rowd; rud; rude. Tadpoles – toddies; tommy-toad; tom-todies; tom-totties.*

The Common Frog is the frog of many people's childhood, the frog of school biology dissections and the source of most of the spawn that is found in ponds and ditches. It is still common, although not as it once was, partly because its numbers were at one time seriously depleted by the large quantities collected for scientific use in schools and elsewhere; the Nuffield school biology course accounted for a great many of them. More recently, it has declined as a result of pollution and loss of habitat. Nonetheless, the popularity of garden ponds has proved something of a salvation for it, although a disease called 'red leg', caused by a bacterium but possibly also involving a virus, has accounted for large numbers in recent years and gardeners have been distressed to find quantities of bloodied dead bodies.

Common Frogs need little description therefore. They are typical frogs in possessing long hind legs for hopping and their colour varies widely, quite apart from the marked changes that take place in most amphibians in response to the lightness or darkness of the background. They tend to be mostly shades of green, brown, yellow or olive above, with more or less scattered dark blotches, but touches of pink, red and orange are also frequently seen. The underside is usually white, yellow or rather vivid orange and there is a consistent dark blotch behind the eye. Albino frogs have been found but not melanistic ones.

> I have located distinctly reddish Common Frogs on the Derbyshire moorlands and Glen Lyon, Perthshire; the common denominator seems to be acidity and plenty of Sphagnum peat.[1]

Yellow frogs attract special attention and Roy Palmer tells of an old custom in some Warwickshire villages of infants being given yellow frogs to suck for the relief of mouth ulcers.[2]

The Common Frog occurs throughout mainland Britain and reddish coloured individuals are especially frequent in Scotland. It is found on many of the offshore islands and on Guernsey, although its place on Jersey is taken by the closely related, slightly smaller and even longer-legged Agile Frog *(Rana dalmatina)*. There are Common Frogs in Ireland and, although there have been several introductions there, it is generally thought that some populations at least are native.

The name 'frog' is Old English and, over the centuries, has given rise to many other names for things that share some similar characteristic: for instance, 'froghopper' for the group of hopping, sap-sucking insects (see Homoptera, page 72) and 'frogfish' for the Angler Fish, with its wide, frog-like mouth (see page 171).

Common Frogs eat a wide range of invertebrate prey, although slugs and snails always figure highly. In turn, they constitute a major food for many predators, most importantly herons, birds of prey, crows, Grass Snakes, Hedgehogs, Foxes, Otters, Polecats and rats. Large numbers are killed by domestic cats, although they are rarely eaten, and considerable numbers are killed on roads. My cats have always pawed them, apparently uncertain what to do; presumably they find the cold, wet skin unpleasant.

Frogs and their behaviour have often been taken to indicate weather patterns, a bright, healthy skin suggesting fine weather and dullness foretelling rain. If spawn was laid at the edges of ponds it was considered an indication of storms while spawn laid in deep water suggested drought.

Frog mating in general has understandably attracted interest, and the tenacity with which the male frog grips the female during mating is well known. The males can always be distinguished by the pad-like patch on the first finger, which becomes covered with tiny prickles during the breeding season. The spawn, unlike that of toads or newts, is laid in large, frothy masses and the date at which the first spawn appears tends to be fairly constant in particular districts or even in individual ponds.

The mass exodus of baby frogs from their breeding ponds has long been observed and, because the movement of many thousands of baby frogs usually takes place in wet weather, it has given rise to the belief that frogs can rain from the sky. The mass movement of adults, however, is a feature of toads, not of frogs.

> The people of Madingly which is on the outskirts of Cambridge, get up to some bizarre behaviour every year helping the toads to cross the road to reach the local pond safely. Each evening, a group of dedicated individuals go out with an old milk float to catch the toads and help them on their way. There is even a 'Beware Toads Crossing' sign in the village.[3]

Pollution and loss of habitat have recently led to a decline in numbers of the Common Frog *(Rana temporaria)*.

To find a frog in a house was a serious indication of death or some equally dreadful event, the extreme rarity of the occasion no doubt encouraging the believers: toads certainly enter houses but frogs hardly ever do so. In some districts, even meeting a frog outdoors was thought a good omen. Much frog lore, however, is interchangeable with that relating to toads. For instance, like toads, frogs or parts of them were commonly hung around the neck in the belief that this would effect a cure or impart some special power.

The widely recounted story of the bone-charm links the two superstitions. A frog, either alive or killed in a certain way, was placed inside a perforated box in an ants' nest. When the bones had been picked clean of flesh, they were thrown at midnight into a running stream. How anyone was able to see frog bones floating in a stream at midnight beggars belief but, in any event, the observers were looking intently for the one bone that was swept upstream rather than being carried downstream. That bone, if it could be retrieved, endowed its owner with the powers of the toadmen (see Common Toad, page 177).

The expression a 'frog in the throat' to describe a type of cough is familiar enough, by analogy with the croaking sound that the frog itself makes, but it has at times taken on a more literal meaning. Eating a live frog was widely recommended as

In the breeding season, the male Marsh Frog *(Rana ridibunda)* displays to the female by extending its vocal sacs and calling.

a treatment for whooping cough, which was itself known as 'frog in the mouth' in some areas. A Yorkshire variant was to take nine frogs (no more, no less), make a soup from them and then administer it to the whooping-cough sufferer without telling them of the ingredients.

Apart from such ancient and best-forgotten medicinal practices, Britain has never adopted the tradition of eating frogs that prevails in France (see Edible Frog, below), although a most remarkable item appeared in *Country Life* in 1904:

> Many folks will find it difficult to believe that as near London as Ashford, Middlesex, the village lads commonly eat baby frogs, removing the stomach, and toasting what is left with a dab of butter. This curious fact was first brought to my notice sometime ago, but not until recently have I received confirmation of it. Not that it proves very much, as village lads, as a rule eat anything.[4]

Pool Frog *(Rana lessonae)*

OTHER COMMON NAMES: *Boston waite, Cambridgeshire nightingale, Dutch nightingale, Whaddon organ (East Anglia).*

Zoologists have argued for more than a century over the British status of this brown or green frog, which is roughly the same size as the Common Frog, or slightly larger, but with a more pointed nose and no brown patch behind the eye. Its distinctive, loud, rather musical croaking (some of which was probably caused by Natterjacks) had been known in Cambridgeshire and other parts of East Anglia for many years and the animal responsible, even though its exact identity was unknown, was given a number of distinctive regional names. There is no doubt that European frog species were introduced into several parts of East Anglia, and Pool Frogs were known to be among them. Then, recently and unexpectedly, a fossil Pool Frog bone was found, indicating that the animal had, after all, probably been a British native for a very long time.

Unfortunately, the Pool Frogs in East Anglia had been diminishing rapidly and, at about the same time as their ancestry was looking more certain, their future was looking much less so. The last native specimen, named Lucky and rescued from a pond at Thetford in 1993, died in captivity in 1999. Tony Gent of English Nature said:

> It is just possible there is still one lurking somewhere. Lucky did breed with other frogs from the Continent, so we still have British genes.[5]

Marsh Frog *(Rana ridibunda)*

OTHER COMMON NAMES: *laughing frog.*

See Edible Frog below.

Edible Frog *(Rana esculenta)*

These two European frogs have been introduced to Britain on several occasions and have become successfully established. The Marsh Frog is a handsome, olive-green creature, the largest European frog, up to 15 centimetres (6 inches) in length, highly aquatic and seldom found far from water. It has a voracious appetite and eats prey as large as dragonflies. It also has a very loud, cackling voice, hence the alternative name 'laughing frog'.

The British population of Marsh Frogs somewhat bizarrely originated with the playwright Edward Percy Smith (1891–1968), author of *The Misdoings of Charley Peace*, *The Shop at Sly Corner* and other masterpieces now seen and heard far less frequently than his frogs. He imported twelve animals from Hungary in the winter of 1934–5 and released them into a pond in his garden at the village of Stone-in-Oxney, abutting 'upon a tract of land where the flats comprising the Walland, Romney and Denge Marshes usurp the place of the salt-water'.[6] The frogs liked what they found and moved to a nearby mere the following summer. The noise they produced prompted letters to the local MP and even to Sir Kingsley Wood, then Minister of Health, and resulted in considerable coverage in the national press. After this there was no stopping them and, using the various dykes, pools and other waterways, and quite possibly aided by human intervention, they became quite widely distributed in Kent, with isolated populations elsewhere in southeastern England. In little more than half a century, they have established themselves as easily the most successful alien amphibian in Britain.

The Edible Frog is the frog whose legs are eaten by the French. (Contrary to popular belief, the French are referred to collectively as 'frogs' probably because the Paris swamps once contained frogs, or toads, not because of their eating habits. It has also always intrigued me that the French equivalent of having a frog in the throat is '*avoir un chat dans la gorge*'.) The increasing numbers of frogs' legs on English menus, however, belong not to the Edible Frog but to an Indian species.

The Edible Frog *(Rana esculenta)* is highly aquatic and, unlike the Common Frog *(R. temporaria)*, can eat underwater.

Slightly larger than the Common Frog, the Edible Frog superficially looks more like the Marsh Frog, being usually green and spotted. Like the Marsh Frog, it is highly aquatic. Edible Frogs can even eat under water, something that Common Frogs seem unable to do. The first properly recorded introduction of Edible Frogs to Britain was to some ponds near Norwich in 1837, but there have been many others since and isolated populations exist at a number of sites in the Home Counties. Thereby hangs a very complicated story that has only recently been fully unravelled: it appears that a population of Edible Frogs cannot be sustained unless Pool Frogs are also present. It has long been realized that the two are closely related but it is now known that the Edible Frog is in reality a hybrid, but no ordinary hybrid, and the sex life of these frogs beggars belief; even geneticists find it scarcely credible.

Edible Frogs are initially produced when a male Pool Frog mates with a female Marsh Frog. The offspring then preferentially mate with Pool Frogs rather than with other Edible Frogs; the result, nonetheless, is more Edible Frogs. As long as there are always some Pool Frogs mating with each other, thus ensuring that there is a genetically pure Pool Frog population available for mating with the Edible Frogs, the *status quo* continues. The genetics are such, however, that if Edible Frogs do happen to mate with other Edible Frogs, all their offspring will be female Marsh Frogs. With this knowledge, it is now clear that, among the early introductions into Britain of European Edible Frogs, there must have been some Pool Frogs too; alternatively, the Edible Frogs must have been put into ponds where native Pool Frogs already existed.

Toads
(Bufonidae)

Common Toad *(Bufo bufo)*

OTHER COMMON NAMES: *bul-cranag/granasg (male); bulgranack (male, Cornwall); cranag; farmer; gangrel/gangril (northern England); glouton; hornywink; jack; joey; josey (Worcestershire); nalter-jack (Suffolk); natter; natter-jack; paddock; puddock; puddock-rude; puddoke; puddow-rudd (young animals); slug; tade; taed; taid; tiad; tode; tod (Cornwall); todelinge (young animals); tooad (Isle of Wight); wilky; winky.*

Certain animals, and the hare is the most obvious example, have accumulated a mass of lore and legend for no immediately apparent reason. The toad, by contrast, is probably the best instance of a British animal about which a huge number of beliefs and stories have accumulated for reasons that are entirely understandable. Almost everything about the Common Toad is strange, and even the zoologist Thomas

This purpose-built 'toad tunnel' enables toads to pass safely beneath the road in order to get to their spawning grounds.

Pennant described them as being 'the most deformed and hideous of all animals – objects of detestation'.[7]

The Common Toad is a large animal, especially when compared with the Common Frog, of up to 10 centimetres (4 inches) in length, occasionally more. It is usually a uniform buff in colour, although animals that are more red, green, yellow, grey, black or white have been seen. Like other amphibians, the colour changes to a degree in order to blend with the surroundings. Its skin has an irregular, warty appearance, the warts behind the eyes being especially large. These skin swellings secrete a poisonous protective substance that has certainly helped to colour people's judgement of them.

Toads will eat 'all living animals that are susceptible of being swallowed'; their appetite is prodigious and the only requirement seems to be that the prey is moving. Slugs are among their favoured food, so toads are extremely welcome in gardens.

> My father was head gardener and bailiff at the Bristol Mental Hospital. They had several large greenhouses there and my father liked to keep a toad in the greenhouse to eat woodlice. The tomato house, like other tomato greenhouses, would get whitefly. My father used to treat it with cyanide for which he had to have a doctor present when he applied it. The next morning he used to look for the toad and, sure enough, it was alive and well despite being in the house where the cyanide was. This went on for a number of years and the toad must have been a good age when my father retired.[8]

Toads move in a slow and ponderous manner and seem never to give up, whatever obstacle is placed in their way. For amphibians, they are highly intelligent, learn quickly and can be tamed, and they are extremely long-lived (40 years at least). They have a marked homing instinct and will return to the same resting spot in some damp corner time after time, even when they are taken considerable distances away. Sometimes, after emerging from hibernation, many hundreds of animals partake in mass migrations to their breeding grounds and it is on these relentlessly direct marches that huge numbers may be killed on roads. They will walk past what appear to be otherwise suitable ponds, whereas frogs, apparently less intelligent and less discerning, will lay their spawn in the nearest convenient puddle.

Such mass movements led in the past to tales of toad plagues and, by analogy with the Old Testament, it was thought that these must be manifestations of divine retribution. One such event in 1660 even made the august pages of an early volume of the *Philosophical Transactions of the Royal Society*:

> Strange and True Newes from Gloucester of a Perfect Relation of the Wonderful and Miraculous Power of God shewed for injustice at Fairford, betwixt Farrington and Scicester; where an innumerable company of Froggs and Toads (on a sudden) overspread the Ground, Orchards & Houses of the Lord of the Town, and a Justice near adjacent; and how they divided themselves into two distinct Bodies & orderly made up to the house of the said Justice, some climbing up the walls and into the Windows and Chambers; and afterwards how strangely and unexpectedly they vanisht away to the Admiration of All.[9]

Once toads reach their breeding grounds, their passion knows few limits. As with frogs, the mating process entails the males gripping the females extremely tightly. Frequently, one female is embraced by a large number of males and may even be drowned or crushed to death in consequence. The mating instinct is such that anything of roughly the same size as another toad will be grasped with equal enthusiasm; it must have been this that gave rise to the belief that, once you were seized by a toad, it would exercise some malign influence over you until it was forcibly removed with boiling water. Toad spawn is laid in water, but in ropes with a double row of eggs rather than the masses that frogs produce. Other than in spring, toads are much less aquatic than their relatives.

The Common Toad is absent from Ireland (St Patrick presumably banished toads at the same time as he banished snakes) but occurs throughout Britain, except for the far north and west of Scotland. It is also widely distributed across Europe, western Asia and North Africa. The British population is thought to number around 20 million, a figure exceeded by few other vertebrates.

The name 'toad' is from the Old English *tadde*, or *tadige*, but there the trail ends, and its earlier origin is unknown. The best-known and most widely used of the regional names is 'paddock', or 'puddock', which is especially common in the north of England and Scotland and is also used for frogs. It was immortalized by Robert Herrick in *c.*1640:

> Here a little child I stand
> Heaving up my either hand
> Cold as paddocks though they be
> Here I lift them up to Thee
> For a benison to fall
> On our meat and on us all. Amen.
> (Robert Herrick, 'Child's Grace')

It originates in the Middle English word *pad*, meaning 'a toad' (or 'frog'), probably from the Old Norse. 'Puddock', or 'puttock', are also regional names for both the Buzzard and the Kite and it has been suggested, I feel improbably, that the name was transferred to them because they eat toads and frogs (see Red Kite, page 245). *Bufo* is the Latin name for toad and came to mean the black tincture used by alchemists.

The venomous nature of toads has been known for centuries and led largely to the inseparable link between toads and witchcraft. The glandular secretion from the skin contains a wide range of irritants, including the hallucinogen bufotenine and several glycosides allied to the heart stimulant digitalin. It is certainly poisonous to humans, although probably never fatally so. Gilbert White knew that 'a quack, at this village, ate a toad to make the country people stare; afterwards he drank oil',[10] although Frank Buckland, in *Curiosities of Natural History* wrote that:

> Mr Blick, surgeon, of Islip, Oxfordshire, tells me that a man once made a wager, when half drunk in a village public-house, that he would bite a toad's head off; he did so, but in a few hours his lips, tongue, and throat began to swell in a most alarming way, and he was dangerously ill for some time.[11]

There is a story that toad venom brought about the demise of King John (1167–1216). Matthew Paris, a contemporary of the ill-fated monarch, was quoted in William Prynne's *History of King John* (1670) as recording that, at Swinshead Abbey, where the King had repaired after the unfortunate loss of his baggage train in The Wash on 12 October 1216, an aggrieved monk administered toad poison in a cup of ale. As it was a Wassail cup, both men were required to drink from it and both passed away soon afterwards.[12] This seems rather improbable, although the death of the king has never properly been explained and it is widely thought that he was indeed poisoned; that good old standby, a dish of lampreys (see River Lamprey, page 118), has also been implicated. Surely, however, there must have been more reliable substances available to anyone sufficiently malevolent. Clifford Brewer thinks that a perforated ulcer is the more likely cause.[13]

The nineteenth-century verb 'to toady' also has a story to tell about toad venom. Its modern meaning is 'to flatter or attend to with servility' and, although it has never conjured an attractive image, its link with the amphibian and its venomous nature is an old one and no more appealing. The verb comes from the noun, 'toady', a contraction of the early seventeenth-century word 'toad-eater'; literally 'someone who eats toads' and originally 'the attendant of a charlatan, employed to eat toads to enable his master to exhibit his skill in expelling poison'. Toad-eaters were to be found at every country fair throughout the seventeenth and eighteenth centuries, swallowing toads and then falling to the ground in feigned agony, only to be restored by some 'medicine' dispensed by their employer: a quack doctor or mountebank who would then sell his proven cure-all to the astonished audience. They were considered among the most miserable of individuals.

> Be the most scor'd Jack Pudding in all the pack,
> And turn Toad-eater to some foreign quack ...
> (Thomas Browne, 'Satires on Quackery')

People who have attempted to imitate the old toad-eaters have generally come unstuck, finding that the cocktail of skin irritants produces a wide range of symptoms, ranging from

Common Toads *(Bufo bufo)* mate later than frogs and lay their eggs in a long string, which becomes entwined around weed.

numbness in the lips to serious chest pains. The modern explanation is that the toad-eaters ate only small toads and contrived to keep the animals very calm beforehand; an irritated or frightened toad may 'sweat' its venom far more profusely than one that is calm. Nonetheless, as we are so often told, 'Don't try this at home'.

Dogs certainly froth at the mouth after interfering with toads, although there seem to be no reports of deaths in Britain. Nonetheless, the secretions are clearly not uniform in their effect on other creatures. Hedgehogs seem to eat toads with impunity, and so do some Grass Snakes and some birds, such as Buzzards and Stone Curlews, although crows are said to tear away the skin first.

> Stephen Hewitt, a naturalist with Carlisle City Council and Paul Duff, a vet from Culgaith, Cumbria launched an investigation in 1997 after more than 1,000 common toads were discovered dead or dying in southern Scotland and Cumbria. Nearly 40 were found in the same area in the following year. All had their legs expertly removed. Experts ... suspected that humans were responsible, possibly in connection with the restaurant trade. But ... the two men wrote in the Veterinary Record 'the weight of evidence indicates that the deaths were caused by animal or bird attack'.[14]

The fact that a large number of potential predators leave toads alone must assist the toads in achieving their undoubted longevity. There are many historical accounts of toads of immense age (and size) being found, none more picturesque than that relating to one reputedly found at Berkeley Castle in Gloucestershire. An early seventeenth-century account gives the details:

> Out of which dungeon ... was (as tradition tells,) drawne forth a Toad, in the time of Kinge Henry the seventh, of an incredible bignes, which in the deepe dry dust in the bottom thereof, had doubtlesse lived there divers hundred of yeares ... which in bredth was more than a foot, neere 16 inches, and in length more. Of which monstrous and outgrowne beast the inhabitants of this towne, and in the neighbour villages round about, fable many strange and incredible wonders; makinge the greatnes of this toad more than would fill a peck, yea.[15]

Yea, indeed.

This engraving from *A Rehearsall at Winsore* (1579) shows a witch feeding her familiars, which include two toads.

Toads, like hares, were thought to be the familiars of witches and there are countless old accounts of toads being used or implicated in witchcraft. As late as the nineteenth century, country people reportedly treated toads with care in case they were really something, or someone, in disguise. It is a folk belief well entrenched in the psyche and may not entirely have gone, even today:

> One night when our elderly baby-sitter was leaving we opened the back door to find a toad sitting just outside on the doorstep. Our dear lady was most alarmed and refused to pass the toad, leaving the house by the front door instead. She never liked using the back door after that.[16]

> Toad-magic could be used against witches as well as by them. A Devonshire charm to destroy their power was to take three small-necked jars and place in each the heart of a toad studded with thorns and the liver of a frog full of new pins. The jars, carefully corked and sealed, had then to be buried in three separate churchyards, seven inches below ground-surface, and seven feet from the church porch. While this was being done, the Lord's Prayer had to be repeated backwards, an evil and dangerous proceeding which must surely have prevented the charm's use by any but the very determined or the very frightened. The operator was supposed thereafter to be safe from witchcraft for ever.[17]

Despite their venomous secretions, toads found a place in more conventional medicine, or what passed for it in earlier times, and toads, or toad skins, were widely used as a treatment for cancer, because it was thought that the toad would draw out the cancer's poison. As recently as the early twentieth century, 'in the Fens north of Ely', a woman showing the first signs of breast cancer would consult the local 'handywoman', who would take a toad and rub its back until beads of moisture appeared. When the entire skin was exuding droplets, the animal was rubbed over the affected breast until its skin was quite dry and its warts had shrunk to small pimples. The unfortunate animal was then returned to a water-butt and the patient would have a plaster of houseleeks placed over the affected area. Gilbert White knew of something similar 150 years earlier, but even then he was sceptical. In Scotland, farm-workers rubbed live toads onto strained wrists to gain relief. In many places, briefly holding a live toad in the mouth was thought to cure thrush and I have seen suggestions that the secretions do indeed contain some anti-fungal substance that would affect the thrush-causing *Candida*; this is perhaps worthy of further research.

It is surprising how often spitting is suggested as a means of warding off something potentially harmful or unlucky and, in this respect, toads can oblige because spitting at a toad or throwing stones was a common way of avoiding the effects that it might have. Simply keeping out of its way was also considered sensible because a toad could bring bad luck if, by chance, it walked over your foot.

The origin of the once widespread notion that a toad has a precious jewel in its head is as mysterious as the idea is unfounded. Nonetheless, it was familiar to John Bunyan in 1678, and to Shakespeare:

> If that a pearl may in a toad's head dwell,
> And may be found too in an oyster-shell,
> If things that promise nothing do contain
> What better is than gold – who will disdain
> That have an inkling of it there to look
> That they may find it?
> (John Bunyan, *Pilgrim's Progress*)

> Which, like the toad, ugly and venomous,
> Wears yet a precious jewel in his head ...
> (*As You Like It*, Act 2, Scene 1)

The idea was old even in Shakespeare's time; Alexander Neckam (1157–1217) wrote of toad jewels or 'toadstones' and of their value as an antidote to poison, especially when carried about the person. This rather begs the question of what exactly people did carry around with them and where they obtained these wondrous objects. Something must have provided a thriving trade for the charlatans of the time, although 'toadstone' was later used more precisely as a geological term for a type of igneous intrusion in limestone deposits; a stone that became known technically as 'bufonite'.

Shakespeare, like most of his contemporaries, really thought rather little of the animals:

> Here is the babe, as loathsome as a toad
> Amongst the fairest breeders of our clime ...
> (*Titus Andronicus*, Act 4, Scene 2)

Time and again in folk medicine, the practice of hanging an object around someone's neck with a view to either curing or preventing something unfortunate is encountered. Toads fill the bill admirably and live toads or parts of toads were commonly placed in a small bag and suspended on a necklace to this end. Any relief that might be obtained from the old affliction of scrofula, or king's evil, was always to be cherished and there were people who made this a full-time occupation. Toad-doctors travelled the country, charging handsomely for the provision of two hind legs in a small bag. The gatherings at which they performed were known as 'toad fairs'. A Herefordshire story told of a farm-worker as late as the end of the nineteenth century who wore a toad's heart around his neck in the belief that he was thereby protected from anyone discovering his petty pilfering activities. The travelling toadman was a familiar person in rural England until relatively recent times. He had:

> ... power over horses and pigs and, in some cases, over human beings, especially women ... he could do anything he liked with horses, and his services were always in demand on any farm where they were kept. Toadmen seem to have flourished particularly in eastern England, and faith in their powers existed until a very late date in the Isle of Ely, South Lincolnshire, and the Soke of Peterborough. A contributor to 'Folk-Lore' says that in 1950 he was told in the March, Peterborough and Stamford districts that they were well known between the two wars. One man said he had worked on the same farm with one, and others spoke of them either from their own personal knowledge, or from the information of friends and relatives. All agreed that the charm, though potent with animals and living beings, would not work with tractors and machinery.[18]

An old Fenland rhyme offered:

> Make a black spit on mutton fat
> Then rub it inside a horse's hat.
> Scrape it off within a week
> Then go outside a toad to seek
> And make it sweat into a pot.
> With a wooden spoon mix the lot,
> And you'll have a healing balm
> To keep the body free from harm.[19]

> An uncle who lived just over the border [Hertfordshire] in Essex used to liken something to 'a toad under a harrow' but we're not sure what it was.[20]

> There was a local expression [Berkshire] 'Like a twod (toad) under an 'arrow', presumably meaning jolly uncomfortable.[21]

There are many folklore links between toads and certain birds. Yellowhammers were bracketed with toads in Scotland as things of evil (see Yellowhammer, page 379). Wheatears' eggs were believed to hatch into toads and Stonechats' eggs were believed to be incubated by toads (see Stonechat, page 339) while, oddest of all, larks and toads were thought to exchange eyes (see Skylark, page 318).

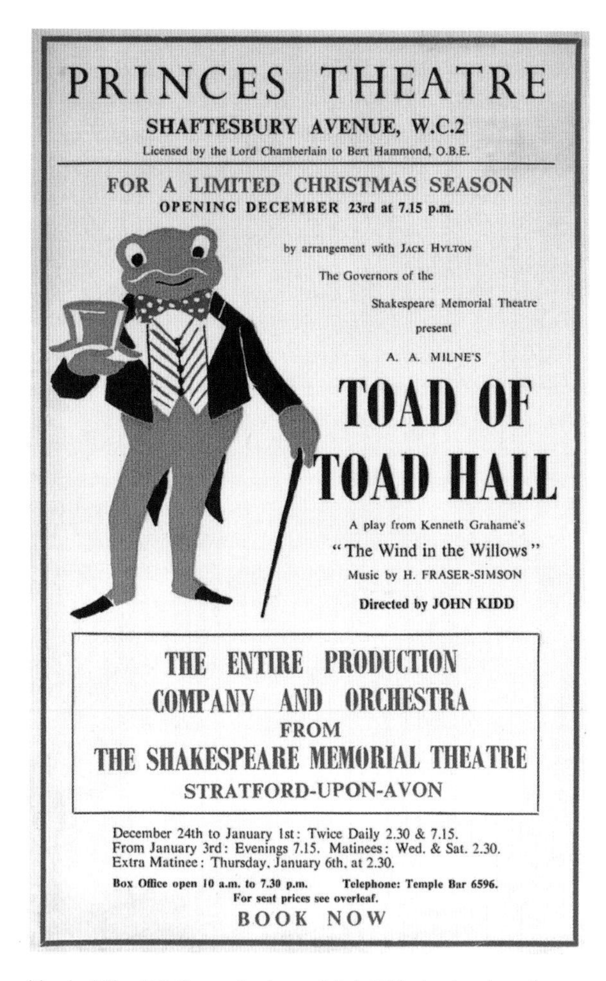

Toad of Toad Hall was the hero of A.A. Milne's play, based on Kenneth Grahame's *The Wind in the Willows*.

The association of toads with fungi (toadstools) is fascinating. Most people have concluded that the link between the names is merely a reflection of the fact that both are likely to be found in cool, damp places and that someone, at some time, saw a toad sitting on top of a fungus. I have certainly seen a toad apparently licking a toadstool as it deftly removed a slug that was feeding on the cap. Adrian Morgan thought that both represented 'the dark and evil threads of nature's tapestry'.[22] One of the most distinguished British mycologists, Geoffrey Ainsworth, simply gave up: 'the superficial derivation of "toadstool" is as apparent as the reason for the association of toads with agarics is obscure'.[23] Another mycologist, Peter Findlay, commented on the occurrence of the word 'todestole' in John Gerard's *Herball* of 1597.

> The spelling for 'tode' here seems significant, 'tode' being the German word for death. The similarity of this word to the amphibian 'toad' might easily have led to the confusion of the two meanings especially as the common toad was always thought of as a venomous beast and associated with witchcraft.[24]

Taking this reasoning further is perhaps the most ingenious explanation, quoted by Trevor Beebee:

> This word is clearly derived from the German 'tod' and 'stuhl,' meaning 'death-stool,' in reference to the poisonous nature of the same, and the supposed dangerous character of others, of the stalked fungi, and ... has nothing whatever to do with toads, as commonly supposed.[25]

It should come as no surprise that a creature so generally maligned as the toad has not made its mark as a fictional character of any great endearment; with one exception. Kenneth Grahame, who could find charm in almost every kind of creature, immortalized the toad in *The Wind in the Willows*.

> It's never the wrong time to call on Toad. Early or late he's always the same fellow. Always good-tempered, always glad to see you, always sorry when you go ... He is indeed the best of animals ... so simple, so good-natured, and so affectionate. Perhaps he's not very clever – we can't all be geniuses; and it may be that he is both boastful and conceited. But he has got some great qualities, has Toady.[26]

It was 'Toad of Toad Hall' who later became the eponymous hero of the story in A.A. Milne's stage adaptation of the book, first performed in 1930.

It is impossible to be sure that Taddiford in Hampshire really was 'toad ford' or that Tadmarton in Oxfordshire was 'toad mere' because, although *tadde* or *taddige* were the old English words for toad, they were also applied to the frog, and present-day residents might prefer the latter explanation. The shape and form of the toad, especially its mouth, have given rise to a number of derived names for other living things: the plants Toadflax and Toad-rush, the Toadfish and, occasionally, Toad-spit as an alternative to Cuckoo-spit for the secretion of the froghopper nymph (*Philaenus spumarius*; see page 72). But 'Toad in the Hole' for a sausage baked in batter? The name dates from 1787 but whoever first saw a resemblance in appearance or taste must have had either a warped sense of humour or highly defective faculties.

Natterjack *(Bufo calamita)*

OTHER COMMON NAMES: *Birkdale nightingale (Lancashire); golden-back; jar-bob (Hampshire); running toad; Thursley thrush (Surrey); walking toad.*

Natterjacks are smaller than Common Toads, very much more locally distributed and rather different in behaviour (the Natterjack is much more gregarious) and habitat. The two animals are superficially similar in colour but the Natterjack is most readily distinguished by the pronounced thin, yellow line down the centre of the back. It was not recognized as a distinct British species until the late eighteenth century, when it was found by Sir Joseph Banks near his family seat at Revesby Abbey in Lincolnshire, where it was known locally as the 'natter jack'. It had clearly been there for a long time and may well have contributed to much earlier toad lore and tradition by being confused with its larger relative.

The local name 'running toad' is appropriate as it runs in short bursts on its short legs, rather than crawling like a Common Toad or hopping like a frog. The name 'Natterjack', like its regional names 'Birkdale nightingale' and 'Thursley thrush', is a reference to its ratchet-like call, which was once very familiar throughout much of south and eastern England.

In Britain today, the Natterjack is mainly a coastal animal of East Anglia, Lancashire, Cumbria and southwest Scotland. A few occur in Hampshire, in some of the same places as the Sand Lizard and Smooth Snake, and there is also a small population in southwest Ireland. The scientific name *calamita* is taken from the Greek *kalamos*, 'a reed', and refers to the vegetation among which it lives. The Natterjack has declined greatly since its original discovery and has disappeared from many of its old habitats as sandy scrub and heathland everywhere has been built on or 'improved'. However, it has been sometimes been successfully reintroduced to former sites.

It is very much a burrowing beast that prefers to do its burrowing in sand and it is found more frequently where there are temporary pools of water rather than permanent ponds or lakes. Like the Common Toad, it tends to limit its aquatic activities to the breeding season; its spawn ropes can be distinguished by having only a single row of eggs.

The Natterjack Toad *(Bufo calamita)* has a distinctive yellow line down the centre of its back.

Newts
(Urodela, or Caudata)

SALAMANDERS AND NEWTS
(Salamandridae)

OTHER COMMON NAMES: *ask (Scotland, northern England); askel/aspel (Shropshire); asker (Lancashire); eft; evvet.*

One newt surpasses all others in making a literary mark: the newt whose eye contributed to the most celebrated of all witches' brews, the one that was concocted in the fourth act of Shakespeare's *Macbeth.* It was very probably a Palmate Newt *(Triturus helveticus)*, the commonest of the three species in Scotland today and quite possibly in the eleventh century as well.

Newts are tailed amphibians (Order Urodela), the only types of tailed amphibian in Britain, and thus they contrast with the tail-less frogs and toads. Elsewhere in the world there are creatures closely related to newts called salamanders; among these is the largest of all living amphibians, the Japanese Giant Salamander *(Andrias japonicus)*, which is over 1 metre (3¼ feet) long.

In general, all newts return to the water to breed and have a clearly developed aquatic adult stage, whereas salamanders do not, but there are exceptions and it is impossible to draw a hard-and-fast line between them. Newts are most usually seen when they are in water and are shy and secretive on land, which has led to the widely held belief that they are fully aquatic.

The three British species are closely related and, for much of the year, are not easily distinguished; the males in the breeding season are the easiest to identify. In general form and function, they are typical amphibians and, apart from their tails, differ little from frogs and toads. However, they characteristically lay their eggs singly, attached to water plants, and not in masses.

One very unusual and distinct characteristic displayed by newts, but not by frogs and toads, is neotony, in which the immature tadpole fails to metamorphose but instead continues to grow and develop to sexual maturity while still a larva. In some exotic species, most famously the Mexican Axolotl *(Ambystoma mexicanum)* and its relatives, the neotonous state is the norm but it is rare in British species. It occurs occasionally in the Smooth Newt *(Triturus vulgaris)* and can be induced artificially by aerating the water.

Historically, newts have often been confused with lizards because they have roughly the same overall shape but, biologically, they are very different. Nonetheless, while lizards readily part with their tails and grow new ones, newts have a remarkable ability to regenerate toes, or even complete legs, if

The Smooth Newt *(Triturus vulgaris)* is the most numerous of British newts.

they are lost or damaged. Occasionally, an incompletely severed limb stimulates the growth of a new one so that newts may be found with five legs or even more.

The name 'newt' has a comparable origin to 'Adder' (page 195), in being derived from the combination of a noun and an indefinite article. In Old English, the animal was an *ewt* but, by Late Middle English, an 'ewt' had charmingly become a 'newt'. The old name survives today in the regional name 'eft'.

Newts feed on a wide range of tiny animal life, such as zooplankton, freshwater shrimps, small molluscs and similar creatures, although the Great Crested Newt *(Triturus cristatus)* eats prey as large as water beetles, leeches and even small dragonflies. In turn, newts themselves are eaten by a range of predators, including large fish, water birds and Grass Snakes. The tadpoles are eaten extensively by fish.

Like frogs and toads, newts sometimes move in large numbers to and from their breeding ponds and this behaviour has long aroused interest and mystery. Also like frogs and toads, they hibernate in winter but, even now, little is known of their terrestrial behaviour because they tend to move only at night and spend the day hidden in damp places beneath stones and in holes. For no better reason, I suppose, than that they creep (when on land), newts have come in for a fair share of mistrust and association with evil. In 1590, Edmund Spenser described:

> ... These marishes and myrie bogs,
> In which the fearfull ewftes do build their bowres.
> (Edmund Spenser, *The Faerie Queen*)

There is a widely held belief in Ireland that relates to what are there called 'man-keepers', 'man-creepers' or similar names: creatures that will enter the mouth of a sleeping person and stay alive inside him until he dies. 'Man-keeping' has been attributed to lizards (see Viviparous Lizard, page 188) and to snakes but most frequently to newts, where it must apply to the Smooth Newt, the only species found in Ireland. I have been unable to determine its origin and there seems to be no equivalent belief in England, although, in some ways, it is paralleled by the idea that earwigs enter your ears when you are asleep. A related notion is that anyone who licks such a creature will have the power of curing burns and scalds.

I have never really fathomed the expression 'drunk as a newt' (or its more vulgar equivalent). The best explanation I have found is that it can be attributed to 'the somewhat tipsy-looking side-to-side movement of the newt's relatively big head, produced by its sigmoid swimming movements'.[27] I cannot say that I am convinced.

Smooth Newt *(Triturus vulgaris)*

OTHER COMMON NAMES: *common newt; ebbet; eft; evat; evvet; spotted newt.*

Probably the commonest newt, and certainly the most widely distributed, the Smooth Newt is no smoother than the Palmate Newt, which it so closely resembles. It is rarely more than 10 centimetres (4 inches) in length, and is olive-green-brown in overall colour, with whitish flanks and an orange underside, more intense in males than females. It is more or less spotted all over and, in the breeding season, the males develop a crest that superficially resembles that of the very much larger Great Crested Newt. However, the Smooth Newt's crest is lower and more rounded and there is often a hint of blue beneath the tail of the breeding male. In females, the spots often merge to form indefinite black lines and this can usefully distinguish Smooth Newts from Palmate Newts.

This is the commonest newt in garden ponds and can occur in large numbers. When I recently drained my smallest pond, which has a surface area of only 4 square metres (about 5 square yards), I rescued 26 individuals of varying sizes, indicating that they were breeding, despite the presence of Goldfish and Golden Orfe. Within a few weeks of refilling the pond, the newts were back; their homing instincts are strong and they have been known to travel a kilometre or more.

Smooth Newts occur in ditches and natural and artificial ponds throughout England, Wales and Ireland, less so in Scotland, and are absent from the Isle of Man and Jersey. Despite the numbers in my garden pond, Smooth Newts, like other newts, are more usually found where there are no fish to eat their larvae.

Palmate Newt *(Triturus helveticus)*

The Palmate Newt is so called because the males develop webbed feet in the breeding season. They also develop a crest but it is much lower and less pronounced than in the Smooth Newt. Otherwise they are similar, although the Palmate Newt rarely exceeds 9 centimetres (3½ inches) in length, making it the smallest British amphibian. Both sexes have a slightly pink and unspotted throat, which is a useful identifying feature, as is the filament at the end of the tail.

Palmate and Smooth Newts often occur in the same ponds and their overall behaviour and biology are very similar. Apart from being the commonest species in Scotland, and possibly therefore being Macbeth's newt, the Palmate Newt tends to be absent from much of the English Midlands and East Anglia. They do not occur in Ireland but are present on Jersey and on some of the Scottish islands.

Great Crested Newt *(Triturus cristatus)*

OTHER COMMON NAMES: *eft; great newt; warty newt.*

This is unquestionably the most splendid British amphibian, a really large animal, up to 16 centimetres (6 inches) long, dark brown or almost black overall, with a warty texture and spots that are hard to see against the dark background colour. The underside is bright orange, with black spots, and the female has an orange line running along the base of the tail. In the breeding season, the male develops a large, jagged crest, which, apart from its size, is very different from the much smoother crests of the other British species. In behaviour, they are typical newts, although they stay in the water longer after the breeding season than the other species.

Great Crested Newts are less common in garden ponds, preferring large expanses of weedy water with an absence of fish. Some ponds hold very large populations (one site was estimated to contain 30,000) but the species has nonetheless declined dramatically in recent years, presumably through loss of suitable habitats and pollution. It is fairly widely distributed throughout England but scarce in the north and west and in Scotland. I recently revisited three sites in Derbyshire where 40 years ago I knew I could find them: two of the ponds had gone completely and, in the third, I could find no amphibians of any sort. It is the only newt to have full protection under the Wildlife and Countryside Act 1981, something it shares with the Natterjack.

The terrestrial and hibernating behaviour of Great Crested Newts is still poorly understood, although it is believed that they will venture considerable distances underground.

> In the garden of the house we have recently bought [Suffolk] there seems to be a large breeding colony of them. Consequently I decided to build a wildlife area for them. During the digging, about eight or nine inches down beneath well compacted clay I found an old oak threshold from a long since disappeared outside toilet. I heaved it out, and there, in what seemed a sealed chamber, I found a mature and very plump newt, how did it get there, and, had it matured there? Later, while planting a small yew hedge across a well established lawn I came across another newt at least two feet down, also quite healthy and also in closely compacted soil. Since then, while digging the vegetable patch, I have unearthed several more, but closer to the surface.[28]

During the breeding season, the male of the Great Crested Newt *(Triturus cristatus)* develops a jagged crest and an orange underside.

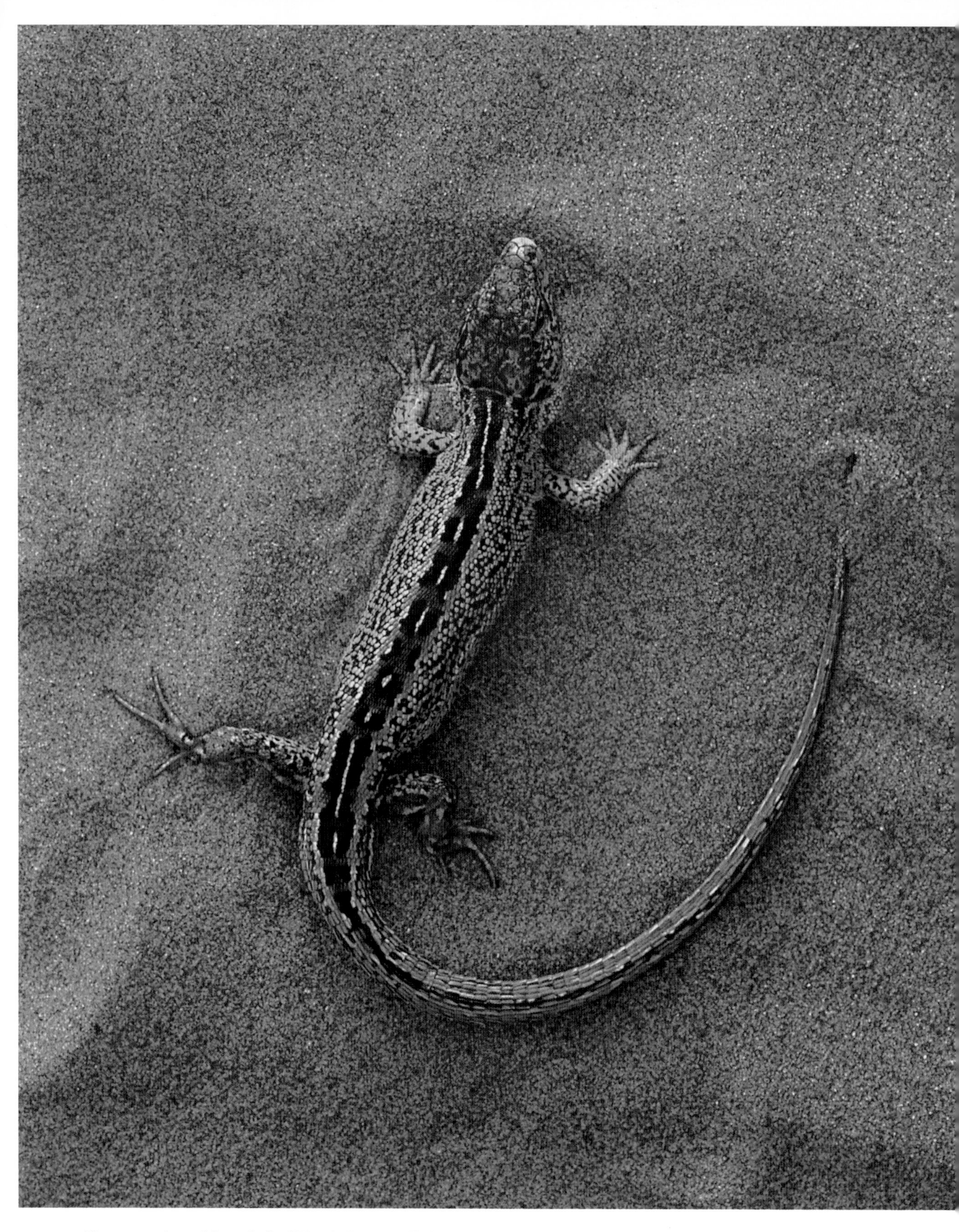

The green colour of the male Sand Lizard *(Lacerta agilis)* is much more vivid in the breeding season.

Reptiles (Reptilia)

Reptiles are cold-blooded vertebrates with dry skins. Most have overlapping scales but some have scutes (from the Latin for 'shield'), arranged edge to edge. Unlike amphibians, they do not need to enter water to breed and although some species, and indeed some groups, have adopted an aquatic habitat, they have close relatives that are entirely terrestrial.

Today, reptiles are relatively minor members of the global fauna, although one ancient group, the dinosaurs, was once supreme. The principal groups of modern reptiles are lizards, snakes, tortoises and their relatives, and a peculiar, distinct group, the worm lizards, or amphisbaenians.

Most reptiles lay eggs that hatch into perfect miniatures of the adults, although some appear to produce living young because the eggs have a thin membrane that ruptures as they are being laid. There are native or introduced members of all reptile groups in Britain except the amphisbaenians, although, as with amphibians, because they are mainly creatures of warm climates, the British species list is not long.

The name 'reptile' is late Middle English, originating in the Latin *repere*, meaning 'to creep'; in the perception of many people, this is their most outstanding characteristic.

Lizards

(Sauria)

> If on any warm day when you ramble around
> Among moss and dead leaves, you should happen to see
> A quick trembling thing dart and hide on the ground,
> And you search in the leaves, you would uncover me.
> (Thomas Hardy, 'The Lizard')

Like other reptiles, lizards feature in relatively minor ways in both the natural history and lore of Britain; the climate is simply too cold for them ever to have become sufficiently common. Of a world total of some 3,000 species, there are only three in Britain: two fairly typical, one not.

The typical lizard is carnivorous, has an elongated body with skin like fine leather (regrettably, some large species have been put to just that use), a long tail, eyes that have lids and, unlike those of snakes, can therefore blink, four legs, and an ability to run very fast. The Viviparous Lizard *(Lacerta vivipara)* and the Sand Lizard *(Lacerta agilis)* meet all these criteria. So does their close relative the large and beautiful European Green Lizard *(Lacerta viridis)*, which is native to Jersey. The Slow Worm *(Anguis fragilis)*, however, really is much slower and belongs to a small group of lizards that lack legs and are therefore commonly mistaken for snakes.

The most unexpected lizard attribute, and one that snakes certainly lack, is an ability to shed the tail, which they do when attacked. This curious facility arises because of the presence of fracture planes within some of the tail vertebrae. The tail will regrow, although never as perfectly as before, but this attribute more than anything has conferred some aura of mystery on what are otherwise relatively straightforward little animals.

Most people who live and garden in the country have probably seen a Slow Worm and possibly a Viviparous Lizard too, although Sand Lizards are rare and localized. Apart from Slow Worms, lizards in Britain, unlike those in warmer climates, tend to be uncommon in urban areas, partly because Blackbirds and domestic cats catch them. There is still a widespread confusion between lizards and newts and many people believe they have seen a lizard when, in truth, it has been a newt:

> We have lizards in our garden [Bromsgrove, Worcestershire] ... every evening when we went out with our torch to collect slugs we were lucky enough to see them on many occasions. A lady I work with has a farm a couple of miles down the road and she too has seen lizards ... we feel very privileged to have them in our garden.[1]

With the exception of geckoes, which do not occur in Britain, lizards are not nocturnal.

Lacertids

(Lacertidae)

Viviparous Lizard *(Lacerta vivipara)*

OTHER COMMON NAMES: *ask; awsk; common lizard; eft; four-legged cripple (Cornwall); furze evvet (New Forest); harriman (Shropshire); heather ask; henete; longscripple (Cornwall); maneater/mankeeper (Ireland); pajerpaw; stellion; swift (Cheshire).*

Viviparous Lizards are not truly viviparous in giving birth to live young, but, like Adders and many other reptiles, they appear to do so, the eggs hatching as they are deposited. The animal is very variable in colour but usually brown overall, with pale streaks and sometimes light and dark spots, although some animals are more olive-coloured or grey. The males are usually yellow or orange beneath. Melanistic individuals occur from time to time.

> Between 1964 and 1967 ... I spent much of my leisure time walking in the country around Cobhay Farm, in the parish of Bathealton near Wiveliscombe, Somerset ... I discovered that Cobhay was the epicentre of a large lizard population ... on any warm May day, I could hope to come across a dozen or so of these enterprising reptiles. I recorded in my diary that on one morning walk on 28 May 1967, I saw twenty seven lizards on a two mile stretch of roadside ... The ones that intrigued me most, instead of being mottled browns, had squamous coats as black as coal and during those years I estimate that I must have come across twenty such specimens, averaging at between five and ten per cent of the local population. Without an insight into the future, I set myself the task of trying to breed from a pair of these melanic specimens. They thrived on a diet of spiders and flies and produced seven ebony babies in the summer of sixty five ... by the May of 1970, I could no longer find even one lizard sunning itself on the roadside walls or the speedwell banks.[2]

The Viviparous Lizard is more slender and smaller overall than the Sand Lizard and, in Britain, the two are most simply distinguished by the Sand Lizard's much more restricted range. The Viviparous Lizard occurs in sunny, undisturbed habitats of many kinds throughout Britain and it is the only lizard in Ireland. It is found on some of the Scottish Islands but, like all other terrestrial reptiles, is absent from Orkney and Shetland.

It tends to be more common in coastal areas (sometimes on the sea shore itself) and there are large swathes of central England where lizards are rarely seen; it was one of my regrets, when growing up in Derbyshire, that we lived in an area particularly low in numbers. Outside Britain, the Viviparous Lizard is one of the most widely distributed

The Viviparous Lizard *(Lacerta vivipara)* prefers quiet, sunny habitats and tends to be more common in coastal areas.

vertebrates in the world, occurring across Europe and Asia; it also lives further north than any other species of reptile in the world.

Viviparous Lizards eat a wide range of invertebrate prey but are especially partial to spiders. The lizards in turn are eaten by a wide range of predators, particularly Hedgehogs, Weasels, Adders, Smooth Snakes, Buzzards, Kestrels and, once upon a time, Red-backed Shrikes, which impaled them on their 'larders' (see Shrikes, page 356).

The lizards tend to begin hibernation fairly early, sometimes in late summer, when there appears to be plenty of food still available. Little is known of their hibernation sites but they probably usually enter small underground holes, although I have found them in small groups beneath log piles. The numerous ancient accounts of lizards (and of other hibernating creatures such as toads and bats) being found entombed within solid trees and rocks merely emphasize how torpid the hibernating animal becomes. Perhaps it was the discovery of lizards in such places that led to the unwholesome practice of placing a lizard (albeit usually a dead one) in the foundations of a new field wall.

The confusion that long existed between lizards and newts is evident from the names 'eft' and 'evvet' being used at different times for both. There are several explanations for the odd name 'maneater' (or sometimes 'mankeeper'), which exists in Ireland and must therefore refer to either the Viviparous Lizard or the Smooth Newt (see Newts, page 183). It alludes in part to a widely quoted and presumably therefore genuine belief that physical contact with a lizard, preferably by licking it (or vice versa), means you will never suffer from burns and can heal someone who does by your touch. Frankly, I cannot see the connection. The name 'lizard' itself is much more straightforward; it is a Middle English word, originating in Old French.

Sand Lizard *(Lacerta agilis)*

This is a rare and lovely lizard, the larger and more beautifully marked of the two British species of *Lacerta.* With its bright green flanks and brown-striped or brown-spotted back, the male is especially handsome, the more so in the breeding season when the green colour intensifies. The female is an overall pale brown but, as with the Viviparous Lizard, the colours and patterns are very variable.

They emerge in spring from their hibernating quarters beneath the sand but it may be some time before they begin feeding on a wide range of invertebrates. The courtship display and rituals are long and involved, with much running around, arching of the body and threatening attitudes between males. Eggs are laid in the sand during the summer.

Sand Lizards have a very restricted distribution in Britain, being more or less confined to sandy heaths in Dorset and Surrey, where they form the most important prey of the similarly rare Smooth Snake, and on the coastal sand dunes near Southport on Merseyside.

As a boy in Dorset, I could always find Sand Lizards
on the heaps of gravel swept up by road workers.[3]

The Sand Lizard was once much more widespread and, in recent years, some success has been obtained with artificial introductions to apparently suitable habitats close to its former territories

'Sand Lizard' is a uniquely British name; elsewhere in Europe it often occurs in high mountain grasslands.

The beautifully marked Sand Lizard *(Lacerta agilis)* is now a rare and local British species.

Slow Worms and Glass Lizards
(Anguidae)

Slow Worm *(Anguis fragilis)*

OTHER COMMON NAMES: *blind worm; deaf-adder.*

The Slow Worm is unlikely to be mistaken for any other lizard but, with its elongated, brown, legless body, it is commonly mistaken for a snake. It therefore inspires the same illogical behaviour in people and is often killed. As I have indicated on page 188, the simplest way to distinguish a Slow Worm from a snake is the presence of eyelids; although quite why this should have given rise to the very widely used name 'blind worm' has always mystified me. By comparison with a real worm, the Slow Worm has extremely obvious eyes but, nonetheless, an old rhyme is found in many parts of the country:

If the Slow-worm could see and the Viper could hear,
Then England from Serpents would never be clear.[4]

However, it is the similarity to snakes that has really coloured the Slow Worm's reputation; Shakespeare was far from alone in thinking them not only serpents but poisonous ones at that. The witches in *Macbeth* added 'blind-worm's sting' to their brew and the fairies implored them to keep away from Titania:

You spotted snakes with double tongue,
Thorny hedge-hogs, be not seen;
Newts, and blind-worms, do no wrong,
Come not near our fairy queen.
(*A Midsummer Night's Dream*, Act 2, Scene 2)

For Robert Herrick, it was simply yet another declaration of love to ask that:

No will-o'-th'-wispe mis-light thee;
Nor snake or slow-worme bite thee.
(Robert Herrick, 'The Night Piece, to Julia')

But Walter Scott seems to have confused it with a slug:

... there the slow blindworm left his slime
On the fleet limbs that mocked at time.
(Sir Walter Scott, *The Bridal of Triermain*)

Slow Worms are probably the most familiar of British reptiles today because they are commonly found in gardens where they perform the invaluable role of dining on slugs. They occur throughout mainland Britain and also on most of the islands but are absent from Ireland. Slow Worms are more tolerant than any other native reptile of the presence and actions of man. Unlike other British lizards and snakes, they do not

The Slow Worm *(Anguis fragilis)* may look like a snake but is really a legless lizard. It will shed its tail to avoid capture.

bask in the sun but share with amphibians a fondness for cool, damp places. When handled carefully, they wriggle violently for a while then settle down. However, if seized quickly or roughly, the origin of the scientific name *fragilis* becomes apparent because they very rapidly part company with their tails, which remain to continue wriggling.

The young are born alive by ovovivipary in late summer and baby Slow Worms are remarkable little things, with a dark stripe and a metallic iridescence that resembles the colour of a young eel as much as anything. They are thought to be the longest-lived of all lizards; the remarkable age of 54 years has been reliably recorded.

Snakes
(Serpentes)

Within recorded history, there have been only three British species of snake; all are still present, one of them, the Smooth Snake *(Coronella austriaca)*, albeit only just. All have a more restricted distribution now than in the past, although the very severe decline in numbers brought about by the spread of intensive agriculture in the 1950s and 1960s seems to have slowed and modern habitat conservation work is now paying dividends. Snakes in Britain have never been so common as to be encountered every day but the Grass Snake *(Natrix natrix)* and Adder *(Vipera berus)* have long been numerous enough for most people to see them during the course of a lifetime in the countryside. Nonetheless, it remains true that the majority of town-dwellers in Britain have probably never seen a wild snake.

The name 'snake' is Old English and, over the centuries, it gradually came to be applied to things with snake-like habits or actions. The notion that snakes were sly and untrustworthy, presumably because no-one could ever really tell what they would do next (their mode of moving certainly gives no clues), resulted in untrustworthy and sly people being known as 'snakes' from the late sixteenth century. Even today, 'snake in the grass' is a fairly serious expression of opprobrium. The verb 'to snake', meaning 'to move in a snake-like, twisting manner', however, is much more recent and nineteenth century.

While there is a large amount of classical snake folklore therefore, the indigenous beliefs, language and literature on snakes is relatively impoverished. No snakes occur in Ireland, the popular and fanciful belief being that they were banished from there by St Patrick.

His exorcism of the venomous reptiles has a similar
significance as the dragon stories of the East,

St Patrick is credited with banishing snakes from Ireland and is therefore often depicted with a serpent at his feet.

symbolising the conquest of good over evil, the triumph of Christianity over paganism. In Christian Art he is always represented with a serpent at his feet.[5]

Snakes are legless reptiles, although some non-British species do retain vestigial external legs in the form of small claws. They differ most obviously from the other British legless reptile, the Slow Worm, in having no lids to the eyes, which therefore never close (snakes do not blink), and in being coloured other than uniformly grey-brown. When adult, they are also much larger than Slow Worms.

Snakes are remarkable, fascinating and almost hypnotically compelling creatures. Even otherwise learned and rational people have been seduced by an interest in snakes into making fatuous remarks. The antiquary John Aubrey (1626–97) was educated at Trinity College, Oxford, entered Middle Temple, excavated Avebury and was among the earliest Fellows of the Royal Society. Yet in his work *Remaines of Gentilisme and Judaisme* (1686–7), there is the following pearl:

> Take on Mid-summer night at xii, when all the planets are above the earth, a serpent, and kill him, and skinne him, and dry him in the shade, and bring it to a powder. Hold it in your hand and you will be invisible.[6]

Deer were long thought to eat snakes during the summer months, thus rendering their flesh inedible, although surely no-one can ever have seen a deer do such a thing. However, deer, and goats, will trample on snakes. It was also widely believed that, no matter how hard anyone tried, a snake could not be killed before sunset. Presumably, no-one ever told this to Harry 'Brusher' Mills, most famous of an extraordinary group of nineteenth-century individuals who earned their living by catching snakes in the New Forest.

Harry Mills was born near Lyndhurst in 1840 and apparently gained the nickname 'Brusher' either because he used to brush the wickets of local cricket clubs between innings or, more probably, because one hard winter, he brushed the ice for the skaters on Foxlease Park lake, close to where he then lived. He realized that a living could be made by catching the snakes of the Forest (all three species) and selling them either to London Zoo, at 1s (5p) each, where they were fed to other snakes, Secretary Birds and the like, or for the production of Adder-fat ointment, which was believed to cure snake bites. In his forties, he took up a solitary existence in a small hut, deep in the forest near Brockenhurst, where he remained until it was destroyed by vandals shortly before his death from a heart attack in July 1905. In his time, he is thought to have caught over 20,000 snakes, simply using a forked stick to pin them down before he picked them up to slip into his sack.

Was he ever bitten? It seems highly probable and it has been suggested that the very frequency of the bites gave him some immunity, or that he was protected by the effects of the rum that he drank at the old Railway Inn in Brockenhurst, now renamed The Snake Catcher in his memory. An elaborate marble headstone with a charming picture of him at his task marks his grave in the beautiful old churchyard of St Nicholas in Brockenhurst. 'Brusher' Mills was succeeded as New Forest snake-catcher by George Wateridge (1869–1948), who kept his snakes in a dustbin at the end of his garden and periodically despatched them by train to London from Lyndhurst Road station. After Wateridge's death, the old tools of 'Brusher' Mills then passed to Douglas Bessant, who kept up the craft for some time but then emigrated, his departure marking the end of New Forest snake-catching.

As snakes grow, they slough, or shed, their old skin, which peels away as a thin, papery, translucent cast of the animal's body. It is hardly surprising that this curious object has not only attracted attention but also has been endowed with some extraordinary powers. There is an old and very widespread belief that it would cure a headache and countless country-dwellers with hangovers and similar afflictions have accordingly stuck an old snakeskin inside their hats. There was also a widely held view that:

> ... the skin cast by a snake is very useful in extracting thorns, &c., from the body, but unlike other remedies,

> it is repellent, not attractive; hence it must always be applied on the opposite side to which the thorn entered. In some cases where the skin had been applied on the same side, it has forced the thorn completely through the hand.[7]

In some districts, snakeskin was hung around the chimney of a house to bring good luck but how this belief could have originated is anyone's guess.

Snakestones feature frequently in ancient British snake lore. These were stones endowed with magical and medicinal properties; the cure of snake bite being one of them. There is not much consistency, however, in the prerequisites of a snakestone: they were variously small stones with holes in them, glass-like pebbles, pieces of coral – almost anything it seems – that was curiously and mysteriously shaped. They were believed in some way to have been produced by the snakes themselves.

Perhaps the most understandable of the old snake beliefs related to those snakestones that really looked the part. Before the development and appreciation of palaeontology as a scientific discipline in the nineteenth century, and later its role in endorsing Darwinian evolutionary theory, people quite understandably believed that fossil ammonites were snakes turned to stone. There were two main centres of the belief in England. The first was Whitby, where the local Blue Lias beds abound with Jurassic ammonites. Heads were commonly carved onto the fossils by the quarrymen, who sold them as charms, and they were often worn as amulets. The underlying belief was that the local Saint, Hild (or Hilda; AD 614–680) had turned the snakes to stone when she founded her abbey in AD 675.

'Brusher' Mills, the famous New Forest snake-catcher, outside the charcoal-burner's hut, near Brockenhurst, where he lived.

> When Whitby's nuns exulting told ...
> ... of thousand snakes, each one
> Was changed into a coil of stone,
> When Holy Hilda pray'd;
> Themselves, within their holy bound,
> Their stony folds had often found.
> (Sir Walter Scott, *Marmion*)

As Hilda of Whitby was one of the most influential women of her time, it seems appropriate that the women's college founded at Oxford in 1893 should have taken her name. Today, St Hilda's is the only Oxford college still restricted to women and it bears three ammonites on its coat of arms (and one on its tableware), although they are officially described as 'coiled serpents argent'. The magazine for the College old girls is, nonetheless, called *The Ammonite.*

In fairness, I should add that an alternative version, told in Whitby itself, is that Hilda put a stop to a plague not of snakes but of snails by commanding them to move over the cliffs, fall onto the beaches below and then turn into ammonites. Hild, at least, was real enough, the daughter of King Oswy of Northumbria, but her West Country counterpart in the turning-snakes-to-fossils business, the 'fifth-century virgin' St Keyna (after whom the town of Keynsham is sometimes erroneously said to have been named) seems to have been purely the figment of someone's imagination.[8]

Britain has no tradition of eating snakes but, inevitably, many found their way into folk medicine in a wide variety of recipes; and not simply as a cure for snake bite itself.

> You may also finely mince the heads and tails of Serpents, and feed therewith Chickens or Geese, being mingled with crums of Bread or Oates, and these Geese or Chickens being eaten, they help all to take away the Leprosie, and other foulness in Mans body ... The bloud of a Serpent is more precious that Balsamum, and if you anoint your lips with a little of it, they will look passing red: and if the face be anointed therewith, it will receive no spot or fleck, but causeth to have an orient or beautifull hew. It represseth all scabbiness of the body, stinking in the teeth gums, if they be therewith anointed.[9]

Typical snakes
(Colubridae)

Grass Snake *(Natrix natrix)*

OTHER COMMON NAMES: *collared snake; common snake; green snake; hedge snake; ringed snake; water snake.*

Any snake found wild in Britain and measuring longer than about 60 or 70 centimetres (24 or 28 inches) will be a Grass Snake; any longer than about 90 centimetres (36 inches) will be a female Grass Snake; the record for Britain was one over 1.8 metres (6 feet) in length.

> During a long career in forestry I spent many years in the New Forest. One day, in the early 50s, a colleague and I were measuring felled oak timber. The process involved the use of a girthing tape and a timber 'sword' – a flexible metal strip with a hook to engage in the buckle of the tape. As I passed the sword under the fallen log there was a very audible hiss as an enormous grass snake emerged. Seeing its remarkable length – and having the equipment available, we managed to gently seize the reptile, and with some difficulty, hold it down on top of the log for accurate measurement. It was 4 feet 8 inches (Give or take a few wriggles). When released it took off unharmed into the surrounding bracken in a few seconds, to the relief of my dog who was barking defiance at a safe distance.[10]

The Grass Snake is a large and very handsome creature, although rather variable in colour. Generally a lovely olive-green above, it can also tend to brown or grey but almost always has a characteristic collar (the 'ring' of its alternative name), which usually comprises two crescent-shaped yellow marks behind the head with a black blotch to their rear. In older animals, the ring sometimes fades and can be paler yellow, orange or even reddish; I was as delighted as our cat was puzzled a short while ago when he presented me with an exquisitely orange-ringed and rather annoyed female Grass Snake that he had caught in the garden. The Grass Snake usually has vertical black streaks or spots along the flanks but, unlike Adders, melanistic or albino Grass Snakes are rare in Britain. Apart from the occasional introduced or escaped animals, Grass Snakes do not occur in Scotland.

Like others in their genus, Grass Snakes are typical water snakes and are rarely found far from places where their preferred prey of frogs, newts and toads occur (snakes are not affected by toad venom). They are excellent swimmers. I often find Grass Snakes in my garden pond but I think the most ambitious specimen I ever saw was one making a very fast crossing of the Thames in Oxford during Eights Week as a

The Grass Snake *(Natrix natrix)* is a highly capable swimmer and seldom found far from water.

flotilla of lustily propelled rowing eights bore down on it. As befits an aquatic creature, they will also eat fish and, in winter, when Minnows were scarce, I used to spend my pocket money on Goldfish to keep my Grass Snakes going through the winter; captive snakes in Britain are best not allowed to hibernate as they rarely survive. In the wild, however, they hibernate from approximately October to March; compost heaps and leafmould piles are good places for them and care should be taken when disturbing garden compost in the winter. The eggs are laid among rotting vegetation in summer.

When caught, Grass Snakes may very convincingly feign death. This is often said to be a fairly uncommon reaction but, in my experience, most do it. If this fails, and if you pick them up, they almost invariably hiss loudly, may try to bite and may squirt a rather unpleasant, fishy-smelling liquid from the cloaca. That said, they soon quieten down and can become remarkably tame. Grass Snakes are eaten by a number of birds, including herons, and also by Badgers, Hedgehogs and other predators.

Smooth Snake *(Coronella austriaca)*

Always much the rarest and most locally distributed British snake, the Smooth Snake is now probably confined to the New Forest, parts of Dorset and northern Surrey, where there may be no more than about 2,000 animals in total. It was long dismissed as a variant of the Grass Snake and was only recognized as a British species by Dr J.E. Gray, who, in 1859, described a specimen that had been caught near the flagstaff in Bournemouth and sent to him by a local resident.

> The species, however, had been caught 6 years earlier by Mr F. Bond between Ringwood and Wimborne in the New Forest, and ... was put into a bottle and forgotten ... .[11]

At that time, it was fairly widespread in the south of England but has since vanished from its old haunts in East Sussex, Berkshire and Wiltshire.

Its characteristic habitat is essentially the same as that of the Sand Lizard, which is its preferred prey, although its distribution is even more restricted. Superficially, it could be mistaken for an Adder, being of similar length, but its markings never assume the diamond or criss-cross pattern of Adders. It differs from the Grass Snake in seldom being more than about 70 centimetres (28 inches) in length and lacking the coloured collar. Overall it is grey or brown, sometimes with a pinkish hue and usually with black spots on the back and a dark line through the eye.

In addition to Sand Lizards, it will eat other lizards, small rodents and also other snakes, including young Adders. It is unique among British snakes in using one or two coils to restrain its prey, although it is not a true constrictor like a python.

It typically hides during the day in some warm shelter and your best chance of seeing a Smooth Snake is probably by visiting a sandy heath in central Dorset on a warm summer's day and carefully lifting any piece of debris that you find; and then replacing it just as carefully because it is illegal knowingly to disturb protected animals.

Although not a true constrictor, the Smooth Snake *(Coronella austriaca)* restrains its prey with a couple of coils of its body.

> I recently found a Smooth Snake on a heath close to Wareham after lifting a sheet of corrugated iron and was able to video it. They are extremely smooth to feel, quite unlike a Grass Snake; more like a Slow Worm.[12]

It is a fairly slow-moving, very shy and secretive creature. Relatively little is known of its behaviour in Britain although, like the Grass Snake, it hibernates, almost certainly below ground. Like several other reptiles, Smooth Snakes apparently give birth to living young, the membranous eggs rupturing immediately they leave the mother's body. Because of its rarity and secretive habits, little is known of its natural enemies in Britain but some birds of prey undoubtedly account for a quantity.

Vipers
(Viperidae)

Adder *(Vipera berus)*

OTHER COMMON NAMES: *ether (Worcestershire); hog-worm; viper.*

The Adder has always had a special place in the lives, beliefs and traditions of Britain, not just because it is the only poisonous species of snake in the country, but because it is probably the only wild British animal that can individually inflict death on a normal adult human. Nonetheless, this seldom happens. A recent study showed that, although an average of about 90 people are bitten by Adders in Britain each year, there has been no death for over 20 years. The only really vulnerable people are the very young, the very old and those already in poor health. Even then, they need a 'full-frontal' bite, with both fangs injecting venom; the study suggested that about 70 per cent of Adder bites are 'dry', with less than the maximum volume of venom being injected. The venom is intended to kill the animals on which the Adder feeds and it will certainly dispose of a lizard in under a minute, affecting the heart via the bloodstream. It was long thought that the animal's flicking, forked tongue was the source of the poison:

> She hath ... struck me with her tongue,
> Most serpent-like, upon the very heart.
> (*King Lear*, Act 2, Scene 4)

The treatment for an Adder bite (still annoyingly often referred to as a sting) is to obtain medical assistance as soon as possible and most certainly not to try to suck out the venom or apply a tourniquet. Only in serious cases will an antivenin will be used. However, in times past, things were different. Among the traditional cures for an Adder bite was the fat from (presumably) another Adder that had recently been deep-fried; not something that most people tend to have available at a moment's notice. In Scotland, pieces of live pigeon were recommended (see Wood Pigeon, page 297).

Folk remedies are legion; countless herbal brews and incantations have been used; and apparently used more than once, no doubt simply because Adder bites are rarely fatal and the local medicine man or woman, being the only person who knew the reality, could be fairly certain of emerging with their reputation enhanced. The 'bag of heads', literally a bag containing the heads of an Adder, a toad and a newt, was still being used as a general aid to healing as recently as the late nineteenth or early twentieth century in wilder places. The bag was dipped in water and the water that ran out when it was lifted up was applied to the affected part.

The name 'Adder', like 'newt', is one of those delightful words that is a compound of a noun and an indefinite article. It originated in the Old English *naedre*, 'a snake' or 'creeping thing', the 'n' being lost in Middle English when a *naedre* became an 'adder'. 'Viper' is sixteenth century, from Old English and then Latin; and perhaps originally from *vivus parere*, 'to bring forth alive', an allusion to the ovoviviparous habit.

There is an old and peculiar link between snakes, especially Adders, and trees, particularly ash trees but sometimes hazel as well. The snakes were believed to avoid and dread them. Striking an Adder with an ash branch was thought to kill it outright and a West Country cure for snake bite was not simply (and foolishly) to suck and spit, but also to intone a rhyme: 'Ashing tree, ashing tree, take this bite away from me.' This belief led to the rather dangerous notion that, if you walk through a snake-infested area carrying an ash branch, you are protected. It is, I suppose, true that the Adder's natural habitat is not especially the same as that of the ash but I cannot believe that no-one has ever seen an Adder in an ash wood; and Adders certainly occur in plenty of places where hazel is to be found.

The legend of St Leonard slaying a dragon in the forest near Horsham that now bears his name – and silencing the Nightingales in consequence (see Nightingale, page 337) – was one of several snake stories whose telling was elaborated over the years. A very convincing and plausible explanation is that these dragon myths were encouraged by landowners seeking to deter intruders: an old version of the modern 'Danger – Adders' notice which is a far better deterrent than any 'Keep Out' sign.

Once people began to take an interest in wild animals more for their own sake than in terms of the benefit or harm they could bring, natural history was born. Nonetheless, this did not immediately result in anything profound or sensible. The belief that female Adders swallow their young is a classic instance of a little observation being a dangerous thing. It was first recorded in 1577 in *The Description of England* by William Harrison, published jointly with that favourite chronicler of Shakespeare's, Raphael Holinshed. It is a charming story:

> I did see an Adder once myselfe that laie (as I
> thought) sleaping on a moulehill, out of whose mouth
> came eleven young adders of twelve or thirteen inches
> in length apiece, which plaied to and fro in the grass
> one wyth another, tyll some of them espyed me. So
> soone therefore as they sawe me they ran againe into
> the mouth of theyr damme whome I killed, and then
> found eache of them shrowded in a distinct celle or
> pannicle in hyr belly ... .[13]

There is no scientific evidence for this whatsoever and, moreover, they would not fit. The belief seems to have arisen because people have seen baby Adders wriggle beneath their mother's body for protection.

There is also a notion that Adders are deaf. 'Deaf as an Adder' is an expression used almost as regularly as 'deaf as a post' to describe the hard of hearing. In a more elaborate version, the Adder lays one ear to the ground, using its tail to close the other ear. It seems to stem from the Bible and the Psalms of David, 'the deaf adder stoppeth her ears, and will not heed the voice of the charmer', but where this idea came from is anyone's guess. In any event, *Vipera berus* does not occur in the Holy Land so it must have been another species.

Adders are relatively short, stout snakes, identifiable by their distinctive coloration. Overall, they are usually up to around 65 centimetres (26 inches) in length and have a zigzag black pattern down the back, although this occasionally appears simply as a single broad stripe. Black adders are surprisingly common and tend to be especially so in certain localities. (The surname

A male Adder *(Vipera berus)* basking on a rock, its characteristic zig-zag pattern very obvious.

Blackadder, made famous in recent times by the popular comedy television series of that name, seems to be derived from the River Blackadder in Berwickshire, which may or may not originally have had anything to do with snakes. A River Whiteadder is nearby.) Like other British snakes, Adders hibernate, emerging from underground in March, by which time their preferred food, the Viviparous Lizard, has also emerged.

Adders occur in a wider variety of habitat than any other British snake and may be found in coastal sand dunes, on chalk hills, moorland, woodlands, almost anywhere that is relatively sunny and undisturbed. Their intolerance of disturbance means that they are most unlikely to be encountered in gardens or close to human habitation. The Adder is also the only snake to occur in Scotland; it has a wide distribution and is one of the very few reptiles to occur naturally north of the Arctic Circle.

They are surprisingly good swimmers. Apart from human beings, who seem bent on destroying them on sight, the Adder has few natural enemies, although they are eaten by some birds of prey and also by Hedgehogs, although it is arguable whether they are immune from the venom, as is sometimes suggested.

> A friend of mine was gutting a Pheasant and found inside an Adder. She washed it and placed it in cling film in the freezer.[14]

While Adders must always have attracted a considerable degree of human attention because of their poisonous nature, they also have a peculiar habit that has added to the aura of mystery: Adders dance, or at least they appear to do so. Among persons privileged enough to have seen the performance will undoubtedly have been those who have not only wanted to spread the story but also to elaborate it. As with much other peculiar animal behaviour, this is a courtship ritual, and a reasonably common one among this group of snakes world-wide. Two or more male Adders face each other and half raise their bodies from the ground in attempts to push their rival away. As their bodies twist and intertwine, they do indeed appear to be dancing.

> My 21 years in the [New] Forest were full of memorable, vivid incidents whilst walking quietly about my duties ... A curious gathering of adders (some six or eight) which involved a slow dancing movement in a circle. Probably a mating prelude.[15]

Tortoises, Terrapins and Turtles

(Chelonia)

Tortoises and their allies, characterized by their hard shell, or carapace, are familiar enough to everyone. They are evolutionarily a very ancient group: superficially identical animals coexisted with the dinosaurs and the fact that there are around 250 living species is testimony to a very practical and enduring design.

In Britain, a 'tortoise' is a mainly vegetarian, terrestrial animal, a 'terrapin' a mainly carnivorous freshwater creature, and a 'turtle' a mainly carnivorous marine one. Strangely, the name 'tortoise' has nothing to do with the characteristic shell but comes from the Latin *tortus*, meaning 'twisted', from which the word 'torture' is also derived. It refers to the relatively obscure fact that the southern European tortoises, which would have been familiar to the Romans, have twisted feet. 'Terrapin', although it has a different meaning in North America from Britain, is nonetheless a North American word, originating, like 'musquash' (see Muskrat, page 469) from one of the languages of the Algonquin people. 'Turtle' is thought to have been a sailor's corruption of the French word *tortue*, which was also derived from *tortus*.

Although there are several European species, no tortoise or terrapin has been native to Britain within recorded history. The European Pond Terrapin (*Emys orbicularis*, often called the European Pond Tortoise) was the last, surviving after the final ice retreated but disappearing around 5,000 years ago. Consequently, while tortoises feature rather extensively in classical lore (Aesop, for instance, played on the tortoise's legendary slowness), they have played no part in British traditions.

Nonetheless, tortoises especially are very familiar as pets and, when I was young, every year, street-corner petshops had for sale numerous specimens of the three Mediterranean species, mainly the Spur-thighed or Greek Tortoise *(Testudo graeca)*, with rather smaller numbers of Hermann's Tortoise *(Testudo hermanni)* and occasionally the Marginated Tortoise *(Testudo marginata)*. Sometimes, the European Pond Terrapin would be offered too.

The keeping of a pet tortoise was a long-established practice and part of every child's growing up. Gilbert White, in his *Naturalist's Journal*, refers to a tortoise called Timothy, and his aunt and uncle, Henry and Rebecca Snooke, certainly had one:

> Mrs Snooke's tortoise, after it had been buried more than a month, came forth, & wandered round the garden in a disconsolate state, not knowing where to fix on a spot for it's retreat.[16]

To supply the pet trade, countless thousands of tortoises were brought to Britain every year, often as ballast in ships, and countless thousands died, most failing to survive hibernation in the first winter. Today, the trade in tortoises is rightly very severely restricted.

In recent times, tortoises and terrapins have only occasionally survived for short periods in Britain and any seen for sale in pet shops are likely to be captive-bred, and very expensive. While forty years ago, I could buy a Greek Tortoise for half a crown (12½p), I was recently quoted £240 for one. Some concern has been expressed about the American Red-eared Terrapin *(Trachemys scripta elegans)*, of which at least one small group already exists and which might just manage to spread. The North American Snapping Turtle *(Chelydra serpentina)* also appears from time to time in reservoirs and waterways (even in the Serpentine), presumably liberated after outgrowing someone's aquarium. It could almost certainly breed because it occurs naturally as far north as Canada, So far, however, none have ever established self-sustaining populations, although, from time to time, stories have appeared in the press of someone's pet turning up again after several years absence.

Turtles, however, are another matter. One of the highlights of my life, as it must have been for many other people, was the first time I watched turtles laying their eggs on a tropical beach at night. Through the medium of television, it is now a very familiar activity, yet it remains extraordinary and moving. However, because turtles are popularly associated with tropical beaches and seas, the notion that they might be part of the British fauna is unexpected. Certainly all seven species of marine turtle breed only on tropical or warm temperate shores (two at least in the Mediterranean), but four or five species have been seen in the North Atlantic and have sometimes been stranded on the British coast.

There were around 600 recorded strandings of marine turtles on British shores in the twentieth century. Around 20 per cent were of the Loggerhead *(Caretta caretta*; despite the name, its head is really no more substantial than that of the other species), which is the commonest turtle in the Mediterranean. There were fewer numbers of the smaller, extremely rare and oddly named Kemp's Ridley *(Lepidochelys kempii)*, which breeds almost entirely on only one beach in Mexico. Identified by an American called Kemp, this is a slightly different form of the Ridley Turtle, which was intriguingly named not after a zoologist but after a distinguished English botanist Henry Nicholas Ridley, who played a major part in introducing rubber to Malaya. On very rare occasions, two other turtles have been found: the Green Turtle *(Chelonia mydas)* and the Hawksbill *(Eretmochelys imbricata)*, with its beak like a bird of prey and overlapping horny plates, which were the source of tortoiseshell, once used extensively for decorative goods.

However, the majority of strandings are of a species justifiably considered British, the Leathery Turtle *(Dermochelys imbricata)*, also known as the 'leatherback', or 'luth'. This remarkable beast is the largest of all sea turtles and quite unlike any other in appearance. Instead of a carapace of interlocking horny plates, it has a mass of small, bony platelets embedded in its thick skin and very prominent ridges running the length of the back. The name 'Leathery Turtle' appropriately describes the appearance. The name 'luth' originates in an old Arabic word for the lute; the appearance of the carapace bears a remarkable resemblance to the underside of the musical instrument.

Although Leathery Turtles breed on warm shores, they have long been regularly seen in the North Atlantic, where they pursue their very specific food source: large species of jellyfish. Uniquely among modern reptiles, they are able to generate some of their own body heat and can maintain a temperature of around 18°C (64°F) above their surroundings. This means that, in summer, they can routinely follow their food into the cold waters off the coasts of northern Europe. Ten Leathery Turtles have been seen in Shetland since 1955 and sightings and strandings further south are almost routine. For this reason:

> ... there is no doubt that this animal merits recognition as a true member of the British fauna, every bit as much as migrant birds or butterflies.[17]

Much the most celebrated stranding took place in 1988, when considerable numbers were seen on the beaches of Wales and southwestern England. As commonly happens, some were caught in fishermen's lines. One suffocated after suffering such a fate off the Welsh coast and was washed ashore on 23 September on Harlech Beach, Merionethshire. It was a male, probably of very great age, 2.88 metres (9½ feet) long, 2.74 metres (9 feet) across the flippers and weighing around 916 kilograms (212 lb); it proved to be the largest marine turtle ever recorded anywhere. It now takes pride of place, mounted and slightly shrunken, in an exhibit at the National Museum of Wales in Cardiff.

Mock Turtle soup provided the inspiration for Lewis Carroll's Mock Turtle, shown here with the Gryphon and Alice.

Although relatively few people in Britain have seen turtles, a great many more have tasted them, at least in times past. Among the first to be blamed for this must be Hannah Glasse who, in *The Art of Cookery Made Plain and Easy* (1747), gave a recipe for a soup that included, not the turtle meat (which was eaten separately), but the mass of gelatinous matter from close to the upper and lower shells.[18] Thus the infamous Turtle Soup of grand banquets was born. Over the years, the Green Turtle is the species that has been slaughtered most commonly to supply the basis for this concoction. The addition of sherry to turtle soup is said to have been the invention of the American Mary Leiter, wife of George Nathaniel Curzon, Viceroy of India ('No gentleman has soup at luncheon'). She was arranging a grand dinner (not a lunch) in 1905 that included an American guest who could not drink alcohol. The remaining guests, being English, could not possibly be served dinner without alcohol, so the compromise was to add alcohol to the soup.

However, even in the eighteenth century, turtle soup was costly and so a variant began to appear: the less expensive (and more zoologically acceptable) Mock Turtle Soup, which was based on calf's head or ox-tail. I once tasted the real thing and found it indistinguishable from the substitute, which is yet further evidence, if it be needed, of the inexcusable nature of the slaughter. Lewis Carroll, in *Alice's Adventures in Wonderland*, most memorably brought the Mock Turtle to life as a creature in its own right:

> Then the Queen left off, quite out of breath, and said to Alice, 'Have you seen the Mock Turtle yet?'
> 'No', said Alice. 'I don't even know what a Mock Turtle is.'
> 'It's the thing Mock Turtle soup is made from,' said the Queen.[19]

The Leathery Turtle has featured little in culinary extravagance, mainly because the flesh is so unpalatable, perhaps because of its jellyfish diet. Thomas Pennant reported an experience that could well be attributed to someone having just obtained a copy of Hannah Glasse's then new book:

> The late Bishop of Carlisle informed me that a tortoise was taken off the coast off Scarborough in 1748 or 1749. It was purchased by a family at that time there, and a good deal of company invited to partake of it. A gentleman who was one of the guests told them it was a Mediterranean turtle, and not wholesome; only one of the company ate of it, and it almost killed him, being seized with a dreadful vomiting and purging.[20]

The rearing and shooting of Pheasants *(Phasianus colchicus)* is a major rural industry in Britain – around 20 million are shot annually.

Birds
(Aves)

Everyone knows what a bird looks like and no-one is likely to confuse one with anything else. Birds have not really resembled any other creatures since they evolved from an offshoot of the dinosaurs around 150 million years ago. They are warm-blooded vertebrates whose bodies are covered with feathers – although there are usually also some scales present, especially on the hind limbs – and whose forelimbs are modified as wings. Egg-layers, they usually show a considerable degree of parental care of their offspring. They have attracted more language, lore and literature than any other group of creatures, which is a testimony to the fascination that they hold for human beings, and their role in human life for good and ill, and, I believe, because almost everyone in Britain stands an extremely good chance of seeing one every day of their lives.

Divers
(Gaviidae)

Red-throated Diver *(Gavia stellata)*

OTHER COMMON NAMES: *Arran ake (Dumbarton); burrian (Scotland); cobble; galrush (Ireland – Dublin Bay); loom (Shetland); loon; lune (Ireland, Devon); mag loon (Norfolk); rain goose (Shetland, northern Scotland); red-throated loon; silver grebe (Kent); speckled diver; speckled loon; sprat borer (Essex); sprat loon; spratoon (East Lothian, Norfolk).*

This sleek, grey-brown bird has a red throat in the breeding season. It is a winter visitor to coasts around Britain and, although it feeds on sea fish, it breeds only on small freshwater lochs in the far north of Scotland and Ireland.

The name 'rain goose', in one form or another, is probably very old and is sometimes applied to divers generally. A Shetland rhyme tells how these birds are believed to be especially noisy before the onset of bad weather:

> If the rain gose flees to da hill,
> Ye can gang to da heof when ye will;
> But when sho gangs to da sea,
> Ye maun draw yir boats an flie.[1]

The Red-throated Diver is considered the best weather indicator of all Shetland birds and, to the initiated, considerable detail can be read into its calls. When it makes sharp, short notes, fine weather may be expected to continue; when the call is long and plaintive, a prolonged spell of wet weather is likely. This information can usefully be combined with the direction in which the bird is flying: inland, towards the hills, suggests fine weather whereas flight out to sea portends rain and wind.

The old name 'loon' is a general one for divers, although it is most usually applied to the Red-throated Diver and was also sometimes used for grebes. It was originally Old Norse and imitates the moaning, wailing call that, once heard, echoing across some remote highland loch, will never be forgotten. At one time, the name came to mean 'cowardly', perhaps in reference to the way the birds (both divers and grebes) dive rather than face the foe. Shakespeare certainly thought this:

> The devil damn thee black, thou cream-fac'd loon!
> Where gott'st thou that goose look?
> (*Macbeth*, Act 5, Scene 3)

However, this reference to 'cream-faced' may mean that Shakespeare had the Great Crested Grebe in mind. The popular use of the term 'loony' for someone particularly stupid is nineteenth century in origin and stems from the word 'lunatic', not from the name of the bird.

The Red-throated Diver *(Gavia stellata)* takes its name from the red patch that develops on its neck during the breeding season.

Black-throated Diver *(Gavia arctica)*
OTHER COMMON NAMES: *black-throated loon; herring bar (Sussex); lesser imber; lumme; northern doucker; speckled loon.*
This is another winter visitor to British coasts, with some birds breeding in northern Scotland, although more usually on large freshwater lochs rather than the small lochs preferred by the Red-throated Diver. It is a very handsome, grey-and-black bird with a speckled back and with a grey-black throat when breeding.

Among its regional names is the lovely 'northern doucker'. The name 'doucker', in different spellings, has been used in many places for most of the divers and has also at various times been applied to almost any bird (and much else) that dives. Its etymology is wonderful and convoluted. 'Douk' and 'dook' are Scottish forms of 'duck' (the verb). The Scottish word 'jouk' has the same meaning: after the Appin incident in Robert Louis Stevenson's *Kidnapped*, Alan Breck calls to David Balfour: 'Jouk in here among the trees.' 'Douk' also finds an outlet in Duiker, a small African antelope which 'ducks' into the bush, 'duiker' being an Afrikaans word derived from the Dutch *duiken*, meaning 'to dive'.[2] From thence we come to that haunt of low life, a 'dive', where patrons used to 'jouk' out of the way when the law arrived; and where they listened to music on a 'juke' box.

Great Northern Diver *(Gavia immer)*
OTHER COMMON NAMES: *Arran hawk; cobble; ember (Orkney); great doucker; great northern loon; gunner; Holland hawk (Scotland); imber diver (Ireland); immer, immer goose (Orkney); loom; loon; naak (Scotland); nauk (Northumberland); ring-necked loon (Cork, Scotland).*
This utterly magnificent creature is similar to the Black-throated Diver but larger and, although a regular winter visitor to British coasts, it has bred in Scotland on only a handful of occasions.

The various regional names reflect its loud, wailing call, but it is most unfortunate that a bird with so distinguished a British name as Great Northern Diver, should, once it crosses the Atlantic to North America, become known as the Common Loon.

Grebes
(Podicipedidae)

What a wonderful name 'grebe' is, although it conceals a sad history. The word was originally French and harks back to a time, in the eighteenth century and later, when ladies of fashion wore grebe-feather muffs, boas and collars, the dried skins of the birds being known as 'grebe fur'.

It is especially sad that grebes should have been persecuted in this way, because they are fascinating creatures which belong to a side-branch of the ornithological tree and are apparently unrelated to any other type of bird, living or dead.

Some of the earliest bells, made in Aldbourne, bear an engraving of a dabchick, or Little Grebe *(Tachybaptus ruficollis)*.

They are streamlined, almost tail-less birds with paddle-like feet situated at the rear of the body (the colloquial name 'arse-foot' pays tribute to this anatomical peculiarity). This enables them to swim very effectively underwater, where they catch fish and other aquatic animal life. Grebes are often exquisitely coloured and engage in elaborate and beautiful courtship rituals.

Little Grebe *(Tachybaptus ruficollis)*
OTHER COMMON NAMES: *arsefoot; black chin; black-chinned grebe (Berkshire); bonnetie (Angus – Forfar); dabber (Buckinghamshire, Berkshire); dabchick; dapchick (Somerset); didapper, diedapper (Norfolk, Hampshire, Dorset); divedapper, divedop (Lincolnshire); diver (Renfrewshire); dobchick; doucker (Perth, Shropshire); jack doucker (Shropshire); little doucker (East Lothian); loon; mither o' the mawkins (Stirlingshire); penny bird (Scotland); small doucker (East Lothian); spider diver; tom pudding (Northern Ireland, Shropshire).*
The smallest of all grebes, this is a neat, short-billed, brown species with a chestnut breast. It seldom flies, preferring to dive for cover, as many of its names suggest. Although fairly widespread and common, Little Grebes are seen infrequently because they live on waterways where there is dense aquatic vegetation. However, they are commonly heard making a

This painting by John Gould and H.C. Richter, 1873, shows a pair of Little Grebe *(Tachybaptus ruficollis)* with their offspring.

shrill, bubbling call, especially during spring. They have long been common but now appear to be expanding northwards, although they are vulnerable in very cold winters.

The lovely seventeenth-century name 'dabchick' and its numerous regional variants are widely used and are indicative of a bird that is held in much affection. A delightful Wiltshire story comes from the village of Aldbourne, to the northeast of Marlborough:

> Long ago, on the little pond in the middle of Aldbourne village a dabchick was found. None of the villagers had ever seen such a bird before, which is not surprising, for Aldbourne is a downland village unlikely to attract a dabchick. They sent for the oldest inhabitant, an ancient invalid who had to be fetched in a wheelbarrow ... After he had been wheeled round the pond three times he declared that the bird was a dabchick. Ever afterwards, the inhabitants of Aldbourne were known as 'Aldbourne dabchicks'. There was great rivalry between Aldbourne and the neighbouring parish of Ramsbury. The traditional Aldbourne Feast, held on or near the feast day of St Mary Magdalen, to whom the parish church is dedicated, generally ended in a free fight between Aldbourne men and intruders from Ramsbury. Ramsbury folk, too, enjoyed playing tricks on the Aldbourne carrier, who served both villages. One of the greatest insults was to tie a dead dabchick to the tail of the carrier's cart and to run behind shouting, 'Yah! Aldbourne dabchick! Yah! Aldbourne dabchick!' Ramsbury villagers, incidentally, would be in a better position than those of Aldbourne to secure a supply of dead dabchicks, for the river Kennet flows nearby. It sounds odd and trivial, but the association between Aldbourne and the dabchick is evidently old and important. An extraordinary fact is that some of the earliest bells, made in Aldbourne – and Aldbourne was noted for them – bear the engraving of a small, long-necked bird which is undoubtedly supposed to be a dabchick. These bells date from the reign of James I. Aldbourne also possessed a ducking ceremony. Before anyone could call himself an Aldbourne dabchick he had to be ducked in the same tiny pool.[3]

Great Crested Grebe *(Podiceps cristatus)*

OTHER COMMON NAMES: *arsefoot; ash-coloured loon; ash-coloured swan; car goose; crested doucker (Scotland); gaunt (Lincolnshire); great copped doucker; great diver (Cheshire); greater copped doucker; greater dabchick; greater dobchick; greater loon (western Ireland, Norfolk); grey loon; horned doucker; molrooken (Northern Ireland – Lough Neagh); satin grebe; tippet grebe.*

A marvellously handsome bird, the Great Crested Grebe has a long neck, white cheeks, a chestnut frill around its neck and a small dark crest. The young are strikingly striped on the head. This species was once mercilessly collected for its feathers but now has complete protection. Consequently, it has been steadily expanding its range northwards throughout Britain, although it is still absent from the north of Scotland, where its place is taken by the divers. The massive expanse of Lough Neagh in Northern Ireland, the largest lake in the British Isles, is now a stronghold of this species.

The persecution of the Great Crested Grebe has had one very positive outcome: it led directly to the formation of the Royal Society for the Protection of Birds (RSPB) in 1904. This was the result of a strong reaction from a group of women at the end of the nineteenth century to the grebe-fur industry that had provided the wherewithal for the creation of their fashionable wardrobes. This was not, however, the first time in its history that the Great Crested Grebe had been regarded as a choice bird: its bones were among those found in the prehistoric lake village of Glastonbury where presumably, along with herons, Cranes, Mallard and other water birds, it had been eaten.

Opposition to the hunting of the Great Crested Grebe *(Podiceps cristatus)* for its feathers led to the formation of the RSPB in 1904.

Red-necked Grebe *(Podiceps grisegena)*

OTHER COMMON NAMES: *greve (Yorkshire); rolling pin (Sussex).*
This large grebe is thicker set than the Great Crested Grebe and has a striking chestnut-coloured neck. It is a summer visitor to the eastern part of the country, where it seems likely to breed in future.

Slavonian Grebe *(Podiceps auritus)*

OTHER COMMON NAMES: *black-and-white dabchick; dusky grebe; hell diver; horned grebe; magpie diver, rolling pin (Sussex); small diver (eastern England).*
Chestnut overall, with a dark head and a broad, yellowish eye-stripe, this beautiful bird has long been a winter visitor, breeding occasionally in Scotland.

'Slavonia' is an old name for part of northern Russia, where it is common.

Black-necked Grebe *(Podiceps nigricollis)*

OTHER COMMON NAMES: *black grebe; eared grebe; lesser crested grebe; rolling pin (Sussex).*
A darker, smaller bird than the Slavonian Grebe, this is another winter visitor but a much more regular breeding species, mainly in the north of the country.

Shearwaters
(Procellariidae)

Shearwaters, like petrels and albatrosses, are sometimes called 'tubenoses' because they possess a tubular nose, the nostrils not being buried within the beak, as in most birds, but formed in a horny tube along the upper mandible. They are usually fairly large, oceanic birds, coming ashore to breed and superficially resembling gulls, although generally darker in colour. They feed on fish, squid, crustaceans and sometimes carrion and are also extraordinarily long lived. It is thought that some Fulmars may be over a hundred years old.

Fulmar *(Fulmarus glacialis)*

OTHER COMMON NAMES: *mallemuck (numerous spellings).*
This gull-like bird, white with a grey back, is the only pale-coloured European shearwater. It has acquired almost legendary status in British ornithology, partly because it was the favourite bird of the popular and pioneering naturalist James Fisher, whose huge monograph, *The Fulmar*, now fetches a small fortune in second-hand book shops. It also has a special place because of its longevity, its vast increase in population and spread southwards during the last century, and its historic role in being so closely tied in to the economy of small islands, especially St Kilda. James Fisher's book begins with the following words:

> The hero of the story which I have to tell is that primitive, rather stupid, but also successful bird which

> is more likely to be met with in the middle of the northern oceans than any other. It appears to observers in different guises, according to their viewpoint; to some it is a ghost, to some the conqueror of the sea, to others foul, or a fool, or food ... Its name is fulmar, which means foul gull; it has many other names, but this is the oldest – at least a thousand years old – and the best, even though the bird is not a gull.[4]

'Fulmar' was originally the Norse name for the bird, which became adopted by the Gaelic-speaking St Kildan islanders and then entered English in 1698 in a book entitled *A Late Voyage to St Kilda*. This was written by an extraordinary man named Martin Martin, tutor and land steward to the MacLeod family who owned the island. Martin was the first to reveal to an unsuspecting outside world the way in which the islanders collected Fulmar vomit, an oily substance forcibly ejected from the beak:

> ... sometimes of a reddish, sometimes of a yellow colour, and the inhabitants and other islanders put great value upon it, and use it as a catholicon for diseases, especially for pains in the bones, stitches, etc.[5]

It was exported and formed a valuable part of the St Kildan economy, being delivered to London and Edinburgh, where it was used, apparently successfully, in the treatment of dental abscesses, sprains, rheumatism and boils, although whether its properties have ever been scientifically examined is uncertain.

Martin's account hid a much wider truth: that a more systematic 'harvesting' of the oil, the flesh, the feathers and the eggs of the Fulmars was taking place, the whole operation being known as 'sea-fowling'. It was pivotal to the island's economy and continued until the evacuation of St Kilda on 29 August 1930, when the last 36 islanders left. This effectively meant the end of sea-fowling in the North Atlantic, a pursuit that in its entirety had embraced not only the St Kilda Fulmars but also Manx Shearwaters, Gannets, Puffins and numerous other sea birds, including the legendary Great Auk. In spring on St Kilda, a few adult birds were taken, as well as eggs, but it was the taking of young Fulmars that was most important. It is impossible to find a better account of the collection of young birds than that given by the Reverend Neil MacKenzie, minister on the island from 1830 until 1844:

> On an average twelve thousand will be killed as they are about ready to leave the nest ... About the beginning of

Three generations of one family of islanders outside their cottage on St Kilda, *c*.1910.

> August, the young Fulmars are about fully fledged and ready for killing. During the preceding week an unusual excitement and alertness pervades the village. Every possible preparation is being made. The women bring the cattle home from the sheilings, grind sufficient meal to last the killing time, while the men test the ropes, make good deficiencies, and provide barrels and salt. Other fowling is really hardly anything more than amusement, and may almost be prosecuted or not as you like, but for ten days at this time it is quite different ... A large and valued portion of the winter's food must now be provided or you have to do without it. The breeding places have all been carefully examined some time before, and an estimate made of the young birds which they respectively contain. They are now divided into as many portions as there are groups of four or five men who are to work together. These portions are now assigned by lot to each group and all is ready. When the day considered most suitable comes all move off to the rocks, and the men either climb down to the breeding-places or get lowered by rope if necessary. The birds must all be caught by hand, and skilfully too, or much of the valuable oil will be lost. They must be caught suddenly and in such a way as to prevent their being able to draw their wings forward or they will squirt the oil. It cannot do this easily while you hold the lower joints of its wings back against each other. Caught in the right way its neck is speedily twisted and broken and the head passed under the girdle. When the man has got strung about him as many as he can conveniently carry, they are passed up to the women who are waiting above. At once they are divided into as many shares as there are men in the group, when the women-kind and children seize upon their shares and begin to drain out the oil into receptacles, which are generally made of the blown-out and dryed stomachs of the Gannet. This they do by the very simple means of holding the bird bill downwards and gently pressing, when about a gill of oil flows out by the bill. This oil is much valued, some used by themselves for various purposes and the surplus sold. When all are got home, plucking off the feathers, disposing of the internal fat, and salting the carcasses for winter use, goes on till far in the night. Early the next morning the same round begins, and so on from day to day till all the accessible breeding places are visited. All this time there is nothing but birds, fat, and feathers everywhere. Their clothes are literally soaked in oil, and everywhere inside and outside their houses nothing but feathers; often it looks as if it were snowing.[6]

This begs two questions. First, what effect did the repeated cropping have on the population of the birds? To which the answer must be none because the Fulmar was expanding in numbers both during and after the fowling took place; it is now reckoned to be running at 3–4 per cent per annum. Second, what judgement should be placed on what seems today to be a barbaric practice? I believe that the whole exercise can be closely equated with the rearing of chickens for slaughter, with the more agreeable proviso that it represented a means of survival for a human population that had no real alternative.

The Fulmar *(Fulmarus glacialis)* was once essential to the economy of small islands and was hunted for its oil and flesh.

There are a number of theories to explain the remarkable success and expansion of the Fulmar over the past century or so.

> Within my lifetime, this species' southernmost nesting site was the Farne Islands, then it became the low and soft cliffs at Hunstanton, Norfolk; now it is the chalk cliffs of the South Coast.[7]

It has been suggested that a mutation has occurred in the Icelandic population of the birds that has favoured their expansion southwards; much as a comparable mutation is believed to have led to the northwesterly spread of the Collared Dove (see page 299). However, it is believed that an increased use by the birds of the waste discarded at sea by fishing boats is also significant in aiding their survival away from land. Whatever the reason, in future years more and more people in the southern half of Britain will become as familiar with this remarkable bird as those in the north are now.

Almost all the world's population of Manx Shearwaters *(Puffinus puffinus)* nests around Britain.

Manx Shearwater *(Puffinus puffinus)*
OTHER COMMON NAMES: *baakie craa, booty (Shetland); cleaver (Cornwall); cocklolly (Pembrokeshire); cockthroden, crew (Scillies); cuckee, cuckles (Pembrokeshire); lyre, lyrie (Shetland, Orkney, Scotland); mackerel bird (Cornwall); mackerel cock; night bird; puffin (Scillies); scraber (Hebrides); sea swift; skidden; skimmer; skinner.*
How the Puffin and the Manx Shearwater exchanged names is described on page 293 and it is regrettable that, today, the Manx epithet no longer has its former significance. The colony that once nested on the Isle of Man, described as 'probably the largest ever known', was destroyed by rats about 200 years ago, although it is gratifying that shearwaters are now nesting there again.

Nonetheless, even without the Manx contribution, almost all the world population of these large, dark-brown-and-white birds breeds in the British Isles, migrating to this country annually from the South Atlantic to Skomer and Skokholm off Pembrokeshire, to Rhum in the Hebrides and to some of the small Irish islands. There are also a few nesting on the Isles of Scilly and Lundy. Like Fulmars, young Manx Shearwaters were harvested for their flesh and oil, but they were never as important for the Manx economy as Fulmars were for the much more remote St Kilda; this too is now an activity of the past.

The Shetland and Orkney names 'lyre' or 'lyrie' originate from Old Norse and simply mean 'fat'.

Petrels
(Hydrobatidae)

Storm Petrel *(Hydrobates pelagicus)*
OTHER COMMON NAMES: *alamonti (Orkney); alamouti (Shetland); allamotti; assilag (Western Isles/Hebrides); gourdal (Kerry); gourder; hornfinch; little Peter; Martin oil (western Ireland); mitey; mitty; Mother Carey's chicken; oily-mootie (Shetland – Foula); pinnick (Cornwall); sea swallow; spence, spencie (Western Isles/Hebrides); storm finch (Orkney, Cornwall); stormy petrel; swallow (Shetland); Tom tailor; waterwitch; witch.*
See Leach's Petrel below.

Leach's Petrel *(Oceanodroma leucorhoa)*
OTHER COMMON NAMES: *fork-tailed petrel; Leach's storm petrel.*
Other than to sailors, petrels must be the least known of all common and numerous British birds, simply because they are birds of the open ocean, coming close to land only at night and coming ashore only to nest. They are also very small and dark, superficially like martins when on the wing and flying low over the water to catch, not insects, but plankton and tiny fish.

Their nesting sites are mainly in the north and west, often on remote rocky cliffs. It is therefore small wonder that they are little known and their real numbers largely a matter of guesswork. Perhaps because they are so enigmatic, they have acquired a characteristic lore.

Their association with storms and the name 'storm petrel' are genuine enough because, in extreme weather, they can be blown ashore and inland; to the sailor, their presence presages storms, which is one reason why they were disliked and known as 'witches'. 'Petrel' is a seventeenth-century name, probably a reflection of the pitter-patter of their feet on the waves, although an alternative explanation links them with St Peter, who walked upon the water. Names including a reference to oil are allusions to the plump, fat young birds that were once collected for their oil. The name 'Mother Carey's chicken' is still being used surprisingly widely; its origin is as recent as the eighteenth century, and it is generally thought to be a corruption of 'Mother Mary's chicken'. 'Leach' was W.E. Leach, who first identified this species in 1817. The Cornish name 'pinnick' was 'a dialect name generally used for a puny child or person of small appetite'.[8]

Gannets
(Sulidae)

Gannet *(Morus bassanus)*
OTHER COMMON NAMES: *basser, bass goose (Angus – Forfar); channel goose (Devon); gan (Wales); herring gant (Norfolk); mackerel gant (Yorkshire); saithor, sethor (Cornwall); soland (Scotland); solan goose; spectacled goose; sula (Scotland, northern England); zethar (Cornwall).*
Gannets are enormous sea birds, the size of a decent goose, and they occur in great numbers, around half a million nesting in Britain every year. Nonetheless, many people have never seen one because they have visited the wrong part of the coast. Those whose marine experience is limited to sandy beaches and who never go anywhere near the great rocky cliffs of the north and the west will have missed one of the most staggering sights that British wildlife has to offer. The vast colonies nesting on the rocky coasts of Scotland, and, most famously of all, on Bass Rock in the Firth of Forth, defy belief. The furthest south that you will see a Gannet colony close to the mainland is on the island of Grassholm off the Pembrokeshire coast, although a hundred years ago they were also on Lundy and there is every hope that, at their current rate of expansion, they may return there.

There is some mystery about the name 'Gannet'. It is traceable back at least to Old English, when it simply meant 'a goose', any goose, but by the thirteenth century, Lundy was being referred to as 'Gannets' Rock' and it is not clear why this distinctive sea bird should have been called a goose. William Lockwood points out that Lundy had been a Viking stronghold ('Lundy' itself being the Old Norse name *lunde* for the Puffin, see page 293), so logically the northern name 'sula', or 'solan', would have been more appropriate. In all older references to Bass Rock, Gannets are called geese.[9]

Bass Rock lies in the parish of North Berwick, and the local priest is still called Minister of St Andrew Blackadder and Vicar of the Bass. As part of his *teind* (a northern word for tithe), he is entitled to receive 'twelve solan geese with their feathers' each year. In the time of Reverend Matthew Murray (1763–72), these provided welcome additions to the family's normal fare of 'cabbage and fruit from the garden, eggs and chickens from the poultry yard, and milk and butter from the manse cows'. Even in the nineteenth century, the Laird of the Bass was still paying 12 geese to the minister, but the present incumbent expresses no sadness that the practice has now ceased.[10] Even at its peak in the eighteenth century, Gannet-harvesting was a precarious trade, with payments having to be made to the Laird (the proprietor of the Bass), the climber, the men who caught the birds and the carrier who took them to Edinburgh.

> In the nineteenth century, the Laird of the Bass was still paying twelve geese to the minister at North Berwick as part of his stipend. The eighteenth century accounts show how little profit was gained from much effort in making the geese a commercial venture ... eighty gannets

About half a million Gannets *(Morus bassanus)* nest in vast colonies on the coasts of northern and western Britain every year.

> were needed to produce a stone of feathers, and three stone were required for a feather bed. In 1767, a stone of gannet's feathers at the Bass cost 10s sterling and in 1874 had gone up to 18s. In the 17th century the birds fetched 1s 8d apiece, plucked and prepared for the market ... By the nineteenth century, the value had dropped but at its end cart-loads of young gannets were still being sold and the oil was used for oiling boots to keep out the damp.[11]

Remarkably, the Dutch call the bird John O'Gaunt.[12] The gannet is a very imposing bird, being white with black legs and black wing-tips. It is one of the few web-footed birds to have all four toes united by a web, which betrays its relationship, along with the cormorants, to pelicans. It dives dramatically into the sea to catch fish and squid.

> The Gannet was known in Cornwall as saithor, derived from saith or saeth, Cornish for arrow; a reference to the bird's fishing technique when it dives into the water like an arrow.[13]

The word 'gannet' has become a slang expression for a greedy person, although Gannets only appear greedy: they are big birds with big bodies to feed. Like most British birds they are legally protected, with one interesting and historic exception. Along with Fulmars (see page 205), Gannets were among the birds for which legal allowance was made because of their ancient role in the economy of small Scottish islands. Sea-fowling on St Kilda, where Gannets were 'harvested' along with the Fulmars, and the historical collecting of Gannets from Bass Rock have already been described (see page 209), but young Gannets are still taken as food from the colony at Sulasgeir, 25 kilometres (15 miles) off the coast of Lewis. This lump of rock is only about 800 x 200 metres (875 x 220 yards) in area but it is host to around 20,000 Gannets in the nesting season; its name means, simply, 'Gannet rock'. It has been uninhabited since 1549, but the men of Ness, from the Ness district of Lewis, make annual visits there to collect young birds, or 'gugas'. Traditionally they rowed across the open sea, staying on the rock for about a week and returning with a boatful of dried birds and feathers. Today, they use motorboats, but the crossing and landing at Geodha Phuill Bhain are still hazardous.

The flesh is clearly an acquired taste but, whereas the locals ate it out of little alternative, it did find a role as an upper-class delicacy. When Gannets from Bass Rock were being sold in Edinburgh at 1s 8d apiece, the flesh was served at the dining tables of the Scottish kings and 'the Edinburgh elite', generally as an appetizer:

> It is very good flesh, but it is eaten in the forme as we eate oysters, standing at the sideboard, a little before dinner.[14]

Charles II is often said to have tasted solan goose and been unimpressed, but the source of this story is elusive. On St Kilda, a pudding was made of guga fat, which was stuffed into the bird's stomach and boiled in a watery gruel. The liquid was taken as a cough medicine for coughs and the fat was used to heal wounds.

Cormorants
(Phalacrocoracidae)

Cormorants have long necks, long, strong beaks and a powerful build, ideally equipping them for their role as fish-catching, diving sea birds. There are two species in Britain: the almost cosmopolitan Cormorant, or Great Cormorant, and its smaller and more numerous relative, the Shag. Unlike other aquatic birds, their plumage is not waterproof, which is why they so characteristically dry themselves with outstretched wings.

Cormorant *(Phalacrocorax carbo)*

OTHER COMMON NAMES: *brongie (Shetland); coal goose (Kent); cowe'en elders (Kirkcudbrightshire); gorma; gormer (Northumberland); hiblin (Orkney); jack cocking, Jan cockin' (Cornwall); lairblade (Orkney); locring, lorin (Shetland); mochrum elders (Wigtownshire); morvran (Cornwall); palmer; parson (southern England); scarf (Orkney); scart (Northern Ireland, Orkney, Lancashire); sea crow; spilgarn (Cornwall).*

The specific name *carbo* means 'black' and it is their blackness that has endowed Cormorants with some of the same dark lore associated with crows. 'Sea crow' is not an alternative name for nothing and the name 'Cormorant', which dates from the fourteenth century, owes its origin to the Latin word for 'sea raven'. The Cornish name 'morvran' also has 'sea crow' as its origin.

> A dialect name at Mousehole [Cornwall] was spilgarn which probably means 'carn-hoverer' from the habit of both Cormorants and Shags standing on rocks with their wings outstretched.[15]

The northerly names like 'lorin' have a Norse origin and mean 'white thighs', a reference to the conspicuous white patches that develop during the breeding season. An old tale tells that, originally, all birds were naked and, when the decorating artist reached the Cormorant, all his store of colours, with the exception of charcoal, had gone. This overwhelming blackness has resulted in Cormorants frequently being identified with Satan and, even if not exactly satanic, they have certainly been considered bad news:

> On Sunday, September 9th, 1860, a cormorant took up its position on the steeple of Boston church, in Lincolnshire, much to the alarm of the superstitious. There it remained, with the exception of two hours' absence, till early on Monday morning, when it was shot by the caretaker of the church. The fears of the credulous were singularly confirmed when the news arrived of the loss of the Lady Elgin at sea, with three

The plumage of the Cormorant *(Phalacrocorax carbo)* is not waterproof and birds are often seen stretching out their wings to dry.

hundred passengers, amongst whom were Mr Ingram, member for Boston, with his son, on the very morning when the bird was first seen.[16]

The Cormorants' habit of swallowing whole really very large fish as they bring them to the surface, and of doing so in full view of anyone watching, has led to it being synonymous with greed. In Shakespeare's play, John of Gaunt speaks none too flatteringly of Richard II:

> With eager feeding food doth choke the feeder;
> Light vanity, insatiate cormorant,
> Consuming means, soon preys upon itself.
> (*Richard II*, Act 2, Scene 1)

Like many other people, I was greatly surprised when I first saw Cormorants some distance from the sea; they were perched with a heron in a tree overhanging a river. They are more usually seen from a cliff top, as dark shapes a few 100 metres (110 yards) out to sea, and we try to guess where they will surface from their dives. Because they are fishing birds *par excellence*, they have been bitterly persecuted. Although they are now fully protected, their real impact on fisheries is both uncertain and a matter of genuine concern and interest, because they not only perch in riverside trees but are also nesting increasingly far inland.

Their efficiency as fishers led to Cormorants being listed among the 'noyfull fowls and vermin' in Elizabeth I's edict of destruction, although it is an efficiency that has also been turned to good effect. It is well known that the Oriental Cormorant *(Phalacrocorax sinensis)* is used to catch fish for its human owners in parts of the Far East, but it is less widely appreciated that both James I and Charles I employed a Master of the Cormorants, who 'at the royal bidding, travelled widely to display the skill of his charges'.[17]

> The Flesh pulverized given at Night going to Bed, is good to cure the Elephantiasis, or Leprosy, Epilepsy, and Opulations of the Spleen. One Dram of the Liver made into Powder, mixt with seven Grains of the Livers, and Galls of Eels in Powder, taken inwardly, provokes the Terms, facilitates Labour in Women, and is good against the Bitings of mad Dogs. A Dram of the Blood pulverized, taken in any proper Vehicle, is useful against Poison, Plague, and malignant Fevers. The Eggs strengthen the Stomach, and help to cure the Dysentery. A Dram of the Ashes of young Ravens, or Cormorants, given every Morning for three Mornings, is accounted good against the Epilepsy, Gout and Morphew. The Fat and Blood make the Hair black. It is reported, that the Dung being hung about the Necks of Infants, mitigates their Cough, and Pain in breeding of Teeth.[18]

The Shag *(Phalacrocorax aristotelis)* is smaller than the Cormorant *(P. carbo)* and develops a crest in the breeding season.

Shag *(Phalacrocorax aristotelis)*

OTHER COMMON NAMES: *cole goose; colmorn; crane (Northumberland); crested cormorant; green cormorant (Ireland); morvran (Cornwall); scarf (Shetland); shagga (Cornwall); skart (Orkney); spilgarn, trainygoat (Cornwall); tufted skart.*

Shags are considerably smaller than Cormorants but have a small, forward-pointing crest, which develops during the breeding season. Otherwise they are very similar and have similarly been persecuted as fish-eaters. They are common and widespread, although the numbers fluctuate considerably for reasons that are not fully understood.

The name, despite its modern sexual connotation, simply means 'tufted', a reference to the crest, as it has done since the sixteenth century. The northern names, such as 'scarf', come from Old Norse and are imitative of the harsh, croaking call.

Bitterns and Herons
(Ardeidae)

Herons are large, mainly fish-eating waders: long-legged, long-beaked and generally slender of body. Their close relatives, the bitterns, are usually more squat and decidedly more noisy. There are three British species in this family: the Grey Heron, a magnificent although dreadfully persecuted bird; the shy and rarely seen Bittern; and its smaller and even more rarely seen relative, the Little Bittern. Improbable as it may seem, their closest relatives lie among the hawks.

One other heron, and a newcomer at least within recorded history, was added to the British breeding list in 1996, and, with the gradual warming of the climate, its expansion seems assured. This is the Little Egret *(Egretta garzetta)*, a small, white heron found in many parts of the world, including many other regions of Europe, which is always a delight to watch.

Two other related birds have also featured in relatively recent British bird history. The first is one of the most notable birds of Europe and European folklore. The White Stork (*Ciconia ciconia*, Family Ciconiidae) has been a British breeding bird within recorded history but clearly not a common one because it has never really entered the country's traditions. A pair nested most famously in 1416 on St Giles' Cathedral in Edinburgh, although, as Clinton Keeling points out, this must have been a pretty rare event for it to have been noted and recorded in such a materialistic age as the fifteenth century.[19]

That bizarre white creature, the Spoonbill (*Platalea leucorodia*, Family Threskiornithidae), with its extraordinary spoon-shaped beak, was certainly once more common; Chris Mead reminds us that it disappeared from most of Britain as recently as 1605, persisting in East Anglia until about 50 years later.[20] It is tempting to suggest that it is absent from British folklore because, historically, East Anglia was a remote and relatively uninhabited region, but the facts may not bear this out: Oliver Rackham points out that, in the thirteenth century, east Norfolk and east Suffolk were the most populous parts of the British Isles; although they were certainly still remote.[21] Whatever the explanation, there may well be an opportunity to put things right. A few tens of birds are now likely to be seen in Britain in most years and it can surely be only a matter of time before they breed again.

Bittern *(Botaurus stellaris)*

OTHER COMMON NAMES: *bitter bum; bitterurne (Surrey); bittour (Northumberland); bog blutter; bog bumper (Scotland); bog drum (Ireland, Scotland); bog jumper; boom bird (Wales); bottle bump (Yorkshire); bull of the bog (Roxburghshire); bull of the mire (Northumberland); bumble; bump cors (Wales); bumpy coss; buttal (Norfolk); butter bump; clabitter (Cornwall); heather bluiter, heather blutter (Scotland); miredrum (Northumberland).*

Very few people have seen a Bittern in the wild but considerably more have heard one. The extraordinary, deep booming call made by the male bird carries for at least a kilometre and differs recognizably from one bird to another; consequently, by recording and analysing the sound by computer, the specific call of every male Bittern in Britain can be recognized. This has enabled ornithologists to state with some certainty that, in 1999, there were 19 male Bitterns in England. This is a slight improvement on previous years and certainly an improvement on the end of the nineteenth century, when there were none. Historically, however, when reed beds were much more widespread, the Bittern was a more common and widespread species, nesting in Scotland and Ireland as well as many other parts of England.

They are small, brown, rather stout birds, all but invisible as they sit motionless among the reed stalks, with head stretched upwards and the speckled neck pattern blending perfectly with the surroundings.

However, it is the extraordinary call that has given the Bittern its many regional names and endowed it with some extraordinary lore. I can think of only one other bird, the Emu, that makes such a weird and far-carrying sound. To the Romany people, hearing the sound was an augur of death. Most famously, Chaucer, to whom the bird was a 'bitore', recounted one of the once widespread beliefs in 'The Wife of Bath's Tale', in which the eponymous lady imitated what was imagined to be the way the Bittern produced the sound by pushing her own mouth into the water to impart a secret to the marsh (mire) rather than tell it to another person:

And as a bitore bombleth in the myre
She leyde hir mouth unto the water doun ...

When motionless among reeds, the Bittern *(Botaurus stellaris)* is almost invisible, but its booming call is unmistakable.

The alternative version, in which the bird pushed its beak into a reed stalk rather than the water, using it as a trumpet or sounding tube, was equally widely believed.

Like Grey Herons, Bitterns were once eaten and, today, with only 19 males in the entire country, it seems remarkable that, when George Neville was enthroned as Archbishop of York in 1465, the banquet included 200 of them, as well as an incredible array of other creatures. The original description of this event, which has become known as 'the Neville Feast', survives in the Public Record Office and was given wider audience by a printed version published in 1645. Even allowing for the possibility of exaggeration in quantity (although medieval feasts could certainly last for days and leave their host in debt for years), it seems improbable that the list of species itself was other than correct. To anyone interested in the history of British animal life, the account is so appalling that it has a grotesque fascination because, in addition to ale, wine and domestic livestock such as geese, pigs, peacocks, pigeons, 'conyes', kids, chickens and capons, the bill of fare included:

Swanns, foure hundred.
Plovers, foure hundred.
Quailes, one hundred dozen.
Mallards and Teales, foure thousand.
Cranes, two hundred and foure.
Bytternes, two hundred.
Hernshawes, foure hundred.
Pheasants, two hundred.
Partridges, five hundred.
Woodcocks, foure hundred.
Curlews, one hundred.
Egritts, one thousand.
Stags, Bucks and Rose, five hundred and four.
Pasties of Venison cold, one hundred and three.
Pasties of Venison hot, one thousand five hundred.
Pikes and Breames, five hundred and eight.
Porpisses and Seales, twelve.

Small wonder that there were:

Cookes in the Kitchen, 60.[22]

In *A Propre New Booke of Cokery*, published in 1545, an account is given of the optimum seasons for various meats. It tells that: 'Herons/curlue/crane/bitture/bustarde be at all tymes good but best in winter ... and should be served with a sauce gallentyne.'[23]

Little Bittern *(Ixobrychus minutus)*

This is a truly beautiful but very rarely seen bird, pink-buff overall, with a black cap and back, that has bred occasionally in the east of England.

Grey Heron *(Ardea cinerea)*

OTHER COMMON NAMES: *brancher; crane (West Yorkshire); diddleton frank (Fens); ern (Somerset); frank (Suffolk); frank hanser (Fens); frog-eater; harn (Norfolk); harnser (Suffolk); harnsey (Norfolk); hegrie, hegril's skip (Shetland); herald (Angus – Forfar); hernser (Suffolk); heronseugh (Berwickshire, Yorkshire); heronshaw (Nottinghamshire); hersew; hershaw; huron (Roxburghshire); jack hern (Sussex); jemmy crane (Lancashire); jemmy heron (Kirkcudbrightshire); jemmy lang legs; jemmy lang neck (Lancashire); Jenny crow (north); Joan-na-ma-crank (Cumberland); lang-necket haaran (Berwickshire); long neck (Yorkshire); long necky (Dumfries-shire); long-necked heron (Ireland); longie crane (Pembroke); longnix (Cheshire); moll hern (Midlands); ooze bird (Cornwall); skip hegrie (Shetland); tammie herl (Perth); varn, yarn, yern (Cheshire).*

The curse of fish-farmers and koi-carp enthusiasts, the Grey Heron is a very common and very big bird. It is only when you see one at close quarters, perhaps as it takes off after emptying your garden pond, that you realize just how enormous they are. Although their numbers drop in hard winters when waters are frozen and fish in general difficult to come by, they seem to recover very quickly and there are probably over 20,000 Grey Herons in Britain today.

They cannot be mistaken for anything else, with their long legs, long, more or less white and very flexible neck (which is bent when they fly), grey back and long, spiky crest. Perhaps their most surprising trait is in nesting not among waterside vegetation, as might be expected, but usually in tall trees and in large colonies, very often close to human habitation.

There has long been a curious belief that a heron's nest has two holes in the bottom through which its legs dangle, or that it incubates its eggs by straddling the nest. It seems odd that this story has survived, despite being so easy to check.

> An old belief prevalent in Norfolk was that two broods were usually reared in the same nest each spring, and an even stranger notion was prevalent amongst the country people that the second set of eggs was incubated by the young of the first brood ... This myth may have arisen from the fact that herons start incubating with the first egg, with the inevitable result that eggs and young must be in the nest together.[24]

One constant problem that arises when investigating the history of the heron in Britain is its confusion with the Crane *(Grus grus)*, which was also a British bird until the end of the seventeenth century (see page 269). The name 'Crane' was used in some areas for the Grey Heron, but Eckwall's *Concise Oxford Dictionary of English Place Names* suggests:

> There is no reason to assume any other meaning for the word than crane, such as heron. The two birds are always kept well apart in early records.[25]

It seems scarcely credible that the plethora of 'crane' names, from Cranborne in Dorset ('crane's stream'), through Cranfield in Bedfordshire ('feld frequented by cranes'), to Cranwell in Lincolnshire ('crane's spring'), could all refer to *Grus grus*. The fact that there are very few place names including a reference to herons appears to support this. Nonetheless, a recent study at the University of Manchester seems to endorse Eckwall's conclusion. By using the more detailed publications of the English Place Name Society, which embrace minor places and fields, the authors identified nearly 300 names related to Cranes. They then satisfied themselves that *Grus grus* was once widespread and common, and likely to be the bird in question, by the fact that clearly identifiable Crane bones have been found in at least 78 archaeological excavations, dating from the Mesolithic to the Middle Ages and including the well-known Iron Age village site at Glastonbury. They concluded that:

> Although the names Crane and heron have been confused in East Anglia and elsewhere, this surely post-dates the local extinction there of the Crane; with no choice available for distinguishing the two species, it seems likely that the name Crane was transferred to the most similar bird type.[26]

They also quote a statute of Henry VIII, passed in 1534, which specifically protected the eggs of the Crane, and, considering this with other evidence, believe that the Crane was not only a common winter visitor but also bred in this country. The presence of both herons and Cranes (as well as bitterns) on ancient menus such as that of George Neville's Feast provides further evidence for the fact the birds were always recognized as distinct.

The Grey Heron *(Ardea cinerea)*, unpopular with fish-farmers and keepers of ornamental fish, is now a protected species.

The Manchester argument is compelling but goes no way towards explaining why there are so few places commemorating the heron. This name first appeared around the end of the thirteenth century, having been introduced from France, and the name and its derivatives have a multiplicity of spelling variations: heronshaw, hernshaw and so forth. Many of them appear in recognizable form in personal surnames and were perhaps applied to people who were good fishermen – or who simply had long necks or even long, thin legs.

The most widespread and varied of heron lore relates to a notion that the birds have some special attraction to fish, which is perhaps not so outlandish in view of their efficiency at fishing. It was said that a heron's foot exuded a scent that lured fish within striking distance and, in some places, fishermen carried a heron's foot in their pocket in the hope of emulating this. Another version was recounted by that otherwise shrewd observer Izaak Walton:

> ... some affirm, that any bait anointed with the marrow of the thigh-bone of a heron, is a great temptation to any fish[27]

although, as Frank Lowe pointed out,

> ... unfortunately for that theory there is no marrow in a heron's thigh bone ...

Lowe continued:

> ... Within my own experience I have met anglers who believe that certain oils extracted from dead herons possess properties attractive to fish, and this belief has existed for more than 200 years.[28]

In 1740, John Williamson wrote that:

> ... some anoint their bait with the marrow cut out of a heron's thigh-bone; and some use the fat and grease of a heron ...

Anglers were advised to smear the line with the following concoction, the receipt for which is given in Williamson's own words:

> ... Take of Man's Fat, Cat's Fat, Heron's Fat, and the best Assa-Foetida, of each two Drams; Mummy finely powdered, two scruples; Camphire, Galbanum, and Venice Turpentine, of each one Dram; Civet, two grains; make them, all according to art, into an indifferent thin ointment, with the chymical oils Lavender, Annise and Camomile, of each an equal quantity. This ointment, which for its excellency some call Unguentum Piscatorum Mirabile, prodigiously causes fish to bite. The man's Fat you may get of any surgeons who are concerned in Anatomy, and the Heron's Fat from the Poulterers in London.
>
> Heron's bowels cut in pieces, and put into a phial and buried in horse-dung, will turn to oil in fifteen days; an ounce of asafoetida is then mixed, when it will be the consistency of honey.[29]

Fishing lines anointed with this mixture were said to make wonderful catches of fish in the west of Ireland.

Today, the notion of eating herons will strike most people as odd, if not revolting, because the flesh is presumed to have an unavoidable fishy taste. However, the fact that herons were once such choice subjects for the greatest of banquets, like the Neville Feast, must belie this. Heron-eating goes back a very long way:

> Excavations of British and early Roman settlements near Colchester were started by the Colchester Excavation Committee in 1930. Among avian bones obtained there were two of the heron, and these could be definitely dated, as they came from a sealed pit in association with objects of a Belgic-British culture of the half century which ended at the latest in A.D. 43, in the reign of Cunobelinus (Cymbeline).[30]

Like swans, herons were once royal birds, their fate the prerogative of the Crown, and there are numerous references to the nobility dining on herons in royal and other accounts throughout the Middle Ages. Even when the Crown did not claim them all, the right to kill herons was limited to Freemen and there were severe penalties for anyone else. Fines of 6s 8d for killing an old bird and 10s for a young one were among the lesser punishments; in Scotland, you could lose your right hand. Young herons were the more highly prized for eating and ownership of a heronry from which they could be taken was therefore a desirable asset. Frank Lowe reports that, in 1517, the woods belonging to an abbot in Somerset contained a heronry that gave an annual return of about a hundred birds. Commonly, these young birds were fattened for the table in special aviaries. In the thirteenth century, when the authorities of the City of London were obliged to issue a formal tariff of food prices to counter the inflated rates being charged by some dealers, while 1½d would buy the best hen and 4d a Pheasant, 6d was the price of a heron.[31]

Herons are now protected under the Wildlife and Countryside Act 1981 and it is illegal to kill them. Therefore, probably no-one in Britain still eats heron but it seems reasonable to assume that, in the past, they were always simply roasted; on 18 May 1812, for instance, a feast at the Hall of the Stationers' Company in London listed six herons among the joints and game roast. There is, however, a relatively recent recipe for Heron Pudding and an interesting earlier variant said they were larded with 'swynefat and eaten with ginger'.[32]

It was partly their desirability for the dining table that led to herons being such choice subjects for hawking. The wish to keep fisheries free from them was a later and additional factor, although herons caught by falcons were often released to provide further sport or were kept for training purposes. Peregrines especially were flown against herons with remarkable success (see page 258). The sport was popular in Shakespeare's time and an enigmatic reference has a bearing on this and an equally intriguing explanation. Hamlet says to Guildenstern:

> I am but mad north-north-west; when the wind is
> southerly I know a hawk from a handsaw.
> (*Hamlet*, Act 2, Scene 2)

'Handsaw' is a corruption of the old heron name 'heronshaw', or 'hernshaw', and because the morning used to be the favourite time for hawking, when the wind blew from the northwest the birds often flew so that any person watching them had the sun in their eyes. It was therefore impossible to distinguish the quarry from its pursuer. When the wind was southerly, however, the birds flew away from the sun and one could then easily 'know a hawk from a handsaw'.

Sad to relate, however, herons gained the reputation of being cowardly birds because they flew high to try and avoid the falcon rather than face it and fight. John Swan, in his *Speculum Mundi* of 1635, referred to another of those odd notions that could surely have been proved fallacious if only someone had looked:

> When they fight above in the air they labour both
> especially for the one thing, that they may ascend and
> be above the other. Now if the hawk getteth the upper
> place he overthroweth and vanquisheth the heron with
> a marvellous earnest flight ...

If the heron rises above the hawk, it:

> ... then with his dunghe defileth the hawk, rotting and
> putrifying his feathers.[33]

There is a belief, from places as far apart as the Isle of Man and Guernsey, that when herons fly low, or fly repeatedly up and down, rain is to be expected.

Swans, Geese and Ducks
(Anatidae)

Commonly called 'wildfowl' (in Britain) or 'waterfowl' (in the USA), these are hugely familiar birds everywhere. There is no area of open water in the country that does not have a resident, visiting or migrant population of one or more species. However, in no other group of British birds do the boundaries between wild, tame, introduced and feral become so blurred, because of the many exotic and often very attractive species that have become established here and that have sometimes hybridized with native species.

This section concentrates mainly on the 'traditional' and well-established species that have been in Britain long enough to have attracted serious attention from the general public and consequently have acquired some lore, or at least long-standing English names, although much duck and goose lore and custom must have originated from domestic breeds rather than their wild ancestors.

Apart from their fondness for water (generally freshwater), wildfowl have in common beaks that are more or less flattened and relatively broad, a generally gregarious, flocking nature, often a simultaneous moult of the flight feathers (which leaves them grounded until they have grown a new set), generally webbed feet and, an unusual feature among birds, a penis. They feed in one of three characteristic ways – diving, dabbling or grazing – but these do not correlate with the three subdivisions of the family, which are much more vague: generally swans have long necks, geese are big, and ducks are small.

Although the names are not interchangeable, the word 'goose' is certainly attached as part of a variant name to many of the species more strictly called 'ducks' and, indeed, to other birds as well; most notable is the Gannet, which is often called the solan goose, while the local names of small water birds, such as Coots and divers, rather commonly include the word 'duck', although nothing but a swan is ever called a swan.

SWANS

Mute Swan *(Cygnus olor)*

Foreign visitors to this country at the beginning of June 2000 could have been forgiven for thinking they had entered a time warp. Picking up their daily paper, they would have read the following news item:

> July Start for Swan Upping
> The dates for this year's swan upping ... were
> announced yesterday by David Barber, the Queen's
> Swan Marker ...[34]

What they were discovering were the bare facts about no more strange a custom than the countless others that so colour and enliven the life of the nation: the Ceremony of the Keys, the Changing of the Guard, Trooping the Colour, the State Opening of Parliament, the Distribution of the Royal Maundy; the only truly unusual aspect about Swan Upping is that it is a ceremony involving wild birds.

Had these foreign visitors made their way to the River Thames at Sunbury Lock at the start of the third week of July, they would have encountered a most colourful spectacle as six traditional Thames rowing skiffs, bearing standards and carrying men in bright costumes, started to make their way up-river towards Abingdon Bridge, picking up and examining cygnets as they did so. The journey takes five days, with stops for refreshment and toasting of the Monarch *en route*.

The precise date of the beginning of Swan Upping seems to be lost in history; even the Official Web Site of the British Monarchy is vague on the matter but suggests the twelfth century. Others have indicated 'at least as early as 1186'. The custom stems from a period when swans were a most important component of banquets and feasts and when, not surprisingly, the Crown claimed them as royal birds, notionally

Swan Upping, the marking of swans to determine ownership, began in the twelfth century and has continued in recent centuries.

as a means of conserving them but in practice to prevent the common people gaining access to an easy luxury food.

However, there were always rival claimants and the Act of Swans of 1482, which formalized the ownership of swans, was one of several official proclamations and ordinances relating to their ownership and, most importantly, its proof. It introduced a right of 'possession by prescription' and only landowners were allowed a 'swan mark', which signified ownership. The swans were marked by nicking the upper mandible with a sharp knife. Swan Motes, or Swanning Courts, enforced the laws and the details were recorded on Swan Rolls.

The entire country was divided into Swan Areas of considerable size, each with a Swan Master to whom the Monarch's rights were delegated. Within these areas, the flock was known as a Game of Swans. The Swan Masters were generally men of some standing: in the sixteenth century for instance, the Swan Masters in the Whittlesea Mere district were the Cecils of Stamford and Burghley, the role passing from father to son. Swan marks and the rights attached to them could be transferred and sold or leased; Christ's College, Cambridge (see below), obtained its right by a conveyance by Robert Collet of the parish of St James, Clerkenwell.

The national flock is known collectively as the Queen's Game of Swans. The Crown retains the right to ownership of all unmarked swans on open water, although the Abbotsbury swannery in Dorset claims an exception granted by Queen Elizabeth I. The present Queen exercises her right only on certain stretches of the Thames and its tributaries, where she shares ownership with two of the ancient livery companies: the Vintners and the Dyers. The Dyers claim their rights from a grant by the Crown in 1473, the Dyers from a grant made sometime before 1483. The Royal swans were once marked with five nicks – two lengthways and three across the beak – but the practice was discontinued at the request of King George V and royal swans are now unmarked. Swans belonging to the Vintners have a mark on each side of the mandible (see The Swan with Two Nicks, page 222); those of the Dyers are marked on one side only. It is worth adding that some conservation and animal welfare bodies consider the whole exercise archaic, barbaric and unnecessary. The Queen's prerogative continued when most other royal privileges to wild animals were abolished in 1971; it was pointed out that the prerogative 'enabled Her Majesty to make highly cherished gifts to foreign heads of state'.[35]

Abbotsbury swannery, Dorset, is home to the largest gathering of swans in the country and attracts a great number of visitors.

The word 'upping' dates from 1560 and refers to the upturning of the swans as they are lifted out of the water; it has also been called 'swan hopping'. The ceremony was formerly a simple matter of recording and marking but became more of a spectacle in the eighteenth century. Today, the boats are attractively decorated and the oarsmen wear red-white-and-blue jerseys. The Royal Swan Marker wears a scarlet jacket with a distinctive arm band, the Vintners' Swan Marker wears green and the Dyers' Swan Marker wears red. At one time, the birds were both marked and pinioned by having their wing feathers clipped, but this was discontinued in 1978. Today, the cygnets are carefully weighed and measured to gain information about the health and viability of the populations.

Swan populations throughout Britain have not been particularly healthy in recent years. I can recall the sorry state of affairs at that symbolic swan home of Stratford-upon-Avon in the early 1970s, when the once-extensive flock (swan flocks are sometimes called herds) had been reduced to one or two birds. Today, swans are present again in large numbers, as they are elsewhere, a situation largely attributable to local bans on the use of lead fishing weights and then their national prohibition in 1987; the poisonous weights were picked up by the birds from the river bed and banks as they fed on water plants. Nonetheless, in August 2001, it was being reported that swans at Stratford still had quantities of lead in their bodies, a reflection of the fact that only the use of lead shot above a certain size had been prohibited. Swan populations have increased so much, however, that they now vie with geese as being among the commonest causes of electricity failure in rural areas.

The largest gathering of British swans has probably always been at Abbotsbury in Dorset, just behind the western end of Chesil Bank. It is often claimed that the swannery began in 1393 (certainly the Post Office issued 600th anniversary stamps in 1993) but in fact this date simply reflects the first record of the swannery, when the resident swanherd, William Squilor, was fined 7d for opening a sluice and flooding the nesting site. The swans had been there long before, probably attracted by the eel grass *(Zostera marina)* which grows in the Fleet and on which they feed. The local Benedictine monks, prohibited from eating four-footed beasts, took advantage of this local supply of meat and eggs. The feathers too were of commercial value and, even today, Abbotsbury still sends quill feathers to Lloyds of London, where they are used to enter details of ships lost at sea in official records. The Abbotsbury swannery is now a major tourist attraction where between 600 and 1,000 Mute Swans can reliably be seen.

Reference has been made to swans as wild birds, which they probably are, although there is a widely quoted belief that they were introduced to Britain, probably from Cyprus and possibly by Richard the Lionheart after the Third Crusade. However, the fact that swan bones were among those unearthed at the lake village of Glastonbury, which dates from 100 BC to 60 BC, makes a nonsense of this.

Mute Swans are the biggest of British swans: huge, heavy white birds, easily distinguished in flight by the sound of their wing-beats and at close quarters by the orange bill, which, in the male, has a large black basal knob.

The Mute Swan is the British swan, although there are three, possibly four, other breeding species. 'Swan' is an old name, unchanged since Old English, from whence it is traceable back to other ancient languages, and means 'sound'. 'Mute Swan' therefore appears to be an apparent contradiction.

In reality, *Cygnus olor* has been the 'Mute Swan' only since 1785, when the name replaced the earlier 'tame swan'. Although Mute Swans are usually silent, they certainly hiss and whistle when threatened and may utter squealing noises too, but anyone who has heard them fly will realize that the deep, singing throb of their wing-beats, which carries for a very long distance, is a good enough reason for them to have been the original 'sound' bird and the inspiration for the name. The juvenile name 'cygnet' is Middle English; the name 'cob', for the male, meaning 'big' or 'stout', dates from 1570; 'pen', for the female, dates from 1550 but has no obvious origin although it might reasonably be connected with the word 'pen' meaning 'a feather, plume or quill' (and hence, a writing implement).

Swans, perhaps the most distinctive of all British birds, are unlikely to be confused with anything else, although, conceivably, one could just be convinced that a very young cygnet might be mistaken for an ugly duckling. The idea that swans are superior to their relatives occurs in several versions, as in the old saying 'all his geese are swans'. This is most usually attributed to Horace Walpole (1717–97), when writing of Sir Joshua Reynolds, and is commonly used 'of one who, unable to grace his board with the royal bird, sought to make the humble goose masquerade in that guise'.[36]

Since men have recorded anything about birds, they have had things to say about swans. These birds have inspired countless tales, poems and music, although, rather strangely, they seem to have featured less in the literature and musical composition of Britain than of many other nations. Shakespeare was the 'Sweet Swan of Avon', an epithet coined by Ben Jonson on the strength of the Ancient Greek belief that the soul of Apollo, the god of music, passed into a swan, hence the Pythagorean fable that the souls of all good poets passed into swans. But although Shakespeare made 11 references to the birds, he created nothing comparable with Tchaikovsky's *Swan Lake*, Sibelius's *The Swan of Tuonela* or even Hans Andersen's *The Ugly Duckling*.[37]

Where British literature is swan-inspired, it is very often in reference to that curious belief of the 'swan song', the fanciful notion that swans sing only once, just before they die:

> I will play the swan,
> And die in music
> (*Othello*, Act 5, Scene 2)

> Like some full-breasted swan
> That, fluting a wild carol ere her death,
> Ruffles her pure cold plume and takes the flood
> With swarthy webs.
> (Tennyson, 'The Passing of Arthur')

And, most memorably of all:

> Swans sing before they die – 'twere no bad thing,
> Did certain persons die before they sing.
> (Coleridge, 'Epigram on a Volunteer Singer')

This belief is unfounded and its origin obscure, although it appeared in the writings of Aristotle, Plato, Euripides, Cicero and other classical authors. Aristotle believed that, as death approached, swans flew out to sea. Pliny, however, was sceptical. There is an Irish belief that swans also sing at other times and a story tells of a flock of them, wearing silver chains and gold coronets, alighting on a lough called Bel Dragon (location unknown), where they sang so sweetly that all who heard the sound fell into a deep sleep for three days and nights.[38]

An old and widespread belief was that swans are hatched only during thunderstorms, the thunder and lightning breaking the eggs. As their eggs are the largest of any British bird, it was surely reasonable to assume that something better than body heat would be required. There is a quotation attributed to an Earl of Northampton – possibly Henry Howard (1540–1614), the First Earl – in a work entitled *Defensative against the Pyson of Supposed Superstition*:

> It chanceth sometimes to thunder about that time and
> season of the yeare when Swannes hatch their young;
> and yet no doubt it is a paradox of simple men to think
> that a Swanne cannot hatch without a crack of thunder.[39]

The association with storms has other manifestations: Swans flying against the wind foretold a hurricane and their nests were built high before heavy rains, while, when the 'white swan' visits Orkney, a severe winter will follow.

Swans have long been considered excellent meat, which was the original motivation behind the Swan Upping ceremony. The swan bones found at Glastonbury suggest that they were eaten from 100 BC to 60 BC, and their remains have also been found at the Roman settlement of Silchester. In *The Canterbury Tales*, Chaucer said famously of his worldly monk 'A fat swan loved he best of any roast', while the poet Matthew Prior (1664–1721) wrote in 1717:

> If you dine with my Lord May'r, Roast-Beef and Ven'son is your fare; Thence you proceed to Swan and Bustard.

In addition to the Crown and the Dyers' and Vintners' livery companies, the Cambridge colleges of St John's, Christ's, King's, Trinity and Jesus had a long tradition of eating swan on special occasions. At Christ's, the earliest reference was in 1542:

> ... ten shillings for a couple of swans to Mr Coke [the master cook] against his reconing.[40]

The accounts of Jesus College from 1601 to 1620 have many references to 'upping', 'fatting' and 'wintering' of swans in considerable numbers; there were 83 in 1601–02. There are also scattered references to swans in the archives of St John's. These date back to 1637 when there was:

> Paid to the Master for two breeding swans and 9 young ones, to stock the College swan park as he had paid before to Arthur Jordan, £2 2s. 9d.[41]

By 1691, there was payment for 'The carriage of two swans at Christmas' and by the nineteenth century swan-eating seems to have become a regular activity. Spot-checks in the archives of the accounts that the cook had with the senior bursar from 1828 to 1871 reveal that cygnet and swan were on the menu around Christmas time in the 1850s and 1860s; items included swan's giblet soup, while the college Feast Books record 'jeunes cygnes' on 27 December between 1879 and 1894. There is no twentieth-century reference however and:

> ... certainly by about 1950, chicken or goose or perhaps another fowl was being eaten at the May Ball, with swan wings and necks of wax added.[42]

Francesca Greenoak gave a description of Mock Swans at the St John's May Ball suppers:

> The swan's necks are made of wax which is first melted and poured into a special mould; the eyes and bills are painted black and orange. The 'bodies', contrived of cold turkey, 'float' on a 'lake' of green aspic decorated with mock water lilies ...[43]

The college itself has no record of this but the college chef said that swan was eaten as a gallantyne sauce at the May Balls up to 1986, when it became difficult to obtain, partly because of animal rights protests.[44]

The association of St John's College with swan-eating gave rise to one of the better known and more repeatable of Cambridge limericks:

> There was a young man of St John's
> Who thought to make love to the swans
> The college porter
> Said 'Please take my daughter,
> The swans are reserved for the dons'.

The custom of dining on swan was wonderfully parodied in Tom Sharpe's wickedly black 1974 comedy of Cambridge college life, *Porterhouse Blue*,[45] and there are numerous regional recipes for real roast swan (or, more usually, cygnet). A Norwich version is:

> 1 cygnet; 3 lb. rump steak; 3 onions; a little nutmeg; salt.
>
> Stuff the cygnet, trussed the same way as a goose, with minced rump steak well seasoned with grated nutmeg, salt and pepper and finely chopped onions. Sew up, cover the bird with greased paper and roast for about 2 hours, according to size [an adult swan can weigh 12 kilogrammes (26 pounds)], basting frequently. Serve with gravy and port wine sauce. In former days, the birds were wrapped up in a flour and water paste and roasted for 4 hours.[46]

Nonetheless, Cambridge does not have a monopoly on academic swan-eating. The archives of Magdalen College, Oxford, record charges made in 1491 'for the keeping of two swans, one of which belonged to the president, 10d'. The last entry for swan-keeping at Magdalen was in 1697, although 'a gift of two black swans for ornamental purposes was accepted in 1904' and, even more bizarrely, in March 1884, 'an offer of two emus was accepted; only one came, the other died on the way'.[47] The Emu became a great favourite and there was dismay when ultimately it was found dead as a result of a surfeit of the college's rich fruit cake.

Swan-eating was popular for a long time in the Norwich area and the local hotels were formerly supplied with swan meat by the Norwich Swan Pit. Historically, the City Corporation, the Great Hospital, the bishop and the dean all had their own marks and associated rights to the birds on the River Wensum and nearby Broads. Regrettably, it is Norfolk, once the home of so many swans and so much swan-eating, that has now become synonymous with its successor on the nation's dining tables, the turkey. When you have tired of swan roasts, however, the village of Budby in Nottinghamshire can offer you its Swan Pie:

> Swan meat; 2 onions; a little sugar; a few mixed herbs; salt and pepper to taste; suet pastry.
>
> Only the finest swan meat should be chosen. Cut up into small pieces and stew gently for 3 hours in water, seasoned with pepper and salt, sugar, chopped onions and herbs. Strain and put the meat into a pie-dish, lined with pastry. Add the pastry, cover the pie and bake for 30 minutes.[48]

The following account recalls a personal experience of swan-eating:

> I remember the incident well. I was taken by my mother as a child, probably aged about twelve to Whiteley's department store in London, where they had a restaurant ... After roast swan, which tasted rather like duck, we had stewed dates. So the date was around 1950.[49]

Although the English are historic swan-eaters, in many places it is thought very unlucky to kill swans; in parts of Scotland and Ireland, this was because they were believed to embody human souls and a person committing the act, or someone from the same parish, would die within a year. There are also several versions of an old Celtic legend that tells of children being turned into swans for periods varying up to 900 years before being released on hearing the sound of a certain consecrated bell. Swans occur repeatedly in ancient pagan Celtic beliefs and there is an appealing tale from Shetland, said to have 'been handed down since Norse times', to explain why the swan has a coat of down while the heron, or haigrie, has merely feathers:

> The two birds were once ordered to keep watch all night, and the one that first heralded the dawn would be rewarded with the 'double down'. The swan became drowsy and fell asleep, while the haigrie remained awake until as daybreak approached she could hardly keep her eyes open. Then the swan, refreshed after a good night's sleep, opened her eyes and noticed the first lightening of the heather and called out to the heron: 'Haigrie, haigrie, daylight in the heather. I hae da double doon and du da single fedder.'[50]

It is in Celtic and Gaelic tales that the story of the swan maiden, in which virgins and virgins alone are turned into swans after death, most commonly occurs:

> A number of the variants of the 'swan maiden' stories involve an aversion to iron, as for instance in the Welsh story where Wastin of Wastinog watched the swan maidens for three moonlit nights, one after the other. He captured one of them and she agreed to marry and stay with him unless he struck her with a bridle and when, later, he accidentally did so, she flew off to the lake with their children and plunged in ...
> One bold theory about the origins of these stories is that they were mythic interpretations of the Bronze Age peoples of the time when they were overruled and dominated by Iron Age Celtic invaders.[51]

There are other popular misconceptions about swans. If you ask any member of the public as they stand admiring the swans in front of the theatre at Stratford-upon-Avon what

It is not unusual to see a young cygnet of the Mute Swan *(Cygnus olor)* being transported on the back of one of its parents.

they know about the birds, the chances are high that you will be offered two notions: first, that they mate for life and second, that a single blow of its wing will break a man's arm (or leg). Mating for life is generally true, although hardly unusual among birds; breaking anyone's limb is pure fancy.

While at the river's edge in Stratford, you need walk only a short distance to discover one of the commonest manifestations of the national fascination with swans in the shape of The Black Swan/Dirty Duck inn. More inns are named after Swans than after any other bird: Black Swans, White Swans, Old Swans, Old White Swans, Four Swans, Cygnets, even Railway Swans; some towns have several. Among other variants are The Swan and Falcon, Swan and Bottle, Swan and Rummer, Swan and Salmon, Swan and Horseshoe, Swan and Sugarloaf; the list goes on. The seemingly inexplicable Swan with Two Necks is probably The Swan with Two Nicks and an allusion to the marks of the Vintners' Company (see page 217). I am told that there was once a Swan with Three Necks in Lad Lane, London, although neither the inn nor the lane still exist. The inn and livery stables called The Swan and Hoop, at The Pavement in Moorgate, was very probably the birthplace of John Keats in 1795. The Swan Tavern in Change Alley in the City of London was the meeting place of the Aurelian Society, the first body in this country to exist for the study of butterflies (see page 75), and a meeting was taking place there when the great Cornhill fire of 25 March 1748 destroyed the premises.

St Hugh of Lincoln (*c.*1140–1200) is reputed to have had a tame Whooper Swan *(Cygnus cygnus)*.

The notion of swans being black is an old one and curiously there are Black Swan inns that pre-date the discovery of Australia, the only place where the Black Swan *(Cygnus atratus)* occurs naturally. Black Swan inn signs nonetheless tend to depict black-coloured Mute Swans rather than true Black Swans. There are also Swan Theatres – a famous historical one in London and a modern one in Stratford – and swans are common as heraldic birds.

The word 'swan' crops up regularly in place names – Swanscombe in Kent, Swanmore in Hampshire and Swanage in Dorset being among the more obvious – but it should not be assumed that these are necessarily connected with the bird because, while in Old English *swan* meant 'swan', *swān* meant 'herd', 'swineherd', 'young man' or 'servant'. Surprisingly, Swansea in Wales has no avian connection but commemorates the tenth-century Swedish King Svend Tveskaeg (Sweyne Forkbeard). Unexpectedly, however, Elterwater in Cumbria is 'Swan Lake' because *elpt* meant 'swan' in Old Norse. 'Swan', or 'Swann', is a fairly common surname but is more likely to relate to occupations than birds. It is obviously the gliding movement of swans on the water that has given rise to the expression 'swanning around', which gained widespread use in the Second World War in relation to tanks moving aimlessly on the battlefield.

Bewick's Swan *(Cygnus columbianus)*

OTHER COMMON NAMES: *tame swan; tundra swan.*

The smaller of the two swans that reach Britain as winter visitors, this yellow-beaked bird can be seen and heard, honking and whooping, on fields and marshes. It was not until 1830 that it was recognized as distinct from the Whooper Swan and named in honour of the naturalist and engraver Thomas Bewick, best known for his *History of British Birds*,[52] who had died two years earlier.

Whooper Swan *(Cygnus cygnus)*

OTHER COMMON NAMES: *elerech (Cornwall); elk (Northern Ireland; northern England, East Anglia); whistling swan; wild swan (Sussex).*

A winter visitor, which occasionally breeds in Ireland and Scotland, the Whooper Swan is larger than Bewick's Swan but otherwise very similar and very noisy, with a wide range of loud, croaking or bugle-like sounds.

St Hugh of Lincoln (*c.*1140–1200) is said to have had a tame Whooper Swan and he is commonly depicted in its company, although how he acquired this unusual bird or why he took a fancy to it is unclear. The regional name 'elk' recalls the Old Norse swan name *elpt.*

A Whooper Swan *(Cygnus cygnus)* displaying. The Whooper Swan is larger than Bewick's Swan *(C. columbianus)*, but otherwise alike.

GEESE

The name 'goose' has changed little since Old English and its origin is certainly very much older. William Lockwood traces it back to an Indo-European word that 'must have come into existence by 3000 BC, at the latest, making goose arguably the most ancient bird name in our vocabulary'.[53] When distinguishing the sexes, the male becomes a 'gander', which is *gandra* in Old English and similarly may be very ancient. The term 'gosling' for the young bird is Middle English. No-one seems to know when the plural of 'goose' became 'geese' and there are alternative uses of the word 'goose', for instance to mean a tailor's smoothing iron, where the plural 'gooses' is acceptable.

An odd feature of the word 'goose' and to some extent 'duck', is why, unlike other bird names, they so commonly have an 'ey' or 'y' ending when used familiarly. 'Goosey-Goosey Gander' must be the best-known instance of this, but the Tavistock Goosey Fair is another. The children's story character Ducky-Lucky has an echo in my own childhood: a small island in the river that ran through the Derbyshire village where I once lived was never 'Duck Island' but always 'Ducky Island'.

There are so many everyday words and expressions in which the word 'goose' appears ('goose-flesh', 'goose-pimple', 'goose-neck', 'gooseberry', 'goose-step', 'goose-month') and countless sayings ('He's cooked his goose', 'He can't say boo to a goose', 'He killed the goose that laid the golden egg', 'What's sauce for the goose is sauce for the gander') that it can be no surprise that the goose is among the oldest of domesticated creatures. It has been kept for centuries for its meat, eggs, feathers and fat, and, as the story of Ancient Rome reminds us, its ability to act as a guard and warning. The farmyard goose is derived from the Greylag Goose *(Anser anser)* and has been domesticated for at least 4,000 years; Egyptian depictions show a recognizably similar bird that had clearly already been subjected to breeding and selection. But it has always been the wild goose that was chased:

> Nay, if our wits run the wild-goose chase, I am done;
> for thou hast more of the wild-goose in one of thy
> wits, than, I am sure, I have in my whole five: was I
> with you there for the goose?
> (*Romeo and Juliet*, Act 2, Scene 4)

The white front that distinguishes the White-fronted Goose *(Anser albifrons)* is on the face rather than the breast.

British geese can be conveniently divided into the grey geese (greyish above, brownish below) of the genus *Anser* and the black geese (dark greyish above, whitish below, with black head and neck) of the genus *Branta.*

Bean Goose *(Anser fabalis)*

OTHER COMMON NAMES: *corn goose; grey goose (Sussex); small grey goose; wild goose (Ireland, Scotland).*

This large goose, with orange-yellow beak and legs, is an uncommon winter visitor to farmland where it forages among grain, potatoes and, of course, beans.

The name was originally used in Lincolnshire and the naturalist Thomas Pennant, when he first mentioned it in 1750, thought it derived from the shape of the beak tip, which was similar to that of a horse bean *(Vicia faba)*; its diet seems a much more obvious source.[54] The Bean Goose was not recognized as being different from the Pink-footed Goose until 1833, but it is now generally considered to be a form of that species, which is a pity because its distinctive English name may therefore disappear.

Pink-footed Goose *(Anser brachyrhynchus)*

OTHER COMMON NAMES: *grey goose (Sussex); long-billed goose (Yorkshire).*

Like a Bean Goose but with pink legs and beak is a reasonably accurate description of this winter visitor to the north of Britain, which has bred here on an odd occasion.

White-fronted Goose *(Anser albifrons)*

OTHER COMMON NAMES: *bald goose; laughing goose; tortoise-shell goose (Ireland); white-faced goose.*

This goose has a pale brown rather than a white breast; the white front is on its face. Together with its orange legs, these serve as its obvious visible distinguishing features, although the local name 'laughing goose' is the best identifying clue as it has a high-pitched and very distinctive cackle, yapping as opposed to barking, as Geoff Sample says.[55] It is a winter visitor and a rare breeder, grazing more on grass than crop residues.

Greylag Goose *(Anser anser)*

OTHER COMMON NAMES: *fen goose; grey goose; marsh goose; stubble goose; wild goose.*

This is the true British resident native goose, although in numbers it now takes second place to the introduced Canada Goose *(Branta canadensis).* In many areas it is also outnumbered by feral populations of its descendant, the farmyard goose, and together they constitute the 'grey goose' of countless stories and the goose that is so often associated in folk tales and on inn signs with foxes. 'The Fox's Foray', or at least the refrain, is a familiar folk song:

> He took the grey goose by the neck,
> And swung him right across his back;
> The grey goose cried out, Quack, quack, quack,
> With his legs hanging dangling down O!
> Down O! Down O!
> The grey goose cried out, Quack, quack, quack,
> With his legs hanging dangling down O![56]

Truly wild populations of the Greylag Goose are gradually spreading out again from their recent restriction to the north of Scotland and the Western Isles and, like many other geese and ducks, are hybridizing with other species.

The name 'Greylag' intrigues people but the otherwise reliable Francesca Greenoak is among many writers to have fallen into the trap of taking the name 'lag' literally, to mean the goose that lagged behind as a resident bird when others flew off on migration, with the fall-back explanation of it being a goose that grazed leas, 'leas' in turn becoming 'lags'.[57] The reality is that 'lag-lag-lag', or 'lag 'em, lag 'em', was the old farmyard call for geese and a 'lag' is simply a goose and possibly a name of great age. It does have many other meanings, among them its use in 'old lag' and also a sixteenth-century word meaning 'urinate', but how many of these have anything to do with geese has yet to be investigated.

Paralleling the situation with swans, the Celtic areas of Britain have long had a taboo against eating goose, although it was kept for ritual sacrifice. In England, it was very different and goose-eating was in its prime in the days before the American custom of eating turkeys, which began at the second Thanksgiving in 1621, was introduced and slowly gained a following. In Charles Dickens's *A Christmas Carol*, the Cratchits took their goose to the bakery for roasting:

> There never was such a goose ... Its tenderness and flavour, size and cheapness, were the themes of universal admiration. Eked out by apple-sauce and mashed potatoes, it was a sufficient dinner for the whole family ...[58]

Even today, it is at Christmas that goose is most likely to be found on the nation's dining tables, but in earlier times it was different. Old Martinmas (originally 12 November but now 13 April) was an occasion for roast goose in some districts, although at Farndale in Yorkshire (where Old Martinmas for some reason was two days earlier), the goose eaten on Martinmas Sunday was traditionally cooked in a pie. Nonetheless, it was Michaelmas (29 September) that for centuries was synonymous with the goose:

> The story goes that Queen Elizabeth was having goose for dinner on Michaelmas Day when the news of the total defeat of the Spanish Armada was brought to her; in remembrance of which she resolved always to eat goose on that day. There is, however, a little

The native Greylag Goose *(Anser anser)* is commonly seen in stubble fields in autumn, hence the regional name 'stubble goose'.

> discrepancy here. Considering that the armada was literally swept off the seas between July 21st and 25th, it is most unlikely that the good tidings must have required two months and four days to reach England. At all events, a goose was regularly served up for dinner long, long before Queen Elizabeth made her appearance in the world. The custom arose out of the practice of the rural tenantry bringing a stubble-goose to their landlords when paying their rent, or to make them lenient if on occasion an excuse had to be tendered in place of the money. At this particular time of the year geese were plentiful and at their best, owing to the benefits they derived from stubble feeding. And since the landlords received many more such presents than they could themselves consume, they forthwith passed them over to their friends or acquaintances. In this way the Michaelmas goose became a standing dinner-dish.[59]

Thomas Blount, in his *Fragmenta Antiquitatis* of 1679, wrote of the widespread practice of goose-eating at Michaelmas:

> Probably no other reason can be given for this custom but that Michaelmas Day was a great feast, and geese at that time most plentiful.[60]

A cookery book of 1709 commented:

> So stubble geese at Michaelmas are seen
> Upon the spit; next May produces green.[61]

'Stubble goose' is a regional name for the Greylag and a 'green goose' was one fed on green pasture. In many places it was thought unlucky not to eat a stubble goose at Michaelmas because hard times for the family would otherwise ensue. An appealing rhyme from the early eighteenth century elaborated:

> Q. Yet my wife would persuade me (as I am a sinner),
> To have a fat Goose on St Michael for dinner:
> And then all the year round, I pray you would mind it,
> I shall not want money – oh! Grant I may find it.
> Now several there are that believe this is true,
> Yet the reason of this is desired from you.

> A. We thinke you're so far from having more,
> That the price of the Goose you have less than before:
> The custom came up from the tenants presenting
> Their landlords with geese, to incline their relenting
> On following payments.[62]

Also:

> ... as early as the tenth year of Edward IV, we read that John de la Hay was bound, among other services, to render to William Barnaby, Lord of Lastres, in the county of Hereford, for a parcel of the demesne lands, one goose fit for the lord's dinner on the feast of St Michael the Archangel.[63]

Today, St Michael's goose is more likely to be a goose obtained from a well-known supermarket but it remains wonderful eating. Mrs Beeton's observation is well worth noting:

> A teaspoonful of made mustard, a saltspoonful of salt, a few grains of cayenne, mixed with a glass of port wine are sometimes poured into the goose by a slit ...[64]

There is a Lincolnshire dish called Mock Goose where, Heaven help them, people force a piece of pork into the shape of a goose and pretend ...

The period around Michaelmas Day was certainly the time when geese were plentiful. Tavistock Goosey Fair, which dates from 1105 and was once a week-long festival for the buying and selling of geese, takes place during the second week of October – although today without its geese. A Roast Goose Fair is held at Redruth on 12 October, while Nottingham Goose Fair, the biggest fair in Britain, takes place on the last three days of the first week of October; the latter achieved immortality when Alan Sillitoe featured it in his 1958 novel *Saturday Night and Sunday Morning*.[65] By then, however, it had fallen a long way from its peak in the Middle Ages when 20,000 geese changed hands. The geese were commonly driven long distances to the city and were often persuaded to walk first through warm tar and then sand so their feet were well shod.

'Wayzgoose' was an old name for the stubble goose (*wayz* being an old word for 'stubble') and also the name of a feast at which goose was served. It was especially common in the printing trade. A Yorkshire feast called Tuning Goose was held to celebrate the safe gathering of the harvest, while in Shropshire the last shearing was called 'cutting the gander's neck'. On the Isle of Man, the old name for the daffodil translates as 'goose-leek' and farmers thought it unlucky to bring daffodils into the farmhouse before the goslings had hatched.

Not all geese have met such an agreeable end as the dining table. Several parts of the country have had variants of a weird and barbaric ritual that entailed horse-riders attempting to pull the greased head from a goose tied to a pole, and, as long ago as 1756, a writer to the *Gentleman's Magazine* was complaining of a practice that he had encountered in the West Country:

> A friend told me lately, that in a cold winter a year or two ago, as he was riding over the moors near Bridgwater, in Somersetshire, he saw a great number of geese dead upon the moors; and upon enquiring into the cause of it, he was informed that it was the custom of the people there, every year, to pick the down off the geese while they were alive, in order to sell it, and then to send the naked geese upon the moors again, where, if the weather grew cold before their feathers grew again, they languished and died. It was some time before I could believe such barbarity was practised in a country which calls itself Christian, and where the principles of humanity are carefully taught. But as I am well assured it is a fact, I beg the favour of you to publish this account in your next magazine, that some humane and generous persons, who live about these places, may do all they can to put a stop to it.[66]

Unfortunately, a seemingly promising rural custom, Cornish Geese Dancing on Plough Monday, which involves much good humour and alcohol, boys dressing as girls and vice versa, and behaving 'in such an unruly manner that women and children were afraid to venture out' has proved misleading.[67] The name is merely a corruption of 'guise-dancing', or 'dancing in disguise'.

The Tavistock Goosey Fair has been held in Devon since 1105. Buyers at this 1955 fair are inspecting the geese.

The flying V-formation of Greylag Geese *(Anser anser)* has been used to predict a variety of events, including frosty weather.

Geese have long been credited with being good indicators of weather changes: the goose 'hath a shrewd guesse of rainie weather' runs a seventeenth-century comment and there is a widespread country saying that, when snow falls, 'the old lady is plucking her geese'. Not surprisingly, the characteristic and extraordinary V-formation in which geese and other wildfowl fly has been used to predict events, including the numbers of weeks of frosty weather.

Many children's stories and rhymes have involved geese, but none more famously than 'Goosey-Goosey Gander':

Goosey Goosey Gander where shall we wander
Up stairs and down stairs and in my Lady's chamber.
Old father long legs will not say his Prayers,
Take him by the left leg and throw him downstairs.(68)

There is an obvious anti-Roman Catholic sentiment here, stemming from the old notion that Catholics were left-handed, but Iona and Peter Opie have studied the rhyme at some length and they conclude that the modern version is an amalgam of two older rhymes. The earliest version they found is from 1784:

Goose-a, goose-a, gander,
Where shall I wander?
Up stairs, down stairs,
In my lady's chamber;
There you'll find a cup of sack
And a race of ginger.(69)

It has been suggested that this nursery rhyme may not be as innocent as it seems and may hark back to the widespread use of the goose as a sexual symbol. Geese were associated with Aphrodite and, in parts of Europe, a dead goose was traditionally carried in bridegrooms' processions. Its wandering into ladies' chambers might have been no mere accident. The notion of 'Mother Goose' (which first appeared in print in French in 1697 as Perrault's *Contes de Ma Mère L'Oye*)(70), with which children will be familiar through rhyme, story and pantomime, may have the same far-off link.

Yet another use of geese to promote a belief comes in an eighteenth-century rhyme protesting at the Enclosure Acts:

Hang the man and flog the woman
Who steals the goose from off the Common,

But leave the far worse villain loose,
Who steals the Common off the goose.[71]

Another well-known goose rhyme of unknown origin but charming imagery runs:

Gray goose and gander,
Waft your wings together,
And carry the good king's daughter
Over the one-strand river.[72]

No-one seems to know who the good king was, or what is meant by the 'one-strand river', because, although the piece is probably very old, there is no record of it before 1844.

'The Fox's Foray' rhyme has already been mentioned (see Greylag Goose, page 225) and another of its many verses runs:

The fox and his wife, without any strife,
Said they never ate a better goose in all their life:
They did very well without fork or knife,
And the little ones picked the bones O!
Bones O! Bones O!
They did very well without fork or knife,
And the little ones picked the bones O![73]

This may have other echoes because goose and duck bones, especially breastbones, have had traditional significance. In many places, country sages claimed to be able to foresee the weather by close examination of the breastbone, although any connection with the modern practice of pulling wishbones, as Francesca Greenoak suggests, seems debatable (see also Mallard, page 235).

Even in this material age, many children are probably familiar with one of the numerous regional variants of the association between the goose and that well-known and traditional playground practice of hair-pulling:

> If a girl has made herself obnoxious, or is so rash as to announce that it is her birthday, her hair is pulled ceremonially, once for each year of her age, plus very often 'one for luck', or 'one to be a good girl in future'. In Manchester they then ask: 'A bull or a goose?' If she replies 'goose' she is let loose, but if she replies 'bull' she receives a further pull. Alternatively they ask: 'A hen or a goose?' if she replies 'goose' she is let loose, but if she replies 'hen' her hair is given another ten pulls.
>
> In Blackburn, Lancashire, the child is asked: 'Cock, hen, goose, or gander?' If she replies 'cock' they say 'Give a good knock'; if 'hen' – 'Pull again'; if 'goose' – 'Let loose'; if 'gander' – 'Whither do I wander'. (Information from boy aged 10, and girl aged 11.)
>
> In Swansea they hope for even further sport. The Tormentors ask 'Hen or goose?' If the child replies 'goose' she is let loose, but if she says 'hen' they sing out 'Do it again', and pull her hair again, a tug for each year, and keep on holding. She is then asked: 'Brick or a stone?' If she says 'stone' they say 'Leave her alone', and her hair is released, but if she replies 'brick' they 'Give her a kick' and still hold her hair. They then ask 'Duck or a feather?' If she chooses 'duck' that is 'Bad luck', but if she chooses 'a feather' she can be consoled with the thought that it is now 'Good luck for ever'.[74]

Children have always had particularly endearing ways of building their relationships, but there is another association with geese that they perhaps found less attractive. Goose grease has acquired an almost legendary status for its lubricating and insulating properties. I can remember my old geography master swearing blind that, in his childhood, he had known people

'Goosey Goosey Gander' is one of the many well-known nursery rhymes to have been illustrated by F.D. Bedford (1897).

The Canada Goose *(Branta canadensis)* has spread, now outnumbering the native Greylag Goose *(Anser anser)*.

who, when they were children, had been liberally smeared with goose grease in the autumn and then sewn into their clothes for the duration of the winter. It has other values too:

> The fat of a Goose tempered and mixed with a Spider and Oyl of Roses together, being used as an Ointment upon the breasts, preserveth them safely, as that no milk will coagulate or curdle in them after any birth.[75]

It has also been used (not necessarily simultaneously) as a cure for croup, sore throats and:

> ... pains, aches or numbness from a cold cause, it softens hard tumors, and contractions of the nerve and sinews, it is exceeding good against a cold gout, it also cures the alopecia, and chaps of the lips, being applied to the grieved parts ...

If you continue to ail, then hold on to your goose because:

> ... Half a dram of the powder made of the skin of the feet drank, is commended to cure the mensium profluvium, outwardly it is applied to kibey heels, the ashes made of the feathers stop bleeding; the heart and lungs pulverized and drank, are good against pectoral disorders; the gall mixt with honey cures most disorders of the eyes ...

And if you are really serious about goose medicine and are made of stern stuff, then note that:

> ... The dung is a specific against the jaundice, it helps to cure the scurvy, dropsy, gout, and green sickness, it also opens all manner of obstructions in the womb, liver, and spleen ... the dung which is green, and gathered in the spring is best.[76]

The rather obscure St Werburga (died *c.*700), traditionally held to be the daughter of Wulfthere, King of Mercia and his wife Ermengild, is often associated with a goose and depicted in its company. This stems from a belief that she restored a goose to life, although the association seems spurious because the story was borrowed by the eleventh-century saint/biographer Goscelin of Saint-Bertin from one associated with a quite different personage: the Flemish St Amelburga.

Gosport in Hampshire may be a market place where geese were sold, Goscote in Leicestershire is 'a hut for geese', Gosbeck in Suffolk is 'goose stream', Gosford in Devon is 'goose ford' and Goswick in Northumberland is 'goose farm' and has been since 1202: this is perhaps evidence of specialist goose-keeping at an early date, in contrast with Gosfield in Essex, which is simply 'a feld frequented by wild geese'.

Canada Goose *(Branta canadensis)*

OTHER COMMON NAMES: *cravat goose (southern Scotland).*

This large, noisy, honking bird has become the most familiar goose to many people since its dramatic spread in recent years. It was introduced as an ornamental bird to country estates sometime in the seventeenth century (it was present in St James's Park in London in 1678) where it more or less remained until, for no very apparent reason, it began to escape and spread in large numbers during the twentieth century. There are believed to be well over 50,000 birds in Britain now and they dominate many ornamental lakes, fouling the surrounding grassland.

They are typical black geese but happen to be the largest and, like swans, they have become major factors in bringing down power lines and so cutting off electricity to rural areas. Various measures aimed at limiting their spread have been suggested and attempted, but none has proved very effective.

> Pilots have been warned to be on guard against potentially catastrophic bird strikes following an upsurge in Canada geese activity around Heathrow ... The geese present a particular risk to aircraft because of their size, typically up to 16 lb, and their tendency to fly in large formations ... A senior British Airways pilot said ... 'I have never known the airport under siege from birds in the way it is at the moment.'[77]
>
> A woman stopped breathing after being knocked out when a low-flying Canada Goose hit her in the face ...

she was treated for a head wound and taken to hospital, but was discharged later.[78]

Barnacle Goose *(Branta leucopsis)*

OTHER COMMON NAMES: *bar goose (Essex); claik, clakis (Scotland); Norway barnacle (Ireland); rood goose; routhecock (Orkney); tree goose; white-faced barnacle.*

To all intents and purposes, the Barnacle Goose is just another species of black goose, considerably smaller than the Canada Goose and the only goose with an all-white face. It is a local winter visitor from the Arctic tundra to the north and west of Britain. Behind this unimpressive *curriculum vitae* lies one of the oddest and weirdest snippets of British animal lore: the Barnacle Goose was once believed to grow from something very unusual. William Turner, in *On Birds* (1554), wrote:

> None have seen the Barnacle Goose's nest or egg: nor is this surprising, since such geese are said to have spontaneous generation in the following way. When the firwood masts or planks of ships have rotted in the sea, then a kind of fungus breaks out upon them: in which after a time the plain form of birds may be seen, and these become clothed in feathers, and eventually come to life and fly away as Barnacle Geese.[79]

For centuries it was believed that Goose Barnacles *(Lepas anatifera)* became Barnacle Geese *(Branta leucopsis)*.

Yes, it was for centuries thought that Goose Barnacles became Barnacle Geese.

> And all be turn'd to barnacles ...
> (*The Tempest*, Act 4, Scene 1)

Even their alternative name, 'tree goose', harks back to a Scottish belief that the fruits of certain trees fall into the sea and grow into geese. The last entry in John Gerard's *Herball* of 1597 is the 'Goose tree, Barnacle tree, or the tree bearing Geese':

> They are found in the North parts of Scotland and the islands adjacent Called Orchades, certaine trees whereon do grow certaine shells of a white colour tending to russet, wherein are contained little living creatures: which shells in time of maturity doe open, and out of them grow those little living things, which falling into the water do become fowles, which we call Barnacles; in the North of England, brant Geese; and in Lancashire, tree Geese ...[80]

What is most odd is that no-one has been able to determine how this very peculiar belief originated and otherwise knowledgeable people have certainly been taken in by it. Giraldus Cambrensis, writing in 1185, said:

> I have frequently seen with my own eyes, more than a thousand of these small birds, hanging down on the sea-shore from a piece of timber, enclosed in their shells and already formed.[81]

Admittedly, with a fair bit of imagination, the pendulous form of the Goose Barnacle could just be taken for the neck and head of a goose, although someone might have taken the trouble to wait and see what happened to them. Nonetheless, as recently as the twentieth century, Barnacle Goose was still eaten in parts of Ireland on Fridays and during Lent because of a belief that these birds were not born of the flesh. None of this was of concern to the Romans; Pliny thought roast Barnacle Goose 'the most sumptuous dish known to the Britons'.

Brent Goose *(Branta bernicla)*

OTHER COMMON NAMES: *barnacle (Ireland); black goose (Essex); brabt (Norfolk); brand goose; clatter goose (East Lothian); crocker; horie goose; horra goose (Shetland); quink goose; rat goose; road goose; rood goose; rott goose; ware goose (Durham).*

The Brent Goose is very similar to the Barnacle Goose and possibly belongs to the same species, but it is a visitor to more southern parts of Britain and has a softer voice. The name 'brent' comes via 'brant' from the Old Norse *brandgads*, meaning 'burned [or black] goose'. The names 'rood goose' and 'horra goose' also come from the same source and were perhaps originally imitative of the call. 'Quink' is also imitative and 'it is possible that the brand of ink of the same name was called after the goose-quills which were used as pens'.[82]

DUCKS

Ducks are smaller than geese and most have rather broader beaks; they are also more numerous in terms of both individuals and species. In most species, the males are easily identified and much more strikingly coloured than the females, which can be confusingly similar to each other. Like geese, ducks have been domesticated for centuries, if not millennia. Almost all of the many varieties of domestic duck, whether bred for functional or ornamental purposes, are derived from the native Mallard *(Anas platyrhynchos)*, one of the most adaptable of all birds, and it is in relation to the domestic breeds that most lore has originated, just as with geese.

The name 'duck' has always meant 'duck' in the sense of 'diving' or 'bowing down low', as it does in the modern verb 'to duck', but there is some mystery about its use for birds that dive down in the water. The ever-reliable William Lockwood says that it has been used in this way only since the fifteenth century, although Eckwall, in his *Concise Oxford Dictionary of English Place Names*, says the village of Dukinfield in Cheshire which dates from 1285, is possibly 'a field frequented by ducks'.[83] Before the fifteenth century, ducks were generally called *ende* ('last recorded *c.*1475'), which came via the Old English *ened* from the same source as the Latin name *Anas*; hence Andwell in Hampshire ('duck stream'), Anmer in Norfolk ('duck mere'), Enborne in Berkshire ('duck stream'), Enford in Wiltshire ('duck ford') and Enmore in Somerset ('duck mere').

The word 'duckling' for the young bird was first recorded in 1440. The word 'drake' for the male apparently pre-dates 'duck' and was first used for the bird around 1300, although the name Sheldrake, which incorporates it, is considerably older (see page 234). William Lockwood believes that the name arose first for domesticated birds and was an imitation of the coarser call of the male; in fact, it is the female that utters the harsh 'quack', the male's call being a softer 'queek'.[84] The game of skimming a stone over water has been called Ducks and Drakes since 1583.

The now very familiar imitative word 'quack' is as recent as 1617, although there is an earlier form, 'quackle', from 1564. 'Quack' was once also used in relation to the call of Frogs. 'Quack-quack' as a 'nursery' name for ducks is nineteenth century. 'Quack' meaning 'a medical pretender' comes from an old word 'quacksalver' and may have nothing to do with birds.

Ducks in general have long been thought of as rather stupid and comical. It is no coincidence that Donald Duck was a duck, rather than a goose, or even a chicken, which were both thought more sensible birds. It is often forgotten that, when Donald Duck first appeared in 1934, it was in a Walt Disney cartoon called *The Wise Little Hen*. And it was at the height of McCarthyism in the USA that the Trade Union leader Walter Reuther devised his infamous test for communist sympathies: 'If it looks like a duck, walks like a duck and quacks like a duck, it's a duck'.

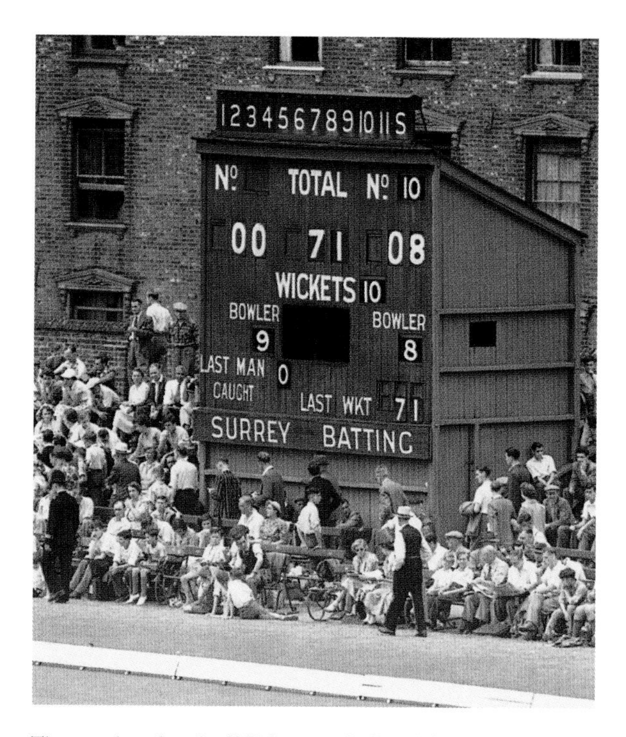

The scoreboard at the 1952 Surrey v. India cricket match at the Oval shows the last batsman out for 'a duck'.

Just like 'goose', the word 'duck' has come to have numerous other connotations, all indicative of a bird that has been close to humans and human affections for a very long time. One of the commonest in modern English is to 'make a duck', meaning a score of nought in cricket, which was originally 'a duck's egg', for no better reason than that someone decided that a duck's egg looks vaguely like the figure zero. It soon became a 'duck', 'a pardonable contraction from duck-egg' (*Daily News*, 1868)'.[85] There is a record of the future King Edward VII, then Prince of Wales, having scored 'a duck's egg' in the 1860s, but the earliest of all references is of some unfortunate batsman who, on 7 July 1856, scored 'another duck, so making a pair'.[86] At least the connection with ornithology has not been totally lost; the former England batsman Michael Atherton has said that he never ate duck on the eve of a match.[87]

Among other popular modern 'duck' uses are the expressions 'water off a duck's back' and 'lame duck'. 'Ducky' as a term of endearment, greatly beloved of bus conductors and camp comedians, in fact dates from 1819, although 'duck' itself used in this way is rather older:

> Will you buy any tape,
> Or lace for your cape,
> My dainty duck, my dear-a?
> (*The Winter's Tale,* Act 4, Scene 4)

Ducks have long been popular sporting quarry, especially since the development of sporting firearms (fowling pieces) over the past two or three centuries, the punt-gun being a particularly formidable item of hardware that can have a barrel over 3 metres (10 feet) long. Did you ever sing the slightly macabre nursery rhyme 'Sam, The Sportsman'?

> There was a little man, and he had a little gun,
> And his bullets were made of lead, lead, lead;
> He went to the brook, and shot a little duck,
> Right through the middle of the head, head, head.
>
> He carried it home to his old wife Joan,
> And bade her a fire for to make, make, make,
> To roast the little duck he had shot in the brook,
> And he'd go and fetch her the drake, drake, drake.[88]

Nowadays most species are protected from people with guns but many have suffered particularly badly in recent years because they form very vulnerable prey for feral Mink (see page 416).

Egyptian Goose *(Alopochen aegyptiacus)*

An escape from ornamental wildfowl collections, this medium-sized, multicoloured bird from Africa has become well established in East Anglia, whence it has begun to spread to other parts of the country. Despite its name, it is related to the Shelduck, not to geese.

It is grey beneath, with a pink beak and legs, a grey-green back, pale brown head and neck, white wings with black tips and a large patch of green primary feathers. It tends to nest in old holes in trees and its continued expansion has been described as 'probably unstoppable'.

Shelduck *(Tadorna tadorna)*

OTHER COMMON NAMES: *bar duck (Ireland); bar gander (Essex); bar goose; barrow duck (Gloucestershire, Somerset); bay duck (Norfolk); bergander; burranet (Cornwall); burrow duck; links goose (Orkney); pirennet; St George's duck; scale duck, sheld fowl (Orkney); sheldrake; shell (Yorkshire); skeel duck (Scotland); skeel goose; skeldrake; sly duck (Yorkshire); sly goose (Orkney); stock annet (eastern Scotland).*

If ever a bird was neither goose nor duck, this is it: a large, strikingly coloured creature, mainly white but with a dark green head and a chestnut band across its chest. Shelducks have traditionally been thought of as mainly coastal birds that feed on invertebrates and characteristically nest in old Rabbit burrows. In recent times, however, they have tended to be found further inland, a habit that has been correlated with an overall increase in their numbers. They have suffered in the past through being shot for sport and were also once turned out of Rabbit warrens when Rabbits were an important food source.

> The rabbit burrows had another usage at Brean [Somerset] where they became nesting places for Shelduck which in Somerset are called Burrow Duck

On calm, shallow waters the punt is an ideal craft for wild-fowling. The punt-gun could be as long as 3 metres (10 feet).

> or Barrow Duck due to their nesting habits. This is an area of the Bristol Channel well known as a nesting ground since they fed on a certain type of worm found in the mud banks.[89]

The male was called the Sheldrake before the species became known as the Shelduck; this first occurs as a bird name in about 1325, although there is a record of a man named Adam le Sceldrake in Suffolk in 1275. It has nothing to do with shells, although it is sometimes written 'Shell Duck'. In truth, it is 'sheld duck', a name meaning 'variegated duck' and a reference to its coloration.

Muscovy Duck *(Cairina moschata)*

When I was a boy, we kept Muscovies with our domestic ducks, and they were once most likely to be seen in smallholdings, farmyards and ornamental wildfowl collections. Feral populations have increased in recent years, however, and they are now well established in many areas. They are ungainly green-black and white birds with a large red outgrowth, known as a caruncle, on the beak.

They come from South America, not from Moscow, and the misleading name 'Muscovy' is a corruption of 'musk duck'; they emit a musk-like odour from a small pouch beneath the beak. Their adaptation to domestication is nothing new; they were kept for meat and eggs in Peru and Colombia long before the Europeans arrived there.

Mandarin *(Aix galericulata)*

Despite their name, most Mandarins occur naturally in Japan rather than in China, but a very great number, perhaps 7,000, now live in Britain, having escaped from collections. There are now 'more wild Mandarin ducks in Surrey than in China'.[90]

Their popularity is no surprise because they are small and exceptionally pretty birds, the male having a delightful crest and sail-like orange feathers that stand proud from its back. They nest in holes in trees and seem likely to increase their range in this country.

Wigeon *(Anas penelope)*

OTHER COMMON NAMES: *bald pate; black wigeon; cock winder (Norfolk); easterlings; golden head (Ireland); grass whew (Yorkshire); half duck; lady duck (Lancashire); lady fowl; pandle whew, smee duck (Norfolk); whew duck (Northumberland); whewer; whim; whistler; winder (eastern England); yellow poll (Ireland).*

A chestnut head on a more or less grey-and-white whistling duck identifies the male of the Wigeon, a bird that is often thought of

The Mandarin *(Aix galericulata)* male has a striking crest and sail-like orange feathers when in breeding plumage.

simply as a winter visitor, although considerable numbers do breed in this country, nesting on the ground close to fresh water. They are predominantly northern and eastern birds in Britain.

They have been called 'Wigeons' (sometimes spelled 'Widgeon') since the early sixteenth century, the name probably being originally imitative of their call. Probably because they were easy prey for duck-hunters, the name also came to be used in the eighteenth century to mean a stupid person. 'Whew duck' and 'whewer' are imitative of the call; a 'pandle' is a Winkle (see page 29), which the ducks collect as they feed on bottom-living plant life.

Gadwall *(Anas strepera)*

OTHER COMMON NAMES: *gray; grey duck; rodge; sand wigeon (Essex).*

Gadwalls are betrayed by their sound rather than their appearance; on the ground, the black bottom of the male is the only obvious feature to distinguish it from a small female Mallard. In flight, a white flash at the rear of the wing stands out.

It is the chattering call that gave rise to the imitative name 'Gadwall', which dates from the seventeenth century. The scientific name *strepera* arose from a belief that the birds were particularly obstreperous.

Originally visitors to the north of Britain, Gadwalls were introduced to Norfolk in the nineteenth century and have spread very successfully to many parts of the country.

Teal *(Anas crecca)*

OTHER COMMON NAMES: *jay teal (Kirkcudbrightshire); speckled teal; tael duik (Scotland); throstle teal (Lancashire).*

My first contact with Teal came, like that of many another country boy, through the Teal, Blue and Silver, one of the great series of nineteenth-century Scottish trout and sea-trout flies. Even if anglers rarely had the chance to use them, the name was part of angling folklore. Sadly, a good many Teal and a good many other ducks have been sacrificed over the years to satisfy fishermen's demand for their feathers.

A great many more have fallen prey to other sportsmen because this little whistling bird has long been regarded as excellent for the table; along with Mallard and Wigeon, their bones were found at the Roman settlement of Silchester.

They are very pretty birds – the drake with a chestnut head and pronounced dark green eye-stripe – but they are small. Mrs Beeton pointed out that two were required for a dish and cooked them for only 9 to 15 minutes; she thought them generally better flavoured 'after the frost has set in' and the cost was 1s 6d each.[91]

There is a breeding population of Teal in the more northern and upland areas of Britain, where increased afforestation seems to have worked against them, although they are supplemented by large numbers of winter visitors.

The male Teal *(Anas crecca)* has a pretty and distinctive green eye-stripe.

'Uniquely among British birds' says William Lockwood, 'the Teal is not known to have been called by any other name'.[92] Can this really be true? The name was first recorded in 1314 and was evidently originally imitative of the call.

Mallard *(Anas platyrhynchos)*

OTHER COMMON NAMES: *common duck; grey duck (Dumfries-shire, Lancashire); mire duck (Angus – Forfar); moss duck (Aberdeenshire, Renfrewshire); muir duck (Stirlingshire); stock duck (Shetland, Orkney); wild duck.*

The vast numbers of Mallard resident in Britain are supplemented by winter visitors, thus raising the total still higher. They are present on almost every waterway, where they outnumber other ducks and generally all other wildfowl too. Not least among the reasons for their success is that they are simply very adaptable and nest in all kinds of improbable places, often well away from their traditional sites on the ground close to water. Mallard nests can be found in trees and on buildings, often very close to human habitation and, although they are strong fliers, they are neither fast nor graceful in the air. It was therefore ironic that when one of Sir Nigel Gresley's streamlined A4 Pacific locomotives set the world record for steam traction of just over 200 kilometres (126 miles) per hour in 1938, it was not the 'Merlin', 'Falcon' or 'Sparrowhawk' engine but the 'Mallard'.

Mallard are also the largest of British inland ducks and, despite being so common, they are attractive, the drake, with its glossy dark green head, white neck ring and chestnut breast, being a very handsome creature. Look closely and you will see a tiny, curled-up twist of feathers on top of its tail; then look closely at the various strangely coloured ducks that may accompany Mallard on the local pond or in the local farmyard and you

Ducks feature in inn names throughout the country, including The Dog and Duck in London's Soho.

will see that same tuft, which betrays their relationship. Most varieties of domestic duck are indeed derived from Mallard, and, like the domestic goose, ducks have had a long and close association with mankind; the Chinese were keeping Mallard 2,000 years ago.

The name 'Mallard' came from Norman French but was originally used to denote the male only, the species as a whole later being called 'wild duck'. It is only since the mid-eighteenth century that it has been more usually employed in its modern sense of the entire species.

Among the breeds derived from Mallard are the strikingly pure white 'decoy' ducks used by wildfowlers to attract wild birds. Decoying was the principle means of catching ducks, both for sport and for eating, in the days before sporting guns became available. Old decoy ponds, a number of which still exist in one form or another, were equipped with tunnel nets into which the birds were driven. These were in use as early as the reign of King John (1199–1216), although, by the sixteenth century, England saw the introduction of:

> ... the more sophisticated Dutch technique, whereby dogs were employed to appear from behind a series of screens to arouse the curiosity of the ducks and so entice them up the pipes.[93]

In Shakespeare's time, duck-hunting often meant hunting a tame bird in the water and allowing it to be chased and caught by spaniels; the many inns still named The Dog and Duck, or The Duck in the Pond, are reminders of this crude pursuit.

Mallard drakes appear to be particularly promiscuous and many a female has fallen foul of several of them in rapid succession. In reality, the drakes abandon their mate as soon as they have settled her onto the eggs and proceed immediately to find another, a practice called by ornithologists, believe it or not, 'serial monogamy'. Nonetheless, female Mallards do have the reputation of being good and caring mothers, carefully fostering their offspring. It was their fussing that led Shakespeare to write of Antony and his affection for Cleopatra:

> ... Antony,
> Claps on his sea-wing, and, like a doting mallard,
> Leaving the fight in height, flies after her ...
> (*Antony and Cleopatra*, Act 3, Scene 10)

All Souls' College, which has some of the finest buildings in Oxford and whose peaceful existence stems from an absence of students, has long been regarded as an institution blessed with a measure of eccentricity. In its glorification of the Mallard, however, there is no doubt that the college is something rather special. As James Morris, in his *Oxford* (1965) says:

> As for a preposterous ritual, no event in Europe can be much sillier, not the most footling country frolic or pointless Anatolian orgy, than the Ceremony of the Mallard at All Souls.[94]

All Souls is replete with Mallard iconography and the bird appears in a multitude of guises: there is a model mallard of some sort in most rooms; the nineteenth-century chairs in the Hall are carved with the college crest and a Mallard; there is a Mallard-shaped telephone; and there is a splendid and very accurately painted plaster Mallard drake on the wall of Hawksmoor's college buttery, which came from a Gothic pavilion built in the garden in 1753. All of which begs the question why; to which the answer is not entirely clear. Conventional wisdom says that, when the foundations for the front quadrangle were being dug in the 1440s, a large Mallard was discovered hiding in a drain and was taken by Archbishop Henry Chicheley (?1362–1443), the College's founder, as 'a sure token of the future prosperity of his good work'.[95]

In fact, there is no evidence that such a bird ever existed and the story did not appear until the seventeenth century. The truth is probably similar to that behind much of British animal lore: it is a story that has become warped in the telling over the years. It appears that a thirteenth-century seal, bearing the image of a mythical griffin and the name of a clerk called Malard, was discovered while digging a drain in the sixteenth century. From then on, it was fertile imagination and embroidery all the way.

None of which has stopped the dons of All Souls from enjoying themselves, sometimes to excess. When the Mallard itself was first mentioned in the seventeenth century, the College Visitor, Archbishop Abbott, warned that 'Civil men should never so far forget themselves, under pretence of a foolish mallard, as to do things barbarously unseeming'.[96]

The Fellows elect a Lord Mallard, who undertakes a number of ceremonial duties and, twice a year, at an appropriate dinner, composes and delivers an additional verse to the song, recounting some recent college news and activity.

Matters really become exciting on 14 January, the College Foundation Day. In times past, this seems to have been an annual celebration of the Mallard but, since 1801, the full ritual has taken place only in the first year of every century, when the Fellows set off in nocturnal pursuit of the mythical bird, a journey that takes them over the college roofs. For the college itself to admit that, some 400 years ago, 'it is possible that during this period the Mallard was generally associated with debauchery and revelry' is compelling. Unfortunately, 'since the Mallard ceremony happens only once every hundred years, no Fellow can take part in it twice: there are no experts or custodians of the ceremonial, and our records of it are thin'.[97]

It is known nonetheless that, in 1801, the Fellows hunted a live Mallard tethered to a pole and drank its blood; its head remains today, stuffed and with a slightly surprised look, mounted in a glass case in the Coffee Room. In 1901, this was considered barbarous and they substituted a stuffed Mallard, which was carried on a pole in front of the Lord Mallard as the Fellows processed. A passer-by witnessed the 1901 ceremony:

> [There] came a few men with torches, one carrying before him on a long pole the stuffed Mallard, then the Lord Mallard carried on a chair shoulder high by four men. At his side walked two others carrying wands in one hand and with the other ready to steady his lordship should it be needful.[98]

The Lord Mallard on this occasion was Cosmo Gordon Lang, subsequently Archbishop of Canterbury. With 35 other Fellows, he later sat down to the Mallard Dinner, which surprisingly, appeared to include just about everything except Mallard.

On 14 January 2001 the ritual was revived, with Dr Martin West as the Lord Mallard, and the Fellows sang, as Fellows have sung down the years:

The Griffin, Bustard, Turkey & Capon,
Lett other hungry Mortalls gape on
And on theire bones with Stomacks fall hard,
But lett All Souls' Men have ye Mallard.

Hough by the bloud of King Edward
By ye bloud of King Edward,
It was a swapping, swapping mallard!

Some storys strange are told I trow
By Baker, Holinshead & Stow
Of Cocks & Bulls, & other queire things
That happen'd in ye reignes of theire Kings.

Hough the bloud, &c.

The Romans once admir'd a gander
More than they did theire best Commander,
Because hee saved, if some don't foolle us,
The place called from ye Scull of Tolus.

Hough the bloud, &c.

The poets fain'd Jove turn'd a Swan,
But let them prove it if they can.
To mak't appeare it's not atall hard:
Hee was a swapping, swapping mallard.

Hough the bloud, &c.

Hee was swapping from bill to eye,
Hee was swapping from wing to Thigh;
His swapping tool of Generation
Oute swapped all ye winged Nation.

Hough the bloud, &c.

Then lett us drink and dance a Galliard
In ye Remembrance of ye Mallard,
And as ye Mallard doth in Poole,
Lett's dabble, dive and duck in Boule.

Hough the bloud, &c.[99]

Verse five, known as the indecent verse and omitted for some years, was restored in 2001 when the Fellows sat down to the Mallard Centennial Dinner, which did at least include a duck:

Salade du ho**M**ard
Terrine de foie gr**A**s
Fi**L**et de turbot
Canard rôti à **L**a cocotte
Sorbet de ch**A**mpagne
Souffle de f**R**omage
Tarte **D**e chocolat et de fruits[100]

I understand that an edict was issued in advance of the 2001 event, forbidding any Fellow from speaking to the press about the matter. Perhaps the dons felt that their ancient and perfectly harmless ritual might be subject to public ridicule, and subsequently the correspondence columns did include some strongly held views, not least from another Oxford college that was suspected of harbouring not a Mallard but sour grapes.

Ritual or not, for many people duck flesh is an unsurpassed meat, even today, when the carcasses of intensively farmed birds are on every supermarket shelf. Commercial duck-farming has long been associated with Aylesbury in Buckinghamshire and with the county of Norfolk, but it is not apparent why they have become known there as 'ducklings' rather than 'ducks'. 'A Mallarde is good after a frost' said *A Propre New Booke of Cokery* in 1545,[101] and roast duck undoubtedly represents the apogee, but Gervase Markham (*c.*1568–1637), in *The English House-wife*, offered an interesting variant:

> To boil a mallard curiously. Take the mallard when it is fair dressed, washed and trussed, and put it on a spit and roast it until you get the fat and gravy out of it. Then take it from the spit and boil it, and take the best of the broth into a pan: put to it the gravy which you saved, with a piece of sweet butter, currants, vinegar, pepper and grated bread. Thus boil all these together, and when the duck is boiled sufficiently, lay it on a dish with the broth upon it, and so serve it forth.[102]

Wales too has a recipe that is uncommon today:

> Boiled Salted Duck
>
> 1 duck; 1 lb. onions; a little butter, flour and milk; salt and pepper.
>
> Salt the duck the day before. Put in a saucepan with enough water to cover and simmer for one and a half hours. Make onion sauce by boiling the onions in milk and water, melting the butter in another saucepan, stirring in flour and gradually adding boiled milk and onions. Dish up the duck and pour the sauce over it.[103

The lore of duck-eating, however, does not end with picking over the carcass. Francesca Greenoak believed that:

> The 'picking of the wishbone' and a smattering of weather lore may be the last living tokens of early beliefs which carried much more important meaning. Nowadays, most people simply break the wishbone and wish on the larger piece but in the past, the bone was carefully examined for signs which, among other things, would forecast the weather. Colour was an important feature, as may be seen in this quotation from Notes and Queries (at as late a date as 1875): In Richmondshire, some persons say that the breastbone of ducks after being cooked are observed to be dark-coloured before a severe winter, and much lighter coloured before a mild winter.[104]

Ducks have also been linked with weather forecasting in more plausible ways:

> When November's ice is strong enough
> to bear the weight of a duck,
> The winter will be nothing more than
> mud and bloody muck![105]

An alternative and slightly more decorous version runs:

> If there's ice at Martinmas will bear a duck
> There'll be nothing after but sludge and muck.[106]

Also:

> If ducks fly backwards and forwards, and continually plunge in water and wash themselves incessantly, wet weather will ensue.[107]
>
> If ducks dart through the pond they are swimming in and flap their wings, it is looked upon as a coming breeze.[108]

Or, as they say in Scotland:

> When ducks are driving through the burn,
> that night the weather takes a turn.[109]

The belief is apparently widespread that taking duck's eggs into the house for setting after sunset is bad luck because they will never hatch into ducklings.

For some seriously dotty duck lore, you need look no further than the ancient belief of disease transference:

> Powers beyond the reach of man were able to give a particular disease to every sufferer; and since those powers settled in the person of a witch or a medicine man, it enabled them to transfer one creature's distemper to another. So was it not possible that an ordinary human being should be able at least to transfer disease to a slave, a dog, or a horse? Pliny speaks of pains in the stomach being cured by transferring the ailment into a puppy or a duck ...

From thence, it is but a short step to an old English cure for thrush:

> ... Capture the nearest duck that can be met with, and place its mouth, wide open, within the mouth of the sufferer. The cold breath of the duck will be inhaled by the child, and the disease will gradually, and as I have been informed, not the less surely take its departure.[110]

The special properties attached to goose grease have already been described (see Greylag Goose, page 225), but Mallard too are not to be left out of the reckoning:

> If that your hearing fail, an old disease,
> Is cur'd with Earth-worms boyled with Ducks grease.[111]

Friends who are hard of hearing have enquired what is to be done with the unholy blend: do you put it in your ears or eat it?

The male Mallard *(Anas platyrhynchos)* is notoriously promiscuous while the female is reputed to be a devoted mother.

Pintail *(Anas acuta)*

OTHER COMMON NAMES: *cracker; harlan (Wexford); lady bird (Ireland); sea pheasant (Hampshire); sprig tail; thin neck (north); winter duck.*

A handful of these very individual birds breed in Scotland and East Anglia, but they are also frequent winter visitors. The drake is a pretty, distinctive bird with its eponymous long black tail, a brown head and long white neck. It is a rather gently whistling species, nesting on the ground and feeding omnivorously.

They tend to be coastal birds in winter and the name 'sea pheasant' comes from the Solent, where the swimming flocks look superficially like Pheasants and wildfowl in general have the local name of 'spratters': birds that catch Sprats.

Garganey *(Anas querquedula)*

OTHER COMMON NAMES: *cricket teal (Hampshire); garganey teal; pied wigeon; pied wiggon; summer duck; summer teal (Norfolk, Somerset).*

'Garganey' is unusual among British animal names in having come via Switzerland from Italy, where the name *garganello* is imitative of the call.

They are small birds, barely bigger than Teal, but with a distinctive white eye-stripe. Garganey are omnivorous, ground-nesting ducks that have spread significantly in recent years from their old stronghold in the east of England.

Shoveler *(Anas clypeata)*

OTHER COMMON NAMES: *beck (Norfolk); blue-winged shoveler; blue-winged stint; broad bill; kertlutock; kirk tullock; maiden duck (Ireland); scopperbill (Norfolk); sheldrake (Waterford); shovelard, spoon beak, spoon bill (Norfolk); shovel bill; whinyard (Waterford).*

Shovelers are unmistakable: their beaks are long, flattened at the end, and work just like shovels, sieving out small particles of plant material and animal life from the surface of the water. The drake's green head and chestnut flanks complete a very striking picture. At one time a popular target for sportsmen, they have recovered and spread considerably over the past century. 'Shoveler' (sometimes 'Shoveller') is a fifteenth-century name, although 'shovel' itself is Old English.

Pochard *(Aythya ferina)*

OTHER COMMON NAMES: *blue poker; bull-headed wigeon (Northern Ireland); diver; doucker (Roxburghshire); dun bird (Ireland, Dumfriesshire; Essex); dun air, dun curre, dun poker (Yorkshire); freshwater wigeon, gold head (Northern Ireland); great-headed wigeon; poker (Lincolnshire, Hampshire); poker duck; red-eyed poker; red-headed curre; red-headed poker; red-headed wigeon (north); red neck (Cheshire); smee duck (Norfolk); snuff-headed wigeon; vare-headed wigeon; well plum; whinyard (Wexford); wigeon diver (Cork).*

Pochards are on the increase, seeming in recent times to have taken advantage of artificial reservoirs and flooded gravel workings, although they have been British breeding birds for only around 200 years. All earlier references to them are simply as winter visitors. They are very smart, neatly turned-out birds, the drake with a chestnut head, black neck and rump on a grey body. Diving ducks, they feed on submerged plant life and small animals.

'Pochard' is a mystery name; even its pronunciation is arguable. It may relate to the bird's feeding habit, but whether or not it derives from the words 'poking' or 'poaching', it all amounts to much the same thing.

Tufted Duck *(Aythya fuligula)*

OTHER COMMON NAMES: *black curre (Hampshire); black poker (Norfolk); black topping duck (Yorkshire); black wigeon; blue-billed curre (west); blue neb (Northumberland); crested diver (Ireland); curre wigeon (Somerset); doucker (Argyllshire – Islay); dovver (east); gold-eye duck (Wexford); magpie diver; white-sided diver (Armagh); white-sided duck.*

'Crested duck' might seem a more obvious name for these black-and-white diving ducks but 'tufted' they have been since the seventeenth century. The Yorkshire name 'black topping duck' is delightful and these ducks have until recently been mainly northern and western birds; Scotland is their stronghold, although it has been only in the last 150 years that they have colonized Britain and the breeding birds are still supplemented by winter visitors.

The freshwater Zebra Mussel *(Dreissensia polymorpha)* (see page 39), which arrived from continental Europe, presumably attached to ships, and was discovered in the London Docks in

The Tufted Duck *(Aythya fuligula)* is easily recognized by its black-and-white colouring, drooping crest and vivid yellow eye.

1824, may have aided their spread, because small molluscs are their main food, along with other aquatic invertebrates and some plant life.

The Northumberland name 'blue neb' embraces the Old English word 'nebb', which is 'the oldest word in our language for a bird's beak'.

Scaup *(Aythya marila)*

OTHER COMMON NAMES: *black duck (Somerset); black-headed diver; black-headed wigeon; blue neb (Northumberland); bridle duck (Dublin); covie duck (Northumberland); dun bird (Essex); frosty-back wigeon; golden-eyed diver; green-headed diver (Belfast); grey-backed curre (Scotland); Holland duck (Angus – Forfar); mule (Wexford); mussel duck (Norfolk); Norway duck (Belfast); Norwegian duck (Banffshire); scaup duck; silver pochard (Yorkshire); spoonbill duck (East Lothian); white-faced duck.*

Scaup are winter visitors to the sea shore, mainly to the north of Britain, and occasionally breed on freshwater sites in Scotland. They are superficially very similar to their close relative, the Tufted Duck, but lack the tuft. 'Scaup' sounds like a Nordic word and probably is originally, although it applies not to the bird but to the mussel bed on which it feeds; the names 'scaup duck' and 'mussel duck' are more appropriate.

Eider *(Somateria mollissima)*

OTHER COMMON NAMES: *colk; coo-doos; crattick (Scotland); cudberduce, culverts (Northumberland); dunter duck; dunter, dunter goose (Shetland, Orkney); dusky duck; edder; eider; great black-and-white duck; St Cuthbert's duck (Northumberland).*

Like many people, I grew up believing that an eiderdown was something put on a bed, and that it was one word. In fact, 'eiderdown' has been used for the down itself since 1744 and subsequently for the quilt that was stuffed with it. Yet the connection was one reason for the decline in numbers of these birds up to the end of the nineteenth century – a decline that also led to the extinction of the Great Auk (see page 291) as its feathers, in turn, became used as a substitute for the increasingly scarce down of the Eider. Today, Eider are protected, nesting sites are provided and the down is carefully collected without disturbing the birds. In the past, it was different; nests, down and eggs tended to be collected together and repeat layings were also harvested.

Like the Great Auk itself, the Eider is both northern and almost exclusively marine. It is a sea duck in every sense, feeding on molluscs, and has a remarkable ability to swallow whole cockles and digest them. It is a sleek creature, mainly black above and white below, and has a soft, cooing call. In Orkney, they are called 'lazy birds' because the drakes tend to leave the ducks to incubate the eggs; although why they should be singled out in this way is not obvious because this is a common habit among ducks.

A female Eider Duck *(Somateria mollissima)*. Over the centuries, the Eider has been much persecuted for its down.

The name 'St Cuthbert's duck' recalls the sojourn of St Cuthbert (*c.*634–87) on Lindisfarne in the Farne Islands.

> It was he, according to a narrative of a monk of the 13th century, who inspired these ducks with a hereditary trust in mankind by taking them as companions of his solitude when for several years he resided alone on Lindisfarne. There is good reason to accept this and similar traditions as largely true, for a like ability in 'gentling' birds and other wild animals is manifested today by some persons of a calm and kindly sort.[112]

Also:

> ... as long ago as 1397 the Bursar's roll of the Monastery of Durham, which contained the shrine of St Cuthbert, mentions the use of eider-down for stuffing and cushions.[113]

Eider are now becoming more common southwards and their further spread around the English coasts is confidently predicted. *Eider* is a Nordic word, reaching English via Danish from Icelandic.

Two male and one female Long-tailed Ducks *(Clangula hyemalis)* feeding.

Long-tailed Duck *(Clangula hyemalis)*

OTHER COMMON NAMES: *calaw (Shetland, Orkney); caloo; coal-and-candlelight (Orkney); col-candle wick (Fife); coldie (Angus – Forfar); darcall; ice duck (Northumberland); long-tailed hareld; mealy bird (Norfolk); northern hareld (Aberdeenshire); sea pheasant (northern England); sharp-tailed duck; swallow-tailed sheldrake.*

The only really long-tailed British duck, this is an imposing black-and-white winter visitor to northern coasts, where it feeds on molluscs and small crustaceans.

Although the name 'Long-tailed Duck' is self-justifying, the plethora of delightful local names certainly bears some comment. 'Calaw', 'caloo' and 'coal-and-candlelight', among others, all imitate a call that has been described as yodelling and certainly carries a long way. Some writers have attempted to extend the explanation for 'coal-and-candlelight' to include the fact that the bird is charcoal-coloured and resembles an old candle-holder, but it is charming enough not to need this. The American name is 'old squaw' and stems from a belief that its cries, which really are most untypical of a duck, sound like old women gabbling.

Common Scoter *(Melanitta nigra)*

OTHER COMMON NAMES: *black diver (Ireland, Northumberland); black duck (Ireland, Norfolk, Essex); Doucker (Westmorland, Lancashire); scoter; scoter duck; sea duck (Norfolk); sea hen (Northumberland); surf duck (Scotland); whilk.*

Another of the winter-visiting sea ducks, the Common Scoter breeds on a few freshwater lochs in Scotland but is a much scarcer bird and its range does not spread southwards. It is a truly beautiful bird: the drake is the only duck anywhere that is all-black, except for an orange top to the beak.

'Scoter' was originally a Yorkshire word and, since the seventeenth century, Yorkshire ornithologists have had to live with the fact that one of their forebears must have had bad handwriting: 'Scoter' must surely have been intended to be 'Sooter', the 'soot-coloured duck'.

Velvet Scoter *(Melanitta fusca)*

OTHER COMMON NAMES: *astracannet (Northumberland); black diver; double scoter; great black duck; velvet duck.*

This bird is almost identical to the Common Scoter, but has white wing-bars in both sexes. It is another winter visitor but one that has never bred in this country. Can the Northumberland name 'astracannet' possibly have arisen because someone thought the birds looked like an astrakhan hat?

Goldeneye *(Bucephala clangula)*

OTHER COMMON NAMES: *brown-headed duck; buffe-headed garrot; curre; diver; diving duck (Shetland); doucker (Scotland, Roxburghshire); freshwater wigeon (Ireland); gingling curre (west of England); golden-eyed garrot; gowdy duck (Orkney, East Lothian); grey-headed duck; morillon; mussel cracker (Lancashire); pied curre (southern England); pied wigeon; popping wigeon; rattlewings (Norfolk); whistler; whiteside (Westmorland).*

No duck ever had less need for any other name, yet the Goldeneye has many. The drake is black above and white below, with a tiny, piercingly golden eye that has identified it since the seventeenth century. It is a winter visitor, mainly to coastal areas. A small number of birds nest inland, generally in holes in trees, and it is one of the few ducks that will readily take to nesting boxes.

Smew *(Mergellus albellus)*

OTHER COMMON NAMES: *lough diver; magpie diver (Ireland, Kent); pied diver; red-headed smew (female); small herring bar (Sussex); smee; smee duck (Norfolk); vare wigeon; weasel coot (Norfolk); weasel duck (Northumberland, Norfolk); white merganser; white nun (Ireland); white wigeon (Devon); white-headed goosander.*

No bird has had teeth since the likes of *Hesperornis* in the Upper Cretaceous, around 80 million years ago, but the Smew, and the Red-breasted Merganser and Goosander (see below) are the best that are currently available. They belong to a group of ducks called 'sawbills', because of the serrated, tooth-like edge to their slender beaks, which enables them to capture slippery fish.

The Smew is a small, almost white bird with a few black markings. It is a winter visitor to the southern half of the country, where it can be seen in estuaries, sheltered coastal places and fresh waters inland. 'Smew', 'smee' and similar names are imitative of the call and, although these forms come from Norfolk, they have Dutch origins.

Red-breasted Merganser *(Mergus serrator)*

OTHER COMMON NAMES: *bardrake (County Down); earl duck (Scotland); grey diver (Argyllshire – Islay); harle (Orkney); herald (Shetland); herald duck (Shetland, Angus – Forfar); herle (Northumberland); land harlan*

(Ireland); lesser toothed diver; popping wigeon (Ireland); rodge; sandbill (Sussex); sawbill (Galway, Aberdeenshire, Stirlingshire); sawbill wigeon (Galway); sawneb (Aberdeenshire); sawyer (Suffolk); spear drake, spear wigeon, tuke (Sussex); yearel (Northumberland).

This is an impressive bird and another fish-eater. The drake has a dark green head, a rough crest, white throat and chestnut neck. The female is also unusually colourful, being grey but with a pretty chestnut head.

The Red-breasted Merganser is a winter visitor to the coast but breeds on inland waters and, over the past century, has gradually spread southwards from Scotland and now breeds as far south as central England.

The name 'Merganser' is a hybrid eighteenth-century word, obtained from the Latin root *merg-*, meaning 'diving' and *anser*, meaning 'goose'.

Goosander *(Mergus merganser)*

OTHER COMMON NAMES: *dun diver (Sussex); green-headed goosander; harle; jacksaw (Yorkshire); land cormorant (Dublin); pied wigeon (Shropshire); rantock (Orkney); sawbill (Stirlingshire); sawneb (Aberdeenshire); sawyer (Suffolk); shell duck (Stirlingshire); spear duck (Sussex); spear wigeon (Kerry).*

Confusingly, the scientific name of the Goosander is *merganser*, but then the history of the two birds is also confused. They look very similar, although the Goosander is paler coloured and slightly larger. The Goosander was once known as the 'gossander' and, although the name may have had some connection with 'goose', its real origin is obscure and has provided both ornithologists and philologists with fruit for endless speculation.

Unlike mergansers, Goosanders tend to remain on fresh water all year round and so eat fish that fisherman would rather they left alone – a habit that has worked very much against them. Despite widespread persecution, they have gradually spread southwards after colonizing Scotland in the latter part of the nineteenth century and further expansion is to be expected.

Ruddy Duck *(Oxyura jamaicensis)*

This is a pretty little American duck with a stiffly upright tail, a chestnut body, a white face and black cap and a strikingly blue beak. A few escaped from the Wildfowl Trust at Slimbridge in Gloucestershire in the late 1950s, but they have since spread and pose a serious threat to existing duck species.

The Government is planning a duck pogrom. Bureaucrats in Whitehall are drawing up plans for a final solution to the Ruddy Duck – genocide. This will be expensive, at a total cost of £800,000, or £200 per duck.[114]

The Red-breasted Merganser *(Mergus serrator)* is another sawbill. The youngsters diligently follow their parent – or hitch a ride.

Hawks and Eagles
(Accipitridae)

This huge family embraces all the British diurnal raptors (birds of prey that, unlike owls, feed in daylight) except falcons and the Osprey. They share the same familiar hooked beak, extraordinary eyesight, feet with long, curved claws and also the less than endearing habit of defecating a powerful jet of white liquid for a considerable distance (which is why their nests are often visible at long range, looking as if they have been splashed with white paint). They differ from falcons mainly in the structure of their eggs.

As far as British representatives are concerned, the family can be usefully subdivided into the Honey Buzzard and kites, the sea eagles, the harriers, the Goshawk and Sparrowhawk, and the Buzzard and true eagles. They tend to nest in tall trees or on rocky ledges and take a wide variety of prey, largely dependent on local availability and the size of the bird.

Such imposing creatures have attracted the attention of men over the centuries; most have been particularly evident in heraldry and, globally, eagles must have attracted almost more attention, lore and symbolism than any other bird, although relatively few people in Britain have ever seen one in the wild. Many of these birds, or at least their names, have entered everyday language in different ways: 'hawk-eyed', 'spread-eagled' and so on.

Hawking was a hobby of Queen Elizabeth I, as shown in this illustration from George Turbevile's *Booke of Falconrie* (1575).

Together with the true falcons, some of these so-called 'short-winged' hawks have been used in Britain for the sport of falconry. The Goshawk and Sparrowhawk were historically the most important, although you are likely to see almost any of them in modern demonstration or entertainment falconry. The word 'hawking', which has been more or less interchangeable with 'falconry' over the centuries, is now also applied to the activities of other creatures, such as dragonflies and bats, that hunt and catch aerial prey. Falconry came to Britain from the East (it is still a national sport in Syria, Iran and parts of India) and, from the reign of Ethelbert (around AD 860) until the middle of the seventeenth century, it is traditionally said that: 'it was the favourite amusement of all, from the monarch to the page; and one in which the fair sex took particular delight.'[115]

Falconry was for centuries inseparable from the amusements of royalty and the Dukes of St Albans were the hereditary Grand Falconers of England. It was only the development and widespread availability of sporting firearms that really presaged its decline, although there has been a modern application of this ancient sport that its former adherents could scarcely have foreseen:

> Trained falcons and hawks continue to be used to control birds on relatively small numbers of aerodromes in the UK ...

But there are drawbacks in using this technique to diminish 'bird strike' damage to aircraft:

> ... Several hours of maintenance are needed every day of the year, irrespective of whether falcons are to be flown for bird control ... falcons must be flow in pursuit of birds regularly to sustain their effectiveness ... media attention, public interest and the demands of looking after the falcons can distract the bird control organisation from its task.[116]

Honey Buzzard *(Pernis apivorus)*

This wonderful but shy and secretive bird has always been a rare summer visitor to Britain, although it has at times been confused with other species because of its very variable plumage. It exhibits a wide colour range from pale grey to dark brown and is most commonly mistaken, in its dark form, for a Buzzard. As many as 50 pairs now breed here, mainly in the south of England, although the breeding sites are invariably enveloped in security by ornithologists.

'Honey Buzzard' is a recent name, coined by Francis Willughby in the seventeenth century after he found the combs of wasps in its nest.[117] It is not, however, a true Buzzard; nor was it honey that the bird was collecting but wasp larvae, which, together with bee larvae, are its principal food.

The Red Kite *(Milvus milvus)* is becoming established in Britain again after successful introductions from Spain and Sweden.

Unexpectedly for a bird of prey, it digs these larvae from the ground, although it will also occasionally eat other insects, frogs and small birds. This species may become more common as global warming results in more bees and wasps for longer periods, although its spread will always be limited by the availability of woodland habitat.

Red Kite *(Milvus milvus)*

OTHER COMMON NAMES: *baegez (Cornwall); crotch tail (Essex); fork tail (Yorkshire); forky-tailed kite (Sussex); glead (Scotland, northern England); glead hawk (Cheshire); gled, greedy gled (Scotland, northern England); puttock; scoul (Cornwall).*

The dubious delights of motorway driving have been greatly enhanced over the past few years. As I travel to London along the M40 from my home in the Midlands, I can almost guarantee that, to the south of Oxford, one or more of these magnificent birds with its striking, deeply forked, chestnut tail will soar over head. Their presence is due to a deliberate and successful policy of introducing Spanish Red Kites to several parts of England. Comparable introductions of Swedish Red Kites have taken place in Scotland. Yet, for the past 200 or so years, as a plethora of Welsh names reveals, these were birds of the Welsh mountains although, by the 1930s, they had been reduced to a single breeding female.

> I moved to Wales in the mid-nineteen seventies when red kites were scarce and great efforts were being made to encourage them. Large areas were out of bounds during the nesting season. A bulldozer driver working nearby told me that he was joined daily at lunch-time by a kite which collected remnants of his sandwiches from the bonnet of his machine.[118]

Some natural increase was already underway by the time the introduction policy was begun in 1989, although, successful as it has been, it seems unlikely that the Red Kite will ever achieve quite the widespread distribution that it enjoyed historically.

In the Middle Ages, it was a common scavenger in London (see Raven, page 367) and seems to have been as familiar over communal rubbish tips as gulls are today; Shakespeare refers to kites no fewer than 15 times. Although a Red Kite was again seen flying above Brick Lane in central London in December 2000, it was the gradual improvements in cleanliness and sanitation in the towns of England that drove it increasingly further into the countryside and, finally, to its Welsh mountain refuges. Kites feed on small rodents, other birds and insects, as well as carrion and human refuse, and their habit of collecting old rags to incorporate in their nests has long been known:

> My traffic is sheets; when the kite builds, look to lesser linen.
> (*The Winter's Tale*, Act 4, Scene 3)

The kite's scavenging habits led to it and its name becoming terms of opprobrium:

An engraving, *c*.1826, of Kitley House in Devon. Kitley means 'kite wood'.

> Ah, you kite! – Now, gods and devils!
> (*Antony and Cleopatra*, Act 3, Scene 13)
>
> Detested kite! thou liest:
> (*King Lear*, Act 1, Scene 4)

The name 'Kite' was originally imitative of the long, whistling call and is sometimes used for the Buzzard too. The old Middle English name 'puttock', or 'puttok', dating from *c*.1450, has similarly been applied to both birds and, according to William Lockwood, may well have the same root as 'piddle', or 'pickle', which are both names used for the Marsh Harrier and which originated in an Old English word meaning 'swooper'.[119] John Clare, in 'The Flitting', wrote of 'The sailing puddock's shrill "peelew"'. There is another possible explanation. 'Puddock', or 'paddock', are regional names for both the Toad (see page 177) and the Frog (see page 174); Francesca Greenoak, and others, have suggested that it came to be applied to these birds, which are relatively unusual in eating Toads.[120]

Kitley in Devon is 'kite wood', as is Putley (from 'puttok') in Herefordshire; Kitnor in Somerset (now called Culbone) was 'a hill frequented by kites' and Putford in Devon may also have an origin in 'puttok'. Pudleston in Herefordshire is 'the hill of the mouse-hawk'. There are Kite and Kite's Nest inns and kites on coats of arms. The word has been used for the flying toy since the mid-seventeenth century and for the geometrical shape that it mimics. There is no obvious reason for the name being used for a bill of exchange or a means of obtaining credit, but from thence has come the expression 'flying a kite'. The word 'kite' is also used for the highest sails of a sailing ship.

As is usual with birds of prey, there seems to be no evidence of anyone having eaten kite, but parts of it were certainly used medicinally:

> The powder, or ashes of the flesh, or head, and liver ... prevail against the gout, and epilepsy. The testicles pulverized being taken for a considerable time, cause fruitfulness, and restores in consumptions, dose two drams. The dung, blood and grease are proper to be applied to arthritic pains, such as the gout, &c.[121]

White-tailed Eagle *(Haliaeetus albicilla)*

OTHER COMMON NAMES: *erne (Scotland, Orkney); sea eagle.*

White-tailed Eagles are more closely related to kites than to the more familiar Golden Eagle and other true eagles, but they are bigger; indeed, they are the biggest northern European raptor. They are hugely impressive, especially in flight, when their wing-tip primaries are extended and their fan-shaped, white-tipped tail is spread.

They have never been very widespread British birds but they may once have been more common than the Golden Eagle, certainly along the west coast of Scotland, in the Western Isles, Orkney and Shetland. Skye was one of their strongholds. Essentially coastal birds, they only occasionally nested inland, as at Rannoch Moor and perhaps even at Windermere. Their diet is certainly not confined to fish. Large numbers of gulls are taken and an analysis of Norwegian bird kills revealed that a very wide range of mammals 'from red deer calves to water voles' were also included in their prey.[122]

The Highland clearances of the nineteenth century spelled the beginning of their end and sheep-farmers were merciless in their persecution of the White-tailed Eagle. Charles St John added to his *curriculum vitae* in the extermination of Ospreys (see page 253) by shooting scores of White-tailed Eagles, claiming that shepherds had asked him to do it.

There is no real evidence of them nesting in England after 1794 and the last to breed in Scotland is believed to have been a pair in Skye in about 1916. Norwegian birds were introduced to the island of Rhum in the Inner Hebrides in 1975 and a chick was first reared by a wild pair in 1985; since then they have spread gradually and there may now be about ten breeding pairs. The future of this magnificent bird as a British species is probably secure.

Marsh Harrier *(Circus aeruginosus)*

OTHER COMMON NAMES: *bald buzzard (Essex); bog gled (East Lothian); brown hawk (Ireland); duck hawk; dun pickle (Wiltshire); kite (Ireland); marsh hawk; moor buzzard; moor hawk; puttock; snipe hawk (southern Ireland); white-headed harpy.*

In the bookshop at the bird reserve of Minsmere in Suffolk, it was once possible to buy a print of a most attractive and evocative painting by John Reaney of a bird of prey flying low over the Minsmere reed beds. That bird was a Marsh Harrier and the association is appropriate because, by the early 1970s, the British population was reduced to one pair breeding at Minsmere. Once fairly widespread, they had almost vanished through shooting by the beginning of the twentieth century, recovered slightly over succeeding decades, then declined again as pesticides took their toll. In the last quarter of the century, however, they have made a spectacular recovery, not only in their traditional reed-bed home but also on farmland, and Marsh Harriers are nesting again in Wales and Scotland, and Ireland too. Some are residents, some summer visitors, but all are spectacularly beautiful: brown overall, the male has chestnut underparts and a grey tail, while the female has a dark eye-stripe on a pale head.

Harriers are slender, graceful birds, typically gliding low to drop onto small birds and mammals. Unlike the dogs

Marsh Harrier at Minsmere **by John Reaney reflects a time when this reserve was home to the last British pair of Marsh Harriers.**

known as harriers, which chase hares (see page 482), the bird name 'harrier' originates from an Old English word meaning 'army', so a 'harrier' was someone who made raids. It was used for birds in the sixteenth century and was originally applied only to the Hen Harrier, meaning a bird that 'made raids on', or 'harried', hens. It was later transferred to other species, where it is meaningless: Marsh Harriers do not harry marshes.

Hen Harrier *(Circus cyaneus)*

OTHER COMMON NAMES: *blue gled (Scotland); blue hawk (Wicklow, East Lothian); blue kite (Scotland); blue sleeves; bod-glas (Wales); brown gled (Scotland); brown kite; dove hawk; faller; flapper; furze kite; gorse harrier (Sussex); grey buzzard (Hampshire); hen driver; hen harrow; katabella (Orkney); miller; moor hawk; ringtail (East Lothian); seagull hawk (Galway – Connemara); vuzz kite (Devon); white aboon gled (Stirlingshire); white hawk; white kite (Donegal).*

The hawk that harries hens is beautiful and graceful, with striking differences between the sexes. The female is a tawny brown overall, but the male is a wonderful blue-grey bird with black wing-tips. It is a bird of moorlands and coniferous forests, and of open, wild places, where it is still widely persecuted by gamekeepers, who perceive it as an enemy of the Red Grouse.

> The UK Raptor Working Group, representing land-owning, sporting and conservation interests, said that with only 570 pairs of hen harriers in Britain, and 1,283 pairs of peregrines, more drastic action could not be justified. It said diversionary feeding – placing dovecotes and dead rats and rabbits near harrier nests during the breeding season – should be used to reduce the predation of red grouse.[123]

Recent studies have also shown that there is a habitat conflict because, if the moors are managed suitably for Red Grouse, the populations of birds such as Meadow Pipits, on which Hen Harriers feed, tend to fall.

They were once very common birds on the Scottish islands; on the Hebrides it was said that anyone experiencing singular good luck must have seen a Hen Harrier, but there too they have now declined.

Montagu's Harrier *(Circus pygargus)*

This is the smallest harrier, superficially very similar to the Hen Harrier, and a summer visitor that has begun to nest again in East Anglia and also around the New Forest in recent years. It was an uncommon but regular British species for many years, although it all but vanished from Britain around the middle of the twentieth century.

The eponymous Montagu was the man who, in 1802, first recognized it as being distinct from the Hen Harrier, although he called it the 'Ash-coloured Falcon'.[124]

Goshawk *(Accipiter gentilis)*

OTHER COMMON NAMES: *goosehawk (Scotland); great hawk.*

There is a great deal of mystery about the Goshawk. Even its name may be incorrect because, while its meaning, 'goose hawk', is fairly evidently from the Old English, it seems highly improbable that, whether as wild or falconers' birds, they ever caught geese. In places, Peregrines, which have been flown against geese, are called Goshawks and it may well be that this was one of their original names.

There is also a mystery about the history of this bird, because no-one seems to know how common it has been over the years. Chris Mead describes its history as 'blanketed by a fog of disinformation',[125] while Ron Murton said 'the evidence that it ever was a regular British breeding bird is flimsy'.[126]

However, it is undeniably a woodland species, requiring fairly extensive areas of woodland to establish itself. It may well move into maturing conifer plantations, where there is certainly no shortage of prey; Wood Pigeons, Rabbits, crows and many other species of bird and mammal are all taken. The consensus seems to be that, in Britain, there are now probably over 300 pairs of this striking grey hawk with speckled underparts.

Undoubtedly, escaped falconers' birds have contributed to the resurgence in Goshawk numbers because it has long been a

A woodland species, the Goshawk *(Accipiter gentilis)* preys mainly on crows and Rabbits *(Oryctolagus cuniculus)*.

sporting favourite. In the hierarchy of falconry birds, it was the yeoman's hawk. Like the Sparrowhawk, it was flown from the fist and required skill on the part of the falconer to approach the quarry as closely as possible. It was certainly flown against a wide variety of targets: Rabbits, game, Moorhens and Coots, ducks – almost anything it seems, except a goose.

Sparrowhawk *(Accipiter nisus)*

OTHER COMMON NAMES: *blue hawk (Stirlingshire, East Lothian, West Yorkshire, Oxfordshire, Buckinghamshire, Berkshire); blue merlin (Perth); gleg hawk (Renfrewshire); gold tip (Yorkshire); hedge hawk; maalin (Shetland); pigeon hawk (Yorkshire); spar hawk (Scotland, Surrey); spare hawk, spur hawk, stannin hawk (Scotland, Yorkshire); stone falcon (Sussex).*

The Sparrowhawk has been so called since people in the British Isles spoke Old English but, at the present rate of decline in sparrows, that might have to change. There are plenty of other possible candidates: 'finchhawk', 'larkhawk', 'tithawk' would all suffice because this bird really is a scourge of small feathered things. The local name 'hedge hawk' is also pretty accurate, for this hawk materializes as if by magic from hedgerows to pounce on its prey. Its increasing prevalence in gardens can lead to its downfall, however. The local Scottish name 'gleg hawk' means 'sharp-eyed hawk' but seems to be a derivation from the old kite name 'gled hawk' or, in Cheshire, 'glead hawk', rather than a comment on its eyesight. There is an old saying 'as gleg as a gled' ('as sharp-eyed as a kite'), which has not helped matters.

Sparrowhawks are now second in number only to Kestrels among British birds of prey but it is remarkable that they have endured so well because gamekeepers, who believe they take game chicks, and organochlorine pesticides, which entered the food chain, have at times reduced their numbers drastically. Historically, they were probably always fairly common and were popular with falconers, who treated them much like small Goshawks, flying them from the fist. They are the only British bird of prey to feed exclusively on other birds and they have been flown against almost anything smaller that flies.

The antiquary John Aubrey (1626–97) wrote in his *Miscellanies Upon Various Subjects* of 1696 that:

> Not long before the Death of King Charles II a Sparrow-Hawk escaped from the Perch, and pitched on one of the Iron Crowns of the White Tower, and entangling its string in the Crown, hung by the heels and died. Twas considered very ominous, and so it proved.[127]

A delightful story is told of the Great Exhibition of 1851, where Joseph Paxton's monumental glasshouse, the Crystal Palace, was apparently plagued with an invasion of sparrows. It is said that the matter so concerned Queen Victoria that she sent for Wellington, possibly imagining that a regiment of cavalry would see them off. The Iron Duke, only a year before he

The Sparrowhawk *(Accipiter nisus)* builds a nest of sticks, down and twigs, into which the female lays three to six eggs.

died, had another, and better, solution: 'Sparrowhawks, Ma'am,' he is reported to have said. There appears to be no contemporary confirmation of the story, which may have been the invention of a provincial newspaper, although it is reliably said of the Great Exhibition that:

> No man delighted in it more than the Duke of Wellington. He made several visits to the palace during its construction and many after its opening.[128]

Buzzard *(Buteo buteo)*

OTHER COMMON NAMES: *bald kite; barcutan (north Wales); bascud; buzzard hawk (Angus – Forfar); gled (northern Scotland); goshawk, kite (Ireland); puddock; puttock (Scotland); shreak (Gloucestershire).*

When I was a boy, it was a real red-letter day to spot a Buzzard, its scarcity being due largely to the disappearance of Rabbits after the myxomatosis outbreak of 1952–55. This also represented a dip in a gradual recovery from its almost complete extinction in the early years of the twentieth century, following years of persecution. Now this magnificent large bird of prey is, after the Kestrel, one of the most commonly seen hawks.

Historically, they were once extremely common, but their downfall was to be foreseen when, in 1457, James I of Scotland ordered their destruction. In the sixteenth century an order of Henry VIII forbidding egg-stealing specifically excluded Buzzard eggs and then, in the nineteenth century,

A bird of open hillsides and wooded valleys, the Buzzard *(Buteo buteo)* is often seen wheeling high, searching for small mammals.

the desire to preserve game (there is no doubt that Buzzards do take a few game-birds), coupled with the spread of urban and industrial development and gradual habitat destruction, led to their fast disappearance from many regions of Britain. Buzzards take a wide variety of prey – Rabbits especially, but also rodents, small birds, insects and frogs – but, unlike kites, do not generally scavenge.

The name 'Buzzard' is fourteenth century and comes, via French, from Latin and the classical name *Buteo*, which is imitative of the drawn-out mewing call. It has passed into some curious common use:

> The term 'Femme Buzzard' was used by my wife's family from Leeds to describe a woman (usually of mature years) with big hair, or even to mean mutton dressed as lamb.[129]

Buzzards have been thought untrustworthy. There is an Irish saying:

> Do not credit the buzzard, nor the raven, nor the word of a woman.[130]

They have also been considered birds of ill omen:

> I chose an eagle, and did avoid a puttock.
> (*Cymbeline*, Act 1, Scene 1)

Its call was believed to foretell rain, as John Clare, in his *Village Minstrel*, obviously knew:

> Slow o'er the wood the puttock sails;
> And mournful, as the storms arise,
> His feeble hole of sorrow wails,
> To the unpitying, frowning skies.

There is an everyday saying 'as blind as a buzzard', but its connection with the bird is obscure, as is the use of the word, common since the fourteenth century, to mean a stupid person. Neither does the town of Leighton Buzzard have any direct connection; it is derived from a family of that name, although how they came by it is not apparent.

Golden Eagle *(Aquila chrysaetos)*

OTHER COMMON NAMES: *black eagle; erne; mountain eagle; ring-tailed eagle.*

One's first sight of a captive Golden Eagle is unforgettable and the relatively few people who have seen one in the wild have experienced something altogether very special. The Golden Eagle is not the biggest British bird, nor even the biggest raptor. It is not the rarest bird nor the most spectacularly coloured (it is not even golden). However, it has a majesty unlike anything else and British writers down the ages have fallen under its spell.

Golden Eagles are tawny overall, rather than golden, with massive yellow talons. They have never been common in Britain, not least because of the huge territory that each pair needs.

Leslie Brown believes that there have probably never been many more than 500 pairs of British Golden Eagles and there are not many short of that number today.[131] Although they once occurred as far south as the southern Pennines (they nested at Ashopton in Derbyshire during the 1660s), Golden Eagles have always been essentially birds of the Scottish Highlands. They have been heavily persecuted by shepherds and gamekeepers in the past but have held their own for the past 25 years or so and have even begun to stray south again to England; they have attempted recently to nest in the Lake District after an absence of almost two centuries. There seems to be no reason why they should not maintain this population level. Food is plentiful: they eat Rabbits and other small mammals, as well as birds and carrion but, contrary to common belief, do not carry off lambs or other livestock – or babies. No eagle can carry more than its own weight and the heaviest that most can lift is probably a Brown Hare.

It is because the Golden Eagle has never been an everyday sight that British eagle lore and literature is nothing like as extensive as that of many other countries, or of classical times, and most of the numerous eagle references by authors such as Shakespeare, Spenser and Chaucer are to classical rather than British beliefs.

Eagles are not part of the national coat of arms, as they are in many countries of continental Europe, although they are very familiar on family crests. The eagle was famously a Napoleonic symbol and the form of the heraldic eagle with wings and legs displayed led to the sixteenth-century expression 'spread-eagle'. The word 'eagle' seldom occurs in British place names (even Eagle in Lincolnshire has nothing to do with the bird, being derived from the Old English for 'oak wood'), but if we turn back to the Old English name 'erne', which the Norman French 'eagle' later replaced, we find a number of examples: Yarnfield in Wiltshire ('eagle feld') and Yarnscombe in Devon ('eagle valley'). Nonetheless the word has been used only rarely in English in anything other than its ornithological sense.

One exception is the golfer's use of 'eagle' to mean a score of two strokes under par, but then golf is a Scottish game.

> The term eagle seems to have been coined as an analogy with 'birdie' and was ... in use by the 1920s. The term 'birdie' seems to have originated from the phrase 'a bird of a shot'. In US slang, a 'bird' was used to denote anything excellent or wonderful. When used by golfers, the phrase may also have implied that the ball 'flew like a bird'. This term was in use by around 1910. Although nobody seems to have a definitive answer as to the reasons why certain birds were used as golfing terms, one would presume that the rarity of the bird coincides with the rarity of achieving each shot. For example, an eagle shot, being a two under par score for one hole, is a less rare bird than an albatross, used to describe a three under par shot for one hole.[132]

Oddly enough, although they have not bred there since about 1750, there is probably more eagle lore in Wales than in any other part of Britain. The Welsh name for Snowdonia is *eryri*, but it is not clear whether this is derived from the word for eagle (*eryr*) or the word for snow (*eira*). (If this seems familiar, an eagle's nest is widely called an 'eyrie' but this has a different etymology, probably coming, via French, from Latin.) Although eating eagles has not been widely recommended, a Welsh belief holds that the descendants of people who have eaten eagles can cure shingles, and the Welsh name for shingles is also *eryr*.

Undoubtedly, when Golden Eagles did occur in Wales, Snowdonia was their principal home and hearing their cries was said to foretell some momentous event, such as the birth of a national hero or the rise or fall of some great family. As they are usually silent away from the nest, it is understandable that hearing them could be thought rather special. If a solitary eagle was seen on a Welsh crag, it was thought to be a sentry, indicating the arrival of an enemy; if several were seen together (which is extremely unlikely), it was a sign of peace.

In Ireland, where Golden Eagles were breeding until 1912, there is a tradition that Adam and Eve live on in the form of a pair of eagles on an island off the west coast of Galway. It is usually said to lie in Killary Bay, although there is an Eagle

The Golden Eagle *(Aquila chrysaetos)* inhabits the Scottish Highlands, where it preys on small mammals, birds and carrion.

A Golden Eagle *(Aquila chrysaetos)* perched on a pine stump overlooking a Highland glen.

Rock further south, overlooking the Aran island of Inishmore. Eagles also undoubtedly once occurred on the Isle of Man, where the old Manx name for them was *Dreain*, or 'the Druid's bird', and it has been suggested that the Druids were the origin of the 'King of the Birds' tag that eagles have acquired.

There are a great many pubs called The Eagle and in W.R. Mandales' familiar old rhyme 'Pop Goes The Weasel' (see Weasel, page 415), 'in and out of the Eagle' was 'the way the money went'. There are some especially puzzling Eagle inns, however:

> I live in the village of Bispham Green in Lancashire which was once part of the Earls of Stanley's estates ... the pub across the green from my home is the 'Eagle and Child'.[133]

Another variant is The Old Eagle and Child, found in Derbyshire. The names recall a story of the local family of the Stanleys, the Earls of Derby, whose crest is an eagle with wings extended preying on an infant in a cradle.

> In the reign of Edward III, Sir Thomas Latham, ancestor of the house of Stanley and Derby, had only one legitimate child, a daughter, but at the same time he had an illegitimate son by a certain Mary Oscatell. This child he ordered to be laid at the foot of a tree on which an eagle had laid its nest. Taking a walk with his lady over the estate, he contrived to bring her past this place, pretended to find the boy, took him home, and finally prevailed upon her to adopt him as their son. This boy was afterwards called Sir Oscatell Latham, and considered the heir to the estates. Compunction or other motive, however, made the old nobleman alter his mind and confess his fraud.[134]

Pieces of eagle were once prized for their medicinal properties, although acquiring them in any quantity in Britain must always have been a difficult task:

> The flesh pulverized helps to cure the Gout ...
>
> The bones filed, or calcined, are good against the Megrim, and other pains of the head ...
>
> The brain reduced into powder is proper to be given to them, who are afflicted with the Jaundice, Scotoma, or any sort of vertigo ...
>
> The powder of the liver, and blood drank with oxymel for ten days cures the falling sickness as Dioscorides testifies. The powder of the testicles drank provokes lust. The gall mixt with white wine of honey cures most disorders of the eyes. The dung outwardly applied ripens tumours and aposthumes. The Aetities or a stone found in the nest of an Eagle, about the bigness of a peach, or apricock, being bound to the left arm, or stomach, prevents abortion, or miscarriage, being tied to the thigh it facilitates the delivery, but must be presently removed after the birth, lest it draw the womb out.[135]

Ospreys
(Pandionidae)

Osprey *(Pandion haliaetus)*

OTHER COMMON NAMES: *bald buzzard; eagle fisher (Scotland); fish hawk (Shetland, Scotland); fishing eagle (Northumberland); mullet hawk (Hampshire, Sussex, Kent).*

Thousands of people, myself included, have seen their first wild Osprey at the famous nesting site near Loch Garten on Speyside. On 24 July 1969, I saw a female brooding her young and the male arriving with a fish just as a third adult bird attempted to land on the same nest. It was a thrilling experience. Twenty-five years later, I watched an Osprey catching trout in the Athabasca River, where it descends from the Columbia Icefield in the Rocky Mountains of Alberta. The distance between these locations emphasizes how widespread and successful a bird the Osprey is; it breeds throughout much of the northern hemisphere, and also in Australia, yet its British history is a patchwork of ornithological and human triumph and failure.

There is only one species in the Osprey family and it is perfectly adapted as a fish hawk, with bare legs and specially roughened feet for gripping the fish, which it holds in a characteristic fore-and-aft position. There was an old belief that Ospreys had one foot formed like that of a hawk, to grip their prey, and one like that of a goose, to swim. Shakespeare may not have been aware of this, but he certainly knew that they were birds designed for a very specific purpose:

> I think he'll be to Rome,
> As is the osprey to the fish, who takes it
> By sovereignty of nature.
> (*Coriolanus*, Act 4, Scene 7)

Ospreys are big birds, with a wing-span greater than any other resident British species except the Golden Eagle. They are heavily built, dark above, pale below and with a grey head, a small crest and a dark eye-stripe that endows them with a scowling, sinister appearance.

Ospreys must have once been fairly widespread in Britain, as evidenced by the regional names from southern England and fairly frequent literary mentions. They declined gradually, however, and by the nineteenth century were effectively confined to Scotland. The persecution that they suffered there, largely at the hands of a band of astonishingly intrepid, albeit wholly misguided egg-collectors, has entered ornithological legend. Apart from their other attributes, Ospreys have suffered the disadvantage of laying remarkably beautiful and very varied eggs, and collectors were not satisfied with one egg; they wanted a representative of each of the almost infinite variants.

During the eighteenth and nineteenth centuries, much the most famous and romantic Osprey nesting site was on top of the crumbling tower of the ruined island castle at Loch an Eilein, in the old Rothiemurchus Forest in Perth. The first report of their presence was in 1808 but, year after year, attempts were made to rob the nest, usually by professional egg-collectors, the most remarkable being Lewis Dunbar. From 1848 to 1852, Dunbar made five raids on the nest, while his brother William and his companion Charles St John were acting similarly, nest-robbing and shooting adult Ospreys in Sutherland. Unfortunately, St John has gained a totally unjustified reputation as a sportsman and 'good fellow'. A collection of his writings was reprinted in 1982, endorsed with a Foreword by Robert Dougall, sometime President of the RSPB. Yet this was a man who set about methodically to collect Scottish Ospreys to extinction. In a letter to the egg-collector John Wolley, dated 29 June 1850, St John wrote:

> I believe, at this moment, there is only one osprey's nest in this country [Sutherland] and that has been taken by Lord Grosvenor's keeper. I am afraid that Mr St John, yourself and your humble servant, have finally done for the Ospreys.[136]

In 1851 came Lewis Dunbar's most daring exploit and the most incredible nest raid in the history of British birds. The eggs he obtained were supplied to John Wolley, who recounted the story in his extraordinary book *Ootheca Wolleyana*, published in 1864:

> These two beautiful eggs I obtained from a correspondent [Lewis Dunbar] on 3rd May 1851, they being the only result of a ten days' nesting expedition undertaken by him. He took them at three o'clock in the morning of 29th April 1851, at the ruins of an old castle on an island in an Inverness-shire loch. After walking nearly all night he reached the spot in the midst of a snowstorm and having tied a cord to his life preserver, he swam off, leaving the other end in charge of a man on the shore. On the island, he tied the rope to a stone and climbed up the ruins, slipping about in six inches of snow. Having found two eggs in the nest, he discovered that he had left his cap behind him. He tried one egg in his mouth, but could not breathe with it; and at last he swam ashore on his back with an egg in each hand. He blew the eggs in the boathouse, washing out the inside with whisky.[137]

Dunbar emigrated to Australia in 1853 but Ospreys nested intermittently at Loch an Eilein and elsewhere in Scotland until the very early years of the twentieth century. Migrating Swedish birds then visited Scotland from time to time with increasing regularity, until a pair finally nested, probably in 1954. Over the next few years, a pair certainly nested close to Loch Garten but were disturbed. In 1958, the nest, despite being guarded, was robbed, but in 1959 three young were raised and this success

The famous nesting site of the Osprey *(Pandion haliaetus)* in the Scottish RSPB reserve of Loch Garten is rigorously protected.

The Osprey *(Pandion haliaetus)* is well adapted for fishing. It typically carries its prey in the fore-and-aft position.

was repeated annually thereafter. By 1969, public access had been well organized. The nest was spectacularly robbed in 1971 but, since then, it has been upwards almost all the way and now around 100 pairs of Ospreys nest in Scotland. In the summer of 2001, birds artificially introduced to Rutland Water in England lived up to expectations and mated for the first time. Rather more unexpectedly, so did a pair of Scottish Ospreys that had migrated naturally south to the English Lake District.

The name 'Osprey' was first recorded in English as 'ospray' in *c.*1460 and came via Old French from a Latin word meaning 'bone-breaker'. The local name 'mullet hawk' alludes to the fact that, in some areas, Ospreys routinely fish in the sea, Mullet being a favoured prey. The word 'osprey' also became familiar as a term for the various types of feathers and plumes used for trimming hats, but this may have been a rather non-specific usage:

> ... the plume of feathers on a hat was named an osprey
> by late 19th-century milliners. It is possible that they
> meant to call it a spray and that another bird-name
> came in and confused them ...[138]

Ospreys have not always been such prized birds. They were among the 'noyfull fowls and vermin' condemned in the reign of Elizabeth I and were also believed somehow to transfix fish into surrendering. The poet Michael Drayton (1563–1631), a Warwickshire contemporary and friend of Shakespeare, alluded to this in his monumental and extraordinary 'patriotic poetical topography' *Poly-Olbion*:

> The ospray oft here seen, though seldom here it breeds,
> Which over them the fish no sooner do espy,
> But betwixt him and them by an antipathy
> Turning their bellies up as though their deaths they saw,
> They at his pleasure lie, to stuff his gluttonous maw.[139]

In similar vein, George Peele (*c.*1558–96) wrote:

> I will provide thee of a princely osprey,
> That, as he flieth over fish in pools,
> The fish shall turn their glistening bellies up,
> And thou shalt take thy liberal choice of all.[140]

It has been claimed, somewhat improbably, that James IV of Scotland kept a trained Osprey to catch fish, although more plausibly they are said to have been:

> ... kept, with cormorants and otters, in special buildings
> on the Thames at Westminster at least as late as 1618.[141]

In this 1931 photograph a falcon is released from a moving open car when a Rook is sighted.

Falcons
(Falconidae)

The sport of falconry appears to have obtained its name from this family of birds, but in reality it may be the other way around. The name 'falcon' is very old, originally Norman French, and first occurs, as 'faucon', in *c.*1250. Then, a 'faucon', or 'falcon', was any bird employed for falconry and, since the mid-seventeenth century, the term has been restricted to this family. The original thirteenth-century spelling betrays the old pronunciation, and today survives only in the surname 'Faulkner'. 'Falcon' itself appears very occasionally as a personal name, never more famously or heroically than in the Polar explorer, Captain Robert Falcon Scott, who was the father of the ornithologist Peter Scott and always known as 'Con' to his family.

Falcons have long, pointed wings (in falconry circles they are often referred to as 'long-winged' hawks) and tend to nest in holes or, rather commonly, in the abandoned nests of other birds, such as crows. They are grouped together on the strength of a nebulous range of characters, including the colour and structure of their eggshells. However, their most striking attribute is their almost complete mastery of the air, a feature that they share with Swifts. The family includes some of the fastest flying of all birds, if not the fastest; they reach astonishing speeds as they stoop (dive downwards) to take their prey.

Kestrel *(Falco tinnunculus)*

OTHER COMMON NAMES: *blood hawk (Oxfordshire); creshawk (Cornwall); cristel hawk; field hawk (Surrey); hoverhawk (Buckinghamshire, Berkshire); keelie (Edinburgh); kite (Shropshire); maalin (Shetland); mouse falcon (Orkney, Yorkshire); mouse hawk; red hawk (Stirlingshire, Lancashire); sparrow hawk (Ireland); stanchel; stand hawk (West Yorkshire); stanniel; stannel; stannel hawk; steingale; stonegall; vanner hawk, wind bivver (Sussex); windbibber (Kent); wind cuffer (Orkney); wind fanner (Surrey); windhover (western and southern England); windsucker (Kent).*

Despite their familiarity, few people have seen a Kestrel, or any other bird of prey, at close quarters and so are oblivious to the beauty of its grey head (brown in the female) and beautiful red-brown back. 'When it comes to seeing an animal properly, the wild is no substitute for a zoological garden.'[142]

Leslie Brown wrote that the Kestrel is probably 'the only British diurnal raptor that the average person could see almost every day, summer or winter, when walking or driving through the countryside'.[143] That was written in 1976, and today I would hazard a guess that one of the best ways to see Kestrels is from a car. They have become hugely familiar as they hover above motorway verges, seeking Field Voles (*Microtus agrestis*), their favoured prey. Strangely, there seems to be no satisfactory explanation for the wide variation in the height at which they hover; it does not seem to be related to the ease of sighting prey, nor, for instance, the length of the grass.

So many of its charming regional names, such as 'hoverhawk', 'stand hawk', 'wind cuffer' and the common 'windhover', relate to this characteristic and impressive activity. There is an old and rather more unexpected name for the Kestrel that alludes to the same behaviour and that the etymological detective work of William Lockwood has unearthed: in 1599, it was called a 'wind fucker'. The name survived at least to the nineteenth century in northern England as 'fuckwind', the literal interpretation being 'wind beater', an allusion to the beating wings of the hovering bird. As Lockwood pointed out, however, with the detached triumphalism that only a true academic philologist can muster: 'The present bird name thus uniquely preserves fuck in its hitherto unrecognised, primary sense'.[144]

Nonetheless, modern sensitivities are unlikely to see such ancient linguistic relics oust the name 'Kestrel', which is also sixteenth century and sometimes appeared as 'kastrel', or 'castrel'. It was originally French and probably imitative of the 'ke-ke-ke' call, although, in general usage, it long ago replaced the oldest English name for the bird, the fifteenth-century 'staniel'.

Kestrels feature little in falconry, mainly because they do not naturally feed on other birds or large mammals, and historically, when they were mentioned, they were at the bottom of the falconry hierarchy, merely being the birds of knaves or servants.

There are probably 60,000 breeding pairs of Kestrels in the British Isles, but they are in decline. As with other birds of prey that feed on small mammals, this may simply be one of the periodic population changes that reflect the rise or fall in the numbers of the mammals (Field Vole populations rise and crash every four years or so), although it also seems to be due partly to a shortage of nesting sites.

Merlin *(Falco columbarius)*

OTHER COMMON NAMES: *blue hawk (North Yorkshire); hawk kestrel, hobby (Shetland); jack; maalin (Shetland); rock falcon; rock kestrel (northeast England); small blue hawk (Stirlingshire); sparrow hawk (Scotland); stone falcon (Scotland, north Wales); stone hawk (Yorkshire, Cheshire); tweedler (Lancashire).*

Merlin was a magician but I can see no resemblance between the old, grey-bearded sage and this bird, the smallest of all the falcons, except the greyness, and that is only in the male (the female is dark brown); both sexes have pale underparts. It is typically a bird of the upland moors, where it found a ready-made enemy in the gamekeeper, who persecuted it mercilessly. It also suffered badly from the chemical pesticide pollutions of the 1950s and 1960s, and, to a degree, from the afforestation of its habitats. In recent times, Merlins have begun to recover, to some extent by changing their habits and abandoning ground-nesting in favour of old crows' nests in trees.

The Merlin *(Falco columbarius)* was favoured as a hunting bird by noblewomen in the Middle Ages.

The persecution was unjustified because, while Merlins are bird-hunters, they take small perching birds, such as pipits, larks and small waders, rather then game-bird chicks.

The Merlin is an important falconry species. It was the ladies' hawk, being lighter on the wrist than other falcons, and, although too small to hawk game, it was used, and in some places still is, for catching larks. In 1904, a contributor to *Country Life* wrote:

> It is quite unnecessary to say to our readers that the principal use to which falconers put the merlin now is that of lark-hawking. In days when this pastime was generally indulged in, it was held in repute as being more particularly the ladies' hawk; but the days long are gone by which a country dame was accustomed to sally forth with her hawk on her wrist, and when a merlin was a most suitable present to give her.[145]

Skylarks were kept in captivity especially for hawking:

> A supply of live larks should be always kept in an aviary for this sport, to encourage the Merlins, by throwing one up from hand, after unsuccessful attempts with wild ones.[146]

When the Merlin was released, both it and the lark prey rose ever higher in the sky, quite often out of sight of the ladies and their gentlemen on the ground. Lacking field-glasses, they would see nothing for minutes on end, until either the lark escaped and the Merlin returned or both were spotted heading earthwards at high speed, the stooping falcon then usually catching the lark.

The name 'Merlin' seems to have no connection with the Merlin of Arthurian legend and, although it appears to have originated in Old French, its meaning is obscure; it may have something to do with smallness.

During the last century, the bird has become synonymous with speed, strangely even more so than its faster flying relative; after all the Spitfire aeroplane was powered by the Rolls Royce 'Merlin' engine, not the 'Peregrine'.

Hobby *(Falco subbuteo)*

OTHER COMMON NAMES: *jack; merlin (Berkshire); riphook; robin; tree falcon; van-winged hawk (Hampshire).*

The Hobby is a summer visitor to Britain, roughly the same size as the Kestrel, dark slate-grey above, paler beneath, and with a rather short tail. The male bird has striking chestnut-coloured thighs. In flight, it has been likened to a giant Swift. Its natural habitat requirement for open countryside close to woodlands is not easily satisfied, although recently it appears to have adapted to farmland, using the old nests of crows or similar birds. Although not a common bird (Leslie Brown in *British Birds of Prey* called it 'Britain's least known bird of prey'), it appears to be slightly on the increase.

Hobbies feed on large insects, such as dragonflies and beetles, and also on small birds, especially Swallows and martins, and they have been employed by falconers to take larks and other small species. The ease with which birds of prey could be obtained for sport in 1855 will appal people today:

> Leadenhall Market is, perhaps, the most likely spot in which to meet with young Hobbies, as during the season young Hawks, as also other birds and beasts of various denomination, may be seen there.[147]

The name 'Hobby', as 'hoby', first appeared in the mid-fifteenth century and seems to have originated in an Old French word meaning to 'jump about', evidently a reference to the bird's agility when hawking its prey. The scientific name *subbuteo* may seem familiar because it was used for a once-popular game of table-top football. 'Subbuteo' means 'smaller than Buteo' (the Buzzard) and someone must have decided that, as table football was a hobby ... .

Peregrine *(Falco peregrinus)*

OTHER COMMON NAMES: *blue-backed falcon (northern England); blue hawk (Ireland, mid-Scotland); duck hawk (Gloucestershire, Sussex); faakin hawk (Aberdeenshire); game hawk (Scotland); goshawk (Ireland); hunting hawk (northern England, Somerset); perry hawk (Yorkshire); spotted falcon; stock hawk (Shetland); tiercel gentle (Western Isles/Hebrides).*

It seems to be generally agreed that the Peregrine is something very special. Leslie Brown called it 'a hot contender for the honour of the world's most successful bird', although it would have to compete for that title with the Herring Gull. Nonetheless, the comment was based on good evidence. It occurs as a resident species on all continents and is probably the most widely distributed of all birds: it will migrate if climate or other circumstances so dictate; it adapts to almost any type of habitat; and, above all, it is a matchless flier, wholly adapted to aerial life and 'killing all its prey with a stoop of dazzling velocity and accuracy'.[148] There has inevitably been much scope for speculation on the speed attained by a stooping Peregrine and it seems clear that in excess of 240 kilometres (150 miles) per hour is highly probable.

As a British bird, the Peregrine has a chequered history. It has long been the supreme bird for falconry but, during the late nineteenth century, came to be persecuted on grouse moors. It held its own but the population was halved during the Second World War, when adults were killed and nests destroyed deliberately to protect carrier pigeons. It recovered again but then, as a top predator, fell foul of organochlorine pesticide usage in the 1950s and 1960s. When J.A. Baker so poetically described his personal observations of Peregrine behaviour in 1967, the bird was at a seriously low ebb.[149]

With the ban on the use of the worst of the pesticides, it again recovered and there are now over 1,000 pairs nesting in Britain, but this does not necessarily mean that the Peregrine is safe because its preferred prey in many areas is pigeons. Baker found that Wood Pigeons accounted for 38 per cent of the 619

A bird of kings, the Peregrine *(Falco peregrinus)* was, and remains, the falconer's top choice.

Peregrine kills that he analysed over ten winters, with little or no evidence of domestic or feral pigeons being killed. Others, however, have felt differently and, in June 2000, the RSPB reported increased poisoning of Peregrines by 'disgruntled pigeon fanciers who mistakenly believe these birds of prey are responsible for the high mortality among racing pigeons'.[150] However, the evidence does seem weighed against this:

> The Raptor Study Group ... said that only 7.5% of homing pigeons were killed by raptors out of 52% that failed to find their way home. The rest got lost or were killed by flying into pylons.[151]

Traditionally, the Peregrine has been a bird of mountains and sea cliffs (to see wild Peregrines stooping from high cliffs onto gulls is a sight unparalleled in British bird-watching), but its adaptability has been evident in the way that, in lowland areas, it is now to be found nesting on pylons and tall buildings. It is a big, heavily built bird, mainly dark grey above and paler beneath, but with pronounced streaking.

The Peregrine was, and still is, top of the falconer's choices; it was the bird of kings and princes and was used to hawk a wide range of prey. To the falconer, the female Peregrine is a 'falcon' and the male a 'tiercel' (from the Old French meaning 'one third', because it is about a third smaller than the female), although the name 'Peregrine' itself also has origins in falconry. Young birds, trapped while making their first long-distance flights from some unknown foreign origin, are thought better than birds taken from the nest and are called 'passage hawks'. To Albertus Magnus of Cologne (1206–80), they were 'foreign falcons', the literal translation of the Latin *Falco peregrinus*. The name has come into English usage as 'peregrinatory', meaning 'travelled', or 'wandering'.

Peregrines have been used to hawk a wide range of prey, from Rooks, pigeons and Magpies to the choicest of game birds. A list of game killed by the Duke of Leeds' hawks in the year 1830 includes a tiercel that killed 11 grouse, 64 partridges, 3 Woodcocks, 1 Land Rail and 1 Magpie, and a falcon that killed 15 grouse, 63 partridges, 1 Woodcock and 2 Rooks.[152] The most spectacular prey was the heron, which, at that time, was still a choice table bird (see Grey Heron, page 214). Heron hawks were first flown at Rooks before being turned against herons, with sometimes astonishing results. The last English heron-hawking club, the High Ash Club at Didlington in Norfolk, closed in 1838, but clearly herons were still hawked in England for some time afterwards, using birds based at Loo in Holland, because there are reports of a pair of passage hawks in Norfolk killing 57 herons in 1844.

Like 'Merlin', 'Peregrine' is sometimes encountered as a personal name but there seems to be no logical reason why the best-known Merlin (albeit as 'Merlyn') of recent times, Merlyn Rees b.1920 and Peregrine Worsthorne b.1923, have both been politically motivated men.

The Red Grouse *(Lagopus lagopus)* has achieved fame as an emblem of 'The Famous Grouse' whisky.

Grouse
(Tetraonidae)

Grouse are typical game-birds, many being fairly small and of stocky, well-fleshed appearance, with feathered legs. They are also fairly sedentary in the sense that they are mostly resident species which undertake few migratory movements. They fall into two main groups, open-country grouse (the Red Grouse and Ptarmigan) and woodland grouse (the Black Grouse and Capercaillie), although all are principally vegetarian, buds and shoots forming their typical diet. Most are ground-nesting and fly relatively little.

However, among the four British grouse is one that has attracted a great following and added immensely to the market value of otherwise fairly featureless tracts of countryside. The grouse, the Red Grouse, has been the most famous British game-bird for around 150 years (there is even a brand of whisky to prove it) and the annual shooting, which begins on the 'Glorious Twelfth' of August, is a pivotal date in some people's social calendar.

The name 'grouse' has replaced the more charming sixteenth-century 'grows', which originated from French and Latin and is probably imitative of the call of the Black Grouse (not the Red Grouse). The name is the same in both singular and plural. The use of the word 'grouse' to mean 'grumble' is relatively recent and seems to have originated in the nineteenth century, although it appears to have no connection with the birds.

***Grouse Moor at Balmoral* by Archibald Thorburn, 1911, underlines the significance of grouse-shooting in the social calendar.**

Willow Grouse (Red Grouse) *(Lagopus lagopus)*

OTHER COMMON NAMES: *brown ptarmigan; gor cock; gor hen (Scotland); heath cock; heather cock; moor bird; moor fowl, moorfowl (Scotland, northern England); moor game (Yorkshire, Cheshire); moor poot; moor pout, more cock; more hen, moss hen (Yorkshire); red game; red ptarmigan; willow ptarmigan.*

The significance of the Red Grouse has declined in two ways in recent years. As a sporting bird, it has never really recovered from the aftermath of the two world wars, when management of the moorlands and control of predators was severely neglected. This has been compounded by its susceptibility to nematode parasites, the so-called 'grouse disease' that tends to occur in 4–8-year cycles and severely reduces its numbers. Scientifically too, the Red Grouse has lost its former status as the only species of endemic British bird: *Lagopus scoticus* has now been reduced to a subspecies of the widespread circumpolar Willow Grouse *(Lagopus lagopus)*.

Willow Grouse are fairly small birds, about the same size as a bantam, and the old name 'Red Grouse' is reasonably appropriate because the plumage of British birds is typically a dark red-brown overall, although there are local colour variations. Grouse in Britain feed very largely on the young shoots of heather and, like some other grouse species, have digestive systems especially adapted to cope with this very fibrous material. The many regional names that include 'moor' tells us most of what we need to know about its habitat. 'Moorfowl' appears very commonly in northern literature, for example in Robert Louis Stevenson's *Kidnapped*:

> Never a word passed between us; each set his mouth and kept his eyes in front of him ... with the moorfowl crying 'peep!' in the heather.[153]

The choice of 12 August for the start of grouse-shooting was dictated by the end of the bird's breeding season but, in turn, influenced matters of much greater moment: Parliament in London made sure that it was always in recess in good time for the Members to make their way to Euston and head north.

> ... in 1915, a bill was passed in the Lords authorising the shooting of grouse on 5th August, so that those for whom the war gave less spare time could be ensured their full share of enjoyment. It is to the credit of the Commons that they rejected the bill, amidst shouts of 'we want to shoot Germans not grouse'.[154]

It was on 12 August in the same year, 1915, that eight guns shot a record bag of nearly 3,000 birds on the Littledale and Abbeystead beat in Lancashire. In reality, however, the Red Grouse is not among the great game quarry of English history. It was not until access to the moors became easier and the birds could be driven over guns that grouse-shooting became so important. Brian Vesey-Fitzgerald gave an admirably succinct summary of its history:

> Grouse were shot, of course – generally over dogs – but grouse shooting as a sport that attracted high rentals did not commence until the advent of the breech-loader and the railway. Lang gave the shooting world the breech loader in 1853 ... And from that date sheep farming declined. Owners found tenants who could afford to pay much more to shoot grouse than farmers could pay to graze sheep.[155]

Grouse-shooting and its associated social activities have formed the background to a number of stirring novels, most famously John Buchan's *The Thirty-Nine Steps*, although, more recently, Isabel Colegate's novel *The Shooting Party*, set just before the First World War, achieved widespread notice through the all-star 1984 film version.[156] Nonetheless, the Red Grouse's relatively late arrival on the sporting scene meant that the great classical writers, such as Shakespeare, ignored it totally.

In order to retain its texture and flavour, grouse must be plucked very carefully and should never be washed or skinned but wiped with a damp cloth. Grouse conventionally roasted is pretty dry fare but my adaptation of an old Scottish recipe called, would you believe it, 'Grouse à La Rob Roy' is recommended:

> Grouse; a little butter; a little lemon juice; sprigs of heather; rashers of bacon; salt, pepper and cayenne to taste; bilberries.
>
> Stuff the grouse with the bilberries and keep the liver. Rub in the salt, pepper, cayenne and lemon juice then wrap the grouse totally in bacon rashers, interspersed with sprigs of heather, and then cover with greaseproof paper. Roast for 20 to 30 minutes according to the size of the grouse. Take off the paper, bacon and heather 10 minutes before serving to brown the bird. Boil the livers for about 10 minutes and chop very finely. Spread them over a piece of wholemeal toast and place under the grouse a minute or two before serving. Only a very little dripping or butter should be left in the roasting tin when doing this to prevent the toast becoming soggy.
>
> Serve with fried breadcrumbs. Decorate with watercress and serve with chanterelle mushrooms, runner beans, creamed potato and game chips. The gravy should be clear melted butter; rowan (or failing that, cranberry) jelly should be served with it.[157]

Ptarmigan *(Lagopus mutus)*

OTHER COMMON NAMES: *cairn bird; grey ptarmigan; rock grouse; rock ptarmigan; snow chick; snow grouse; white game; white grouse; white partridge.*

'Ptarmigan' is a crudely anglicized version of a Scottish Gaelic name that imitates the bird's croaking call and this close relative of the Red Grouse is certainly a bird of Scotland, and of the far north at that. It is an Arctic–Alpine species, inhabiting higher ground than the Red Grouse, where it feeds on the shoots and fruits of bilberries and related plants; 'rock ptarmigan', like 'cairn bird', is an alternative name that reflects its rough mountain habitat.

The Ptarmigan is among the species that could be lost to the British fauna as global warming proceeds and pushes its range further northwards. There is considerable evidence that debris left by tourists on the Scottish mountains has aided its recent decline by attracting crows and other predators, which have then turned their attention to the Ptarmigan.

Like many other creatures of the far north, the Ptarmigan changes from a summer grey-brown to a winter snow-white in order to ensure year-round camouflage. In autumn, as the changeover takes place, the birds take on a wonderfully mottled appearance in which they blend perfectly with lichen-covered rocks. In past times, this colour change was presumably unfamiliar to anyone outside Scotland but whether or not the home-loving Tennyson ever saw a Ptarmigan, when writing 'The Marriage of Geraint', he clearly knew sufficient to draw the obvious conclusion that:

> The ptarmigan that whitens ere his hour
> Woos his own end.

The male Ptarmigan *(Lagopus mutus)* has more prominent red wattles than the female, which is paler in appearance.

Mrs Beeton thought that Ptarmigan, when young and tender, were 'exceedingly fine eating, and should be kept as long as possible to be good'. They were best, she thought, 'roasted before a brisk fire'.[158]

Black Grouse *(Tetrao tetrix)*

OTHER COMMON NAMES: *birch hen (female); black cock (male); black game; brown hen (female); grey hen (female); grigear (female); hasel/hazel hen (female); hazel grouse (female); heath bird; heath cock (male); heath fowl; heath hen (female); heath poult (Hampshire); killockdoe (Scotland); moor pout (female); muir pout (female).*

The Black Grouse and the Capercaillie are the two British woodland grouse species. The Black Grouse is roughly the size of a domestic fowl, with marked differences between the sexes. The cock is a handsome bird, glossy black, with white wing-bars and a beautiful lyre-shaped tail, while the hen is smaller, grey-brown and with a slight tail fork. In consequence they join the handful of British birds in which males and females have different names, 'black cock' and 'hasel/hazel hen' being the commonest.

The British Black Grouse population has plummeted over the past century or so, the blame being apportioned to egg-collecting, shooting and the disappearance of their woodland edge habitat. They were once relatively common almost everywhere, right down to the south coast of England, and were still being shot in the Home Counties after the First World War. However, they have since dramatically retreated northwards and Scotland is now the place to see them, although, even there, numbers are declining. One unexpected factor in this, for both Black Grouse and Capercaillie, has been the widespread use of high deer fences in woodland; the low-flying birds have simply collided with the fences and been killed in hundreds. Grants are now available for the removal of the fences.

During its courtship, the male Black Grouse *(Tetrao tetrix)* displays to the female with fanned tail and drooping wings.

Like other grouse, Black Grouse indulge in fairly elaborate courtship displays that involve a special courting ground to which the females come for mating, a dominant male defending the central position. These courting areas are called 'leks' (possibly from the Swedish *leka*, 'to play') and were once familiar features of native woodlands.

Capercaillie *(Tetrao urogallus)*

OTHER COMMON NAMES: *auer-calze (Scotland); capercailzie; cock of the wood (Ireland); cock of the mountain; great grouse; wood grouse.*

The Capercaillie, or 'great grouse', is a bird of impressive dimensions, impressive appearance and equally impressive history. The males are the size of a goose, more or less dark grey overall and with a wide-spreading, forked tail; the females are much smaller, browner and rather like a female Black Grouse but with a reddish chest.

Capercaillies are indigenous British birds that have been caught and eaten for millennia, as evidenced by the bones found on prehistoric kitchen sites. Curiously, they have never been legally classed as game because, by the time the Game Act of 1831 was passed, they had become extinct in Britain. Although they were once widespread, the destruction of the native coniferous woodlands of Scotland in the seventeenth and eighteenth centuries led to their demise and by the late eighteenth century they had gone. As ever, there is argument about when and where the last bird was seen shot, but apart from 'a rather doubtful record' of an old cock killed near Fort William in 1815, it seems that the last native birds were killed in Ireland *c.*1760.

Capercaillies were reintroduced at Taymouth in 1837 and 1838 by Lord Breadalbane, Sir Thomas Fowell Buxton and others, and they soon began to re-establish themselves, colonizing the new conifer plantations into the early years of the twentieth century. Now, like Black Grouse, they are again in decline, and for similar reasons (see Black Grouse, above).

Also like Black Grouse, Capercaillies use a display ground (less usually called a 'lek') and the relatively few people to observe the 'formalized' proceedings have evidently been very impressed. The display takes place early in the morning:

> The neck is stretched up, the tail is fanned and held vertical, and the wings are drooped and the song uttered. Occasional 'rogue' males are prone to display at any time away from the lek and can even attack humans and vehicles.[159]

I have seen two versions of the origin of the name 'Capercaillie', or 'Capercailzie', which is fairly evidently Gaelic. One says that it is derived from *cabhar coille*, meaning 'cock of the woods'; the other, from William Lockwood, says that the peculiar sound made by the male is derived from *capull coile*, 'the horse of the

Still Life with Game Birds **by James Hardy Jr (1832–89).**

woods': 'the shrill, frenzied climax of the male's nuptial song [being] for all the world like the whinnying of a horse'.[160]

The general advice for cooking both Black Grouse and Capercaillie seems to have followed that for Red Grouse, although it is often said that the Capercaillie's diet of pine shoots gives the meat a distinct turpentine flavour.

Partridges, Quails and Pheasants (Phasianidae)

Although pheasants are now familiar and common almost everywhere, they are, as Chris Mead picturesquely put it, 'no more part of the natural avifauna than battery chickens'.[161] They are semi-domesticated and also introduced. All the related and spectacularly beautiful ornamental pheasant species – particularly the Golden Pheasant *(Chrysolophus pictus)*, Silver Pheasant *(Lophura nycthemera)* and Reeves' Pheasant *(Syrmaticus reevesi)* – that so amaze people who encounter them in the wild are also introduced, and so are Red-legged Partridges. Quails are migratory; which leaves the Partridge, now called the Grey Partridge, as the only resident native British gamebird in this very familiar and hugely important family that also includes the Peacock and the domestic fowl.

All are heavily built birds with short wings and strong, short beaks. They feed mainly on the ground, fly relatively infrequently and usually only for short distances (the migrating Quail being an obvious exception), and make very simple nests of a few twigs or leaves, usually on the ground. Why domestic fowl should have remained so totally domesticated is a mystery. They seem never to have established themselves in the wild, other than for short periods in the New Forest. Why are there not more feral fowl? Perhaps the best explanation is 'a combination of their stupidity, Foxes, Stoats and gypsies'.[162]

Red-legged Partridge *(Alectoris rufa)*

OTHER COMMON NAMES: *Frenchman; French partridge; green partridge; red leg; red partridge.*

The most distinctive feature of this bird is not the red legs but the white face with a black stripe. It has an interesting history as a British species. It originated in southwestern Europe (not exclusively France, despite the alternative names) and was first brought to Britain as a sporting bird in 1673, when some were

The Grey Partridge *(Perdix perdix)* is considered to be among the finest of game-birds for the table.

liberated at Richmond and Windsor. The introduction was unsuccessful but, after several further attempts, a release near Orford in Suffolk led to a colony being established, since when many further introductions have been made and it is now fairly widespread.

Grey Partridge *(Perdix perdix)*

OTHER COMMON NAMES: *brown partridge; common partridge; English partridge; girgirick (Cornwall); grey bird (Suffolk); grey partridge (Sussex); mountain partridge (Kent); pairtrick (East Lothian); paitrick (Ayrshire); patrick (West Yorkshire, Lancashire); partrig (Yorkshire); petrick (Aberdeenshire); stumpey (Sussex).*

The name 'Partridge' is descended, through French, from Latin, but it is odd that so important a game-bird seems never to have had an original English name. There have been numerous spelling variants and the modern form dates from 1579.

It was then, and for centuries, a familiar bird throughout Britain, its horseshoe-marked breast, especially evident in the male, and its dumpy appearance being its trademarks. Sadly, it is now a bird in retreat. It has all but disappeared from Ireland, large areas of southwest England and Wales and parts of Scotland. Although, after harvest, less grain is now left among the stubble for the birds to feed on (adult partridges are mainly grain-eaters), this does not appear to be the main cause of its demise. Largely to blame is the loss of insect food for the young as a result of modern agricultural practices. 'Good farming and partridges go hand in hand' was an old saying; but not, it seems, by modern definitions. Never again will 1,671 birds be shot in a single day, as happened at Holkham in Norfolk on 7 November 1905.[163]

Partridges are particularly endearing birds, both to look at and to eat, as many people will agree. It was once said:

> If the partridge had but the woodcock's thigh
> 'T'would be the best bird that e'er did fly.[164]

John Clare (1793–1864), in his earnest desire to immortalize birds' nests, found a special appeal in the extraordinarily large clutch that the partridge produces:

> The partridge makes no nest but on the ground
> Lays many eggs and I have often found
> Sixteen or eighteen in a beaten seat
> When tracing o'er the fields or weeding wheat
> They lay in furrows or an old land rig
> Brown as the pheasant only not so big
> They're often found by pasture boys at play
> And by the weeders often ta'en away
> The boys will often throw the eggs abroad
> And stay and play at blind eggs on the road
> They lay in any hole without a nest
> And oft a horse's footing pleases best
> And there they safely like till weeders come
> When boys half fill their hats and take them home.
> (John Clare, 'The Partridge's Nest')

In practice, 16 or 18 eggs is by no means the limit. I once saw 23 eggs in a partridge nest, and no doubt some particularly fecund bird has produced yet more.

Partridges have long been considered among the finest of all game-birds for the table; a belief with which it is difficult to argue. Although roast partridge is perhaps the more common dish, an interesting traditional Sussex alternative is Ashdown Partridge Pudding:

> 1 brace of partridge (boilers); 4 oz mushrooms; a few mixed herbs; salt and pepper to taste; some wet crust; 8 oz suet; 4 oz rump steak; a little chopped parsley; 1 wineglass claret; 1 pint stock; ½ lb flour; water to mix.
>
> Line a greased quart pudding basin with the crust, leaving enough to cover the pudding. Place a thin slice of steak at the bottom. Cut the partridges into neat little joints and season with salt and pepper, mix with mushrooms, parsley and herbs, and add to the meat. Pour stock and claret over the whole and cover with suet crust, pinch the edges together, tie a cloth over and boil for 3 hours.[165]

Many people have shared this fondness for partridge. In 1536 Henry VIII issued a proclamation in order to preserve the partridges, Pheasants, and herons over an area from his palace at Westminster to St-Giles-in-the-Fields:

> ... and from thence to Islington, Hampstead, Highgate, and Hornsey Park. Any person, of whatever rank, who should presume to kill, or in any wise molest these birds, was to be thrown in prison, and visited by such other punishments as to the king should seem meet.[166]

Partridge has found medicinal value too:

> The potestates of the flesh cure consumptions, the broth of it is good against the jaundice, elephatiatis, and hectic fevers: The fume of the feather helps to cure hysteric fits, the apoplexy, and vertigo ...

The liver and spleen were once used to treat epilepsy while:

> ... The gall being rubbed on the belly facilitates the birth.[167]

The partridge, almost invariably accompanied by a spaniel, is one of the most common birds found on inn signs; most parts of the country have a Dog and Partridge somewhere. It is also a fairly common surname, although its origin is vague. Perhaps it signifies someone who shares one of the bird's traits, such as dumpiness?

Quail *(Coturnix coturnix)*

OTHER COMMON NAMES: *but-for-but (Cheshire); corncrake (Sussex); deadchick (Shetland); quailzie (Scotland); quick-me-dick (Oxfordshire); rine (Cornwall); throsher; wandering quail; weet-my-feet (Northern Ireland, East Lothian); wet weather (Sussex); wet-my-feet; wet-my-lip (west Norfolk).*

Quail, dressed ready for the table, have become rather common on the supermarket shelf of late. They are probably still in the gourmet category; like Pheasant, they are better braised than roasted as they then become too dry. Commercial production, of both birds and eggs, has become a big business, although the modern supermarket Quail is a farmed, if 'free-range' bird, commonly the Japanese species *(Coturnix japonica).*

The British population of the wild migratory European Quail fluctuates enormously and there are undoubtedly 'Quail years', when large numbers reach Britain from the Mediterranean and tropical Africa and the more usual breeding number of around 300 pairs may multiply tenfold. Nonetheless, the sparse references to Quail in old shooting and wildfowling records suggest that it is has never been a common British bird. Superficially, Quails are like small, delicate partridges and, also like partridges, are birds of open, rough grassland.

The Old English names (and the modern Welsh) meant 'stubble-field hen' and 'Quail' (as 'quayle') dates from the fourteenth century and was originally imitative of its call, as are most of the curious regional names. The Quail is an endearing little bird but, nonetheless, its supposed (but purely imaginary) amorousness is said to have led to the name being used in the seventeenth and eighteenth centuries to mean a whore. It may be coincidence, but the collective noun for Quails (and partridges) is a 'bevy', a word of unknown etymology, related to the Old French for drinking *(bevée)*. This word was used by Milton in *Paradise Lost*: 'A Beavie of fair Women, richly gay.'

Pheasant *(Phasianus colchicus)*

Pheasants are Pheasants, give or take a spelling variation, and so it has been since 'feasaund' appeared in *c.*1299. The name came, through French, from Latin and Greek and meant a bird from the River Phasis (now renamed the River Rhioni), which enters the Black Sea to the south of the town of Poti in what is now Georgia. In practice, Pheasants originate from a rather wide area of southeastern Europe, not just the eponymous river.

When they were first introduced to Britain, and by whom, has never been discovered, but there have been several introductions since. The Romans are rather commonly and probably justifiably held responsible and it is certain that the birds were present before the Normans because an ordinance of Waltham Abbey, dated 1059, mentions them. 'Thomas à

In this early twentieth-century illustration by Joyce Merier, Maid Marian joins Robin Hood, with Pheasant's feather in his hat.

Becket dined on Pheasant the night before he was murdered.'[168] Today, Pheasants occur almost everywhere in Britain, although they have never been nearly as common in Ireland. The original introductions, of Caucasian birds which had a dark neck and no neck-ring, became known as the 'Old English' type. Subsequently, other types were introduced from the Far East; these had a conspicuous white neck-ring and became known as 'Ring-necked Pheasants'. Later still, other forms were introduced from other parts of Asia, including dark, melanistic forms ('Black Pheasants').

The British population is now a mixture of these and many other different variants. Moreover, the chances are that any Pheasant seen in Britain today has been reared and released onto managed land for shooting. Around 20 million birds are shot annually and the rearing and shooting of Pheasants has become a major rural industry, despite the fact that the birds are subject to a range of debilitating diseases. They are shot for eating, although there is a long tradition of the cock bird's tail feathers being stuck into hats – for instance, they feature in almost every representation of the legendary outlaw Robin Hood.

In his speech prior to the Haxey Hood Game, the Fool, wearing a Pheasant's feather in his hat, is ceremonially 'smoked'.

> The Haxey Hood game: on the twelfth day of Christmas, the villagers of Haxey and Westwoodside [Lincolnshire] will celebrate by playing the Hood, a centuries old tradition. The object is to carry a three-foot length of leather-covered rope, the Hood, to a pub. Order is maintained in the melée, which can last for hours, by red-shirted referees known as Boggins, who are led by the Lord of the Hood. He wears hunting pink and a top hat decorated with flowers and Pheasant feathers.[169]

While Pheasants have become an integral part of the countryside, country economy and headgear, they have never truly entered the national psyche. Creative artists of most kinds have neglected them, although they have always been choice subjects for painters specializing in sporting subjects, generally being depicted dead and hanging over the side of a table. Shakespeare mentioned them only once, John Clare did not write a poem about their nest, and I cannot remember anyone ever singing a song about them. There are not even very many Pheasant inns. It is all a rather curious paradox.

However, there must be room for speculation in the use of the word 'pheasant' to mean 'a wanton, low person' (as it did from the seventeenth to the nineteenth centuries), for a 'pheasantry' to be 'a brothel' (nineteenth and early twentieth century) or, in an interesting zoological cross-fertilization, for a 'Billingsgate pheasant' to mean a 'red herring'.

Rails, Crakes and Gallinules
(Rallidae)

The rails and their relatives are a motley bunch, rather like domestic fowls in overall appearance but with a wide range in size, from birds the size of a finch to others as big as a goose. They have in common short tails, short wings, short beaks, long legs and big feet, and males and females tend to be similar.

In Britain, there are two very common and very distinctive species (the Moorhen and Coot), and three rare ones (the Water Rail and two kinds of crake); the latter are very similar, speckled brown above and greyish below. They nest on the ground among aquatic vegetation and feed on a range of aquatic plant and animal life.

Water Rail *(Rallus aquaticus)*

OTHER COMMON NAMES: *bilcock; brook ouzel; brook runner; brown hen; darcock; grey hen; grey skit (Devon); gutter cock (Cornwall); moorhen; rail (Suffolk); rat hen (Yorkshire); scarragrise (Lancashire); sharming (Norfolk); skitty (Somerset); skitty cock (Devon, Cornwall); skitty coot.*

Descriptions of Water Rails invite the adjective 'secretive', and with good reason: they are probably the shyest and most difficult to observe of all British birds. I have only ever glimpsed three of them in the wild. As a result, no-one seems to know how rare or common they really are, because they live in places that are difficult to penetrate: among dense marshland vegetation from where they rarely venture into the open. They are also crepuscular.

Exploring wet marshes at night trying to spot Water Rails, or even to hear their distinctive squealing, grunting and piglet-like call, is not something to be undertaken by the faint-hearted. Nonetheless, when they were much more common and widespread, a good many people did just that because Water Rail eggs were collected and sold for eating. The birds themselves are residents and feed on insects, other small creatures and some plant matter, and they nest among the densest of vegetation.

The name 'Rail' (as 'rale') first occurs in the middle of the fifteenth century but it is impossible to decide whether this and other old references relate to this bird or the very similar Corncrake. Indeed, they are so similar that Corncrakes were once believed to turn into Water Rails for the winter, although the Water Rail has a longer beak. The word 'rail' comes from old regional French and is related to the word 'rattle', a reference to the call. The delightful name 'bilcock' (which is also used for the Moorhen) is allied to the verb 'to bolt', an allusion to the bird's habit of slipping away from view. The Norfolk word 'sharming' usually refers to the call rather than the bird itself.

The Corncrake *(Crex crex)* owes its scientific name to the male's cry, thought to be a portent of death in parts of Scotland.

Spotted Crake *(Porzana porzana)*

OTHER COMMON NAMES: *skitty, skitty cock, skitty coot (Devon); spotted rail; spotted skitty (Devon); spotted water hen; water crake.*

The Spotted Crake is a rare summer visitor, with barely a handful of birds living and nesting in similar impenetrable places as the Water Rail (see above). Even if you see one, you need to be very sharp-eyed to detect the shorter beak that betrays its identity. The name 'Crake' is onomatopoeic and certainly very old.

Corncrake *(Crex crex)*

OTHER COMMON NAMES: *bean cracker, bean crake (south Pembrokeshire); corn drake (North Yorkshire); corn rake (Yorkshire); corn scrack (Aberdeenshire); cracker, craker, creck (northern England, Shropshire); daker (Surrey); daker hen (Westmorland); gallinule crake; gallwell drake; gorse duck; grass drake (West Yorkshire); grass quail (Cheshire); hay crake (Yorkshire); land drake (Shropshire); land hen; land rail; meadow crake; meadow drake (Nottinghamshire); rape-scrape (West Country).*

The Corncrake is among the great lost birds of British ornithology and the distinctive, unbelievably repetitive nocturnal call of the male (which gave rise to the scientific name *crex*) is among the lost sounds of the British countryside. The reasons for its disappearance are self-evident as it is a species of the long, lank grass that typified the ancient hay meadows and corn (wheat) fields; since mechanized cutting began in the nineteenth century, its decline has been inevitable. Careful management of meadows in selected areas has led to a very slight recovery but it will never again be a common bird.

Unusually for a British bird name, 'Corncrake' was originally Scottish. However, Scotland has always been a stronghold of the species (the Isle of Skye is now the place to see them) and it is in Scotland that the belief arose that hearing its call was a portent of death. In Shetland, the Corncrake was said to make its distinctive sound when lying on its back to prevent the heavens from falling, an odd notion that may have arisen because it does indeed lie on its back when feigning death to avoid its predators.

Moorhen *(Gallinula chloropus)*

OTHER COMMON NAMES: *bilcock; bilter (North Country); cuddy (Scotland); dabchick (Shropshire); kitty coot; marsh hen; mere hen; moat hen; moorcock; moor coot; morant (Shropshire); nightbird (Sussex); pond hen; skitty (Somerset); stank hen; stankie (East Lothian); water hen; water rail.*

The name 'water hen' seems more accurate and appropriate than 'Moorhen' because there are still relatively few areas of river or lake that I knew growing up where this bird's distinctive purring sound cannot be heard; it was certainly the name with which I grew up in Derbyshire. However, 'moor' is here used in its old meaning of 'marsh', or 'mere' (as in 'mere hen'), rather than in its modern connotation of an extensive dry region of heather.

The Moorhen is a rather tame, gangly, long-legged and very distinctive creature, very dark brown overall, with a white tail and striking red-and-yellow beak. Nonetheless, it has largely been ignored in literature and there are surprisingly few waterside inns named after it, although there is a Three Moorhens somewhere.

Coot *(Fulica atra)*

OTHER COMMON NAMES: *bald coot (Yorkshire, Gloucestershire, Somerset, Sussex); bald duck; bald-faced coot (Yorkshire); bel poot (East Lothian); bell kite (Scotland); black diver (Ireland); black hen (Shetland); cute; queet; smuth; snaith (Orkney); snythe; water crow (Dumfries-shire); water hen (Somerset); whistling duck (Scotland); white-faced diver (Ireland).*

The expression 'bald as a Coot' is perhaps understandable because the very striking, featherless white front of the face is certainly a form of baldness. However, the mid-nineteenth-century American expression 'daft as a Coot' seems a little uncharitable and why American Coots should be thought of as simpletons remains a mystery. One suggestion is that the bird's apparent baldness means that its brains are inadequately protected, but it is interesting that, when applied to people, it is generally used for the elderly, 'silly old coot' being a commonly heard expression.

Coots are very common birds almost everywhere and are frequently encountered with ducks on ornamental lakes and ponds, where they respond in similar fashion to being fed by the public, although they are generally rather more vegetarian than most of their relatives. They are also somewhat aggressive, especially towards each other.

The Coot, rather than the duck, has personified watery places for some notable writers, and never better than in these lines by Tennyson:

> I come from haunts of coot and hern,
> I make a sudden sally
> And sparkle out among the fern,
> To bicker down a valley.
> (Tennyson, 'The Brook')

The 'hern' referred to here is not the moor-hern but the heron (see page 214), although it is also appropriate that Tennyson chose the old poetic verb 'to bicker' meaning 'to move lightly and quickly'. The etymology is said to be unknown, but one wonders if there is not some common root with 'bilcock' (an alternative name for the Moorhen). Perhaps the most famous literary coot, however, belongs not to Tennyson but to Arthur Ransome, and the children who belonged to the Coot Club in his 1934 story about sailing on the Norfolk Broads.[170]

The name 'Coot' is ancient, certainly Middle English, and originally imitative of its sharp, repetitive call. At Horsey in Norfolk, a Coot Custard Fair was held certainly until the nineteenth century, where items made from Coot eggs were on

The bare white skin on the head of the Coot *(Fulica atra)* has given rise to the expression 'bald as a Coot'.

sale. Indeed, Norfolk was once a pretty dangerous place if you happened to be a Coot because, for many years during the nineteenth century, a public Coot drive was held on Hickling Broad on St George's Day (23 April):

> It was indeed regarded in the nature of a public holiday. The guns were divided into two parties. One party manned the boats, and it was their duty to sweep the water in line. The rest lined the shore of some bay or whatever stretch of the Broad it had been decided to drive the birds into. The coot will not rise from the water (which it can only do with difficulty after the manner of a diving duck) unless it is forced to do so, and the birds would come swimming along in front of the boats. Then finding themselves trapped between two lines, they would rise and fly wildly round in all directions. The shooting I understand, was inclined to be dangerous. But the bags were enormous. The record for a public shoot at Hickling seems to be 910 coot in 1901, but on 25th February, 1927, 1,175 coot were killed by twenty guns.[171]

Similar Coot massacres were held at Slapton Ley in Devon.

Both Moorhens and Coots were once served at table – 'young moorhen are very good eating'[172] – and a Herefordshire recipe says that Coots are best in pies or stews and suggests that they and Moorhens should be skinned, like Rooks.

Cranes
(Gruidae)

Crane *(Grus grus)*
Cranes are very large, long-legged birds with an omnivorous diet, a famously dramatic courtship behaviour and often a long migratory flight. The European Crane was a fairly frequent British bird until the end of the seventeenth century and has been seen again in East Anglia over the past 20 years. Its prominence in British bird lore is largely due to its historical confusion with the Grey Heron (see page 214).

Bustards
(Otididae)

Great Bustard *(Otis tarda)*
If Queen Victoria had visited Salisbury Plain soon after her coronation, she might have caught sight of one of the last Great Bustards, which ceased to be a British breeding bird a year later, in 1838. According to Edward Jesse:

> The last Bustard known to have been killed in England, was shot in the spring of 1843, in Cornwall. It was a female, and had been seen in a turnip field for several days. This is the only instance of the Bustard being found in Cornwall.[173]

What a loss. This was a big, heavy bird of open grasslands, superficially like a long-legged game-bird, but with the male dramatically dwarfing the female in size. An unsuccessful attempt was made to reintroduce the bird during the twentieth century, although enthusiasts have not given up.

Oddly for such a huge, conspicuous creature, it entered British lore and literature only sparsely, although its memory survives in the Bustard inns that can still be found in parts of Wiltshire and Lincolnshire. According to Herbert Edlin, Gilbert White thought that, from a distance, they could easily be mistaken for Roe Deer.[174] Clinton Keeling put this to the test at Whipsnade, and pronounced the comment valid.[175]

Oystercatchers
(Haematopodidae)

Oystercatcher *(Haematopus ostralegus)*
OTHER COMMON NAMES: *chalder, chaldrick (Orkney); dickie bird (Norfolk); gilliebride (Scotland); krocket (Aberdeenshire); mere pie (Suffolk); mussel cracker (northern England); mussel picker (Ireland); olive (Essex); oyster picker (Somerset); oyster plover; pienet; pynot (Northumberland); scolder (Orkney); sea nanpie (Yorkshire); sea pie (Lancashire, Norfolk, Gloucestershire, Cornwall); sea piet (Northumberland); sea pilot; sea pyot; shalder, shelder (Shetland); sheldro, skeldrake (Orkney); tirma; trillichan (Hebrides).*

I have always had a soft spot for these very common, unsubtle, noisy waders. I think it is the sheer audacity of the bright orange beak in combination with the striking black-and-white plumage. Oystercatchers are resident birds of the sea shore, where they nest in a small scrape among shingle and in short vegetation close to the sea, although they are increasingly spreading inland. They do not really 'catch' anything; they may smash a few oysters in the course of a day's work, but mussels and other shellfish, along with crustaceans, seem to provide their staple fare.

I much prefer the names 'mussel cracker' and 'sea pie' (using 'pie' in its Magpie sense of 'black-and-white'), which are much more attractive than the American 'Oystercatcher'. A much older English name is 'olive', first found as 'oliff' in 1541 and evidently imitative of the call. The Scottish and Gaelic name 'gilliebride' is sometimes said to mean 'servants of the Bride', but 'Bride' in this instance probably refers to the mysterious St Bridget of Ireland (also known as St Bride, Abbess of Kildare, who died *c.*AD 525), who, according to some beliefs, was the patron saint of birds and carried an Oystercatcher in each hand. Nonetheless, in all the references I have found, St Bridget is said to be the patron saint of healers, blacksmiths and poets, and is usually depicted not with an Oystercatcher but a cow. There is, however, a further religious link because, according to another Gaelic tradition, an Oystercatcher:

> ... covered Jesus with seaweed when his enemies appeared in hot pursuit. The oystercatcher was therefore blessed, and still shows, as it flies, the form of a cross on its plumage.[176]

Although Oystercatchers *(Haematopus ostralegus)* are sea shore birds, they are increasingly being seen further inland.

Stilts and Avocets
(Recurvirostridae)

Avocet *(Recurvirostra avosetta)*

OTHER COMMON NAMES: *awl-bird (Suffolk); butterflip; clinker (Norfolk); cobble's awl; cobble's awl duck; crooked bill; picarini; scooper; scooping avocet; shoe awl; shoeing horn; yarwhelp, yaup (Norfolk); yelper (Lincolnshire).*

As its generic name suggests, the outstanding feature of the Avocet is its upward-curving beak, an adaptation for feeding on small creatures on the surface of water, which it scoops up by swinging its head from side to side. In this and its overall appearance, it cannot be mistaken for any other wader and, with its long, graceful legs and striking black-and-white plumage, it is undeniably elegant. It is also an evocative and symbolic bird for British ornithology: in recognition of its success at the important bird reserve of Minsmere, the Avocet was selected as the logo of the RSPB.

Avocets are summer visitors to Britain, although some overwinter. They nested in modest numbers in the marshes of eastern England in the early nineteenth century but, for some reason, they left and the last pair nested on Romney Marsh in Kent *c.*1843. They continued to visit and, inevitably, be shot; around 18 were killed in the Thames estuary between 1858 and 1865. Apart from an odd couple that nested in Ireland in 1938, however, nothing further significant happened until the East Anglian marshes were flooded as part of the coastal defences of the Second World War. Avocets liked the result and began to breed there in small numbers and, in 1947, eight pairs bred in Suffolk. Since then, it has been upwards all the way; there are now over 600 pairs and their range is gradually extending.

The common name is clearly not English, although it has been in use since the seventeenth century. It originated in Italy, *avosetta* being the name used for the bird on the Venetian coast, where they occur in quantity.

The Avocet *(Recurvirostra avosetta)* is the logo of the Royal Society for the Protection of Birds.

Stone Curlews
(Burhinidae)

Stone Curlew *(Burhinus oedicnemus)*

OTHER COMMON NAMES: *clew (Surrey); collier, collier-jack (Cheshire); cullew (Suffolk); great plover (Suffolk, Surrey, Sussex); land curlew (Kent); little bustard, night curlew (Sussex); night hawk; Norfolk plover (Norfolk); stone plover; thick knee (Norfolk); thick-kneed bustard (Sussex); whistling plover; Willie reeve (East Anglia).*

The Stone Curlew is a decidedly odd summer visitor, and although it has the same pale brown coloration with darker speckling as the Curlew, it lacks the distinctive curved beak. 'Thick knee' is an appropriate name because it does have long, yellow, knobbly legs, although the few people who have ever seen the bird are most struck by its bright, staring, yellow eyes.

They have never been common British birds and are really a southern European species, at home on dry heaths and grassland, where they nest in a small hollow on bare ground (among stones) and feed mainly on insects, although they will take young birds and mice. They are also among the rather few birds that eat toads, with evident immunity to the venom. They are thought to have been affected by myxomatosis, which, in reducing the number of Rabbits, also reduced the area of closely grazed grassland that they prefer. Nonetheless, there are still a few hundred Stone Curlews nesting in the south and east of England.

Plovers
(Charadriidae)

Thirty years ago, flocks of Lapwings were among the most characteristic sights and sounds of the British countryside. Sadly, the Lapwing has suffered an enormous decline in numbers for several reasons (see Lapwing, page 273). Its disappearance, however, means that Britain is becoming seriously short of the distinctive group of birds called plovers.

These birds are waders but with shorter beaks than the sandpipers, fairly long legs, long wings and an odd way of scratching their heads, by lifting one foot over a wing. They use their short beaks to probe in mud and soil for insects and other small animals and, like other waders, usually nest in the open on the ground.

There has been a good deal of discussion about the origin of the name 'plover' and many people have tried to link the birds with rain. *Plovere* is the Latin verb 'to rain' and *pleuvoir* the French. Rather succinctly, William Lockwood has pointed out however that 'the many attempts to find a rational link between the Plover and rain, by writers ancient and modern, have necessarily been in vain'.[177] They have been in vain because 'plover' has nothing to do with *plovere* but comes, via the Old French *plovier*, from the Late Latin word *plovarius*, which was simply imitative of the bird's very clear and far-reaching call. For no very obvious reason, 'plover', like 'pigeon', 'Pheasant' and 'Quail', is among the bird names that have been used at various times to mean a low or wanton person, Ben Jonson being among the culprits.

Little Ringed Plover *(Charadrius dubius)*

OTHER COMMON NAMES: *little ringed dotterel.*

A common summer visitor, this bird has gradually been extending its breeding base in Britain since it first arrived in 1938. It has a dark breast-band, a dark eye-stripe, pale brown back and white underparts. In most respects it is therefore similar to the more common Ringed Plover, but it is smaller, more slender and with clay-coloured not orange legs. It is most likely to be seen close to inland waterways, at lakesides, on sewage farms, gravel pits and similar places.

Ringed Plover *(Charadrius hiaticula)*

OTHER COMMON NAMES: *bull's-eye (Ireland); dulwilly; grundling (Lancashire); knot (Belfast); ring dotterel; ringlestone; ring-neck (Yorkshire); sand lark (Northumberland); sand tripper (County Down); sandy laverock (Shetland, Orkney); sandy loo (Orkney); sea lark; shell-turner (Sussex); stonehatch; stone plover; stone runner (Norfolk); tullet, tullot (Lancashire); wideawake (Somerset).*

A larger version of the Little Ringed Plover, but with orange legs, this plover is also much more common and much more of a coastal bird, as many of its regional names testify, although it is being found breeding inland with increasing frequency.

Most British Ringed Plovers are residents, but their numbers are supplemented by winter visitors from Scandinavia and elsewhere.

Kentish Plover *(Charadrius alexandrinus)*

Very similar to the two Ringed Plovers, the Kentish Plover can be distinguished by its black breast-band, which has the centre part missing, and black legs. It was once a rare British breeding bird and, today, is an even rarer migrant.

The naturalist John Latham, who was sent specimens from Sandwich in Kent in 1802, was responsible for naming it, just as he was for naming the Sandwich Tern some years previously.

Dotterel *(Charadrius morinellus)*

OTHER COMMON NAMES: *dot plover (Norfolk); land dotterel, moor dotterel, spring dotterel (Yorkshire); stone runner (Norfolk); wind (southern England).*

If there was ever a case of giving a bird a bad name, this inoffensive and pretty wader is a prime example. Rather like 'plover', 'dotterel' is another word whose meaning has been confused and misunderstood, the general assumption being that 'dotterel' ('dotrelle' in the fifteenth century) comes from the old word 'dotard', meaning 'a fool', and is therefore of the same origin as the modern word 'dotty'. Another story links the association between the bird and stupidity with its extreme tameness and the relative ease with which it can be caught. Sadly for the bird, this is all fallacious because the name has nothing to do with stupidity: 'Dotterel' originated in the word 'dot', which still survives in the Norfolk name 'dot plover', and is simply an imitation of part of its plaintive call.

The Kentish Plover *(Charadrius alexandrinus)* is distinguished by its black legs, eye-stripe and breast-band.

Dotterels have a narrow neck-band and white stripe above the eye and, in summer, a very distinctive, rich russet breast. It is their habitat, however, that is even more distinctive, for they are northern birds of high mountain moors and were once a favourite quarry for wildfowlers, being regarded as excellent eating.

The ornithologist Francis Willughby who, with John Ray, was responsible for many of the modern names of British birds, wrote in 1676:

> The dotterel is a very foolish bird, but excellent meat, and with us accounted a great delicacy. It is taken in the night time by the light of candle, by imitating the gestures of the fowler: for if he stretches out an arm, that also stretches out of a wing: if he a foot, that likewise a foot: in brief, whatever the fowler doth, the same doth the bird and so being intent upon men's gestures, it is deceived and covered with the net spread for it.[178]

As is the case with other popular wildfowl, inns called The Dotterel are still to be found in places where hunters once congregated. Less expectedly, there is at least one Dotterel inn on the Yorkshire Wolds and also Dotterel farms in eastern England; both places where the birds paused during their migration, only to be shot. Dotterels were clearly once seen in southern England because, according to an old Wiltshire rhyme:

> When dotterel do first appear,
> It showes that frost is very near:
> But when that dotterel do go,
> Then you may look for heavy snow.[179]

The Dotterel *(Charadrius morinellus)* is a northern bird, although folklore suggests that it was once common in the south.

The Dotterel seems to be the same bird as the Mountain Plover (the 'red bird of the black turf ground') of the Isle of Man, which features in an old Manx lullaby sung to the tune of 'Here We Go Round the Mulberry Bush'. Translated from its original Manx, this runs:

Little red bird of the black turf ground,
Where did you sleep last night?
I slept last night on the top of the briar,
And oh! what a wretched sleep!

Little red bird of the black turf ground,
Where did you sleep last night?
I slept last night on the top of the bush,
And oh! what a wretched sleep!

Little red bird of the black turf ground,
Where did you sleep last night?
I slept last night on the ridge of the roof,
And oh! what a wretched sleep!

Little red bird of the black turf ground,
Where did you sleep last night?
I slept last night between two leaves
As a babe ' twixt two blankets quite at ease,
And oh! what a peaceful sleep![180]

A related and delightful Manx story links the Mountain Plover *(Ushag-reaisht)* with the Blackbird *(Lhondoo)*:

It is said that once upon a time the haunts of Lhondoo were confined to the mountains, and those of the Ushag-reaisht to the lowlands. One day, however, the two birds met on the border of their respective territories, and, after some conversation, it was arranged to change places for a while, the Ushag-reaisht remaining in the mountains, till the Lhondoo should return. The Lhondoo, finding the new quarters much more congenial than the old, conveniently forgot his promise to go back. Consequently the poor Ushag-reaisht was left to bewail his folly in making the exchange, and has ever since been giving expression to his woes in the following plaintive querulous pipe: Lhondoo vel oo cheet, vel oo cheet? 'Black-bird are you coming, are you coming?' The Lhondoo, now plump and flourishing, replies 'Cha-nel dy bragh, cha-nel dy bragh!' 'No never, no never!' The poor Ushag-reaisht, shivering says 'T eh feer feayr t' eh feer feayr' 'It's very cold, it's very cold!'[181]

Golden Plover *(Pluvialis apricaria)*

OTHER COMMON NAMES: *black-breasted plover, grey plover (Ireland); hill plover (Scotland); plover (Roxburghshire, Surrey); sheep's guide (Cheshire); whistling plover (Scotland, Cheshire, Norfolk, Somerset); yellow plover (East Lothian).*

Not really golden but speckled olive-yellow with some black on the underparts, the Golden Plover is resident in Britain, breeding mainly in northern uplands during the summer but moving to lower ground in autumn (where it is threatened by more intensive agriculture). It is sometimes found on the coast.

The Cheshire name 'sheep's guide' arises from a belief that it gives warning to sheep of impending danger by its plaintive call. In Aberdeenshire, its cry is said to be giving friendly advice to the ploughman: 'Plough weel, shave weel, harrow weel.' In other areas, such as north Wales, it is the Golden Plover rather than the Lapwing, Curlew or other waders that has been linked with the legend of the Seven Whistlers (seven birds, flying together by night, whose cries forbode disaster). In common with other birds with plaintive calls, Golden Plovers have also sometimes been regarded as lost souls.

Grey Plover *(Pluvialis squatarola)*

OTHER COMMON NAMES: *bull head; bull-headed plover (Suffolk); mud plover; rock plover (Ireland); sea cock (Waterford); sea pigeon (Yorkshire); sea plover; silver plover (Scotland, Yorkshire, Cheshire); strand plover (Cork); whistling plover.*

The Grey Plover is not really grey and, despite its regional names, is not exclusively coastal. However, it certainly lacks the olive-yellow of the Golden Plover and is much more common on the coast in Britain, where it is a winter visitor. Some young birds remain in the country throughout the summer.

Lapwing *(Vanellus vanellus)*

OTHER COMMON NAMES: *chewit (Perth, Lancashire); cornwillen (Cornwall); flopwing; green plover; horneywink (Cornwall); hornpie (Norfolk, East Suffolk); lappie (Fife); lappin (Northumberland); lappinch (Cheshire); lipwingle (Bedfordshire); lymptwigg (Exmoor); old maid (Worcestershire); peesieweep, peesweep, peeweep (Scotland); peewit; piewip, peweep (Norfolk); phillipene (Ireland); plover; scochad, shouchad, shuchad, sochad (Caithness, Sutherland); tee whip, tee whippo, tee wup (Orkney); teuchit (Angus – Forfar); teufit (North Yorkshire – Cleveland); teewheep, tewhit, (Kirkcudbrightshire); tewet (Northumberland); tewit (Lancashire); tieves' nacket (Shetland); tuet (Lancashire Westmorland, West Yorkshire); wallock, wallop, wallopie, weep (Morayshire); weep, wype (Northumberland).*

The huge list of regional names is certainly not exhaustive but what does it tell us about this, the largest British plover? First, that it makes a distinctive cry, of which the common name 'peewit' is as good an imitation as any. Second, that it seems predominantly northern (although this is not especially true). Third, and less obviously, that it has the largest crest of any common British bird. The first part of the name 'Lapwing' originates in an old northern European name for a crest and the name as a whole is found in an Old English form recognisably similar to 'Lapwing' as early as the eighth century.

Lapwings are familiar because they are mainly birds of farmland and the recent dramatic decline in their numbers in Britain is certainly not new. They suffered a similar decline throughout the nineteenth century, although for a quite different reason: they were persecuted because their eggs were considered a delicacy. The culinary 'plover's eggs' were Lapwing eggs and Ron Murton cites some appalling figures:

> ... on one estate near Thetford, about 280 dozen lapwing eggs were taken annually in the 1860s but by the 1880s the annual take had fallen to 60 dozen ... in view of such intensive exploitation, it is impossible to isolate the harmful effects of a changing agriculture.[182]

The egg-collecting trend continued into the early years of the twentieth century; P. Weaver noted that over 1,900 eggs were taken by one person on the Norfolk marches in 1921.[183] Nonetheless, there was a spectacular increase in the numbers

The Lapwing *(Vanellus vanellus)*, with its distinctive crest, was once a common sight on ploughed fields and farmland.

of Lapwings after the passing of a specific piece of protective legislation: the Lapwing Act of 1926. Their recent decline seems to be related to the change in farming practice from spring to autumn sowing, which has entailed the loss of the winter fallow that the birds require.

Lapwings seem to have accumulated good and bad press in similar measure. Much of the bad stems from that distinctive, call, which is said to cry 'Bewitched, bewitched' and so bring evil on all who hear it. There is a contemporary reminiscence in Edwin Lees' remarkably evocative *Pictures of Nature in the Silurian Region Around the Malvern Hills*, written in 1856, about a boy in the village of Colwall in Herefordshire, who caught a young Lapwing and showed it to a woman in the village who said that he must let it go or evil would befall him.[184] The Lapwing's call is associated with the legend of the Seven Whistlers (see Golden Plover, page 273) and is particularly frequently linked with mining disasters. Christina Hole describes several instances:

> They were heard before the terrible explosion at Hartley Colliery in 1862. A writer in Notes and Queries relates how on September 6th, 1871, he saw immense flocks of birds flying through the storm and crying; his servant said they were the Seven Whistlers, and that a calamity would follow. The next day there was a colliery explosion at Wigan in which several men were killed. Most miners refused to go down the pit at all after such a warning. The Leicester Chronicle for March 24th, 1855, records that a collier was asked on March 16th why he had not gone to work. He replied that all the men had stayed away that day on account of the Seven Whistlers, and added that on two previous occasions some collier had been foolish enough to disregard the warning. Two lives had been lost in each instance.[185]

A writer in *Country Life* in 1904 spoke of others:

> In September, 1874, large numbers of the miners employed at the Bedworth Collieries in North Warwickshire ... refused to descend the pits in which they are employed. The men are credulous enough to believe that certain nocturnal sounds, which are doubtless produced by flocks of night-birds in their passage across the country, are harbingers of some impending colliery disaster. During Sunday night it was stated that these sounds, which have been designated 'Seven Whistlers', had been distinctly heard in the neighbourhood of Bedworth and the result was that on the following morning many of the men positively refused to descend the pits.[186]

Also:

> The Morfa Colliery, in south Wales, is notorious for its uncanny traditions. The 'Seven Whistlers' were heard there before a great explosion in the sixties, and before another in 1890, when nearly a hundred miners were entombed ... Innumerable instances could be cited of the alleged appearance of the 'Whistlers' just prior to the death of a country gentleman, especially in the case of those who had been great preservers of wildfowl.[187]

Robert Burns in 'Flow Gently Sweet Afton', anxious to avoid the disturbance of Mary's sleep, wrote:

> Thou green-crested lapwing, thy screaming forbear,
> I charge you disturb not my slumbering fair.

Sir Walter Scott, when he wrote, in 'Helvellyn', of the 'gentle lover of nature' who was found dead there, said:

> And more stately thy couch by this desert lake lying,
> Thy obsequies sung by the grey plover flying.

An often-repeated Scottish story tells of the seventeenth-century Covenanters, who met in secret to uphold the Covenants of 1638 and 1643 between England and Scotland to establish and defend Presbyterianism. Once such meeting:

> ... in some hidden, heathery glen of the misty hills was discovered and roughly dispersed because of the hovering, bewailing plovers, fearful for their young, clamouring overhead'.[188]

Later, the Scottish poet and scholar John Leyden (1775–1811) made telling reference to this long-held dislike of Lapwings:

> The Lapwing's clamorous whoop attends their flight,
> Pursues their steps where'er the wanderers go,
> Till the shrill scream betrays them to the foe.
> (John Leyden, *Wandering Covenanter*)

The ingenious wiles which the Lapwing employs to protect her nest have not enhanced her reputation. The characteristic limping and calling that she uses to draw potential predators away from the nest led to the Scottish rhyme:

> Pease weep, pease weep
> Harry my nest and gar me greet.[189]

Chaucer in his *Parlement of Foules* had thought the 'false lapwyng', to be 'ful of trecherye' for this subterfuge. Nonetheless, Shakespeare could appreciate its parental affection:

> Far from her nest, the lapwing cries away:
> My heart prays for him, though my tongue do curse.
> (*The Comedy of Errors*, Act 4, Scene 2)

The precocity of Lapwing chicks, which run while they still have pieces of eggshell attached to their feathers, did not pass unnoticed and brash, young people were likened to them:

Knots *(Calidris canutus)* and Dunlins *(Calidris alpina)* in flight.

This lapwing runs away with the shell on his head [says Horatio about Osric].
(*Hamlet*, Act 5, Scene 2)

For centuries, Lapwing eggs were collected and the birds were killed for the table. Most old 'plover' recipes relate to them, although Mrs Beeton pointed out that:

... there are 3 sorts – the grey, green, and bastard plover, or lapwing. They will keep good for some time, but if very stale, the feet will be very dry. Plovers are scarcely fit for anything but roasting, though they are sometimes stewed, or made into a ragoût.[190]

One nineteenth-century recipe for Plover Pie required six birds and a pint of rich, cold beef gravy, presumably to impart some taste. It is a tragic thought that so many birds were killed when they were not even considered a worthy meal.

Sandpipers and Related Birds
(Scolopacidae)

This family brings together a number of rather different birds, many popularly called 'waders' and all blessed with relatively long beaks, but with little else in common. Some are shore birds while others are typically species of high moorland and mountains; some move from one habitat to the other according to season. The only reasonably common feature on which ornithologists seem to be agreed is that very few of these birds breed further south than latitude 50°N.

Most nest on the ground and feed by probing in mud, sand or soil for insects, worms, crustaceans or other small creatures. Many waders characteristically feed both at night and during the day, their habits being dictated not by day length but by the state of the tides. Some species congregate in mixed flocks and many are difficult to tell apart, which is reflected by several sharing the same regional names. Others, however, are more distinctive and have attracted a disproportionate amount of attention by virtue of being the targets of people with guns and appetites.

Knot *(Calidris canutus)*

OTHER COMMON NAMES: *black sandpiper; dun (Cheshire); dunne (Belfast); ebb cock (Shetland); gnat; gnat snap; grey plover (Scotland); grisled sandpiper; howster; knat; knet; male (Essex); red sandpiper (Ireland); red knot; sea snipe (Dublin); silver plover (Scotland); spease (Northumberland – Lindisfarne).*

The largest species in a genus of very similar-looking waders, Knots are speckled greyish brown birds with a scale-like patterning on the back. They are typically seen in large, close flocks on the winter sea shore; they appear to be huddling together for warmth, although this seems improbable as they nest in the high Arctic, and British winters must be balmy in comparison.

They also stay close together when they take off and this has tempted wildfowlers to try to shoot several with a single shot. Brian Vesey-Fitzgerald quoted the record bag from a shoulder gun ('an old 4-bore I think') as 80 birds, a dramatic improvement on the conventional 'two birds with one stone'.[191] They were shot on both a large and small scale because they were considered very good eating.

The name 'Knot' imitates the rather low, grunt-like call and dates back to the fifteenth century; 'knat' and 'gnat' are related words ('gnat' has no connection with the insect of that name). There is a poorly founded belief that the name 'Knot' has some connection with King Cnut and that Knots were his favourite birds (hence *Calidris canutus*), presumably on the basis that they behave much as he did, foolishly playing around the tide-line.

Sanderling *(Calidris alba)*

OTHER COMMON NAMES: *curwillet (Cornwall); ebb cock (Shetland); ox-eye (Essex, Kent); peep (Northumberland); ruddy plover; sand lark; sandling; sand runner (East Yorkshire); sea lark (Ireland); snent; stint; towillee (Cornwall); towilly; tweeky (Northumberland).*
Smaller than a Knot but bigger than a Dunlin sums up the Sanderling, another sea-shore wader and, as befits its name, one most likely to be seen on sandy shores, running through the water's edge and feeding on small sea creatures.

The Cornish name 'towillee' is 'from the noise made by the wings in flight'.[192]

Little Stint *(Calidris minuta)*

OTHER COMMON NAMES: *brown sandpiper; little sandpiper; ox-bird; purre; stint; wagtail (Sussex).*
A rather uncommon small wader, the Little Stint is usually seen in Britain during its autumn migration and only very occasionally on the winter sea shore. It is most readily distinguished from the Dunlin by its considerably smaller size and shorter beak.

The lovely name 'Stint' first appeared as the even more charming 'styntis' in 1452 and is obviously related to other regional words for small waders, such as 'stent' and 'snent'.

Temminck's Stint *(Calidris temminckii)*

This bird is similar to the Little Stint but with clay-buff rather than black legs. It is always uncommon in Britain, although a few breed in northern Scotland.

As a somewhat bizarre justification for the pursuit of birds' eggs (fancifully dressed up in the term 'oölogy'), the first record of the British nesting of Temminck's Stint in 1951 was made by egg-collectors rather than by 'armchair ornithologists, or even bird-watchers and photographers'.[193]

Purple Sandpiper *(Calidris maritima)*

With a little imagination, the Purple Sandpiper could be described as purple-brown. It is certainly darker than most other small waders but at close quarters is better distinguished by its yellow legs and the yellow base to the beak. Small numbers are seen in Britain every year and a few odd pairs nest in Scotland. It is more likely to be seen on rocky coasts among seaweed than on sandy beaches.

Purple Sandpipers *(Calidris maritima)* are more likely to be found on rocky coasts than on sandy beaches.

Dunlin *(Calidris alpina)*

OTHER COMMON NAMES: *bundie (Orkney); churre (Norfolk); dorbie (Banffshire); ebb cock (Shetland); ebb sleeper; jack plover (Yorkshire); jack snipe (Shetland); ox bird; ox-eye (Sussex, Kent, Essex); peewee (Northamptonshire); pickerel (Scotland); plover's page (west Scotland); purre; sand mouse (Westmorland); sand snipe (Gloucestershire); sandy (Northumberland); sea lark (Northern Ireland, East Lothian, Cheshire, Gloucestershire); sea mouse (Dumfries-shire, Lancashire); sea peek (Angus – Forfar); sea snipe (northern England); stint (Northumberland).*
Statistically, the chances are that any small wader on a winter shore is likely to be a Dunlin because it is the commonest winter visitor to Britain. The name 'sea mouse' is particularly apt because it looks just like a mouse with long legs as it scurries about its business. Dunlins breed in some quantity in Britain but their numbers are declining; feral mink predation is one reason but so, more surprisingly, is the fact that their eggs are being eaten by Hedgehogs (see page 385).

The name 'Dunlin', from the sixteenth-century 'dunling', simply means 'dun-coloured bird', although they tend to be grey-white rather than brown. This is because the name was originally applied to the birds as seen in their summer plumage; the summer and winter birds were thought to be two separate species until the early nineteenth century.

Ruff *(Philomachus pugnax)*
OTHER COMMON NAMES: *fighting ruff (male); gambet; oxen-and-kine; reeve (female).*
This remarkable wader is a most unusual bird in many ways, not least in being among the very few British birds in which the male and female have different names. The male is considerably larger but, in winter, both look like sandpipers with extra long legs. A male Ruff in its summer breeding plumage is something quite extraordinary, however, with a huge neck ruff in a wide range of colours, ranging from black, through red-brown, to white. Perhaps the oddest feature is that males with wholly white ruffs are non-breeding birds, serving solely to stimulate the other males.

Unfortunately, the Ruff is a very rare British bird and even rarer as a breeding species. Its British home is in the Lincolnshire and East Anglian marshes, where it was at one time common enough to be netted regularly by wildfowlers, fattened and sold for eating.

The origin of the name has been the subject of much debate. The conclusion seems to be that the bird's original name was 'reeve', or 'ree', which can be traced back to Old English 'ree', meaning 'frenzied' or 'aggressive'. William Lockwood suggests that:

> At some later date, when the original meaning of Ree and Reeve had been forgotten, a new term Ruff came into use for the male, so that the [other] terms became attached to the nondescript female.[194]

What a loss this bird is to British ornithology. A great many naturalists would give a hundred Avocets to have the Ruff back as a regular breeding bird.

Jack Snipe *(Lymnocryptes minimus)*
OTHER COMMON NAMES: *gaverhale (Devon); half snipe (Norfolk); jedcock; jid; juddock; plover's page (Orkney); St Martin's snipe.*
See Snipe below.

Snipe *(Gallinago gallinago)*
OTHER COMMON NAMES: *blitter; blutter; bog bleater (Ireland); common snipe; ern bleater; full snipe (Somerset); hatter-flitter (Cornwall); heather bleater (Ireland, Scotland); horse gawk, horse gowk (Shetland, Orkney); Jill snipe (Ireland); lady snipe (Cheshire); mire snipe (Aberdeenshire); snippack; snippick (Shetland, Orkney); whole snipe (Somerset, Sussex); snite.*
'Snipe' is among the most wonderful of bird names and Britain is blessed with two species: the Snipe itself, which is still a fairly common, although declining, resident, to be seen all year round on marshes, mud flats, sewage farms and other wet, muddy places; and the Jack Snipe, about half the size, which is mainly a winter visitor. Their fondness for mud stems from their fondness for worms and insects, which they obtain with their long,

The Ruff *(Philomachus pugnax)* performs a dramatic courtship display, defending its mating area, or 'lek', against all rivals.

probing beaks, curious organs that are unexpectedly pliable. A snipe's eyes are set far back on its head so that it can still see when the beak is plunged full length into mud.

In Old English, the word 'snipe' was originally the even more wonderful 'snite' and meant 'a long, thin object', so it is hardly surprising that it is sometimes used for other related birds, such as the Woodcock, which have similar beaks (there are still people in the southwest of England who talk of 'snites'). The name is commemorated in the village of Snitterfield, Warwickshire, called 'Snitefeld' in *Domesday Book* and originally 'a feld frequented by snipes'. Snydale in West Yorkshire ('a haugh frequented by snipes') is comparable. The modern verb 'to snipe', meaning 'to shoot at a single person' or 'to single out a person for criticism', appears to come directly from the shape of the bird's beak.

Other names refer to the birds' curious bleating sound: Robert Burns quite splendidly referred to the Snipe as 'the Blitter frae the Boggie' and both Celtic and Welsh names allude to goats, which the sound really does imitate. This bleating noise is known as 'drumming', although it sounds little like a drum, and how it is produced has been disputed and debated for years. It is formed not in the throat but externally; the birds fly upwards, twist sideways and then swoop downwards with their tail spread, fan-like, and it is now generally believed that the noise is produced by the outer tail feathers. However, there are still countrymen who will you tell you otherwise – and it still does not sound remotely like a drum. The reason for the performance, like so much extraordinary behaviour, is courtship. In parts of Scotland, it is said that Snipe drumming indicates cold, frosty nights and dry weather and also that Snipe have no peace at night because one once rose to begin its drumming in front of the ass that carried Christ, causing it to stumble.

The Snipe *(Gallinago gallinago)* is noted for the length of its bill, which has a sensitive end for searching out food in mud.

Snipe have long been prized by wildfowlers. The Jack Snipe, although now protected, was once the preferred quarry because its smaller size made it a more difficult target. It was said that: 'For a bit of good sport, all you want is an acre of land, a bag of cartridges, and a Jack Snipe.'[195] Today, however, the Snipe and the Woodcock are the only two waders that may legally be killed and sold.

There are probably as many inns with shooting connections called The Snipe or The Woodcock as there are The Pheasant or The Dog and Partridge.

In my copy of Mrs Beeton's cookery book, a sentence in the recipe for Dressed Snipes has been underlined by an earlier owner, presumably on the basis of some experience:

> They should be sent to the table very hot or they will not be worth eating.[196]

So why were they eaten?:

> I once heard a man at Worsbrough reservoir [West Yorkshire] tell me that a Snipe 'has not a single bone in its body' and that is the reason why they are eaten.[197]

Woodcock *(Scolopax rusticola)*

OTHER COMMON NAMES: *cock; great snipe; kevelek, kyvellack (Cornwall); quis (Wiltshire).*

The Woodcock looks like a large Snipe, with the same unusually long, probing beak, although, as its name implies, it is more of a woodland bird and finds its food not only in the mud of woodland pools and stream-sides but also among leaf litter. British Woodcocks are mainly residents, although the numbers are supplemented by some winter visitors and the arrival of migrating birds is reflected in several local sayings and beliefs, for example, the Yorkshire name of 'woodcock pilot' for the Goldcrest (see page 350).

It was long believed that Woodcocks were among the birds that went to the moon in autumn. John Gay (1685–1732) was merely reflecting this widespread belief when he wrote the strange lines:

> He sung where Woodcocks in the Summer feed,
> And in what Climates they renew their Breed;
> Some think to Northern Coasts their Flight they tend
> Or, to the Moon in Midnight Hours ascend.
> (John Gay, 'The Shepherd's Week: Saturday or The Flights')

He would have done better to stick to the northern coasts. Interestingly too, John Gay was writing at a time in the early eighteenth century when all Woodcock genuinely went elsewhere to 'renew their breed'. They took to breeding in Britain

The Woodcock *(Scolopax rusticola)* is a woodland bird that was once, incredibly, thought to go to the moon in autumn.

before the early nineteenth century, but a variety of practices in forestry and agriculture have led to their recent decline.

Woodcocks are handsome birds, even when dead on a poulterer's slab, which is where they are most likely to be seen. They are popular and challenging wildfowling quarry because they generally rise silently and fly with an irregular course that makes them difficult to shoot. Partly for this reason, they were often caught by netting and, like other birds that can relatively easily be trapped in this way, came to be thought of as stupid. The word 'woodcock' was once used for a stupid person, something instantly denied by those who bear it as surname. Shakespeare, however, did not mince his words:

> O this woodcock! What an ass it is!
> (*The Taming of the Shrew*, Act 1, Scene 2)

Woodcocks have another odd flight characteristic in the territorial activity called 'roding' ('1768, origin obscure'). The male flies in a circuit at dusk and dawn with 'an owlish flight with interrupted wing beats', uttering two distinctive cries: a repeated croak followed by a short squeak.

> If you come across a roding Woodcock, it is worth waiting at the spot for 10 minutes; there is a good chance the bird will come round the same circuit again.[198]

If you are extremely fortunate, you might see a Woodcock carrying its young between its legs. This was once thought fanciful but has now been authenticated and is unique among birds:

> In our village in Kent we have established a tiny community wood – only one hectare in extent, about 300 metres from the Village Hall and Pub. While clearing an access path through the young trees, I was suddenly aware of a pair of extended wings coming towards me. A sudden scuffle at my feet, and turned to see the bird flying back up the path I had just cut. Between her thighs dangled the little legs of her chick! A woodcock exercising extreme Mothercare![199]

A delightful old pun on the bird's beak appeared in the eighteenth-century slang use of the word 'woodcock' for a tailor presenting a long bill. The word 'snipe' for a lawyer presenting a long bill followed a century later.

> Near Hindhead in Surrey is ... the comically named Woodcock Bottom (and the 'Woodcock' public house). I have never seen anything like a woodcock there.[200]

However, when in season, Woodcock is still found on the menus of specialist game restaurants, although it can come as a surprise to guests ordering it for the first time; they may be forgiven for thinking that they have been served a bird which has, a few minutes earlier, lain down on a piece of toast and died. Traditionally, Woodcock and Snipe are both cooked very rare, complete with head, feet and entrails, and served on toast. You may also be surprised if you order Scotch Woodcock, a nineteenth-century dish of egg yolks, cream and anchovies, whose name is a play on the English belief that the Scottish are cautious with their money and that this offers a substitute for real Woodcock, which is rather choice and costly.

Black-tailed Godwit *(Limosa limosa)*

OTHER COMMON NAMES: *barker (Suffolk); jadreka snipe; red godwit (Ireland); shrieker (Norfolk); small curlew; whelp; whelpmoo; yarwhelp (Suffolk).*

See Bar-tailed Godwit below.

Bar-tailed Godwit *(Limosa lapponica)*

OTHER COMMON NAMES: *godwin (Ireland); half curlew (Norfolk); half whaup (Angus – Forfar); pick (Norfolk); poor Willie (East Lothian); prine (Essex); scammel (Blakeney); sea woodcock (Shetland); set hammer (northern England); shrieker (Shropshire, Norfolk); speethe (Northumberland – Lindisfarne); stone plover; yardkeep; yarwhelp; yarwhip.*

The two godwits are remarkably similar, both being long-legged, long-beaked waders, greyish overall in winter but with more or less orange-red colouring on the neck and breast in summer. The beak is black at the tip but increasingly pink towards the base. Both are winter visitors, but the Bar-tailed Godwit is most likely to be seen on mud flats, sandy beaches and estuaries, whereas the Black-tailed Godwit will be seen on marshes and flooded grasslands.

The existence of such names as 'yarwhelp' and 'whelpmoo', which imitate the clamorous call that the Black-tailed Godwit makes only around its nesting sites, reveals that it was once a much more common breeding bird than it is today. It is

The Bar-tailed Godwit *(Limosa lapponica)* occurs in large flocks along the tide-line in winter, mainly on mud flats and beaches.

nonetheless odd that the same names are used for the Bar-tailed Godwit, which has never bred in Britain and is generally fairly silent, so it can only be assumed that the names have become interchangeable because of the birds' similarity.

The basis of the name 'godwit' has been argued about for centuries. One suggestion is that it originated in the Old English word meaning 'good white', but the *Oxford English Dictionary* is totally noncommittal on the matter. William Lockwood's view, that it originated in Lincolnshire and is 'an example of onomatopoeia modified by folk etymology', is more convincing.[201] Certainly the cry of godwits does encompass a very wide range of different sounds, shrieks and whelps. Godwits also share some names with Whimbrels (such as 'half curlew' and 'half whaup'), which refer to their superficial resemblance to Curlews, although they are smaller and have straight, not curved beaks.

Like Curlews, godwits were once considered choice birds for the table. All sources, including the *Oxford English Dictionary*, quote Sir Thomas Browne (1605–82) on this matter: 'Godwyts ... accounted the daintiest dish in England; and I think, for the bigness, of the biggest price,' which, presumably, comes from his *Account of Birds ... Found in Norfolk* (1682).[202] Nonetheless, Sir Thomas, 'originator of the grand style in English literature', did not seem to be a man given to gastronomic excesses. I have been unable to trace a recipe for the godwit but imagine they were cooked underdone, after the manner of Snipe and Woodcock.

Whimbrel *(Numenius phaeopus)*

OTHER COMMON NAMES: *brame (Suffolk); chequer bird; chickerel (Dorset); corpse hound; curlew jack (Yorkshire); curlew knave (Cumberland); curlew knot (Lincolnshire – Spalding); half curlew (Norfolk, Suffolk); little whaup (Scotland); May bird (East Anglia, Gloucestershire, Sussex, Cornwall); May curlew; May fowl, May whaap (Ireland); peerie whaup (Shetland); summer whaup; tang whaup (Shetland, Orkney); titterel (Dorset, Sussex); whimbrel curlew, winderel (Northumberland).*

Whimbrels are similar to Curlews but smaller and rarer. Summer visitors throughout much of Britain rather than residents, they breed in the north of Scotland and names like 'May curlew' and 'May fowl' reflect their spring-time arrival. Both have the same distinctive long, curved beak, brown back and paler underparts, but the Whimbrel does not have quite the same evocative, eerie call, its own cry being imitated by the name 'Whimbrel'; this was first used by John Ray in 1678, who said that it had been supplied to him by a correspondent in North Yorkshire.[203] A similar old name from Durham was 'whimperel'. The name 'corpse hound' is fascinating and has the same origin as the wild-hunt beliefs surrounding the Curlew (see below). There are several old names for the Whimbrel (like 'spowe', used especially in Norfolk) that are evidently based on imitative Norse words for 'Curlew'. It is under 'spowe' and similar names that Whimbrels were revealed as table birds in the household accounts of the Norfolk Lestrange family, spanning the years 1519 to 1578.

Curlew *(Numenius arquata)*

OTHER COMMON NAMES: *awp; calloo; collier, collier-jack (Cheshire); courlie (Sussex); crithane (Ireland); curlew-help (Lancashire); full curlew; gelvinak (Cornwall); great curlew; great whaup (Orkney); guilbinn (Western Isles/Hebrides); marsh hen (Suffolk); seven whistler (Sussex); stock whaup (Shetland); whaup (Shetland, Scotland, north of England); whistling duck (Somerset); whitterick (East Lothian).*

For naturalists of my generation, even those who had never been near a live Curlew, the bird's glorious, haunting call was one of the most familiar of sounds because it was used to introduce that marvellous BBC wildlife programme, *The Naturalist*; this brought us not only birds but also such pioneer natural-history broadcasters as Peter Scott and James Fisher.

Almost all of the bird's names represent attempts (and, apart perhaps from 'whaup', not very satisfactory attempts) to imitate their call. 'Curlew' itself, the most generally preferred name, dates from *c.*1340. In a phrase that I suspect must have been coined by Robert Burns, Curlews are delightfully described as 'lang-leggity beasties'.

Always fairly vocal, they can be heard calling on the coastal mud flats, marshes and estuaries where they are usually found in winter. However, the best place to hear a Curlew is on the high summer moorlands, where the melancholy whistle drifts across the heather. I am far from alone in believing that the call of a Curlew in such a place is probably the most wonderful wild sound in Britain. Although Curlews have spread from the moorlands to breed on lower ground, and even on farmland, they are now declining as British breeding birds because of changes in farming practices and the drainage of wetlands.

As recently as 1950, it was said that: 'The autumn is the best time for shooting curlew; they are then very good eating and give excellent sport.' September birds, it was said, should not be skinned but merely drawn; 'winter birds – they taste a bit kippery – should be skinned' but 'whether in the autumn or winter the Curlew provides good nourishing fare for the humble folk of the shore cottages'. Nonetheless, they were never considered easy birds to shoot, although they can be lured by imitative calls or simply by objects left lying on the ground; an old Irish saying was 'He's a good hunter who kills a curlew'.

In early spring, the male Curlew *(Numenius arquata)* establishes its breeding territory, from where it displays to females.

Curlew-eating has a long history and the bird's value was embodied in the old rhyme:

> The curloo be she white, be she black
> She carries twelve pence on her back.[204]

A Lincolnshire version runs:

> A curlew lean, or a curlew fat,
> Carries twelve pence upon her back.[205]

The black-and-white references are to the colour change from summer to winter. Francesca Greenoak cited a poulterer's price list, issued in the year 1275 by order of Edward I, which listed Curlews at 3d.[206] By 1384, the price had increased to 6d. They were served among other game at the coronation banquet for Henry VI in 1422 and also, almost inevitably, at the 'The Great Feast At the Inthronization of the Reverend Father in God, George Neavill Archbishop of Yorke, Chancellor of England, in the sixth yeere of Edward the Fourth' (see Bittern, page 212). A fifteenth-century rhyming bill of fare offered:

> What fishe is of savor swete and delicious,
> Rosted or sodden in swete herbes or wine,
> Or fried in oil, most saporous and fine
> The pasties of a hart –
> The crane, the fessaunt, the peacock, and curlewe,
> Seasoned so well in licquor redolent,
> That the hall is full of pleasant smell and scent.[207]

Curlews were never eaten in Shetland, where they were thought weird and uncanny, an allusion to the belief that Curlews were supernatural. The name 'Seven Whistlers' is applied to different birds, including the Curlew, and, when a flock of Curlews flew overhead, uttering their strange cries, they were believed to be lost souls. Other parts of the country also take the sight of a flock of Curlews as a presage of death; they were called 'wisht hounds' in Devon and 'Gabriel's hounds' in northern England. Edward Armstrong has linked Whimbrels and Curlews with the belief of the Wild Hunt 'when the tempest rages and the wind shrieks on winter nights, a ghostly company of hunters rides across the sky', although he concedes that geese probably have a better claim because their call resembles the baying of hounds;

he points out, however, that 'whaup', the northern name for the Curlew, is also 'the name of a goblin with a long beak which moves about under house-eaves at night'.[208]

Curlew are also among the very few birds in which the female is larger than the male.

Spotted Redshank *(Tringa erythropus)*

OTHER COMMON NAMES: *black-headed snipe; Cambridge snipe; couland snipe; dusky godwit; dusky redshank; spotted snipe.*

See Redshank below.

Redshank *(Tringa totanus)*

OTHER COMMON NAMES: *clee (Aberdeenshire); ebb cock (Shetland); pellile (Aberdeenshire); pool snipe; red leg (Norfolk); red-legged horseman; red legs (Sussex); redshank tattler; sandcock; shake (Galway – Connemara); swat (north of England); teuk (Essex); took; tuke (Sussex); warden of the marshes; watchdog of the marshes; watery pleeps (Orkney); whistling plover (Cheshire).*

As their name suggests, Redshanks have red legs, a characteristic that distinguishes them from most otherwise similarly coloured, brown, speckled waders. The Spotted Redshank also has a strikingly different, overall dark, almost black, summer colouring, whereas the Redshank changes little between seasons. Most British Redshanks are residents, breeding on moorland, wet coastal meadows and on wetlands inland, but their numbers are declining because of land drainage and predators. The Spotted Redshank is an irregular winter visitor and migrant.

Most of the regional names refer to the leg colour ('Redshank' itself is sixteenth century), the very characteristic call, or the habitat; 'pool snipe' was for many years a popular variant. Redshanks have never been considered serious wildfowling quarry but their eggs were once collected for eating; along with the eggs of Snipe, Ruff and Black Tern, they were a substitute for the more desirable plovers' eggs, which they closely resemble. Redshanks are now among the ground-nesting birds thought to suffer seriously from egg-predation by Hedgehogs (see page 385).

Greenshank *(Tringa nebularia)*

OTHER COMMON NAMES: *barker; greater plover; green-legged horseman; green-legged long shank.*

Greenshanks, not surprisingly, have green legs; they also have a slightly upturned beak and are the largest species among their close relatives, the Redshanks and the larger sandpipers. In Britain, Greenshanks are mainly summer visitors in small numbers, nesting in the far north on upland moors, generally near trees. Some are to be found on the coastal marshes in winter.

There is clearly something compelling about the Greenshank because the experienced ornithologist, Desmond Nethersole-Thompson devoted an entire book to it.[209] Like many ornithologists of earlier generations, he came to the subject through egg-collecting, from which stemmed his fascination with the Greenshan ... From his many accounts of egg-hunting expeditions comes this evocative description of a passion that has now, fortunately, passed away:

The red legs of the Redshank *(Tringa totanus)* distinguish it from most other waders.

> The tradition of greenshank-hunting has also been enriched by the memory of the almost legendary egg-hunting Highland laird who used to order out his gillies and stalkers with their telescopes. They have been described to me as grim men in kilts and tweeds, cursing profusely on wet hillocks, while they waited for the greenshank whose nest they had been ordered to locate upon pain of instant dismissal. Whenever the angry line of moist gillies had successfully done its work, the head-keeper was ordered to report to his master, who always consumed, as a libation, a bottle of the best in the cellar. Then, the old Chief ... with his eagle feathered bonnet cocked on the side of his head, bowled and jolted in a coachman-driven dog cart, along the track nearest the nest. In the evening, inspired by natural bonhomie, or something better, he is said to have rolled priceless clutches of eggs all over his billiard table, while his personal piper, like a punch-drunk Blackcock, strutted round the room to the tune of 'The Cock o' the North'![210]

The eggs were collected for their own sake, not for eating, but Greenshanks have been eaten in the past. The Eighth Duke of Argyll (1823–1900), who found time for a little ornithology among his numerous other distractions (Keeper of the Great Seal of Scotland, Secretary of State for India, owner of 170,000 acres and the 'sole complete copy of the first book printed in Gaelic' among others), wrote to the artist John Gould that, 'Its flesh is excellent, far superior to that of the redshank,' and John Latham, writing in 1785, said: 'they are sometimes sent up to the London markets in which I have bought them and thought their flesh to be well-flavoured.'[211]

Green Sandpiper *(Tringa ochropus)*

OTHER COMMON NAMES: *black sandpiper (Suffolk); drain swallow (Yorkshire); horse gowk (Shetland); martin snipe (Norfolk); whistling sandpiper.*

See Common Sandpiper below.

Wood Sandpiper *(Tringa glareola)*

See Common Sandpiper below.

Common Sandpiper *(Actitis hypoleucos)*

OTHER COMMON NAMES: *dickie-di-dee (Lancashire); heather peeper (Aberdeenshire); killieleepsie (East Lothian); kittie needie (Kirkcudbrightshire); land laverock (Scotland); sand lark (Ireland; Scotland); sand snipe (Cheshire); sandie laverock (northern England); sanny (Aberdeenshire); shad bird (Shropshire – Shrewsbury); shore snipe (Perth); skittery deacon (Stirlingshire); steenie pouter (Orkney); summer snipe; tatler; water junket; water laverock (Roxburghshire); waterypleeps (Orkney); weet weet; willy wicket (north of England).*

'Sandpiper' is an appropriate name for birds that make a piping call on sandy beaches. Originally, in the seventeenth century, it was applied only to the Common Sandpiper, but it gradually became extended to related British species and also the more distantly related Purple Sandpiper; it occurs in regional names for several other birds in the Scolopacidae as well.

The Common Sandpiper *(Actitis hypoleucos)* is often seen near inland waters, probing the edges of streams with its long bill.

The Common Sandpiper is fairly common and is often the only wader to be seen on inland waters. The Green Sandpiper (green only on its legs) is usually seen as a winter visitor; the Wood Sandpiper is a rare winter visitor and occasional in the north of Scotland. All tend to differ in their nesting habits from other waders, the Common Sandpiper building a nest in vegetation on the ground; other sandpipers may use old nests of other species in trees.

Turnstone *(Arenaria interpres)*

OTHER COMMON NAMES: *bracket (Northumberland – Lindisfarne); ebb pecker (Shetland); sea dotterel (Norfolk); skirl crake (Shetland, East Lothian); stanepecker (Shetland); stone raw (Armagh); tangle picker (Norfolk).*

Turnstones also turn other things, such as shells and pieces of seaweed, in their search for insects and other small creatures. However, it is the almost disdainful way in which they flick aside small stones that gave them their common name in the seventeenth century.

They are pretty little waders with short beaks and, or so it appears from a distance, a black-and-white patterning, although they are mainly brown above and white below. 'Their coloration always looks tortoiseshell to me; and I think their orange limbs are exquisite.'[212] They are common winter visitors, although a few juvenile birds are likely to be seen in summer.

The Blackburn 'Skua' aeroplane of the Fleet Air Arm is seen here flying over a convoy of British ships *c*.1940.

Phalaropes
(Phalaropidae)

Red-necked Phalarope *(Phalaropus lobatus)*
See Red Phalarope below.

Red Phalarope *(Phalaropus fulicarius)*
OTHER COMMON NAMES: *deargan (Red-necked Phalarope only); glasan (Red Phalarope only); brown phalarope; coot-foot; grey phalarope (Red Phalarope only); half-web red phalarope; hyperborean phalarope; jacu.*
The phalaropes are a small family comprising three species of wader, of which one is a rare vagrant but the remaining two, the Red-necked and the Red (sometimes confusingly called the Grey), regularly visit Britian in small numbers, mainly in the autumn, although a few Red-necked breed in Scotland. Many of the common names fail to distinguish between the two. In truth, 'wader' is a bit of a misnomer for, unlike other superficially similar wading birds, they spend most of their time swimming and the most frequently seen species, the Red-necked Phalarope, draws attention to itself by spinning around on the water to disturb plankton. It is a strikingly pretty and strangely tame, grey, brown and white bird with a slender black beak and, in summer, an orange-red throat.

The rarer Grey is only likely to be seen in its summer plumage, when it may be distinguished by its russet-brown underparts.

The curious name 'phalarope' is originally French but is derived from the Greek *phalaris*, meaning 'coot'. The birds appear never to have acquired other English names.

Skuas
(Stercorariidae)

These extraordinarily aggressive, dark-coloured relatives of the gulls are summer visitors now found only in the north of Scotland. Skuas are nature's dive-bombers, harrying other birds in flight, especially terns, gulls and auks, in order to persuade them to drop their prey. Appropriately, the name was adopted for the single-engined Blackburn 'Skua' aeroplane employed by the Fleet Air Arm at the beginning of the Second World War and, though less appropriately, for a modern ground-launched missile, the 'Sea Skua'.

Because of this behaviour, skuas are usually described as parasitic, and one species even has the scientific name parasiticus, although this is probably stretching the definition. 'Piratical', another adjective that is often used, seems preferable. Skuas are not entirely dependent on other birds' food and will also catch and eat live fish, other birds and their eggs.

'Skua' is a northern word originating, via Faroese, from Old Norse and originally imitative of the bird's aggressive call note.

Arctic Skua *(Stercorarius parasiticus)*
OTHER COMMON NAMES: *Arctic bird; Arctic gull; boatswain (north Scotland, Orkney); chaser (Yorkshire); dirt bird (Scotland); dirty Allan; dirty aulin (Orkney); dung bird; dung hunter; dung teaser (Berwickshire); fasceddar; feaser; kepshite (Northumberland); labbe; man-of-war bird; scoutie aulin (Shetland, Orkney); shite scouter (Scotland); shooi (Shetland); skait bird; trumpie (Orkney); wagel (Cornwall).*
Strikingly graceful, even beautiful, with long dark wings and a very long, slender, dark tail, this bird is unmistakable in outline, although variable in colour and both pale and dark forms exist.

Despite its grace, this skua has accumulated a fair list of frankly excremental names: 'kepshite', 'fasceddar', 'shite scouter', 'dung bird', 'dirty Allan' and 'dirt bird' leave little to the imagination. However, they all stem from an old belief that the skua, rather than persuading its quarry to drop its food, was persuading it to defecate. This was perhaps an understandable mistake for observers at ground level with no field-glasses.

Arctic Skuas have suffered a population decline in recent years, possibly due to a shortage of Sand Eels and predation by Great Skuas.

Great Skua *(Stercorarius skua)*
OTHER COMMON NAMES: *allan (Scotland); boatswain (Yorkshire); bonxie (Shetland, Orkney); brown gull, black gull (Kerry – Tralee); hagden (Cornwall – south coast ports such as Looe); herdsman (Orkney); jager; lords and captains (Cornwall); morrel hen (Yorkshire); robber bird; sea crow; tod bird (Yorkshire); Tom Harry (Cornwall, Scilly); tuliac; wease-alley (Sussex); wagel (Cornwall – Mount's Bay and St Ives).*
This, the largest skua, is a big, heavy bird with short wings and tail. In flight, its most distinguishing features are the white flashes on its wing-tips.

'Bonxie', meaning 'dumpy', is a widely used name, originally from Shetland, for all the large, heavy species of skua that are probably biologically distinct from the more slender types. The origin of the name 'robber bird' is self-evident, while 'Tom Harry' may well reflect its harrying nature; it seems to have an origin in a Cornish word meaning 'dirty' and so may be related to the many regional names for the Arctic Skua. 'Herdsman', an Orkney name, is said by Francesca Greenoak to recall a time when Great Skuas were welcomed by shepherds because they drove White-tailed Eagles away from the lambing grounds.[213] It is undeniable that, after the eagles departed from Orkney, Arctic Skuas were persecuted by crofters, who collected eggs and birds from the nesting sites for food. They are now expanding their range south along the coasts of mainland Scotland.

Gulls
(Laridae)

The popular BBC radio programme *Desert Island Discs* has an instantly recognizable signature tune over which is played a sound familiar to everyone in Britain: the squawking cry of the Herring Gull. As has often been pointed out, while gulls are familiar in Britain, by and large they do not live and breed on tropical islands and the nasal cry of the Sooty Tern *(Sterna fusca)* might have been more appropriate.

Nonetheless, gulls remain the archetypal British sea birds and, because they are residents, and our companions during cold, out-of-season walks along the pier or promenade when the terns have departed south, they have become much more intimately linked with our life and lore.

Gulls and terns are closely related, gulls being larger and heavier, and lacking the long, forked tail. Like terns, they are difficult to identify, partly because of the often rather marked differences between summer and winter plumage and also because of the difference between adults and juveniles. Gulls are highly gregarious, communal-nesting birds with long wings and webbed feet. They are also fairly good swimmers, although they rarely dive. Most eat fish, but many are also scavengers.

They have been called 'gulls' since the fifteenth century, the word meaning 'wailer' in medieval Cornish; it was spread widely, especially during the Tudor period, when Cornishmen provided much of the manpower of the English navy. The use of 'sea gull' (or 'seagull') dates from the sixteenth century and is prevalent in everyday usage, although less in ornithological circles.

Working fishing boats are often attended by vast flocks of hungry gulls, hoping for rich pickings in the form of discarded fish.

Most of the folklore surrounding gulls fails to distinguish between the species, and gulls in general have long been thought to represent the souls of the departed (especially dead fishermen and sailors who have drowned at sea); consequently, it was considered ill luck to kill one. Their sometimes plaintive calls may have helped to create this illusion but, as recently as the late nineteenth century, there was a firm belief in some coastal villages that old fishermen turned into gulls when they died. A less frequently heard but related story is that a gull seen flying in a straight line over the sea is following the drift of an unseen corpse, rolling along the sea bed. The gull is the soul, attending the body of the dead man who cannot rest.

Three gulls (and only three) seen flying together overhead spell death for the observer or someone close, while a gull flying against the window of a house serves as a warning to any member of the family at sea. It is still commonly said in fishing villages that noisy flocks of gulls around a sailor's home presage a death and Marie de Garis tells of an incident on Guernsey, just before the Second World War, when everyone remarked on the remarkably noisy and restless flocks around the home of a well-known local boat-owner who lay inside dying.[214]

However, the most widespread and logical of the beliefs about gulls is surely one that I heard my mother saying countless times when I was a child: that gulls seen inland spell bad weather at sea. As we lived in Derbyshire, about as far from the sea as it is possible to get, the belief argued that, for gulls to fly such a distance, it was a very bad sign indeed. There is a Yorkshire saying on a similar theme:

> Seagull, seagull, get thi on t'sand.
> It'll nivver be fine while thou'rt on t'land.[215]

There was a time when to see gulls inland was something of a novelty. In Oakfield, near Rugby, it is said that schoolboys who spotted two or more gulls in the month of May could claim a half-day holiday. However, gulls have become much more familiar away from the coast in recent times and no-one now seems to remark on the contradiction of so characteristic a bird of the sea being so typically the one that wheels in flocks following the farmer's plough. Also, wherever you live, it always seems to be a gull, and a large one at that, that deposits its excrement on your newly washed car, and the origin of the ludicrous idea that a bird, especially a gull, defecating on you is some sort of good omen is a mystery.

The name 'gull', like 'Rook' and 'pigeon', has unaccountably come to be associated with cheating and swindling (no-one seems sure if there is any connection with the word 'gullible'). Shakespeare certainly knew the meaning:

> Why, 'tis a gull, a fool, a rogue ...
> (*Henry V*, Act 3, Scene 6)

There can be no better instance of gullibility than a serious report, produced during the First World War, to the effect that gulls might be trained to spot submarines, by tying fish to a submarine and thus encouraging them to follow it.

In times past, gulls have been eaten, out of necessity in coastal communities, but generally as something exotic to serve at a regal banquet alongside Lapwings, larks and thrushes. Their eggs have been much more widely used however.

> A late sixteenth century interrogatory refers to the payment of a dozen gulls and other birds yearly as rent for 'the Gull Rocke next the lands called Penhale, in the parish of St Peran'. The annual tithe in 1611 for Padstow's Gull Rock was two gulls.[216]

Predictably, folk medicine also has its recommendations for pieces of gull: 'Aurelianus saith, that the brain of this bird being dried, and smelt to by children helps to cure them of the epilepsy.' At one time, gull gall was once used to treat disorders of sight (it was probably no bad thing that the patient could not see what he was eating) and its ventricle to cure (not cause) 'weak stomachs', while the whole bird, prepared in an undisclosed manner, was used for 'stoppage of urine, stone and gravel'.[217]

Little Gull *(Larus minutus)*

A local, mainly southern bird in Britain, the Little Gull has no local names. It is a small gull with, in summer, an all-black head (unlike the Black-headed Gull), red legs and beak.

The Little Gull *(Larus minutus)*. In winter, the black head is reduced to face patches and a dark spot on the crown.

The Black-headed Gull *(Larus ridibundus)*. The black of the head is reduced to a dark spot behind the eye in winter.

Black-headed Gull *(Larus ridibundus)*

OTHER COMMON NAMES: *bakie (Shetland); black cap; black head; brown-headed gull; carr swallow (East Yorkshire); cob; collochan gull (Kirkcudbrightshire); crocker; hooded crow, hooded mew (Orkney, East Lothian); maddrick gull (Cornwall); masked gull; mire crow (Cornwall); patch; peewit gull; pewit (Cornwall); pickie buret, pickmire (Roxburghshire); pick sea; pictarn; pigeon gull (Yorkshire); pine; pine maw (Antrim); Potterton hen (Aberdeenshire); puit gull; red-legged gull (Ireland); red-legged pigeon mew (Norfolk); redshank gull (Ireland); rittock (Orkney); Scoulton peewit, Scoulton pie (Norfolk); sea crow (Yorkshire, Cheshire); skarraweet, skirriweet (Cornwall); sprat mew; tarrock (Cornwall); tumbler (Yorkshire); tumbling gull; white crow.*

Considerably bigger and much more common than the Little Gull, the Black-headed Gull is black only at the front of the head in its summer plumage, the neck being white; the legs and beak are red. This is the bird most frequently seen inland and following the plough and it is also among the more vocal, often seen in large wheeling and screaming flocks.

The old name 'mew', which is used for this as well as other gulls, was in widespread use as the normal English name for these birds until the seventeenth century, when it was generally replaced by 'gull'. It seems to have been imitative in its original form. These gulls were also once called 'peewits' or 'peewit gulls' in some areas, a name presumably also imitative, as it is for the true peewit, or Lapwing.

> In 1946, while temporarily living in Essex, I specifically recorded in my journal on the frequency with which one saw flocks of Peewits and Black-headed Gulls on the ground together.[218]

One of the really endearing folk stories about gulls relates to this bird and is delightfully told by Ernest Ingersoll:

> One day about AD 550 the blackheaded gulls, flying as usual along the coast of Wales, and scanning the sea sharply for food or anything else interesting to a gull, found floating in a coracle – a round, wickerwork canoe – a human baby a day or two old, contentedly asleep on a pallet made of a folded purple cloth. Several gulls seized the corners of this cloth and so carried the child to the ledge of the Welsh cliff where they nested, plucked feathers from their breasts to make a soft bed, laid the baby on it, then hastened to fly inland and bring a doe to provide it with milk, for which an angel offered a brazen bell as a cup. There the waif lived for several months; but one day, in the absence of the gulls, a shepherd discovered the infant and took him down to his hut and his kind wife. The gulls, returning from the sea, heard of this act from the doe. They rushed to the shepherd's cottage, again lifted the babe by the corners of its purple blanket, and bore him back to the ledge of their sea-fronting crag. There he stayed until he had grown to manhood – a man full of laughter and singing and kind words; and the Welsh peasants of the Gower Peninsula revered him and called him Saint Kenneth.[219]

Common Gull *(Larus canus)*

OTHER COMMON NAMES: *annet (Northumberland); barley bird (south Devon); blue maa (Shetland); cob (Norfolk, Suffolk, Essex, Kent); coddy moddy; gow (Aberdeenshire); green-billed gull; maa, mar (Kirkcudbrightshire); mew (Scotland); sea cobb (Norfolk, Suffolk, Essex, Kent); sea mall, sea maw, sea mew (Scotland); see bird (Roxburghshire, Scottish Borders – Teviot Dale); small maa (Shetland); white maa (Orkney); winter bonnet; winter gull; winter mew.*

A yellow beak and green-yellow legs are the distinguishing features of the Common Gull, which is not the commonest gull but is a frequent companion of the Black-headed Gull inland. The species is now much less common than previously but does not seem seriously threatened as a British breeding bird.

'Common' means simply 'ordinary' rather than 'abundant'. The unexpected name 'barley bird' is also used for such unrelated species as the Greenfinch, Nightingale and Yellow Wagtail and merely alludes to some real or imagined activity coinciding with a particular stage of the development of the barley crop.

Lesser Black-backed Gull *(Larus fuscus)*

OTHER COMMON NAMES: *blackback; caudy maudy, coddy moddy (Northamptonshire); gray cob; gray gull; said fool (Shetland); saith fowl; yellow-legged gull.*

This is a large gull with a black back, but not as large as the Great Black-backed Gull, as might be expected. Any black-backed gull seen inland, over farmland and in any quantity, is likely to be the Lesser Black-backed Gull and a significant number now nest not only inland but also on buildings, although it is officially considered a threatened species. It is now generally thought to be no more than a subspecies of the Herring Gull.

Herring Gull *(Larus argentatus)*
OTHER COMMON NAMES: *cat gull (Kirkcudbrightshire); laughing gull (Belfast); silver back; silvery gull (Ireland); white maa (Shetland, Orkney); Willie gow (Aberdeenshire, East Lothian).*
The commonest British gull, the Herring Gull has a yellow beak and pink legs. It is as likely to be seen inland as on the coast and is the gull most often seen squawking in groups around municipal refuse sites. It does eat Herrings but not in preference to anything else and certainly no more than, say, the Lesser Black-backed Gull, but the name has been used since the seventeenth century. Despite persecution, diseases and predator damage to the nests, the species seems extremely robust.

Great Black-backed Gull *(Larus marinus)*
OTHER COMMON NAMES: *baagie, baakie, baugie (Shetland, Orkney); black-and-white gull; black back; black-backed hannock (Yorkshire); carrion gull (Ireland); cobb (Galway, Wales, Essex, Kent, north Devon); goose gull, gray gull, greater saddleback (Ireland); gull maw (East Lothian); parson gull, parson mew (Galway, Sussex); saddleback (Lancashire, Norfolk, Cornwall); swarbie (Shetland, Orkney); swart back (Orkney); wagell gull (Yorkshire).*
This very large, usually solitary bird has a big yellow beak, pink legs and a black back. When seen at close quarters, the Great Black-backed Gull is a large, menacing creature, quite capable of killing and eating other birds as large as a full-grown duck. This is exemplified by an old Somerset tradition about gulls that suggests you should never look a gull straight in the eye or even try feeding it because, if you do, one day it will return while you are swimming and peck out your eyes.

The Great Black-backed Gull, together with the Lesser Black-backed Gull and the Herring Gull, are the three sea birds that are excluded from protection under the Wildlife and Countryside Act 1981; they may be killed and their eggs taken at all times by any authorized person.

Kittiwake *(Rissa tridactyla)*
OTHER COMMON NAMES: *annet; cackreen; craa maa (Shetland); keltie (Aberdeenshire); kiff (Cornwall); kishiefaik (Orkney); kittiake; kittick (Orkney); kittie (east coast, Banffshire); kittiwaako (Orkney); petrel (North Yorkshire – Flamborough Head); sea kittie (Norfolk, Suffolk); tarrock (young birds); waeg.*
The most seafaring of British gulls, the Kittiwake is confusingly like the Common Gull but with black legs and all-black, rather than black-and-white, wing-tips. It is likely to be seen on land only in the nesting season, when huge, noisy colonies assemble on the cliffs.

The beautiful name 'Kittiwake' is originally Scottish and, unlike so many supposedly imitative bird names, quite closely resembles the three-note cry. Because of its rather plaintive call, the legend that gulls are the souls of the departed finds a variation here: Kittiwakes are believed to be the souls of dead children. The scientific name *tridactyla* refers to the fact that it possesses only three toes; the hind toe is lacking.

Kittiwakes *(Rissa tridactyla)* come ashore in the breeding season, forming vast colonies, like this one on the Farne Islands.

Tragically, this most lovely bird suffered more than other gulls because of the millinery trade's demand for its feathers. Ron Murton told of the large industry that once existed at Clovelly in Devon and elsewhere in the nineteenth century, citing a report of 9,000 Kittiwakes being killed on Lundy for this purpose in the space of a fortnight.[220]

Terns
(Sternidae)

On a summer's day on the shingle beach at Blakeney Point in Norfolk, and at countless other places on the coast of the British Isles, terns just cannot be avoided. They dive at intruders in the most menacing fashion, their needle-sharp beaks aiming directly at the head as they try to drive them away from their nesting grounds, which typically are in the open, on bare shingle.

All our terns are summer visitors to Britain. They are closely related to gulls, sometimes being placed in the same family, but are smaller and much more streamlined fliers, with webbed feet, deeply forked tails and proportionately longer, more slender wings. Adults of all British species have black-capped heads when in their summer plumage and, although their beaks and leg colours vary, they are mostly difficult to identify. They tend to hover before diving spectacularly to catch fish.

The name 'tern' is thought to be derived from an imitation of the screeching call and it is a pity that the alternative and highly appropriate 'sea swallow' is not used more widely. There are some regional names for terns in general, especially in Cornwall:

> The dialect scarraweet ... and Miret ... possibly responsible for the names of Great Merrick Lodge off St Martin's and Merrick Island in Tresco Channel in the Isles of Scilly.[221]

All British tern species have been persecuted for their eggs and for their feathers, which were once extensively used in the millinery trade. Sometimes, for reasons beyond our control, they also suffer on their winter migrations (see Roseate Tern, below).

Sandwich Tern *(Sterna sandvicensis)*

OTHER COMMON NAMES: *boatswain; crocker kip (Sussex); screecher (Kent); surf tern.*

Particularly pale in colour, the Sandwich Tern has a yellow-tipped black beak and black legs. Its association with the town of Sandwich on the Kent coast dates back to 1785, when the naturalist John Latham was sent specimens collected from there by some local boys.

Like most British terns, Sandwich Terns rarely stray inland away from the coast and, like most other terns, they were exploited in the past for their eggs. The populations have since recovered but still fluctuate considerably.

Roseate Tern *(Sterna dougallii)*

The Roseate Tern has a black beak with a touch of red at its base and red legs. The British breeding population is precarious. In common with other migrating terns that overwinter in Africa, they suffer badly at the hands of the local people, especially children, who catch them and wreak a terrible toll.

Common Tern *(Sterna hirundo)*

OTHER COMMON NAMES: *cherrick (Cornwall); clett; darr, dippurl, great purl (Norfolk); great tern; gull teaser (south Devon); kingfisher (Northern Ireland – Lough Neagh); kip; kirrmew; miret (Cornwall); pease crow; piccatarrie (Shetland); picket-a (Orkney); pictarnie (Fife, East Lothian); pirre (Northern Ireland); rittock; rippock; scraye; sea swallow; shear tail (Orkney); skirr (Dublin – Lambay Island); sparling (Lancashire); speikintares (Ross and Cromarty); spurling (Lancashire); starn (Norfolk); taring (Shetland); tarnie; tarret; tarrock; Willie fisher.*

With the scientific name *hirundo* and its deeply forked tail, this is the archetypal sea swallow. Its beak is red with a black tip and its legs are orange-red.

Common Terns are being increasingly seen inland, although most still remain at the coast; in England they are accurately named because they are the commonest species.

As part of the courtship ritual, the male Sandwich Tern *(Sterna sandvicensis)* feeds fish to its mate.

Terns wheel around the lighthouse at Rockabill, County Dublin, on the east coast of Ireland.

Arctic Tern *(Sterna paradisaea)*

OTHER COMMON NAMES: *pickieterno, ritto (Orkney); rittock; sea swallow, skirr (Ireland); sparling (Lancashire); tarrock (Shetland); tarry (Northumberland).*

This is a bird of breathtaking romance, with an entry in the *Guinness Book of Records* for the longest of all migratory flights. The annual journey of the Arctic Tern is one of the world's great wonders: each of these modest little birds, weighing not much over 100 grams (3½ ounces), escapes the northern winter by flying from the Arctic to the Antarctic pack ice, where it spends the summer, returning to the Arctic again in spring. Some call in to the British Isles to breed and, in the north, take the place of the Common Tern, although their numbers have declined considerably in recent years and they are now among the most threatened species. They are similar to Common Terns, but with an all-red beak, and they often nest in mixed colonies.

Little Tern *(Sterna albifrons)*

OTHER COMMON NAMES: *chite perl; dip ears (Norfolk); fairy bird (Galway); hooded tern; little darr (Norfolk); little pickie (Angus – Forfar); richel bird; sea mouse; sea swallow; shrimp catcher (Norfolk); skirr (Ireland); small purl (Norfolk); sparling (west Lancashire).*

The smallest of the terns, with a voice out of all proportion to its size, this is the only British species with a yellow, black-tipped beak. Today, it is found mainly in East Anglia, where it is the subject of protection, to which it seems to be responding.

Black Tern *(Chlidonias niger)*

OTHER COMMON NAMES: *black kip (Sussex); blue darr; blue daw (Norfolk); carr crow; carr-swallow (Cambridgeshire); cloven-footed gull; clover-footed gull; scare crow; starn (Norfolk); stern.*

The Black Tern is not all black, but it is certainly dark above, with an all-black head and beak and only a shallowly forked tail. It occurs irregularly in Britain and may be seen away from the coast, around fresh water; hence the references to 'carr' (boggy ground) in its regional names.

They were once much more common and their eggs were among those that were substituted for plover's eggs by Norfolk collectors who were sending supplies to London in the 1800s.

Auks
(Alcidae)

Sadly, since the demise of the Great Auk, there are only five British auk species: the Guillemot, Razorbill, Black Guillemot, Puffin and Little Auk. All are short-winged, diving sea birds, feeding mainly on fish; fairly poor fliers, they are wonderful swimmers, using their wings for propulsion. Despite being very vulnerable victims of oil pollution, they are among the most successful and adaptable of modern British birds.

The name 'auk' tended to be used for the Razorbill until the late eighteenth century and can be traced back to the Old Norse word *alka*, meaning 'neck', an allusion to the way that Razorbills stretch their necks when they return to roost. Auks were originally known as ' penguins', one of the few English words with a Welsh origin; it means 'white head' and refers to the large white patch on the face of the Great Auk (*Pinguinis impennis*, see below). When the flightless southern hemisphere birds now known as penguins were first seen, they were given their name because of their superficial similarity to the Great Auk.

Great Auk *(Pinguinus impennis)*

What surprised me most was the smallness of the Great Auk. In my mind's eye, I had always imagined it to be some sort of giant, the size of an Emperor Penguin, but the slightly fragile-looking museum specimen that I saw was only about 1 metre (3¼ feet) tall. This extraordinary creature was the only flightless European bird and also the only British bird to have become extinct within recorded history. It has acquired almost legendary status since the last two living individuals in the world were found on 4 June 1844, on a ledge beneath the sheer 75-metre (245-foot) cliffs of a volcanic rock called Eldey Island, off the southwest coast of Iceland, where they were clubbed to death by three Icelandic fishermen. Over a century later, these actions were to cost Iceland's Natural History Museum dear when, wanting a stuffed specimen, it was obliged to purchase one of the handful that still exist for a world-record £9,000 at Sotheby's in London.

It ceased to be a British breeding species when a nesting pair was killed on Papa Westray on Orkney: the female in 1812 and the male, which stayed on the same ledge until the following season, in 1813. The last British Great Auk was found on St Kilda *c.*1840 and beaten to death by two local men named M'Kinnon and MacQueen, who believed it to be a witch. Great Auks had once been regular breeding birds on St Kilda, but clearly, by 1840, they had been unfamiliar to a whole generation of islanders. The MacQueen family seems to have a fair bit to answer for in the extermination business; it was an earlier MacQueen who reputedly disposed of the last Scottish Wolf (see page 402). The appalling tale of the way that the population of the Great Auk was reduced from tens of millions to nought in a few centuries has been told many times, notably by Symington Grieve, in his 1885 book *The Great Auk or Garefowl*, and then, even more comprehensively, by Errol Fuller in *The Great Auk* (1999). It has even spawned a curious novel, a semi-factual, anthropomorphic tale called *The Last Great Auk* by Allan Eckert (1963).[222] The birds were killed initially for their flesh, oil, eggs (one collecting trip recorded taking a barely believable 100,000 eggs in a single day) and, when supplies of Eider down had been depleted, feathers.

Its old name, 'garefowl', or 'gairfowl', dates from the late seventeenth century in English but originated in Old Norse. It was applied with more than a touch of irony for the Norsemen gave this ponderous, flightless creature the same name as they used for the bird known today as the 'gyrfalcon', one of the most perfect of all flying creatures. 'Garefowls' became 'Great Auks' in the mid-eighteenth century.

Guillemot *(Uria aalge)*

OTHER COMMON NAMES: *auk (Orkney); bridled guillemot; bridled marrot (western Scotland); eligny (Pembrokeshire); foolish guillemot; frowl (Scotland); kidda, kiddaw (Cornwall); lavy; lamy (Hebrides); Maggie (Angus – Forfar); marrot (Firth of Forth); meere (Gloucestershire); morrot (Firth of Forth); muir-eun (Donegal); murre, muse (Cork; Devon, Cornwall); quet (Aberdeenshire); ring-eyed scout (western Scotland); ringed guillemot (Yorkshire); scout (Orkney, Angus – Forfar, Yorkshire); sea hen (East Lothian, Northumberland, Durham); silver-eyed scout (western Scotland); skiddaw, skuttock (East Lothian); spratter (Hampshire); strany; tarrock; tinkershire; tinkershue; wil-duck (Suffolk); Will (Sussex); willock (Orkney, Norfolk); Willy (Norfolk).*

Guillemots are among the true birds of the sea and, like many auks, only come ashore to breed. They are fish-eaters, dark grey above and white below, with an elongated, pointed beak, and, since the demise of the Great Auk, they are the largest of British auks. They are also extremely numerous and isolated oil spillages, population fluctuations and the 'sport' of the nineteenth-century shooting fraternity have made little impression on them. They must be among the most successful modern British birds, capitalizing on the large quantities of small fish available in coastal waters.

Despite their relatively brief sojourns on land, they have acquired a multitude of regional names, most of which are onomatopoeic, based on the call of either the juvenile or adult birds.

In spring, Guillemots *(Uria aalge)* assemble in vast breeding colonies on the rocky ledges of offshore islands.

'Skiddaw' and 'kiddaw', according to William Lockwood, are imitative of the young birds' call.[223] Francesca Greenoak argues that they are variants of 'skite', in turn derived from 'shite', and an allusion to the excrement-covered cliffs on which the birds breed.[224] 'Guillemot' sounds, and is, French, and comes from the old Christian name Guillaume. 'Murre' and similar words have also been used for other auks, especially the Razorbill.

Guillemots were important to people who lived on remote sea coasts. Their eggs were collected as food and the adults were killed, to some extent for eating (although little compared with Puffins) but more importantly for their feathers and as a source of fertilizer. 'Sea-fowling' by the inhabitants of the remote Atlantic island of St Kilda has already been described (see Fulmar, page 205), but the following description by J.M and I.L. Boyd, relates particularly to Guillemots and is especially vivid:

> Among the most intrepid St Kildan seafowling expeditions were the ascent of Stac Biorach (73 m), the rock-fang in the Soay Sound, for guillemots, and the nocturnal visits to the great stacks to kill adult gannets in spring, driven on by winter starvation. The St Kildans had none of the boating and climbing aids of the present-day, only sail and oars, bare feet and horse-hair rope. Who today will row 6 km to Biorach, climb it without boots and rope, and row the 6 km home, as did the young men of old St Kilda with their harvest of guillemots?[225]

Guillemot eggs were clearly an acquired taste. The Reverend Neil MacKenzie, writing at the beginning of the twentieth century, said:

> The eggs are very good eating when fresh. After they are incubated for a few days most of the egg appears, when boiled, to be changed into a thick, rich cream, and in this condition they are also relished. Sometimes eggs, not only of this species but of some others which have not been hatched, are found late in the season. Some of these when cooked look like a piece of sponge cake, have a high gamey flavour and are esteemed a great delicacy. Others are as bad as the most vivid imagination can depict.[226]

Because of dependence on wildfowl for food, St Kilda was exempt from the Bird Protection Act of 1880. In 1918, towards the end of the First World War, it was announced that:

> In order to allow the collection of certain sea-birds' eggs for food, the protection of eggs of the Herring-gull and Black-headed Gull, and of Puffin, Guillemot and Razorbill is suspended in certain counties until given dates. This does not authorise the taking of any eggs except those specified; and it is generally arranged that the collection shall be done by properly accredited persons.[227]

On the Isle of Arran, it was said that the local inhabitants justified eating Guillemots on Friday (traditionally a meatless day), because they never flew over land and ate only seafood.

Razorbill *(Alca torda)*

OTHER COMMON NAMES: *alk; auk (northern England); bawkie (Orkney); faik (Hebrides); falk; gurfel; helligog (Shetland); helljay; marrot (Aberdeenshire, East Lothian); murr, murre (Cornwall); puffin (Antrim; Sussex); raow, row (Cornwall); scout (Scotland); sea blackbird (Pembroke); sea craa (Shetland); sea crow (Orkney).*

When this bird's beak is seen at close quarters, it becomes obvious why the eighteenth-century name 'auk' was replaced by 'Razorbill'. In all other respects, however, it looks like a slightly darker Guillemot and the two birds often nest in close proximity. Razorbills have never been very important as food and neither birds nor eggs were taken in any quantity by the seafowlers. Like Guillemots, they are extremely successful in the present coastal and marine environment.

Black Guillemot *(Cepphus grylle)*

OTHER COMMON NAMES: *diving pigeon (Northumberland – Farne Islands); doveky; Greenland dove (Orkney, Cornwall); Greenland turtle; puffinet (Northumberland – Lindisfarne); rock dove (Ireland); scraber (Hebrides, East Lothian); sea dovie (Angus – Forfar); sea pigeon (Ireland); sea turtle; spotted guillemot; taister; tinkershere; toyste; turtle (Northumberland); turtle dove (Northumberland – Lindisfarne); turtur (Firth of Forth – Bass Rock); tystie (Shetland, Orkney).*

Although not closely related to the Guillemot, this bird is more or less true to its name in summer, when it is black, apart from a striking white wing-patch. In winter, however, it is grey and white, although still with vividly red legs and feet.

Perhaps the oddest feature of Black Guillemots is the way in which so many of the regional names allude to them as pigeons or doves; this can only be because of the comparable displays of affection shown by the pairs at their nests on rocky shores. They are yet another successful sea bird, principally of the northern and western coasts of Scotland and Ireland.

Puffin *(Fratercula arctica)*

OTHER COMMON NAMES: *Ailsa cock (Scotland); Ailsa parrot (Antrim; Scotland); Bass cock (Scotland); Bill (Galway); bottlenose (Wales); bouger; brille (Cornwall); brilly (Cornwall – Mousehole); bulker (Hebrides); bully; cliaheen (Galway); Coulter neb (Northumberland – Farne Islands); Flamborough Head pilot (Yorkshire); guldenhead (Wales); Londoner, Lundy parrot (Cornwall); marrot; cockandy (Fife); mullet (Scarborough); nath (Cornwall); pal (Wales); parrot-billed Willy (Sussex); pipe (Cornwall); pope (Cornwall – St Ives); pope, Scilly parrot (Cornwall); scout (Northumberland – Farne Islands); tammie norie (Orkney, Shetland); tommy; tom nodd (Northumberland – Farne Islands); Welsh parrot (Cornwall); willock (Kent); Willy (Sussex).*

The Puffin most famously appears on the stamps issued by the Bristol Channel island of Lundy of which it is the symbol; even the island's unofficial 'currency' is the puffin. This is, indeed, for good reason because the island is named after the bird: the Puffin is called *Lunde* in both Norwegian and Icelandic.

The island of Lundy, in the Bristol Channel, is named after the Puffin *(Fratercula arctica)*, which features on its postage stamps.

The name 'Puffin' has a peculiar history, however. It is English in origin and was used initially for the cured carcasses of nestling shearwaters, which were a highly prized delicacy until the late eighteenth century, birds being collected mainly in the Isles of Scilly and on the Calf of Man: 'The young are fit to take in August when great numbers are killed and barrelled with salt, which the inhabitants boil and eat with potatoes' (George Montagu, 1802).[228] 'Puffin' is related to 'puffing' and is a reference to the fat young birds; *Puffinus* is still the scientific generic name for shearwaters.

This exchange of names seems to have arisen from a confusion between shearwaters and Puffins, which both nest in burrows, and, by the late nineteenth century, 'Puffin' had taken on its present-day meaning. Its need for grassy banks in which to nest sets the Puffin apart from other British auks, which prefer rocky ledges and crevices. The charming name 'Londoner' has a lovely origin 'in West Cornwall and Porthleven, possibly because of its visitor-like custom of standing upon cliffs in crowds and gazing vacantly seawards'.[229]

In winter, Puffins are little different from their black-and-white auk relatives. They are smaller than Guillemots but larger than Little Auks. In summer, however, there is nothing else like them, apart from the tropical parrots, because the beak, triangular in outline, becomes bright red and yellow (due to 'plates' that are moulted at the end of the season); when the beak is dripping with a mouthful of small fish, the bird assumes a most endearing and comical appearance. It was clearly an inspired decision by the publishers at Penguin to call their first children's book imprint Puffin.

At its most colourful when in breeding plumage, the Puffin *(Fratercula arctica)* bears more than a passing resemblance to a parrot.

The attractive northern name for the Puffin, 'tammie norie', finds another meaning in 'a shy person' and turns up in a charming Scottish rhyme:

> Tammie Norie o' the Bass
> Canna kiss a bonny lass....[230]

Puffins have been touched by one of folklore's odder beliefs: the notion of the bird-fish (see also Barnacle Goose, page 231, and Little Auk, below). The sixteenth-century Swiss zoologist, Konrad von Gesner (1516–65), 'the father of modern zoology', wrote of the Puffin:

> It is eaten in Lent because in a measure it seems related to the fishes, in that it is cold-blooded ... The English make the puffin a bird and no bird, or bird-fish.[231]

In the latter remark, he was drawing on a letter that he had received from no less an authority than the English physician John Caius (1510–73), who said:

> It is used as fish among us during the solemn fast of Lent: being in substance and looke not unlike a seal.[232]

The view was repeated by numerous other writers. Thomas Mouffet, for instance, said that Puffins were:

> ... birds and no birds, that is to say birds in shew and fish in substance ... permitted by Popes to be eaten in Lent.[233]

George Owen, however, in his *Description of Pembrokeshire* (1603) was at a loss to explain it:

> The Puffin ... is reputed to be fishe, the reason I cannot learne.[234]

In modern times, Edward Armstrong thought the most plausible argument for regarding Puffin as fish was simply that it feeds on them.

Puffins were among the main food birds of the St Kildan islanders, the adult birds generally being roasted; the eggs were little prized, but the feathers were sold and commanded a high price. It is now impossible to know just how many birds were taken, but one account records 89,600 Puffins being killed in 1876 alone. The Reverend Neil MacKenzie may have been closer to the mark when, in the 1830s, he suggested that between 18,000 and 20,000 were taken each year. After the beginning of the twentieth century, the numbers taken dropped sharply, although some Puffins were still being collected for eating when St Kilda was evacuated in 1930.

An often-quoted and wonderful story relates to the Isles of Scilly and the Duchy of Cornwall, which owned, and still owns, most of them. When Edward the Black Prince was created the first Duke of Cornwall in 1337 the Scillies were listed under 'foreign rents' and the islands of Agnes and Ennor (now called St Mary's) paid their annual dues in Puffins:

> ... in return for finding twelve armed men, keeping the peace and paying yearly at the gates of Launceston Castle at Michaelmas [29 September] a rent called waiternfee [watching fee] of three hundred puffins ...

By 1440, this had been reduced to 50 Puffins, or 6s 8d.

> ... On the mainland rent was paid partly in Puffins in at least one instance: Stephen Hoskyn of Penzance paid Thomas Bouryng in 1494 or 1495: xij pofyns sufficiently savyed for manys mete or ij d. For every pofyn ... payable at the feast of the purification of our lady yearly. And that the said Stephen and hys heyrs shall cary the said pofyns to the said Thomas and hys heyrs to Salcomb or Bourynggelegh in the county of Devonshire.[235]

According to Welsh legend, King Arthur appears from his mountain cave as a Puffin, a Raven or a Chough.

Little Auk *(Alle alle)*

The smallest of the British auks, this diminutive bird is characterized at close quarters by its apparent absence of a beak. It is a regular winter visitor to the north of Britain, being a more northerly species than its relatives and breeding principally in the Arctic. It is a true seafarer, spending winter at sea among the storms, which sometimes drive it closer to the shore.

It is these storms that have led to the Little Auk, like the Puffin, becoming involved in the bird-fish belief described previously (see Barnacle Goose, page 231), but in an individual and peculiar way. Throughout history, tales exist of small, black diving birds originating from floating driftwood, their bodies being washed ashore on the Irish coast. So-called 'wrecks' of Little Auks are fact; large numbers of dead birds are, from time to time, blown ashore during especially violent storms and it is not unreasonable to assume that the combination of barnacle-covered wood and tiny birds on the beach drift-line led people to believe that the barnacles were indeed the birds' early stages.

Pigeons and Doves
(Columbidae)

There is no real ornithological difference between a pigeon and a dove, except that, today, the word 'dove' tends to be reserved for the smaller species. The word 'pigeon' comes originally from Latin via French and once meant a young bird only. 'Dove' is almost certainly Old English and of Germanic origin, and is an attempt to imitate the characteristic soft, cooing call. William Lockwood has made an intriguing analysis of the preference for one name over the other. He points out that 'dove' in traditional Germanic belief was a pagan symbol of death and so did not endear itself to the Anglo-Saxon church, which preferred the name 'culver', which is still used in some areas. 'Culver' is derived from the Latin *columbula*, the diminutive of the classical name *Columba* and a prominent Christian symbol (not least because of Saint Columba).[236] Culver Cliff on the Isle of Wight is said to be named from the large number of pigeons that once roosted there.

Although these birds tend to be thought of as white or grey, some tropical species, especially the fruit pigeons, are very brightly coloured. British pigeons and doves are more or less omnivorous, although most tend to eat more seeds than insects. They are odd birds in some ways nonetheless: they store their food temporarily in a distended crop (anyone preparing a Wood Pigeon for the table will probably find a considerable dollop of undigested grain deposited on the table); they feed their newly hatched young on 'pigeon's milk', a curd-like substance secreted by the cells lining the crop; and they drink in a very unusual way, not by tipping the head backwards like other birds but by sucking. Doves have long been associated with love and fidelity, a reflection of their apparently endearing and tender courtship, during which they rub necks together while cooing softly.

Because of its pale colour and apparently gentle demeanour, the dove has also long been a symbol of peace and purity. It was a dove that Noah sent out from the Ark and in this story, as in many others, its whiteness and purity was contrasted with the blackness and evil of the Raven.

Rock Dove/Feral or City Pigeon *(Columba livia)*

The true wild Rock Dove occurs only on the far western coasts of Britain, but, even there, it is impossible to identify it for certain because it cohabits with its very common and widespread descendant, the Feral or City Pigeon, the pigeon of Trafalgar Square and the curse of many an inner city. The characteristic feature which distinguishes these birds from Stock Doves and Wood Pigeons is the presence of black wing-bars. However, the many domesticated breeds include brown-and-white, all-white and all-black birds – and almost every combination in between.

At one time, every visitor to London went to Trafalgar Square to feed the pigeons *(Columba livia)*, an activity now discouraged.

Huge sums of money are spent by inner city authorities to discourage pigeons from roosting and fouling buildings: slippery jellies or carefully tensioned thin wires are applied to ledges and other potential roosting sites and tourists are implored not to feed and encourage them, but to almost no avail.

> Ken Livingstone, London's mayor, was accused of signing a 'death warrant' for the capital's pigeons yesterday after he paid off the last bird seed seller in Trafalgar Square. Bernard Rayner, age 45, whose family had occupied the pitch for half a century, agreed to give up his fight against eviction just hours before a High Court hearing ... Mr Livingstone, who described the 6,000 or so pigeons around Nelson's Column as 'rats with wings' and a health hazard, announced plans last year to disperse them to make room for cultural events.[237]

> In a ruling that could affect all landowners who allow wildlife on their property to cause a nuisance, Mr Justice Gibbs said Railtrack was to blame for the actions of the pigeons on a 150-year-old Victorian railway bridge across the Balham High Road in south London. Wandsworth council which has had to pay around £12,000 a year to clear droppings from pavements, said they amounted to a public nuisance for which Railtrack was liable ... Mr Justice Gibbs cited legal argument over the activities of pigeons dating back to 1356. 'As early as the fourteenth century Londoners, irate with the pigeons at St Paul's Cathedral, were throwing stones at them, thereby breaking windows, much to the Bishop's consternation,' he said.[238]

Despite their huge numbers, little notice has hitherto been taken of Feral Pigeons in ornithological censuses.

Pigeons have been domesticated since *c.*4000 BC and were probably the first birds to be tamed by man. They have certainly been kept in Britain for centuries and the many, often very beautiful, pigeon lofts and dovecotes ('doocots' in Scotland) found in country-house gardens are testimony to their past importance as sources of food. Over the past 200 years, homing and racing pigeons have been bred for sport, a pastime that originated in Belgium, where every village once had a Société Colombophile.

For centuries, they have also been used as message-carriers. The Sultan of Baghdad set up a pigeon post *c.*1150 and, during the Second World War, my uncle, too old for active service,

Carrier pigeons were used to carry messages during the Second World War. This bird is being released from a flying boat.

volunteered and enlisted in the Royal Air Force, where his pigeon-keeping knowledge was turned to good use. He was sent to the Middle East and put in charge of a squadron of pigeons used for carrying messages, silently and unlikely to be detected by the enemy. *The Times* of 19 August 1943 reported the successful rescue of a downed bomber crew in the Mediterranean who released their carrier pigeon, which took message of their plight back to their base. The Germans had similar ideas about the use of pigeons:

> The reason we were to operate hawks in the Scillies was because reports had been received by Intelligence from the coastguards of pigeons flying over the islands. With the help of the Army and Air Force Pigeon Service, Intelligence tried to check up on these flights but without success. They were accordingly regarded as 'suspect' and it was our responsibility to catch these pigeons and deliver any messages they might be carrying ... During our four months' stay, the hawks had ten flights at suspect pigeons, out of which they killed seven ... Of the seven pigeons caught only two were carrying messages.[239]

As a testament to their service:

> In Beach House Park in Worthing [there] is a drinking pool and stone memorial to 'warrior birds'; carrier pigeons killed in the war.[240]

There is a sadder side to this: the British population of wild Peregrines was effectively halved during the Second World War when many were killed as it was felt that they posed a threat to the 'pigeon post' (see Peregrine, page 258).

Stock Dove *(Columba oenas)*

Other common names: *blue pouter (Suffolk); burrow pigeon (Yorkshire); bush dove; craig doo (Northumberland); cushat (Northamptonshire); hill pigeon, sand pigeon (Cheshire); Scotch cushat; Scotch queest (Cheshire); stoggie (Yorkshire); wood dover (Scotland); wood pigeon (Lincolnshire).*

A bird resembling a small Wood Pigeon but with no white on its neck could be a Feral or City Pigeon, although in the open countryside, particularly in wooded areas of the English Midlands, it is more likely to be a Stock Dove, especially if it has no obvious black wing-bars. It is called a Stock Dove because it commonly nests in 'stocks' (tree stumps), although it is also content to use holes, such as old rabbit burrows (hence the names 'burrow pigeon' and 'sand pigeon'). It was thought to be among the unlucky birds, as an old rhyme translated from Scottish Gaelic tells:

> I heard the cuckoo with no food in my stomach, I heard the stock-dove on top of the tree, I heard the sweet singer in the copse beyond, And I heard the screech of the owl in the night. I saw the wheatear on the dyke of holes, I saw the snipe while sitting bent, And I foresaw the year would not go well with me.[241]

Pigeons were once kept in dovecotes as a source of food. These white Stock Doves *(Columba oenas)* are purely ornamental.

Wood Pigeon *(Columba palumbus)*

Other common names: *clatter dove (Yorkshire); cooscot (Durham - Teesdale, North Yorkshire – Craven); cooshat (Yorkshire); cruchet (northern England); culver (Dorset); cusha (Roxburghshire); cushat (Westmorland, North Yorkshire – Craven, Buckinghamshire, Berkshire); cushie, cushie doo (Scotland); doo, dow (Norfolk, Suffolk); queece (Cheshire); queest; quest; quice, quease (Shropshire, Gloucestershire); quist (West Midlands, Wiltshire); quisty (Suffolk); ring dove; timmer doo (Scotland); too-zoo (Gloucestershire); wood quest (Ireland; Dorset).*

The Wood Pigeon is one of the only truly wild British birds to have found its way onto the meat counters of the supermarket, a testimony to the fact that it is one of the two species (the Feral Pigeon is the other) that is permitted under the Wildlife and Countryside Act 1981 to be both 'killed by authorised persons at all times' and 'sold at all times'. It is much the largest British pigeon and provides a good meal. However, despite the pigeon family's association with love and endearment, you are warned against thinking that dining on Wood Pigeon will endow you similarly. Richard Martial (d.1563), Dean of Christ Church and a man who oscillated between Protestantism and Rome, and who perhaps should therefore have had his mind on other things, wrote:

> Ringed doves make a man's loins slow and dull; who would be lusty should not eat this bird.[242]

Nonetheless, lustiness or not, the following is a very simple old recipe that will ensure that even fairly dry birds are rendered succulent:

Steamed and Browned Wood Pigeon
2 pigeons
4 rashers of streaky bacon with rinds removed
1 large onion
Salt and pepper
Rub the prepared birds with pepper and salt and tie two rashers of bacon over each. Steam them together with the onion until tender then brown them lightly under the grill or in the oven. Use the onion and juices from the steamer to make a rich gravy. Serve with mashed potato and creamed, mashed swede.[243]

Apart from their size, the most obvious distinguishing feature of Wood Pigeons is the presence of white flashes on the side of the neck of the adult bird. They are much the most widespread pigeon and also much the most numerous, with around 3.5 million pairs, although Collared Doves may be more common locally. They are excluded from legal protection because they are pest species, causing considerable damage to farm crops. They eat seeds, fruits and nuts of many kinds and strip foliage, such as that of brassicas, and there has been a steady increase in their numbers since oil-seed rape has provided them with a fresh and novel food source. In autumn, they are often to be seen feeding on acorns and beech mast and Gilbert White, writing in the late eighteenth century, said:

I have consulted a sportsman, now in his seventy-eighth year, who tells me that fifty or sixty years back, when the beechen woods were much more extensive than at present, the number of wood-pigeons was astonishing that he has often killed near twenty a day; and that with a long wild-fowl piece he has shot seven or eight at a time on the wing as they came wheeling over his head.[244]

The species has officially been called the Wood Pigeon since 1912, when the name superseded the earlier 'ring dove', and most of the sometimes quaint regional names are based on various people's interpretations of the deep, and I think rather attractive, cooing call. It is often the first bird that I hear in the morning as it usually starts the dawn chorus. The cooing sound has been interpreted as 'what shall I do, I can scarce maintain two?', a reference to pigeons usually having only two eggs, in contrast to most other birds which lay more. The endearing name 'quist' finds its way into a Wiltshire saying 'thee bist a queer quist', meaning a person slightly dim and easily fooled. This meaning was also attached in various ways to 'pigeon' itself in the sixteenth century and later: for instance, 'to pigeon' someone was to fool or dupe them and 'to fly a blue pigeon' was to steal lead from a roof. A 'pigeon hole' has now come to mean a small compartment, like that in a pigeon loft, although earlier meanings include the stocks in which miscreants were placed and a wide gap between printed words.

Apart from their obvious and widespread use as food, Wood Pigeons have played some bizarre roles in both medicine and English history. There is a peculiar statement in Pepys' diary pertaining to the illness of Catherine of Braganza, Queen to Charles II: 'the Queene ... was so ill as to be shaved, and pidgeons put to her feet, and to have the extreme unction given her by the priests.'[245] This puzzling reference to the Queen's close encounter with both death and pigeons is explained in John Keogh's *Zoologia Medicinalis Hibernica* of 1739:

The anus of a live Pigeon, applied to the soles of the feet in a fever, mitigates it. Pigeons being cut up alive, and applied to any place pained, they ease the Pain, and draw away the malignity, if applied to the head they help to cure the megrim, vertigo, phrenzy, head-ach &c. they also give ease in the gout, being applied to the parts afflicted therewith ...

Pigeon products have also found other applications. The blood was used for blood-shot eyes and:

... the dung pulverized ... is good against chronic diseases, such as the gout, vertigo, megrim, haemicrania, collic, pains in the sides, shoulders, &c ... The said dung discusses scrophula's, angina's, struma's, oedema's and all hard swellings, a poultis being made of it with barley meal, and vinegar ... Pliny useth them for the stone and difficulty of urine thus: Take Mice and Pigeons dung, of each half a dram; two Hog-lice or Chisleps bruised, drink it, and the pain will cease, and you shall void either the Stone, or much smal sand.[246]

The unfortunate belief that cutting up live pigeons is of medicinal use found one of its oddest manifestations in Scotland, where pieces of the unfortunate bird were applied in an attempt to effect a cure for adder bites and where the heart, liver and lungs were believed to be an effective laxative for cattle.

The feathers of pigeon and doves were once used as filling for pillows and it was very widely believed that a peaceful death was impossible while a sick person lay on them. To ease a dying person's passing therefore, a custom known as Drawing the Pillow took place, in which the pillow was pulled backwards from under him.

In Guernsey, a girl could find out whom she was to marry by killing two pigeons and removing their hearts, which she roasted on skewers between midnight and one o'clock while remaining totally silent. If she was to marry, she would then see her future husband walk in. There is also a rather widespread belief that the heart of a pigeon stuck with pins could be used by lovesick or love-crossed girls to summon their straying lovers. A heart stuck with pins also echoes the ancient use of doves as innocents in sacrifice, a practice that originated with Abraham (Genesis 15:9).

Collared Dove *(Streptopelia decaocto)*
Not long ago, I was asked to advise a film company about creating an accurate rural and garden setting for a production based in the 1930s because the producers were concerned that the garden should not include plant varieties not then in existence. Their biggest problem, however, proved to be not the plants but the penetrating call of the Collared Dove, because the species did not exist as a British bird until 1953. Yet in many areas of the British Isles, it is now much the commonest pigeon. In half a century, the British breeding population has gone from nought to around half a million birds. It was originally an Asian species but irrupted and spread north and west in a expansion that has been unequalled by any other European bird in the last 50 years or so. This is generally attributed to a genetic mutation that affected the birds' sense of direction and encouraged them to move north and west. However, as the British Trust for Ornithology pointed out:

> We would now expect natural selection to have started reducing this tendency in our population as birds that do not move will have a better chance of survival than those heading out over the Atlantic.[247]

The success of the Collared Dove at the expense of the Turtle Dove has been magnified by the fact that it regularly produces more broods each year. It is particularly common in rural areas and it is also very pretty, a slim, long-tailed buff-coloured bird with a characteristic black neck-band.

In this illustration by Kate Greenaway, from *The Quiver of Love*, both Cupid and the doves in the tree are symbols of love.

Turtle Dove *(Streptopelia turtur)*
OTHER COMMON NAMES: *summer dove (Surrey); troet (Cornwall); turtle; turtur (Wales); wood dove (Sussex); Wrekin dove (Shropshire).*
The Turtle Dove is the original symbol of love and also the species most likely to be confused with the Collared Dove, the most obvious differences being its slightly smaller size and lack of a collar. It is a summer visitor and a bird of woodland edges and hedgerows rather than of towns and villages, but it has decreased in recent years and is now officially a species of conservation concern.

Its old name was 'turtle', originating in the Latin *turtur* and imitative of its soft, rippling call. The word 'dove' was first added *c.*1300 but 'turtle' as a single word continued throughout the literature until the eighteenth century. The bird's role as a mark of affection is old, witness Chaucer's *Parlement of Foules*:

> Nay, God forbede a lover shulde change!
> The turtel seyde, and wex for shame red, ...
> For though she dyede, I wol non other make
> I wol bin hires til that the deth me take!

Shakespeare also remarked on their supposed faithfulness:

> When arm in arm they both came swiftly running,
> Like to a pair of loving turtle-doves,
> That could not live asunder, day or night.
> (*Henry VI Part 1*, Act 2, Scene 2)

Izaak Walton was another writer who was very familiar with beliefs about the bird:

> Why then, sir, I will take a liberty to tell, or rather to remember you what is said of turtle-doves: first, that they silently plight their troth, and marry; and that then the survivor scorns ... to outlive his or her mate, and this taken for a truth; and if the survivor shall ever couple with another, then not only the living but the dead (be it either the he or the she) is denied the name and honour of a true turtle-dove.[248]

Walton also knew one particular Dove very well: his beloved river that forms the boundary between Derbyshire and Staffordshire. However, the river name, and comparable place names like Dover, Dovaston and Dovenby, have nothing to do with the bird; most are derived from something very different, the old name 'dubro', meaning 'stream'. Dufton in Westmorland, however, is 'dove farm', and my native village of Duffield in Derbyshire is 'a feld frequented by doves'.

Parrots
(Psittacidae)

Rose-ringed Parakeet *(Psittacula krameri)*

OTHER COMMON NAME: *ring-necked parakeet*

There is little British folklore about parrots because these remarkably beautiful, long-lived and intelligent birds originate from far warmer climates; but things could change. One species, the bright green Eurasian Rose-ringed Parakeet, has established itself in Britain after numerous escapes from captivity over the past century and, some 30 years ago, began breeding in the southeast of England, where it has since spread dramatically.

> Ring-necked Parakeets have arrived in the area. Mother's stepsister talked about them invading Benfleet more than thirty years ago. As pretty as they are, the noise they make is so rotten that I pray they won't nest too near. Zig and Carl watched a flock of around twenty fly from a tree in a neighbour's garden and I had a pair visit our spruce tree one morning. They were mobbed by the Magpies and flew away.[249]

There are now more breeding pairs of Rose-ringed Parakeets in Britain than there are of Crested Tits, and probably of Woodlarks too.

Long John Silver's parrot, from Robert Louis Stevenson's *Treasure Island*, is one of the few to have featured in British literature.

Cuckoos
(Cuculidae)

If there is one ornithological signpost each year more widely recognized than the arrival and departure of Swallows, it is the hearing of the first Cuckoo in the spring. For decades, this singular event has been a matter for the correspondence columns of the national press and the jottings of writers on country affairs. Kenneth Gregory's 1976 anthology of 'Letters to *The Times* 1900–1975' was even called *The First Cuckoo* and, over the centuries, the bird has inspired both poets and authors.[250] Yet, while most people have heard a Cuckoo at sometime in their lives, few have ever seen one; or could even describe the colour of the only species in Britain. Sadly, their chances of doing so may become fewer, as the Cuckoo is clearly arriving in fewer numbers, possibly because of a fall in the populations of their host birds.

Cuckoo *(Cuculus canorus)*

OTHER COMMON NAMES: *gawk gawy, (Dorset); geck, gog, gok (Cornwall); gowk (Scotland, Yorkshire); hobby (Norfolk); Welsh ambassador.*

Because its call is such an obvious inspiration for the name, Cuckoos have seldom been called anything else, although the widespread use of 'gowk', especially in Scotland and the north of England, is interesting. 'Gowk' is Old Norse in origin and an early use of it in *c.*1450 even offered the alternative: 'The goke ansered hur and sayd V times, Cukkow.' It is also supposedly onomatopoeic, although the birds seldom seem to say anything other than '*cuckoo*'. It may also comes as a surprise to learn that, in France, Cuckoos utter a more drawn-out Gallic '*coucou*', while in Germany they say '*kuck-kuck*', in Spain and Portugal '*cuco*', in Italy '*cucu*', in Russia '*kukushka*' and in Poland '*kukutka*'; it is only as far north as Sweden that they begin to say '*gök*'. It is evidently all in the ear of the beholder.

> Cuckoos don't sing this month ... 'They only sing "Cuckoo"'.[251]

The first use of the name 'Cuckoo' in English is in the earliest English canon, that delightful and celebrated composition 'Summer Is Icumen In', which was probably written in Reading by the monk John of Fornsette in *c.*1240 and includes the line 'Lhude sing cuccu'. The name is clearly not English but came originally with the Normans from the old French *cucu*, which must therefore rule out the earlier name of Cuckfield, in Sussex, from being 'cuckoo's feld', although Gokewell in Lincolnshire is 'gauk's stream'.

The Cuckoo is commonly mistaken for a Sparrowhawk, which is similar in overall outline and also has a barred breast with mainly grey upper parts. Generally greyer and also more reddish forms of the Cuckoo occur quite commonly too. The resemblance to a Sparrowhawk might help explain the old belief (said to be still prevalent within living memory) that Cuckoos turn into hawks in autumn.

In most people's experience, they are fairly solitary birds and are seldom seen in groups, although a report from Northamptonshire says: 'We have cuckoos in these particular fields every year, the most I have ever seen together was seven.'[252]

Cuckoos call principally in early summer, when they are most frequently to be seen as they perch in a prominent position looking for caterpillars (including the hairy caterpillars that most other birds reject) and other small creatures on which they feed.

> As a child, I was very privileged, although I didn't realise it at the time. My grandfather was stockman for the horticulturist E.A. Bowles and I had free access to his estate ... We lived in Edmonton and I remember Cuckoos calling loudly every year. There was also abundant insect life in urban gardens – plenty of woolly caterpillars.[253]

Later in the season, Cuckoos tend to become shy and secretive, reflected in a Welsh belief that they then disappear completely ('The first week of May frights the cuckoo away'), but they only truly depart for tropical Africa from late summer onwards. A Lancashire saying is closer to the mark: 'The first cock of hay frights the cuckoo away.'[254] There are many variants of one of the best known of all bird rhymes:

> In April come he will
> In May he sings all day
> In June he changes tune
> In July he prepares to fly
> In August go he must [255]

A neat continuation in Sussex is:

> If he stay until September.
> 'Tis as much as the oldest man can remember.[256]

At Tinwell in Rutland, the rhyme was part of a delightful May Day procession recalled by a Mrs Harvey:

> The boys had a separate garland to the girls. The children moved off at eight o'clock or eight thirty in the morning with the girls leading and the boys following in a separate group. The procession stopped at each house in the village, with first the girls, then the boys singing:
> 'Good morning, lords and ladies, it is the first of May,
> We hope you'll view our garland, it is so very gay.
> The cuckoo sings in April, the cuckoo sings in May,
> The cuckoo sings in June, in July she flies away.'[257]

It has long been said that the first Cuckoos arrive in Sussex on 14 April and are generally heard in North Yorkshire a week later. There are certainly Cuckoo Days in many places, but they seem to be only vaguely related to the bird's arrival: in Hertfordshire, as in Sussex, it is 14 April, the date of the Heathfield

This Reed Warbler *(Acrocephalus scirpaceus)* is feeding a parasitic young Cuckoo *(Cuculus canorus)* many times its own size.

(or 'Heffle') Fair; in Hampshire it is 15 April, the date of Beaulieu Fair (where it was said the Cuckoo was going to buy an overcoat); in Northamptonshire, it is 15 April; in Guernsey 19 April; while at Towednack, near St Ives in Cornwall, a Cuckoo Feast is held on the Sunday nearest to the 28 April, commemorating the day when the bird first brought spring to the county. There is a charming and often-told Sussex belief that an old woman is responsible for all Cuckoos and fills her apron with them in spring. If she is in a good mood, she will allow several to escape at Heathfield Fair, but if she is feeling angry, only one or two will be allowed their freedom.

There are many other delightful tales linking Cuckoos with rural fairs. A Worcestershire story says that the Cuckoo comes in time for Tenbury Fair and that:

> The cuckoo comes in April, and sings his song in May,
> He buys a horse at Pershore Fair, and then he rides away.[258]

Pershore Fair is held on 26 June and, if the Cuckoo continues to be heard after this date, the suggestion is that he could not find a suitable horse there and had to wait until he could buy one at the nearby Stow-on-the-Wold Fair. There is a version of this in Herefordshire, linking Orleton Fair on 23 April, where the Cuckoo first buys his horse, and Brampton Fair on 11 June, where he sells it, while in Hertfordshire, the Cuckoo is looked for after the Spring Fair at Sawbridgeworth on 23 April. Unfortunately, these stories are all confounded because of the confusions between historical and modern fairs, and between today's dates and those in the pre-1752 calendar, when 11 days were lost. The old date for Pershore Fair, for instance, was not 26 June but 15 June, although it still followed Stow-on-the-Wold Fair, which is in mid-May. The Brampton and Sawbridge dates are old ones, but the Tenbury Fair referred to must be modern because its old fair was on 7 July and died out before 1635.

The facts about the Cuckoo's arrival are rather different, and may become even more so as the impact of global warming becomes increasingly evident. A calculation of the mean arrival date of the Cuckoo, as revealed by the Sussex Bird Reports for the 30 years 1966–96, gave a date of 7 April; interestingly, Gilbert White in eighteenth-century Hampshire noted the earliest and latest dates as 7 and 26 April.[259]

Nonetheless, no-one will argue that the arrival of the Cuckoo signifies spring, as evidenced by the names of many other things. The Cuckoo Flower *(Cardamine pratensis)* usually blooms around this time, although Cuckoo spit, the frothy secretion of the froghopper nymph *(Philaenus spumarius)*, appears rather later. By mid-April, the Cuckoo Pint *(Arum maculatum)* has produced its leaves and also most of its flowers, with their curious, club-like spadix, which is fancifully imagined to resemble a penis and thus reflects in some way the imagined promiscuity of Cuckoos.

Almost every part of the country has its sayings and beliefs related to the hearing of the first Cuckoo. In Cornwall, if you hear a Cuckoo calling from the right, it is believed to signify prosperity and, if you have money in your pocket when you hear the Cuckoo, you will never be in need of it; similar tales exist in the Warwickshire village where I live. At Wooler in Northumberland, it is said that, if you are walking on a hard road when the Cuckoo first calls, the ensuing season will be full of calamity, but if you are standing on soft ground it is a lucky omen. In Scotland it is considered good fortune to be walking when the Cuckoo is first heard; you should then sit down on a bank, and pull your stocking or sock off your right leg, saying 'May this to me, Now lucky be'. A curious variant of this exists in the west of Scotland, where you should pull off your shoes and socks when hearing the Cuckoo and, if you find a hair on the sole of your left foot, it will be the exact colour of the hair of your future spouse; if there is no hair, you will still be single the next year.

In Shropshire, miners took a holiday, called a Gaudy Day, or Cuckoo Morning, when the Cuckoo was first heard, while in Somerset, boys would run away on first hearing the call, so that they would not be lazy through the coming year. In Norfolk and in Sussex, whatever you are doing when you hear the Cuckoo, you will do all year; in Berwickshire, Midlothian and Cornwall, the direction you are looking when you hear its call is where you will be within a year. If you are looking at the ground, it means death while, only a little less worryingly, in Norfolk, Herefordshire and Sussex, hearing the Cuckoo while in bed means illness or death in the family. In Sussex, too, hearing a Cuckoo after 24 June was another sign of an impending family death. A more widespread belief is that hearing the Cuckoo after midsummer portends illness or death. A Cuckoo landing on a rotten branch was a similarly bad sign in Norfolk but a crow, a Cuckoo and a hawk on the same roof at the same time meant three deaths were absolutely certain (and equally absolutely improbable).

The phallus-like spadix of the Cuckoo Pint *(Arum maculatum)* is linked with the Cuckoo's imagined promiscuity.

Hearing the Cuckoo has not surprisingly been a token for gardeners or farmers to turn their attention to particular tasks. In the West Country, it is said 'When you hear the Cuckoo shout, 'Tis time to plant your tetties out,' which suggests that Cuckoos arrived much earlier in times past than they do now. (Mid- or even early April is a bit late to be planting potatoes in Dorset and Devon.) Closer to the mark might be a story recounted by a Miss Trump of Broadclyst in Devon, who told of an old man in the village who would stop whatever he was doing when the first Cuckoo was heard and start to sow his French Beans.[260]

Farmers were said to believe:

When the cuckoo comes to the bare thorn,
Sell your cow and buy your corn;
But when she comes to the full bit,
Sell your corn and buy your sheep.[261]

In the Scottish Borders, oats that were sown after 1 April in years when there had been bad weather in March were called 'gowk's oats' and farmers took a dim view of this:

> Cuckoo oats and Woodcock hay
> Make a farmer run away.[262]

Cuckoos have often been regarded as foolish birds. In Scotland, northern England and apparently Cornwall, sending someone on a Fool's Errand on 1 April is called 'a gowk's errand' or 'hunting or hounding the gowk'. In Scotland the day is known as Huntingowk Day. The victim is sent from one fruitless destination to another, asking the whereabouts of the gowk until he or she realizes that they themselves are the object of the search. The jokers would put this message in a sealed envelope:

> Don't you laugh and don't you smile
> Hunt the gowk another mile.[263]

A similar saying runs:

> The first and second day of April
> Hound the gowk another mile.[264]

To which the response is:

> April gowks are past and gone,
> You're a fool and I am none.[265]

Since 1596 calling someone 'cuckoo' has meant they were stupid. 'Cloud cuckoo land' is an alternative name for 'cloudland', a nineteenth-century poetic allusion to a land of fantasy.

For no obvious reason, there are customs associating Cuckoos with cherry trees. In Yorkshire, children would sing around the tree:

> Cuckoo, cherry tree,
> Come down and tell me
> How many years afore I dee.[266]

And in Northamptonshire:

> Cuckoo, cherry tree,
> How many years am I to live?
> One, two, three.[267]

They then shook the tree in turn and the number of cherries that fell indicated how long the children would live. There are other beliefs associating Cuckoos with foretelling life expectancy; a west of Scotland saying ran:

> ... the cuckoo, the first time you hear it in spring, cries once for every year you have yet to live.

A variation was:

> Cuckoo, cherry tree,
> Good bird, tell me
> How many years I shall be
> Before I get married.[268]

In many places in England, people have long believed that, by imprisoning a Cuckoo, they could capture eternal spring. At Madely in Gloucestershire a story is told of men forming a ring around a Cuckoo to entrap it; they are remembered for their abject failure as 'The Wise Men of Madely'. There are comparable tales from Zennor in Cornwall, where a wall was built around a bird; from Borrowdale in Cumberland; and at both Wing in Rutland and Gotham in Nottinghamshire, where a hedge was planted around the entire village in the hope that the Cuckoo would stay inside. Some years ago, this belief was studied in the Lake District by John Edward Field, who found numerous places called 'Cuckoo Pen'; he believed that the name was associated with ancient earthworks, indicating that the idea was an extremely old one and well entrenched in the national psyche. He published his findings in 1913 in an extraordinary book called *The Myth of the Pent Cuckoo.*[269] Similar stories of the 'Pent Cuckoo' exist in other places, such as the Chilterns and the South Downs, associating Cuckoos with ancient ritual and the retaining of spring fertility.

> Every April the people of Marsden in West Yorkshire hold a Cuckoo Day. The event is based around a time-old legend in which the people of Marsden longed for eternal spring. A cuckoo was captured and put into a tower built by the villagers, but the bird flew away as the tower had no roof, thus taking eternal springtime with it. The National Trust jointly organises a Cuckoo Day with Marsden Community Association ...[270]

While most folklore relates to the Cuckoo's diary, to the ornithologist it is merely another migrating bird and the fact that many members of the Cuckoo family, including the British species, are parasitic on other birds is of much greater interest. Cuckoos build no nest of their own but lay their eggs singly in the nests of another species. Cuckoos lay up to 20 eggs, but individual females tend to lay only one type, which resembles the eggs of just one of the numerous host species. Over a hundred different hosts have been recorded for Cuckoos in Europe, but the commonest in Britain are the Meadow Pipit, Dunnock, Reed Warbler and Pied Wagtail. The bright blue egg of the Dunnock is the one that Cuckoos seem to have the greatest difficulty in matching and studies of birds eggs collected a century or so ago have revealed that the colour matching generally is now less good than it once was;

Cuckoos *(Cuculus canorus)* lay their eggs singly in the nests of other species.

this may be due to the fragmentation of habitats, which means that each female Cuckoo now finds it difficult to locate sufficient nests of her 'own' host species. Where the baby Cuckoo is of similar size to the host's offspring, they tend to be reared together but, where the Cuckoo baby is much larger, it usually ousts the other youngsters from the nest.

It may be this curious behaviour that lies behind the notion that Cuckoos are promiscuous. The odd connection between the appearance of the Cuckoo Pint flower, a penis and the Cuckoo's morals has already been mentioned and the word 'cuckold' also seems to be related to this. John Brand discussed it:

> It is not easy to understand how it has come about that this word, which is generally derived from cuculus (a Cuckoo), has been assigned to the injured husband, for it seems more properly to belong to the adulterer, the Cuckoo being well known to be a bird that deposits its eggs in other birds' nests. The Romans apparently used cuculus in its proper sense of adulterer, with equal propriety calling the cuckold himself curruca, or hedge-sparrow, which bird is well known to adopt the other's spurious offspring. The Cuckoo, says Johnson in his Dictionary, is said to suck the eggs of other birds, and lay her own to be hatched in their place; from which practice it was usual to alarm a husband at the approach of an adulterer by calling 'cuckoo', which by mistake was in time applied to the husband.[271]

Chaucer did not help matters, characteristically calling the Cuckoo 'a leud bird', and even Milton said it was 'a rude bird of hate'. However, it was Shakespeare (in one of his numerous Cuckoo references) who encapsulated it most perfectly in the closing song of *Love's Labour's Lost* (Act 5, Scene 2):

> When daisies pied and violets blue,
> And lady-smocks all silver white,
> And cuckoo-buds of yellow hue
> Do paint the meadows with delight,
> The cuckoo then on every tree
> Mocks married men; for thus sings he,
> Cuckoo;
> Cuckoo, cuckoo: O word of fear,
> Unpleasing to a married ear!

The last word on the Cuckoo is a rhyme which not only echoes the odd belief that it sucks eggs and more certainly follows the spring rains, but also summarizes the common feeling about it.

> The cuckoo's a bonny bird,
> She sings as she flies,
> She brings us good tidings,
> And tells us no lies.
>
> She sucks little birds' eggs
> To make her voice clear,
> And never cries Cuckoo!
> Till spring-time of the year.[272]

Barn Owl and Owls
(Tytonidae and Strigidae)

Why owls in children's stories should always be 'old and wise' has never been properly explained, although it may be associated with the owl being an emblem of Athene, the Greek goddess of wisdom. This, of course, begs the question of which came first: the association with Athene or the association with wisdom? Perhaps it is due to a combination of their piercing gaze, their quiet, often motionless stance and the fact that owls are almost always seen from below and so seem to look down in a supercilious manner.

The large, forward-staring eyes that endow owls with such stunningly powerful night vision, and the haloes that surround them, certainly give the impression that the birds are wearing spectacles. Indeed, in some stories, the owl is even illustrated with spectacles, 'Ernest Owl' of Jane Pilgrim's *Blackberry Farm* tales, who is obviously a Short-eared Owl, being a classic instance of this. The identity of two other famous literary owls is less certain: Old Brown in Beatrix Potter's *Tale of Squirrel Nutkin*, who lived in a hollow oak, certainly looks like a Tawny Owl, as does Owl ('wise though he was in many ways, able to read and write and spell his name WOL') in A.A. Milne's

Winnie-the-Pooh, although it is difficult to tell exactly from Ernest Shepard's drawings.[273]

Nonetheless, the owl was wise in Shakespeare's time; it was the bird with 'five wits' (*King Lear*, Act 3, Scene 4) long before people associated wisdom with wearing glasses. Shakespeare mentioned owls over 20 times, possibly more than any other bird, and, like countless other people, he was clearly fascinated by them and their largely unseen, nocturnal life style which has endowed them with a mystery matched by few other birds.

On a global scale, the British owl fauna is modest, consisting of two fairly common resident species (the Barn Owl and Tawny Owl), two less common ones (the Long-eared and Short-eared Owls), the very successful introduced Little Owl and one newcomer, the Arctic Snowy Owl.

Ernest Shepard's charming illustration of A.A. Milne's Owl and Winnie-the-Pooh.

Barn Owl *(Tyto alba;* Family Tytonidae*)*

OTHER COMMON NAMES: *berthuan (Cornwall); Billy wix, Billy wise (Norfolk); cherubim; church owl (northern England); gil-hooter (Cheshire); gill howter (Cheshire, Norfolk); gillihowlet (Scotland); gilly owlets (Shropshire); hissing owl (Yorkshire); hobby owl (Northamptonshire); hoolet (Scottish Lowlands); hullart (Cheshire); Jenny howlet (North Yorkshire); Jenny owl (Northumberland); Madge howlet (Norfolk); moggy (Sussex); oolert (Shropshire); owlerd (Shropshire); owlud (Worcestershire); padge; pudge, pudge owl (Leicestershire); roarer (Scottish Borders); screaming owl (Yorkshire); screech owl; scritch owl; silver owl (Angus – Forfar); ullat (Yorkshire); ullet (Cheshire); ullert (Shropshire); white hoolet; white owl; woolert (Shropshire); yellow owl.*

This is the owl with the ghostly white face and shrieking call, the screech owl, the owl that screamed when 'the deed' was done (*Macbeth*, Act 2, Scene 2). It is one of the archetypal farmland birds but also one that has declined dramatically in recent years, having fallen by some 70 per cent from an estimated total of around 12,000 pairs at the beginning of the twentieth century. The British Trust for Ornithology's 'Project Barn Owl' was established to try to understand the reasons behind this drop, which seems to be the result of a combination of factors associated with alteration in farming practices, loss of nesting sites and climate change. The tide may have turned because, at the British Association meeting in 2000, the Barn Owl was said to have been 'pulled back from the brink of extinction'.[274]

The name 'owl' dates back to Middle English and this species has been called the Barn Owl since the late seventeenth century, partly because it nests in barns (although it also nests commonly in hollow trees) and because it hunts its prey around barns. Most of the multitude of regional names allude to its call or its colour, but the name 'howlet' may be the most familiar. It has usually been used to mean a small owl and 'howlet's wing' was among the ingredients of the witches' brew (*Macbeth*, Act 4, Scene 1). This is but one instance in which owls have been associated with death and the foretelling of death. For Keats, its cry was 'the gloom bird's hated screech'; for Ben Jonson it was 'the shrieks of luckless owls'; and for Byron, it was 'the owl whose notes the dark fiend of night deplores'.[275]

A screech owl calling constantly foretells death and three nights running was enough in Oxfordshire, although in Wales an owl calling continually around a village meant simply that a local girl would shortly lose her virginity.

> Our nearest neighbour was a good old countryman who believed 'the old white owl' brought news of a death, no less.[276]

> My maternal grandmother who lived with an aunt in the then small village of Weaverham in Cheshire (near Northwich) died in 1944 or 1945. My aunt told my mother that the day before my grandmother died she had observed an owl sitting on a post at the end of the garden staring intently into my grandmother's bedroom during daylight hours. Apparently local folklore has it that this is a sign of imminent death in the household. My aunt whom I had not known to be of a superstitious nature, forever after was convinced of the truth of this occurrence.[277]

> My grandfather was very ill in 1962 and as I walked up to the house in the twilight I saw something that I had never seen before, an owl sitting on the thatch over his

An archetypal farmland bird, the Barn Owl *(Tyto alba)* can be seen systematically patrolling fields in search of small rodents.

> bedroom. This was once thought to be a sign of death and my grandfather died two weeks later. This was the only time I ever saw an owl on the roof. It was many, many years before I mentioned it to anyone and certainly not to my mother who was very anti-superstitious.[278]

Chaucer thought it bad luck even to dream about owls and, in many countries, individual families have acquired superstitions of warning birds, those in Britain being especially associated with owls. The Wardour family of Arundel are said to be forewarned of death when two white owls appear on their roof and, similarly, the Oxenham family of the village of South Tawton in Devon knew that a family member would die when a ghostly white owl appeared in a bedroom. Since at least the seventeenth century, the Bishops of Salisbury have also kept a weather eye open for the appearance of large white birds, often thought to be owls. The appearance of two in 1885 was said to presage the death of Bishop Moberley and, on 16 August 1911, a Miss Olivier said that she saw them while returning home, only to be told that the incumbent Bishop Wordsworth had just passed away.

This is a very depressing association for a creature of such astonishing beauty. Undeniably, owls are predators and they kill to eat, but the Barn Owl's principal food in Britain is the Bank Vole, followed closely by mice and shrews; this should surely have made them farmers' friends, rather than fearsome fiends. There is, however, an explanation for the widespread association of owls with the foretelling of death: owls become more evident in autumn, when they tend to hunt closer to human habitation; and autumn is followed by winter, a time when more deaths occur in human populations, which was especially the case in the days of rural poverty. Also, anything so inextricably linked with the night and darkness already has the odds stacked against it, whether or not it rids the farm of vermin. In consequence, owls have been widely persecuted and, in many parts of Britain, and elsewhere, it was customary to nail an owl, or part of one, to a barn door as protection against evil.

The appearance of an owl at childbirth has long been thought a most serious omen. In Shakespeare's *Henry VI Part 3* (Act 5, Scene 6), as Richard of Gloucester prepared to stab Henry VI, he was warned 'The owl shriekt at thy birth, – an evil sign' and it certainly did not bestow much of a charmed life on Richard thereafter.

Presumably because they appear to lead a fairly solitary, lonely life, owls have sometimes been believed to be doleful and melancholy. In Cheshire, it was said that anyone looking into an owl's nest would be melancholy forever after. Robert Burns certainly thought owls pretty miserable:

Shut out, lone bird! from all the feather'd train,
To tell they sorrows to th'unheeding gloom:
No friend to pity when thou dost complain,
Grief all they thought, and solitude thy home.
(Robert Burns, 'To the Owle')

So did Gray, who must have seen plenty of them as he contemplated his 'Elegy Written in a Country Churchyard':

Save that from yonder ivy-mantled tower,
The moping owl doth to the moon complain
Of such as, wand'ring near her secret bow'r,
Molest her ancient, solitary reign.

Owls in general, and Barn Owls in particular, have often been associated with ivy. Michael Drayton wrote:

And, like an owl, by night to go abroad,
Rooster all day within an ivy tod.
(Michael Drayton, 'The Owl')

In James Shirley's *Triumph of Peace* (1633), a scene is set in a woodland where 'in the further part was scene an ivy bush, out of which came an owle'.[279] There are even inns called The Owl in the Ivy Bush and the expression 'like an owl in an ivy bush' is used for someone having a distant, vacant look, as when they stare into a drinking cup.

On Guernsey, it was said that owls were forced to hunt at night as punishment for refusing to give a feather to Robin Redbreast when he first brought fire to the island and children would always run past a tree or other place where owls were roosting. Owls have also been believed to herald bad weather: hailstorms in England and snow in south Wales.

Nonetheless, places have been named for their owl associations: Ouseden in Suffolk is 'owl valley' and Outchester in Northumberland was 'a Roman fort inhabited by owls'. Owls also occur very commonly in heraldry, both on family and municipal crests, as in Dewsbury and Oldham, and in Leeds, where they are a tribute to the first Alderman Sir John Savile, whose arms contained three owls argent. Also, in Yorkshire, the Sheffield Wednesday football team are known as 'The Owls'.

No-one appears to dine regularly on owl (Tawny Owl is reported to be 'unexpectedly oily'), although their bones were among those found in the prehistoric lake village at Glastonbury, where presumably they were habitually eaten. They have, however, certainly found medicinal use. John Swan in his *Speculum Mundi* (1635) said that an habitual drunkard, if given owl's eggs, would henceforth detest strong drink as much as he had formerly loved it. Owl's eggs, charred and powdered, have been recommended as a remedy for poor eyesight and, in Yorkshire, owl broth was once used to treat whooping cough. Swan, among others, also gave us a memorable remedy for gout, which was to eat salted owl; the recipe ran something like:

Take an owl, pull off her feathers, salt her well for a week, then put her into an oven, that so she may be brought into a mummy.[280]

This unholy blend should then be beaten into a powder, mixed with boar's grease, and thoroughly anointed onto the afflicted part. At other times, both epilepsy and madness have been treated with various bits of owl.

The West Country appears to be an owl-rich region if custom is anything to go by. A strange pastime known from both Dorset and Wiltshire was the so-called 'duck hunt'. An owl was somehow attached to the back of a duck, which was then released onto the village pond and, in the ensuing mêlée, both birds were generally drowned.

From Dorset comes a St Valentine's Day practice:

Three single young men went out together before daylight on St. Valentine's Day, with a clapnet to catch an old owl and two sparrows in a neighbouring barn. If they were successful and could bring the birds to the inn without injury before the females of the house had risen they were rewarded by the hostess with three pots of purl in honour of St. Valentine, and enjoyed the privilege of demanding at any house in the neighbourhood a similar boon. This was done, it is said, as an emblem that the owl, being the bird of wisdom, could influence the feathered race to enter the net of love as mates on that day, whereon both single lads and maidens should be reminded that happiness could alone be secured by an early union.[281]

John Udal, the authority on Dorset folklore and the source of this story also claims, with no apparent reason, that the residents of the village of Houghton, near Blandford, have always been known as Owls.

Sheffield Wednesday ('The Owls') features its distinctive owl badge on match programmes.

Snowy Owl *(Nyctea scandiaca)*
A huge, almost pure white owl of the far north, this is arguably the most imposing species on the list of British nesting birds. It is a regular visitor to the far north of the British Isles and bred on the Shetland island of Fetlar from 1967 to 1975.

Its overall population size and the extent to which it moves south from its traditional breeding and feeding grounds seem largely dependent on the fluctuations in the populations of rodents, especially lemmings, on which it feeds.

It is unique among the 130 species of owl in exhibiting sexual dimorphism, the female being larger and more spotted.

Little Owl *(Athene noctua)*
OTHER COMMON NAMES: *Belgian owl, Dutch owl, French owl, Indian owl (Sussex); Lilford owl (Northamptonshire); little Dutch owl (Somerset); little grey owl; little night owl; little spotted owl, Spanish owl (Somerset); sparrow owl.*
After the Tawny Owl, the Little Owl is now the commonest and most widespread British owl and certainly the smallest. A dumpy little bird, it is usually seen in daylight, flying or hopping on the ground, feeding on insects and worms and occasionally a few small mammals and birds.

It is a bird of parks, other wooded places and rocky areas, which it has colonized actively throughout England and Wales and southern Scotland since it was introduced from Europe in the nineteenth century by the amateur ornithologist Lord Lilford, among others. This is betrayed by its local name 'Lilford owl'. Lilford was a noted naturalist and writer, best known today as the author of one of the most sought-after bird books, *Coloured Figures of the Birds of the British Islands* (seven volumes, 1885–97), illustrated by Archibald Thorburn and others.[282]

Its success in the cool, damp climate of Britain is surprising because, in its native habitat, it is a bird of dry, almost semi-desert areas. Curiously, one of its main natural predators in Britain is the Tawny Owl.

Young owls have an innocent, fluffy-toy appeal that disguises their predatory life style. The eggs hatch over a long period, resulting in nestlings of very different ages and appearances:

> I also see a little owl which has raised young for several years in the same trees – two summers ago I was lucky enough to see them the day they fledged – three chicks in various states of plumage, one was completely white, one was half soft and half grey, the other had all its new feathers.[283]

Tawny Owl *(Strix aluco)*
OTHER COMMON NAMES: *beech owl; Billy hooter; brown hoolet; brown ivey owl; brown owl; brown ullert; ferny hoolet; gilly hooter (Shropshire); golden owl; grey owl; hill hooter (Cheshire); hollering owl (Somerset, Sussex); hoolet; ivy owl; Jenny hoolet; Jenny houlet (Yorkshire); Jinny yewlet; screech owl (Sussex); tawny hooting owl (Shropshire); ullet; wood owl; wood ullat.*
This is a medium-sized, brown or greyish owl with black eyes; the brown and grey birds representing two distinct colour

The Snowy Owl *(Nyctea scandiaca)* is unusual in exhibiting sexual dimorphism. The female is larger and more spotted than the male.

The Tawny Owl *(Strix aluco)* nests in tree-holes, preying on small mammals and birds, which it brings back to its young.

phases. This owl became officially known as the Tawny Owl, which superseded the much older 'Brown Owl' or 'Grey Owl', only in the eighteenth century, and it is the species mainly responsible in Britain for the 'wise old' image.

It has lent one of its names to the Brown Owls who lead the Brownies, the junior branch of the Girl Guides, and who are supposed to give wise counsel to their young followers. Or has it all been a big mistake? When Robert Baden-Powell invented the Brownies in 1918, he took the name from the Victorian children's book *The Brownies and Other Tales* by Juliana Horatia Ewing.[284] In this story, the Brownies, a type of primitive domestic help, took guidance from Old Owl. Old Owl then became Brown Owl, but did anyone really check? In the book, only one feature of Old Owl's appearance is described:

> When he got in, no bird was to be seen, but he heard a crunching sound from above, and looking up, there sat the Old Owl, pecking and tearing and munching at some shapeless object, and blinking at him – Tommy – with yellow eyes.[285]

No Tawny Owl has yellow eyes, which is a feature of the Long-eared Owl, and so generations of Brownies should probably have been calling their leader 'Long Ears'. Perhaps it is just as well that, since 1968, officially at least, Brownie leaders have been called 'Brownie Guiders'.

Tawny Owls are most likely to be seen hunting at dusk or resting, well camouflaged, during the daytime in a tree. The familiar '*tu-whit-tu-whoo*' is heard when a pair are calling to each other; only the male hoots, the female having a sharper '*whit-whit*' call. The calls of Tawny Owls fascinated Gilbert White:

> From what follows, it will appear that neither owls nor cuckoos keep to one note. A friend remarks that many (most) of his owls hoot in B flat; but that one went almost half a note below A. The pipe he tried their notes by was a common half-crown pitch-pipe, such as masters use for tuning of harpsichords; it was the common London pitch.
>
> A neighbour of mine, who is said to have a nice ear, remarks that the owls about this village hoot in three different keys, in G flat, or F sharp, in B flat and A flat. He heard two hooting to each other, the one in A flat, and the other in B flat. Query – Do these different notes proceed from different species, or only from various individuals?[286]

Like the Barn Owl, the Tawny Owl feeds on small mammals but also takes such varied livestock as worms, frogs, small birds and even snakes. In rural localities, Starlings and Grey Squirrels form a significant part of the diet and Tawny Owls are frequently seen in gardens, especially in late autumn. As with other owls, its diet can be investigated by dissecting the pellets of undigested bone, feathers and other matter that it regurgitates.

Long-eared Owl *(Asio otus)*

OTHER COMMON NAMES: *cat owl; horn coot; horned owl (Somerset); hornie hoolet (Scotland); long ears (Berkshire); long-horned ullat (Yorkshire); tufted owl (Sussex).*

Superficially similar to the Tawny Owl when in flight, the Long-eared Owl can be distinguished by its red-orange eyes and long ear-tufts; these are raised when it is alert and are then striking and unmistakable, although they are only false ears.

This is widespread but uncommon in Britain, although the commonest owl in Ireland, and is usually found in coniferous woods. It is largely a summer visitor although some overwinter. Like the Barn Owl, it probably feeds mainly on voles.

Short-eared Owl *(Asio flammeus)*

OTHER COMMON NAMES: *brown yogle, cat ool (Shetland); day owl; fern owl (Ireland); grey hullet (Lancashire); grey yogle (Shetland); hawk owl; moor owl; marsh owl, march owl; moss owl (Yorkshire); mouse hawk; pilot owl (Suffolk); red owl (Dartmoor); sea owl (Kent); woodcock owl (Norfolk, Berkshire, Kent).*

The 'ears' of the Short-eared Owl are so short as to be hardly visible, its main distinguishing feature being the very obviously barred brown-and-white plumage pattern, which gives away its identity in the *Blackberry Farm* books (see Owls, page 304).

They are usually active during the daytime, and early morning and later evening are the times when you may both see them and hear the short '*hoo-hoo*' call of the male. They feed on a wide variety of rodents and other small animals, and are typically birds of the open countryside, although they prefer to nest among young conifers; as these grow to maturity, so the population seems to decline. The resident population is supplemented each year by summer visitors from Scandinavia.

Nightjars
(Caprimulgidae)

Nightjar *(Caprimulgus europaeus)*

OTHER COMMON NAMES: *dor hawk (east Suffolk, Cornwall); eve churr (Hampshire); evejar (Surrey); fern hawk, fern owl (Gloucestershire); flying toad (Lancashire); gabble ratch (Yorkshire); gabble ratchet; gnat hawk (Hampshire); goat chaffer (Scotland); goat owl (East Lothian, Shropshire); goat sucker (Somerset, Surrey, Sussex); heath-jar (Surrey); heave-jar; jar owl; Jenny spinner (Cheshire); lich fowl (Shropshire, Cheshire); moth hawk (Angus – Forfar); moth owl (Cheshire); night churr (Cornwall); night crow (Northamptonshire, Cornwall); night hawk (Lancashire, Shropshire, Norfolk, Hampshire, Cornwall); night swallow (Devon, Surrey); puck bird (Sussex); puckeridge (Hampshire, Surrey, Sussex); razor grinder (Norfolk); scissors grinder (Norfolk, Suffolk); screech hawk (Berkshire, Buckinghamshire); spinner (Wexford); wheel bird (Stirlingshire).*

Like many other naturalists, I found the Nightjar was the last of the fairly common British birds that I encountered. This was because Nightjars are not designed to be seen and, when I was a boy, my parents did not encourage me to venture alone into remote woods at night. Nightjars are shy, secretive summer visitors to Britain and are most likely to be found in open areas in pine or other coniferous woodland. They are usually described as nocturnal although, in their feeding habits, they are more usually crepuscular, using their huge, gaping mouth to catch moths and other flying insects. They are also extremely well camouflaged when at rest among twigs on the ground and merge perfectly into the bark of a thick branch, where they tend to sit.

They have probably benefited from the increase in conifer plantations within the past century, but their relative invisibility makes them difficult subjects for study, although the curious low purring song that the males produce often betrays their presence, as well as puzzling many a late-evening dog-walker and dog. This call is reflected in many of the regional names and also in the name 'Nightjar' itself, which dates from 1630 and echoes the jarring song. Although few people today will appreciate the name 'Jenny spinner', the sound really does resemble that of an old spinning wheel. The inclusion of 'owl' in some names is a reference not only to the bird's night-time activity but also to its silent, owl-like flight.

Much the most fascinating of Nightjar names is 'goat sucker'; this does not occur until 1611 but reflects a much more ancient notion that the birds suck the udders of goats (and cows) to get at the milk and, so it is sometimes said, infect them with blindness and disease. This is a widespread belief wherever Nightjars are found but its origins are obscure, although Edward Armstrong says that it is mentioned by Aristotle. Perhaps it is because of the enormous size of its open mouth.

The names 'puck bird' and 'puckeridge', from southern England, relate to a link with that mischievous and evil character of folk legend Puck, also known as Robin Goodfellow, or Hobgoblin, and a being whose existence is deeply entrenched in the country psyche. An affliction of cattle called 'puckeridge' was thought to be the result of Nightjar activity. The birds have also been associated with human death: the Shropshire name 'lich fowl' (or 'lych fowl') is derived from the local word 'lych' meaning 'a corpse', while in the north, Nightjars were believed to be the souls of children who died before being baptized. In Somerset the birds were thought to be witches in disguise, who could be destroyed only by a gun loaded with a silver sixpence.

Both Robin Goodfellow (illustration by Charles Folkard) and the Nightjar *(Caprimulgus europaeus)* are known as 'Puck'.

Swifts
(Apodidae)

Swift *(Apus apus)*

OTHER COMMON NAMES: *black martin (Scotland, Hampshire); black swift (Kirkcudbrightshire); brown swallow (Renfrewshire); bucharet; cran swallow, crane swallow (East Lothian); devil (Berkshire); devil bird (West Yorkshire); devil shrieker (North Yorkshire – Craven); devil's bitch, devil squeaker, devil swallow (Yorkshire); devil's screamer (Yorkshire, Hampshire); devil's screecher (Devon); deviling (Westmorland, Lancashire, East Anglia); devilton; diverton (Suffolk); harley (Angus – Forfar); hawk swallow; jack squealer (Shropshire); screamer; screech martin; screecher (Hampshire); screek (Gloucestershire); skreek owl (Worcestershire); shriek owl; skeer devil (Somerset, Devon); squealer (Sussex); swing devil (Northumberland); tile swallow (Yorkshire); whip (West Yorkshire).*

The Swift and the Swallow are not related, despite their superficially similar appearance and habits, and the fact that they have so often been linked ('The swallow and the swift are God Almighty's gifts' says a Herefordshire rhyme). Swifts are not hirundines but belong to the family Apodidae and their closest relatives are the tropical humming birds. Apodidae means 'no feet' and therein lies a widespread belief. Swifts do of course have feet but, because their legs are very short and relatively useless, they use them only for hanging onto vertical surfaces and crawling into their nests in the roofs of buildings or holes in trees. Once a Swift is in the air, however, it is a strong acrobatic flier, totally adapted to life in the sky. No bird has a greater mastery of the third dimension.

The Swift has been so called since the mid-seventeenth century, the name coming from the much older adjective. Many of the regional names are still extensively used and, given the bird's dramatic, all-black coloration, arrow-like appearance and behaviour, especially the piercing, shrieking cries that accompany the flocks as they wheel and twist overhead, it is easy to understand how they came to be associated with the devil. Indeed, no other British bird has so many 'devil' names. In Lincolnshire, the Swifts that dive and circle around a church tower are said to be lost souls, bemoaning their missed opportunities for redemption.

There are several places where it is thought unlucky to kill a Swift and an old story from Hampshire tells of a farmer who, for some reason, killed not one but 17 Swifts and then, within seven weeks, lost 17 of his cows.

To see Swifts flying high as a sign of good weather is common among countrymen. A Shakespearean actor friend who often performs in outdoor productions, always looks for them high above before going on stage in the certain knowledge that the evening will then pass without the cast being soaked: 'If they are flying low, then it pays to be doing a play like *Pericles* whose plot positively benefits from bad weather.'

Swifts arrive later and leave earlier on migration than Swallows and are in Britain for only about three months between May and August. Like Swallows, they were once thought to hibernate in improbable places (see page 321).

The martlet is an heraldic bird, probably based on the Swift *(Apus apus)*. The coat of arms of East Sussex features six martlets and a Saxon crown.

The martlet is an heraldic bird and, although this name is also used for the House Martin, the fact that, in heraldry, it is usually depicted without legs suggests that it is based on the Swift. There is an interesting story behind the adoption of the martlet as a mark of cadency (relative status) for the fourth son, the reasoning being that, by the time the older sons have had their share of the family fortunes, there is little left for the fourth, who must fly away and seek his fortune elsewhere. Occasionally, a martlet is depicted with feet, when it becomes a true heraldic martin.

Kingfishers
(Alcedinidae)

Kingfisher *(Alcedo atthis)*

Other common names: *fisher (Yorkshire).*

One of the most evocative landscapes by the nineteenth-century artistic genius Sir John Everett Millais is an oil painted in 1892 on a backwater near Stanley in Perth, where the artist was staying. It is called *Halcyon Weather* and when Millais exhibited it at the Royal Academy in the same year, the year of his presidency, it was accompanied by a quotation: 'Expect St Martin's summer, halcyon days' (*Henry VI Part 1*, Act 1, Scene 2); and thereby hangs a tale of great ancestry and romance.

The word 'halcyon', in the sense of 'calm, peaceful, undisturbed', has been in use since 1578, but is of much older derivation. The Halcyon was the Greek Kingfisher, possessed of the power to keep the water calm while it built its nest upon the surface. The 'halcyon days' were the seven days before and after the winter solstice when the bird was brooding and the tale has been beautifully told by Frank Gibson:

> Ceyx the King of Trachyn, being disturbed in his mind respecting the fate of his brother, determines on consulting the oracle at Claros, and makes known his intention to his wife, Halcyone. She pleads with tears that he will take her with him and let her share by his side the dangers of the long and hazardous voyage; but Ceyx cannot think of his beloved Halcyone being placed in such a direful position, and he gently refuses her request. Then Halcyone seeing that her request will not be granted, and feeling a presentiment of the fate which is to overtake her husband, falls at his feet in a swoon. When she raises her weeping eyes the ship is on the seas and Ceyx is waving his hand to her in farewell; and, watching it as the sails disappear from view, Halcyone returns to her home and weeps for her beloved husband. True to the presentiment of Halcyone, a great storm overtakes the ship on which Ceyx is journeying to Claros, and, unable to battle with the waves, it sinks to the bottom of the ocean. Ceyx, with

a few of the sailors who do not go down with the sunken vessel, clings to a spar and for a while keeps upon the surface of the water, but presently a great dark billow overwhelms him, and he is engulfed. All this while, faithful Halcyone is praying for the safety of her husband, and making every preparation for his return before the end of two months, as he had promised. But the fate of her husband is made known to her. Morpheus flies through the dark to her bedside, and taking upon him the form of the ill-fated Ceyx, tells her that the prayers she has offered in his behalf have availed him nothing, for he has perished on his voyage. Halcyone, on hearing this dire news, is overcome with anguish. She wildly beats her face and breast, and rushing to the sea-shore, where she had last seen her husband when he set sail to Claros, she declares that she cannot live without him, and resolves to go to him in the sea. But while she is standing by the side of the sea, she sees a body floating upon the water, and, as she watches it, the waves bring it nearer and nearer to her. Then she recognises in this body the form of her unfortunate husband, and she springs upon a mole in the sea to clasp him. But, even as she leaps upon the stones, wings bear her upon the surface of the water, and she becomes transformed into a kingfisher. With her bill she kisses the dead face of her beloved Ceyx; with her wings she embraces him, and the gods having compassion upon her, change her husband also into a kingfisher so that as birds their love might remain unchanged. And thereafter, when they rested upon the face of the waters, the sea remained calm, and the passage of the deep was safe.(287)

Milton refers to this story in his ode 'On the Morning of Christ's Nativity':

The winds with wonder whist
Smoothly the waters kist,
Whispering new joys to the mild Ocean,
Who hath quite forgot to rave,
While birds of calm sit brooding on the charmed wave.

In reality, Kingfishers do not brood on the waves, and the British birds build their nest in a deep hole in a river bank from whence they emerge as a blur of electric blue and red.

Occasionally I see a kingfisher flash along a brook near the village in the area of the carp ponds which were kept when Silverstone was cut off from the surrounding areas during the Winters.(288)

This is a fairly typical observation because few people are fortunate enough to see a Kingfisher at rest, usually perched on an overhanging branch as it waits to pounce on some unsuspecting small fish or other aquatic creature, when its long beak, vivid blue back and red underparts are readily seen. Interestingly, whereas the beautiful red breast is the result of pigmentation in the feathers, the blue of the back is due solely to light refraction.

Kingfishers seldom eat their catch immediately but usually beat the fish on a branch or stone to stun it before swallowing it headfirst. They take especial care to kill those with sharp spines, such as Sticklebacks and Bullheads, so that the spines cannot be raised to stick in the gullet.

The Kingfisher was once known just as 'fisher' (a name that still survives, especially in Yorkshire); this became 'King's fisher' in 1318 and then 'Kingfisher' in the eighteenth century but, strangely for so incredibly coloured a bird, it seems to have acquired no other names. The name 'Kingfisher' now crops up in many contexts: it is often used to describe a particular shade of blue and is common as a name for waterside inns.

Others were killed for a stranger reason: the widespread but inexplicable belief that kingfishers do not moult, and that their skins never decay and have the power of preserving any material with which they came into contact. Housewives once kept a skin among their linen to preserve it and Giraldus Cambrensis, writing in the twelfth century, knew the story well:

It is remarkable in these little birds that, if they are preserved in a dry place when dead, they never decay; and if they are put among clothes and other articles, they preserve them from the moth and give them a pleasant odour.(289)

Presumably because of their strong fishy flavour, Kingfishers seem never to have formed part of the diet although:

The postestates of the Flesh cure consumptions, being taken to the quantity of three ounces for a dose. The flesh pulverized taken in any proper vehicle, to the quantity of two drams at a time is good against the epilepsy. Also the Heart, being dried, and hung about the neck of an infant, is said to drive away the said disorder.(290)

Hoopoes
(Upupidae)

Hoopoe *(Upupa epops)*

The first sighting of a Hoopoe is a triumphal moment for any British bird-watcher. Such happenings are most likely to be in spring but they could occur in almost any part of the country because the birds, although rare, are nonetheless regular visitors and have been known to nest; there have been 20 attempts in the past 200 years according to Chris Mead.(291)

The Hoopoe is unmistakable in flight, when it has been likened to a huge, black, white and tawny pink butterfly with rounded wings and a 'bouncing' motion. On the ground, it is much less visible, until it raises its huge, black-tipped, pink crest. It feeds on insects which it obtains from the ground with its long, probing beak and it is often associated with cattle, seeking out the beetles that live in the dung.

The Kingfisher *(Alcedo atthis)* is one of the most brilliantly coloured of British birds.

The Hoopoe *(Upupa epops)*, although rare, is a regular visitor to Britain.

The name 'Hoopoe' is imitative of its booming and far-carrying call. The history of the Hoopoe as a British bird has been described as 'long and disgraceful' because, whenever it has attempted to establish itself, collectors have turned their guns on it. Nonetheless, it may have another opportunity as global warming makes British summers more to its liking. Because of its scarcity, the Hoopoe has attracted little British folklore but elsewhere it has been credited with the same talismanic powers that are attributed to woodpeckers, Ravens and Swallows.

Woodpeckers
(Picidae)

Generations of children must have recited the playground rhyme 'Why did the owl 'owl? Because the woodpecker would peck 'er,' although the name 'woodpecker' is far from universal, even today. It was first used in 1530, but the alternative name 'yaffle', for the Green Woodpecker, is probably used most widely.

Those same schoolchildren, like many an adult, must also have pondered how a bird could spend so much time hammering its head against a tree without acquiring a gigantic headache. The answer lies in the structure of the skull, which is modified to absorb the shock and is a common feature of this very large family. There are four British representatives (although none occurs in Ireland): three residents (the Green Woodpecker and the Great Spotted and Lesser Spotted Woodpeckers, which are two of a larger group of confusingly similar black-and-white species) and that rare and strange summer visitor, the Wryneck.

All woodpeckers feed mainly on insects and drill holes in tree trunks in which to nest, but the familiar drumming sound is not, however, the sound of holes being made but a means of communication. Even more extraordinary is the woodpeckers' tongue: to enable the bird to seek out insects deep within a rotten tree, the tongue is unusually long, the Wryneck's being five times the length of its beak. This creates a logistical problem: where do you put a tongue far longer than your beak? The answer is to tuck it around the back of your head, but on the inside. Woodpeckers' skulls are modified, not only to withstand hammering, but also to accommodate their incredible tongues.

Wryneck *(Jynx torquilla)*

OTHER COMMON NAMES: *barley bird, barley snake-bird (Hampshire); cuckoo's fool (Gloucestershire); cuckoo's footman; cuckoo's knave (Wales); cuckoo's leader (Norfolk); cuckoo's marrow (central England); cuckoo's mate (Shropshire, East Anglia, Gloucestershire, Somerset, Hampshire, Surrey, Sussex); cuckoo's messenger; cuckoo waker (Somerset); dinnick (Devon); emmet (Surrey); emmet hunter (Somerset); long tongue; mackerel bird (Channel Islands); pee bird (Surrey); peel bird (Sussex); peet bird (Somerset); rind bird; rinding bird; rine bird (Surrey); slab (northern England); snake bird (Gloucestershire, Somerset, Hampshire); summer bird (Northumberland); tongue bird (Somerset); turkey bird (Suffolk); twister; weet bird (Hampshire); writhe neck.*

The Wryneck was once much more widespread – how else could it have accumulated so many regional names? – but it has suffered a huge decline not only in Britain but also in a large part of western Europe. When it does occur in Britain today, it is as a summer visitor to the extreme southeast of England and yet there are local names for it as far north as Northumberland.

It is a dull grey-and-brown bird, only distantly related to the true woodpeckers, and is commonly mistaken for a large treecreeper. Like true treecreepers and true woodpeckers, but unlike nuthatches, it works its way up a tree trunk, not down. Unlike true woodpeckers, however, it nests in existing holes (even in nest boxes) and does not drill its own.

The odd name 'Wryneck' dates from 1585 and, as with regional names such as 'writhe neck' and 'twister', takes its inspiration from the bird's ability to twist its head through 180 degrees. When combined with its hiss-like call, this serves to disarm its enemies. References to 'Cuckoo' in some of the regional names, which it shares with a few other summer visitors, allude to the time of its arrival. 'Cuckoo's marrow' means simply 'Cuckoo's mate', just as 'marrer' in northeast Derbyshire is a work colleague.

The Wryneck *(Jynx torquilla)* is often mistaken for a large treecreeper.

Witches are said to have tied a Wryneck to a wheel (further evidence surely that they were once easier to come by), which was then turned, either to act as a 'charm on faithless lovers' or to work a spell against enemies.

Green Woodpecker *(Picus viridis)*

OTHER COMMON NAMES: *awl-bird; bee-bird; coit (Cornwall); eaqual; ecall (Shropshire); eccle (Oxfordshire); eekle (Worcestershire), green ile (Gloucestershire); green peek (Lincolnshire); hecco; heffalk (Yorkshire); hefful; hew hole (Somerset); high hoe (Shropshire); hood-awl (Cornwall); icwell (Northamptonshire); jack eikle (Worcestershire); jar peg (Northamptonshire); kazek (Cornwall); laughing Betsey (Gloucestershire); laughing bird (Shropshire); nickle (Nottinghamshire); nicker pecker; pick-a-tree (Yorkshire); popinjay; rain-bird (Sussex); rain-fowl (Northumberland); rain-pie; snapper; sprite (Suffolk); stock-eikle (Staffordshire); storm cock (Shropshire); weather cock; wet bird, wet-wet (Somerset); whetile (Essex; Hertfordshire); whitewall; whittle (Somerset); wood-awl; wood hack (Lincolnshire); wood knacker (Hampshire); wood pie (Somerset); wood spack (Norfolk, Suffolk); wood sprite (Norfolk, Gloucestershire); woodwale; woodwall (Somerset); wood yaffle (Suffolk); yaffingale (Wiltshire, Hampshire); yaffle; yaffler (Herefordshire); yappingale (Somerset); yockel (Shropshire); yuckel (Gloucestershire).*

There cannot be many birds with such a huge accumulation of names and probably very few that have so many still in regular use. Many of those that include 'laughing', or words like 'hefful' and 'yaffle', are imitative of its curious, harsh call; others refer to its supposed role as a bringer of rain. 'Hew hole' tells us its nesting habits, while 'popinjay' has been borrowed from parrots and 'bee-bird' reminds us of its fondness for bees. It was 'papingay' to Chaucer in *The Romaunt of the Rose*, but whether it or the Golden Oriole was the medieval 'woodwele' is a matter of dispute (see Golden Oriole, page 355). Nonetheless, as is evident from the length of the list, it is a fairly common bird, although it never fails to attract attention whenever it arrives to pick ants from the lawn. Certainly no other large bird likely to be seen in a garden is so brightly and strikingly coloured, its yellow-green back contrasting with a bright red cap and long, dark beak. They are widely distributed, although absent from the north of Scotland, Ireland and the Isle of Man.

Why Green Woodpeckers should be credited with foretelling rain is not immediately obvious (in Scotland, their drumming is believed to forecast dry weather with frost at night), although there are several European tales and a French story told by Charles Swainson is relevant:

> At the beginning of all things, when God had finished creating the earth, He ordered the birds to excavate with their beaks the hollows that were destined, when filled with water, to become seas, rivers, and pools. All obeyed except the woodpecker, who in sullen obstinacy sat still and refused to move. What was the result? Why, that when all was completed the good

Both male and female Green Woodpeckers *(Picus viridis)* incubate their young and feed them by regurgitating their food.

> God declared that, as she was unwilling to peck up the earth, her lot would be to be ever pecking at wood; and moreover that, as she had nothing to do with making the cavities in which water was to be stored, she should drink nothing but the rain, and get that as she could! Hence it is that the wretched bird is ever calling to the clouds 'Plui-plui,' and that she ever keeps an upward, climbing attitude, in order to receive in her open beak the drops which fall from the sky.[292]

Edward Armstrong has made an extensive study of world-wide beliefs relating to woodpeckers and their association, at least in part, with the legend of the thunderbird, although many of the European beliefs may relate to the very large, non-British Black Woodpecker *(Dryocopos martius)*. (This species now occurs in France, just across the Channel, and, with the continuance of global warming, may well appear in southern England before long.) Nonetheless, in localities spread widely across Europe, Asia, North America and Africa, woodpeckers of one sort or another have been credited with special powers. It is often overlooked that Romulus and Remus, the legendary founders of Rome, were cared for not only by a she-wolf but also by a woodpecker. Armstrong's account concludes that the link between woodpeckers and oak trees led initially to a woodpecker cult among Neolithic woodland people, who farmed with crude implements that included ploughs. The fact that woodpeckers dig when feeding on the ground has reinforced an historical link between the birds and plough-based agriculture and, naturally, if the bird could truly bring rain, its importance to such a culture can readily be envisaged. The story is fascinating although, in fairness, not all authorities find the evidence for it as compelling as Edward Armstrong did.[293]

Woodpeckers have long been considered pests because of the damage they cause to timber, formerly to the fine oak trees needed for ship-building but of late, and more importantly, to telegraph poles. The damage to the wood may not in itself be extensive, but it can admit wood-destroying fungi and so woodpeckers were among the 'noyfull fowls and vermin' condemned to destruction in an Act of Elizabeth I. John Clare had the damage perfectly assessed:

> A noise in oaks above the head
> Keep on throughout the day
> Wood peckers nests are neerly made
> And natures carpenters are they
> Through hardest oaks their whimbles go
> And thick the sawdust lies below.
> (John Clare, 'Walks in the Woods')

Woodpeckers of all kinds suffered a serious decline when the formerly extensive forests were cleared and it is only relatively recently that they seem to have found new habitats in plantation forests, parks and gardens and become familiar again. There may be no inn called The Yaffle but there are several called The Woodpecker, sometimes with a Green Woodpecker and sometimes with one of the spotted species on the sign.

Andrew Marvell was among many who committed the bird to verse, but his account is especially praised in Charles Swainson's *Folklore and Provincial Names of British Birds* for having 'discerned so well, three hundred years ago, the ecological niche of the Green Woodpecker':

> He walks still upright from the Root
> Meas'ring the Timber with his Foot;
> And all the way, to keep it clean
> Doth from the Bark, the Wood-moths glean ...
>
> But where he, tinkling with his beak
> Does find the hollow oak to speak
> That for his building he designs
> And through the tainted side he mines
>
> Who could have thought the tallest oak,
> Should fall by such a feeble stroke
> Nor would it, had the tree not fed,
> A traitor worm within it bred.
> (Andrew Marvell, 'Upton Appleton House')

There is also a sinister belief connected with Peonies:

> When I was a child, in Kent, my mother warned me that if I picked peonies in the daytime my eyes would be pecked out by a woodpecker, a rather revolting idea for a small child! I had never heard this anywhere else until a year or two ago when I bought a copy of 'Curtis' Flower Garden Displayed' in which, in talking of the medicinal properties of the peony, it says that Theophrastus, Pliny and Apuleius believed the same thing. John Gerard had heard it as he wrote 'of necessitie it must be gathered in the night, for if any man shall pluck off the fruit (of the peony) in the daytime, being sene of a Woodpecker, he is in danger to lose his eie'. I have no idea where my mother heard this superstition as she was no Classics scholar nor for that matter would she have read Curtis. She came from South Wales – is that the connection?[294]

Strangely, the only place that Peonies grow wild in the British Isles is on the island of Steepholm in the Bristol Channel, not a million miles from south Wales.

Great Spotted Woodpecker *(Dendrocopos major)*

OTHER COMMON NAMES: *black-and-white woodpecker (Norfolk); eckle (Gloucestershire); French pie (Leicestershire); hickwall, magpie-ile (Gloucestershire); pied woodpecker (Surrey); witwall (Gloucestershire, Surrey); wood pie (Staffordshire, Hampshire); woodwall (Hampshire).*

The Great Spotted Woodpecker is the most widespread British woodpecker. It is also currently the most familiar because its population has increased over the past 30 years, partly because of the large amount of dead elm wood available after the Dutch elm disease epidemic of the 1970s and 1980s. It is also much the most omnivorous of the three species, having the greatest proportion of vegetable matter in its diet. It is a fairly frequent visitor to gardens and bird tables, especially in early summer, and seems to be particularly attracted by peanuts, which resemble its natural pine-kernel food. Being omnivorous, it is not averse to the occasional young bird and quite commonly raids tit nests.

In the seventeenth century, it was called 'Greater Spotted', to distinguish it from the 'Lesser Spotted', but the more logical comparative word 'greater' has now been supplanted by 'great'. However, it is now sometimes called simply the 'Spotted Woodpecker', in which case, to avoid confusion, the Lesser Spotted should be called the 'Barred Woodpecker'. The two are distinguished most easily by size, although the Great Spotted has a black back with white shoulders – the Lesser Spotted is barred black and white – and is also crimson beneath its tail.

Lesser Spotted Woodpecker *(Dendrocopos minor)*

OTHER COMMON NAMES: *barred woodpecker; crank bird, French magpie (Gloucestershire); French pie; hickwall (Gloucestershire); lesser galley bird (Sussex); little wood pie (Hampshire); tabberer, tapperer (Leicestershire); wood tapper (Shropshire).*

A shy bird of open woodland, copses, parkland, orchards and similar places in the southern half of England, this is the smallest European woodpecker, no bigger than a Greenfinch. (See also Great Spotted Woodpecker above.)

The Great Spotted Woodpecker *(Dendrocopos major)* is widespread and a fairly regular visitor to bird-feeders.

Larks
(Alaudidae)

Mere mention of a bird called a lark immediately brings to mind the Skylark, a species that vies with the Nightingale as the most celebrated of avian voices. There is nonetheless one other British resident lark, the Woodlark, and one winter visitor, the Shore Lark (written as one and two words respectively). They are brown birds in which both sexes are similar and they are characteristic species of open, treeless countryside, where they feed on insects and seeds.

Despite the Skylark's particular fame, all are good songsters, typically singing on the wing, although it is the special glory of the Skylark's delivery, as it rises ever higher, that has always given it a unique appeal. The average length of its performance seems to be about 3 minutes, although I have heard a bird in Derbyshire that must have managed 10; Eric Simms has said that the longest he ever tape-recorded was 6 minutes but full-scale oratorios of an hour or more have been known.[295]

'Lark' is an ancient name, traceable to Old English; it simply means 'little song' and it has been used for other birds quite unrelated to the true larks ('sea lark', for instance, is used for the Dunlin and Ringed Plover).

Woodlark *(Lullula arborea)*

OTHER COMMON NAMES: *shore lark; skylark.*

Woodlarks are smaller than Skylarks, brown above with a flecked breast and in appearance much like a pipit, with which they have often been confused. They have a shorter tail than the Skylark and a much smaller and less obvious crest. Woodlarks also have a much more restricted distribution. They bred in most parts of England and Wales 200 years ago, since when there have been considerable fluctuations, but they are now rapidly expanding their range from the southern quarter of England. They are much shyer, less sociable and less obvious birds than Skylarks, although there has been sufficient confusion for them to be called Skylarks and, in some areas, even Shore Larks.

The Woodlark *(Lullula arborea)* builds a nest of grass and moss on the ground, into which three or four eggs are laid.

However, John Clare knew the difference:

The woodlark rises from the coppice tree
Time after time untired she upward springs
Silent while up then coming down she sings
A pleasant song of varied melody
Repeated often till some sudden check
The sweet toned impulse of her rapture stops.
(John Clare, 'The Woodlark')

So did the nineteenth-century poet Gerard Manley Hopkins, who tried with less than dramatic success to capture its song in verse:

Teevo cheevo cheevio chee:
O where, what can that be?
Weedio-weedio: there again!
So tiny a trickle of song-strain.
(Gerald Manley Hopkins, 'The Woodlark')

Skylark *(Alauda arvensis)*

OTHER COMMON NAMES: *field lark (Surrey); ground lark; lady hen (Shetland); laverock, lavrock (Scotland); learock (Lancashire); lerruck (Orkney); lintwhite (Suffolk); melhuez (Cornwall); rising lark (Northamptonshire); short-heeled lark (Scotland); sky-flapper (Somerset); sky laverock (Northamptonshire).*

As recently as the early 1970s, the Skylark was not only a very common bird but was also being taken seriously as a farm pest because of winter damage to sugar beet, rape and other crops. How circumstances change. Along with the three resident British thrushes, it is now the bird whose populations have most dramatically collapsed. The huge drop in numbers seems to be a response to several changes in farming practice, including a decline in mixed farming and rotational cropping and an increase in autumn sowing. Certainly, by 2000, numbers had fallen to about one-third of the level 30 years previously, and so a generation of people is growing up, even in the country, for whom a song of exquisitely lyrical beauty is now relatively unfamiliar.

Skylarks are slightly larger than Woodlarks and have a more distinctive crest, although this is not evident at a distance. They have been called Skylarks as opposed to simply larks since John Ray, in 1678, referred to them as 'The common or skie-lark',[296] while 'laverock' and similar regional names are merely literary equivalents of the word 'lark'. Like Woodlarks, they nest on the ground, characteristically in the side of a large grass tussock.

Poet after poet has eulogized the Skylark's song. For John Donne, it was 'the Lirique Lark'; for Shakespeare, 'the herald of the morn' (*Romeo and Juliet*, Act 2, Scene 5); and for Wordsworth:

There is madness about thee, and joy divine
In that song of thine;
Lift me, guide me, high and high
To thy banqueting place in the sky.
(William Wordsworth, 'To a Skylark')

Shakespeare referred to the belief that larks and toads exchanged eyes and it was Juliet's wish that they had exchanged voices too, so that the toad's call would not herald Romeo's inevitable departure at daybreak. It is the fact that larks begin to sing early in the morning, that has given rise to the familiar saying 'rising with the lark'. Inevitably, it took a composer to capture it in its full glory: Vaughan Williams' romance 'The Lark Ascending' was inspired by both the bird and George Meredith's poem of the same name; as arranged for violin and orchestra it is as close to evoking real bird-song as anything composed before or since.

The beauty of the Skylark's song has meant that, like so many other singing birds, it has been captured and kept in a cage, sometimes being cruelly blinded in the belief that this would enhance the sound. To satisfy this craving, larks have been caught in various ways, perhaps the oddest being the one to which Shakespeare alluded:

To be thus jaded by a piece of scarlet,
Farewell nobility; let his Grace [Cardinal Wolsey]
go forward
And dare us with his cap like larks.
(*Henry VIII*, Act 3, Scene 2)

The birds were attracted by a mirror, or 'larking-glass', and a small piece of red cloth (like a cardinal's cap) and, when they came to investigate, they were trapped in a net. Larks have also been caught in large numbers by a quite different method: Merlins were used by falconers for hawking larks, large numbers of which were kept in captivity and released for the Merlins to pursue (see Merlin, page 257).

Large numbers of larks, no matter how caught, have also been used to satisfy a centuries-old fondness for lark pie:

Pie of Larks or Sparrows
Time to bake, one hour and a half.
A dozen small birds; a rumpsteak; a small bunch of savoury herbs; the peel of half a lemon; a slice of stale bread; half a cupful of milk; six eggs; pepper and salt; two ounces of butter.
Make a forcemeat with the slice of bread soaked in milk, and beaten up, a small bunch of savoury herbs chopped fine, and the peel of half a lemon minced, a seasoning of pepper and salt, a piece of butter, and the yolks of six eggs; mix all together, put it in a stewpan and stir it over the fire for a few minutes until it becomes very stiff, then fill the inside of each bird. Line a pie-dish with the rumpsteak, seasoned with pepper and salt and fried lightly; place the birds on it, cover them with the yolks of the hard-boiled eggs cut into slices, and pour in sufficient quantity of gravy. Put a paste round the edge of the dish and cover it over, glaze it with the yolk of an egg brushed over it, make a hole in the top, and bake it.[297]

Skylarks *(Alauda arvensis)* were once caged for their song (J.M. Bechstein's *Natural History of Cage Birds*, 1885).

A lark, roasted or boiled 'eased the pain of the collic', although presumably not in parts of Scotland, where anyone eating a lark would receive three curses, one for each of the black spots on its tongue. In Shetland, the lark was a sacred bird and, if it or its nest were disturbed, it would sing, not a beautiful song, but a curse. Quite the most peculiar belief associated with larks, however, must be that from Lincolnshire, where adders were said to spit poison into the air, killing larks stone dead and causing them to fall into the snake's open mouth. Can this have any connection with the saying from nearby Norfolk that, if you want to hear what the lark is saying as it rises heavenwards, you should lie down on your back in a field and listen?

There are a good many inns called The Lark or The Skylark, and The Lamb and Lark is an occasional, although inexplicable, pairing. Skylarks appear in heraldry, often as a pun, for instance in the arms of the Clarke family. But what of the more common use of the name, in the sense of 'frolicking' or 'larking about'? The connection is as obscure as why small Delphiniums should be called 'Larkspurs'.

The Shore Lark *(Eremophila alpestris)* has distinctive black and yellow face markings and two black 'horns'.

Shore Lark *(Eremophila alpestris)*
OTHER COMMON NAMES: *horned lark; sea lark (Yorkshire); snowbird, snowflake (Lancashire).*
Shore Larks are distinctly different from their two relatives, having a white, unspeckled breast, a yellow face with black markings and two small black 'horns' on the head. They are uncommon winter visitors from northern Europe, unknown in Britain before 1830, and very rare breeders. They are seen occasionally on shingle beaches, salt marshes and sometimes adjoining stubble fields, especially along the east coast.

Swallows and Martins
(Hirundinidae)

Uniquely graceful, streamlined and masters of the air, swallows and martins occupy a special place in animal folklore and human affection. They are small birds, characterized by their long, pointed wings, forked tails and wide, gaping mouths that catch insects in flight as they wheel and twist. All three British species are summer visitors throughout the British Isles, their arrival and departure being of especial significance and interest, while the fact that two of them nest typically on buildings and are rather infrequently found on natural structures reinforces their close affinity with man.

'Swallow' is an ancient name, traceable back at least to Middle English and possibly originating in an earlier term for a 'cleft stick', a reference to the forked tail. 'Martin' too is old and has a more complex history, occurring as 'martoune' in about 1450 and variously spelled 'martin', 'marten', 'martinet' and 'martelet'; these are all derived originally from the personal Norman French name 'Martin' at a time when giving human Christian names to animals was a common practice. It has also been applied to several other birds, most notably the superficially similar but unrelated Swift.

The name 'martlet', which has been specifically used for the House Martin, and sometimes the Swift, may be of similar origin but is of particular interest because it survives in heraldry in the form of a bird superficially resembling a Swallow but obviously based on the Swift as it has no legs (see also Swift, page 310).

Sand Martin *(Riparia riparia)*
OTHER COMMON NAMES: *bank martin (Sussex); bank swallow (North Yorkshire – Craven); bitter bank, bitterie (Roxburghshire); pit martin (Yorkshire); river swallow; sand backie (Angus – Forfar); sand swallow; sandy swallow (Stirlingshire, Roxburghshire); shore bird; witchuk (Orkney).*
This is the smallest British swallow, dull brown above and

white beneath, with a brown band across the breast. Sand Martins are highly gregarious and sometimes occur in very large numbers, usually flying and feeding over water and nesting communally in holes in river banks and similar places. In recent times, sand and gravel workings have been the places to see them and they are clearly very efficient at making use of any opportunities with which they are presented.

The name 'Sand Martin' seems to have replaced the earlier 'sand swallow' in the late seventeenth century. It is visually the least distinguished of the swallows and, being the only one that nests away from human habitation, has attracted less attention that its relatives. However, characteristically, John Clare did not miss an opportunity:

Thou hermit, hunter of the lonely glen
And common wild and heath – the desolate face
Of rude, waste landscapes far away from men,
Where frequent quarries give thee dwelling place,
With strangest taste and labour undeterred
Drilling small holes along the quarry's side,
More like the hants of vermin than a bird,
And seldom by the nesting boy descried –
I've seen thee far away from all thy tribe,
Flirting about the unfrequented sky,
And felt a feeling that I cant describe
Of lone seclusion, and a hermit joy
To see thee circle round nor go beyond
That lone heath and its melancholy pond.
(John Clare, 'The Sand-Martin')

Although still common, Sand Martins are among the migrating birds that have suffered because of drought in the areas of Africa south of the Sahara over which they fly.

Swallow *(Hirundo rustica)*

OTHER COMMON NAMES: *barn swallow; chimney swallow (Northumberland, Sussex); house swallow; red-fronted swallow; swallie (Lincolnshire); tsi-kuk (Cornwall).*

'One swallow does not make a summer' is one of the best known of all English proverbs and the Swallow's abnormally long tail-streamers, steel-blue back, white underparts, black head and red chin combine to make it one of the best-known British birds. The saying seems to date from 1539 and a translation of Erasmus's *Adagia*: 'It is not one swalowe that bryngeth in somer. It is not one good qualitie that maketh a man good,' although it is clearly of classical origin and 'One swallow does not make a spring,' in its classical Greek version, appears in the Nicomachæan *Ethics of Aristotle.*[298]

The Swallow's open, cup-shaped mud nest, lined with feathers, is typically attached to the rafters beneath the roof of a house or barn.

You know how, if you come out of sunlight into an unlit room, you stand there blinking for a bit ... well at a previous home, I was always amazed at the way in which a Swallow would fly at 40 mph through the open doorway of our barn, right up to its nest on a beam.[299]

The dates of arrival and departure of migrating Swallows are two of the most notable British sign-posts of the passing of the seasons and there are several regional customs and traditions based around them. In the south of Scotland, for instance, many a gardener will not sow peas until the Swallow has arrived and a Scottish saying, associating Swallows and Cuckoos, runs:

Gang and hear the gowk yell
Sit and see the swallow flee
'Twill be a happy year with thee.[300]

According to many a country sage, once they are here, and if they fly high, fine weather is assured. There may be more truth in this than in much other ornithological weather lore because insects, the Swallow's food, are more likely to be carried aloft on thermals when the weather is warm and settled.

To destroy a Swallow's nest or kill the bird has been widely thought to bring ill luck; in North Yorkshire, it was believed that rain would fall for a month on the crops (not unheard of in that part of the world), and that the cows would run dry or their milk become blooded. Other regions have their own local versions of bad luck: also in Yorkshire, a Swallow falling down a chimney portends death; a Swallow flying under your arm (most unlikely) means that the arm will be paralysed; and a Swallow removing a certain hair from your head (a hair that we all apparently have, although there seems to be no means of identifying it) means that you are doomed to eternal perdition. There is a Norfolk belief that, if a large flock of Swallows gathers around a house, someone will die and their spirit will leave with the birds, while a note from Sheringham more specifically suggests that, if Swallows sit in long rows on the church before departing, they are deciding who is to die before they return. By contrast, having Swallows nesting on your house is good fortune indeed.

Nonetheless, in Ireland, the Swallow is one of the devil's birds and, in parts of Scotland, the Swallow, like the Magpie, is believed to have a drop of devil's blood under its tongue and has at times been persecuted in consequence. It has also been one of the birds credited with talismanic powers, able to find a stone with marvellous properties that can bring sight to the blind and, like other birds with a red breast or red head, it was believed to be a bringer of fire.

The Swallow was among the birds long thought to hibernate or fly to the moon in autumn and even Gilbert White, who realized that some migrated, was unsure whether others remained, hidden in holes. He declared:

I am more and more induced to believe that many of the swallow kind do not depart from this island, but lay themselves up in holes and caverns; and do, insect-like and bat-like, come forth at mild times, and then retire again to their latebrae.[301]

The notion clearly worried him, however, as he also wrote: 'But if hirundines hide in rocks and caverns, how do they, while torpid, avoid being eaten by weasels and other vermin?' Nonetheless, it seems to have been a widely held view that some flew away but others remained. The seventeenth-century poet Thomas Carew, for instance, wrote:

Like swallowes, when your summer's done,
They'le flye and seeke some warmer Sun.
(Thomas Carew, 'To A.L. Persuasions to Love')

He also, in reference to a supposed emergence from hibernation (or even 'temporary' death), said:

But the warm sun thaws the benumb'd earth,
And makes it tender; gives a sacred birth
To the dead swallow.
(Thomas Carew, 'The Spring')

Dr Johnson with a wonderful turn of phrase, neologism and imagery said that Swallows certainly slept all winter and wrote:

The tumbling waters of Swallow Falls, near Betws-y-Coed in Wales, have been likened to the swooping flight of the Swallow.

A number of them conglobulate together by flying round and round and then, all in a heap, throw themselves under water and lie in the bed of a river.[302]

Among the poetic and literary references to the Swallow is one from Thomas Hardy's poem 'Overlooking the River Stour':

The swallows flew in the curves of an eight
Above the river-gleam
In the wet June's last beam:
Like little crossbows animate
The swallows flew in the curves of an eight
Above the river-gleam.

There are inns called The Swallow, there are a handful of places arguably named for an association with Swallows, probably the most likely being Swalcliffe in Oxfordshire ('a cliff where Swallows nested'), and the name has come to be used as an adjective for anything that swoops like the bird in flight (Swallow Falls near Betws-y-Coed).

Swallows have also been used in folk medicine and the following is a particularly depressing recipe from 'How to make my Aunt Markam's Swallow-water' in *Mistress Jane Hussey's Still-Room Book* of 1692:

Take 40 or 50 Swallows when they are ready to fly, bruise them to pieces in a mortar, feathers and all together you should put them alive into the mortar. Add to them one ounce of castorum in powder, put all these in a still with white wine vinegar. Distill it as any other water ... You may give two or three spoonfuls at a time.

The gruesome blend was said to be:

very good for the passion of the heart, for the passion of the mother, for the falling sickness, for sudden sounding fitts ... for the dead palsie, for apoplexies, lethargies and any other distemper of the head. It comforteth the brains ...[303]

House Martin *(Delichon urbica)*

OTHER COMMON NAMES: *black martin (Northumberland); easin swallow (North Yorkshire – Craven); eaves swallow; martin swallow (East Lothian); martlet; river swallow (Yorkshire); swallow (Roxborough, West Yorkshire); window martin; window swallow (Northumberland).*

The House Martin is a conspicuous bird, black above, with white underparts and a white rump. Rather like Sand Martins, they have a less swooping flight than Swallows and tend to be the higher fliers. Max Nicholson recounted how tolerant they are of heat after watching about a hundred young birds in the full August sun on a rooftop where the temperature was around 27°C (80°F).[304]

Few birds are so closely associated with human habitation, their nests under house eaves being neat, enclosed structures

Swallows *(Hirundo rustica)*, with forked tails, and House Martins *(Delichon urbica)* frequently rest on telegraph wires.

with an entrance hole at the top. Shakespeare knew the House Martin and its nesting habits well:

> This guest of summer
> The temple-haunting martlet, does approve
> By his lov'd mansionry, that the heaven's breath
> Smells wooingly here; no jutty, frieze,
> Buttress, nor coign of vantage, but this bird
> Hath made his pendent bed and procreant cradle:
> Where they most breed and haunt, I have observed
> The air is delicate.
> (*Macbeth*, Act 1, Scene 6)

As with Swallows, it is deemed very good luck if House Martins nest on your house and very bad luck if you harm them. C.I. Paton, writing in 1938, said:

> One of my brothers was at Shelfanger in Norfolk, and the house of a local farmer was being painted. The farmer's wife wanted some house-martins' nests removed, as they were soiling the window-sills, etc. The man employed in painting, however, said that his master had warned him not to injure the nests as it was very unlucky, and he had known a man break his leg after having done so.[305]

Pipits and Wagtails
(Motacillidae)

These small, mainly ground-living, relatively long-legged birds feed chiefly on insects and seeds. Pipits are more or less brown with fleck-like markings but are difficult to distinguish from each other. Wagtails are larger but more slender, generally with striking patterns or colours and long tails. There are two British residents and one summer visitor in each group. The name 'pipit' is imitative of the call but, until the end of the eighteenth century, all pipits were believed to be types of lark. By the early nineteenth century, however, it had been decided that there were three distinct British species, named approximately according to their habitat, although the Tree Pipit was more specifically named because it almost invariably lands on a tree before descending to the ground. The difficult of distinguishing one pipit from another was well illustrated by a note some years ago in the *Bulletin of the British Museum*:

> It's a pity pipits have
> No diagnostic features
> Specifically they are the least
> Distinctive of God's creatures.[306]

Tree Pipit *(Anthus trivialis)*

OTHER COMMON NAMES: *blood lark (Cheshire); field lark; lesser field lark; pipit lark; short-heeled field lark (Scotland); titman (Sussex); tree lark (Yorkshire, Nottinghamshire, Somerset); tree pipit; wood lark (southern Scotland, north of England, Cheshire).*

The Tree Pipit is the only non-resident British pipit and it typically sings while descending to its perch after a short flight. It is a bird that has spread to take advantage of the increase in plantation forests during the twentieth century, although it is subject to considerable fluctuations in numbers for no obvious reason.

Meadow Pipit *(Anthus pratensis)*

OTHER COMMON NAMES: *banks teetick (Sheltland – Bressay); butty lark (Hampshire); cheeper; chitty (Lancashire); cuckoo's sandie; cuckoo's titling (Durham); earth titling (East Lothian); field titling; gowk's fool (north of England); ground lark (Yorkshire); heather lintie (Cumberland, Westmorland); hill sparrow (Shetland, Orkney); hill teetick (Shetland); ling bird (Cumberland, West Yorkshire); meadow lark (Hampshire); meadow titling; moor tit; moor titling (North Yorkshire – Craven); moss cheeper (Scotland); moss cheepuck (Northern Ireland); peep (Angus – Forfar); teetan (Orkney); tietick (Shetland); tit; titlark; titling; wekeen (Kerry).*

The Cuckoo often lays its eggs in the nest of the Meadow Pipit, hence the references to 'Cuckoo' in the regional names for this bird. An old Irish tale conjures up an image of a tiny parent pipit feeding a huge nestling Cuckoo and says that, although the pipit is constantly trying to enter the Cuckoo's beak, if it should ever succeed, the world will end. The Hampshire name 'butty lark' means 'companion bird' and similarly refers to the association with the Cuckoo, as does the Welsh *gwas y gog*, meaning 'Cuckoo's knave'.

Many of the Meadow Pipit's *(Anthus pratensis)* common names reflect its association with the Cuckoo.

That notable politician and amateur ornithologist Viscount Grey of Falloden (1862–1933) found much pleasure in the Meadow Pipit's modest, repetitious, trilling song, which he thought made 'a minute but imperceptible contribution to the happiness of the day'.[307] Fortunately, Meadow Pipits are still very common birds on the uplands that Lord Grey knew so well.

Rock Pipit *(Anthus petrosus)*

OTHER COMMON NAMES: *dusky lark; gutter teetan; pipit lark; rock lintie (Aberdeenshire); rock pipit; sea lark (East Yorkshire); sea lintie (Scotland); sea mouse; sea titling; shore teetan (Orkney); tangle sparrow (Shetland, Orkney); teetan, teetuck (Shetland).*

The Rock Pipit is a widespread and still fairly common resident, most readily recognized because of the rocky-coast habitat where it breeds. Eric Simms, who wrote the definitive book on the family, described watching them on the towering rocky cliffs of Buchan, north of Aberdeen:

> Through the roar of the breaking sea and the cacophony from the sea-bird city I hear the plaintive 'feest' of a Rock Pipit followed shortly by a song like that of a Meadow Pipit but harsher, more metallic and capable of penetrating the noise around me. A small bird rises strongly above the cliff-face, singing as it climbs and then descending at an angle before dropping silently down to the rocks on the shore. A hundred metres away another mounts up and parachutes down to a projecting rock buttress halfway down the cliff-face. This then is the world of the Rock Pipit.[308]

Yellow Wagtail *(Motacilla flava)*

OTHER COMMON NAMES: *barley bird (Nottinghamshire, Surrey, Sussex); barley-seed bird (Yorkshire); cow-bird (Sussex); cow-klit; dishwasher (Sussex); golden dishwasher (Somerset); gypsy bird; ladybird, maw-daw (Sussex); oat-ear; oatseed bird; potato dropper (Cheshire); potato setter; quaketail; spring wagtail (Yorkshire); summer wagtail; sunshine bird; tater setter; waterie wagtail (Aberdeenshire); water wagtail; yaller wagster (Cheshire); yellow Molly (Hampshire).*

The Yellow Wagtail is a summer visitor to England and Wales and is biologically a complex species comprising several different, rather distinctive races. The race seen in Britain is lemon-yellow beneath and olive-yellow above, with an olive-yellow head. Elsewhere in Europe, the head colour varies and some forms are distinctly blue. Like other wagtails, it is usually seen along fast-flowing streams, where it feeds on insects and other small aquatic invertebrates and nests in holes in the banks and similar places. The decline in wet, marshy habitats has resulted in a drop in its numbers.

A summer visitor, the Yellow Wagtail *(Motacilla flava)* is a shy, slender bird, usually seen beside fast-flowing streams.

'Wagtail' (as 'wag tayle') replaced the older name 'wag-start' around the beginning of the sixteenth century and all three species are commonly referred to as 'water wagtails'. 'Cow-bird' and similar names refer to the fact that the Yellow Wagtail commonly feeds on insects associated with cattle. Max Nicholson listed it alongside Jackdaws and Starlings as a bird with a particular dependence on grazing animals.[309] References to oats, barley, potatoes, and indeed sunshine, all allude to seasonal happenings when it arrives in the spring. 'Potato dropper' and 'potato setter', for instance, indicate that the first sighting of the bird served as a sign to farmers and gardeners that potatoes were due to be planted. This belief was frequently heard in Cheshire, where the bird is particularly common, and where the name 'yaller wagster' is probably a survival of the old 'wag-start'. Cheshire farmers also turned over the money in their pockets on seeing the bird in the hope that the Yellow Wagtail would change it to gold.

Grey Wagtail *(Motacilla cinerea)*

OTHER COMMON NAMES: *barley bird (general); barley-seed bird (Yorkshire); dishwasher, dun wagtail (Sussex); gypsy bird; oatseed bird (Yorkshire); water wagtail; winter wagtail (southern England); yellow dishwasher (Somerset); yellow wagtail (Ireland, Somerset, Sussex).*

The name 'Grey Wagtail' implies a dull-coloured bird, although it is only grey above, being as yellow as the Yellow Wagtail beneath; it also has a strikingly long black tail. It is an increasingly common resident throughout much of the British Isles and is most frequently seen in the usual stream-side habitat favoured by wagtails.

Pied Wagtail *(Motacilla alba)*

OTHER COMMON NAMES: *devil's bird; deviling (Ireland); dish-dasher (Somerset); dishlick (Sussex); dishwasher (Shropshire, Gloucestershire, Somerset, southeast England); dishwipe; ditchwatcher (Surrey); gypsy bird; lady dishwasher (Gloucestershire); lady wagtail (Somerset); Molly washdish (East Anglia, Somerset, Dorset); nannie wagtail (Nottinghamshire); nanny washtail; pegy wagtail; Peggy wash-dish; piedie wagtail (Cheshire); quaketail; scullery maids (Wiltshire); seed bird (Yorkshire); seed lady (Peebles); waggie (East Lothian); wagtail; washdish; washerwoman; washtail; water wagtail; watitty (Cheshire); wattertiwagtail (Cumberland); wattie; wattie wagtail (Westmorland); white wagtail; Willie wagtail (Orkney).*

The following poem must describe a Pied Wagtail, the commonest of the three species in Britain:

> Little trotty wagtail, he went in the rain
> And, tittering tottering sideways, he ne'er got straight again.
> He stooped to get a worm and look'd up to catch a fly
> And then he flew away ere his feathers they were dry.
>
> Little trotty wagtail, he waddled in the mud
> And left his little foot-marks, trample where he would.
> He waddled in the water pudge and waggle went his tail
> And chirrupt up his wings to dry upon the garden rail.
>
> Little trotty wagtail, you nimble all about
> And in the dimpling water pudge you waddle in and out.
> Your home is nigh at hand and in the warm pigsty,
> So little Master Wagtail I'll bid you a 'Good bye'.
> (John Clare, 'Little Trotty Wagtail')

This wagtail tends to occur frequently away from water and is the species most likely to be seen in parks and gardens. In common with the other wagtails, it is sometimes called the 'gypsy bird' because of an old belief, prevalent in Warwickshire and probably other places, that, if you see a wagtail, a gypsy will be somewhere near. The Romany people themselves have a saying that, if a wagtail is seen while they are travelling along a road, this is a fortunate sign; if it stays in the road as they approach, they will meet strangers; if it flies away, they will encounter relatives.

In many districts, a wagtail tapping on a window foretells death, just as it does with Robins and some other birds. The belief is particularly strong in Ireland, where it is considered a bird of serious ill-omen; the devil's bird indeed. In Cornwall, however, a window-tapping wagtail merely tells that a stranger will visit, an interesting echo of the Romany association of the bird with strangers. A belief from Dartmoor is

The Pied Wagtail *(Motacilla alba)* is thought to resemble the washerwomen of yesteryear.

that no cat will touch a Pied Wagtail, which suggests that it must therefore possess unnatural powers.

People are often puzzled by such names as 'dishwasher', 'washerwoman', 'Peggy wagtail' and 'washdish' that are sometimes used for all three wagtail species, although especially the Pied. The explanation generally offered is that these birds, in their black-and-white garb, are like washerwomen of old, and their bobbing jerking actions imitate those of a woman scrubbing clothes or washing dishes.

Waxwings
(Bombycillidae)

Waxwing *(Bombycilla garrulus)*
This small, crested bird is predominantly chestnut brown, with a paler, pink-brown breast, a grey rump, and dark wings with yellow-and-white patterning and wax-like, bright red tips to the feathers. A beautiful and mainly Arctic species, it is a fairly regular winter visitor, especially to the eastern and northern parts of Britain, where the sometimes very large flocks attract immediate attention by their acrobatic feeding on ornamental fruit trees in parks and gardens.

> One winter's day I looked out and rushed around until I found my glasses. Waxwings! Like painted toy birds they swarmed over crab apples, cotoneasters and ivy, at the back of the houses and in the road. I rang everybody I could think of and nobody was in! No-one to show or tell![310]

Despite the excitement they engender, their appearance in large numbers was once believed to be a sure sign of profound problems in the form of very bad weather, war and pestilence. 'Waxwing' is a fairly recent but appropriate description of a bird that has acquired no other English names. The specific name *garrula* is singularly inappropriate, however, as it is a remarkably silent bird.

Dippers
(Cinclidae)

Dipper *(Cinclus cinclus)*
OTHER COMMON NAMES: *Bessie ducker; Bobby; brook ouzel; colley (Cheshire); ess cock (Aberdeenshire); piet (Scotland); river pie (Ireland); water blackbird (Ireland; Scotland); water colly (Somerset); water crow (Scottish Lowlands, Westmorland); water ouzel; water peggie (Dumfries-shire); water piet (Scotland); water thrush (Cornwall); white-throated dipper.*
One of the many pleasures of stream-fishing, even when the fish are not biting, is to watch a Dipper as it busies itself about its extraordinary life style. It is a common resident, the only British member of the most aquatic family of all perching birds and the only one regularly to submerge itself when searching for the aquatic invertebrates on which it feeds.

> I'm always astounded that this small bird can stand on a glass-smooth stone while water pours over and around it.[311]

For many years, there was great discussion about the possibility of Dippers literally flying under water and many people only believed it when it was proved on film. Dippers are associated with streams throughout much of the British Isles, although they are absent from the south and east of England and occur in fewer numbers overall than they once did; perhaps acid rain has played a part in reducing the aquatic life on which they feed. Their habitat and habit, bobbing up and down on a rock before diving, are unique enough, but these little Wren-like birds are also distinguished by a vivid white breast and dark brown back.

The bird has been called the Dipper since 1678, although the name has been used generally since 1388 for birds that dip into water; early references may apply to either this species or the Dabchick. In some places too, it has been confused, at least in name, with the Kingfisher, and it has occasionally been confused with the female of that brilliantly coloured and very different species. Names including 'pie', or 'piet', refer to its black-and-white appearance (see also Magpie, page 358). In Gaelic, its names mean 'blacksmith', presumably in reference to its dark-coloured back.

Wrens
(Troglodytidae)

Wren *(Troglodytes troglodytes)*

OTHER COMMON NAMES: *Bobby wren (Norfolk, Suffolk); chitty (northern England); chitty wren (Ireland); crackadee; cracket; crackeys, crackil (Devon); cuddy (Somerset); cut (north and west England); cutteley wren; cutty; gilliver wren (Lincolnshire); guradnan (Cornwall); Jenny (Yorkshire, Lancashire, Lincolnshire); Jenny wren (general); jitty (Cheshire); juggy wren (Surrey); Kitty-me-wren (Scotland, northern England); Kitty wren; Our Lady's hen (Scotland); puffy wren; puggie wren (Surrey); Sally (Ireland); scutty (Somerset, Hampshire, Sussex); skiddy (Gloucestershire); stag (Norfolk, Cornwall); stumpit (Lancashire); stumpy toddy (Cheshire); tiddy wren, tidley wren (Essex); tintie (Nottinghamshire); titmeg; titty todger (Devon); titty wren (Suffolk, Gloucestershire, Wiltshire); tope; tom tit (North Yorkshire – Craven, Norfolk, Suffolk,); two fingers (Cheshire); wirann (Orkney); wran (Ireland, Scotland); wrannock; wranny (Cornwall).*

The name 'Wren' comes from the Middle English 'wrenne' and means 'a little tail'. Almost all additions and variants ('titty', 'chitty', 'puggy', 'two fingers' and so forth) are references to the bird's smallness. Those whose memories extend back before 1961 will remember the farthing, which was equal to one-quarter of an old English penny and the smallest coin then in circulation; a Wren featured on its reverse.

The Wren is a familiar bird in gardens, woods, hedges – indeed almost anywhere that offers the dense cover that it needs, and, in recent years, it has become the commonest resident bird in Britain, nesting in all manner of unlikely places.

> A Wren nested for several years running in Challock Church, Kent, on a shelf on the lectern beneath the wings of the great brass eagle. Quite recently, during the course of a funeral hymn, a mummified Wren was blown out of one of the organ pipes, landing in the lap of the squeaking lady organist who thought it was a mouse. The little bird was perfectly shaped and was used as the model in a painting.[312]

The male builds several nests and the one that the female chooses is then lined with feathers and used.

The Wren is brown above with a speckled brown breast and a short, cocked tail and, in the way it works over the bushes, picking off aphids, may be likened to an avian vacuum cleaner. Perhaps it is its size, its relative secretiveness or simply its familiarity that has attracted such profound and deep passions. No-one seems to know, but it has featured abundantly in folklore and legend since before recorded history. An exquisitely beautiful and accurate painting of a Wren appears on the Sherborne Missal of *c.*1400.

A Wren *(Troglodytes troglodytes)* features on the reverse side of a farthing.

Wren folklore centres on the extraordinary ritual of Hunting the Wren. It was most comprehensively reviewed by Edward Armstrong, 'the last in England's line of distinguished parson-naturalists', who wrote the definitive scientific monograph on it.[313] He called the Wren Hunt 'among the most elaborate bird rituals surviving in Europe' and it is one that has always been important in the British Isles.

In essence, the ceremony involves boys and sometimes (especially in Ireland) young men, known as 'Wren Boys', often dressed in curious costume, who catch a Wren, kill it or place it alive in a small decorated 'Wren House', and then carry it in procession. The event usually took place on St Stephen's Day (26 December), although it also occurred on New Year's Day and at other times during the winter. In Wales, it was generally associated with Twelfth Night. There are few accounts from Scotland, but it has been recorded from many parts of England and Wales and has always been particularly strong in Ireland and the Isle of Man: 'In the Isle of Man ... the tradition is alive and well ... Peel is one of the last bastions of 'Hun'tha wran'.[314] Songs may accompany the ritual.

> In Co. Kerry where we live the Wren boys still turn out on St Stephen's Day purportedly gathering money to bury the poor bird. Thirty years ago they still dressed in the traditional costume of straw woven into a cape and covering the head: a very frightening sight at the door for two English people unaware of the custom. Nowadays any kind of costume seems acceptable and face paints add to the colour. Mostly they give value for the money they collect, singing and playing musical instruments. This is particularly true in Co. Tipperary where they call at houses and all the pubs in the area. A famous song written in the twentieth century by a Kerryman, Sigerson Clifford, is called The Boys of Barna Sraide recalling boyhood days and each verse ending 'when the boys of Barna Sraide went hunting for the wran'.[315]

Iona and Peter Opie quote the following, recited by 'a 10-year old urchin in the heart of Dublin' in the mid-twentieth century:

> The wran, the wran, the king of all birds,
> On Stephen's day was caught in the furze.
> We chased her up, we chased her down,
> Till one of our little boys knocked her down.
> We drowned her in a barrel of beer,
> A happy Christmas and a merry New Year.
> Up with the kettles and down with the pan,
> A penny or twopence to bury the wran.[316]

Traditionally the bird was killed with a bow and arrow and sometimes its corpse was dismembered and its feathers put to various uses. In some places, as in parts of Wales, the whole ceremony was associated with weddings and marriage, the homes of recently married couples being visited by the procession. In recent times, both public opinion and the Law have protected birds from cruelty and killing, and artificial Wrens have been substituted. In 1999, Conor O'Brian, who was a Wren Boy in the early 1940s in west Cork, and anxious to dispel the notion that any harm befell the birds, said in *The Countryman* that the purpose could be satisfied by 'a ball of wool, plucked from a friendly sheep and suitably dyed'.[317]

This is a much simplified summary of a complex and varied ritual. In parts of Ireland in particular, the Wren Hunt was inextricably associated with the traditional Folk Play, in which horns and animal-skin costumes were worn by the male characters, and other odd clothes by men masquerading as women. In some areas, what Edward Armstrong sensitively describes as 'pantomimic obscenities' and representations of sexual congress were involved. Similarly complex ceremonies existed in other parts of Europe, the whole merging imperceptibly with the Morris Dance, the Sword Dance and ancient Dionysiac ritual.

The custom survived in many parts of England and Wales until well into the twentieth century and is clearly still performed on the Isle of Man. Its origin is very old and Armstrong examined this in some depth. Parallels are found in the rituals of other cultures and with other birds in many parts of the world. In Edward Armstrong's own summary:

> We may conclude that the Wren Cult reached the British Isles during the Bronze Age and was carried by megalith builders whose cultural inspiration came from the Mediterranean region. Probably these folk cherished mainly solar magico-religious beliefs. The Wren Hunt represents New Year ceremonial having as its purpose the defeat of the dark earth-powers and identification with the hoped-for triumph of light and life.[318]

The Wren Bush and the Wren Boys parade in the Irish countryside on St Stephen's Day (*Illustrated London News*, 1850).

The female Wren *(Troglodytes troglodytes)* chooses one nest out of several built by the male and lays five to eight eggs.

Why the Wren? This is difficult to answer because other birds feature in what are clearly related ceremonies elsewhere. Perhaps we need look no further than the fact that it has always attracted attention and curiosity because of its diminutive size.

Despite such ceremonial persecution of the Wren, it has been thought in many places ill luck to hurt one and, certainly, it seems to have been used little if at all as a source of folk medicine. It was thought to be the bird that brought water to the island of Guernsey and, for that reason, was sacred and not to be harmed.

On another island, however, Wrens have suffered greatly. Several distinct island races have been described by ornithologists, the most celebrated being the St Kilda Wren *(Troglodytes troglodytes hirtensis)*, which lives on the island cliffs with Fulmars and Puffins. This proved to be almost its undoing; it was persecuted mercilessly by egg- and skin-collectors wanting to add this rarest of creatures to their collections until the Wild Birds Protection (St Kilda) Act of 1904 came to its rescue as it did for Leach's Petrel.

In some areas, the sight of Wrens congregating presaged bad weather, but they generally seem to feature little in weather lore. Throughout the centuries, the Wren has often been associated with that other familiar small bird, the Robin, and the very odd notion that one is married to the other is discussed elsewhere (see Robin, page 330).

Accentors/Dunnocks
(Prunellidae)

Dunnock *(Prunella modularis)*

OTHER COMMON NAMES: *Billy (Oxfordshire); black wren (Ireland); blind dunnock; blue dickie (Renfrewshire); blue dunnock; blue Isaac (Gloucestershire); blue jannie; blue jig; blue sparrow (Scotland); blue tom; bush sparrow (Stirlingshire); creepie (Scotland); cuddy (North Yorkshire – Craven); cudgie (Nottinghamshire); dickie, doney (Lancashire); dyke sparrow; dykesmowler; dykie; fieldie; field sparrow (Roxburghshire); foolish sparrow; (Worcestershire) hedge accentor; hedge Betsy (Suffolk); hedge Betty (Warwickshire); hedge chanter; hedge creeper (Yorkshire); hedge chat (Northamptonshire); hedge dunny (Lancashire); hedge mike (Sussex); hedge scrubber (Suffolk); hedge spick; hedge spurgie (Aberdeenshire); hedge warbler; hempie (Yorkshire, Scotland); Isaac, hazock (Worcestershire); philip; phip; pinnock; reefouge (Ireland); segge (Devon); shufflewing (North Yorkshire – Craven); smokey (Northumberland); sparve (Cornwall); titlene; titling (northern England); whin sparrow (East Lothian); winter fauvette.*

The Dunnock is among the commonest small birds in the British Isles, although, like so many others, it is now rather less common, perhaps because of the loss of hedgerows. It is a pretty if undistinguished resident with a brown back and grey head:

> The tame hedge sparrow in its russet dress
> Is half a robin for its gentle ways.
> (John Clare, 'The Hedge Sparrow')

It is more slender than a true sparrow and is a common garden visitor, at least in Britain; elsewhere in Europe, it tends to be a woodland bird. It nests everywhere in low bushes and hedges, hence the constant reference to hedges in its many regional names; indeed, it is probably generally still better known as the 'hedge sparrow'. 'Dunnock' is the older name, however, occurring as 'donek' in the second half of the fifteenth century, and simply means 'dun [grey-brown] bird'. The name 'hedge sparrow' was not adopted until 1570.

Dunnocks are probably best known as being the most favoured host for Cuckoo eggs. So tolerant are they of this intrusion that their benign attitude has found expression in such names as 'foolish sparrow' and 'blind dunnock'. To Chaucer, the Dunnock was the 'heysoge' and he took a dim view of the Cuckoo parasitizing it:

> Thow mortherere of the heysoge on the braunche
> That broughte the forth, thow reufullest glotoun!
> Lyve thow soleyn, wormes corupcioun,
> For no fors is of lak of thy nature!
> Go, lewed be thow whil the world may dure!
> (Chaucer, *The Parlement of Foules*)

Thrushes and Related Birds
(Turdidae)

The thrushes are a very large family of perching birds that embraces some of the commonest and best-loved British species, including some with an extensive folklore. These can be conveniently subdivided into the enormous and mainly exotic group of robins and chats (the Robin, Nightingale, Black Redstart, Redstart, Whinchat, Stonechat and Wheatear) and the true thrushes (the Ring Ouzel, Blackbird, Fieldfare, Song Thrush, Redwing and Mistle Thrush).

Robin *(Erithacus rubecula)*

OTHER COMMON NAMES: *Bob (Nottinghamshire); Bob robin (Stirlingshire); Bobby (Suffolk, Hampshire – New Forest); ploughman's bird (Yorkshire); redbreast; reddock (Dorset); robinet (West Yorkshire, Lancashire, Derbyshire); robin redbreast; robin ruck, robin ruddock (northern England); ruddock (northern England); Thames gierdet; tommy-liden.*

No British bird is less in need of an introduction than the Robin and perhaps the only real surprise is that it should be listed here, for few people realize that it is a close relative of the thrushes. Many years ago, when *The Times* polled its readers for the title of 'national bird', the Robin won easily. It is familiar mainly through living contentedly in close proximity to human beings, often nesting in garden sheds, cars, old saucepans, and even an unmade bed while the owners were having breakfast (recounted by Max Nicholson).[319] According to Richard Kearton, it also nested 'in the hole made by a cannon ball through the mizzen mast against which Lord Nelson was standing when he received his death wound on HMS *Victory*', although, good bird man and pioneer photographer as Kearton was, this might be mere fancy because the plaque on the deck saying 'Here Nelson Fell' is nowhere near the mizzen mast.[320]

When a gardener begins to dig, a Robin *(Erithacus rubecula)* will often appear in search of food turned up by the spade.

Robins have endeared themselves in many other ways as well, particularly through being such regular and constant visitors to bird tables and perhaps, above all, through being especially evident in winter. Few gardeners will be winter-digging for long before a Robin appears, investigating the livestock turned up with the spade.

> I recall once, as his general practitioner, visiting a retired farm worker, aged 90 years, in a Gloucestershire village. It was a freezing January day, with the ground solid and rock-like. I eventually found the patient in his garden wielding a pick-axe: he was turning the soil so that 'my Robin can find a worm'.[321]

The Robin has also become as much a popular symbol of Christmas as snow, stagecoaches and Norway spruce.

> The north wind doth blow,
> And we shall have snow,
> And what will poor Robin do then?
> Poor thing![322]

The Romany people believe that the Robin feels the cold in winter more than any other bird and this is why it seeks shelter in wagons and houses in search of food.

> When the snow is on the ground,
> Little Robin Red-breast grieves;
> For no berries can be found,
> And on the trees there are no leaves.
>
> The air is cold, the worms are hid,
> For this poor bird what can be done?
> We'll strew him here some crumbs of bread,
> And then he'll live till the snow is gone.[323]

Robins are also uniquely conspicuous; no other British bird of comparable size has such an obvious red breast and this has endeared them to the extent that there is 'hardly a corner of the world in which the English have not managed to find

The single Robin on this nineteenth-century Christmas card demonstrates the bird's territorial nature.

some red-breasted bird which they could call a robin'.[324] Unusually coloured Robins have naturally enough attracted attention and white or yellowish birds, although still with red breasts, have been reported. The sexes are indistinguishable, which is not in itself particularly startling, although it is rare for the female of any bird species to share such bright coloration with the male.

Common residents throughout the British Isles, Robins have inspired a vast literature. They have also attracted some classic ornithological investigation; David Lack's seminal work, *The Life of the Robin*, published in 1943, set the standard for modern bird study. They are not the sweet creatures that might be imagined, however.

They are extremely aggressive and territorial and their appeal certainly does not lie in their being seen in flocks. David Lack himself commented on the popular Christmas cards that show 'four or even more robins perching happily together on a holly branch', suggesting that 'no more inappropriate symbol could be devised for the season of peace and goodwill. Should the depicted incident occur in nature, furious conflicts would arise'.[325]

The Robin was originally known as the 'ruddock' ('the ruddock ... with charitable bill'; *Cymbeline*, Act 4, Scene 2), traceable to Old English and from the same source as 'ruddy', meaning 'red'. 'Redbreast' arose around the end of the fourteenth century and 'Robin', the familiar and popular form of the Christian name Robert (which both originated from Norman French), was appended to it in the mid-sixteenth century. The expression 'cock robin', which occurs far more frequently in written language than in speech, is no more than 300 years old and is comparable with 'cock sparrow' and other terms of familiarity. For many people, however, the bird is still 'robin redbreast' and, although the redbreast part had been largely dropped by ornithologists by the late nineteenth century, it was still almost invariably used by Wordsworth, who included it in numerous poems:

> Art though the bird whom Man loves best
> The pious bird with the scarlet breast.
> (Wordsworth, 'The Redbreast Chasing the Butterfly')

Perhaps the oddest, least explicable and most widespread of beliefs is that the Robin is married to the Wren:

> The robin and the wren
> Are God Almighty's cock and hen.

This ancient rhyming couplet, of unknown origin, occurs in a number of versions and is echoed many times in popular literature and verse, most famously in 'The Happy Courtship, Merry Marriage and Pic-nic Dinner, of Cock Robin and Jenny Wren', published in 1806, of which the best-known and most frequently quoted lines run:

> Cock Robin got up early
> At the break of day,
> And went to Jenny's window
> To sing a roundelay.
> He sang Cock Robin's love
> To little Jenny Wren,
> And when he got unto the end
> Then he began again.

Modern writers have also drawn on the myth:

> When Robin's not a beggar,
> And Jenny Wren's a bride ...
> (Christina Rossetti, 'Summer')

Apart from the marriage story, tradition renders Robins and Wrens almost inseparable. Thomas Evans' poem, published a few years earlier than 'The Happy Courtship', suggests a slightly different relationship:

***The Death of Cock Robin* by John Anster Fitzgerald (1832–1906).**

Jenny Wren fell sick
Upon a merry time,
In came Robin Redbreast
And brought her sops and wine.

Eat well of the sop, Jenny,
Drink well of the wine.
Thank you, Robin, kindly,
You shall be mine.

Jenny Wren got well,
And stood upon her feet;
And told Robin plainly,
She loved him not a bit.

Robin he got angry,
And hopped upon a twig,
Saying, Out upon you, fie upon you,
Bold faced jig!
(Thomas Evans, 'The Life and Death of Jenny Wren')

An old Essex verse runs:

The Robin and the Redbreast,
The Robin and the Wren,
If ye take out of the nest
Ye'll never thrive again.[326]

An old Cornish saying is:

Hunt a robin or a wran,
Never prosper, boy nor man.[327]

Even rhymes about spiders endorse the association:

If 'twere not for the robin and the wran
A spider would overcome a man.[328]

Robins are also associated with the ancient and mysterious ritual of the Wren Hunt (see Wren, page 327), occasionally serving as substitute Wrens.

The most famous of all Robin rhymes, however, is also perhaps the most mysterious. 'An Elegy on the Death and Burial of Cock Robin' was published in its now familiar form *c.*1744 and may be a complex political commentary relating to the downfall of Robert Walpole's Government in 1742. There are grounds for thinking that the basis of this rhyme is much older as there are comparable ancient rhymes existing elsewhere in Europe, and it has been claimed to be the story depicted in a fifteenth-century stained glass window in the rectory at Buckland in Gloucestershire.

Who killed Cock Robin?
I said the Sparrow,
With my bow and arrow,
I killed Cock Robin.

Who saw him die?
I said the Fly,
With my little eye,
I saw him die.

Who caught his blood?
I, said the Fish,
With my little dish,
I caught his blood.

Who'll make his shroud?
I, said the Beetle,
With my thread and needle,
I'll make the shroud.

Who'll dig his grave?
I, said the Owl,
With my pick and shovel,
I''ll dig his grave.

Who'll be the parson?
I, said the Rook
With my little book,
I'll be the parson.

Who'll be the clerk?
I said the Lark,
If it's not in the dark,
I'll be the clerk.

Who'll carry the link?
I, said the Linnet,
I'll fetch it in a minute,
I'll carry the link.

Who'll be chief mourner?
I, said the Dove,
I mourn for my love,
I'll be chief mourner.

Who'll carry the coffin?
I, said the Kite,
If it's not through the night,
I'll carry the coffin.

Who'll bear the pall?
We, said the Wren,
Both the cock and the hen,
We'll bear the pall.

Who'll sing a psalm?
I, said the Thrush,
As she sat on a bush,
I'll sing a psalm.

Who'll toll the bell?
I, said the Bull,
Because I can pull,
So Cock Robin, farewell.

All the birds of the air
Fell a-sighing and a-sobbing,
When they heard the bell toll
For poor Cock Robin.

Children through the ages have loved Robins, pictures of Robins and songs, stories and rhymes about Robins. A typical child's Robin song dating from *c.*1800 runs:

Little Robin Redbreast sat upon a tree,
Up went pussy cat, and down went he;
Down came pussy, and away Robin ran;
Says little Robin Redbreast, Catch me if you can.
Little Robin Redbreast jumped upon a wall,
Pussy cat jumped after him, and almost got a fall;
Little Robin chirped and sang, and what did pussy say?
Pussy cat said, Mew, and Robin jumped away.[329]

There are also several versions (some decidedly rude) of a popular eighteenth-century rhyme:

Little Robin Redbreast
Sat upon a rail;
Niddle noddle went his head,
Wiggle waggle went his tail.[330]

Another children's story, or at least a story about children, reveals another side to Robin folklore. The tale once known simply as *The Children*, but now generally called *Babes in the Wood*, is encountered most often today as a pantomime, but it was first published as a melodrama in 1601 and then appeared as a ballad in Thomas Percy's *Reliques of Ancient English Poetry* of 1765.

In this remarkably gruesome story, two children are abandoned in a wood overnight to die and Robin Redbreast covers their bodies with leaves. This tale has resonances in other references to Robins caring for the dead and the belief is found in at least three significant sixteenth-century literary works. The dramatist John Webster's tragedy *The White Devil* (*c.*1608) includes the sombre request:

> Call for the robin-red-breast and the wren,
> Since o'er shady groves they hover,
> and with leaves and flowers do cover
> The friendless bodies of unburied men.

Similarly, in Thomas Lupton's *A Thousand Notable Things of Sundrie Sortes* (1579):

> A Robbyn red breast, fynding the dead body of a man or woman, wyll cover the face of the same with mosse. And as some holdes opinion, he wyll cover also the whole body.[331]

And in 'The Owl' (1604), Michael Drayton wrote:

> Covering with moss the dead's unclosed eye
> The little redbreast teacheth charity.

In *Babes in the Wood*, illustrated here by Randolph Caldecott, two children left in the woods are covered with leaves by Robins.

However, there is another, strange and even more sinister side to Robin folklore. Robins have long been associated not just with caring for the dead but with the foretelling of sadness or death.

> A Robin or a butterfly in the house was considered unlucky. This has been passed down in our family living in this village since 1730.[332]

While there was, and still is, a widespread belief that it is unlucky for a Robin to enter a house, in other places, such an event presages death.

> It was unlucky for a Robin to enter the house, someone would die.[333]

> In the Yorkshire Dales, it is still half-believed that a Robin entering the house foretells a death'.[334]

> I have heard that a Robin in the house foretold death or bad luck.[335]

> A neighbour told us once, when we had found a Robin's nest by a house wall, that it was unlucky and there would be a death in the family.[336]

> The Robin being a sign of death is very much prevalent in Suffolk. Both my mother and mother-in-law were very scared of a Robin even approaching the back yard and if one came into the house it was a foregone conclusion that death was near. In fact all birds in the house were deemed unlucky.[337]

> Robins entering the house have always portended woe if not death. A recent confirmation of this was when my sister-in-law found one in her bedroom the day my niece nearly died. The same day, 140 miles away, I found one, for the first time, in my tool-shed.
> I had not heard the news that my niece, who is also my God-daughter, was so gravely ill.[338]

> Since a child over sixty five years ago I have always been terrified of a bird entering the house as my mother said it would be followed by a death ... in the autumn, a Robin that we fed in the garden kept trying to come into the house and succeeded twice. After the first time, my sister-in-law died but it had been expected all year; and after the second time my greatest friend died quite unexpectedly. I can give no explanation of this as the Robin had never tried to come in before and has not again.[339]

> My grandfather, William Wallis, was an agricultural labourer in Essex in the eighteen eighties. When he was a child his youngest sister, a baby, was gravely ill.

'We were sitting downstairs' he said 'and a Robin hopped in through the door. Mother began to cry. "Molly will be dead before morning" she said. And that was true. The baby died in the night. Always a bad sign when a Robin comes indoors.'[340]

In fairness it should be added that, in some localities, such a happening was taken as a good omen.

I seem to remember that in Hampshire it was thought unlucky for a Robin to come into the house but we felt it a privilege ... my mother tamed a Robin which used to peck on the sitting room window and come into the room. On one occasion it sat on a flower decoration and sang whilst she hoovered and on another, I was sitting reading when she opened the window and it perched upon my book.[341]

My mother seems to have been unusual in believing that a Robin in the house foretells a birth, and with good reason. My sister tells me that there were great shrieks and flappings of aprons when Mummy and her 'daily' went into her bedroom and found a Robin sitting on top of the wardrobe. A few months later, I arrived![342]

Romanies are pleased when a Robin taps on their wagon window and, in Oxfordshire and Gloucestershire, there is a curious hybrid variant of the belief. There, it is said that someone who is ill will not recover if a Robin comes indoors, unless this happens in November, when it is a good omen, although November itself is traditionally the month of the dead. In Oxfordshire too, a Robin singing close to a house is bad news and if a Robin alights on a chair in which someone is sitting, the occupant will be dead within the year. In some places a Robin tapping on a window is sufficient to portend misery and there are people today who take the receipt of a Christmas card bearing a Robin as an indication of death or misfortune within the year. Clearly, the continued sale of Robin-bedecked cards must mean that this can have little credence.

In Wales, a Robin singing on the threshold of a house was similarly a death sign and a Robin in or near a coal mine a sure sign of tragedy. A report in the *South Wales Weekly News* of 14 September 1901 recounted that a Robin had nested in the pump house of the colliery at Llanbradach and that an explosion had followed. In Buckinghamshire, a Robin singing plaintively also presaged death and there is said to be a tradition at Hurstpierpoint School in West Sussex that, if a Robin sings on the chapel altar, one of the boys will die. This may be supported by fact: the cleric and luminary Sabine Baring-Gould (1834–1924) taught at Hurstpierpoint between 1857 and 1864 and has been quoted as saying that he saw a Robin enter the chapel through the open east window, land on the altar and begin to chirp; and that shortly afterwards, the bell began to toll for a boy who had just died. Uunfortunately, there seems to be no direct contemporary corroboration of this although 'during his time here there were, sadly, a number of deaths of pupils, and also two masters'.[343] Nonetheless, in 1945, in answer to a query on the matter in the school magazine, an interesting reminiscence emerged, originally related by H.L. Johnston, a master at Hurstpierpoint from 1896 to 1906:

In the depths of winter, a red-breasted Robin *(Erithacus rubecula)* hopping in the snow is a cheering sight.

William Pratt ... who spent all his working life at the school ... told me that once in the 60s, after a death in the school, a sparrow (not a robin) was seen flying about the Chapel, and from this doubtless the legend arose. During my time, when there was a death in the school, no bird appeared, although I remember, in another term, a robin attending the service, alighting on the head of the Captain of the School during the sermon. Beyond rousing him suddenly from what was evidently deep contemplation, no untoward result occurred.[344]

Keeping Robins in cages has widely been thought to bring ill-luck and, perhaps for this reason, they have never been popular as cage birds, even in times past, when almost anything with feathers that could sing was imprisoned. William Blake, in *Auguries of Innocence*, was certainly aware of it:

A Robin Redbreast in a Cage
Puts all Heaven in a Rage.

There is also a comparable and widespread view that no good will come to anyone who harms a Robin. Romanies believe that harming a Robin will result in them having a crooked finger. In Wales it is believed that ill will befall the culprit's family, perhaps even that their house will burn down; in Ireland that a swelling will appear on the hand that performed the act; and in Lincolnshire that the miscreant will break an arm or leg or, if the Robin was killed wantonly, that it will cause the death of the person's mother. From Scotland and Yorkshire comes the belief that a cow belonging to the family of anyone killing a Robin would in future produce bloody milk, while in Suffolk it is said that trembling of the hands will result. There is a first-hand account from Cornwall of a groom who, as a boy, shot a Robin and then held it in his hands, which shook thereafter.

Kingsley Palmer tells of a West Country Wassailing ceremony that involves Robins.

> [The ceremony] has taken place on Old Twelfth Night (17 January) for very many years at Carhampton, a Somerset village on the A37 near Minehead. Behind the Butchers Arms Inn a tree is addressed in song ... Shotguns are fired to ward off evil spirits and pieces of cider-soaked toast are lodged in the tree branches for robins, which are said to be good spirits in disguise. A bonfire is lit in the orchard, and the participants stand round it drinking freely of hot cider themselves. Fears that the grubbing up of trees might jeopardise the ceremony were ended when the publican bought the orchard to preserve it.[345]

Stealing Robin's eggs was at least as sinful as harming the bird itself and, in Wales, the thief would fall into the hands of the devil and witches. English tradition is slightly less harsh: in Devon, the household crockery would be smashed and, in Yorkshire, your friends would boo and hiss and point and shout 'Robin takker, robin takker, Sin, sin, sin!' Children in Dorset were told that interfering with a Robin's nest would cause their fingers to grow crooked but, in Guernsey, it was simply considered beneath a boy's manhood to rob a Robin's nest. The Essex rhyme linking Robins with Wrens, quoted above, also prophesies ill to egg-thieves. There is a similar Scottish notion:

Since then no wanton boy disturbs her nest:
Weasel nor wild-cat will her young molest,
All sacred deem the bird of ruddy breast.[346]

This has its origin in another widespread Robin tradition, and is related to other birds bearing red markings, for it was thought that the Robin had a drop of God's blood in its veins. A similar religious significance stems from the story of the two birds that perched on Christ's cross, one of which was the Magpie (see page 358) and the other a small brown bird. The Magpie was scornful of the Lord and was condemned in consequence, but the brown bird wiped away Christ's tears with its wings and used its beak to pull thorns from His brow. A drop of blood fell on its breast and God rewarded the Robin for sharing in His suffering: it would henceforth be the bird of God and its eggs as blue as heaven.

Another tradition associated with red-coloured birds is that they are bringers of fire and, in one version, the Robin takes the fire from a burning Wren, scorching his breast in the process. According to a Welsh story the bird's breast was scorched as it dropped water to quench a fiery pit, while a Guernsey tradition holds that there was no fire on the island until the Robin brought it there by carrying a lighted torch in its beak, burning its breast feathers in the process. Other island birds, except for some reason the owl, took pity on the Robin and each gave him a feather.

A game is played in Cornwall and Scotland called Robin's Alight, in which the participants sit around a fire and pass a whirling, flaming stick from one to another; a forfeit is due from whomsoever is holding the stick when the flame goes out.

In the Lake District, a Robin singing in the morning indicates rain before nightfall, whereas singing in the evening presages a fine day to follow. In Guernsey, a Robin singing very loudly from the top of a hedge is a promising sign; one chirping while hidden in the hedge bottom is not. In the southeast of Ireland, a Robin entering a house signifies not death, as in other places, but impending frost, while a Suffolk rhyme offers:

If the robin sings in the bush,
Then the weather will be coarse;
But if the robin sings on the barn,
Then the weather will be warm.[347]

The Robin appears as an inn name although it is unexpectedly rare, perhaps due to its presence in buildings being thought unwelcome. It is also occasionally seen in heraldry, most famously on the coat of arms of the City of Glasgow, an association that commemorates the piety of St Kentigern, also known as St Mungo (another St Kentigern tale is responsible for Glasgow's arms also sporting a salmon – see page 130). The saint seems to have been a seventh-century monk and bishop of Strathclyde and there are many legends about him. It is said that, when a boy, he was a pupil of St Serf at Culross, where he was greatly in favour. In a fit of jealousy, other students killed St Serf's pet Robin and then blamed Kentigern for the deed, although the saint's piety subsequently restored the bird to life.

Nightingale *(Luscinia megarhynchos)*

OTHER COMMON NAMES: *barley bird (East Anglia).*

For one of the most celebrated of birds, immortalized in music, poetry and literature, the Nightingale has surprisingly few regional names. For centuries, it has been known as the Nightingale and the sole regional variant, the East Anglian name 'barley bird', is also applied to several other species; in any event, there is no special reason for associating Nightingales with the barley crop, although Edward Armstrong suggests that it arose because the bird, a summer visitor, is first heard when the barley crop is being sown.[348]

The name 'Nightingale' can be traced directly to the Old English 'nihtegale' and has become so inseparable from night singing that even totally unrelated creatures that make night-time noises (no matter how unmusical) have been graced with it: in parts of East Anglia, frogs are called ' fen nightingales' and, in the Southport area, even Natterjack Toads have been called 'Birkdale nightingales' (see page 182). Probably the most celebrated human instance of this was 'The Swedish Nightingale', the coloratura soprano Jenny Lind (1820–87). However, it is certainly not the only night-singing bird; Robins do so and the inspiration for the immortal bird of 'A Nightingale Sang in Berkeley Square' was probably a Sedge Warbler.

Nightingales are rather undistinguished, medium-sized, brown birds with a slightly red-brown tail. They are mainly species of deciduous woodland, coppice and dense undergrowth, from whence they tend to pour forth their song, although not exclusively in the hours of darkness.They are now in decline largely because of the loss of this dense ground cover. They are no longer found in Wales and have retreated to the east and southeast of England. In 1998, the British Trust for Ornithology launched a Nightingale Appeal in an attempt to reverse this trend and supported it with a wonderful CD of quite extraordinary historic recordings of Nightingales.

It has been said that the Nightingale's legendary state is almost entirely due to poetry and that it has no traditional folklore. This is is not altogether true, although it has certainly provided poets with a remarkable muse. To the classical poet, the bird is Philomel, after Philomela, a pivotal character who was turned into a Nightingale in the course of a Greek legend entailing much murder, mutilation and cannibalism. When Shakespeare, in a similarly unsavoury tale of murder, mutilation and cannibalism, wrote that 'Philomel must lose her tongue today' (*Titus Andronicus*, Act 2, Scene 5), he was simply repeating a common mistake that stems from an erroneous version of the story by Ovid; in the original legend, it was Philomela's sister Procne, not Philomela herself, whose tongue was cut out. Nonetheless, it was the unfortunate Philomela's fate that seems to have resulted in the Nightingale being endowed with so much pathos and sorrow, although, unlike Philomela and many another poetic heroine (and Jenny Lind), the Nightingale that sings is always male.

A typical story of Nightingale pathos is one of uncertain origin that has been repeated several times by such luminaries as the seventeenth-century dramatist John Ford, as well as by later writers, including Cowper and Moore. It tells of a contest between a Nightingale and a lute-playing shepherd to determine who had the richest and finest song. They exchanged chord after chord until the shepherd finally broke forth into 'a full-mouthed diapason' and the poor bird, straining to imitate the rich, deep sound, fell exhausted and dead upon the lute.

Milton, Shakespeare, John Clare, Shelley, Wordsworth, Chaucer, Coleridge, Marvel, Keats – no poet worthy of the name has failed to write about the Nightingale, although their eulogies are not always supported by accurate ornithology. Cowper, for instance, wrote six verses to the Nightingale in which he claimed, without a shred of evidence, to have heard the bird singing on New Year's Day 1792. W.H. Hudson wrote of him: 'He was as bad a naturalist as any singer before him or after him, and as a true poet has a perfect right to be.'[349] Even Shakespeare, a real countryman and no mean observer of nature, was happy to add to the widespread and quite erroneous belief that Nightingales never sing in the daytime:

> The nightingale, if she should sing by day,
> When every goose is cackling, would be thought
> No better a musician than the wren.
> (*The Merchant of Venice*, Act 5, Scene 1)

Despite its undistinguished appearance, the Nightingale *(Luscinia megarhynchos)* is one of nature's finest songsters.

Another long-standing and rather foolish belief, echoed in much of the Nightingale poetry, is that the birds sing only when their breast is pressed against a thorn.

> A Nightingal gan sing; but woe the lucke;
> The branch so neare her breast, while she did quicke her
> To turn her head, on sodaine gan to pricke her.[350]

Much local Nightingale folklore relates to the time of its arrival in spring. In many places it is heard at approximately the same time as the first Cuckoo: 'On the third of April, Come in the Cuckoo and the Nightingale.' There are also many beliefs attached to its very uneven distribution (it is, for instance, rare in the West Country and absent from Ireland), which has given rise to such myths as that it only occurs where cowslips are found, or where hops are grown. Edward Armstrong quotes a belief that Nightingales are never heard at the village of Havering-atte-Bower in Essex because its singing there once interrupted Edward the Confessor during his devotions.[351]

Nightingales are among the summer-visiting birds that, until the early nineteenth century, and before migration was understood, were believed either to fly to the moon or to hibernate in caves, hollow trees and even the mud at the bottom of ponds.

They have sometimes been credited with talismanic powers and they have also on occasion been considered creatures of ill-omen. A Shropshire belief was that death would follow their singing. An endearing Sussex tale, relating to the slaying of a dragon in the forest near Horsham by St Leonard, tells how the Nightingales, who had sought to distract the beast with their singing, then fell silent.

Like other singing birds, Nightingales have been caught and caged; the London naturalist and writer J.E. Harting knew of a Middlesex gamekeeper in the nineteenth century who paid for his rent by catching them. In one year alone, he sold 180 birds at 18s a dozen.

Nightingale flesh has been used medicinally and was thought:

> ... exceeding good for consumptive Persons, and such as are afflicted with a Cachexia, or an evil Disposition of body: The Gall ... is an excellent Collyrium for most Disorders of the Eyes.[352]

Strangely, the bird does not seem to feature in place names, nor in heraldry, although it is does occur as a surname; Florence Nightingale is a particularly well-known example.

The last word on the Nightingale rightly belongs to one of the greatest of countrymen, naturalists and writers. No-one has ever eulogized it more wonderfully than Izaak Walton:

> But the nightingale (another of my airy creatures) breathes such sweet loud music out of her little instrumental throat, that it might make mankind to think miracles are not ceased. He that at midnight, when the very labourer sleeps securely, should hear, as I have very often, the clear airs, the sweet descants, the natural rising and falling, the doubling and redoubling of her voice, might well be lifted above earth, and say, 'Lord, what music has Thou provided for the saints in heaven, when Thou affordest bad men such music on earth?'[353]

Black Redstart *(Phoenicurus ochruros)*

OTHER COMMON NAMES: *black redtail; blackstart; Tithy's redstart.*

The Black Redstart is a much scarcer summer visitor than its close relative the Redstart and differs in its black breast and head, which serve to enhance the striking red tail. Historically, it was a bird of rocky places rather than woodlands, but in England it is now unusual in being among the few species to have moved into industrial places and large buildings, where it often nests on ledges. Francesca Greenoak records one that nested in a jumbo-jet hangar at Heathrow in 1973.[354]

Redstart *(Phoenicurus phoenicurus)*

OTHER COMMON NAMES: *Bessy brantail; Fanny redtail; fiery brantail, fiery redtail (Shropshire); fire flirt; fire tail (Scotland, Nottinghamshire, Norfolk, Warwickshire, Somerset, Hampshire); flirt tail, Jenny redtail (Yorkshire); Katies brantail (Shropshire); nanny redtail; redrump (Yorkshire); redstare; redster; redtail (Scotland, Norfolk, Oxfordshire); white cap (Shropshire); woh snatch (Cheshire).*

Redstarts are common summer visitors, most frequent in the north of Britain. They are Robin-sized birds with a strikingly obvious rust-red tail which flicks constantly up and down, a rust-red or orange breast and a slate-blue back. It has been

The male Redstart *(Phoenicurus phoenicurus)* has a distinctive rusty-red tail with a reddish belly and a black throat.

Each of these Stonechats *(Saxicola torquata)* has a beakful of food. They feed mainly on insects, worms and spiders.

said that only their migratory habit has prevented them from becoming as familiar and much loved as the Robin itself. They are essentially birds of woodland, parks, heaths and other areas with large old trees and they nest in holes in trees, stone walls or sometimes buildings.

The name 'Redstart' literally means 'red tail' and is almost certainly much older than its first recorded use, as 'redstarte', in 1570. Most of the regional variants allude to this very striking character. It was known as the 'firetail' in the seventeenth century:

> Around the rotten tree the firetail mourns
> As the old hedger to his toil returns,
> Chopping the grain to stop the gap close by
> The hole where her blue eggs in safety lie.
> Of everything that stirs she dremeth wrong
> And pipes her 'tweet tut' fears the whole day long.
> (John Clare, 'The Firetail's Nest')

Whinchat *(Saxicola rubetra)*

OTHER COMMON NAMES: *bush chat; dock topper (Somerset); fern lintie (Aberdeenshire); furr chuck (Norfolk); furze chat (Worcestershire); furze hacker (Hampshire); gorse chat (Westmorland); gorse hatch; gorse hopper (Cheshire); grass chat (West Yorkshire); tick (Nottinghamshire, Shropshire); uthage (Shropshire); utick; whinchacker (North Yorkshire – Craven); whincheck (Lancashire); whin clocharet (Angus – Forfar); whin lintie (Aberdeenshire); yutick.*

Whinchats are small summer visitors with a fairly local distribution in Britain, typically found among gorse (whin) and bracken on heathland and in other open areas, where they nest on the ground. They have declined significantly with the loss of the dense grass in which they nest. They are small, pretty birds with a buff-coloured breast and a black eye-stripe.

The name 'Whinchat' is seventeenth century ('winchat' is sometimes seen today); the word 'chat' occurs in the regional names of many birds and evidently relates to their call. The unusual 'utick', or 'yutick' are onomatopoeic and interestingly occur as an inn name, which baffles visitors and locals alike.

> One local name I still have to fathom is a Yutick.
> It is on a pub sign in Blackburn. The pub is called
> The Yutick's Nest. It could possibly be a local name
> for the Linnet.[355]

Stonechat *(Saxicola torquata)*

OTHER COMMON NAMES: *blackcap; blacky cap (Ireland); blacky top; bushchat; chickstone (Yorkshire); furze chitter (Cornwall); furze hacker (Hampshire); gorse chat; gorse jack (Gloucestershire); heath tit (Sussex); moretetter (Yorkshire); moor titling; stanechacker (North Yorkshire – Craven); stane chapper; stone chatter; stone clink; stone clocharet (Angus – Forfar); stonepricker (Cheshire); stonesmith.*

Stonechats are the resident equivalents of Whinchats and are very similar birds, although with a more intensely orange breast. They were once found in similar places as the Whinchat but are now largely restricted to some of the heathlands of the eastern and southern coasts of England and in the New Forest, where they can find the vegetation to construct their remarkable tunnelled nest, deep in a dense, low-growing bush.

The name 'Stonechat' is eighteenth century, but there seems no obvious reason for this bird to be more associated with stones than the Whinchat.

The Stonechat, for no obvious reason, was one of seven creatures in the Isle of Man believed to pass the winter in a torpid state and there are peculiar beliefs linking both Stonechats and Wheatears with toads. In Scotland, it was thought that Stonechat eggs were incubated by toads and that toads hatched from Wheatear eggs. Stonechats were consequently thought to contain a drop of the devil's blood and, for all these reasons, both birds have at times been persecuted. An old rhyme from Galloway sounds a caution to anyone interfering with a Stonechat nest:

Stone chock!
Devil tak'!
They who harry my nest
Will never rest,
Will meet the pest!
De'il break their long back
Wha my eggs wad tak! tak![356]

Interest in the arrival of the Wheatear *(Oenanthe oenanthe)*, one of the earliest summer migrants, vies with that of the Cuckoo.

Wheatear *(Oenanthe oenanthe)*

OTHER COMMON NAMES: *chack; chacks (Orkney); chat (Northamptonshire); check bird; chickell (Devon); chock; chuck (Norfolk); clocharet (Angus – Forfar); clodhopper (Northamptonshire); coney chuck (Norfolk); cooper (south Pembrokeshire); dyke hopper (Stirlingshire); fallow chat; fallow finch; fallow lunch; fallow smirch; fallow smiters (Warwickshire); fallow smith; furze chat, fuzz chat (Sussex); hedge chicker (Cheshire); horse masher (Cornwall); horse musher (Hampshire); horse smatch; jobbler, snorter (Dorset); stanechacker (Northern Ireland, Aberdeenshire, Lancashire); steinkle (Shetland); stonechat (Westmorland, West Yorkshire, Northamptonshire); straw mouse (Cheshire); wheatears; white ass (Cornwall); white rump (Norfolk); white tail; wittol (Cornwall).*

The Wheatear is a particularly beautiful and graceful small thrush, slightly larger than the Stonechat and the Robin, with a buff breast, a black eye-stripe, grey back and white rump. It is still a fairly common summer visitor to open grassy areas of the north and west, where it feeds on insects, nesting in a crevice, Rabbit hole or similar place.

The name 'Wheatear' has an interesting origin, first occurring in approximately this form *c.*1600 and often said to relate to the bird fattening itself on the ripening wheat. In reality, its etymology is older, and rather more vulgar: it appears to originate in a Middle English word meaning 'white arse'. Many of the regional names are evidently onomatopoeic or habitat-related, but 'horse smatch' and related variants are most intriguing. There are three theories: the first, from Oxfordshire, holds that it arose because the bird flew along the road beside horse-drawn traps and carriages, apparently 'matching' or racing them; the second that it is attracted by the insects that accumulate on horse droppings; and the third that its call seeks to imitate or 'match', the clicking sound that people make when talking to horses.

Although Wheatears have been persecuted in the north of England and in Scotland (see Stonechat, page 339), in some parts, interest in their arrival vies with that of the Cuckoo; it is held that when first spotted in their normal habitat on short turf, good fortune will follow, but if they are first seen standing on a stone, the death of the observer will ensue. In County Kerry in Ireland, Wheatears are called 'the cunning little man under the stone'.

Roasted Wheatear was once highly recommended, although the size of the birds and the difficulty of obtaining them in large numbers meant that it was always considered a delicacy. Mrs Beeton included an illustration of a Wheatear hanging in a poulterer's next to an Ortolan.[357]

Ring Ouzel *(Turdus torquatus)*

OTHER COMMON NAMES: *blackbird (Shropshire); cowboy (Tipperary); crag ouzel (North Yorkshire – Craven); ditch blackie (East Lothian); flitterchack (Orkney); heath throstle; hill chack (Orkney); Michaelmas blackbird (Dorset); moor blackbird, mountain blackbird (Scotland, North Yorkshire); mountain colley (Somerset); mountain ouzel; mountain thrush (Kirkcudbrightshire); ring blackbird; ring thrush; rock blackbird (Stirlingshire); rock ouzel; rock starling (Roxburghshire); round-berry bird (Galway – Connemara); tor ouzel (Devon); whistler (Wicklow).*

A once common summer visitor, the Ring Ouzel resembles a Blackbird, but with a black beak and a white band across the breast. It is typically a bird of high moorland areas, where increased afforestation and general 'improvements' have resulted in a loss of nesting sites. Even when they were much more common, they were unfamiliar to people who did not venture to such places. They have been notoriously loath to come close to houses, which led to the Orkney belief, reported by J.W.H. Trail, that if a flitterchack is seen near to a dwelling, one of the occupants would leave or die.[358] Nonetheless, the fact that they were once much more widespread is very apparent; Gilbert White commented that:

> A young man at Lewes, in Sussex, assured me that about seven years ago ring-ousels abounded so about that town in the autumn that he killed sixteen himself in one afternoon.[359]

'Ouzel', or 'ousel', is the oldest name for the Blackbird *(Turdus merula)* and is traceable back at least to Old English. It was replaced by 'Blackbird' in the seventeenth century but is still in use in the official name for *Turdus torquatus* and also appears in some regional Blackbird names. The generic name *Turdus* for the true thrushes derives from the Latin word for 'thrush'.

Blackbird *(Turdus merula)*

OTHER COMMON NAMES: *amsel; blackie (Scotland, North Yorkshire); black uzzle (North Yorkshire – Craven); colly/ colley (Gloucestershire); merle (Ireland, Scotland); ouzel; ouzel cock; woofell; Zulu (Somerset).*

One of the commonest British breeding birds, the Blackbird is familiar in town and country alike, and is a frequent garden visitor with a beautiful song, the epitome of the dawn chorus.

> There was a saying when I was young, that if the Blackbird sings when he is going to bed, he will rise in the morning with a wet head.[360]

There has been some decline in its numbers in recent years, although for the British Trust for Ornithology's magazine *BTO News* to emblazon its Winter 1998 front cover with 'Bye-bye Blackbird?' may have been over-alarmist.

It is surprising that so familiar a species has not acquired more regional names and also strange that, of the all-black birds, it is this bird rather than one of the crows that has become known as the Blackbird. Nonetheless, it was still the ouzel for Shakespeare:

> Alas a black ousel, cousin Shallow!
> (*King Henry IV Part 2*, Act 3, Scene 2)

> The ousel-cock so black of hue,
> With orange-tawny bill ...
> (*A Midsummer Night's Dream*, Act 3, Scene 1)

The old name 'colly', or 'colley', which survives as a dialect name, especially in the West Country, originates in the old word 'coaly', meaning 'coal-black' or 'sooty'. It also survives as the 'four colly birds' in the familiar Christmas song 'The Twelve Days of Christmas'.

Albino or partially albino Blackbirds are surprisingly common and sometimes lead to them being mistaken for Ring

The adult male Blackbird *(Turdus merula)* is all black, with an orange bill and yellow eye ring.

The female Blackbird *(Turdus merula)* has less distinguished brown plumage and is spotted.

Ouzels but, surprisingly, albinism is rare, if not unknown, in the closely related Song Thrush. The female of the species is a much less imposing brown bird and is sometimes called the 'French blackbird' in Cheshire, 'French' being used in the sense of something 'foreign' or 'unusual' rather than pertaining to France.

There seem to be very few English places named after this now common bird. The River Ouse appears to be an obvious candidate but the name has a different origin from the name 'ousel', although this may not be true of Ouslethwaite in West Yorkshire.[361] A number do occur in Ireland, however, derived from the Gaelic name *lon*: for instance, *Coill-nan-lon* ('the wood of the Blackbirds'), now called Kilnalun, near Donegal.

A long-standing affection for Blackbirds exists in Ireland, where it is thought that, if they sing especially loudly, rain is on its way. (The ever-soft Emerald Isle must surely therefore have a large population of very noisy Blackbirds.) In County Meath, there is a saying, 'When the Blackbird sings before Christmas, she will cry before Candlemas', probably a reference to the fact that, when the birds nest very early, the mortality of the nestlings is high. Ireland is also the origin of the delightful story of St Kevin, who was praying in the temple of the rock at the sanctuary of Glendalough in County Wicklow. Suddenly a Blackbird flew down and laid an egg in one of his outstretched hands. The appreciative (and evidently very patient) saint remained with his arms outstretched until the eggs were hatched and the nestlings had flown.

Interestingly, in view of the Irish love of Blackbirds, in Scotland it is a Jacobite symbol, probably because Charles II, with his swarthy complexion, was called 'Black Boy' when he was young. An old Scottish song runs:

> Upon a fair morning for soft recreation
> I heard a fair lady was making her moan,
> With sighing and sobbing and sad lamentation,
> Saying, 'My blackbird most royal is flown.
> My thoughts they deceive me,
> Reflections do grieve me,
> And I am o'erburdened with sad misery.
> Yet if death should blind me,
> As true love inclines me,
> My blackbird I'll seek out wherever he be.'[362]

The first Blackbird rhyme that many people hear is probably the well-known 'Sing a Song of Sixpence', which refers to 'four and twenty' of them baked in a pie. Theories on its real meaning are legion, but it is probably sixteenth century and probably has political overtones. That Blackbirds should have been chosen for their symbolism is nonetheless revealing because, until recent times, they seem to have been fairly elusive birds of woodland and thickets, with little human contact. Gilbert White, Thomas Bewick and other older naturalists scarcely mention them.

An early mid-eighteenth-century version of another familiar nursery rhyme also includes Blackbirds:

There were two blackbirds
Sat upon a hill,
The one was nam'd Jack,
The other nam'd Gill;
Fly away Jack,
Fly away Gill,
Come again Jack,
Come again Gill.[363]

The pagan names 'Jack' and 'Gill' were replaced by the more socially acceptable ones of two apostles in the mid-nineteenth century while, at the same time, the 'Blackbirds' became 'dicky birds' (although the origin of the word 'dicky' applied to birds is obscure) to produce the version known today:

Two little dicky birds,
Sitting on a wall;
One named Peter,
The other named Paul.
Fly away, Peter!
Fly away, Paul!
Come back, Peter!
Come back, Paul![364]

A few of the 'four and twenty blackbirds' from the rhyme 'Sing a Song of Sixpence' (*Songs for the Nursery*, 1818).

Not surprisingly, poets through the centuries have also admired Blackbirds:

Down Wessex way, when spring's a-shine,
The blackbird's 'pret-ty de-urr!'
In Wessex accents marked as mine
Is heard afar and near.
(Thomas Hardy, 'The Spring Call')

The Blackbird arises occasionally as an inn name (although The Three Blackbirds in Hertfordshire probably derives its name from the arms of the local family of Seabright and should really be three martlets/martins).

Like thrushes, Blackbirds seem to have attracted medicinal interest, their flesh being:

... exceeding good for them who are afflicted with the Consumption, Colic, or Dysentery, it being constantly eaten. The Dung mixt with the Juice of Lemons and Camphire in fine Powder, applied to the Skin cures Lentils, Freckles, and other Defedations of the fame.[365]

They have also at times been served at the meal table (possibly in pies): Andrew Boorde remarked in 1547 that 'of all small birds the lark is best, then is praised the blackbird and thrush'.[366]

Fieldfare *(Turdus pilaris)*

OTHER COMMON NAMES: *big felt (Ireland); blue bird (Devon, west Cornwall); blue black (northwest England); blue felt; blue tail (central England); cock felt (Northamptonshire); feldefare (Ireland, central England); felfar (northern and central England); felfaw (Yorkshire); felfer (Yorkshire, Lancashire); felfit; fellfare (Northamptonshire); fellfoot (East Anglia); fellfor (Cumberland, West Yorkshire, Lancashire); felt (Oxfordshire); felter (Surrey); feltifer, feltiflier (Scotland); feltyfare; fildefare (central England); fildifire (Shropshire); flirty fleer (Scotland); fullfit, fullfor (East Anglia); grey thrush (Scotland); hill bird; jack bird; pigeon felt (Oxfordshire, Buckinghamshire, Berkshire); screech bird, screech thrush (Stirlingshire); snow bird (Shropshire); storm bird (Norfolk); storm cock (Scotland, Shropshire); veldiver (southwest England); velly bird; velverd (Wiltshire).*

This is a winter visitor to Britain although the few numbers that stay to breed are continually increasing. Larger than the Song Thrush and Redwing but slightly smaller than the Mistle Thrush, it is distinguished by its grey head and chestnut back. It is a bird of open countryside and fields and evidently always has been, judging by its name. This dates back to the Middle English 'feldefare' and has survived in countless regional variants; the list given here is not exhaustive.

The name means field wanderer or traveller; as in wayfarer.[367]

It is also a markedly noisy, screeching, chuckling or even grunting bird; the Welsh name *socen llwyd* means 'grey pig'. The characteristic image of Fieldfares is of a flock searching for food on a cold winter's day over a snow-covered field, an image that has endured since Chaucer wrote of them as the 'frosty feldefares'. They also epitomized winter for John Clare in *The Shepherd's Calendar*:

> Flocking fieldfares, speckled like the thrush,
> Picking the red haw from the sweeing bush
> That come and go one winters chilling wing.
> And seem to share no sympathy with Spring.
> (John Clare, 'March')

Song Thrush *(Turdus philomelos)*
OTHER COMMON NAMES: *dirsh (Somerset); drush (Dorset); garden thrush; grey bird (Sussex, Devon, Cornwall); mavie (southwest Scotland); mavis (Ireland, Scotland, East Anglia); thirstle (Shropshire, Devon, Cornwall); throggie (Cheshire); throstle (Ireland, northern and central England); thrush; thrusher (Buckinghamshire, Berkshire); thrushfield (Shropshire); trush drush (Somerset); whistling Dick (southern England); whistling thrush.*
The Song Thrush is among the common songbirds that have recently suffered a serious decline in numbers, especially in farmland and gardens. The reasons are not fully understood but seem to be related to the loss of damp ditches and field margins where the birds found much of their food. These are the birds that most famously utilize so-called 'thrushes' anvils' – stones against which they smash snail shells – to the satisfaction of gardeners everywhere.

The Song Thrush has a beautifully speckled breast but can be immediately distinguished from the similarly adorned Mistle Thrush by its smaller size and, when in flight, by its pale orange, not white, underwing areas.

> I leant upon a coppice gate
> When frost was spectre-grey,
> And Winter's dregs made desolate
> The weakening eye of day.
> The tangled bine-stems scored the sky
> Like strings of broken lyres,
> And all mankind that haunted nigh
> Had sought their household fires.
>
> The land's sharp features seemed to be
> The Century's corpse outleant,
> His crypt the cloudy canopy
> His wind his death-lament.
> The ancient pulse of germ and birth
> Was shrunken hard and dry,
> And every spirit upon earth
> Seemed fervourless as I.

The Song Thrush *(Turdus philomelos)* smashes snails against stones in order to feed on the contents.

At once a voice arose among
The bleak twigs overhead
In a full-hearted evensong
Of joy unlimited;
An aged thrush, frail, gaunt, and small,
In blast-beruffled plume,
Had chosen thus to fling his soul
Upon the growing gloom.

So little cause for carolings
Of such ecstatic sound
Was written on terrestrial things
Afar or nigh around,
That I could think there trembled through
His happy goodnight air
Some blessed Hope, whereof he knew
And I was unaware.
(Thomas Hardy, 'The Darkling Thrush')

Its characteristic mud-lined nest has been familiar to generations of schoolchildren and also to John Clare:

How true she warped the moss to form her nest
And modelled it within with wood and clay.
(John Clare, 'The Thrush's Nest')

This is also the thrush that pulls worms from lawns:

Terrifying are the attent sleek thrushes on the lawn,
More coiled steel than living-a poised
Dark deadly eye, those delicate legs
Triggered to stirrings beyond sense – with a start, a bounce, a stab
Overtake the instant and drag out some writhing thing ...
(Ted Hughes, 'Thrushes')

The name 'thrush' originates in Old English and over the centuries it has alternated in popularity with 'throstle', which, although superficially similar, has a different derivation. (The word 'thrush', as applied to the human disease, is of unknown origin, but logic suggests that the symptoms of the disease resemble the speckled appearance of the bird's breast.) The curious name 'mavis' is very old, probably originally French *(mauvis)* or Spanish. The specific common name 'Song Thrush' dates from the late seventeenth century, but other names seem to have been used indiscriminately for both the Mistle Thrush and Song Thrush.

Thrushes have accrued some odd superstitions: it is widely thought that they dispose of their old legs and acquire new ones when about 10 years old and that they are deaf. Unfortunately for the birds, their flesh was considered a specific against 'convulsions and the falling sickness'.

Redwing *(Turdus iliacus)*

OTHER COMMON NAMES: *felt (Northamptonshire); little feltyfare (East Lothian); pop; red thrush (central England); redwing mavis (Angus – Forfar); redwing throlly (Yorkshire); redwing thrush; swine pipe; wind throstle; wind thrush (Warwickshire, Gloucestershire, Somerset, Cornwall); whin thrush (Gloucestershire); whindle thrush, windle thrush (Devon); windle; wing thrush; winnard (Cornwall).*

The smallest British thrush, this is a common winter visitor to much of the country and breeds in Scotland, flocks arriving in autumn and often associating with Fieldfares. The adults are readily identified by the striking orange red patches on their flanks and underwing areas.

The name 'Redwing' is an obvious reference to the plumage, although it does not seem to have been used before the seventeenth century. The interesting 'swine pipe' probably has nothing to do with swineherds or their pipes but is related to the bird's somewhat whining note and to the names 'wind thrush', 'whin thrush' or 'whining thrush'.

Charles Swainson tells of the 'herring spear' or 'herring piece',[368] a rushing sound heard by Kentish fisherman on dark winter nights and caused by the flight of migrating Redwings crossing the Channel; Frank Buckland commented on the fear that this caused.[369]

Mistle Thrush *(Turdus viscivorus)*

OTHER COMMON NAMES: *big felt (Ireland); big mavis (East Lothian); bull thrush (Hampshire); bunting thrush; butcher bird; char cock; chercock (Westmorland); corney keevor (Antrim); crakle; felfit (east Suffolk); fen thrush (Northamptonshire); fulfer (Norfolk); gaw thrush (Northamptonshire); hillan piet (Aberdeenshire); hollin cock (Yorkshire); holm cock; holm screech (Dorset, Devon, Cornwall); holm thrush; horse thrush (Northamptonshire); jay (Northern Ireland); jay pie (Wiltshire); jercock; Jeremy joy; marble thrush (Northamptonshire); mistle bird (Shropshire); mizzly Dick (Northumberland); muzzel thrush (Roxburghshire); Norman gizer, Norman thrush (Yorkshire, Oxfordshire); screech; shirleycock (Lancashire); shrike; skirl cock; skirlock (Derbyshire); skrike; skrite (southern England); squawking thrush (Isle of Wight); stone thrush (Dorset); storm cock; thrice cock (Warwickshire); wood thrush (Dumfries-shire).*

The largest British thrush, this bird is readily identified by its size and can be distinguished from the Song Thrush in flight by its white, not pale orange, underwing areas. It tends to be a species of open parkland and similar places and is less of a garden bird than the Song Thrush. It is, however, becoming less common everywhere, for reasons that are not obvious. Traditionally, and by name, it feeds on Mistletoe berries but, like its relatives, it is omnivorous and eats a great many other things and the decline in the availability of Mistletoe in Britain means that these berries now form a minor part of the diet.

The West Country name 'holm cock' is a reference to its widespread fondness not for the Holm Oak but for holly.

The Mistletoe berries that formed the traditional diet of the Mistle Thrush *(Turdus viscivorus)* are now in short supply.

'Storm cock' is a wonderfully evocative and still fairly widely used name that aptly describes the fact that, unlike many other birds, Mistle Thrushes seem positively to revel in the approach of storms and shun shelter. The use of the name 'Norman' is thought to symbolize aggressiveness; 'gizer' is from the Old French *guis*, meaning 'mistletoe'.

Warblers
(Sylviidae)

Warblers are small, mainly insect-eating birds with typically short, thin beaks and a wide range of habitats that to some degree characterize the main groups: the swamp warblers, scrub warblers, leaf warblers and tree warblers. Males and females are similar in most species. Only Cetti's Warbler, the Dartford Warbler and the Goldcrest are normally resident in Britain, but there are several common summer visitors, some with a small number of overwintering individuals, and also many occasional visitors and vagrants.

Warblers are difficult to identify, many of them being shy or secretive, and several are remarkably similar in appearance: predominantly brown above and paler below. Some warblers, however, have very distinctive calls, the name 'warbler' being derived from the verb 'to warble' or 'to sing in a vibrating or trilling manner'.

Cetti's Warbler *(Cettia cetti)*

A rare, shy resident, usually found close to water or reed beds, this warbler is generally revealed and identified by its distinctive, loud, repeated song. Eric Simms said that it was the most difficult of bird songs to record because of its unpredictable occurrence, adding that, once heard, it could never be mistaken for any other.[370] George Yeates attempted none too successfully to put its song into words:

> What yer ... what yer ... what yer ... come-and-see-me-bet-you-don't ... bet-you-don't.[371]

Cetti's Warbler first appeared in Britain in 1961 and was breeding in Kent 10 years later, since when it has gradually spread westwards, something that no doubt would have pleased the eponymous Cetti, an eighteenth-century Jesuit naturalist.

Grasshopper Warbler *(Locustella naevia)*

OTHER COMMON NAMES: *brake hopper; brake locustell; chate, chut (Worcestershire); cricket bird (Norfolk); grasshopper lark; reeler.*

A once very common summer visitor, this is also a secretive bird and a species most likely to be revealed by its call, which is remarkably like that of a stridulating grasshopper or cricket. It has suffered a large decline in numbers in recent years, possibly, as with Whitethroats, because of drought in its winter quarters in the southern Sahara, although loss of suitable breeding sites in Britain may also be a factor.

Savi's Warbler *(Locustella luscinioides)*

OTHER COMMON NAMES: *night reeler (East Anglia).*

A relative of the Grasshopper Warbler, the Savi's Warbler has a similar but lower pitched call. It was identified in Britain by ornithologists in the East Anglian fens in 1840 and named in honour of Paolo Savi, the Italian naturalist who first described it in 1824. This came as nothing new to the people of the fens, who had long known it by the traditional name of 'night reeler'. Shortly afterwards, it had all but vanished and, although it made a comeback in the mid-twentieth century, it now seems to be disappearing again.

Sedge Warbler *(Acrocephalus schoenobaenus)*

OTHER COMMON NAMES: *chamcider (Hampshire); chat; chattering Billy (Cheshire); chitter chat (Northumberland); fantail warbler; grey, grey bird (Yorkshire); hay tit (Oxfordshire); Irish nightingale (Ireland); leg bird; lesser reed warbler; night singer (Ireland); night sparrow; pit sparrow; razor grinder (Cheshire); reed fauvette; sally pecker, sally picker (Ireland); Scotch nightingale (Stirlingshire, Roxburghshire); sedge bird (Yorkshire); sedge marine (Norfolk); sedge reedling; sedge wren, seg bird, thorn warbler (Yorkshire); water sparrow (Shropshire); willow lark.*

A widespread summer visitor, the Sedge Warbler has a characteristic white eye-stripe and inhabits reed beds and similar places near water. It is yet another bird that has declined because of overwintering problems in the southern Sahara.

The Sedge Warbler *(Acrocephalus schoenobaenus)* lives near water, where the insects on which it feeds flourish.

Their habit of nocturnal singing is reflected in several local names. Gilbert White said that they would start to sing when a stone or clod was thrown into the bushes and they are often said to imitate the calls and songs of other birds.[372] References to 'sally' in the Irish names relate to willows *(Salix)*. Edward Thomas wrote of Sedge Warblers as 'small brown birds, wisely reiterating endlessly what no man learnt yet, in or out of school'.[373]

Marsh Warbler *(Acrocephalus palustris)*

This species was first distinguished from the Reed Warbler in 1871 when the name was coined. It is an outstanding mimic whose calls copy those of many other birds, most notably the House Sparrow, Linnet, Swallow and Blue Tit. It has declined greatly in numbers with the reduction in the area of osier beds.

Reed Warbler *(Acrocephalus scirpaceus)*

OTHER COMMON NAMES: *babbler; fen reedling; marsh reedling; night warbler; reed chucker; reed sparrow; reed tit; reed wren; rush warbler; small straw (Yorkshire); smastray (Cheshire); water sparrow (Shropshire).*

This common summer visitor is usually, but not invariably, associated with reed beds and it has benefited from such sites as flooded gravel workings. The odd name 'smastray' is a corruption of 'small straw' and is sometimes applied to other warblers that use small pieces of straw or hay to build their nests.

Dartford Warbler *(Sylvia undata)*

OTHER COMMON NAMES: *furze wren (Surrey).*

A very local resident, this bird is confined to gorse and heather in the extreme south of England and therefore suffers badly in heath fires. As one of the few entirely insectivorous resident British birds, it is also very vulnerable in cold winters.

It is a truly beautiful little warbler, dark brown above and with a deep brown-purple breast. It takes its famous common name from a report by the naturalist Thomas Pennant in 1776, who wrote that 'A pair were shot on a common [Bexley Heath] near Dartford in April 1773'.[374] There is hope that it will again become a regular nesting bird in its home county of Kent.

Lesser Whitethroat *(Sylvia curruca)*

OTHER COMMON NAMES: *babbling whitethroat; haychat; hazel linnet; mealy mouth (Yorkshire); white-breasted warbler.*

A summer visitor, nesting in hedgerows and open, scrubby areas, the Lesser Whitethroat is characterized by a dark eye-stripe, greyish upper parts and a white breast. However, it lacks the chestnut wing-bars of the Whitethroat. It was first described from Britain in 1787 but probably earlier and independently by Gilbert White in his *Naturalist's Journal*: 'A rare and I think a new little bird frequents my garden.'[375] It has spread slowly northwards and westwards but has shown some decline in recent seasons.

Whitethroat *(Sylvia communis)*

OTHER COMMON NAMES: *beardie (Scotland); bee bird (Devon); Billy whitethroat (Shropshire); blethering tam (Scotland); caperlinty; Charlie muftie; cut-straw; feather bird (Northamptonshire); flax (Shropshire); great Peggy (Leicestershire); hay jack (Norfolk, Suffolk); haysucker (Devon); hay tit (Shropshire, Oxfordshire); hazeck (Worcestershire); hedge chicken; jack straw (Shropshire); Jenie, Meg cut-throat (Roxburghshire); meggie (northern England); muffit (Stirlingshire); muggy (northern England); nettle bird (Leicestershire); nettle creeper; nettle monger (North Yorkshire, Hampshire); Peggy (Nottinghamshire); strawsmall (West Yorkshire); strawsmear (Westmorland); winnel straw; wheetie whey beard; wheetie why; whey beard; white lintie (Angus – Forfar); whitecap; whittie beard.*

The multiplicity of local names reflects the familiarity of this distinctive and widespread summer visitor. It is a bird that is closely associated with farmland, shrubby areas and hedgerows.

The most widely used name, 'Whitethroat', was first recorded in the seventeenth century but is generally believed to be much older. John Clare was among its literary admirers:

> The happy whitethroat on the sweeing bough
> Swayed by the impulse of the gadding wind
> That ushers in the showers of April – now
> Singeth right joyously ...
> (John Clare, 'The Happy Bird')

The song of the male Whitethroat *(Sylvia communis)* was once common but, since 1968, populations have fallen in numbers.

There was a dramatic fall in numbers of Whitethroats after 1968, which was traced to severe drought in the southern Sahara and the death of large numbers of migrating birds which used up their fat reserves on their journey from further south in Africa. Over the past 15 or so years, there seems to have been some recovery but the situation underlines the fact that, no matter how much good work is done in Britain in improving habitats and nesting sites, the clock can be turned back overnight by factors beyond our control.

Garden Warbler *(Sylvia borin)*

OTHER COMMON NAMES: *Billy whitethroat (East Lothian); garden whitethroat; greater pettychaps (Yorkshire, Lancashire, Northamptonshire); haychat; juggler (Surrey); Peggy (Yorkshire); small straw (West Yorkshire); smastray, strawsmear (Cheshire); streasmear (Westmorland).*

Perhaps the least distinctive of the warblers, this is the archetypal 'greyish above, paler below' bird with a song remarkably like that of its close relative, the Blackcap. Historically a bird of much of England and Wales, it has spread into Scotland while largely retreating from Devon and Cornwall.

On the basis that a bird would not be called the Garden Warbler until most people had gardens, it is no surprise that the name dates from as recently as 1832. The older and much prettier name 'pettychaps', which is imitative of the call, has also been used for the Chiffchaff. Nonetheless, it was to the garden warbler that John Clare's poem was dedicated:

> Hard to discover – that snug entrance wins,
> Scarcely admitting e'en two fingers in
> And lined with feathers warm as silken stole
> And soft as seats of down for painless ease
> And full of eggs scarce bigger e'en than peas.
> (John Clare, 'The Pettichap's Nest')

Blackcap *(Sylvia atricapilla)*

OTHER COMMON NAMES: *black-headed hay-jack (Norfolk); black-headed Peggy; coal hoodie (North Yorkshire); hay bird, hay chat, hay jack (Northamptonshire); jack straw (Somerset); King Harry black cap (Norfolk); mock nightingale; nettle creeper; nettle monger; northern nightingale.*

Although several British birds (such as the Coal Tit, Black-headed Gull and Reed Bunting) have a black cap, at least this feature readily distinguishes this common summer visitor from other warblers, although the cap in the female is chestnut rather than black.

This pretty, tuneful little bird has formally had the name 'Blackcap' since the mid-eighteenth century, although at one time it tended to be called the 'blackcap warbler'.

It is probably the most familiar warbler today, having adapted to life in urban areas better than others and consequently being

This male Blackcap *(Sylvia atricapilla)* feeds on Sea Buckthorn berries. Males have a black crown; the female's is brown.

seen regularly in gardens and at bird tables. This familiarity is enhanced because some birds frequently overwinter in Britain.

> Under the twigs the blackcap hangs in vain
> With snowwhite patch streaked over either eye
> This way and that he turns and peeps again
> As wont where silk-cased insects used to lie
> But summer leaves are gone the day is bye
> For happy holidays and now he fares
> But cloudy like the weather yet to view
> He flirts a happy wing and inly wears
> Content in gleaning what the orchard spares
> And like his little couzin capped in blue
> Domesticates the lonely winter through
> In homestead plots and gardens where he wears
> Familiar pertness – yet but seldom comes
> With the tame robin to the door for crumbs.
> (John Clare, 'The Blackcap')

Wood Warbler *(Phylloscopus sibilatrix)*

OTHER COMMON NAMES: *green wren; hay bird (West Yorkshire); linty white; wood wren; yellow wren.*

This warbler, the Chiffchaff and the Willow Warbler (see below) belong to the group called 'leaf warblers'. They are common summer visitors and similar in appearance, although the Wood Warbler has a striking yellow throat. A bird of deciduous woods, it nests on the ground in undergrowth and recently has become mainly a species of Wales and the west.

Edward Thomas was a lyrical lover of the bird:

> High up among the beeches, the invisible wood wrens sang, and their songs were as if, overheard in the stainless air, little waves of pearls dropped and scattered and shivered on a shore of pearls ...[376]

Bird Songs and Calls describes its song the 'most accomplished ecstatic trill' of any European bird.

Chiffchaff *(Phylloscopus collybita)*

OTHER COMMON NAMES: *bank-bottle; bank jug (Bedfordshire); chip chip; choice and cheep (Devon); feather bed (Oxfordshire); feather pokel; huck muck; least whitethroat; least willow wren; lesser pettychaps; Peggy (West Yorkshire); sally picker (Ireland); thummie; wood oven.*

The markedly two-note song of the male Chiffchaff as it hops and flies among the trees feeding has given this common warbler several of its regional names. 'Chiffchaff' itself seems to be late eighteenth century in origin.

It tends to nest above ground, but in brambles and among evergreen shrubs rather than trees; names such as 'bank-bottle' and 'bank jug' reflect this. The Chiffchaff and the Blackcap are probably the two most successful warblers in Britain at present and both are being seen more and more as overwintering birds.

Willow Warbler *(Phylloscopus trochilus)*

OTHER COMMON NAMES: *bank jug (Bedfordshire); bee bird; feather bed (Oxfordshire); feather poke (West Yorkshire); fell Peggy (Lancashire); golden wren (Ireland); grass mumruffin (Worcestershire); ground Isaac (Devon); ground oven; ground wren; hay bird; huck muck; mealy mouth (North Yorkshire – Craven); miller's thumb; muffie wren (Renfrewshire); oven bird (Norfolk); oven tit; Peggy, Peggy whitethroat (West Yorkshire, Shropshire); sally picker (Ireland); smeu; smeuth (Stirlingshire); smooth; strawsmeer; sweet Billy (Nottinghamshire); Tom Thumb (Roxburghshire); white wren (Scotland); Willie muftie (Scotland); willow sparrow; willow wren; yellow wren.*

Although almost identical to the Chiffchaff in appearance, the song of this common and widespread summer visitor, a long, rippling warble, reveals its true identity. Like the Wood Warbler, it is a tree-living but ground-nesting bird and has always been predominantly a northern species. Eric Simms said of it:

> The Willow Warbler is quite likely to be the one species of warbler that attracts the attention of the layman as he walks through the woods due to the persistence and volume of its delicate and charming warblings. Thomas Bewick noted that many naturalists called it 'the Liquid-noted willow wren'.[377]

Its beautiful, domed, feather-lined nest, hidden and almost buried on the ground has inspired several of its regional names.

The Goldcrest *(Regulus regulus)* is the tiniest bird in Europe. It is most commonly found in coniferous forest.

Goldcrest *(Regulus regulus)*
OTHER COMMON NAMES: *golden-crested wren; golden cutty (Hampshire); golden wren (Stirlingshire); herring spink (east Suffolk); kinglet; marigold finch; miller's thumb (Roxburghshire); moon; moonie; muin (Roxburghshire); thumb bird (Hampshire); tidley goldfinch (Devon); tot o'er the seas (East Anglia); woodcock pilot (Yorkshire – coastal areas); wood titmouse (Cornwall).*
See Firecrest below.

Firecrest *(Regulus ignicapillus)*
OTHER COMMON NAMES: *fire-crested kinglet; fire-crested wren; fire-crowned kinglet.*
The Goldcrest and Firecrest belong to a distinct group of tiny and extremely beautiful warblers. The Goldcrest is the smallest European bird, a resident species distinguished by a yellow crown edged with black. The Firecrest is a winter visitor, sometimes breeding, and is very similar but with a white stripe above the eye. Both (but especially the Goldcrest) are most commonly found in coniferous woodland, where they build exquisite little suspended nests of moss, cobweb and feathers. Outside the breeding season, the Goldcrest is often seen in the company of tits. Like the Dartford Warbler, it is entirely insectivorous and so suffers severely in cold winters. Although officially there is said to be concern about the Goldcrest, it certainly appears to have become much more common in many gardens, where it seems to have adapted to ornamental conifers. In this respect it appears to be at variance with official policy in finding Leyland Cypress an agreeable habitat.

Curious names like 'herring spink' and 'tot o'er the seas' allude to the way that migrating birds are (or were) often trapped in large numbers among the rigging of North Sea fishing boats. Charles Swainson was told that Goldcrests sometimes clustered, exhausted, like bees along the hedges near the Norfolk coastal town of Caister and could be picked off by hand.[378] Eric Simms recounted similar stories and reported a flock of some 2,000 birds near Saltfleetby in Lincolnshire in 1983.[379]

The name 'woodcock pilot' derives from an old Yorkshire belief that their arrival just preceded that of the Woodcock. There have long been stories of Scandinavian Goldcrests hitching a ride to Britain on other birds' backs and this is now more or less accepted by ornithologists.

Flycatchers
(Muscicapidae)

Flycatchers do indeed catch flies, and other insects too, in very characteristic fashion: by sitting on a branch or other prominent perch and suddenly darting away to obtain their prey, returning again to the same spot. It is a charming habit and perhaps in consequence the birds have been looked on kindly.

Spotted Flycatcher

(Muscicapa striata)

OTHER COMMON NAMES: *beam bird; bee bird (Norfolk); chait (Worcestershire); chancider; cherry chopper; cherry snipe; cherry sucker; cobweb (Northamptonshire); grey flycatcher; lead-coloured flycatcher (Northamptonshire); miller (Shropshire); post bird (Kent); rafter; rafter bird; spider catcher; wall bird (East Anglia, Buckinghamshire, Berkshire, Hampshire); wall chat (Yorkshire); wall plat (Devon); wall robin (Cheshire); white baker; white wall (Northamptonshire).*

There is a Somerset tradition that flycatchers bring good luck:

> If you scare the fly-catcher away,
> No good luck will with you stay.

As well as catching insects in flight, the Pied Flycatcher *(Ficedula hypoleuca)* also collects caterpillars from nearby foliage.

It is a bird that could do with a little luck of its own because it was once among the commonest small summer visitors throughout the British Isles, familiar from its habit of catching flies.

> Spotted Flycatchers will commonly feed well into dusk in their nesting areas. But where there is artificial garden lighting, fly catching may be extended into late evening, especially during warm weather conditions. I have seen this behaviour on several occasions in Somerset in recent years.[380]

It is a pale brown bird with a white breast that is characteristically spotted. The many references to walls, beams and similar places in its regional names allude to some of its favoured nesting sites. Regrettably, the species as a whole is fast becoming a fading memory for many people; it has declined dramatically in numbers for some reason, possibly connected with its fate once it migrates from British shores.

Pied Flycatcher *(Ficedula hypoleuca)*

OTHER COMMON NAMES: *coldfinch; colefinch (Northumberland, Cumberland, Westmorland).*

The Pied Flycatcher is a much more local summer visitor than the Spotted Flycatcher, being mainly restricted to the north and west of Britain. It is a shy bird, not commonly seen. Evidence suggests that its numbers fluctuate and it appears to be less common than it once was. The male is strikingly coloured, mainly black above and white below, while the female is grey-brown above.

The strange name 'coldfinch' must be a corruption of 'colefinch' (black finch).

Tits
(Paridae), including Timaliidae and Aegithalidae

Tits are among the most familiar of all British birds and three species of these small, plump creatures have been voted among the 'top 20' garden visitors. Most are attractively, even garishly coloured (both sexes are similar) and they have a special appeal as they swing acrobatically on branches or bird tables. Tits are omnivorous feeders and all the six true species in Britain, as well as their two relatives (the Long-tailed Tit and Bearded Tit), which are included here out of convenience and convention, are resident in Britain, although some have a very restricted distribution.

'Tit' is an abbreviation, although now almost universally used, for the fourteenth-century 'titmouse', or 'titmose', 'tit' (in one of its several senses) meaning any small creature and 'mose' being the ancient name for these birds. It therefore has no linguistic connection with 'mouse', although the sometimes mouse-like movements of the birds probably encouraged its adoption. The plural 'titmice' has presumably resulted from this fortuitous connection and 'tit' sometimes occurs in the regional names of other small birds, such as the Wren and larks. Most traditional beliefs and folklore do not distinguish between the species and most tits, at different times, have been popular as captive cage birds.

Does the well-known children's taunt have any special connection with tits?

> Tell pie tit,
> Your tongue will be slit,
> And all the little dicky birds
> Will have a little bit.[381]

The Derbyshire version is 'tell tale tit'. 'Blab tale tit' is another version, while the Yorkshire variant is:

> Tell Pie Tit
> Laid an egg and couldn't sit.[382]

Bearded Tit (*Panurus biarmicus;* Family Timaliidae)
OTHER COMMON NAMES: *bearded pinnock; beardmanica; least butcher bird; lesser butcher bird; pheasant (Norfolk); reed bunting (Essex); reedling (East Anglia); reed pheasant (Norfolk); water pheasant (Sussex).*
The Bearded Tit is related to such tropical birds as babblers and parrotbills rather than true tits, but it is included here because of its widely used common name. However, 'moustached tit' might be more appropriate, for the male bird has two moustache-like black bands on either side of the face. The charming name ''bearded pinnock' must be related to the thirteenth-century name 'pinnock' for the Dunnock, although there is no obvious reason to connect the two birds.

It is a bird of reed beds and marshes, a habitat accurately reflected in the local East Anglian names, but it is nonetheless one that has gradually been extending its range, although its numbers are subject to fluctuations, largely brought on by cold winters.

Long-tailed Tit (*Aegithalos caudatus*; Family Aegithalidae)
OTHER COMMON NAMES: *bag (Northamptonshire); bellringer (Kirkcudbrightshire); bottle jug (Yorkshire); bottle tit (West Yorkshire, Shropshire, Buckinghamshire, Berkshire); bottle tit; bottle tom; bum barrel (Midlands); bum towel (Somerset, Devon); bush oven (Norfolk); buttermilk can (Warwickshire); can bottle (Shropshire); dog-tail (Cheshire); feather poke; French pie; fuffit (East Lothian); hedge jug; hedge mumruffin (Shropshire, Worcestershire); huckmuck (Wiltshire); jack-in-a-bottle; juffit; jug pot (Nottinghamshire); Kitty longtail; long-pod (Midlands); long-tailed capon (Norfolk, Hampshire); long-tailed chittering; long-tailed mag; long-tailed mufflin; long-tailed pie; miller's thumb; millithrum; mumruffin (Shropshire, Worcestershire); nimble tailor; oven bird, oven builder (Stirlingshire); oven's nest (Northamptonshire); poke bag, poke pudding (Shropshire, Gloucestershire); prinpriddle; pudding bag (Norfolk); puddney poke (Suffolk); ragamuffin.*
Arguably the prettiest and daintiest of the group, the Long-tailed Tit is not closely related to the six species of the family Paridae, but it is a familiar bird in woodlands as well as hedgerows and gardens throughout most of Britain. Few British birds have accumulated a large collection of names and certainly very few have acquired such charming ones.

These tits are almost invariably seen in small flocks and, although in the past they have not been common bird-table visitors, this seems to be changing, perhaps in response to the milder winters that have led to an overall increase in their numbers. The name 'Long-tailed Tit' says it all: the birds are like tiny balls of fluff with very long tails. Names such as 'mufflin', perhaps related to 'muffle', seem to allude to their apparent fluffiness.

They build possibly the most beautiful nest of any British bird: a spherical mass of feathers, moss, fur and other soft material with a tiny hole, all bound together with cobwebs and

The nest of the Long-tailed Tit *(Aegithalos caudatus)* is an elaborate structure of moss, fur, feathers, hair and cobwebs.

covered in lichen. Regional names that include 'bottle', 'oven' and 'barrel' seem to reflect this shape.

> When I was at school in the 1920s at Little Compton, Warwickshire, the boys used to call the nests Bum-Barrel's nests.[383]

Was it the Long-tailed Tit's nest that inspired 'Tommy Tittle-mouse' in *Beatrix Potter's Nursery Rhyme Book* or the nest of a real mouse?

> I've heard that Tommy
> Tittle-mouse
> Lived in a tiny little house,
> Thatched with a roof of
> rushes brown
> and lined with hay and
> thistle-down.
> Walled with woven grass and
> moss,
> Pegged down with willow
> twigs across.
> Now wasn't that a charming house
> For little Tommy
> Tittle-mouse?

The Coal Tit *(Parus ater)* is a bird of coniferous woodland, although it is also seen in deciduous woods and in hedgerows.

Marsh Tit *(Parus palustris)*

OTHER COMMON NAMES: *black cap (Nottinghamshire); black-headed tit (Somerset); Joe Ben (East Anglia); marsh titmouse; saw whetter (Staffordshire); willow biter (Nottinghamshire).*

A rich, glossy black head distinguishes the Marsh Tit from the very similar Willow Tit. Although it has no association with marshes, it does tend to nest in willows and alders. It is a bird of hedges and woodlands generally but is not a frequent garden visitor.

The name 'Marsh Tit' has been used since the sixteenth century; 'black cap' may be more appropriate but leads to confusion with the warbler of the same name and with the Coal Tit.

Willow Tit *(Parus montanus)*

The Willow Tit was for a long time confused with the Marsh Tit and it was not recognized as a separate species until 1897. Since then it appears to have been declining and its future as a British breeding bird does not seem good. The principal feature distinguishing the two species is that the crown of the Willow Tit is a dull, sooty black, not glossy; also it is much more a bird of marshes.

Crested Tit *(Parus cristatus)*

A distinctive bird with a black-and-white speckled crest (the only crested small British bird), the Crested Tit is seen by very few people because it is virtually restricted to the pine forests of Speyside. As a Scottish bird, it has never acquired an English folk name.

Coal Tit *(Parus ater)*

OTHER COMMON NAMES: *black cap (Stirlingshire, Shropshire); black ox-eye (Angus – Forfar); coal; coal hooden (East Lothian); coalmouse (Ireland); coaly hood (Scotland); little blackcap (Yorkshire); tomtit (Ireland).*

The Coal Tit is the only one of the three familiar garden-visiting tits to have no obvious yellow coloration, the mainly black head with white cheeks giving it its common name. Historically it was called the 'coalmouse', or 'colemouse', although this name was also used for the Marsh Tit, Willow Tit and Great Tit. Despite its frequent appearance in gardens, it is naturally a bird of coniferous woodland.

> After the 1987 storm, Coal Tits disappeared, now they are back (it took two years).[384]

Blue Tit *(Parus caeruleus)*

OTHER COMMON NAMES: *ackimule, ackmaull (West Country); allecampagne (Cornwall); bee bird (Hampshire); Billy biter (North*

Yorkshire, Shropshire); blue bonnet (Scotland, West Yorkshire; Shropshire); blue cap; blue ox-eye (Angus – Forfar); blue spick (north Devon); blue yaup (Scotland); hackmal, hagmal, heckymal, hickmall (west Somerset, Devon, Cornwall); Jenny wren (North Yorkshire – Craven); nun; ox eye (East Lothian); pedn-play; pickcheese (Norfolk); pinchem (Bedfordshire); pridden pral (west Cornwall); stonechat (Ireland); tidife; tinnock; titmal (west Somerset, Devon, Cornwall); tom nouf (Shropshire); tom tit; uckimol (West Country); yaup (Renfrewshire).

The most familiar member of the tit family, this is also among the best loved of all British birds. It is the only tit species with any blue coloration and is a distinctive visitor to gardens and bird tables. Despite its appealing appearance, however, it can be very aggressive and the local name 'Billy biter' is a reference to the way in which it will peck and hiss to protect its nest. 'Hickmall' and similar West Country names seem to be derived from an old verb 'hick', meaning 'to peck'. Henry Williamson, in *Tarka the Otter*, quoted 'an old man whose pea-shucks had been hacked open by Blue Tits', who said: 'Withering things they ackimules: they ought to be kicked to flames!'[385]

Blue Tits, and to a lesser extent Great Tits, displayed a fascinating and well-publicized adaptation to human affairs when they discovered the knack of pecking off milk-bottle tops on people's doorsteps to gain access to the contents. The practice was first observed in Southampton in 1929 and clearly the tits learned from each other. Milk-delivery vans were sometimes followed by flocks of birds and doorsteps were once regularly littered with fragments of cardboard bottle-tops. The large-scale irruptions of Blue Tits in 1949 and 1957 were followed by a considerable increase in bottle-top pecking and also by attacks on washing lines, putty in window frames and paper-tearing. A careful survey in 1957 revealed the following number of incidents involving different 'targets': wallpaper (66), books (27), boxes (20), newspaper (17), lampshades, (13), notices (9), labels (7), and a further 57 involving calendars, silver paper, letters, photograph and picture frames, parcels, blotting paper, toilet rolls, tissue paper and razor-blade packets.[386]

Since the disappearance of cardboard bottle-tops and the decline in milk deliveries, this interesting bird behaviour has gone forever, but paper-tearing has been a long-standing feature of the species. As long ago as 1693, Father Jean Imberdis wrote (translation by Professor Eric Laughton):

> Nor will it easier be – nay, not a whit –
> To keep from your domain the greedy Tit.
> Small is the naughty Fowl, yet it can wreak
> No small Destruction with its claws and beak.
> For, when Paper from afar it spies,
> Straightaway through open Window in it flies,
> Its frequent blows the sheet do quickly tear
> Still sodden, and make Havoc everywhere ...
> To Gin and Snare it grows too soon inured,
> And Carelessness is by Experience cured.
> The Lime untouched, always the naughty Tit,
> So keen its zest, to Paper straight will flit.[387]

Blue Tits *(Parus caeruleus)* were notorious for pecking off the tops of milk bottles left on doorsteps.

Great Tit *(Parus major)*

OTHER COMMON NAMES: *ackimule, ackmaull (West Country); big ox-eye (Angus – Forfar, East Lothian, Roxburghshire); black cap; black-capped lolly (West Yorkshire, Northamptonshire); black-headed Bob (Devon); black-headed tomtit (Stirlingshire, Shropshire); eckmall, hackmall (West Country); heckymal (Dartmoor); hickmal (West Country); Joe Ben (Suffolk); Joe Bent (Gloucestershire); ox-eye (Ireland, North Yorkshire, Midlands, Shropshire); pridden pral (west Cornwall); sawfinch; saw sharpener (Roxburghshire); saw whetter (Norfolk); sharp saw (Norfolk); sit ye down; tom noup (Shropshire); uckimol (West Country).*

This is the largest British tit and a common and familiar bird. Considerably larger than the Blue Tit and Coal Tit, it most closely resembles the Coal Tit but with bright yellow underparts.

The names 'Joe Ben' and 'Joe Bent' are probably onomatopoeic of the bisyllabic call; 'saw sharpener' and other 'saw' references are similarly descriptive of the call. 'Ox-eye', like 'bull's-eye', is an old term for anything small.

Nuthatches
(Sittidae)

Nuthatch *(Sitta europaea)*
OTHER COMMON NAMES: *blue leg (Sussex); jar bird; jobbin (Northamptonshire); mud dabber (Somerset); mud stopper (southern England); nutcracker (Shropshire); nuthack; nut jobber (Berkshire, Somerset, Sussex); nut topper; woodcracker; woodhacker (Surrey); woodjar (Gloucestershire); woodpecker (Surrey).*
A small, beautiful, tree-climbing bird, distinguished from the Treecreeper by its slightly larger size, the Nuthatch has a grey back and buff underparts. Unlike the Treecreeper, it does not use its tail as a support, although it similarly collects insects and also eats seeds, including nuts. It is said to be the only British bird that can (or does) run down a tree trunk. Its nesting habits have attracted as much attention as its appearance: 'mud jabber' and 'mud stopper' both allude to its habit of plastering mud around the nest hole to leave just sufficient room for the bird itself to enter.

Nuthatches seem to be on the increase; they have gradually extended northwards to regain some of their old territories and their future looks secure.

The name 'Nuthatch' is old; it occurred as 'notehache' in the mid-fourteenth century and is related to the word 'hack'.

The Nuthatch *(Sitta europaea)* is a tree-climber, usually nesting in holes in trees or walls but is not averse to nest boxes.

All names refer to the Nuthatch's habit of wedging nuts in the bark of trees and striking them with its beak. 'Jobber' and 'jobbin' are from the obsolete verb 'to job', meaning 'to jab or peck'. The Nuthatch was immortalized in John Clare's poem (see Jay, page 357).

Treecreepers
(Certhiidae)

Treecreeper *(Certhia familiaris)*
OTHER COMMON NAMES: *bark creeper, bark runner (Suffolk); bark speiler (East Lothian); brown treecreeper; brown woodpecker; climb tree (Somerset); creeper; creep tree (Norfolk); creepy tree (Yorkshire); cuddy (Northamptonshire); daddy-ike; eeckle (Gloucestershire); little woodpecker (Yorkshire); ox-eye creeper; tomtit (Ireland); tree climber; tree climmer (Surrey, Sussex); tree clipper (Oxfordshire); tree crawler (Sussex); tree mouse (Somerset); tree speiler (East Lothian); woodpecker (Ireland, Perth).*
The Treecreeper's similarity to a mouse is recognized in the West Country name 'tree mouse', although, coincidentally, the bird commonly associates with tits (titmice) and flies like them. One of its alternative names, 'tomtit', recognizes this. 'Spelier', or 'speeler', is a Scottish word for 'climber', while 'cuddy', a local word for any small creature, is also applied to the Dunnock and the Wren. The name 'Treecreeper' replaced 'creeper', or 'common creeper', in the nineteenth century.

A small bird, mottled brown with buff wing-bars, it has large claws, a stiff, fairly long tail and a long, downcurved bill. It is still a fairly common resident throughout the British Isles, except the far north and west, and, like so many other insectivorous birds, its only really serious threat is severe winters.

Orioles
(Oriolidae)

Golden Oriole *(Oriolus oriolus)*
Arguably the most beautiful British bird, the Golden Oriole is the size of a large thrush, the male being a stunning brilliant yellow. The call of short, sharp notes can be mistaken for that of a Jay, although the male oriole also produces a melodious, clear, far-carrying, flute-like call. This is said to be the origin of its onomatopoeic name, rather than a corruption of the Latin *aureolus*, meaning ' golden'.

In the fourteenth century and earlier, the bird was known as the 'woodwale' but, by the late fifteenth century, this name (variously spelled 'woodwall', 'woodwele' or 'woodweele', among other verions) had been transferred to the Green Woodpecker (see page 315). Presumably this woodpecker was becoming more common as the Golden Oriole became rarer. Both live in similar habitats and the female Golden Oriole is distinctly green. An anonymous medieval poet wrote:

The Woodweele sang and wolde not cease
Sitting upon the spray
So lowde he wakened Robin Hood
In the greenwood where he lay.

Similarly, Chaucer, in *The Romaunt of the Rose*, observed:

In many places were nyghtyngales,
Alpes, fynches and woodwales.

The Golden Oriole is a summer visitor to southeast England and has nested regularly in Britain in the past, although it has always been persecuted by trophy-hunters. It began to disappear from many of its traditional nesting sites in parks and woods, but its colonization of poplar plantations in the east of England has prompted a small resurgence. This has continued despite the fact that an increasing number of smokers are now using disposable cigarette-lighters; this seemingly spurious connection arises because poplars, which were once grown largely for match production, have been felled because of the decreasing demand for matches.

Shrikes
(Laniidae)

Despite their small size, shrikes are aggressive hunters with hooked beaks and a behaviour more reminiscent of a bird of prey. Any victims surplus to their immediate needs are impaled on thorns and, in this manner, they accumulate small larders. 'Butcher bird' is a widely used and most appropriate alternative name.

Red-backed Shrike *(Lanius collurio)*

OTHER COMMON NAMES: *butcher bird; butcher boy (Surrey); cheeter (Somerset); cuckoo's maid (Herefordshire); flesher; flusher strike (Surrey); granfer (Somerset); horse-match (Gloucestershire); jack baker (Hampshire, Surrey); murdering bird; nine killer; pope (Hampshire); weirangle (northern England); wurger; worrier (Yorkshire).*

A very attractive bird with a chestnut back, blue-grey crown, black eye-stripe and white chin, the Red-backed Shrike has steadily retreated southwards and eastwards and its numbers are now so diminished that it counts as a 'lost' British breeding species. Its prey consists mainly of large insects and lizards and

The Golden Oriole *(Oriolus oriolus)* has begun to colonize poplar plantations in eastern England.

The Red-backed Shrike *(Lanius collurio)* impales its surplus prey on thorns or barbed wire for future use.

its dramatic decline has been attributed mainly to loss of habitat, especially the rough pasture and scrub that it favours.

The wide range of generally descriptive regional names is nonetheless testimony to its past importance as a British bird. The odd name 'shrike' has no obvious origin and may simply be borrowed from the other bird to which it has been applied, the Mistle Thrush, where it is probably onomatopoeic. The name 'nine killer' can be traced back to 'nyn murder', which occurred first in 1544 and relates to an old belief that the bird kills nine other birds every day. 'Wierangle' (also 'wirrangle', or 'wariangle') dates to *c.*1386 and the word 'waryangle', which is Germanic in origin but otherwise of elusive meaning.

> This Somnur which that was as ful of jangles
> As ful of vermin been thise wariangles
> And ever enquering up-on every thing ...
> (Chaucer, 'The Friar's Tale')

Great Grey Shrike *(Lanius excubitor)*

OTHER COMMON NAMES: *butcher bird; great shrike; murdering bird; murdering pie; sentinel shrike; shreek; skriek; skrike; white whisky John; wierangle (Derbyshire).*

This is the largest shrike, the size of a Blackbird, grey above, white beneath and with black wings. A summer visitor to the east of Britain, it is found at woodland edges and in open countryside with trees and bushes.

The word 'whisky' in one of the regional names has nothing to do with the alcoholic beverage but relates to the way that the tail whisks from side to side.

Crows
(Corvidae)

There are few British birds as obvious, noisy and imposing as the six characteristic black species of this family, nor are the Magpie and Jay likely to be overlooked. They are highly intelligent and are often thought the most advanced and long-lived of all birds. All eight are residents and their biology and, to a large degree, their extensive folklore are similar so it is difficult to relate many of the older beliefs to any particular species.

The name 'crow' can be traced back to the Old English 'crawe', but there is some confusion because the word is often used to refer to birds, especially troublesome birds, in general: scarecrows are used to scare Starlings and pigeons as readily as true crows and Rooks.

The appearance of any crow is usually thought to portend events both momentous and miserable, and the name has come to be used for anything comparably raucous, brooding, sinister or simply coarse. The plant Crow Garlic *(Allium vineale)* is cruder than the real thing and a 'crow-bar' is a rough, unsophisticated tool with an end like a bird's beak.

Jay *(Garrulus glandarius)*

OTHER COMMON NAMES: *blue jay (Lothian); devil scritch (Somerset); gae (Scotland); jay pie (Midlands, Devon, Cornwall); jay piet (Perth); Jenny jay (North Yorkshire); kae; oak jackdaw; scold (Somerset).*

It is extraordinary that so common, widespread, large, conspicuous and distinctive bird as the Jay has accumulated almost no tradition or folklore. It is a pink-brown crow with a striking white rump, black tail and blue and black bars on the wings.

Jays are common almost everywhere in Britain and, although now rather less frequent on farmland, they have begun to spread even into the far north of Scotland, and are often seen and heard in small noisy groups, especially close to trees in parkland and gardens, where they nest. They are omnivorous and will eat living prey up to the size of small birds.

The name 'Jay' first appeared in the thirteenth century, having arrived via Old French from Latin. It is often said to be derived from the old French word *gai*, meaning 'gay', on account of its colour (which is untypical for a crow), but the *Oxford English Dictionary* states unequivocally 'not identical with French *gai*'. 'Jay pie' and 'jay piet' nonetheless serve as reminders that the Jay and the Magpie were once grouped together. The scientific name *Garrulus* is a reference to its harsh, chattering call.

John Clare was almost alone in his literary commemoration of the Jay, and then only because he mistook it for a Nuthatch:

This adult Jay *(Garrulus glandarius)*, with its begging youngster, is holding an acorn in its beak.

In summer showers a skreeking noise is heard
Deep in the woods of some uncommon bird
It makes a loud and long and loud continued noise
And often stops the speed of men and boys
They think somebody mocks and goes along
And never thinks the nuthatch makes the song
Who always comes along the summer guest
The birdnest hunters never found the nest
The schoolboy hears the noise from day to day
And stoops among the thorns to find a way
And starts the jay bird from the bushes green
He looks and sees a nest he's never seen
And takes the spotted eggs with many joys
And thinks he found the bird that made the noise.
(John Clare, 'The Nuthatch')

The American expression 'jay-walker', for someone who crosses the road (highway) carelessly, stems from a common late-nineteenth-century use of the word 'jay' to mean 'a simpleton', although its connection with the bird is not obvious.

Magpie *(Pica pica)*

OTHER COMMON NAMES: *bush magpie; chattermag; chatternag (Somerset); chatterpie (Norfolk, Somerset); Cornish pheasant (Cornwall); haggister; longtailed nan; madge; mag; Margaret's pie; maggit (Worcestershire); maggot (Lincolnshire, Worcestershire, Gloucestershire); magot pie (eastern England); marget; miggy (northern England); mock-a-pie; nanpie (North Yorkshire – Craven); ninut (Nottinghamshire); pianate; piannot (Cheshire); pie; piet (Westmorland); pyat, pyet, pyot (Scotland); pyenate (West Yorkshire); pynot (Derbyshire); tell pienot, tell piet (North Yorkshire).*

Probably no other piece of ancient bird lore is used on a more regular basis today than the old Magpie rhymes beginning with the words:

One for sorrow, two for joy.

But can you remember the next line:

Three for a wedding, four for a boy.[388]

Even fewer people ('The rhyme only became known to me in my teens ... but a single Magpie was always bad news for my parents'[389]) know the continuation:

Five for silver, six for gold,
Seven for a secret ne'er to be told.[390]

Another version runs:

One for sorrow, two for mirth
Three for a wedding, four for a birth.[391]

In Derbyshire, the third line was 'Three for a girl' and this was obviously prevalent in other counties:

Magpies – one for sorrow, two for joy, three for a girl, four for a boy; handed down in our family in Hertfordshire since 1730.[392]

Yet another version is:

I saw eight magpies in a tree,
Two for you and six for me.
One for sorrow, two for mirth,
Three for a wedding, four for a birth.
Five for England, six for France,
Seven for a fiddler, eight for a dance.[393]

And there are still more.

It is said that the unfortunate effects of seeing one Magpie can be cancelled out by seeing a crow immediately afterwards or, in some places, by crossing yourself.

Whenever as we drove to Beaminster we saw one or two Magpies, an old lady who I used to take to the next village would cross herself. She couldn't tell me why other than to say that Magpies were the devil's birds.[394]

Raising your hat was sometimes thought to help, although sometimes even to see two or more Magpies was thought a bad omen if they were flying from right to left.

One magpie was unlucky. If seen one had to turn money over in one's pocket. If one had any that is. Multiples of magpies were fine.[395]

> My grandfather (Scottish) taught me to either hold my collar when I saw a Magpie until I'd seen four other birds, or bow and wish it good day. I still do that and as there seem to be more Magpies around, when out for a drive, I spend a lot of time nodding my head and greeting them.[396]

> Every time we see a Magpie we say 'Good Morning, Mr Magpie, please give our regards to Mrs Magpie and all her baby Magpies'. We are fairly normal, middle-aged people so just why do I feel ever so slightly panicky if I think I've forgotten to say it?[397]

Even battle-hardened soldiers in the modern British Army will always salute a Magpie. It is an unmistakable large black-and-white crow, whose distinctive appearance has long attracted attention. It is very aggressive and omnivorous and, although heavily persecuted in the past, it has trebled in numbers in the past 40 years and has recently reached serious pest status in many parts of Britain, acquiring much (mainly unjustifiable) blame for the decline in populations of a number of once-common songbirds.

The name 'Mag' is of mysterious origin but stems from the same root as 'Margaret', although perhaps via the French *margot*; it seems to have nothing to do with 'maggot' meaning 'a fly larva'. 'Pie' is a thirteenth-century word originating in the Latin name *Pica* and, because of the bird's highly distinctive coloration, the word 'pied' has been widely used since the fourteenth century to mean anything black and white. Curiously, the word 'pica' is used (mainly in medical circles) to mean 'a perverted craving for substances unfit for food, like soil' and it may possibly be connected with the Magpie's kleptomanic habits. 'Piebald' is of the same derivation as 'pica' and the name 'Magpie' itself is used extensively in the same way; no supporter of the black-and-white-striped Newcastle United football team needs to be told the extent to which it has now entered everyday currency. To the Romanies, the Magpie is still 'magot pie'; they use 'magot' familiarly, 'like Robin to a redbreast, Tom to a titmouse or Philip to a sparrow', and in the west of England still use the name 'magatipie'. 'Pyat' and similar names are widespread in Scotland, for instance, in Robert Louis Stevenson's *Kidnapped*:

> 'Mungo,' said he, 'There's many a man would think this more of a warning than two pyats.'[398]

The Magpie has a chattering call and, in Scotland, it was believed that, if its tongue was scratched (ideally with the sharp edge of an old thin silver sixpence, in the days before they were milled) and a drop of human blood inserted in it, the chattering Magpie would acquire the gift of speech.

In common with other crows, and possibly second only to the Raven, Magpies have attracted pagan or devilish significance. It was said, although with no real Biblical evidence, to be the only bird that refused to enter the Ark, perching instead on

The Magpie's *(Pica pica)* love of bright objects has not escaped Clementina M. Hull in her watercolour *Still Life with Magpie*.

top of the roof and chattering as the world was engulfed by the Deluge. According to a related tale, the Magpie was a hybrid of the two birds, the black Raven and the white dove, that were sent out from the Ark by Noah. Its association with ill luck is also biblical and may date from its being one of two birds claimed to have perched upon Christ's cross. At that time, it was said, the Magpie had bright plumage but, in retribution, it was cursed by the Lord to be sad and sombre for ever more and its nest to be open to the storm (a quite erroneous observation because the nest usually has a characteristically domed cover). In Northern Ireland, it is said that the Magpie seldom keeps still because the devil dropped three spots of blood on its tail.

Seeing Magpies sometimes produces more than just odd verses. The sight of a solitary bird encouraged expectoration, at least in Staffordshire, where it was also customary to raise one's hat while simultaneously spitting at the bird and calling 'Devil, devil, I defy you' (not an easy feat). In Dorset, the advice was not only to raise your hat and spit but to do so three times over your left shoulder; if this failed to dislodge the Magpie, the consequences might be serious because, according to another Dorset saying, if the Magpie remained all day while a man worked in a ploughed field, the man would die soon afterwards. Nonetheless, certain farmers may have considered themselves immune from harm because they sometimes kept tame Magpies as watch-birds to give warning of intruders.

Several regional beliefs hold that merely hearing a Magpie presaged ill-luck but that things would change when it went:

Magpie, magpie, flutter and flee,
Turn up your tail and good luck come to me.[399]

In the days of cock-fighting, fowl eggs were sometimes placed in a Magpie's nest in the belief that the young would absorb their foster-parents' aggression, although there seems to be no evidence that they ever hatched. A Caernarvon custom was to break the eggs of Magpies, crows and Blackbirds on May Day, although for what reason is obscure.

Surprisingly for so distinctive a bird, the Magpie features rarely in heraldry. It is also unusual on inn signs, although The Cock and Pye and a Magpie and Crown exist, but there is one Magpie inn that surpasses all others in having played a major part in shaping British history. Not far north of Chesterfield in Derbyshire is the village of Old Whittington and a pretty little thatched cottage, now a museum, called Revolution House. For many years, this building was an ale house called The Cock and Pynot, 'pynot' being an old Derbyshire name for the Magpie. In 1688, William Cavendish, the fourth Earl (and later the first Duke) of Devonshire, who built his seat at nearby Chatsworth, met other local dignitaries here and together they agreed to support the plan to overthrow the Catholic King James II and offer the throne to William of Orange. Cavendish was one of the seven signatories of the letter of invitation sent to The Hague. They succeeded; William took the throne as William III and, in the fullness of time, met his own end through the actions of a Mole (see page 392). The Cock and Pynot sign continued to

'Five for silver' is the saying, although 'five for dinner' might be more appropriate here. Magpies *(Pica pica)* gather around a rabbit carcass.

hang across the road until 1850, but the building gradually fell into neglect and the plotting parlour itself was lost. The museum is now the property of the Borough of Chesterfield.

There seems to be no English place name inspired by Magpies and they have only infrequently appeared in literature, although Shakespeare evidently knew about them:

> Augurs, and understood relations have
> By maggot-pies and choughs and rooks brought forth
> The secret'st man of blood.
> (*Macbeth*, Act 3, Scene 4)

There are several references to dried and powdered Magpie flesh being used as a treatment for epilepsy (using the chatterer to cure a chatterer) and, among other claims, portions of Magpie were reputed to cure melancholy, vertigo and poor eyesight. Eating a Magpie's leg would help someone bewitched and, in Gaelic legend, the bird was believed to be a witch named Dodiag.

Chough *(Pyrrhocorax pyrrhocorax)*

OTHER COMMON NAMES: *chauk; chauk daw; chawk; cliff daw; Cornish chough; Cornish daw; Cornish jack; Cornish jay; Cornish kae; daw; hermit crow; killigrew (Cornwall); long-billed chough; Market Jew crow (Cornwall); palores; red-legged crow; sea crow (Ireland); tsauha.*

The Chough is much the most locally distributed of British crows, being restricted to coastal cliffs and mountains in the far west, although it has vanished from its traditional Cornish haunts. At present, its main British strongholds are the Isle of Man, parts of Wales, Islay and parts of western Ireland. It is slightly larger than the Jackdaw but is readily distinguished by its all-black plumage and especially its curved red beak and red legs, which differentiate it from all other European crows. Choughs are sociable birds, feeding mainly on worms and insects, which they obtain by probing in short turf; it is the diminution in this type of habitat that has largely prompted their decline. They nest in deep rock crevices.

For a bird with so restricted a distribution, the Chough has achieved significant notice over the centuries, although there is always an element of uncertainty in older references because the name 'Chough' was also applied to the Jackdaw and it was not until the past 300 years or so that clear distinctions have been made. References to the name 'Cornish chough' (which first appeared in 1544), 'Cornyssh daw' (*c.*1575) and 'Cornwall kae' (1684) clearly refer to *Pyrrhocorax*, as does 'red-legged crow', which was used for a time in the late eighteenth century. Not until the nineteenth century, however, did plain, simple 'Chough' become the accepted term. The name originated in the beginning of the fourteenth century as 'chogen', later 'chowgh', or 'choughe', with a guttural 'gh'. The name' Market Jew crow' originates in the Cornish village of Marazion, which was also known as Market Jew and was a place where the birds congregated.

Choughs *(Pyrrhocorax pyrrhocorax)* usually nest on ledges in cliffs, or caves, but this pair is nesting in a roof space.

The Chough (the real Chough, unmistakable by its curved beak and red extremities – 'beaked and legged gules') is a fairly common bird in heraldry, especially in Cornwall. It occurs today on the arms of the County of Cornwall and, since 1380, of the City of Canterbury. The latter is a testimony to the three Choughs that appeared, for no immediately obvious reason, on the arms of Thomas à Becket. In fact, this is one of heraldry's more obscure puns: the word *becqué* in French and heraldry means 'beaked' and, in Old French, Choughs are sometimes called *becquets* because they have large beaks. In practice, the coat of arms was probably created for St Thomas some time after his death because the heraldic system was not in use until the thirteenth century.[400] The Three Choughs, rather than a single one, is a common inn name, presumably commemorating St Thomas.

Understandably, the Chough features most in the folklore of Cornwall and the far west, including the Channel Islands. It signified bad luck on Guernsey and was linked with witchcraft, witches being thought to wear red stockings for identification (although why they should wish to advertise their wickedness in such a way is not recorded). According to Welsh legend, King Arthur assumes the guise of a Chough (or a Raven or Puffin – all conveniently common on the sea cliffs) when he emerges from his sleep in a mountain cave.

The call of the Chough is more drawn-out and melodious than that of the Jackdaw, although clearly not melodious enough for the poet Cowper, who placed it among the 'birds obscene of ominous note'.

The Jackdaw *(Corvus monedula)* sometimes steals other birds' eggs and fledglings.

Jackdaw *(Corvus monedula)*

OTHER COMMON NAMES: *caddaw (East Anglia); cader, caddy (Norfolk); carder (Suffolk); caw (Banffshire, Kirkcudbrightshire); cawdaw (Ireland, northern England, Suffolk); chatterjack, chawk (western England); daw; grey head; grey neck (Midlothian); jack (general); jackie, jaikie (Midlothian, East Lothian, Berwickshire); kya (Wigtownshire); paiet, pate (Kirkcudbrightshire).*

The Jackdaw is the smallest of the mainly black crows and is a common bird almost everywhere, often preferring to nest in old buildings and cliffs. According to a Norwich saying, 'When three daws are seen on St Peter's vane together, then we are sure to have bad weather', something that must happen fairly often as Jackdaws are generally gregarious. They are also omnivorous, although they feed especially on insects and worms, and are notorious for taking other birds' eggs.

The old name for the bird, 'daw', was probably fifteenth century, originating from the German. 'Jack' was added about a century later for no obvious reason, although it has many animal associations and is often used to denote maleness; it occurs in other animals' names, for instance, 'Jack Snipe', 'jack rabbit'. 'Daw' was also widely used in the sixteenth century and later to mean 'a knavish or stupid person':

> But in these nice sharp quillets of the law,
> Good faith, I am no wiser than a daw.
> (*Henry VI Part 1*, Act 2, Scene 4)

Nonetheless, like all crows, the Jackdaw is commonly thought to be intelligent. It has often been kept in captivity and tamed and can be taught to imitate the human voice; like the Magpie, another chattering bird, it was sadly once believed to do this better if its tongue was slit.

Although it was the Magpie that featured in Rossini's opera *The Thieving Magpie*, the Jackdaw has a greater reputation than the Magpie for being a thief, not only of other birds' eggs but also of human possessions; there are many accounts of accumulations of objects, especially shiny ones, being found in Jackdaws' nests. It is the only British bird regularly to nest in chimneys and similar places and drops sticks down the flue to build up a platform. Edward Jesse described an enormous pile:

> ... built in the Bell Tower of Eton College in seventeen days in May 1842. It measured ten feet in height, and formed a solid stack-work of sticks.[401]

The practice continues today:

> We have Jackdaws nesting in two of our chimneys which are unused. We think sometimes baby daws must fall down the chimneys because in summer we have come home to the upstairs rooms full of flies. We wonder if these have hatched on the dead baby birds.[402]

Despite this common habit, there was a belief in the north of England that a Jackdaw in a chimney presages death in the house. Notwithstanding its association with buildings, it is rare in London; Richard Fitter called it 'a lost London bird', although Max Nicholson knew of one in 1949 that sat on the roof of 10 Downing Street, chattering quietly while a Cabinet meeting was held beneath.[403]

The flesh of the Jackdaw, 'fresh and warm', was claimed to dissolve tumours if held against them and was also thought beneficial against scrofula, or 'King's evil' (a type of tuberculosis affecting the lymph glands in the neck, causing swelling; its popular name derives from the notion that the monarch's touch would heal it).

Rook *(Corvus frugilegus)*

OTHER COMMON NAMES: *barefaced crow; brancher; brandre (Cornwall); cra (Westmorland); craw (North Yorkshire, Lancashire); croaker; crow; percher; Scotch craa.*

Rooks are large crows, readily distinguished by their bare face. They feed mainly on insects, worms and grain and typically nest in tall trees; they lost many favoured nesting sites when Dutch elm disease destroyed the native elms, but overall the population does not seem to have suffered. They are the most gregarious of the British crows, rookeries (a name now used for the communal nesting places of other bird species, and some mammals) often including huge numbers of individuals. For Rooks to desert a rookery has long been considered a bad omen for those who owned the land and may presage a food shortage; famine would be confidently expected in Orkney.

The name 'Rook' originates in the Old English 'hroc', via the Middle English 'roc', although they have often informally been called 'crows'. One of my boyhood friends in Derbyshire would always say that Rooks flying back to their nests were 'going home from school'. The expression 'as the crow flies' refers to Rooks, which typically fly in straight lines. Overall, Rooks have had a better reputation and a better press than most of their relatives, largely because they feed on insects rather than carrion and animals, although they are nonetheless treated as pests because they also eat large quantities of grain. Rooks were linked with crows and Jackdaws in the Act of Parliament of 1593 that protected corn crops, and any parishes that failed to set nets for their capture could be fined 10s (50p) per day. However, William Marshall, in his *Rural Economy of Norfolk* (1787), said that Rooks were rarely shot because of their value in picking up worms and grubs, especially 'the grub of the cockchaffer'.[404]

The fact that Rooks themselves have traditionally been eaten in rural areas seems commonly to give rise to much mirth; a celebrated episode of the popular television comedy series *The Two Ronnies* once featured a restaurant that served nothing else but Rooks. Perhaps the menu included this typical recipe for Rook Pie, which comes from Nottinghamshire:

> Young rooks; flaky or short pastry; a little butter, flour and stock; a little salt and pepper.
>
> Pluck, draw and skin the rooks and remove the back bone (it is bitter). Season with salt and pepper. Stew in a little water. Then, place the birds in a pie-dish and cover with stock, thicken with butter and flour. Cover with pastry and bake for 1 hour and 30 minutes in a moderate oven.[405]

Rooks have long had a special association with foretelling weather: a tumbling or very low flight indicated rain, as did flight to the mountains in the Isle of Man, roosting at mid-day in Devon, and feeding and remaining close to the rookery in both Wiltshire and Cornwall. If Rooks fly high, fine weather is assured.

> Rooks (or Crows) building their nests signalled the type of summer due, but unfortunately I cannot remember which way round it is. If they are high it is a sign of a good/bad summer and if the nests were low down in the tree it's the reverse.[406]

The nests are usually fairly high whatever the weather, hence the seventeenth-century expression 'crow's nest' for a fort built in a high place or, from the early nineteenth century, the barrel or other precarious look-out at the top of the mast of sailing ships. Rooks coming down to feed in the streets or sitting in rows or large groups are also widely believed indications of rain. As they are reliably gregarious birds, weather prophets who relied on Rook activity must have gained a good reputation in some areas.

If a death (especially of the head of the household) occurred in a family owning land that supported a rookery, there has been a widespread tradition that the Rooks must be told. This is a belief that has sometimes been extended to other crows and parallels a similar practice with bees (see page 99).

> A large rookery at Round Green [West Yorkshire] is part of local folklore; its Rooks are said to be the reincarnated souls of the Elmhirst family, which has lived in the area for many centuries.[407]

The largest rookery in the world is believed to be that at Turrif Castle in Scotland.

For no very obvious reason, there is more than one literary allusion to Rooks being identified with clergymen. A verse in the well-known 'Cock Robin' rhyme runs:

> Who'll be the parson?
> I, said the Rook
> With my little book,
> I'll be the parson.[408]

The use of the word 'rook' to mean 'a cheat or swindler' dates from the late sixteenth century but its association with the bird is obscure. There are several place names associated with Rooks: 'Rockley (Stainborough) is Rook clearing' [West Yorkshire] and the various Rocklands in Norfolk are 'Rook wood'.[409]

Rooks *(Corvus frugilegus)* nest in colonies, or rookeries, high in trees. In some, there may be several thousand pairs.

Carrion Crow *(Corvus corone corone)*
OTHER COMMON NAMES: *blackbill; black crow (Nottinghamshire); black neb (Westmorland); bran (Cornwall); car crow (Yorkshire); carener crow; carner crow (Norfolk); corbie (East Lothian); corbie crow (northern England); craw (Aberdeenshire); daup (York); dob (Westmorland, North Yorkshire); doup (Lancashire); doupe (Westmorland, North Yorkshire); flesh crow (Yorkshire); ger crow (North Yorkshire – Craven); gor crow (Yorkshire, Oxfordshire); hoddy fraw; huddy craw (south Scotland); ket crow (northern England); land daw (Northamptonshire); midden caw; raven crow (Yorkshire).* See Hooded Crow below.

Hooded Crow *(Corvus corone cornix)*
OTHER COMMON NAMES: *blue-backed crow (Ireland); cawdy mawdy (northern England); corbie (Perth); craa (Shetland); Danish crow; denman (Norfolk); dun crow; dunbilly (North Yorkshire – Craven); Dutch crow (Yorkshire); grey crow (Orkney); greyback (northern England); grey-backed crow (Hampshire); Harry Dutchman (Yorkshire); hoddie, hoodie (Morayshire, Perth); Isle of Wight crow; Kentish crow; Kentishman; Market Jew crow (Cornwall); moor crow (Yorkshire); northern crow (North Yorkshire – Craven, Norfolk); Norway crow; Royston crow; Royston Dick (Midlands); scald crow (Ireland); winter crow.*

These two large crows are very closely related, the Hooded Crow being a subspecies of the Carrion Crow, but they are almost completely separated geographically. The Carrion Crow occurs throughout England and Wales, and roughly as far north as the Great Glen in Scotland, while the Hooded Crow takes over in the far north of Scotland and in Ireland. The two interbreed where their ranges overlap. The Carrion Crow is entirely black; the Hooded is grey but with a black head, wings and tail.

They are the only two British birds that have the word 'crow' as part of their commonly accepted name. The name 'Royston crow' originates in the market town of Royston in Hertfordshire, where the Carrion Crow is the resident species but where migratory Hooded Crows once appeared in large numbers in winter on the local heath (as they do elsewhere, hence 'winter crow' and 'moor crow'). Nonetheless, the heath has largely been replaced by roads and industrial estates; and the only crow I could find was the local newspaper, *The Royston Crow*. The absence of the birds does not seem to deter local enthusiasm, however, and they are not only commemorated by the newspaper. Local sports teams are known as 'The Crows' and play in black-and-white strip and, on the crest of Royston Urban District Council, the bird sits proudly atop the Royse Stone, a boulder that once formed the base of a wayside cross at a local road junction. From thence, the crow finds itself on the crest of North Hertfordshire District Council: 'On a Wreath of the Colours two sprigs of Oak in saltire fructed proper enfiling a Mural Crown Or perched thereon a hooded Crow *(Corvus cornix cornix)* close proper'; regrettably, the bird's scientific name is incorrect.

Carrion and Hooded Crows have similar habits and habitats. They are usually more or less solitary, or occur in small groups, generally nesting in trees. They are omnivorous and, like the Raven, are often guilty of pecking out the eyes of animals, especially sheep and lambs. The *Game Rearer's Annual* of 1903 advised farmers to place poison in the eyes of dead sheep 'which usually provides a fatal lure to the hoodies'.

Thomas Pennant in his *Tour in Scotland*, published in 1771, told of Scottish shepherds making offerings to Hooded Crows and eagles in the hope that their flocks would be spared.[410] A Morayshire saying improbably linked Hooded Crows with obnoxious weeds and the neighbouring Clan Gordon as 'the three worst things Murray ever saw'. Both crows regularly feed on carrion, although it is not apparent why the Carrion Crow should have been singled out for this habit. Nonetheless, the name does date back to 1528 and even features in old rhymes. The following was being sung in Cheshire until recently.

The War Memorial in Royston, Hertfordshire, features a Hooded Crow *(Corvus corone cornix)* as part of the Royston crest.

Scarecrows, like this one at Barnsley House, Gloucestershire, were once commonly used to deter crows from crops.

A carrion crow sat on an oak,
Watching a tailor shape his cloak.
Sing heigh ho, the carrion crow,
Fol de riddle, lol de riddle, hi ding do.

The carrion crow began to rave,
And called the tailor a crooked knave.
Sing heigh ho, the carrion crow,
Fol de riddle, lol de riddle, hi ding do.

Wife, bring me my old bent bow,
That I may shoot yon carrion crow.
Sing heigh ho, the carrion crow,
Fold de riddle, lol de riddle, hi ding do.

The tailor he shot and missed his mark,
And shot his own sow through the heart.
Sing heigh ho, the carrion crow,
Fold de riddle, lol de riddle, hi ding do.

Wife, bring brandy in a spoon,
For our old sow is in a swoon.
Sing heigh ho, the carrion crow,
Fold de riddle, lol de riddle, hi ding do.(411)

An early version, without the carrion reference, was found in a commonplace book dating from the reign of Charles I:

There was an old crow
Sat upon a clod;
That's the end of my song.
That's odd.(412)

An alternative was:

There was a crow sat on a clod,
And now I've finished my sermon, thank God!(413)

For many centuries, young children (and others) were employed as bird-scarers, especially of crows. They used whatever means were available to frighten away the birds, hence the expression 'stone the crows'. The Norfolk Labour MP Sir George Edwards, who founded the National Union of Land Workers, even called his autobiography, written in 1922, *From Crow Scaring to Westminster*:

> On coming out of the workhouse in March 1856, I secured my first job. It consisted of scaring crows from the field of a farmer close to the house. I was then six years of age and I was paid 1s. for a seven-day week.(414)

Unlike the all-black Carrion Crow *(Corvus corone corone)*, the Hooded Crow *(Corvus corone cornix)* has a grey back and belly.

Unfortunately, in the second week, Edwards fell asleep, the crows returned and the farmer gave him 'a severe thrashing' and deducted 2d from his wages.

The word 'scarecrow' dates from 1553 but when children performed the task, they were often called 'crow-keepers' ('Scaring the ladies like a crow-keeper', *Romeo and Juliet*, Act 1, Scene 4) or 'crow-herds'. According to a Sussex rhyme, collected in the mid-nineteenth century, 'God made little boys to tend the rooks and crows'. The children sang songs or ditties in a loud voice to frighten away the birds, such as the following from *The History of Little King Pippin*, dating from *c.*1786:

Away, away, John Carrion Crow!
Your master hath enow
Down in his barley mow.[415]

At Ilmington in Warwickshire, they sang:

Ye pigeons and crows, away! away!
Why do you steal my master's tay?
If he should come with his long gun,
You must fly and I must run.[416]

And nearby, in Shottery:

Shoo-hoo, shoo-hoo!
Away, birds, away,
Tek a corn
And leave a corn
And come no more ter-day.[417]

There are recollections from the neighbouring county of Gloucestershire of crow-keeping (and 'rook-starving', which amounted to the same thing) well into the twentieth century. The going rate of 6d a day was much better than Sir George Edwards found in Norfolk, although, if you were unlucky enough to live in Winchcombe, it was only 1d or 2d plus a swede. Clearly, inflation was modest in the crow-keeping profession because Henry White of Bagendon, also in Gloucestershire, who began crow-keeping in 1832 at the age of 10, was even then paid 3d for a 12-hour shift.[418]

Further variant crow-scaring rhymes have been listed by Iona and Peter Opie, but crows feature in numerous other rhymes, some very familiar and widespread. The gardener's adage on sowing beans runs:

One to rot and one to grow
One for pigeon, one for crow.[419]

Another nursery rhyme tells that:

On the first of March
The crows begin to search;
By the first of April
They are sitting still;

By the first of May
They've all flown away,
Coming greedy back again
With October's wind and rain.[420]

There are many regional tales of the behaviour of crows in general presaging changes in the weather. In Wales, an indication of strong winds is given by Ravens and crows flapping their wings and flying at a great height, while sunshine will follow if they are seen flying towards the sun.

Although not generally as feared as the Raven, crows and their black colour have long been considered indications of bad news. To hear a crow call three times in Sussex foretold death, while children in Yorkshire and Lancashire had a rhyme:

Crow, crow, get out of my sight
Or else I'll eat thy liver and lights.[421]

Roy Palmer recounts an episode that occurred in Gloucestershire as recently as 1990, when a large crow landed on a tree in an old man's garden to be greeted by: 'Yes, and you can bugger off. I ain't ready to come with you awhile yet.' One evening, a fortnight later, the same crow landed on the road nearby, walked down the middle and turned into the man's garden. The old man died the next day. There are similar resonances of this in other parts of the country: an East Yorkshire belief was that a crow perching in a churchyard would mean a funeral within week.[422]

A Herefordshire story, dating from 1903 tells of a farming rivalry on the Black Mountain which resulted in two brothers murdering a local shepherd who had warned that, if they killed him, 'the very crows will cry out and speak it'. Some time later, a flock was seen circling above the brothers, who were overheard by a neighbour relating the coincidence to their evil deed and, as a result, they were eventually charged, found guilty and hanged.[423] Coincidentally, the collective noun for crows is 'a murder' with an origin in the folk belief that crows form tribunals to judge their kin and punish bad behaviour. If the verdict goes against the defendant, it is murdered by the others.

> I have witnessed an event of similar sort. It took place at Brockwells Farm, Caerwent, Gwent (my old home) on 16 May 1986 ... A commotion in the Little Banky Field attracted our attention, a real cacophony of angry crow calls. As we approached we could see that the disturbance revolved around three crows and as we drew near, two of them flew away, the third could not for both of its eyes had been pecked out.[424]

As with Magpies and Ravens, a message was carried by the number of crows seen at any one time. A none-too-musical Essex rhyme runs:

> One's unlucky,
> Two's lucky;
> Three is health,
> Four is wealth;
> Five is sickness
> And six is death.[425]

The prefix 'crow' occurs in a number of place names, sometimes, although not invariably, with bird associations: for instance, Crowborough in Staffordshire and Sussex ('crow hill'), Crowell in Oxfordshire ('crow stream') and Crowmarsh, also in Oxfordshire ('a marsh frequented by crows'), Crowhurst in Surrey ('crow wood'). 'Crow Well (Upper Denby) and Crow Edge (Ingbirchworth) indicate Carrion Crow [West Yorkshire]'.[426] It also exists as a surname and occasionally as an inn name. Parts (unspecified) of crow have also found their way into folk medicine; the ashes of the whole bird, for instance, 'mixt with lard, is proper to be applied to any pain or ache'.

Raven *(Corvus corax)*

OTHER COMMON NAMES: *corbie (Scotland, northern England); croupy craw (northern England); corbie craw (Scotland); fiach (Ireland); marburan (Cornwall); Ralph.*

The Raven is a huge black bird, much bigger than any other British crow and the largest of all perching birds, with whiskery throat feathers and a massive beak. It is now mainly western in distribution, the Scottish population having declined in recent years, and it is usually associated with mountains and cliffs, although it will also nest in trees. The Raven is omnivorous and commonly kills smaller birds and rabbits, although, because it was thought to kill game birds and young farm animals, it has suffered long and persistent persecution.

Because of its size, ominous black appearance and coarse, raucous call, its fondness for wild, lonely places and, in ancient times, its arrival on battlefields to feed off corpses, it has always been associated with foreboding and death. The Raven was the first bird that Noah sent out from the Ark, perhaps because it would feed on the dead bodies of animals killed by the flood. Also, when Isaiah described the state of desolation

The Raven *(Corvus corax)*, with its sombre black plumage, is regarded as a bird of ill omen and a portent of death.

that was to descend on Idumea, he declared that: 'the owl also and the raven shall dwell in it' (Isaiah 34:11). (The link with owls is an interesting one; see below.)

There is no doubt that the Raven is in a league apart and there is probably more folklore attached to it than to any other British bird. Even today, its name engenders awe, loathing, fear, suspicion and admiration in almost equal measure; appropriately, the collective noun for Ravens is 'an unkindness'.

The name 'Raven' is ancient, and is little changed from the Old English 'hræfn' and the Old Norse 'hrafn'. Odin, the supreme creator and war god of Norse legend, was known as Hrafnagud, the Raven god; two Ravens, Huginn and Muninn (Mind and Memory), were his eyes and ears and flew across the world, collecting information which they imparted to him when they returned to perch, one on each shoulder. The Danish standard, called 'Landeyda', which was carried before the army, displayed an image of a Raven that raised its wings if the army was to be victorious and lowered them if defeat was in the offing. It was probably not lowered very frequently, as Sir Walter Scott reminds us:

The eponymous hero of Charles Dickens's novel *Barnaby Rudge* is accompanied everywhere by Grip, his pet Raven.

When Denmark's raven soared on high,
Triumphant through Northumbrian sky.
(Walter Scott, 'Harold the Dauntless')

'Corbie' is a widely used Scottish name for the Raven and is sometimes applied to other black crows. It came into Scottish usage through Middle English and Old French, from the original Latin *Corvus*.

In northern tales, as well as in the name 'corbie', the Raven is sometimes interchangeable with its northern and western relative the Hooded Crow: Scottish legend tells of an evil witch, or hag, called *Cailleach*, who appeared either as a Raven or a Hooded Crow – or sometimes as a gull, Cormorant ('sea raven', see Cormorant, page 210) or heron – and lived on after the coming of Christianity, feasting on men's bodies. 'Lady Day' (25 March) was once the far less agreeable 'Cailleach Day'.

Nonetheless, the Raven has not always been considered wholly bad and Noah may have chosen it because it flew 'swift-winged and strong', while constancy and devotion have also been considered among its virtues. After all, paradoxically, God chose the Ravens to feed Elijah: 'Hide thyself by the brook Cherith, that is before Jordan. And it shall be that thou shalt drink of the brook; and I have commanded the ravens to feed thee there' (I Kings 17:3–4). They also appear in the New Testament: 'Consider the ravens for they neither sow nor reap ... and God feedeth them' (Luke 12:24). In consequence, the Raven as an emblem of God's Providence is common in Christian art. The mysterious Italian abbot St Benedict (*c*.480 – *c*.550) is often depicted with a Raven at his feet, a reference to the bird that removed a cup of poison from him.

In more recent literature, Charles Dickens, in *Barnaby Rudge*, gives Rudge a pet Raven and constant companion in the shape of Grip. 'He takes such care of me ... He watches all the time I sleep,' said Barnaby, although Dickens reminds us that Grip, 'as if exulting his infernal character', persistently declared: 'I'm a devil, I'm a devil.'[427]

Ravens have long been associated both with prophecy and bad luck, probably because of their prompt arrival whenever carcasses are in the offing: 'The obscene ravens, clamorous o'er the dead' (Shelley, 'Adonais'). It was fluttering Ravens that warned Cicero of his death and Macbeth was no less concerned to note that:

The raven himself is hoarse
That croaks the fatal entrance of Duncan
Under my battlements.
(*Macbeth*, Act 1, Scene 5)

while Othello says to Iago:

The Ravens at the Tower of London are cared for by one of the Yeoman warders known as the Ravenmaster.

Thou saids't, – O, it comes o'er my memory,
As doth the raven o'er the infected house,
Boding to all, – he had my handkerchief.
(*Othello*, Act 4, Scene 1)

Among the many other writers to make reference to Ravens as portents of death, especially when flying over a house, was Thomas Hardy who, like Elijah in the Bible, also mentioned them in the company of owls, an association that must surely have arisen because both are birds of solitude, strange calls and a mysterious nature:

The bell went heavy to-day
At afternoon service, they say,
And a screech-owl cried in the boughs,
And a raven flew over the house,
And Betty's old clock with one hand,
That's worn out, as I understand,
And never goes now, never will,
Struck twelve when the night was dead still,
Just as when my last loss came to me
Ah! I wonder who next it will be!
(Thomas Hardy, 'Premonitions')

There was a long-held belief, especially in Cornwall (where it was once thought that young Ravens were always hatched on Good Friday), that King Arthur had been turned into a Raven. The story was evidently widespread enough to have reached Spain, because it is alluded to by Cervantes in *Don Quixote* (1605):

There is an ancient tradition, common in all that kingdom of Great Britain, that this king did not die, but by arts of enchantment, was transformed into a Raven and that in due course of time he will return to reign, and recover his kingdom.[428]

In Castell Coch, near Welshpool, there is a legend that a treasure chest buried beneath the castle is guarded by a giant Raven and comparable stories occur in Yorkshire and elsewhere.

As with Magpies and other crows, the number of birds seen at one time carries a message. While it is advisable to view Magpies in quantity, however, the reverse seems to hold with the Raven. Deer-stalkers in the Scottish Highlands are said to welcome one but take it a sign of bad luck to find two or three:

> To see one raven is lucky, 'tis true,
> But it's certain misfortune to light upon two
> And meeting with three is the devil.[429]

In Somerset, anyone hearing a Raven croak three times would turn the other way and cross their fingers and it was believed that to steal eggs from a Raven would result in the death of a baby. At a time when infant mortality was an everyday occurrence, it must have been easy to find evidence to support this. It was considered unlucky to hear a Raven in Lincolnshire but, as it has probably never been a very common East Anglian bird, this is perhaps understandable.

The Raven's sinister habit of pecking out the eyes of animals, shared by the Carrion Crow and other crows, has understandably led to strange beliefs and much loathing among farmers, although, in Wales, it was said that a blind person who was kind to a Raven might regain his eyesight. There is certainly evidence that Ravens will peck the eyes of living sheep and Ron Murton quoted reliable observations, made during a study by the Ministry of Agriculture, Fisheries and Food, of Ravens killing newborn lambs.

> In the school in the Derbyshire village of Biggin is/was the Parish Register for the latter part of the 18th century, listing scores of local Ravens killed, and the bounties paid.[430]

In certain areas, Ravens, like Hoopoes, Swallows and woodpeckers, have been credited with talismanic powers. There is a story told by A.T. Cluness that, in Shetland, a Raven's nest was once robbed and the eggs boiled as a prank before being returned. The parent birds remained for weeks before realizing that something was amiss, when one flew off and returned with a twig which it placed in the nest; the eggs then hatched. The explanation was that the Raven, God's chosen bird, had visited the Holy Land to collect a piece of the True Cross.[431] There are similar stories in Orkney and the Faeroes, and in some versions the birds returned with a stone that had the power to restore life to all it touched. In Wales, the 'raven stone' was an amulet that conferred the gift of invisibility on its owner. To obtain it, the seeker was obliged to climb a tree, discover a nest of Ravens more than a hundred years old, and kill a male fledgling of under six weeks. When the parent bird returned, it would insert the magic talisman into its dead offspring's throat.

However, the most famous Ravens, at least today, must be those associated with London. Because of the squalid state of London's streets and rivers, certain birds and animals were legally protected in the Middle Ages as scavengers of human detritus, the carcasses of dead dogs and other sordid refuse. Kites and Ravens especially benefited from this. As the city became cleaner, however, this status was gradually eroded and, certainly by 1768, Robert Smith wrote that he was well paid for killing Ravens as vermin.[432] The last pair to nest in central London did so in Hyde Park in 1826, but tame Ravens are still kept at the Tower of London. The belief that, should they depart, the White Tower will crumble and England will fall has been taken seriously and, during the Second World War, Churchill took steps to ensure that the birds were well protected. It is not known when the story originated, although their status and protection by royal decree date from the reign of Charles II. It is said that, when the king was viewing the heavens from the new observatory in the White Tower, his view was obscured by bird droppings on the telescope. He ordered the Ravens to be destroyed but, when it was pointed out to him that this would threaten his throne and kingdom, he relented, ordering the birds to be protected and moving his observatory to Greenwich. There are now eight Ravens at the Tower, distinguished by differently coloured leg rings, and they are cared for by one of the Yeoman warders known as the Ravenmaster. Their offspring are taken each year and released in different parts of the country.

There is an ancient version of the Tower Raven belief in the medieval Celtic *Mabinogion*, where it is recounted that, when Bran the Blessed died, his head was cut off and buried on Tower Hill with his face towards France. So long as it remained there, Britain would be safe from invasion. The name 'Bran' means 'crow' or 'Raven'.

Many place names, especially those of Norse origin in the north and west, allude to Ravens.These include Ravendale in Lincolnshire ('raven valley'), Ravenfield in West Yorkshire ('raven's field'), Ravenscar in North Yorkshire ('raven's rock') and Ravensdale in Derbyshire ('valley of ravens'), although 'hrafn', as well as being the Old Norse name for the Raven, was also a personal name, so the association with places may be at one remove from the bird.

'Raven' is also found occasionally as an English personal name and sometimes in surnames. It appears on inn signs, although not surprisingly, given its ominous overtones, inns called The Raven are not common. Similarly, it appears fairly infrequently in heraldry; although it is sometimes found on the arms of families with Danish origins, no distinction is made between the heraldic depictions of the Raven, Rook and crow.

Ravens have played little serious role in traditional medicine, although one of the merits attributed to its various parts is that the eggs 'strengthen the stomach' and help to cure dysentery; the ashes of young birds, taken every morning for three days are effective against epilepsy and gout; and the dung, hung around a child's neck, eases the pain of teething.[433]

Starlings
(Sturnidae)

Starling *(Sturnus vulgaris)*

OTHER COMMON NAMES: *bird of the snow (Wales); black felt; black starling (East Lothian); black steer (Worcestershire); gyp, gyp starnill (North Yorkshire); Jacob (Northamptonshire, Sussex,); shebbie (Shropshire); sheeprack (Northamptonshire); sheep stare (Somerset); sheppie (Cheshire); shepstare (North Yorkshire – Craven); shepstarling; shepster (Cheshire, Lancashire); shippy (Derbyshire); stannel (Northamptonshire); stare (Ireland, North Yorkshire, Dorset, west Cornwall); starn (Shetland); starnel (Derbyshire); starp (Oxfordshire); staynil.*

There is a tendency to regard the Starling as a fairly recent British bird, a species that has expanded its range in Britain as human habitation has expanded. The old Welsh name, 'bird of the snow', harks back to a time when it was just a winter visitor, but certainly the way that it is has adopted television aerials and chimney pots as perching places gives the impression that it is inseparable from present-day life, and its remarkable ability to mimic even such artificial sounds as car alarms reinforces this view. However, it has been fairly common for a very long while:

> Granny, born in Chipping Norton, Oxfordshire, used to call Starlings starps and Fieldfares felts.[434]

> Starlings are still called Starnels in N.E. Derbyshire.[435]

> My father, born in 1918 in North Derbyshire, still refers to Starlings as Shippys.[436]

Its most usual old name (which survives in some areas) was 'stare', with an origin in Old English and, when Thomas Heywood, in *c.*1600, wrote:

> Blackbird and thrush in every bush
> Stare, linnet and cocksparrow'[437]

it is clear that he was describing an everyday bird. Nonetheless, the Starling, with its familiar and characteristic glossy green-and-purple plumage is now diminishing in numbers and sights such as I saw recently, of a flock of nearly a hundred, noisily demolishing the crop of one of my apple-trees, may soon be a thing of the past. One theory is that, as with some other species, intensive farming has reduced the supply of insect food for young birds.

The name 'Starling', the diminutive form of 'stare', was originally used only for the immature birds, which differ markedly in appearance from the adults; that gorgeous glossiness is acquired only gradually. Names like 'shepstare', or 'shepstarling', that have references to sheep, allude to the bird's habit of settling on the backs of sheep to feed on ticks. Its diet is omnivorous, as anyone who has watched a bird table will verify, although insects are generally its preferred food.

Starlings *(Sturnus vulgaris)* are gregarious, wheeling in vast flocks across the sky, like this one above the West Pier in Brighton.

Starlings are certainly gregarious and sometimes occur in truly vast flocks, which settle on buildings or trees to roost, sometimes with appalling effects on the locality. Ron Murton described a small pine plantation in Norfolk that, after a winter's roosting, was covered with excrement to a depth of 15 centimetres (6 inches).[438]

For the Scottish poet Alexander Hume, writing around the same time as Heywood, the Starling's call was its most notable feature:

> The sterling whistles loud,
> The cushats on the branches green,
> Full quietly they crood.
> (Alexander Hume, *The Sterling*)

It is that imitative call that makes Starlings the curse of anyone trying to recognize native birds by sound. They can effectively copy almost any other species, and all manner of other sounds too. The collective noun 'murmuration' seriously underestimates the noise that they make, either singly or *en masse*.

Starlings are quite commonly eaten in rural areas and their large-scale shooting, which takes place on farmland (traditionally on Boxing Day in some places), provides the wherewithal for a good many Starling pies. In Ireland, where the name 'stare' is still widely used, the flesh was thought 'good to resist poison', although it was also in Ireland that an ounce of Starling dung, mixed with alum and white vitriol (zinc sulphate) was believed to cure afflictions ranging from ringworm to herpes.[439]

The parents of young Starlings *(Sturnus vulgaris)* are hard-pressed to meet their constant demand for food.

Sparrows
(Passeridae)

House Sparrow *(Passer domesticus)*
See Tree Sparrow below.

Tree Sparrow *(Passer montanus)*
OTHER COMMON NAMES: *craff (Northumberland); cuddy; easing sparrow; grey spadger (Antrim); hoosie (Northumberland); lum lintie (East Lothian); roo-doo (Northamptonshire); row-dow; spadger (Lancashire, Suffolk); sparr (Kent, Sussex); speug (Lanarkshire, Ayrshire, Stirlingshire); sproug (Caithness); sprug (East Lothian); spug (Shropshire); spuggie (Northumberland, Cumberland); spurdie (Banffshire, Aberdeenshire); spurgie (Aberdeenshire); thack sparrow (Northamptonshire, Shropshire); thatch sparrow.*

Few birds are so closely associated with human habitation as the pretty chirping sparrow; the House Martin is the only other British bird to have acquired the word 'house' as part of its name and that only because it nests on them. 'Spearwa' was the Old English name for sparrow and the two true sparrows were distinguished in Britain only in the early eighteenth century, when the Tree Sparrow, at first called the 'mountain sparrow', was recognized as a distinct species. No-one seems certain how names like 'spuggie' and 'spurdie' originated but, although they seem and sound Scottish, they are clearly more widespread:

> My grandmother, who was brought up in Chalfont St Peter at the beginning of the twentieth century, always used to call house sparrows spugs. I was confused because she called potatoes spuds.[440]

The so-called 'hedge sparrow' is the Dunnock, an unrelated bird. The word 'sparrow' itself appears as a suffix in the alternative names of many other small perching birds, including some warblers and tree pipits; it also appears in the name 'Sparrowhawk'.

In one sense, the bird has long been held in considerable affection, the familiar prefix 'cock', as in 'cock sparrow', being invariably limited to those species that people have found endearing; 'cock robin' is another common example. The chirpy Cockney 'sparrer' was the inspiration for Joan Littlewood's 1962 film, the East London melodrama *Sparrows Can't Sing*. The fact that sparrows chirp almost constantly makes a nonsense of the Isle of Man tradition that 'when sparrows chirp, rain will follow' – although the Isle of Man can be a very wet place.

Sparrows were for many years among the commonest British birds and among the most frequent visitors to garden bird tables and there is still a perception that they must be the most numerous species in the country. The reality is very different, however. House Sparrow populations have plummeted in recent years, for reasons that are not yet understood but are possibly related to changes in habitat, loss of food sources or loss of nesting sites. The fact remains that House Sparrows are no longer everyday birds. They are certainly among the

The Tree Sparrow *(Passer montanus)* has a distinctive chestnut crown.

boldest, however, happily hopping into houses to peck crumbs from the table or floor.

Oddly enough, for a bird so familiar and so generally inoffensive, sparrows are badly represented in folklore. It was the sparrow who killed Cock Robin in the nursery rhyme, although it may simply have been a convenient rhyme for 'bow and arrow' (which is Cockney rhyming slang for 'sparrow'). There is another old rhyme from Lancashire that seems to bear this out:

> The robin and the wren
> Are God's cock and hen.
> The spink and the sparrow
> Are the Devil's bow and arrow.[441]

There is, however, no denying that sparrows can be serious pests of farmland and it was this that gave rise to the miserable practice of Sparrow Clubs, common in almost all parishes in the eighteenth and nineteenth centuries and formed with the sole objective of killing as many sparrows as possible and paying rewards for quantities of dead birds or eggs.The going rate in the early years was 2d or 3d a dozen, although this later rose to 10d or a shilling (12d); one Guernsey parish paid out £72 5s 8d in 1827. A news item in the *Sussex Express*, dating from the 1860s and concerning the thirteenth anniversary of the Rudgwick Sparrow Club, which was celebrated at the Cricketers' Inn, recorded the sad fact that members had sent in 5,313 birds' heads during the year, the first prize going to Mr W. Wooberry, who had personally contributed 1,363 birds.[442]

In Papa Stour in Shetland, sparrows were always very damaging to the corn and the locals believed that the beadle of the kirk had the power of telling them to go away and never return; for this the good beadle was paid a fee or retainer. Until the early twentieth century, the 'Sparrow Beadle' still existed; although when he was last called on to tell anything to the sparrows is not obvious.

The killing of sparrows as pests merges imperceptibly into the use of sparrows as food. Sparrow Pie was a popular country dish until well into the twentieth century and probably still is in some areas (see Skylark, page 318). In my limited experience, grilled sparrows seem to consist mostly of beak but they were considered highly nutritious, although Andrew Boorde, in his *Dyetary of Health* (1547) thought them 'hard of digestion'.[443]

In the village of Cam, southwest of Stroud in Gloucestershire, the 'sport' of 'bird-batting' was practised well into living memory. A group of men with nets, sticks, lanterns and cowbells crashed and smashed their way around the village, beating hell out of any bush and thatched roof they found. Out flew birds, mainly sparrows but apparently also the odd Pheasant and, confused by the light, these were trapped in nets and later formed the ingredients of 'sparrow pudding'. This curious pursuit (described in one source as a 'traditional nocturnal field sport') was once more widespread and generally went under the name of 'bat-fowling'.

> We would so, and then go a bat-fowling.
> (*The Tempest*, Act 2, Scene 1)

Sometimes fires were lit to drive the birds from the bushes, as the following account reveals:

> It was not always used with nets, but instead, 'a Third-part of the company must attend upon the said Fires with long Poles, rough and bushy at the upper Ends, to knock down the Birds that fly above the Lights, and another Third-part must have long Poles to beat the Bushes and other Places, to cause the Birds to fly about the Lights which they'll do as if amaz'd, not departing from it; so that them may be knocked down at Pleasure. And thus you may spend as much of the Night as is dark and find good Diversion'.
>
> What strikes us about bat-fowling is the size of the crowd, and indeed, there are references in Fielding, and other of the eighteenth-century novelists, which show that a company of bat-fowlers was usually a large one, and must have had an uncanny look prowling about the fields with their lights in the middle of the night.
>
> Even more popular was the plan for catching birds with Low-Bell, Net and Light. This was a sport in season from the middle of October to the end of March, and was for 'Plain and Champion-Countries also in stubble fields'. Those who were engaged in it usually started about nine o'clock at night, the air being mild and the moon not shinning. The Bell used 'must be of

a deep and hollow Sound and of such a reasonable size that a man may carry it well in one hand, which toll just as a sheep useth to do whilst it feedeth.' The next requisite was a box 'much like a large Lanthorn and about a Foot and a half square, big enough for two or three great lights to be set in, and let the Box be lined with Tin and one side open to cast Forth the Light.' This was fixed on the breast, and we suppose it is some tradition concerning that old box which still guides the fish poacher on the Tweed. He, too, uses a box-like lantern, which he can carry against his breast, making it dark when he wishes to conceal his presence from the water-bailiffs, and letting the light flash forth when he is about to spear a salmon.

'Concerning the bell, the author asks us to note 'that the Sound of the Low-Bell doth cause the Birds to lie close, and not dare to stir while you put the Nets over them and the Light is so terrible to them that it amazeth them. And for caution you must use all imaginable Silence, for fear of raising them'.[444]

Arguably the most distinguished feat ever achieved by sparrows was their reputed role in the recapture of Cirencester by the Saxons under the leadership of a mysterious African named Gurmund from the post-Arthurian British king Carric whom they besieged there. Gurmund caught large numbers of sparrows and tied lighted pieces of tinder in nutshells to their feet. As they flew home to their roosts, the town was set ablaze. In the Middle Ages, Cirencester was sometimes called *Urbs Passerum* in commemoration of the event.

Sparrows have commonly been kept as pets and the fifteenth-century satirical poet John Skelton (*c.*1460–1529) in his 'Death of Phylyp Sparowe' eulogized one particular bird:

That cat specially
That slew so cruelly
My litell prety sparowe
That I brought up at Carowe.

'Philip' was a common name for pet sparrows because they cried '*phip, phip*'.

Many other poets, including John Donne and John Clare, have referred to sparrows in their works, although Wordsworth's 'Sparrow's Nest', with its bright blue eggs, obviously described a Dunnock. Perhaps the most famous literary and pet sparrow was not a Philip but a Clarence, the subject of Clare Kipps' best-selling story *Sold for a Farthing*, first published in 1943 and reprinted many times.[445]

Another sparrow that achieved serious immortality did so merely through being in the wrong place at the wrong time. On 3 July 1936, the MCC was playing Cambridge University at Lords. The Cambridge bowler Jehangir Khan (father of the more famous Majid) bowled to the MCC batsman Tom Pearce and the ball hit and despatched a nearby sparrow. Although the match reports in both Wisden's *Cricketers' Almanack* and *The Cricketer Magazine* ignored this momentous event, an account appeared in the 'Miscellany' section of Wisden in 1937, where it was reported:

T.N. Pearce, the batsman, managed to play the ball and the bird fell against the stumps without dislodging the bails. The bird is preserved as a relic in the pavilion at Lord's.[446]

And there it remains to this day.

For so common a bird, sparrows have seldom featured in the names of inns, The Sparrow at Letcombe Regis in Oxfordshire and, appropriately enough for London, The Sparrow's Nest in Drury Lane being notable exceptions.

Finches
(Fringillidae)

The finch family includes some of the most familiar and common small perching birds in Britain, one or more species being likely to be seen in most gardens on any given day. All are predominantly seed-eaters and many are regular visitors to bird tables.

The name 'finch' originates in Old English and was originally restricted to the Chaffinch; it is thought to reflect one of its numerous calls. A number of place names commemorate finches, such as Finchale in Durham ('a haugh frequented by finches'), Fincham in Norfolk ('a ham frequented by finches'), Finchampstead in Berkshire ('a finch homestead') and Finchley in Middlesex ('a finch leah'). The word is fairly common as a surname but, for such everyday birds, the infrequency in which they have made their mark in literature and folklore is surprising. Perhaps they are just too ordinary and too commonplace.

Chaffinch *(Fringilla coelebs)*

OTHER COMMON NAMES: *apple bird (Cornwall); apple sheelie; apple sheller (Northumberland); binkie (Ross and Cromarty); blue cap (Aberdeenshire); boldie (Kincardineshire); brichtie (Wigtownshire); brisk finch; charbob (Derbyshire); chay (Morayshire); chink chaffey (Hampshire); chink chink (Shropshire); chy (Aberdeenshire); copper finch (Devon, Cornwall); flackie; fleckie wings (Lancashire); fleckiwing; pea finch; pied finch; pine finch (Midlands, Shropshire); pinkety (Northamptonshire); pink twink (Shropshire, Somerset, Devon); prink prink (Inverness-shire); scobb; scop; scoppie (Northumberland, Cumberland); sheefla (Scotland); sheely (Kinross, Northamptonshire); shelfie (Scotland); shell apple; sheltie (Fife); shieler applie (Northumberland); shilfa (Scotland); shiltie (Fife); snabbie (Fife, Dumfriesshire); spink (Suffolk); tree lintie (Morayshire); treeack; wet bird (Stirlingshire, Rutland); wet chaff (Angus); wheatsel bird (Norfolk); white finch; white wingie (Lanarkshire); whitewing (Armagh; Donegal).*

The Chaffinch *(Fringilla coelebs)* is common everywhere. The male in particular is easily recognized by his distinctive colouring.

The huge number of local names reflects the familiarity of Chaffinches and the affection in which they are held. They are easily the commonest British finch and are among the commonest of all British breeding birds. The male, with his slate-blue crown and neck, chestnut back and pinkish-brown underparts is a familiar sight. The female, though duller in colour, shares the same distinctive white wing-bar and shoulder patch.

The name 'Chaffinch', like 'finch', is from Old English and is a reminder of the time when these birds frequented barns, looking for seeds among the chaff. References to 'apple', and to 'fleck' and 'pink', seem to refer to the colour of the plumage rather than diet. Linnaeus gave the bird the scientific name *coelebs*, meaning 'unmarried', because of an unusual behavioural trait: in Scandinavia, where he lived, most female chaffinches migrate southwards in winter, leaving the lonely males behind.

Brambling *(Fringilla montifringilla)*

OTHER COMMON NAMES: *bramble; bramble cock (Cheshire); bramble finch; bramlin; cock o' the north (eastern and southern Scotland); flat finch (Cheshire); French linnet (North Yorkshire); furze chirper, furze chucker (Hampshire); goldie-wing; Kate (Kent); mountain finch; tartan back (Scotland); yallawing (Northumberland).*

Superficially similar to the Chaffinch but with a striking white nape, the Brambling is a widespread winter visitor and, despite its name, has no particular association with brambles. Like the Chaffinch, it has a fondness for beech mast.

The name seems most likely to be a corruption of 'brandling', meaning a creature that is 'branded' or 'brindled' in appearance. References to 'furze' in the Hampshire names appear to derive from confusion with Whinchats and Stonechats.

Serin *(Serinus serinus)*

This small, yellowish finch, which has a delightful musical song, has slowly been extending its range north from the Mediterranean and has now reached southern England. It is not yet familiar enough to have an English name, so the French name has been adopted.

Greenfinch *(Carduelis chloris)*

OTHER COMMON NAMES: *barley bird; green bird; green chub; greeney; green grosbeak; greenick; green lennart (Northumberland); green linnet; green olf (Norfolk); greenbull (Lancashire); peasweep.*

A common garden bird throughout the British Isles, except for the far north and west of Scotland, this is the most intensely green of any native bird of its size. It is a familiar visitor to the bird table, especially as the increased flailing of farm hedges is removing more of its supply of winter food on wild roses and other native shrubs.

The names 'green olf' and 'greenbull' are suggestive of a green version of the bullfinch. 'Peasweep' imitates the call.

Goldfinch *(Carduelis carduelis)*

OTHER COMMON NAMES: *brancher (juvenile cage birds); draw bird; draw water; foolscoat; gold linnet (northern England); goldie; gold spink; gool French (Devon); gooldspink (Scotland); gowd spink (northern); grey Kates, grey pates (juveniles, northern England); jack nicker (Cheshire, Northamptonshire, Shropshire); King Harry, King Harry redcap (North Yorkshire, Shropshire, Suffolk); lady with the twelve flounces, linnet (Shropshire); melenek, molnek, moleneck (Cornwall); proud tailor (Derbyshire, Nottinghamshire, Warwickshire, Northamptonhamshire, Somerset); redcap (northern England, Suffolk); seven-coloured linnet; sheriff's man; spotted dick (Shropshire); sweet William; thistle finch (Stirlingshire); thistle warp.*

When seen at close quarters feeding on thistle or teasel heads, this is such an extraordinary beautiful bird that it is difficult to believe it is a common British native and not some exotic escape from an aviary. Sadly, it was indeed once a cage bird of enormous popularity. Ian Newton describes a report of 132,000 being caught each year in the mid-nineteenth century in the vicinity of Worthing, which lies on a major migration route.[447] The RSPB, when it was founded in 1904, made the protection of Goldfinches one of its first priorities. The names 'draw bird' and 'draw water' are references to its abilities to pull up a weight attached to a thread, by looping each length under its foot, and birds kept in cages were often obliged to use this means of obtaining both food and water. The same behaviour is seen in captive Redpolls, although they do not appear to have had the names attached to them.

The collective noun for Goldfinches is 'a charm', by derivation from a sixteenth-century meaning of the word, as 'a blended noise, as of birds, school-children etc'. It perhaps has the same Middle English origin as the alternative meaning of 'incantation', or 'magic spell'. It also reflects the twittering sound that the birds make as they flock together.

Goldfinch nesting was the subject of an unusual study in 1943 by the ornithologist Peter Conder, who spent long periods of his time as a prisoner-of-war making records of their activities, an investigation that formed the subject of a scientific paper when hostilities ended. His paper noted that, 22 days after hatching, the Goldfinches left the prison camp, unlike Conder and his fellow observers.[448]

Most of the Goldfinch's many local names relate to its striking colour or to its food; even 'proud tailor' alludes to its gorgeous coat, while 'King Harry' reflects the costumes sported by Henry VIII. The strange 'moleneck' is derived from Cornish.

Although less in English than European art, the Goldfinch has long been depicted in religious paintings, especially devotional works, where, after the dove, it is the commonest bird; its gold wings are deemed particularly important. Its use as a symbol of the Resurrection is something that it shares with other red-marked birds, such as Swallows and Robins, and the resemblance between thistle heads, which are its food, and the Crown of Thorns is probably also important.

The Goldfinch *(Carduelis carduelis)* tends to be seen on neglected land where its favoured food plants are found.

Unsurprisingly, the Goldfinch has been immortalized in literature, an English tradition that began with Chaucer, who compared the apprentice in 'The Cook's Tale' to the bird:

Gaillard he was as goldfynch in the shawe

Then, in *The Flower and the Leaf*, a work attributed to Chaucer there is some perceptive observation:

I was ware of the the fairest medle-tree
That ever yet in al my lyf I sy,
As full of blossomes as it might be.
Therein a goldfinch leping pretily
Fro bough to bough, and, as him list, he eet
Here and there, of buddes and floures sweet.

Nonetheless, the frequency with which Goldfinches, even in the fourteenth century, ate the buds and flowers of medlars has been debated at some length by modern ornithologists.

The assumption that something coloured so golden must be a symbol of wealth is manifested in rhymes such as that describing the wedding of Cock Robin and Jenny Wren:

Who gives this maid away?
I do says the goldfinch
And her fortune I will pay.[449]

Inevitably for so striking a creature, Goldfinches have also found medicinal use: 'The flesh if it be fat is a great restorative in consumptions' and with other ingredients, it was used, 'against all obstructions of the reins and bladder.'[450]

Siskin *(Carduelis spinus)*

OTHER COMMON NAMES: *aberdavine; barley bird; black-headed thistlefinch; golden wren (Cheshire); tea leaves (Sussex).*

This small, yellow-green finch of coniferous woods and wet places associated with birch, willow and alders was once confined to the old coniferous forests of Scotland, but it has spread southwards, following the new conifer plantations, and is now commonly found in flocks with Redpolls.

The name 'Siskin' dates from the mid-sixteenth century and was probably originally onomatopoeic in some European language and, because it is chiefly a bird of Gaelic-speaking areas, it does not seem to have acquired a true English name. The extraordinary early name 'aberdavine', of which there are several spellings, is still occasionally used in Cheshire and Yorkshire but its origin is mysterious. That the bird has been

A bird of coniferous forests, the Siskin *(Carduelis spinus)* has spread south in the wake of new coniferous plantations.

known for centuries is recalled by Max Nicholson's frustrated observation of the medieval minstrel who:

> ... airily mentions the siskins which sang in the glade where Tristan and Isolde embraced near King Mark's Castle in Cornwall, with an entire disregard for the importance of authenticating in proper detail the occurrence of the siskin as a Cornish breeding bird in the 5th century AD.[451]

Linnet *(Carduelis cannabina)*

OTHER COMMON NAMES: *blood linnet, brown linnet (Norfolk); furze bird (Northamptonshire); furze linnet (Oxfordshire); gorse bird; gorse hatcher (Shropshire); gorse thatcher; gorse linnet (Northamptonshire); greater redpole; grey linnet (southern Scotland, England); grey; grey bird (Northern Ireland; Westmorland); heather lintie (Scotland); lemon bird (West Yorkshire); lennert (northern England); lint white (Orkney); lintie (Scotland); red linnet (West Yorkshire, Hampshire); red-breasted linnet; red-headed finch; rose linnet; rose lintie (northern England); song linnet (Yorkshire); thorn grey (Ireland); thorn linnet (Yorkshire); whin grey (Northern Ireland); whin linnet (Stirlingshire); whin lintie (Scotland).*

The Linnet is much loved, despite its preference for open countryside rather than urban areas and despite the fact that many modern town-dwellers have no personal knowledge of it. Its reputation and the affection in which it is held have evidently been carried over from times when the human population in general was more rural and the birds had an abundance of seeds on which to feed.

Mainly resident, it is widespread throughout the British Isles. The male has a grey head, a red forehead and breast, and chestnut back (grey in the breeding season), while the female is more streaked and lacks the red colouring.

The name 'Linnet' comes from Old French and refers to the bird's fondness for linseed (flax). 'Lintwhite' means 'flax-plucker' while the various references to 'gorse' and 'furze' betray its favoured habitats, an association that poets have been happy to use: 'linnets o'er the flowering furze,' wrote the poet James Thomson (1700–48).

Twite *(Carduelis flavirostris)*

OTHER COMMON NAMES: *grey linnet; heather lintie (Shetland, Orkney, Scottish Borders); hill lintie (Orkney); ling linnet (Yorkshire); lintie (Orkney); little peewit (Yorkshire); moor linnet, moor peep (Cheshire); peepie lannart (Northumberland); rock lintie; rockie (Angus – Forfar); trice finch (Lancashire); twite finch (North Yorkshire).*

The Twite is a resident finch now confined to northern and western coasts, although it is not always recognized because of confusion with its rather more colourful close relative, the Linnet.

The Roaches on the Staffordshire–Derbyshire border, where Red-necked Wallabies once bred (see page 384), is the most southerly place that it nests. It was once called the 'mountain linnet'; 'moor linnet' and 'hill lintie' are still used locally. 'Twite' seems a fairly obvious onomatopoeic name.

Redpoll *(Carduelis flammea)*

OTHER COMMON NAMES: *chaddy (Cheshire); chevy linnet (West Yorkshire); chippet linnet; chitty; Chivey linnet (Yorkshire); French linnet; little redpole linnet; redcap (Yorkshire); red-headed finch; red linnet (West Yorkshire); rose lintie (Scottish Lowlands).*

This resident finch is an associate of Siskins and its population is supplemented by winter visitors, although it is absent from large areas, especially of central England. The striking small red patch on the forehead has given rise to its names.

Crossbill *(Loxia curvirostra) and* Scottish Crossbill *(L. scotica)*

OTHER COMMON NAMES: *crossbeak; shell apple.*

Few people get the opportunity to see this distinctive finch at close quarters because, although a fairly common resident (its place in Scotland is taken by its close Scottish relative), it is restricted to coniferous forests. Periodically, large numbers irrupt from the Continent to the British Isles. The association with conifers explains the curiously crossed beak, which is used to extract seeds, especially from pine or spruce cones. Folklore, however, has another explanation. Like other red-fronted birds (for instance, the Goldcrest and Robin), the Crossbill is traditionally associated with the Crucifixion, being

The Crossbill *(Loxia curvirostra)* owes its name to its peculiarly shaped bill which is crossed at the tip.

stained with Christ's blood; even more specifically, it is credited with having twisted its bill while attempting to wrest the nails from His hands and feet.

Although the Crossbill certainly eats apples if it is given the opportunity, the use of the name 'shell apple' seems to have little to do with Crossbills and is simply 'borrowed' from the Chaffinch.

Crossbills have been credited with helping women in childbirth and rousing the household in the event of fire (presumably when kept as cage birds) and were distinguished by a specific Act of Parliament, passed in 1926, which imposed a fine of £1 on anyone convicted of taking their eggs. The eggs were once thought to hatch at Christmas, the fledglings flying at Easter, a belief clearly based on the accurate observation that Crossbills have an unusual breeding season which varies from area to area but may extend from autumn to spring.

Bullfinch *(Pyrrhula pyrrhula)*

OTHER COMMON NAMES: *billy black cap (Shropshire); black cap (Lincoln); black nob (Shropshire); blood olp (male, Norfolk, Surrey); bud bird, bud finch, bud picker (Devon); bull finch; bull flinch (Yorkshire); bull head; bull spink; bulldog; bully (Yorkshire); hoop; hope (West Country); mawp (Lancashire); monk; mwope (Dorset); nope (Staffordshire, Shropshire); olf (east Suffolk); plum bird, plum budder (Shropshire); pope, red hoop (Dorset); tawny (Somerset); thick bill (Lancashire); tonnihood (female, Somerset); Tony hoop.*

One of the prettiest and most pestilential of finches, the Bullfinch, a common resident, is at the same time no friend to the gardener and fruit-grower because of its predilection for the buds of fruit trees and bushes. Possibly because of this, its population has dropped markedly in the past 30 years. It is principally a bird of deciduous woodlands and, in early autumn, when these provide an adequate food supply, it is seldom seen in gardens. However, woe betide the gardener whose fruit cage is not secure when cold weather arrives in November.

Many of the regional names reflect its bud-picking habit, although it is not a recent one. As long ago as 1566, an Act of Parliament allowed for one penny to be offered for 'everie Bulfynche or other Byrde that devoureth that blowth of Fruite'. Ron Murton reported that the churchwarden's records of Tenterden, in the once hugely important fruit-growing county of Kent, show that many such payments were made in 1628–29, and again in 1678–80, while in 1618, William Lawson, in *A New Orchard and Garden*, wrote:

> Your cherries when they bee ripe, wil draw all the blacke-birds, Thrushes and May-pyes to your Orchard.

> The Bull-finch is a devourer of your fruit in the budde, I have had whole trees shald out with them in Wintertime.[452]

Rather unexpectedly, Kent is among the few counties with an inn named after the bird: The Bullfinch at Riverhead.

The old name for the bird was 'alp', or 'alpe', used by Chaucer in *The Romaunt of the Rose*, from which 'olp', 'olf' and similar names are derived. All are obscure in origin, as is the later use of the word 'bull', from which comes the more modern 'Bullfinch'.

Hawfinch *(Coccothraustes coccothraustes)*

OTHER COMMON NAMES: *berry breaker (Hampshire); black-throated grosbeak; cherry finch (Yorkshire); coble; grosbeak (Yorkshire); grosbeak haw; haw grosbeak; Kate; nutcracker (West Yorkshire); pie-finch.*

Hawfinches are relatively large, very stocky finches with a massive bill that equips them well for cracking open hard seeds such as those of damson and cherry and the hawthorn, hence their name which originated in 1674 and has been current since the mid-nineteenth century.

Although widespread residents, they are very shy, secretive birds and, because their nests were not found until 1833, they were long thought to be winter visitors. They are typically birds of lightly wooded uplands and are largely absent from the north and west of the British Isles.

Buntings
(Emberizidae)

Buntings are small, mainly seed-eating, finch-like birds with short, stubby beaks, nesting in shrubs or trees or on the ground and most with rather drab females but brightly coloured males. There are three more or less widespread resident species (the Corn Bunting, Reed Bunting and the beautiful Yellowhammer), one southern resident (the Cirl Bunting), and two northern winter visitors (the Snow Bunting and the Lapland Bunting, *Calcarius lapponicus*). The numbers of almost all are seriously declining.

The name 'bunting' is Middle English and means 'plump' or 'stocky' (the nursery rhyme 'Bye, Baby Bunting, Daddy's Gone a'Hunting' is probably better known to most people than the birds). It was first recorded as a surname in the thirteenth century and has been used for the Corn Bunting since the fourteenth century. However, its association with birds may be older because the name Buntingford ('a ford frequented by buntings') for a village in Hertfordshire was first recorded in 1185.

A species that crops up frequently in older writings is the Ortolan Bunting (*Emberiza hortulana*, literally 'garden bunting'), which is only a rare migrant to Britain but a bird that was netted in vast quantities in southwest France for eating. Sadly, many of these tiny birds were needed to make a meal and they were exported in large numbers to Britain. Mrs Beeton gives a recipe for roast Ortolans that required six birds, each wrapped in a vine leaf and served on toast.[453]

Snow Bunting *(Plectrophenax nivalis)*

OTHER COMMON NAMES: *brambling; mountain bunting; north cock (Aberdeenshire); oatfowl (Orkney); pied finch; snaw fowl (Shetland, Caithness); snow bird (southern Scotland, northern England); snow flake (Orkney, Scotland); snow fleck (northwest England, Norfolk, Hampshire, Sussex); snow flight; tawny bunting; white lark, white-winged lark (Norfolk).*

This striking, mainly black-and-white bird is chiefly found on bare, high mountains, but also occurs in coastal areas further south. It has been known to breed in Britain since the late nineteenth century, a few on the northern Scottish islands but most on the Cairngorm tops. It is among the species that just might be pushed too far north by global warming to remain a British bird.

'Snow Bunting' is an eighteenth-century name; 'snow flake' and 'snow bird' are much older and more attractive.

Yellowhammer *(Emberiza citrinella)*

OTHER COMMON NAMES: *Bessie blaceling (Westmorland); Bessie bunti (Cumberland); blacksmith (Shropshire); chesser (Northamptonshire); gladdie (Devon, Cornwall); gold spink (northern England); goldfinch; goldie (North Yorkshire – Craven, Nottinghamshire); gouldie (County Down); guler (Norfolk); little-bit-of-bread-and-no-cheese; may-the-devil-take-you (Cumberland); pretty-pretty-creature (Gloucestershire); Scotch canary (Scotland); scribbler, scribblie, scribbling lark (Northumberland,*

The village of Buntingford ('a ford frequented by buntings') in Hertfordshire has been so called since the late twelfth century.

Northamptonshire); urin (Lancashire); writing lark (Northumberland, Northamptonshire); writing master (Shropshire); yedda yeldern (Antrim); yellow amber (Shropshire); yellow bunting (Renfrewshire); yellow omber (Shropshire); yellow ring; yellow yite (Scotland); yellow yoit (Kircudbrightshire, Wigtownshire); yellow yorling (Northern Ireland, Scotland, northern England); yellow yowlie (Northumberland).

Much the most striking bunting, this bright yellow bird has attracted attention over the centuries, hence the dazzling array of local names, many of which are still in use.

> As a boy in Perthshire, I was astonished to find the Yellowhammer referred to by all and sundry as the Yellow Yite.[454]

References to 'writing' or 'scribbling' allude to the patterns on the eggs, while such peculiar names as 'little-bit-of-bread-and-no cheese' and 'may-the devil-take-you' are fanciful onomatopoeic allusions to the call that do not really convince anyone. The less than flattering 'urin' presumably describes the colour. 'Yellowhammer' dates from 1587 but is probably originally Old English; although no-one seems to know the meaning of 'hammer', or 'ammer'. It may be a corruption of the German word *ammer*, meaning 'bunting', or originate from the Old English *amore*, 'an unidentified bird'. In the first half of the seventeenth century, the name was adopted to mean a gold coin and was later used for a boy wearing yellow breeches.

It is sad but curious that so inoffensive yet conspicuous a bird should have become associated with the devil, especially in the north of England, Scotland and Northern Ireland:

> The brock and the toad and the yellow yorling
> Tak a drap of the devil's blood ilka May morning.[455]

Like the Badger and Toad it was severely persecuted in parts of Scotland in consequence. Crowds of youths and boys would seek out its nest and destroy both eggs and nestlings. In Northumberland, the nests met a similar fate at the hands of gangs of children, who sang:

> Half a paddock, half a toad,
> Half a drop o' de'ils blood,
> Horrid yellow yorling.[456]

Two theories have been advanced for this apparently illogical sentiment. The first is that its song mimicked human speech and was thus diabolic: the Scottish version of the onomatopoeia is 'Whetil te, whetil te, whee! Harry my nest and the de'il tak ye' and a Cumberland name for the bird was 'may-the-devil-take-you.' The second theory is that the scribble-like markings on its eggs were thought to be cabalistic. These same serpentine marks have given rise to the Welsh name *gwas y neidr*, 'the servant of the snake', 'servant' because

The Yellowhammer *(Emberiza citrinella)*, once common in grassland areas, has declined dramatically in recent years.

the Yellowhammer is supposed to give snakes warning of approaching enemies. A comparable old belief was that snakes are hatched in its nest.

On the principle of curing like with like, it was a remedy for jaundice, at least in Wales, a Yellowhammer being held in front of the face of the afflicted. At least the birds remained alive, a better fate than those whose flesh was used to treat people afflicted with consumption and epilepsy.

Poets, however, have been more charitable to this charming bird:

> When, towards the summer's close,
> Lanes are dry,
> And unclipt the hedgethorn rows,
> There we fly!
>
> While the harvest waggons pass
> With their load,
> Shedding corn upon the grass
> By the road.
>
> In a flock we follow them,
> On and on,
> Seize a wheat-ear by the stem,
> And are gone ...
>
> With our funny little song,
> Thus you may
> Often see us flit along,
> Day by day.
> (Thomas Hardy, 'The Yellowhammer')

John Clare was also hugely taken with the bird:

> When shall I see the white thorn leaves agen
> And Yellowhammers gath'ring the dry bents
> By the Dyke side on stilly moor or fen
> Feathered wi love and natures good intents
> Rude is the nest this Architect invents
> Rural the place wi cart ruts by dyke side
> Dead grass, horse hair and downy headed bents
> Tied to dead thistles she doth well provide
> Close to a hill o' ants where cowslips bloom
> And shed o'er meadows far their sweet perfume
> In early Spring when winds blow chilly cold
> The yellow hammer trailing grass will come
> To fix a place and choose an early home
> With yellow breast and head of solid gold.
> (John Clare, 'The Yellowhammer')

Unfortunately, this once common if widely persecuted bird is no longer in need of unpleasant local customs to reduce its numbers because it has declined dramatically over the past century and has disappeared from many of its old northern and western haunts. Its future in Britain looks bleak.

Cirl Bunting *(Emberiza cirlus)*

OTHER COMMON NAMES: *French yellowhammer.*

This local, southern bird, a paler version of the Yellowhammer, derives its usual but peculiar English name from its scientific name *cirlus*. This, however, has a contorted pedigree, coming via a Bolognese dialect from the Italian *zirlare*, 'to chirp'.

Modern bird-watchers have always known this as a rare bird, although it was once common throughout much of the southern half of the country; its decline has been due largely to the loss of the weedy stubble-field habitats that it requires. Careful conservation measures seem to have stemmed any further diminution of its numbers.

Reed Bunting *(Emberiza schoeniclus)*

OTHER COMMON NAMES: *black bonnet (Scotland); black cap (North Yorkshire – Craven, Hampshire); black coaly hood (southern Scotland); black-headed bunting; bodkin (Lancashire); bog sparrow; chink, coaly hood (Scotland); Colin blackhead (Renfrewshire); pit sparrow (Cheshire); reed sparrow; ring bird; ring bunting; ring fowl (Aberdeenshire); riverside sparrow; seave cap (Yorkshire); spear sparrow (female, Hampshire); toad snatcher (Yorkshire); water sparrow.*

Almost all the regional names leave no doubt that this remarkable bird, which resembles the House Sparrow, is a species of reed beds and other marshy places, where it builds its beautiful hair-lined nest of grasses on reed stems.

It has declined drastically with the loss of many wetland habitats, although this is not the full story: this loss has encouraged it to explore grasslands and other drier places during the last century. Nonetheless, it is now nothing like as common as it was.

Corn Bunting *(Miliaria calandra)*

OTHER COMMON NAMES: *briar bunting (Northern Ireland); bunting lark, buntling lark (Scotland); bunt lark (Norfolk); bush lark (Ireland); chub lark (Yorkshire); common bunting; corn bird (Ireland); ebb; ground lark (Yorkshire); hornbill bunting (Ireland); horse lark (Cornwall); lark bunting (Somerset); skitter brottie, thistle cock (Orkney); whisker bird.*

The references to 'lark' in this bird's regional names stem from its resemblance to the Skylark. This is the largest bunting and was once a common bird on farms (hence the reference to corn), along hedgerows and similar open areas, where it nested in long grass. The major decline in the growing of cereals has caused a phenomenal drop in its numbers everywhere. It has all but vanished from Ireland and Wales, clings to a few coastal places in Scotland, and is seen far less in both the north and south of England. Chris Mead has said:

> The lack of seed-rich winter stubbles and spring-sown barley have been implicated along with the use of pesticides and the reduction of mixed farming. One has seriously to wonder whether this inoffensive and unspectacular bird will survive in Britain.[457]

As its name suggests, the Corn Bunting *(Miliaria calandra)* was once a familiar sight in cornfields.

Red Deer *(Cervus elaphus)* hinds at dawn, Scotland.

Mammals (Mammalia)

Mammals are warm-blooded, mainly terrestrial vertebrates that give birth to living young, nourish them with milk and generally display considerable parental care. Some have largely abandoned the terrestrial environment and either taken to an aerial existence (bats) or returned partially (seals), or totally (whales), to the water. Although biologically the most advanced of British animals, the mammal fauna is relatively small in numbers and, including the whales, comprises about 75 species.

Most people have seen some of them, but few have seen many and very few have seen them all. This is because most British mammals are small, many are more or less nocturnal, and many live well away from the majority of the human population. Nonetheless, they have had a significant impact on people's lives and attitudes: men have eaten some species, hunted some for sport, farmed a few and killed many because they competed with humans for food. Consequently, the lore and traditions associated with mammals are second only to those associated with birds.

Pouched Mammals
(Marsupialia)

Marsupials are mammals that give birth to comparatively undeveloped young and then nurture them further in an external pouch. They have numerous other features that render them evolutionarily much less advanced than the remaining, vast majority of species, which are collectively called the placental mammals. Most marsupials are Australian, with a very few in the Americas, but, although none occurs naturally in Europe and therefore plays no part in British traditions, one at least has certainly made a modest impact in modern Britain.

Kangaroos and Wallabies
(Macropodidae)

Red-necked Wallaby *(Macropus rufogriseus)*
OTHER COMMON NAMES: *Bennett's Wallaby.*
Anyone encountering a small, long-tailed, grey-brown creature, resembling a dwarf kangaroo, bounding along a remote Derbyshire road late at night sometime during the past 50 or so years might have been sorely tempted to doubt their faculties. However, this would have been no mental aberration, because a small population of Red-necked Wallabies from southeast Australia and Tasmania lived for many years in the area of The Roaches on the western edge of the Peak District, on the boundary of Derbyshire and Staffordshire. The original animals escaped from 'a rather good private collection ... contained Nilgai, Black Buck, Yak, that sort of thing',[1] which was formed before the Second World War by Captain Henry Brocklehurst of Roaches House near Leek in Staffordshire (whose father once shot a Giant Panda that ended its days stuffed in Leek Museum). It was probably war-time neglect of the property that enabled the creatures to escape, something at which they are remarkably adept:

> Everyone in this country who keeps Red-necked Wallabies contrives to lose a few.[2]

They adapted to a diet of heather and associated vegetation and some travelled considerable distances, individuals being found:

> ... on the other side of Stoke on Trent and near Chesterfield.[3]

> The locals used to discourage questions about the animals in order to protect their whereabouts but they were always well known to the local postmen.[4]

The wallabies survived until the late 1990s, when they seem finally to have died out, presumably through a combination of cold winters and inbreeding. Other groups, some escapes and some deliberate introductions, have become established at different times in Ireland, southwest Scotland, Sussex and the

Some of the Red-necked Wallabies *(Macropus rufogriseus)* that were introduced to Britain from Tasmania and southeast Australia before the Second World War escaped to establish a small population in the Peak District.

Channel Islands. The Scottish population is descended from two animals from Whipsnade, which were released onto the undisturbed Inchconnachan Island in Loch Lomond in 1975; in 1999 it was estimated to total 43 animals. There is every possibility therefore that self-sustaining colonies could arise again elsewhere.

The name 'wallaby', as might be expected, is from a native Australian word and, as no Englishman had seen one before 1770, it entered the English language as recently as 1828.

Insect-eaters

(Insectivora)

Not all insectivores exclusively eat insects, although their teeth certainly equip them to do so. They are all relatively small and, among placental mammals, are considered to be some of the most primitive. They tend to have poor eyesight, a good sense of smell and a small brain.

Hedgehogs

(Erinaceidae)

Hedgehog *(Erinaceus europaeus)*

OTHER COMMON NAMES: *British porcupine; butter-bump; erchin (Fife); erison; furze-a-boar; furze-man-pig (Gloucestershire); furzehog; hag-hog; hedgepig; herysson; hirchen; hodgen; hurcheaon; hyrchoune; nertchard; niceple; nisbill; nurchon; nysebill; perpentine; perpynt; pochin (Somerset); porcupig; porkpoint; porpentine; pricky back ochun (Yorkshire); rock (Somerset: young animals); sharp-nails; urcheon; urchin; vuz-peg (western England); western hedgehog.*

Sadly, a generation has grown up thinking of the Hedgehog as a two-dimensional animal: the poor flattened creature impressed into the tarmac has become one of the most familiar sights on roads and, in many areas, traffic must take second place only to winter starvation as a cause of death. The relationship between Hedgehogs and cars has become likened to an innocent David facing his Goliath, something that Dick King-Smith turned to charming effect in his 1987 children's story *The Hodgeheg*, concerning a Hedgehog for whom everything is the wrong way round.[5] The cause of the Hedgehogs' downfall is that, instead of running when caught in a vehicle's headlights, they usually simply roll into a ball.

> In East London and Lancashire, animals seem to have learned the habit of running for safety. This is an extraordinary change in a habit that evolved many thousands of years ago and was seemingly instinctive.[6]

No-one can mistake a Hedgehog for anything else. Its prickly back with around 6,000 spines, which are gradually replaced over a period of about 18 months, is unique, although:

When threatened, the Hedgehog *(Erinaceus europaeus)* rolls itself into a ball, so many are killed attempting to cross busy roads.

> ... it is hardly necessary to deny the popular belief that the animal can shoot out its quills like so many arrows; the notion has arisen from the fact that when the animal erects its spines, loose ones sometimes fall out.[7]

As generations of Hedgehog mothers must have appreciated, however, the spines on the newborn young are soft.

The Hedgehog, together with the dormice and bats, are the only British mammals that hibernate, although they may become active during mild winter periods. They do not dig burrows but their winter home, or hibernaculum, and also their breeding nest, which are both lined with grass and leaves, may sometimes be built in abandoned Rabbit burrows, although they are more usually situated above ground in a sheltered spot. The odd word 'hibernaculum' is sometimes used for the hibernation place of other animals too and was well known (as 'hybernaculum') to Gilbert White, who used it to refer to the winter quarters of 'Mr Snooke's tortoise' (see Tortoises, Terrapins and Turtles, page 197).

Hedgehogs live in a wide range of habitats throughout the British Isles, wherever their food is to be found, and have been introduced to some of the offshore islands, including Orkney, Shetland, the Isle of Man and certain of the Channel

This image from a fourteenth-century bestiary shows a Hedgehog *(Erinaceus europaeus)* collecting apples on its spines.

Islands. They are thought to have been introduced to Ireland in *c.*1700, quite possibly as food. They have probably declined in numbers throughout Britain in recent years because of loss of habitat and they seldom occur in quantity in areas where there are many Badgers (which may compete for some of the same food but also eat them). They are, however, familiar and welcome visitors to gardens because they eat large numbers of slugs.

> I cannot help thinking ... that when it comes to hedgehogs, a snail may well have an advantage over a slug in that the latter would appear easier to deal with. An extraordinary incident that seems to bear this out occurred in a garden that was so overrun with snails the woman who owned it made a practice of walking around the place at dusk crushing snails underfoot. One evening, though, when she was scrunching her way around the garden in an orgy of destruction, she suddenly became aware the noise was not coming from her alone. A similar scrunching was going on behind, and when she peered through the gloom she saw a hedgehog bringing up the rear, and having a whale of a time as it munched away at the leftovers. To satisfy her curiosity, she offered the hedgehog a snail fully intact, but the offer was refused until she broke the shell. Then the hedgehog lapped up the contents with as much enjoyment as it had all the others. By the end of the evening the hedgehog had disposed of well over 30 snails, mainly because they were so easily obtainable. This would seem to prove that as far as a hedgehog is concerned, the shell of a snail does act as a deterrent, even though it may not always be a complete safeguard.[8]

Hedgehogs also eat earthworms, beetles and other insects, and sometimes frogs, small rodents, young birds and birds' eggs. It was their fondness for the eggs of such ground-nesting birds as Dunlin, Redshank and Ringed Plover that led to a decision in 2003, based on the RSPB's proposal,[9] to cull large numbers of Hedgehogs in the Western Isles. Conservationists tried to move the Hedgehogs elsewhere.

A supposed association between Hedgehogs and snakes has mystified people and a belief has grown up that Hedgehogs not only eat snakes but are immune to Adder venom. Neither seems likely. They do, however, eat fruit on occasion, but the ancient belief that they roll onto apples or other fruit in order to impale them on their spines has never fully been explained. This phenomenon, reportedly mentioned by Pliny, was illustrated in an English manuscript dating as far back as the thirteenth century. It is still found in children's stories today but has generally been given little credence. Hedgehogs do not store fruit, so why should they bother collecting it, and, in any event, invertebrates are their preferred food. Presumably, on odd occasions, apples have fallen onto a Hedgehog and stuck to its spines, and the story has become elaborated through the telling. The Renaissance naturalist Edward Topsell (1572–1625), in his *Historie of Four-footed Beasts* (1607), was as much to blame as anyone in perpetuating the idea:

> The hedgehog's meat is apples, worms and grapes: when he findeth them upon the earth, he rolleth on them until he hath filled up all his prickles, and then carrieth them home to his den ... and so forth he goeth, making a noise like a cart wheel. The prickly thorns on their backs will not suffer them to have copulation like Dogs or Swine, and for this cause they are a very little while in copulation, because they cannot stand long belly to belly on their hind legs. With the same skin flayed off, and the prickles, brushes are made for garments, so that they complain ill which affirm that there is no good nor profit from this beast.

Topsell also referred to another widespread belief, that they could predict a change in the wind and weather:

> ... when they hide themselves in their den they have a naturall understanding of the turning of the wind. They have two holes in their cave, the one North, the other South, observing to stop the mouth against the winde, as the skilful mariner to stiere and turn the rudder and sailes, for which some have held opinion that they do naturally foreknow the change of weather.[10]

This has been echoed many times:

> The Urcheon is witty and wise in his knowledge of comming of Winds, North and South, for he changeth his Denne or hole, when he is ware that such windes come.
>
> The Hedge-hog commonly hath two holes or vents in his Den or Cave, the one towards the South, the

other towards the North: look at which of them he stops, thence will great storms and winds follow.[11]

Robert Chester, in 1601, even put it into verse:

The Hedgehog hath a sharpe quicke thorned garment
That on his backe doth serve him for defense:
He can presage the winds incontinent,
And hath good knowledge in the difference
Between the Southerne and the Northern wind.
These vertues are allotted him by kind.
Apples or peares or grapes, such is his meate
Which on his back he caries for to eate.
(Robert Chester, 'Love's Martyr')

In 'Poor Robin's Almanac for the Year of Our Lord 1733', the belief was amplified:

Observe which way the hedge-hog builds her nest,
To front the north or south, or east or west;
For it 'tis true that common people say,
The wind will blow the quite contrary way.
If by some secret art the hedge-hogs know
So long before, which way the winds will blow,
She has an art which many a person lacks,
That thinks himself fit to make almanacks.

How this belief arose, and whether it contains any grain of truth, no-one knows, but Hedgehogs do enough strange things without people inventing any more. They spend an inordinate amount of their time in 'self-anointing', or licking their bodies with a frothy saliva. This peculiar habit has never been satisfactorily explained; it might serve to ward off parasites (Hedgehogs are notoriously flea-ridden) or it might deter potential predators (Foxes, large birds of prey and other carnivores will also eat them). However, it has not helped dispel the widely held belief that Hedgehogs are odd.

On the Isle of Man, Hedgehogs and hares were treated with great circumspection because of an imagined connection with witches and, as recently as the late nineteenth century, the Irish author G.H. Kinahan wrote:

Hedgehogs are considered witches, and called granogoes ... A few years ago, I knew in Co. Wexford of a hedgehog having been thrown into a pool to see if it could swim, and because it swam to the bank was considered to be a witch and burnt.[12]

The most famous witches in literature, gathered on the blasted heath, stir their cauldron while incanting:

Thrice and once the hedge-pig whined.
(*Macbeth*, Act 4, Scene 1)

The Hedgehog *(Erinaceus europaeus)*, so-called because of its pig-like snout, is welcome in gardens as it has an appetite for slugs.

As so often, Shakespeare knew his wildlife. Hedgehogs do utter an eerie scream when they are forced to uncurl, for instance when they are attacked, perhaps by a Badger or a dog:

> The shepherd's lurcher, who, among the crags
> Had to his joy unearthed a hedgehog, teased
> His coiled-up prey with barking turbulent.
> (William Wordsworth, *The Prelude*)

In the often-quoted Preservation of Grain Act of 1566, Hedgehogs were listed among the many creatures to be destroyed, and old country churchwardens' accounts throughout England often list the bounties (generally about 2d or 3d per tail or snout) that, in consequence, were paid to parishioners for killing them. The Act was not repealed until 1863, but, although Hedgehogs were quite innocent, countless numbers must have been killed in the intervening centuries.

For a creature whose strange and somewhat bizarre appearance has attracted many an odd tale, the supposed relationship between Hedgehogs and cows' udders is the most challenging and persistent. The belief is simple: that Hedgehogs suck milk from the udders of cows lying in pastures. The obvious questions have often been asked: Can a Hedgehog reach a cow's udder? Can it open its mouth wide enough to suckle and, if so, will the cow tolerate a Hedgehog's sharp teeth? Has anyone ever seen it happen? To all of which the answer is no, although Hedgehogs have been seen licking milk from the ends of udders. Nonetheless, old beliefs die hard and this one evidently lives on in modern form in the Home Counties:

> When we moved here thirty years ago there seemed very few hedgehogs in the locality so my husband used to pick up the ones he saw when he was driving to and from Luton at night, to release them in the vegetable garden. Our neighbour was extremely upset about this as she believed they would climb into her baby's pram to suck the milk from his bottle and in so doing suffocate him.[13]

The name 'Hedgehog', in reference to the animal with a pig-like snout that lives in hedgerows, dates from 1450 and has generally replaced the older 'urchin', or, in its more original form, the Middle English *hurcheon*, which in turn came with the Normans from the Old French *herichon*. Robert Hendrikson summed up the history admirably:

> The English, who have always been poor at spelling French words, stumbled badly over the synonym for hedgehog ... They spelled herichon a number of wrong ways before finally settling on urchin, which looked something like herichon sounded. The English called the hedgehog the urchin for a time. They also applied the name urchin to a mischievous child, because the urchin or hedgehog was popularly believed to be a mischievous elf in disguise. Eventually people stopped using urchin as a synonym for hedgehog, but continued to use it when referring to an impish child. Neither was the name sea urchin abandoned; this spiny creature was originally named for its resemblance to the urchin or hedgehog and was once called the sea hedgehog [see Sea Urchin, page 21].[14]

Most of the regional names for the Hedgehog, like 'furzehog', allude in some way to its resemblance to a pig, as does the Late Middle English word 'porcupine', which has sometimes been attached to it. Hedgehog place names are not common but 'Urchin Royd, a field name in Worsborough township [West Yorkshire] is Hedgehog clearing'.[15]

There has long been a country tradition of eating Hedgehog. Leo Harrison-Matthews, quoting an unnamed writer 'who has much culinary experience of the hedgehog', said that it is 'the best meat in England' (how often is that said of some outlandish fare?) but also gave very practical information on its preparation:

> The classic way of cooking a hedgehog is to gut and stuff it with sage and onion, sew it up and plaster it over with clay; then suspend it over the fire with a length of twisted worsted as a roasting jack, and when the clay cracks it is done. But this is not the best way because although the spines come away with the clay when it is broken open, the smaller hairs are not completely removed. It is better to singe the prickles and hairs off in the fire after gutting the animal, and then to scrape it with a very sharp knife before roasting it without clay. Another method is to gut and skin the animal, wash it well, and simmer it with seasoning in a little water for several hours; when cold the whole sets to jelly, and the 'pudding' can be cut into slices like pressed meat; this is very good.[16]

Self-anointing, so characteristic of the Hedgehog *(Erinaceus europaeus)*, may be a means of warding off fleas.

Traditionally, the Romanies have been Hedgehog-eaters and, inevitably, the animal has found a widespread application in folk medicine. Hedgehog oil was particularly used – according to Roy Palmer, it was produced at Congerstone in Leicestershire by a hermit and used for earache – and this seems to have been a rather widespread practice.[17] William Ellis, in his *Country Housewife's Family Companion*, told of a man being cured of deafness:

> ... with Dripping of Hedgehogs, 3 Drops into his Ear at Night, the same in the Morning, and so for two Days, when it cured him.[18]

Rather less plausible are the virtues attached to Hedgehogs' eyes. The right eye of a Hedgehog, boiled in oil and preserved in 'a brazen vessel', was widely supposed to enable a person to see in the dark, while the left eye, according to Sir Thomas Browne, among others, should be 'fried in oil to procure sleep'.[19] An often-quoted old remedy for epilepsy, fits or other irrational behaviour, was to give the patient a cooked Hedgehog to eat:

> For a lunatic. Take a hedge-hog and make broth of him, and let the patient eat of the broth and flesh.[20]

The ever-reliable John Keogh had a far wider range of suggestions:

> Two ounces of the ashes of the whole Hedge-hog burnt, mixt with half an ounce of tartar, and two ounces of the juice of onions cure tetters, ringworms, the alopecia, and herpes ... The fat is exceeding good to be applied to ruptures ... The liver pulverized was used against 'the cholic, gravel, dropsy, convulsions, and lethargy' ... The spleen was used against 'opulations of the spleen' ... The ashes of the skin, and head mixt with vinegar and honey, cleanse foul ulcers, and causes hair to grow on parts cicatrized ... The fat and liver were also used against 'the flux of the belly and elephantiasis'.[21]

The association between Hedgehogs and the growth of hair is met rather frequently, perhaps in view of the remarkably strong 'hair' that the animal itself possesses. 'Gypsies also rub hedgehog fat into the scalp to keep the hair plentiful and prevent its going grey.'[22] On a far more practical and believable level and, before the invention of barbed wire, Hedgehog skins were nailed to gateposts to prevent cows from rubbing against them.

So distinctive a creature as the Hedgehog has attracted the attention of many writers, including the poet John Clare:

> The hedgehog hides beneath the rotten hedge
> And makes a great round nest of grass and sedge,
> Or in a bush or in a hollow tree;
> And many often stoop and say they see
> Him roll and fill his prickles full of crabs
> And creep away; and where the magpie dabs
> His wing at muddy dyke, in aged root
> He makes a nest and fills it full of fruit,
> On the hedge-bottom hunts for crabs and sloes
> And whistles like a cricket as he goes.
> It rolls up like a ball or shapeless hog
> When gipsies hunt it with their noisey dog;
> I've seen it in their camps – they call it sweet,
> Though black and bitter and unsavoury meat.
> (John Clare, 'The Hedgehog')

Many modern children's writers have also found the Hedgehog endearing, and characters such as Fuzzypeg, in Alison Uttley's stories of Little Grey Rabbit, and Beatrix Potter's Mrs Tiggy-Winkle have become accepted as part of children's folklore, invariably as friendly, cuddly beasts, despite the spines. As the old nursery rhyme says:

> Old Mr Pricklepin
> has never a cushion to
> stick his pins in,
> His nose is black and his
> beard is grey,
> and he lives in an ash stump
> over the way.

Shrews
(Soricidae)

The shrew family includes the smallest European mammals. Superficially mouse-like, although with longer, pointed snouts (they were once commonly called 'picked-nosed mice') and shorter tails, they are biologically very different from rodents and have a number of unusual features, some of which zoologists consider primitive. They walk on the soles of their feet, produce poisonous secretions from their salivary glands, have a single, combined digestive, urinary and reproductive opening, and some, like bats, are able to echolocate using high-frequency sound. Being small, warm-blooded and yet living in a relatively cold climate means that they have developed a phenomenal turnover of energy and need to eat for a very large proportion of their lives: the smallest British species, the Pygmy Shrew *(Sorex minutus)*, is active both day and night and must consume the equivalent of its own body weight every 24 hours. Such distinctive characteristics mean that shrews have attracted attention, admiration, fear and loathing over the centuries. In Britain, there are three mainland species of red-toothed shrew (they really do have red teeth), with two additional white-toothed species (one or other of them in the Scilly Isles and each of the Channel Islands).

Shrews produce a ball-like nest of grass or other vegetation and are remarkably noisy, squeaky and aggressive creatures.

The derogatory sense of 'shrew' was famously applied by Shakespeare to Katharina, seen here in a scene from *The Taming of the Shrew*.

They are not animals to pick up with your bare hands; a shrew bite is not easily forgotten. Edward Topsell, writing in the seventeenth century, knew this well enough, calling the shrew:

> ... a ravening beast, feigning itself gentle and tame, but, being touched, it biteth deep, and poisoneth deadly.

The 'poisoneth deadly' is an exaggeration, although I can vouch for the rest, but Topsell also summarized much of what else has been thought and believed about shrews when he wrote that:

> It beareth a cruel mind, desiring to hurt anything, neither is there any creature it loveth. They annoy vines, and are seldom taken except in cold; they frequent ox-dung. If they fall into a cart-road, they die and cannot get forth again. They go very slowly, they are fraudulent and take their prey by deceit. Many times they gnaw the ox's hoofs in the stable. They love the rotten flesh of a raven. The shrew being cut and applied in the manner of a plaister doth effectually cure her own bites. The dust of a cart-rut, in which a shrew has died, being taken and sprinkled into the wounds made by her poisonous teeth is a very excellent and present remedy for the curing of the same. If horses or any other labouring creature do feed in that pasture or grass in which a Shrew shall put forth her venom or poison in, they will presently die.[23]

I think the appropriate modern expression is 'give a dog a bad name'.

The name 'shrew' is from the Old English *screawa*, but its meaning is obscure and it seems to have nothing to do with the word 'shrewd'. Its use to mean 'a person especially a woman having the character or disposition of a shrew' is Middle English and the connection is obvious, as is the adjective 'shrewish' which is derived from it and which appears from 1565. The most famous of all shrews, Katharina, the woman who was tamed in Shakespeare's *The Taming of the Shrew*, was a literary character with a familiar and popular lineage that can be traced back to shrewish wives in earlier miracle plays and to Chaucer's 'Wife of Bath', among others.

For many years, and in Shakespeare's time, the creature itself was more usually called a 'shrewmouse' (sometimes hyphenated), a word that was first recorded in 1572. Most colloquial names make no distinction between the various species, although the Water Shrew, because of its specialized habitat, has attracted a few of its own. Such apparent connections as the place names Shrewsbury and Shrewly are sadly quite unrelated.

Shrews are taken by several different predators, although Barn and Tawny Owls account for most, followed by Foxes, Stoats, Kestrels and Weasels. Cats also commonly catch shrews, although I have never owned a cat that would eat one; presumably cats find the liquid produced by certain skin glands to be distasteful. This characteristic may have some bearing on their supposed venomous nature. Historically, as Edward Topsell described so graphically, shrews were thought highly poisonous, with a deadly bite; it is said that the specific name *araneus* for the Common Shrew alludes to it having a bite like a poisonous spider. This was subsequently dismissed as mere fancy, although more recent evidence suggests that there is a toxin present in the saliva.

One of the oddest features of many an English village for centuries was the 'shrew-ash', a tree that was believed to have special healing powers, imparted to it by an unfortunate shrew entombed within. The most celebrated, by virtue of having been described so carefully by one so literate, was the shrew-ash of Gilbert White's Selborne, although the practices associated with it seem to have died out by 1776, when he wrote:

> At the fourth corner of the Plestor, or area, near the church, there stood, about twenty years ago, a very old, grotesque, hollow pollard-ash, which for ages had been looked on with no small veneration as a shrew-ash. Now a shrew-ash is an ash whose twigs or branches, when gently applied to the limbs of cattle, will immediately relieve the pains which a beast suffers from the running of a shrew-mouse over the part affected; for it is supposed that a shrew-mouse is of so baneful and deleterious a nature, that wherever it creeps over a beast, be it a horse, cow, or sheep, the suffering animal is afflicted with cruel anguish, and threatened with the loss of the use of the limb. Against this accident, to which they were continually liable, our provident forefathers always kept a shrew-ash at hand, which, when once medicated, would maintain its virtues for ever. A shrew-ash was made thus: Into the body of the tree a deep hole was bored with an auger, and a poor devoted shrew-mouse was thrust in alive, and plugged in, no doubt, with several quaint incantations long since forgotten. As the ceremonies necessary for such a consecration are no longer understood, all succession is at an end, and no such tree is known to subsist in the manor or hundred.

As to the shrew-ash on the Plestor:

> The late Vicar stubb'd and burnt it, when he was way-warden, regardless of the remonstrances of the bystanders, who interceded in vain for its preservation, urging its power and efficacy, and alleging that it had been 'Religione patrum multos servata per annos'.

White remarked that there were:

> ... several persons now living in the village, who, in their childhood, were supposed to be healed by this superstitious ceremony, derived down perhaps from our Saxon ancestors, who practised it before their conversion to Christianity.[24]

A tree that stood until the end of the nineteenth century near the Sheen Gate in Richmond Park was also commonly thought to have been a shrew-ash.

Apart from their supposed poisonous nature, the most widely held of country beliefs about shrews, and one to which Gilbert White referred, was that ill would befall any man or beast unfortunate enough to have one run over him. It was thought that they would become lame in consequence and in Hampshire, and other places too, shrews were once called 'over-runners'. A related supposition, recounted by an old man from Ruyton in Shropshire, was that 'if you see a shrew-mouse ... you must cross your foot or you will suffer from it'.[25] The affliction could apparently be cured by a visit to the nearest shrew-ash or by passing the victim through the arching stem of a bramble that had rooted at both ends, a well-known country remedy for a number of afflictions.

It was thought bad luck to see a shrew when starting on a journey (but then it was thought bad luck to see a great many other things when starting a journey) and, as Topsell commented, it was thought bad luck for the shrew if it should cross a path because it was widely believed that it would die in consequence. The fact that shrews, and mice, are sometimes found dead at the side of paths was considered evidence for this. No-one seems to have considered that dead shrews are seen at the side of paths because that is where people walk; there are just as many lying dead in other places.

Common Shrew *(Sorex araneus)*

OTHER COMMON NAMES: *artishow; artishrew; artisrobe; artisrow; erd-shrew; hardi-shraow (Worcestershire); hardishrew; hardistraw; hardistrew; hardistrow; hardy-mouse; hardyshrew; hartis-straw; harvest mouse; harvest-row; harvest-shrew; harvest-trow; nussrow; over-runner; ranny, rennie (Suffolk); shirrow; shrew-mouse; skrew; skrow; strawmouse (Morayshire); streaw; strow (Galloway); wight; wreen (Shetland).*

Unlike many animals designated as 'Common', the species known as the Common Shrew really is abundant. It is not only the most frequently found British shrew but, after the Field Vole (*Microtus agrestis*; see page 469), it is also the second commonest British mammal, with a national population estimated at over 40 million. It occurs throughout the British Isles, although it is absent from Ireland and, to be strictly accurate, the shrew that occurs on Jersey is the almost identical French Shrew, or Millet's Shrew *(Sorex coronatus)*.

It is dark brown or nearly black above, almost white beneath, and lives among dense grass and other vegetation in woods and at woodland edges and has adapted very well to roadside verges, often burrowing below ground. It feeds on earthworms, beetles, spiders, slugs, snails and other invertebrates.

Pygmy Shrew *(Sorex minutus)*

OTHER COMMON NAMES: *lesser shrew.*

The Pygmy Shrew is paler than the Common Shrew and is the smallest British mammal, its body being only about 60 millimetres (2¼ inches) long and its weight barely 7.5 grams (¼ ounce). For a long time, Pygmy Shrews were believed simply to be the young of Common Shrews and were not recognized as a distinct species until well into the nineteenth century. Unlike Common Shrews, they do not burrow and occur in more open habitats, although they eat similar food. They are the only shrews found in Ireland.

Water Shrew *(Neomys fodiens)*

OTHER COMMON NAMES: *blind-mouse (Fens); otter-shrew; water-mole; water vole.*

The Water Shrew is dark brown or black above, pale beneath, but is most conspicuous when seen in water, where air bubbles trapped among its fur give it a striking, silvery appearance. It is the largest British shrew, reaching over 90 millimetres (3½ inches) in body length and is an utterly fascinating animal that feeds on aquatic invertebrates, some small fish, amphibians and other animals. It is especially fond of watercress beds, but although I know lengths of river bank where I could once be sure to see both Water Shrews and Water Voles on almost every visit, it is now some years since I have seen either.

The Water Shrew *(Neomys fodiens)* usually lives in a burrow system in the river bank that opens just below the water level.

Lesser White-toothed Shrew *(Crocidura suaveolens)*

OTHER COMMON NAMES: *Scilly shrew.*

See Greater White-toothed Shrew below.

Greater White-toothed Shrew *(Crocidura russula)*

OTHER COMMON NAMES: *house shrew.*

The island-dwelling white-toothed shrews live and behave similarly to their red-toothed mainland relatives, although they tend to be less aggressive. The Lesser White-toothed Shrew occurs on the Scilly Isles and also on Jersey and Sark in the Channel Islands while the Greater White-toothed Shrew occurs on Guernsey, Herm and on Alderney.

> The Scilly Shrew is present in considerable numbers here [St Agnes, Isles of Scilly] and is quite different from its native red-toothed cousin which does not occur here, and was probably a stowaway on boats that traded in the past with France and with the Channel Islands. The Scilly Shrew is among the smallest mammals on earth and is not easily seen, but is in every habitat, from hedgerows and banks to boulder beaches. Its life is short – not usually more than a year in the wild. It pursues its activities with frantic haste ... and its diet consist of beetles and sand hoppers. It produces three or four litters a year. If the female is disturbed when caring for a litter, she will gather her babies from the nest, and the young form a line, each biting the base of the tail of the baby in front with the mother in the lead and running to safety. This is called caravanning and is not often observed by the casual visitor.[26]

Moles
(Talpidae)

Mole *(Talpa europaea)*

OTHER COMMON NAMES: *common mole; crode; heunt (Worcestershire); mouldiwarp (and well over 60 similar names: modywart, moodywarp, moudie, moudwarp, moudywort, mowdie, muddywarp and so on, many tending to have regional associations); oont (Herefordshire); tape; want (or wante) (Wiltshire, Hampshire, Isle of Wight, Cornwall); western mole; wont.*

No animal is Britain is so well known but so infrequently seen as the Mole, no creature so fascinating yet so misunderstood. Moles are despised and loathed by the gardener and greenkeeper but adored by generations of children, to whom the likes of Mole in Kenneth Grahame's *The Wind in the Willows*[27] and Moldy Warp of Alison Uttley's Little Grey

Rabbit stories have been as much a part of growing up as the first bicycle and teddy-bear.

Although few people have seen a live Mole, their appearance is familiar from the countless illustrations that portray their spade-like forefeet, short cylindrical body, pointed pink nose and characteristically short, velvety black fur.

> Diggory Diggory Delvet!
> A little old man in black velvet;
> He digs and he delves –
> You can see for yourselves
> The mounds dug by Diggory Delvet.[28]

The expression 'little gentleman in black velvet' seems to date from 1702, when it was a toast drunk by the Jacobites to celebrate the death of King William III (William of Orange). On 21 February of that year, the king was riding his horse Sorrel at Hampton Court when it stumbled over a molehill and threw him. As a result, he broke his collar-bone and died at Kensington three weeks later:

> ... from the complications of a fractured clavicle; he developed pneumonia, which complicated his pre-existing heart trouble, and probably suffered a terminal pulmonary embolism.[29]

And all because of the activities of a diminutive insectivore.

Moles occur throughout Britain, wherever the soil is easily tunnelled and earthworms are plentiful, although they are absent from Ireland. Low-lying water meadows can often be completely pock-marked with the small mounds of soil, called 'molehills', that betray their subterranean activity.

> We feel at every step
> One foot half sunk in hillocks green and soft,
> Raised by the mole, the miner of the soil.
> (William Cowper, 'The Task')

Molehills are sometimes called 'tumps' ('1589, origin unknown'):

> I was brought up as a child in the forties near Ledbury in Herefordshire where we had plenty of Moles. These were known by the older locals as 'oonts' and their molehills were 'oonty toomps'. The 'toomps' were tumps but I never knew the derivation of the noun 'oont' – though later on it was a useful way of getting rid of a surplus 'O' or two in Scrabble.[30]

Moles seldom come to the surface, which is why they are so rarely seen, but construct elaborate networks of tunnels up to 1 metre (3¼ feet) deep, with sleeping chambers lined with grass, moss and leaves. The familiar small molehills, or tumps, which spelled such misfortune for William III are simply spoil-heaps of soil pushed to the surface as the tunnels are made with the forefeet. There is a fairly widespread but quite

Kenneth Grahame's Mole in *The Wind in the Willows*, illustrated by Arthur Rackham, is a favourite of generations of children.

unfounded belief that the higher and the more numerous the molehills, the greater the likelihood of rain.

> Scarce disappears the deluge, when the mole,
> Close prisoner long in subterranean cell
> Frost-bound, again the miner plays, and heaves,
> With treble industry, the mellow mound
> Along the swarded vale.
> (James Hurdis, 'The Village Curate')

Sometimes much larger mounds, called 'fortresses', are built over a nesting site and may contain stores of worms. The discovery of these caches, sometimes of several hundred worms, was a mystery for many years but it has now been proved that the Moles themselves are responsible:

> ... a Danish zoologist kept moles in captivity and supplied them with far more worms than they needed. He found that, under these circumstances, the moles seized one worm after another, bit off the head, twisted the rest into a knot and pushed it into a cavity in the earth.[31]

The worms can be used for food in times of hardship. If they are not needed, however, the worms regenerate their missing parts and make their escape.

Night and day have little meaning below ground, and Moles are active and feeding intermittently throughout the 24-hour period. Sometimes, in very light soils with abundant food near the surface, they construct shallow, short-term feeding runs where the soil surface is merely raised slightly. Although earthworms are their preferred food, Moles will also take slugs, centipedes, millipedes and insect larvae, especially in summer, but not, I think, Toads or Vipers:

> Other concerns there are of Molls ... as the peculiar formation of their feet, the slender Ossa Fugalia and Dogteeth, and how hard it is to keep them alive out of the Earth. As also the ferity and voracity of these animals; for though they be contented with Roots and stringy parts of Plants or Wormes underground, yet when they are above it will sometimes tear and eat on another, and in a large glass wherein a Moll, a Toad and a Viper were inclosed, we have known the Moll to dispatch them and devour a good part of them both.[32]

> A lonely life for the dark and silent mole! Day is to her night. She glides along her narrow vaults, unconscious of the glad and glorious scenes of earth and air and sea. She was born, as it were, in a grave; and in one long, living sepulchre she dwells and dies. Is not existence to her a kind of doom? Wherefore is she thus a dark, sad exile from the blessed light of day? Hearken![33]

It was widely thought that molehills encircling a house would spell death for the occupants and even a molehill appearing 'suddenly' might result in someone dying. A Mole burrowing under 'a wash-house, dairy or other out-house used for domestic work' was not a good portent for the mistress. Whether washing machines fall into the scope of the belief is not recorded. The familiar expression 'Don't make a mountain out of a molehill' was believed to be Greek in origin and first appeared in English in William Roper's *The Life of Sir Thomas More* in 1557.[34]

Moles are not blind, although their eyes are small and feeble. John Swan, writing in 1635, did not know this:

> ... although the mole be blinde all her lifetime, yet she beginneth to open her eyes in dying: whiche is a prettie embleme. This serveth to decypher the state of a worldly man, who neither seeth heaven nor thinketh of hell in his lifetime, untill he be dying: and then beginning to feel that which before he either not believed or not regarded, he looketh up and seeth. For even against his will he is then compelled to open his eyes and acknowledge his sinnes, although before he could not see them.[35]

Neither did Shakespeare:

> Pray you, tread softly, that the blind mole may not
> Hear a foot fall ...
> (*The Tempest*, Act 4, Scene 1)

But Thomas Browne, writing only 33 years later, did:

> That moles have eyes in their head, is manifest unto anyone that wants them not in his own.[36]

Nor are Moles deaf, although they have no external ears. There are also special delicate organs in the nose that are extremely sensitive and can detect alterations in humidity and temperature, and there is a widespread but erroneous belief that hitting a mole on its nose will prove fatal.

Moles have been thought of as pests as long as men have tilled the land. In 1697, John Worlidge, in his *Systema Agriculturae: The Mystery of Husbandry Discovered*, wrote:

> Moles are a most pernicious Enemy to Husbandry, by loosening the Earth, and destroying the Roots of Corn, Grass, Herbs, Flowers, etc, and also by casting uphills to the great hinderance of Corn, Pastures etc.[37]

Over the centuries, they have been trapped, gassed, threatened with foul smells and electronic vibrators, and supposedly unappealing plants have been grown to deter them. The sixteenth-century gardener Thomas Hyall was reported to advocate encouraging boys to play football in the garden so that the vibrations would frighten away the Moles.

One deterrent, often suggested today, is the Caper Spurge *(Euphorbia lathyris)* but I know one old gardener who once saw a Caper Spurge topple over as a Mole tunnelled regardless beneath it. In times past, almost every strongly smelling vegetation has been tried: 'put Garlick, Leeks or Onyens in their passages, and they will leap out of the ground presently'.[38] However, all has been to little avail and Moles are probably as numerous as ever.

Trapping has consistently proved the only really effective control measure and mole-catching has long been one of the skilled and essential country crafts.

> I was born in the village of Bayston Hill near Shrewsbury on St Valentines Day 1924. The population was only a few hundred mostly scattered houses and farms. The farming was mixed, those that worked elsewhere cycled to work although a neighbour an elderly man walked across fields for about 4 miles ... The village had a man who was a rabbit and mole catcher. Some old people called moles mouldywarps.[39]

> In North Cornwall, particularly, there's an expression 'I'm as giddy as a Want' (Want, 'wa' as in wagon) is a Cornish word for Mole. Observers in Cornwall always believed that Moles tunnelled in circles although in Kenneth Mellanby's opinion, this is not necessarily the case. A friend of mine whose father was horseman on the Pencarrow estate at the edge of Bodmin Moor, before the coming of the tractor, supplemented his income with trapping Moles. In the 1930s and during the Second World War, as many as 50 per day were

taken. He remembers albino animals being secured on a couple of occasions and a few that were piebald. Trapping was always suspended at the end of March when breeding commenced. Dried pelts fetched 6d–1/6d each depending on condition – black marks on the skin indicated bruising. The pelts were separated with newspaper and together with Fox (10/6d each) and Badger (£2) were simply dispatched to Horace Friend Ltd, Wisbech, Cambridgeshire.[40]

When melted snow leaves bare the black-green rings,
And grass begins in freshening hues to shoot,
When thawing dirt to shoes of ploughmen clings,
And silk-haired moles get liberty to root,
An ancient man goes plodding round the fields
Which solitude seems claiming as her own,
Wrapt in greatcoat that from a tempest shields,
Patched thick with every colour but its own.
(John Clare, 'The Mole-Catcher')

Mole-catching was a valued profession and John Vine, seen in a publication of 1804, was a famous Kent and Surrey Mole-catcher.

> ...one solitary and silent coadjutor of the husbandman's labours ... the Isaac Bint, the mole-catcher ... Isaac is a tall, lean, gloomy personage with whom the clock of life seems to stand still. He has looked sixty-five for these last twenty years, although his dark hair and beard, and firm manly stride, almost contradict the evidence of his sunken cheeks and deeply-lined forehead. The stride is awful: he hath the stalk of a ghost. His whole air and demeanour savour of one that comes from under ground. His appearance is 'of the earth, earthy'. His clothes, hands, and face are of the colour of the mould in which he delves. The little round traps which hang behind him over one shoulder, as well as the strings of dead moles which embellish the other, are incrusted with dirt like a tombstone ... His remarkable gift of silence adds much to the impression produced by his remarkable figure. I don't think that I ever heard him speak three words in my life. An approach of that bony hand to that earthy leather cap was the greatest effort of courtesy that my daily salutations could extort from him. For this silence, Isaac has reasons good. He hath a reputation to support. His words are too precious to be wasted. Our mole-catcher, ragged as he looks, is the wise man of the village, the oracle of the village inn, foresees the weather, charms away agues, tells fortunes by the stars, and writes notes upon the almanack – turning and twisting about the predictions after a fashion so ingenious ... .[41]

Along with Hedgehogs and numerous other creatures, Moles were condemned in the infamous Preservation of Grain Act of 1566 and Kenneth Mellanby quotes two instances of its application as revealed in old churchwardens' accounts:

> In one case we read 'for the Heades of Everie Moulewarpe or Wante, one halfpenny'. In 1700, the annual fee of the mole-catcher of the parish of Billingham is one pound, and at Arlesey in 1752 ninepence is paid for the destruction of six moles.[42]

The name 'Mole' is Late Middle English. Its origin is said to be obscure but I suspect that it may be an abbreviated form of the alternative and charming name 'moldwarp', which is also Middle English, although it comes directly from Old English and means simply 'earth-thrower'. Several of the huge range of variants of this name are still found in regional use and the name of Alison Uttley's character Moldy Warp the Mole was no fanciful invention. Other uses of the word 'mole', for instance to mean 'a form of wart', have no etymological connection with the animals. However, by analogy with the burrowing activities of *Talpa europaea*, 'mole' has achieved widespread use in recent times to mean a spy who has inveigled his way into some secret organization. Kim Philby, who betrayed MI5, must be one of the best-known real-life 'moles', but this usage really became established through the fictional writings of John le Carré.

The Mole *(Talpa europaea)* has a sensitive nose for sensing temperature and humidity changes and broad, spade-like hands for digging.

People never seem to have eaten Moles, a fact that has been attributed to their flesh rapidly becoming rancid after death. This I can vouch for, although not because of any odd culinary ambitions on my part. Many years ago, I found a completely undamaged dead Mole and put it aside with the intention of turning my taxidermy skills to it a few days later. Foolishly, I failed to put it in the freezer and the stench after a very short time was like nothing I have experienced before or since and rendered the corpse almost unapproachable. Nonetheless, Foxes, herons and some birds of prey, most notably Buzzards, take them, although most deaths of Moles are probably the result of starvation and flooding.

And the rancid flesh has not prevented the Mole from being used for medicinal purposes:

> A mole skinned, dried in the oven, and then powdered, was held in the fen districts to be a specific for ague. It may still be in vogue – it certainly was in use twenty years ago. The mole must be a male.[43]

Curiously, the requirement for a male Mole in Swan's remedy might not have been easy to satisfy. For much of the year, the sexes are difficult to distinguish, and there was an old country belief that 'all moles are males until early spring, when half the animals turn into females'.

Carrying Mole's feet in your pocket was a widely held practice to prevent cramp until relatively recently; the forefeet to prevent cramp in the arms and the hind-feet, the legs. It would also prevent rheumatism and Kenneth Mellanby, writing in 1971, claimed that this was still done 'in the Fens'. If the feet were cut from a living animal, they were believed to cure toothache too and, incredible as it seems, within the twentieth century the blood of a freshly killed Mole was dripped onto warts to treat them. (Was this, I wonder, due to a confusion of the two meanings of the word 'mole' and an instance of treating 'like with like'?) Sometimes, blood from the nose of a living Mole was used and, if this was placed on a lump of sugar and swallowed, it was believed to control fits.

The similarity of Mole's fur to velvet has been put to use and their skins were used to make waistcoats, trousers and other items of fashionable clothing until well into the twentieth century. The finest quality skins were known as 'best winter clear'. 'Moleskin' trousers and other garments can still be bought; but have never been anywhere near a Mole. Modern moleskin is 'a soft, fine-piled cotton fustian, the surface of which is shaved before dyeing'.[44] Plumbers had a rather more prosaic use for the skin of Moles, and as a child I remember our local plumber using a piece of moleskin to wipe the joints of his pipework.

Although moleskin (real moleskin) is usually and traditionally black, the animals are remarkably variable in colour:

> My orange roan spaniel hunts quite happily and will pull moles from the runs along the ground. Several years ago she found two orange coloured moles which eventually finished up in the Natural History Museum.[45]

Bats

(Chiroptera)

OTHER COMMON NAMES: *aerymouse; airy-mouse (Cornwall); at-bat (Shropshire); athern-bird (Somerset); backe; backie; bak; bakie-bird; bakke; barnmouse; bastat; bathymouse; batmouse; bawkie-bird; bit-bat; black-bear-away (Yorkshire); brere-mus; chipper; flickermouse; fliddermouse; flinder; flindermouse; fliner; flintermouse; flittermouse; flitty; flittymouse; fluttermouse; glaik (Lothian); glitmouse; haddabat; hat-bat; leather-mouse; leathern wing; leather-wings; oagar-triunse (Shetland); raamis; raamouse; raamse; raird; ramished; ramsh; rare; rattlemouse; raw-mil; rawmouse; rawmp; raymouse (Gloucestershire); rearie; rearmouse; ree-raw; reelrallm; reelymouse; reerd; reerie; reermouse; reiny-mouse; rennie-mouse; rere-mouse; reremouse; vlitter.*

> Twinkle, twinkle, little bat!
> How I wonder what you're at!
> Up above the world you fly,
> Like a teatray in the sky.
> (Lewis Carroll, *Alice's Adventures in Wonderland*)[46]

> When vesper trails her gown of grey
> Across the lawn at six or seven
> The diligent observer may
> (Or may not) see, athwart the heaven,
> An aimless rodent on the wing. Well that
> Is (probably) a Bat.
> (A.A. Milne, 'The Day's Play')

Bats are the only warm-blooded creatures except birds to have successfully and completely taken to the air. They have achieved this by an enormous extension of the finger-bones to support thin flight membranes. They also have knees that bend backwards but, despite A.A. Milne's poem, they are not rodents or even closely related to them.

Bats are subdivided into two groups. The Megachiroptera (the large bats) are popularly known as 'flying foxes' and use smell to locate the fruit and flowers on which they feed. None occurs in Europe at the moment, although the Egyptian or Rousette Fruit Bat *(Rousettus aegyptiacus)* of East Africa is heading northwards at an astonishing rate and has already taken Israel and Syria in its stride and is heading for Turkey. If climate change is ever to bring a fruit bat to Britain, the chances are that this is it.

By contrast, the existing British species belong to the very different and mainly insect-eating Microchiroptera (the small bats) and the two groups are thought to have evolved from very different origins. The British bats have retained the essentials of mammalian body structure and function, although their ability to echolocate, by which means they find their way around and detect flying insects (an activity called 'hawking'), is more highly developed than in any other group. They are not 'as blind as a bat'; they have eyes, capable of vision, although the extent to which they use them is not known.

All bats are more or less nocturnal and this has done nothing for their place in public perception; any creatures that are active at night immediately attract an aura of mystery because they are seen infrequently and imperfectly. Even when bats are seen clearly and at close quarters, their appearance and habit do them no favours: when not flying, they are scrambling, ungainly creatures; their wing membranes have a curious, unusual feel; their fur and size remind people of rodents, which are also generally unpopular; they have peculiarly and disproportionately large ears (to pick up echolocation signals); and their facial features make them unlikely candidates for any beauty contest. On balance, they probably share with mole-rats some of the ugliest faces among warm-blooded creatures. Then, as a final condemnation of bat-kind, people are aware that some species of tropical vampire bats are blood-drinkers.

Together with the dormice and the Hedgehog, bats are the only British mammals to hibernate (although they will emerge and fly during warm weather) and their winter hibernating places, or 'roosts', are usually in some elusive underground site, such as a cave or mine-shaft. The summer roosts, by contrast, are much more easily found and tend to differ characteristically from species to species. Many bats roost in summer in buildings, sometimes in very large numbers, and their unexpected discovery in a house can come as a serious surprise, although, sadly, the use of modern building techniques and timber preservatives has led to a huge loss of suitable sites. Conservation bodies now actively encourage private individuals and authorities to take account of this when planning any renovation or new buildings. The provision of bat boxes for those species that will adapt to them has become almost as familiar an activity as the erection of bird boxes in gardens.

Given the fact that many bat species not only roost in trees but also crawl into inaccessible holes, it is easy to see how the old belief arose that bats, along with frogs, Toads and lizards, could survive for centuries locked up in a tree. Gilbert White believed that:

> During the hibernation of animals, a temporary stagnation or suspension of active life ensues ... Heat and air are the only agencies which rouse them from their death-like lethargy. Judging from the circumstances of toads, lizards, and bats being found alive in solid rocks and in the centre of trees, this torpidity may endure the lapse of ages without the extinction of life.[47]

The association of bats with churches is old and well known. Bats are unlikely to distinguish between a church and a barn as a suitable roost. Nonetheless, the more specific association of bats with belfries is as mysterious as it is misleading.

> Belfries are just about the last places in a church that you would expect to find bats. They are just too cold and draughty. In the roof of the south porch is a much more likely spot.[48]

It is surprising that someone as familiar with churches as John Betjeman should have been misled: he called his 1945 anthology *New Bats in Old Belfries*. I recently asked a bat expert if there was a correlation between bat populations in churches and the nature of the liturgy: does the High Church service with its incense deter them? The answer was 'apparently not', but a subsequent correspondence on this very topic in the national press threw up contrasting opinions.

> Having two churches formerly affected by bats and their droppings, I can report almost complete success in their removal by the simple expedient of burning incense.[49]

There is a regional tradition of children throwing their hats when they see a bat. A Shropshire saying is 'Billy bat, come under my 'at' (presumably the origin of the regional names 'hat-bat' and 'at-bat') and John Nicholson tells that in East Yorkshire children throw up their caps for the bats to come under, saying 'Black, black, bearaway, Cum doon bi here a-way'.[50] An alternative version runs:

> Black bat, bear away,
> Fly over here away,
> And come again another day,
> Black bat, bear away.[51]

Another old children's ditty offers:

> Airy mouse, airy mouse, fly over my head,
> And you shall have a crust of bread,
> And when I brew and when I bake,
> You shall have a piece of my wedding cake.[52]

Or, in a version cited by Iona and Peter Opie:

> Bat, bat, come under my hat,
> And I'll give you a slice of bacon;
> And when I bake, I'll give you a cake,
> If I am not mistaken.[53]

These invitations, however, contrast strikingly with the widespread and totally illogical fear that many people have of bats becoming entangled in their hair:

> Children and adults in our village were scared of bats flying as it was said that if they got in your hair, you couldn't get them out.[54]

> Walking at night I feared bats in my hair.[55]

There seems no obvious origin for this belief and, given the efficiency of the creature's echolocation system, the likelihood of a bat accidentally landing on anyone's head seems unlikely, although it can happen:

> An elderly lady told me, apologetically, that she did not like birds. In fact, she continued, she did not like any small flying creatures because as a girl, in Essex, a bat had got entangled in her long hair as she was cycling one summer evening. Apparently, she had such difficulty in ridding herself of the bat that she was unable to sleep for some nights.[56]

Nonetheless:

> In 'The Countryman' (Spring 1960) there is an account of experiments conducted in 1959 by the Earl of Cranbrook with the help of three gallant young women, who allowed him to thrust a bat into their hair. Four different kinds of bat were used, and in each case, the creature escaped quite easily without getting entangled in any way.[57]

There is also an often-quoted belief that a bat flying three times around a house presages death for the occupants, but this too

The association of bats with belfries, as in this nineteenth-century engraving, is a misleading but enduring myth.

The bat carving on the staircase at Hatfield House, Hertfordshire.

is a curious myth: bats fly around countless houses every night in the summer with no obviously unfortunate consequences.

> At Hatfield House there is a mysterious bat carved into the great staircase Queen Elizabeth I rode her horse famously up ... .[58]

There is a long-standing and very widespread connection between bats and witches. It has often been said that bats are witches in disguise and, in the famous scene in *Macbeth*, featuring the three witches around a cauldron, 'wool of bat' was an ingredient of the brew (*Macbeth*, Act 4, Scene 1). The familiar expression 'like a bat out of hell' also draws on the association between bats and evil.

There is also a commonly quoted account dating from as late as the beginning of the nineteenth century that:

> ... when a bat is observed to rise and then fall again, it is the witches' hour, when evil ones have power over every human being.[59]

However, there are more benign and possibly fairly accurate inferences to be drawn from bat behaviour, most notably that bats flying early in the evening are a sure sign of good weather.

Nonetheless, pieces of bat have often been used not only in witchcraft but also in traditional medicine, and there are many examples of this, although rather few from Britain:

> The Flesh of a Bat medicinally taken, is good against a Scirrhus of the Liver, the Gout, Rheumatism, Cancer, and Leprosy, it has greater Virtue to cure the above Disorders, when taken after this Manner. Take of the Flesh pulverized half a Dram, Powder of Hogs-lice one Scruple, mix for a Dose. Take of the Blood one Dram, Honey half a Dram, mix them, being applied to the Eyes, it cures Ophthalmias, Films, Webs, Pearls, &c.
>
> There is an Oil made of this Bird, which is exceeding good against the Gout, saith Avincenna. Take half an Ounce of the said Gall, Vinegar two Drams, Salt one Scruple, mix them, it is good to wash any part bitten by a mad Dog.[60]

All bats in Britain are now strictly protected but most are threatened, by either habitat loss or climate change. Despite the pressures of the modern environment, however, bats as a group are evolutionarily very successful creatures: almost one quarter of all species of mammals are bats and they constitute the largest group of British mammals. There are currently 14 breeding species in the country, not including the Greater Mouse-eared Bat *(Myotis myotis)*. The largest bat in Britain, this was recognized as a British species only in 1958 but had effectively disappeared by 1990. It reappeared in Sussex however in 2002 and may yet still breed here. Britain has, however, gained one fairly reliable occasional vagrant, Nathusius' Pipistrelle *(Pipistrellus nathusii)*.

British bats occur in two families: the Rhinolophidae, containing the horseshoe bats, and the Vespertilionidae, containing the rest. Bat identification is never simple and, apart from a few distinctive species, tends to rely on such features as ear length and shape, and such elusive characteristics as the pattern of the teeth and the shape of the penis. Serious bat students now use electronic detectors that enable them to distinguish between species on the basis of their call frequency, volume and pattern, even when the animals themselves cannot be seen.

> One night, a friend of mine and I walked around the Brent Reservoir in North London. We had two detectors, one tuned to the calls of the Noctule and the other to those of the smaller Pipistrelle. We picked up the sonar of both kinds, even as they flew over the noisy North Circular Road, although nobody had ever reported seeing a bat at the reservoir.[61]

Most bat lore therefore makes no distinction between the various species; a bat flying through the sky at dusk is simply a bat, and the majority of people have neither the opportunity nor the inclination to distinguish one from another.

The name 'bat' dates from 1575 and gradually replaced the Middle English word *bakke*, which was of Scandinavian origin and still occurs in modified form in some regional names. The attractive variant 'flittermouse' and similar names are also sixteenth century and come from the German word *Fledermaus*, made familiar from the opera of that name. The resemblance of bats to mice is also echoed in many of the other local and historic bat names, such as 'bathymouse' and 'leathermouse'.

'Rearmouse' and similar names are still met with regionally and may mean 'raw mouse' or perhaps have an origin in the Old English word *hreran*, meaning 'to move'. 'Airy-mouse' is self-explanatory and charming.

> Some, war with rere-mice for their leathern wings,
> To make my small elves coats ...
> (*A Midsummer Night's Dream*, Act 2, Scene 2)

Philemon Holland, in his charming translation of Pliny, reminded us that the unique characteristics of bats have been known and accurately observed for a very long time:

> The Reremouse or Bat, alone of all creatures that flie, bringeth forth young alive: and none but she of that kind hath wings made of pannicles or thin skins. She is the onely bird that suckleth her little ones with her paps, and giveth them milke; and those she will carrie about her two at once, embracing them as she flieth ... No flying foule hath teeth, save only the bat or winged mousee.[62]

Nowhere in Britain seems to be named after the bat. Such likely candidates as Battersea, Batley and Batcombe have quite different origins, and there appears to be no inn called The Bat. However, bats certainly feature occasionally in heraldry: the Wakefield and Heyworth families, for instance, have rather anonymous-looking bats on their coats of arms.

Horseshoe Bats
(Rhinolophidae)

Lesser Horseshoe Bat *(Rhinolophus hipposideros)*
See Greater Horseshoe Bat below.

Greater Horseshoe Bat *(Rhinolophus ferrumequinum)*
OTHER COMMON NAMES: *larger horseshoe bat.*
Horseshoe bats have a curious and distinctive structure called, very aptly, a 'nose-leaf', a horseshoe-shaped piece of skin around the nostrils. They roost in buildings and in caves, where, in textbook fashion, they hang upside-down, often by one leg. In Britain, both species are seriously endangered as a result of habitat loss and they are now mainly confined to southwest England; Greater Horseshoe Bat populations declined by about 90 per cent during the second half of the twentieth century.

> I have happy memories of visiting with bat expert Michael Blackmore a four hundred strong colony of Greater Horseshoe bats in the roof of a school barn in Dorset. I had a tape recorder with me and captured the sounds of beating wings, the loud, high-pitched squeaks of adults and the calls of the young ... Apart from their scientific interest, the recordings also found their place in horror dramas on BBC radio.[63]

Greater Horseshoe Bats *(Rhinolophus ferrumequinum)* roost communally in roof spaces in summer and in caves in winter.

The Lesser Horseshoe Bat is one of the bat species that take much of their prey of small insects and spiders by picking them off shrubs, trees and rocks rather than catching them in flight. The Greater Horseshoe Bat, by contrast, usually catches and eats large insects while flying, although it will sometimes land to feed; I have seen three or four congregated around cow-pats collecting dung beetles, and Gilbert White presumably witnessed something similar:

> While I amused myself with this wonderful quadruped, I saw it several times confute the vulgar opinion, that bats when down upon a flat surface cannot get on the wing again, by rising with great ease from the floor. It ran, I observed, with more dispatch than I was aware of; but in a most ridiculous and grotesque manner.[64]

Evening Bats
(Vespertilionidae)

Whiskered Bat *(Myotis mystacinus)*
This small, shaggy bat is widespread but sparsely distributed in England and Wales. It is confusingly similar to Brandt's Bat and was distinguished only as a separate species in 1970. It occurs in a wide variety of habitats, including gardens, and often roosts in houses in summer, although in small numbers and usually tucked into crevices. The winter roost is below ground in caves and similar places. It is seldom found far from water and feeds mainly on small flying insects.

Brandt's Bat *(Myotis brandti)*
Very similar to the Whiskered Bat in appearance and distribution, this bat also occurs close to water but is less common

near houses, although its summer roost is usually in unoccupied buildings. The winter roost is in mines, cellars or other places below ground.

Daubenton's Bat *(Myotis daubentoni)*

OTHER COMMON NAMES: *water bat.*

This is a bat of fairly open countryside throughout the British Isles, wherever there is water nearby. It typically flies slowly, very close to the water surface, and has a characteristic fluttering wing-beat when collecting insects. The summer roost is usually in tree or rock crevices; the winter roost is usually underground where the bats congregate in large numbers.

Natterer's Bat *(Myotis nattereri)*

OTHER COMMON NAMES: *red-grey bat.*

Of medium size, these are bats of open farmland throughout the British Isles. In summer, they typically roost in old barns, although they may also be found in trees or under bridges, and have adapted well to bat boxes. The winter roost is usually somewhere damp below ground. The bats feed on small flies, beetles and other insects, usually in flight, although they sometimes pluck them from leaves.

This Whiskered Bat *(Myotis mystacinus)*, one of the evening bats or Vespertilionidae, is searching for prey among heather.

Bechstein's Bat *(Myotis bechsteini)*

A medium-sized, very uncommon, mainly moth-eating bat, this species is likely to be seen only in Wiltshire, Hampshire and Dorset. In summer, it roosts mainly in tree-holes and its dramatic decline is almost certainly due to the loss of old deciduous woodlands.

Common Pipistrelle *(Pipistrellus pipistrellus)*

OTHER COMMON NAMES: *common bat; pipistrelle.*

Most British bats are Pipistrelles and, despite a decline in numbers over the past 20 years and an excessively gloomy Worldwide Fund for Nature prediction that they could be extinct by 2007, they still account for over 90 per cent of individual bats. The Pipistrelle is tiny – the smallest European bat – and occurs in considerable numbers throughout the British Isles. It is the bat that most people see as they stand in their gardens at dusk and has a typically erratic, jerking flight as it goes to and fro along a regular route, catching small flying insects.

> We have Pipistrelles thanks to the vogue for outdoor lighting. They catch insects that are attracted to the beam.[65]

Pipistrelles mainly roost in houses in summer, although usually in small, confined spaces, such as cavity walls. The winter roost is above ground, either in buildings (often in house roofs) or trees, and sometimes vast numbers congregate; up to 100,000 bats have been found together in mainland Europe, although reassuringly (as much as anything because of the powerful smell), numbers in British houses are rather lower.

Leisler's Bat *(Nyctalus leisleri)*

OTHER COMMON NAMES: *hairy-winged bat.*

This medium-sized bat has an odd distribution, occurring throughout Ireland and in isolated places in southern England. It is a woodland bat and flies high and rapidly over the treetops, collecting insects. In summer, it roosts in trees, or sometimes in roofs, and it tends to hibernate in similar places. It will also use bat boxes in summer.

Noctule *(Nyctalus noctula)*

OTHER COMMON NAMES: *great bat.*

A big bat and the typical larger bat of small woodlands and parks, the Noctule will also hunt insects over water, fields and, rather commonly, over urban rubbish tips. It is one of the very few mammals of which the female is larger than the male. Gilbert White was the first to recognize it as a distinct species and mentioned it in his diary: 'The large species of bat appears.'[66]

Noctules are remarkably noisy and their calls can often betray the existence of a roost. They are still fairly common in England and Wales, although declining in numbers. Curiously, as a species, they may benefit from the recent enormous drop in the Starling population because Starlings will actively kill Noctules or turn them out of their roosts.

The name 'Noctule' dates from the late eighteenth century. It comes from the Latin scientific name *noctula*, from the Italian *nottola*, meaning 'bat', which is related in turn to the Latin *nox*, *noctis*, meaning 'night'.

Serotine Bat *(Eptesicus serotinus)*
OTHER COMMON NAMES: *serotine.*
Another bat of southern England and Wales, occurring in parks, light woodland and grassy places, the Serotine Bat feeds on a wide variety of insects, which it catches on the wing. These bats are almost always found close to human habitation, roosting in trees and buildings in summer and in buildings in winter. Like the horseshoe bats, they hang freely by their legs. The winter roosts are somewhat of a puzzle because they are occupied only by odd individuals or small numbers. It has been suggested that this is the one species of British bat that is genuinely on the increase.

The name 'serotine' originates in the Latin *serotinus*, meaning 'late'; the bats always emerge after sunset.

Barbastelle Bat *(Barbastella barbastellus)*
OTHER COMMON NAMES: *barbastelle; bearded bat.*
For no obvious reason, this is a very rare bat in Britain. Of medium size, it occupies a wide range of habitats, including parks and gardens, but also mountainous areas and inaccessible valleys. Barbastelles feed on small insects, which are caught on the wing. The summer roosts are in buildings, roofs, holes in trees and bat boxes. The winter roost is usually below ground in mines and caves, although also in trees. No nursery sites with young have ever been found in Britain.

'Barbastelle' means 'small beard'; there are tufts of dark bristles on the muzzle.

Grey Long-eared Bat *(Plecotus austriacus)*
See Brown Long-eared Bat below.

Brown Long-eared Bat *(Plecotus auritus)*
OTHER COMMON NAMES: *common long-eared bat; long-eared bat.*
It was not until 1960 that these two medium-sized species, the Grey and the Brown Long-eared Bats, were recognized as being different; before then they were simply known as Long-eared Bats. They undoubtedly have long ears with conspicuous folds, the scientific name *Plecotus* meaning 'one fold'.

> Strange revelation! Warm as milk,
> Clean as a flower, smooth as silk!
> O what a piteous face appears,
> What great fine thin translucent ears!
> (Ruth Pitter, 'The Bat')

The Brown Long-eared Bat is much the more widespread and, after the Pipistrelle (admittedly a long way after), is the second commonest British species. A bat of woodlands, parks and gardens, it usually roosts in holes in trees, roofs and bat boxes in summer, and in buildings or below ground in winter. It feeds extensively on moths but also eats other large insects. This bat is said to land on the ground more frequently than any other British species.

The Grey Long-eared Bat is a very much rarer bat in Britain, vying with Bechstein's Bat as the rarest of all, and supposedly with fur that is grey instead of brown at the base. Nonetheless, 'the two species are morphologically so similar as to make them almost indistinguishable'.[67]

Meat-eaters
(Carnivora)

Carnivores are meat-eating creatures; at least in theory. They have teeth designed to deal with meat, although some of the most impressive teeth belong to animals that appear to have little need for them. Some bears, for instance, dine happily on vegetable matter and insects that surely do not demand such teeth. There are two major subdivisions of the carnivores: the dogs, bears, racoons and weasels in one group and the cats, hyenas, civets and mongooses in the other. Modern Western society seems to have little room for meat-eating creatures, apart from humans, and the fact that a large carnivore needs a large territory has worked against them. The weasel family (Mustelidae), which includes many relatively small animals, is much the most numerous group of carnivores in Europe and in Britain today.

Dogs, Wolves and Foxes
(Canidae)

Wolf *(Canis lupus)*
To my mind, one of the greatest differences between life in Britain today and a thousand years ago is not the absence of sanitation, transport or medicine, but the deeply entrenched fear of knowing that you were sharing the countryside with the Wolf. To know that you might encounter a Wolf just beyond your village and almost anywhere you walked in the still-vast native forests would surely colour your life and activities to an unimaginable degree. Even as recently as 400 years ago:

> Passage through the Northern and Central Highlands
> in the 16th century was hazardous enough for hospices
> or 'spittals' to be set up where the benighted traveller
> could rest in safety ... wolves were plentiful and hungry
> enough to cause people in the Highland areas to bury
> their dead on islands off-shore.[68]

A wolf approaches a sheep pen as the shepherd sleeps in this illustration from a thirteenth-century bestiary.

In the year AD 937, the English King Athelstan, 'the third of the great West Saxon kings', faced a widespread Danelaw uprising led by the Scottish King Constantine II and his son-in-law Olaf Sitricson of Dublin, and reinforced by Viking troops from Norway. The rebels were defeated at Brunanburh in one of the pivotal but forgotten battles of Dark Age Britain; seldom in any country's history can five kings have been killed in one day. The whereabouts of Brunanburh is uncertain but is most likely to have been 'somewhere between Derby and Rotherham'.[69] The *Anglo-Saxon Chronicle* recorded that the armies left behind them:

> ... to feast on carrion, the dusty-coated raven with horned beak, the black-coated eagle with white tail, the greedy battle-hawk, and the grey beast, the wolf in the wood.[70]

Athelstan is said subsequently to have imposed an annual tribute of 300 Wolf skins on the Welsh King Howel Dda (d.950), an imposition that James Harting believed was maintained by their successors up to the time of the Norman conquest.[71] Other early sources suggest that the payment lapsed after three years because, by then, there were no Wolves left in Wales to kill, although this seems highly unlikely. Various places in England (with Wolf names) have claimed to be the site of the handing-over of the payment.

Far more than any other extinct British mammal, the Wolf has left its mark on both the country and its lore. A Manchester University research group found 230 place names (including field names) in England alone that were derived from the Old English name *wulf* or the Old Norse name *ulfr*, an astonishing testimony to the animal's once widespread occurrence and abundance. Most of these names are in the north but they extend throughout the country, from Wooler in Northumberland ('Wolf bank') to Carplight in Cornwall ('hillock of Wolves'), and there are many comparable ones in the Irish, Scottish and Welsh languages.

Wolves were not only feared, but also hunted and killed; almost a fifth of the English Wolf place names included the Old English word *pytt* or the Old Norse word *graef*, signifying a Wolf pit or Wolf trap. Moreover, by far the greatest number of British animal tales and legends relate to the site and date of the killing of the last Wolves, and James Harting in the nineteenth century, and others, have painstakingly assembled them. In Lancashire, it is said that Sir John Harington killed the last Wolf in England on the promontory called Humphrey Head, near Grange-over-Sands; John o' Gaunt Inn at Rothwell is said to mark the spot where the last Wolf in Yorkshire was killed (unfortunately, however, not by a Yorkshireman but by John of Gaunt, Duke of Lancaster); and in Cornwall:

> ... it is not generally known that the last native wolf lived in the forests of Ludgvan, near Penzance. The last of his race was a gigantic specimen, and terrible was the havoc made by him on the flocks. Tradition tells us that at last he carried off a child. This could not be endured, so the peasantry all turned out, and this famous wolf was captured at Rospeith, the name of a farm still existing in Ludgvan.[72]

The Yorkshire Wolds are said to be the last place in England where a price was placed on a Wolf's head. However, it appears that the last English Wolf was probably seen around the start of the fourteenth century (too early for John of Gaunt or Sir John Harington to have been involved), because eight cattle were recorded killed by Wolves at Rossendale in Lancashire in 1304–05.

What is not in doubt is that Wolves survived elsewhere in these islands for considerably longer. James I of Scotland passed a law in 1427 to enforce Wolf-hunting, which required three hunts a year between 25 April and 1 August (the cubbing season), and Mary Queen of Scots hunted Wolves in the Forest of Atholl in 1563. Every district in Scotland has its 'last Wolf' tale.

> Stirling has the Wolf on the Crag as its emblem in the belief that the last was killed at the nearby Wolf Craig as early as 1599.[73]

Undoubtedly the most evocative tale, however, is from Sutherland and involves the discovery and killing of a she-Wolf with her cubs in a den:

> ... between Craig-Rhadich and Craig-Voakie, by the narrow Glen of Loth, a place replete with objects connected with traditionary legends.[74]

A strong tradition attributes the killing of the last of all Scottish Wolves to one MacQueen of Pall-à-chrocain, who died in 1797, although Derek Yalden considers it 'a fairy story'.[75] MacQueen was 'of gigantic stature, six feet seven inches in height' and had the best deer-hounds in the country. The Wolf in question had evidently killed two children the previous day on the lands of Mackintosh of Mackintosh in

Inverness-shire, which is admittedly improbable unless it was rabid. However, fairy story or not, there is clearly something sinister about the MacQueen family: in the following century, it was another of their number who despatched the last British Great Auk (see page 291). The Wolf-killing supposedly took place in *c.*1743 but Sir Robert Sibbald, President of the Edinburgh Royal College of Physicians and presumably a reliable source, declared Wolves extinct in Scotland in 1684, 20 years after the last official mention of them. In Ireland, they undoubtedly survived for longer, apparently into the first quarter of the eighteenth century. Shakespeare seems to have known of their existence:

> Pray you, no more of this; 'tis like the howling of Irish wolves against the moon.
> (*As You Like It*, Act 5, Scene 2)

The Wolf is still widely distributed in Europe and elsewhere but is everywhere in retreat as a result of persecution and hybridization with domestic dogs, of which it is the ancestor. In a biological sense, it is a top carnivore. During its time in Britain, it would have fed on almost anything that came its way, apart from the Brown Bear: Wild Boar, deer, beavers and other rodents, birds, carrion and, of course, domestic livestock, the taste for which really spelled its end. It has been suggested that Wolves were important predators of Foxes in Britain and that it is only since the Wolf's demise that the Fox has assumed significance. Suggestions for reintroducing Wolves to Britain have been made at various times but have come to nothing.

Fox *(Vulpes vulpes)*

OTHER COMMON NAMES: *bay-reynolds; faws (northern England); foks; kid-fox (young animals); kliket; laste; laurence; lawrie; loss; lowrie; on-beast; ranald; red fox; reynald; reynard; rinkin (Suffolk); roplaw (Scottish borders–Teviot Dale: young animals); tod (Scotland); tod-lowrie; tod-tyke; vixen (female).*

The Fox is the most widespread carnivore in the world (and I think that's true whether or not the British *Vulpes vulpes* and the North American Red Fox, *Vulpes fulvus*, are considered to be the same) and, by that token, is one of the most successful.

There is hardly any part of the British Isles where a driver is unlikely to see the flash of a Fox's eyes in the headlights as it crosses the road, possibly with a Rabbit or other prey in its mouth, and country walks would not be the same without the strong likelihood of catching its characteristic smell as one passes a well-used but otherwise all but invisible track. As Gilbert White wrote in his journal:

> Foxes begin now to be very rank, & to smell so high that as one rides along of a morning it is easy to distinguish where they have been the night before.[76]

However, Foxes are rarely seen clearly in daylight, other than dead by the roadside, when they are revealed as about the size of an average dog, although readily distinguished by the combination of a rich red-brown coloration, pointed muzzle, erect ears and very bushy tail, or 'brush'. It is the red-brown colour that has been largely responsible for the word 'fox' being used for so many other purposes, such as 'foxing' (the red-brown stains on old paper). In practice, Foxes may be more often heard than seen, and to some effect. The scream, usually from the vixen in the mating season, is unearthly and unbelievably similar to the human voice.

In Continental Europe, the Fox is recognized as a major carrier of wildlife rabies, although widespread Fox-killing campaigns have done nothing to halt the spread of the disease and the Fox itself is still a very common animal. I think it was a BBC wildlife documentary film some 20 years ago that first used the Hollywood term 'Twentieth Century Fox' to epitomize the way that the Fox of today has significantly changed its habits, and many others have since used the rather clever word-play. This is because the Fox, while still an animal of farmlands and wild, uninhabited areas, has also adapted remarkably well to the urban environment, where it scavenges in dustbins and other places where food waste accumulates. This behaviour has been attributed in large measure to the shortage of Rabbits that followed the myxomatosis outbreaks:

> The collapse of the rabbit population ... created the 'commuting fox' which lives on the urban fringe and travels into city centres at night to scavenge for food.[77]

This was a life style seen by that keen observer Brian Vesey-Fitzgerald as long ago as 1965, when he called one of his books *Town Fox, Country Fox.*[78]

> Recently in Leicester, as I was waiting at the traffic lights to cross a very busy road, I became aware of something at the side of me, on looking down I was surprised to see that it was an adult Fox. I spoke to it, but did not touch it, it looked up at me with no sign of fear. When the lights changed and the traffic stopped it crossed the road with me and entered a large local park. I gained the impression that it regularly made use of the traffic lights during its search for food.[79]

Even in the wild, Foxes rely heavily on scavenging but will eat almost anything, dead or alive, animal or vegetable, that comes their way.

> I recall a young fox [in the New Forest] lit by a shaft of warm autumn sunshine carefully selecting the ripest blackberries from a bush.[80]

Curiously, most members of the dog family will eat fruit if they can obtain it: Aesop's fable of 'The Fox and the Grapes' might be based on fact. Foxes are also opportunistic in other respects, which is why they have been so successful. Indeed, they do not even always construct their own den, or

The Fox *(Vulpes vulpes)*, with its red-brown coat, black-tipped ears and bushy tail, is a splendid sight against a snowy background.

'earth', but commonly adapt an old Rabbit burrow or Badger sett, sometimes before the original residents have moved out.

However, it is another act of opportunism that has really given the Fox its notoriety, as well as providing the inspiration for innumerable story-book characterizations. This is the 'surplus killing' that occurs when a Fox enters a poultry-house and creates mayhem. In reality, the surplus killing is not wanton but a means of providing for times of hardship. In the wild, most of the potential prey would escape, but this is impossible in an unnaturally confined space, and so the Fox instinctively kills as many chickens as it can.

> Foxes can be deterred from entering hen houses or small gardens by the generous application of lemonade to small boys who will be enchanted to pee around the perimeter. Foxes will not cross human urine.[81]

If stories and songs are to be believed, however, the Fox's traditional prey is not the chicken but the goose. In 'The Fox's Foray' (see Greylag Goose, page 225), verse after verse recounts this fact, and the association of Foxes and geese occurs commonly in inn names, the straightforward 'Fox and Goose' being the commonest. There is also an old game called Fox and Goose and it has been suggested that inns where the game was played adopted the name.

> One of the party called the Fox takes one end of the room or corner of a field (for the game was equally played indoors or out); all the rest of the children arrange themselves in a line or string according to size – one behind the other, the smallest last – behind the tallest one, called Mother Goose, with their arms securely round the waist of the one in front of them (as in the last game) or sometimes by grasping the dress. The game commences by a parley to this effect:
> Mother Goose (to Fox): 'What are you after this fine morning?'
> Fox: 'Taking a walk.'
> Mother Goose: 'With what object?'
> Fox: 'To get an appetite for a meal.'
> Mother Goose: 'What will your meal consist of?'
> Fox: 'A nice fat goose for my breakfast.'
> Mother Goose: 'Where will you get it?'
> Fox: 'Oh! I shall get a nice morsel somewhere; and as they are so handy, I shall satisfy myself with one of yours.'
> Mother Goose: 'Catch one if you can.'
> A lively scene follows. The Fox and Mother Goose should be pretty evenly matched, the 'mother' with

> extended arms seeking to protect her 'brood', whilst the Fox, who tries to dodge under, right and left, is only allowed in case of a successful foray or grasp to secure the last of the train. Vigorous efforts are made to escape him, the 'brood', of course, supplementing the 'mother's' exertions to elude him as far as they are able but without breaking the link. The game may be continued until all in turn are caught.[82]

The Fox and goose link may, however, all be at one remove from natural history because the association is also an old satire. In the magnificent church of St Mary Redcliffe in Bristol is a misericord carving of two Foxes addressing a flock of geese, in an anticlerical satire, while 'a kind of medieval strip-cartoon' appears on the capitals of the late twelfth-century church of St Peter at Tilton in Leicestershire and also on the fourteenth-century carved capitals ('a most unusual thing', according to Nikolaus Pevsner) at All Saints, Oakham, in Rutland.

> At Tilton a fox is seen running away with a goose held by the neck in his jaws, and half-flung over his back ... The ecclesiastical explanation is that the carvings depict the Abbot of Westminster in the guise of the fox, running away with the great tithes, the goose ... The theory is ingenious, but the more likely origin of the carvings is the story already at least two hundred years old in about 1390 when Chaucer used it in his Nun's Priest's Tale ... as the doings of Reynard the Fox and Chanticleer the Cock. Mediaeval sculptors often substituted a goose, in view of its silliness, for the cockerel ... .[83]

The medieval so-called 'Beast Epic' of Reynard and Chanticleer (from French *chant clair*, meaning 'sing clearly'), has its counterparts in other European cultures and languages (for instance, *Le Roman de Renart*[84]) and was a widely used satire on contemporary life. The name 'Reynard' itself is Middle English and dates from the same time, coming through the French *renart* from the Old High German personal name Reginhart.

'Fox' is an Old English word of very ancient European origin and has been linked with the Sanskrit *puccha*, meaning 'a tail'. Most of the uses or combinations derived from it are fairly self-evident ('foxtail', 'fox-trot', 'fox-terrier' and so on), but an interesting modern development has been the slightly puzzling use of the adjective 'foxy' to mean 'sexually attractive' (as in 'foxy lady'). 'Tod' is an old Middle English word, still widely used in Scotland and the north of England, and it was the obvious choice for Beatrix Potter, with her Lake District background, when she called her principal Fox story *The Tale of Mr Tod*.[85] There is also an old, although relatively uncommon, surname 'Todhunter'. *Vulpes* is the old Latin name for the Fox. 'Vixen' for the female is Middle English, perhaps Old English, and, since 1575, has also been used for what the *Oxford English Dictionary* calls 'an ill-tempered quarrelsome

This illustration from C.C. Clark's *The Riches of Chaucer*, shows Chanticleer and his hens, with the Fox lurking.

woman', although, in everyday usage, it tends to be stronger. From around the same time comes 'dog' for the male animal. 'She-Fox', unlike 'she-Wolf', is unusual but it was King Lear's expressed opinion of his daughters:

> Now you she-foxes!
> (*King Lear*, Act 3, Scene 6)

For as long as anyone has recorded anything about Foxes, they have been labelled 'sly', 'wily' and 'cunning' (Field-Marshal Erwin Rommel was the 'Desert Fox' – presumably *Vulpes vulpes* but just possibly *Vulpes ruppelli*), although children's stories do not necessarily depict them as evil in the same way as the Wolf. Foxes are rogues, a bit caddish and, just occasionally, as in A.A. Milne's 'Three Little Foxes', 'who didn't wear stockings and didn't wear sockses', they are almost lovable.

Surprisingly, although Fox-hunting (see below) has developed its own special lore and language, the Fox has entered rather little into British folklore. It was thought that witches could turn themselves into Foxes and, in Scotland, a Fox's head was sometimes nailed to the stable door to deter them. There are also widespread beliefs that a Fox's bite is fatal, if not immediately then within seven years, and in some places a Fox entering a house was considered a death omen (like countless other creatures). If bright sunshine appears in the middle of a period of unsettled weather, it is called a 'Fox day' on the Isle of Man because it is untrustworthy and, in many places, days when both bright sunshine and rain are experienced are said, for no obvious reason, to be 'a Fox's wedding day'.

No-one seems to have eaten Fox on a regular basis, but I am assured that, in the Second World War, Italian prisoners-of-war working on English farms developed a particular taste for them.

Parts of Fox have certainly turned up in ancient medicine: a Fox's tooth was once carried to help an inflamed leg and:

> ... the tongue was used to draw out a thorn [a remarkably widely held belief], or worn as a cure for cataract, the liver and lights were washed in wine, dried and sugared to make a cough preventive. Ashes of fox flesh taken in wine assisted liver complaints or difficulty in breathing, and ... bathing the legs with water in which a fox had been boiled would ease the pain of gout. Fox fat was applied to the scalp in cases of baldness.[86]

Fox fur has been widely used for fashion clothing, the grotesque Fox stole that incorporated the animal's head being especially popular in the 1920s (as a child, I was always horribly embarrassed that my Mother had an old one), although exotic species were used more extensively than the native Fox.

There are plenty of Fox inns to be found, generally in hunting country and generally called The Fox and Hounds or a name alluding to a classical story, like those of Aesop, who was particularly fond of Fox tales: hence The Fox and Stork, Fox and Crow, Fox and Grapes. There are plenty of places commemorating Foxes: Foxhall in Suffolk ('Foxes' burrow'), Foxley in Norfolk ('Fox wood') and Foxton in Durham ('Fox valley') among others. 'Fox' is also a common surname, originally adopted by people with some Fox-like characteristic.

Any consideration of the Fox in Britain is inseparable from the practice of Fox-hunting. This is not uniquely British (Foxes are hunted with hounds in a few other parts of the world, although the long French tradition of mounted hunting with dogs is for deer, not for Foxes), but it has become associated with rural life in these islands to a particularly intimate degree. It probably divides public opinion more than any other legal pursuit in modern Britain. We all 'ken John Peel', whether or not we agree with Oscar Wilde that his occupation was nothing less than 'the unspeakable in full pursuit of the uneatable'. John Peel (1776–1854) was immortalized in the song written by his friend John Woodcock Graves in 1828, but he was real enough and, for over half a century, maintained a pack of hounds at his own expense at Caldbeck in Cumberland. John Peel had a 'coat so grey' but huntsmen more traditionally wear bright scarlet jackets, somewhat confusingly called 'hunting pink', after a tailor called Thomas Pink who first made them. By the time that John Peel's 'view halloo' was awakening the dead in the first half of the nineteenth century, Foxes had been seriously hunted with hounds in England for some 150 years.

The seventh Earl Spencer, with his wife and daughter. The countess is wearing a Fox stole.

In earlier days, animal-hunting had been pretty indiscriminate. In the New Forest for instance it was said that:

> ... in Charles I's time a certain master keeper had a motley pack that hunted buck, fox, otter, and hare indifferently, but with no indifferent success, for the keeper's room was hung with fox-skins.[87]

After the Restoration, however, things changed. Enclosures had fallen into disrepair during the Civil War, huge numbers of deer had escaped and been illegally killed, and the Wolf and the Wild Boar had disappeared. Otters and Wild Cats were available, but in small numbers, leaving only the hare as a traditional quarry. Serious attention then turned to the Fox for sport: it was widespread and accessible and had a long tradition of being a pest (both of farm stock and game-birds) and was therefore perceived as being in need of control. In reality, it was its pest status that initially worked against it; serious hunting people thought it beneath their dignity to hunt Foxes, although George Turberville (*c.*1540–*c.*1610), in his *Noble Arte of Venerie or Hunting* of 1576, wrote that:

> ... the hunting of the Foxe is pleasant, for he maketh an excellent crye, because his sent is very hot and he never flieth farre before the houndes, but holdeth the strongest coverts.[88]

The practice then was to run the Fox to earth and dig it out; only later were earths stopped to allow it to be hunted to its death above ground. 'Venery', or 'venerie', of Middle English origin, was formerly the widely used word for the chase, for hunting game; *vénerie* is the modern French word for hunting.

In Leicestershire, in Quorn country, they will tell you that:

> The first man in England to have a pack of foxhounds is said to have been Thomas Boothby (1681–1752) of Tooley Park, near Desford, who was known as Old Tom o' Tooley. By the end of the eighteenth century

As traditional quarry declined, sportsmen turned to the Fox. This painting, *The Birton Hunt*, is by John E. Ferneley (1781–1860).

> Leicestershire was by far the most famous fox hunting county, with Melton Mowbray being the centre of the sport and also of the glittering social life which followed it.[89]

The expression 'painting the town red' also originates in Leicestershire and:

> ... entered the language in 1837 as a result of the actions of the Marquis of Waterford – 'The Mad Marquis' – and his friends whose hunt celebrations in that year at Melton Mowbray in Leicestershire included literally coating with red paint the White Swan Inn and other buildings in the town.[90]

In reality, there were packs of Foxhounds before Boothby's career began in 1697; the Duke of Buckingham, for instance, had a pack in Yorkshire in 1685. Gradually, Fox-hunting with specially bred Foxhounds, each dog in a pack of 52 having its own name, took over as the pre-eminent rural hunting pastime. There was even a time in the nineteenth century when Foxes were in such short supply as hunting quarry in some districts that cubs were sold at Leadenhall and other markets and some were even imported from France for restocking. To be an MFH (Master of Foxhounds) has long carried a special rural cachet. Master Lucas, in Jane Austen's *Pride and Prejudice* (1813), says:

> If I were as rich as Mr Darcy ... I would keep a pack of foxhounds, and drink a bottle of wine every day.[91]

Gradually, hunting passed out of the confines of the great parks of the landed gentry to operate over open country with large bands of followers. Even those who could not afford horses were able to follow the hunt on foot. (In some upland areas of Britain, the hounds were always followed entirely unmounted.)

The nineteenth century was the age of hunting art and hunting literature, and the age of R.S. Surtees' character Mr John Jorrocks:

> 'Unting is all that's worth living for – all time is lost wot is not spent in 'unting – it is like the hair we breathe – if we have it not we die – it's the sport of kings, the image of war without its guilt, and only five and twenty per cent of its danger.[92]

Some of it, like Jorrocks, was at least stirring stuff; some of it was frankly appalling:

> They've run twenty minutes as close as a wedge,
> By Jove! They have split – two lines since the hedge:
> Old Regent is right. Up the furrow they rip;
> And round swing the rest with the Galloping Whip
> (Brooksby (Captain Pennel Elmhirst), 'The Galloping Whip')

Much better, though later, was Siegfried Sassoon's biographical *Memoirs of a Fox-hunting Man*, which has never been out of print since it was first published in 1928.[93] Among the literary figures to make their contribution to the hunting scene was John Masefield, a keen foot-follower, whose epic 'Reynard the Fox', or 'The Ghost Heath Run', first appeared in 1919:

There they were coming, mute but swift –
A scarlet smear in the blackthorn rift,
A white horse rising, a dark horse flying,
And the hungry hounds too tense for crying.
Stormcock leading, his stern spear-straight,
Racing as though for a piece of plate,
Little speck horsemen field on field;
Then Dansey viewed him and Robin squealed.

At the 'View Hallo' the hounds went frantic,
Back went Stormcock and up went Antic,
Up went Skylark as Antic sped,
It was zest to blood how they carried head,
Skylark dropped as Maroon drew by,
Their hackles lifted, they scored to cry.

The fox knew well that, before they tore him,
They should try their speed on the downs before him,
There were three more miles to the Wan Dyke Hill,
But his heart was high that he beat them still.
The wind of the downland charmed his bones
So off he went for the Sarsen Stones.

The moan of the three great firs in the wind,
And the 'Ai' of the foxhounds died behind;
Wind-dapples followed the hill-wind's breath
On the Kill Down Gorge where the Danes found
death ...

Then 'Leu Leu Leu' went the soft horn's laughter,
The hounds (they had checked) came romping after;
The clop of the hooves on the road was plain,
Then the crackle of reeds, then cries again.

A whimpering first, then Robin's cheer,
Then the 'Ai, Ai, Ai'; they were all too near,
His swerve had brought but a minute's rest;
Now he ran again, and he ran his best.

With a crackle of dead dry stalks of reed
The hounds came romping at topmost speed;
The redcoats ducked as the great hooves skittered
The Blood Brook's shallows to sheets that glittered;
With a cracking whip and a 'Hoik, Hoik, Hoik,
Forrard', Tom galloped. Bob shouted 'Yoick!'
Like a running fire the dead reeds crackled;
The hounds' heads lifted, their necks were hackled.
Tom cried to Bob as they thundered through,
'He is running short, we shall kill at Tew'.
Bob cried to Tom as they rode in team,
'I was sure, that time, that he turned up-stream.
As the hounds went over the brook in stride
I saw old Daffodil fling to side,
So I guessed at once, when they checked beyond'.

The ducks flew up from the Morton Pond;
The fox looked up at their tailing strings,
He wished (perhaps) that a fox had wings ...
(John Masefield, 'Reynard the Fox')

More recently, David Rook's 1970 novel *The Ballad of the Belstone Fox*, the story of a Fox cub reared by a huntsman with his hounds, was somewhat sentimental but later very cleverly filmed.[94] The most successful modern literary Fox, however, must be Roald Dahl's *Fantastic Mr Fox*, from the same year.[95]

Legislation in 1975 outlawed Otter-hunting, and stag-hunting is now severely restricted. This left Foxes, hares and American Mink as legal prey for hunting with dogs in Britain, but the Fox was far and away the most important of the three. Foxes were still significant rural pests but, for many people, the unwritten rule was that, in hunting country, they were never shot but conserved for the hounds. Only where hunting with

Stirrup cups – such as this Fox-head example – were popular among horsemen in the nineteenth century.

The Fox *(Vulpes vulpes)* mates as early as January. The cubs, born some 50 days later, are independent by the age of four months.

packs of hounds was not practised, such as some of the wild northern uplands, was Fox control with a gun socially acceptable; there the Foxes were often dug out with terriers and then shot, although in Wales, gun packs used dogs to flush the Foxes to the waiting guns. It has been estimated that between 50,000 and 100,000 Foxes were killed each year in Britain, about half by the 200 registered packs of hounds (most of which were Foxhounds, although there were also some harriers). About 40 per cent of the Foxes killed by hounds were killed in the autumn cub-hunting season.

Arguments over the moral, ethical and practical position of Fox-hunting began seriously to surface during the latter part of the twentieth century. However, the vocal pressure groups who had long argued against it on grounds of cruelty and from the view that Foxes could be adequately controlled by other means, began to be seriously opposed by others putting forward reasoning based on the impact on the rural economy if it disappeared. In June 2000 a Committee of Inquiry into Hunting with Dogs, under the chairmanship of Lord Burns, reported to the Government. Among its numerous conclusions were that killing foxes at night with rifles might be preferable to hunting but that killing Foxes in daylight with shotguns was cruel, and that the equivalent of somewhere between 6,000 and 8,000 full-time jobs depended on hunting, although the number of people involved might be significantly higher. It believed that hunting:

> ... has undoubtedly had a beneficial influence in lowland parts of England in conserving and promoting habitat which has helped bio-diversity, although any effect has been in specific localities.[96]

After immense debate and much acrimony, a hunting ban finally entered law in February 2005, with many people believing it unenforceable. What is undeniable is that Fox-hunting has had an eradicable impact on the life, lore, language and literature of Britain. And whatever one's views, it is hard to disagree with Charles Dickens, who, in *Oliver Twist*, wrote: 'There is a passion for hunting something deeply implanted in the human breast.'[97]

Bears
(Ursidae)

Brown Bear *(Ursus arctos)*

The Brown Bear was once one of the most widespread of large carnivores but today retains its European foothold mainly in northern Scandinavia and the north of the former USSR, with odd, scattered, remnant populations elsewhere. It is fairly widely believed to have become extinct in Britain by the tenth century, and the fact that there are still place names commemorating bears and numerous inn signs depicting bears has nothing to do with its ancient British history. Most 'Bear' inns recall not native bears but the barbaric 'sport' of bear-baiting, in which bears and dogs were set together, or other entertainments, such as the display of chained 'dancing' bears. These activities used imported animals and were familiar sights in English towns and villages for centuries. As an article in the *Barnsley Chronicle* of 1871 reports:

> The Bull and Bear baitings, not only on May-Day Green and Barebones [Barnsley] but those in the neighbouring villages, used to be well attended and will be remembered by persons who have passed the middle age of life.[98]

The Bear and Ragged Staff, which appears as an inn name in Warwickshire, is based on the crest of the Neville family, later Earls of Warwick:

> Now, by my father's badge, Old Nevil's crest,
> The rampant bear chain'd to the ragged staff ...
> (*Henry VI Part* 2, Act 5, Scene 1)

There is an old legend that the emblem of the first Earl, Arthgal, one of King Arthur's Round Table companions, was a bear because he had once strangled one. The second Earl, Morvid, presumably unable to match the feat, slew a giant instead, using as a club a tree with its branches stripped away, 'a ragged staff', which thus became attached to the existing bear as the family's emblem.

Many bear place names can also be explained by links not with native bears but with personal names, such as 'beaver', 'barley' or other words that are similar in Old English to the word *bera*, 'a bear'. The conclusion is that:

> ... the place-name evidence for Brown Bears surviving in England in Anglo-Saxon times is at best uncertain; if not already extinct, they were at best rare animals by then. The evidence from Wales and Scotland is equally doubtful ... .[99]

Nonetheless, bear remains are common in archaeological digs dating from Mesolithic times, certainly to the Bronze Age, and there is some evidence that British bears were taken to Rome. There is also a commonly quoted dictum, issued by Archbishop Egbert of York in *c.*AD 750 to the effect that:

> ... if anyone shall hit a deer or other animal with an arrow, and it escapes and is found dead three days afterwards, and if a dog, wolf, a fox or a bear, or any other wild beast hath begun to feed on it, no Christian shall touch it.[100]

Egbert, or Ecgberht, who died in AD 766, was one of the leading luminaries of his age and was unusual among archbishops in issuing his own coins. He was a noted Latin scholar, 'supreme in ecclesiastical matters', and both his brother Eadbert and his cousin Ceolwulf were kings of Northumbria. The logical inference must be that he knew what he was talking about and that bears were still present in the eighth century.

The Brown Bear *(Ursus arctos)* has long been extinct in the British Isles but 'dancing bears' like this one were once a familiar sight.

In its time, the Brown Bear would have been a truly awesome member of the British fauna, reaching almost 3 metres (3¼ yards) in length and, although feeding mainly on vegetable matter, fruits, insects and carrion, answerable to nothing. Until, that is, men began systematically to extend further into its natural territories. Albeit armed only with spear and bow, although presumably making use of pit traps, Anglo-Saxon man gradually became the master and the Brown Bear must have retreated north and west, as the Wolf was to do seven or more centuries later, although with no comparable written record of where the last British animal was finally cornered and slain.

Weasels and Related Mammals
(Mustelidae)

Stoat *(Mustela erminea)*

OTHER COMMON NAMES: *carre; clubster (Yorkshire); clubtail; ermine; fite; futteret; lobster; puttice (Kent); weasel (Ireland); whitterick; whitteret; whitrit (Shetland); whutherit.*

> Weasels – and stoats – and foxes – and so on. They're all right in a way ... but they break out sometimes, there's no denying it, and then – well, you can't really trust them, and that's the fact.[101]

So said Rat in *The Wind in the Willows*, but, like all of their family, Stoats do possess a vitality and a rather deserved reputation for aggressiveness that are out of all proportion to their size:

> Stoats, though not as numerous as weasels, probably do quite as much injury, being larger, swifter, stronger, and very bold sometimes entering sheds close to dwelling-houses. The labouring elder folk declare that they have been known to suck the blood of infants left asleep in a cradle on the floor, biting the child behind the ear.[102]

Most people's encounters with a live Stoat are limited to a glimpse of a sleek, sinuous, brown creature dashing across the road in front of a car to disappear among the roadside vegetation. Seen at close quarters, Stoats are chestnut-brown above, with a sharp demarcation line to pure white underparts. There is always a black tip to the tail. Should you ever have the opportunity to examine a Stoat closely, do not be tempted to look too closely at their teeth: a Stoat bite is unforgettable.

Stoats occur throughout the British Isles in almost every habitat, from sea shore to mountain top. They feed mainly on rodents, especially voles, although the males in particular also take Rabbits and birds, often considerably larger than themselves, and use a ruthlessly efficient killing method involving a bite to the back of the neck.

> The stoat, or whitrit as it is known in Shetland, is said to have been introduced by the king's falconer on one of his visits to Shetland. It was customary for this gentleman to come north each summer to catch wild falcons, and every house in Shetland was compelled by law to give one hen to feed the birds during their journey south. One year the people of a certain parish refused point blank to give the required hens, and to get his own back the king's falconer brought with him on his next visit a pair of whitrits. They have spread to most parts of Shetland and can be a nuisance when they take up residence near a hen-house.[103]

Like other carnivores, including Foxes, Stoats will create havoc in the confined space of a hen-house through 'surplus killing' and have, in consequence, been persecuted by farmers and gamekeepers for centuries. They have maintained their numbers because they are not easily tracked down, although many have been caught in gin traps set for Rabbits (and, despite its illegality, probably still are). Although it is curious to imagine of so small a creature, Stoats have also at times been hunted with dogs.

There has been much speculation about the source of the name 'Stoat'. The *Oxford English Dictionary* gives a date of 1460 and 'origin obscure', although there is an Old English word 'stot' that has been applied to various animals, usually male or castrated, and it is tempting to link them. 'Clubster' may have something to do with the club-shaped tail. 'Whitrit' and similar names are also obscure, as are the charming 'fite' and 'futteret'.

Stoats nest in small burrows or hollow trees, often taking over the former home of their prey and even using rodent fur to line the nest. Their presence is sometimes betrayed by the sharp, high-pitched bark-like sound that they utter when calling their young or when the family is threatened. The breeding behaviour of Stoats, however, is singularly eccentric because the males mate with young females that are still suckling from their mothers. As in some other species of the mustelid family, there is then delayed implantation of the fertilized egg and so young female Stoats grow up already pregnant.

Perhaps because Weasels are absent from Ireland (although Irish Stoats are sometimes called Weasels), Stoats there have received a disproportionate amount of attention and today are legally protected. There are many stories from Ireland and elsewhere of people and their animals being set on by packs of Stoats:

> It is well known that in autumn, when several families of stoats have united (probably for hunting purposes) they sometimes make determined attacks on boys or men who have provoked them.
>
> In several newspapers in the early 'sixties appeared an account of an attack by a pack of stoats on a woman and her collie dog near Burtonport in Co. Donegal. Whether the dog provoked the stoats was not clear, but the woman fled to a nearby house. The collie was killed.

The chestnut fur of the Stoat *(Mustela erminea)* becomes white in winter to camouflage it against snow, when it is known as ermine.

> My grandfather told me that he and a friend were attacked by a pack of stoats. He'd been walking home through the snow, after a day's shooting. He managed to knock a stoat out with the butt of his rifle and the other stoats all jumped on the one he'd knocked out. This was only a few minutes walk from his village in Lowland Scotland.[104]

However, these are tales that must have been elaborated in the telling. Stoats do not hunt in packs, although a female with her litter of up to 12 young can be an impressive sight.

While many people have never seen a live Stoat, a huge number have seen plenty of dead ones, possibly without realizing it. A coronation or comparable state occasion offers as extensive an array of dead Stoats as you are likely to see anywhere. The robes of the peers of the realm are edged or faced with ermine, which is white with small black dots. Ermine is the fur of dead white Stoats, and each dot represents the perpetually black tip to the tail. Ermine was first mentioned in 1138 and, in the reign of Edward III, the privilege of wearing garments lined or faced with ermine was strictly limited to the royal family and nobles with a minimum income of £1,000 per annum. In the reign of Henry IV, it was extended to the nobility in general, down to 'knights bannerets and certain official personages'. There is some confusion in old accounts between 'ermine' (strictly the winter fur of a dead Stoat) and 'miniver' (white fur generally, which may have carried the black dots of the ermine tails). The number of bars of ermine on parliamentary robes indicates a peer's rank; the most, four, denotes a duke. Ermine has always been expensive. In 1760, when the Duke of Marlborough had new robes made, the total cost was £44 6s, of which £16 16s was for the ermine alone. Today, ermine is seldom used; most hereditary peers have old family robes remade and most new peers' robes are of artificial fur.[105]

> From the legend that if the whiteness was soiled, the animal would die, it became a symbol of purity.[106]

The association between ermine and purity was employed most famously in the portrait of Queen Elizabeth I, the Virgin Queen, who was depicted with an ermine on her left arm. The picture is in the collection at Hatfield House and was painted in *c.*1585 by either Nicholas Hilliard or William Segar for Lord Burghley. Roy Strong commented on the portrait: 'What sets apart the "Ermine" portrait at Hatfield is the little creature on the sleeve of her left arm whose neck is encircled with a golden collar.'[107] The late Dame Frances Yates was the first to point out that such an emblem adorned the banner in Petrarch's *Triumph of Chastity*:

A group of peers, wearing ermine-trimmed robes, dash through the rain after the coronation of Queen Elizabeth II in 1953.

The 'Ermine' portrait of Elizabeth I at Hatfield House. The ermine symbolizes purity and its gold collar represents royalty.

> Her vyctoriouse standerde was this:
> In a greene felde a whyte armyne is
> With a chayne of golde about her necke;
> A fayre Topazion therto dyd it decke.[108]

The topaz was 'a jewel symbolizing resistance to all human lasciviousness'.[109] Moreover, in his desire to portray the Queen's purity and innocence, Hilliard, or Segar, would have done well to study the ermine a little more closely: the animal in the painting has a black spot almost everywhere except in the one place that it should have one – at the end of its tail, which is impossibly long.

In the past, the name 'ermine' was often applied to the Stoat itself, irrespective of its colour. It is a Middle English word that can be traced back to Old French and may originally have had some connection with Armenia, from where the fur was originally imported; it was believed to be the skin, not of a Stoat, but of 'the Pontic mouse'.

All Stoats are brown in summer but some do turn white in winter to become ermine; the process was for long a mystery because the transformation, when it does occur, may take only a few days. It is partly hereditary and partly controlled by temperature. In the northern part of Britain, all Stoats turn white; further south the change is usually at best partial.

> On 14 January 1993, I was driving in the late afternoon along a road in South Lincolnshire when I saw an all-white mammal with a black tip to its tail dash across the front of the car. As if aware of its own conspicuousness it flung itself with two violent twists to the body into the shelter of a hedge bottom ... To be all white in a mild winter in Lincolnshire must be a grave disadvantage to an animal trying to hide from its enemies.[110]

The rapidity of the change is because the new white coat grows beneath the old one and becomes apparent only when the old coat is shed. The shortening of the days stimulates the winter coat to grow, while the temperature determines whether or not it is white. It appears that proportionately more females than males turn white and so, paradoxically, while for centuries female peers were a rarity, female Stoats have long outnumbered males in the House of Lords.

Weasel *(Mustela nivalis)*

OTHER COMMON NAMES: *beal (young animals); cane (young animals); fairy (Devon); ferry; fozle; futcat (Banffshire); futrat; kane; keen; ken; kime; kine (Surrey, young animals); lavallan; lavellan (Caithness); marder; marten; marten-cat; marter; martern; martre; matrick; martrik; mertrick; mertrik; mouse-hound; mouse-killer (Cheshire); mouse-weasel (Morayshire); mulere (Somerset); mustela; puttice (Kent); quhitred; quihittret; rezzil (East Yorkshire); veary (Dorset); waesel; water-mole; wessel (Yorkshire); weysyl; whesile; whezle (Lothian); whitneck (Cornwall); whitrack; whitred; whitret; whittret (Scotland); whitruck; whut-throat; wreasel (northern England).*

> There is nothing in this beast more strange, than their conception and generation: for Weasels do not couple in their hinder parts, but at their Ears, and bring forth their young at their mouth. Yet it is certain that they have places of conception under their tails: and there-fore how it should come to pass that their young ones should come out of their mouths, I cannot easily learn.[111]

Whether easily learned or not, the habits of Weasels have fascinated men for centuries, and it is regrettable that their name should have come to be so synonymous with evil, slyness, cunning and mistrust. They are smaller than Stoats and similar in habits and prey but, size for size, probably more aggressive. Any carnivore so tiny must be aggressive in order to kill sufficient food to keep its system functioning. Weasels need to eat roughly one-third of their body weight every 24 hours; there is an old expression 'Catch a Weasel asleep if you can'. Mice and voles account for most of their meals and Weasel populations fluctuate with the rise and fall of rodent numbers. They always seem able to restore their numbers in abundance, however, and are probably the commonest carnivores in Britain. They can also tackle prey as large as a Rabbit and certainly take birds and their eggs when the opportunity arises, something that has long been known:

For once the eagle England being in prey,
To her unguarded nest the weasel Scot
Comes sneaking, and so sucks her princely eggs ...
(*Henry V*, Act 1, Scene 2)

I can suck melancholy out of a song, as a weasel
sucks eggs.
(*As You Like It*, Act 2, Scene 5)

The expression 'weasel-word', originally American, is 'a word that ruins the force of a statement, as a Weasel ruins an egg by sucking out its contents'. 'Weasel' itself derives from the Old English word *wesle* but its origin is obscure. Weasels and Stoats share many regional names and, in older writings, it is not always easy to distinguish the two. The West Country name 'veary', however, seems to be related to 'fairy' and to a belief that the Weasel will turn into a fairy if you try to catch it.

As with Stoats, there are beliefs that Weasels hunt in gangs. Kenneth Grahame was responsible for implanting this myth into the minds of children when, in *The Wind in the Willows*, the Chief Weasel and his tribe caused mayhem on a mass scale, but it has no real foundation and adult Weasels being seen with their numerous offspring is the most likely explanation.[112] Sometimes Weasels are observed behaving in a highly erratic manner which has been called 'dancing'. The reason for this is unknown but one theory suggests that they are suffering from a common infection with a parasitic worm that causes pressure on the brain. Apart from in the works of Kenneth Grahame and other children's authors, the Weasel has never gained a place in fine literature. Phil Drabble, in 1957, called his personal account of keeping wild animals as pets *A Weasel in my Meatsafe* but, although charming, it was not quite in the same class.[113]

Weasels occur throughout the British Isles in almost all habitats, although they are less common than Stoats at altitude and are absent from Ireland and most of the offshore islands. Being smaller, they can follow small rodents into their burrows, which they then take over for their own use; like the Stoat, they will use their prey's fur as a nest lining. Because they are so small, predators such as owls and Foxes probably kill significant numbers simply through failing to distinguish them from mice and voles.

Weasels are similarly coloured to Stoats but have a much less definite line between the chestnut back and the white underparts and they never have a black tip to the tail. In Britain they do not turn white in winter, although a distinct subspecies called the Least Weasel does so in northern Europe.

People have never really trusted Weasels. The Romanies believe that it is bad luck to meet one and the usual clutch of bad luck signs is associated with them: one should not run across your path, especially at the start of a journey; it should not appear close to your house and nor should it squeak in your hearing. Chaucer used another old analogy of the Weasel as a 'bad young woman, afflicted by abnormal sexual desires'.

And the Weasel that went pop? This almost certainly has nothing to do with *Mustela nivalis* and, although it is now generally associated with the nursery rhyme, it was 'in the 1870s and 1880s a proletarian (mostly Cockney) catch phrase. Probably erotic origin'.[114] An alternative explanation is that 'pop' was a synonym for 'pawn' and that something called 'a weasel' had to be pawned to pay for drinking at The Eagle, which was an old-time and somewhat unruly music hall in City Road, London.

American Mink *(Mustela vison)*

Paradoxically, the European Mink *(Mustela lutreola)* has not been a British animal since the end of the last glaciation; today it is predominately an eastern European species, although there are isolated groups in France and Spain. However, mink fur has long been a prime choice for ladies of fashion and, the mink coat was once the ultimate status symbol. The creature that parted with its own coat to provide the fur was principally the American Mink, and it was this species that was first brought to Britain in 1928 for fur-farming. It is roughly the size of a Ferret and normally has a rich brown-black fur, although mutant forms with pale brown or silver-blue coloration are the preferred fashion choices. Almost inevitably, many escaped and then, more recently, misguided 'animal rights' groups deliberately liberated some. They were first confirmed to be breeding wild in Britain in Devon in 1956. The consequences have been locally disastrous for the native fauna because, although the mink feed mostly on Rabbits, they also prey on ground-nesting birds and their eggs, as well as poultry, fish and Water Voles. Moorhens and Coots have suffered especially severely, as have sea birds in coastal areas. Although there was initially some suspicion that Otters were suffering in consequence of the mink expansion, this appears not to be so and, with the resurgence in Otter numbers, it seems that mink tend to give way to them in any particular habitat.

Mink numbers have now been reduced from their initial peak to around 18,000 and continue to fall. Trapping is the main means of control and, although there are 20 packs of Mink hounds, some of which had been used for Otter-hunting until its ban in 1975 (how do the hounds know the difference?), they have had little overall effect, killing only 400–1,400 mink each year between April and October. Mink are good climbers and a hunt often ends with the animal being cornered up a tree. There is little hope, however, that the American Mink may yet go the way of the Coypu (see page 481), although the planned ban on fur-farming in Britain will eliminate any new escapes or liberations and just might, ultimately, spell the beginning of their end in this country.

The word 'mink' first appeared in 1466 as a name for 'a stinking animal in Finland', and there are similar words in

Danish and Swedish; the origin is different from that of the Minke Whale (see page 443). Mink have at times been called 'mink-otters'.

Polecat *(Mustela putorius)*

OTHER COMMON NAMES: *beech-marten; carre; club-tail; fitchew (numerous spellings and variants: fitchal, fitchet, fithowe and so forth); foulmart (numerous spelling and variants: foomurt, foumert, fowmart and so forth); martill; martrick; martrone; mertrick; pine-marten; stote (Somerset); western polecat; wild-cat; wilocat (Lancashire).*

One of the memorable moments in my watching of British wildlife was my first sight of a wild Polecat in a corner of a remote Welsh farm in the 1960s. Its white facial patches gave it the characteristic spectacled appearance that I knew so well from captive animals. Wales had been almost the only place to see Polecats since around the end of the nineteenth century, although they have begun to increase considerably in numbers and range in recent years. Recent studies suggest that their range has doubled in the ten years since 1991 and English animals, at around 20,000, now outnumber Welsh ones. Even more surprisingly, it is estimated that there are now around 500 Polecats living again in Scotland.[115] They were once very much more widespread but were persecuted mercilessly by gamekeepers because of their real or imagined destruction of game-birds and poultry. It is certainly true that Polecats have long preferred to live in the proximity of farms although they also feed on Rabbits, rodents, perching birds, frogs, carrion, large insects and other small creatures. They have a reputation for the wholesale and wanton destruction of frogs but this has never convincingly been proved, although there are accounts of stores of fresh frogs being found alive but paralysed after being bitten through the brain. Unlike their smaller relatives, the Stoat and the Weasel, they do not construct a serious nest, making do with any convenient hollow.

The name 'Polecat' is Middle English and may perhaps have something to do with the Old French *pole* (*poule* in modern French), meaning 'fowl'. Among its alternative names, some with an absolute plethora of spelling variations, are 'fitchet', 'fitchew' and 'foulmart'. 'Foulmart' is also Middle English, meaning 'foul marten', an allusion to the disgusting aroma which the creature exudes from its anal glands and which instantly distinguishes it from the much more socially acceptable sweetmart, or Pine Marten. The scientific name *putorius* also alludes to the odour. 'The country folk used to call the male, hob, and the female, jill.' Young Polecats are called 'kittens'. The alternatives, 'fitchet' and 'fitchew' are also Middle English, originally from Dutch; the derived word 'fitch' was sometimes used for the fur, although the Polecat was seldom trapped for fur because of the accompanying foul smell. The name 'beech-marten', which has been applied to the Polecat, properly belongs to a quite distinct and non-British animal, *Martes foina.*

Its facial markings make the Polecat *(Mustela putorius)* easy to identify. Once rare outside Wales, it is becoming more common.

Many people who have never seen a live Polecat may have encountered its domesticated version, the Ferret, which may be at least partly a hybrid with the eastern Steppe Polecat *(Mustela eversmanni).* Albino Ferrets are especially popular with breeders and there has undoubtedly been interbreeding between feral Ferrets and wild Polecats in Britain. Ferrets have been used at least since Roman times for hunting and, in Britain, have long been employed, especially for catching Rabbits; Sir Geoffrey Luttrell's exquisite Psalter, which was created in East Anglia in *c.*1325–35 for use in the diocese of Lincoln, illustrates a Ferret being placed in a Rabbit warren.[116] In modern practice, all except one of the holes in a Rabbit colony are blocked and the Ferret is then introduced via the remaining exit to flush out the occupants. It is sometimes suggested that Ferrets originated in North Africa but this seems doubtful because the Polecat is not native there; a modern theory is that Polecats were domesticated into Ferrets in Spain, where the Rabbit is native.

The name 'Ferret' dates from 1450 and from it has come the familiar verb meaning 'to seek and clear out'. The decidedly odd 'entertainment' of placing a Ferret inside one's trousers seems fairly recent. There is a very widely quoted but disgusting belief that certain 'debilities', especially whooping cough in children, can be cured by giving the sufferer the remains of milk from which a Ferret has drunk.

The Scottish Highlands are the main stronghold of the Pine Marten *(Martes martes)*, which is an inhabitant of coniferous forest.

Pine Marten *(Martes martes)*

OTHER COMMON NAMES: *martin cat; sweetmart; sweet marten.*

> The pine marten, or marten cat, was formerly a common woodland animal, and in the Middle Ages 'hunting the mart' was almost a national pastime. It was hunted by groups of men on foot with sticks and stones, but the real reduction of its numbers came in the nineteenth century, with the onslaughts of the gamekeeper and the high prices paid for the skins. In addition, the use of the gin-trap for rabbits resulted in something little short of a massacre of the pine marten. The last killed in the London region was shot in Epping Forest in 1883.[117]

This is a sad but accurate indictment of the way that one of the most beautiful of native British mammals has been treated. It is truly a striking animal, rather bigger than a Polecat, more the size of a domestic cat, and with rich chestnut-brown fur and a creamy yellow throat. It is fairly accurately named because its commonest habitat is coniferous forest and it is seldom seen in the open.

Until the late nineteenth century, Pine Martens were still fairly common in remoter areas of much of mainland Britain and also on some of the offshore islands. By the 1920s their range was reduced to a small region of northeast Scotland but they have since begun to make up some ground. They occur today in the Scottish Highlands, where there are probably around 3,500, and some of the Lowlands, and there may be isolated populations in Wales and northeastern England. They are also present in small numbers in Ireland. More widespread planting of conifers may improve their chances of spreading further into more of their former territories. There have been limited experiments in reintroducing Pine Martens to some areas but they seem to move very little from the reintroduction sites, which is strange because naturally established Pine Martens have been followed over great distances.

> My friend Richard Balharry once tracked a marten for twenty miles by its footprints in the snowy tracts of Beinn Eighe.[118]

As with so many of their relatives, Pine Martens have been persecuted by farmers and gamekeepers, although quite misguidely. Pine Martens feed mainly on rodents and Rabbits, although they also take insects, carrion, eggs and a considerable quantity of fruit, fungi and other vegetable matter. They have been unfairly blamed for killing Red Squirrels but there seems scant evidence that they do so in Britain, although squirrels do form an important food in other parts of Europe. Their nests, or dens, are typically among the roots at the base of old or fallen Scots Pines but they will live in rock hollows, old birds' nests or squirrels' dreys (Pine Martens are excellent climbers) and similar places.

In the *Treatise on the Craft of Hunting*, written in *c.*1340 by Guillaume Twici, huntsman to King Edward II, a list of beasts of the chase was supplied:

And for to set young hunterys in the way
To venery I cast me first to go:
Of which four beasts be, that is to say,
The hare, the herte, the wulf and the wild boar.
And there ben other beasts five of the chase;
The buck the first, the second is the do;
The fox the third which hath ever hard grace;
The fourth the martyn, and the last the roe.[119]

In the period from 1837 to 1840 the lessee of a single estate near Glen Feshie caused his keepers to destroy no fewer than 246 Pine Martens. They were also killed for their pelts which 'were much used for lining the gowns of magistrates'.[120]

The name 'marten' is Middle English via Old French; 'Pine Marten' is rather later; 'fir marten' has been used occasionally. Martham in Norfolk is 'a ham frequented by martens' but there are few other place names associated with Pine Martens.

The marten cat, long-shagged, of courage good,
Of weasel shape, a dweller in the wood,
With badger hair, long-shagged, and darting eyes,
And lower then the common cat in size,
Small head, and running on the stoop
Snuffing the ground, and hind-parts shouldered up –
He keeps one track and hides in lonely shade
Where print of human foot is never made,
Save when the woods are cut; the beaten track
The woodman's dog will snuff, cock-tailed and
black,
Red-legged, and spotted over either eye;
Snuffs, barks, and scrats the lice and passes by.
The great brown hornèd owl looks down below,
And sees the shaggy marten come and go.

The marten hurries through the woodland gaps,
And poachers shoot and make his skin for caps,
When any woodmen come and pass the place,
He looks at dogs and scarecely mends his pace.
And gipsies often and bird-nesting boys
Look in the hole and hear a hissing noise.
They climb the tree – such noise they never heard,
And think the great owl is a foreign bird;
When the grey owl her young ones cloaks in down,
Seizes the boldest boy and drives him down.
They try agen and pelt to start the fray.
The grey owl comes and drives them all away,
And leaves the marten twisting round his den,
Left free from boys and dogs and noisy men
(John Clare, 'Marten')

Badger *(Meles meles)*

OTHER COMMON NAMES: *badgerd; badget; bason; baud; baunsey; bauson; bauston; baustone; bawsin; bawson; bawsond; bawsone; bawstone; biter; boreson; bosen; bouson; broc; brock; brofie; burran; caniculus; earth dog (Ireland); Eurasian badger; gray; gre; grey; gripper; guisard; melos; pate (northern England); sow-brock (Fife).*

Like many other children of my generation, the first Badger I encountered was Bill, Rupert Bear's trusty friend in Mary Tourtel and Alfred Bestall's classic tales. Possibly without exception, the Badger in children's stories has been similarly depicted as trustworthy, kind, sometimes wise and always genial. When, in *The Wind in the Willows*, Mole and Ratty rang Mr Badger's door bell after being lost in the Wild Wood, they were greeted with:

'Come along in, both of you, at once. Why you must be perished. Well I never!'[121]

Likewise, the real Badger has seldom caused mankind much harm and hardly been more than a nuisance. Yet over the centuries, countless Badgers have been mercilessly and brutally killed and, in recent years, many hundreds have been officially slaughtered. Quite clearly, there are tales to be told and explanations to be given.

Badgers are very big animals: a fully grown male may be not far off 1 metre (3¼ feet) in length and over 12 kilograms (26½ pounds) in weight, something all too apparent if you are unfortunate enough to hit one while driving, a fate that sadly befalls a great many of them. In the course of one 10-kilometre (6-mile) journey recently, I saw no fewer than four dead at the roadside. Badgers are especially common in the southwest of England but are much thinner on the ground in East Anglia and in Scotland. Nonetheless, despite the fact that there are believed to be around a quarter of a million Badgers in Britain, they are rarely seen alive because they are almost entirely nocturnal, emerging from their dens to forage after dark. A Badger's den is now almost always referred to as a 'sett' (or 'set'), although this is a surprisingly new term, first used as recently as 1898. It is possible that Eden Phillpotts did not know this and was confusing Badgers with Otters:

Brocks snuffle from their holt within
A written root of blackthorn old ...
They stretch and snort and sniff the air,
Then sit to plan the night's affair.
(Eden Phillpotts, 'The Badgers')

Badgers, with their familar black-and-white striped faces, cause little or no harm to livestock, feeding principally on earthworms, although they are omnivorous and will take almost anything else that comes their way, including carrion. They are particularly fond of wasps' and bees' nests, which they dig up principally for the larvae, and sometimes cause a nuisance in gardens when they dig up bulbs. Their underground setts are dug typically in deciduous woodland but may be found in almost any reasonably sheltered spot and the network of tunnels, which many animals may share, can be very extensive and include special lavatory areas. The nest chambers are characteristically lined with leaves and grass and sometimes Badgers, Foxes and even Rabbits may cohabit within the same subterranean colony.

Thus far, I have given no indication of why Badgers have been so persecuted, although they are certainly now well and carefully looked after by a specific piece of legislation, The Protection of Badgers Act 1992, which safeguards both the animals and their setts. This consolidated earlier Acts of 1973 and 1991, which made the digging of Badgers illegal, and it is here that we are getting closer to the sad history of Badger persecution. For many centuries, a pursuit called 'Badger-baiting' was a familiar rural pastime. It was officially rendered illegal as long ago as 1835 but continued, strangely, in urban areas. (I am reliably informed that a dentist in Chesterfield was a passionate follower of the activity within my lifetime.) Ernest Neal, the great champion of Badgers in modern times, when writing in 1948, could still say:

> Badger-baiting, in its original sense, is probably a thing of the past, but a slightly more humane variation of it still exists to some extent in the larger towns and cities as well as in the remoter country districts. A badger is caught alive and placed in a barrel or box to which it is chained, and the locals bring their dogs to test them out on it. Betting is a usual sideline, and the owner of the badger often makes quite a haul from the owners of inexperienced dogs. The terriers are often badly mauled in the process and even killed, and the badger may suffer a lot of ill-treatment. At one time the badger's lower jaw was cut away to give the dogs more chance. There is nothing whatever to be said for badger-baiting.[122]

Roadsigns like this, near Moreton Pinkney, Northamptonshire, are designed to protect heedless Badgers from motorists.

It was largely to provide Badgers for this activity that Badger-digging was necessary and it was also the reason why it became necessary to outlaw it:

> It took us four hours and a half, two spades, a mattock, a beetle and wedges, two pairs of prodigious badger-tongs, and three hours of heroism on the part of Nip, who pinned the badger for three hours until the mattock and the spade, the wedges and the tongs got down to him through root and rock and drew him forth ... Now this digging out of the badger is no ignoble thing. At its best it is a work of philanthropy whose object is to remove unwanted badgers from over-populated regions and transport them to estates where they are wanted. At its eighteenth-century worst it was sheer butchery ... Badger tongs are just like sugar tongs, with the difference that they are 223 times larger and heavier. You grip the badger with them by the scruff of his neck and haul him out. If you do not grip him properly he will grip you – and the badger can bite clean through a man's foot.[123]

The strength of the Badger's jaws is legendary and, as a country boy, I grew up knowing that a Badger's teeth would leave their mark on a steel spade.

> There is a popular fallacy that has been current for many centuries about the jaws of the badger, according to which once a badger has seized anything it cannot be made to let go unless its jaw is dislocated. This yarn has probably arisen from the fact that the jaw-joint of the badger is so constructed that the jaw cannot be dislocated without breaking part of the socket into which it fits ... But there is no mechanism for locking the jaws when the badger bites, and if it does 'hold on like a bull-dog', it does so merely by muscular strength and tenacity.[124]

Badgers are blessed with a great many different names. 'Badger' itself is relatively recent – 1523 – and its origin is uncertain. It is often said to be related to the French *bêcheur*, 'a digger', which was what Ernest Neal believed. The *Oxford English Dictionary*, however, prefers a connection with 'badge' and the badge-like facial markings. The verb 'to badger' is late eighteenth century and is derived from the teasing that was the essence of Badger-baiting. An alternative name, popular in some districts, is 'grey', or 'gray', dating from 1686. 'Bauson' was once widely used and is Middle English, from Old French meaning 'piebald'. 'Boreson' seems different and it may be derived from either 'boar' or 'bear'; certainly male Badgers have long been called 'boars' and the females 'sows'. I have heard the name 'pate' used for Badgers very recently in the north of England:

A strictly nocturnal animal, the Badger *(Meles meles)* feeds mainly on earthworms. A two-month old cub is at the centre of this group.

> In Durham there is a small glen called 'Pate-Priest's Glen' after a refugee priest who came over from France and lived a hermit life in the glen and was much given to badger-hunting. The name is also to be found commonly in the parish registers of some of the Lakeland towns in connection with the paying of head-money for badgers. For example, in the Penrith Parish accounts for 1658 there is an inventory, 'Payed for killing of two Paytes, 2s.'[125]

Ritson Graham comments on this paying of head-money for Badgers in Lakeland as follows:

> With unremitting zeal this paid persecution of the badger prevailed in the hill parishes of the three counties for almost a hundred years, the first payment recorded being the one from Penrith dated 1658, and the last from Ulverston in 1741. During this period the payment varied from a shilling to sixpence a head, and in a few instances to fourpence. The parishes of Penrith, Kendal, Dacre, Barton, Kirkby Lonsdale, Orton and Ulverston are all mentioned as having paid head-money for the killing of badgers, and Kendal heads the list with seventy-three badgers paid for within a period of eight years.[126]

However, much the most common and widespread name until very recently was 'brock', which is Old English from Old Celtic, and it is this that is perpetuated in a great many place names, especially of fields and farms. Brockhall in Northamptonshire ('badger hole'), Brockhurst in Warwickshire ('badger hill or wood') and Broxted in Essex ('badger hill') are among the larger settlements with Badger connections. However, care is needed because, while *brocc* is Old English for 'Badger', *broc* is Old English for 'brook'; Brockhampton in Dorset is therefore 'the tun of the dwellers on the brook' and not some ancient Badger territory. Sadly, the name 'brock' came later to mean 'a stinking fellow', a particularly unfortunate association for a mammal that is one of the cleanest in its habits.

Because the Badger is nocturnal, many people become aware of the presence of Badgers in their gardens only when one activates the security lighting or simply makes a considerable noise.

> In 1951, very little was known about the Badger's language and I set out to tape record its whole vocabulary. I was introduced to a sett in the middle of a Theosophist community in Surrey where the animals were tame enough to accept buns thrown to them and to climb dustbins in order to knock off the lids of others close by. The sett covered an area some 75 metres long and 45 metres wide. There were more than 30 entrances ... I spent over a thousand hours at night over four years making recordings and evaluating the Badger's vocabulary. This included grunts, barks, warning calls, threats, purrs, yarls, screams, play noises, fights between sows and the calls of tiny cubs.[127]

As mentioned before, the Badger is a large animal and has large, strong thighs, so perhaps it is not surprising to find that

The finest sporrans are made from Badger pelts. These late nineteenth-century Highlanders are modelling a selection of sporrans.

Badgers have been eaten, apparently on a regular basis until relatively recently in the remoter parts of Ireland (or so I am assured), although Badger hams are also said to have been a delicacy throughout the British Isles. Most published references suggested they were usually smoked or roasted.

> Besides the medicinal use of its Fat, the flesh, when roasted, is good food, like pig's flesh, and makes a good ham; and the skin is tanned for Breeches, Waistcoats etc. and is sometimes dressed with the hair by Furriers. The Hair makes pencils for Painters.[128]

However, in other places, it may have been different. In the West Country, Badgers have always been particularly plentiful:

> On Boxing day in Somerset, jugged hare or even badger, was the poor man's feast. Flavoured with onions, spices and herbs it was cooked in a deep earthenware pot which stood in a pan of boiling water suspended over the fire.[129]

A friend who once sampled Badger (unsmoked) recalls being 'distinctly disappointed as its flavour was nothing startling – in fact it was as much like mutton as anything'. But opinions do vary:

> I was responsible for the Warwickshire Museum Service from late 1948 to the end of 1977. The Museum received numerous dead Badgers, usually road casualties ... I guess that my first Badger stew was made in the 1950s and that the Badger would have been given to the Museum when we had no taxidermist working for us ... I used the main 'cuts' (legs, shoulders and meatier parts) and cooked them like a normal stew, browning first and cooking slowly with vegetables. So far as I can recall, the meat was fairly dark and, no doubt, slightly gamey. It made a good stew.[130]

> The flesh, blood and grease of the badger are very useful for oils, ointments, salves and powders, for shortness of breath, the cough of the lungs, for the stone, sprained sinews, collachs etc. The skin being well-dressed is very warm and comfortable for ancient people who are troubled with paralytic disorders.[131]

You may still see 'Real Badger' for sale, but not to eat or rub on any painful parts, in places that sell high-class gentlemen's toiletries because Badger hair is still reckoned to make the finest shaving brush. Apparently, most of the brushes sold in Britain are manufactured in Germany but, despite a great deal of objection to the practice by Badger conservation bodies, there is no indication of any illegal activity. In French, the word *blaireau* is used for both the animal and the shaving equipment. It is in Scotland, however, that you are most likely to see dead Badgers put to significant use today because the finest sporrans are made from Badger fur, the men who trap and supply the Badgers traditionally being known as 'brochans'. The Clan MacIvor uses a Badger's head 'as an additional embellishment to their cuarain, buskins, or Highland shoe'.[132] Incredibly, given the hairiness of the animal, there is a Derbyshire expression 'as bald as a Badger'; but then my native county never does fail to surprise me.

The earliest reference I have found to the strange and apparently widely believed notion that Badgers have legs of unequal length to enable them to walk satisfactorily around the outside of a hill comes, like many strange ideas, from Edward Topsell:

> He hath very sharp teeth, and is therefore accounted a deep-biting beast: his legs (as some say) are longer on the right side than on the left, and therefore he runneth best when he getteth to the side of an hill.[133]

This grotesque idea is referred to by no less than Thomas Macaulay, who said of Titus Oates that he was 'as uneven as a badger'. As Ernest Neal has sensibly pointed out, it would be 'presumably rather a nuisance on the home journey'.[134]

There are regular and persistent stories told of so-called 'Badger funerals', of animals almost ritually burying their dead. Even a hardened Badger-watcher like Ernest Neal pondered where Badgers go when they die and found a remarkable first-hand account:

> It would appear ... that on certain occasions at any rate badgers will bury their dead in specially excavated holes away from the set in which they have been living. Brian Vesey-Fitzgerald ... gave an amazing account of how in June he witnessed a badger's funeral. In this instance the sow had lost her mate. She came to the set entrance and let out a weird unearthly cry; then she departed for a rabbit warren not far distant. There she excavated a large hole in preparation for the body of her mate. She worked at this over a long period, the time being broken up at intervals by journeyings between warren and set. After some hours a second badger appeared, a male. The sow stood still with nose lowered to the ground and back ruffling agitatedly, and the male slowly approached with nose also lowered. Then the female moving her head swiftly up and down, uttered a whistling sound 'as though the wind had been sharply expelled through the nostrils'; at the same time she moved forward with two tiny jerky steps. When she stopped the male went through a similar motion, his nose to the ground like the sow's. This was repeated. The ritual over, they both retired down the set. After some time they reappeared, the male dragging the dead badger by a hind leg and the sow somehow helping from behind. They reached the warren, interred the body, and covered it with earth. Then the male departed and the sow returned to her set and disappeared. One wonders if all badgers are buried in this way or whether these rites are characteristic only of such special occasions as when a sow loses her mate.[135]

It is not difficult to see how such beliefs have arisen but others have witnessed happenings remarkably like those described by Brian Vesey-Fitzgerald:

> While staying in the Coquet Valley (near Rothbury) in Northumberland in the early 1990s, a farmer/shepherd told me he had witnessed what he called a Badger funeral. One day while driving along a country track he saw a dead adult Badger at the side of the track. Coming back along the track later that night he saw two adult Badgers dragging the dead Badger from the track verge. The next day there was no sign of the dead Badger.[136]
>
> This story was told to my wife and I whilst walking in Grass Woods near Grassington Yorkshire some years ago by our guide Len Huff. Len had lived most of his life in the area which he knew and loved having made a study of its flora, fauna and geology. At the time of our visit there were no Badgers living in the wood but he showed us a disused Badger sett and told us the following story. Whilst observing the sett one evening he was surprised to see an adult Badger leave the sett and return shortly afterwards with another adult Badger. Both Badgers entered the sett and eventually he saw that a Badger was reversing out of the sett. As he watched he was amazed to see that this Badger was pulling a dead Badger from the sett, his amazement increased as the second Badger emerged pushing the corpse from behind. The Badgers then proceeded to bury the body. The visiting Badger left the scene and the original Badger sat on the grave. Len told us that if the sounds he heard from this Badger had been made by a human he would have said that they were weeping.[137]

Badger-baiting is one reason why so many Badgers were killed in years gone by but does not explain the recent large-scale official slaughter previously mentioned. This has a quite different, but in its way no less controversial, explanation. In 1971, a committee was appointed:

> ... to enquire into the causes of the continued relatively high incidence of bovine tuberculosis in West Cornwall, and to advise on what future action would be feasible.[138]

Despite the tuberculin testing (TT) of cattle herds, which began in 1935, the continued occurrence of the disease was a worry and a mystery. Then:

> In Gloucestershire, evidence soon accumulated from which it was concluded by the Ministry of Agriculture, Fisheries and Food that a significantly high level of

> infection with the bovine strain of tuberculosis existed in the Badger population of the Cotswolds and neighbouring parts of Gloucestershire and Wiltshire.[139]

Careful trials and assessments were carried out and the association between tuberculosis in Badgers and in cattle was proved, although the means of transmission has never properly been demonstrated; badgers urinating on pasture land is the most likely explanation. Once the link was shown, the law was changed to allow Badgers legally to be killed by gassing and very large numbers have been eliminated. However, major practical and ethical arguments continue and:

> Public feeling is outraged by the thought that cattle are tested in order to determine whether they have the disease and are then only slaughtered if positive whereas Badgers are caught and killed in order to establish the presence of tuberculosis by post mortem examination. Most of the animals killed are found not to be diseased and it is thought their lives have been sacrificed for nothing ... .[140]

Otter *(Lutra lutra)*

OTHER COMMON NAMES: *dratsie (Scotland); king-otter; lutria; ote; oter; ottar; otyre; river dog; teak/ tike/ tyk/ tyke (Shetland).*

> Twilight upon Meadow and water, the eve-star shining above the hill, and old Nog the heron crying kra-a-ark! as his slow dark wings carried him down to the estuary.[141]

So begins the most celebrated tale of the most celebrated of all Otters. Henry Williamson's 1927 story of *Tarka the Otter*, which lived and was eventually hunted to its end in 'the country of the two rivers' in north Devon has been called 'a book of transcending beauty and truth'. With C.F. Tunnicliffe's drawings, it probably did more than anything to put the Otter on the twentieth-century map. In a sense, this was the Otter's century, a period when its numbers plummeted but then began to recover with the introduction of conservation measures and the passing of legislation in 1975 that finally rendered illegal the old and grotesquely anachronistic pursuit of Otter-hunting.

The Otter is the largest British member of the weasel family, a long, sleek, streamlined creature with tiny ears, a broad muzzle, a thick, muscular tail, webbed feet and close, dark brown fur that looks charcoal grey and spiky when wet.

Otters were once widespread throughout Britain but declined most especially in the 1950s and 1960s through a combination of habitat loss and organochlorine pesticide usage. The work of river authorities intent on improving water flow by the removal of streamside vegetation deprived Otters of suitable places for their homes. They are shy, secretive creatures, easily disturbed by noise and human activity, and, in most areas, they are largely nocturnal.

Although they now have strongholds in the West Country, Wales, some areas of East Anglia, the northwest of England and in Scotland, Otters are clearly reoccupying some of their former territories and deliberate reintroductions have sometimes been very successful. There is a suspicion, however, that some releases of captive-raised otters have been not of the British species but of the Asiatic Short-clawed Otter *(Aonyx cinerea)*, which is a common species in zoos. However, even introductions of *Lutra lutra* require patience and careful planning.

> Otters released into the wild after being bred in captivity have devastated a trout farm. Tim Small, who runs Lechlade Trout Farm, said 30 per cent of his organically reared trout stocks had been killed by otters ... Dr Simon Pickering is the biodiversity officer for the Cotswold Water Park Society ... He said: 'It is not against the law to release captive-bred otters but it is not good to release several in one place which may have happened here'. The Otter Trust was unavailable for comment.[142]

Otters will routinely travel many kilometres and their sudden appearance in a remote spot can be explained by these long-distance journeys. A friend who owns a salmon beat on the River Wye rang me excitedly a little while ago to say that an Otter had taken up residence there; no-one local could remember the last time they had seen one. Contrary to what is sometimes thought, Otters do not seem to be threatened by the spread of American Mink *(Mustela vison)*, which appear to give way to Otters as territorial competitors.

It is sometimes forgotten that the Otter is as at home on the coast as it is on inland rivers, and my own most exciting Otter experiences have been in a remote corner of Pembrokeshire, where I have watched them run effortlessly down the rough and rocky shore to fish in the rock pools and the surf. Nonetheless, seaside Otters need to return to fresh water periodically in order to maintain the insulating properties of their fur. Otters do not seem especially fussy over their choice of fish species, catching coarse fish, trout, salmon and eels in fresh water and whatever is available in the sea. They are not exclusively fish-eaters, however, and crabs, crayfish, small mammals, frogs and small water birds, such as Coots, are also taken.

Perhaps more than any other British mammal, the Otter, although fairly short of local names, has attracted a vocabulary all of its own: its footprint track is called a 'spur', 'spoor' or 'seal'; its lying-up site a 'hover' or 'couch'; its underground home a 'holt' or 'lair'; its curiously sweet-smelling droppings 'spraint'; and the process of producing them 'sprainting'.

> The spraints are black at first, turning paler as they dry and weather ... you are supposed to tell which way the Otter has gone by the position of the spraint. If it lies on the down-river side of its boulder the animal was travelling upstream.[143]

Most of these odd words are of fifteenth- or sixteenth-century origin, as applied to the Otter, but there seems no special

The Otter *(Lutra lutra)* is not confined to fresh waters. This mother and cub are playing among seaweed on the Scottish coast.

reason why this particular mammal should have been so blessed with attention. The word 'Otter' itself is from Old English and perhaps originally came from a Greek word meaning 'water snake'. Males are known as 'dogs' and females as 'bitches'. The delightful 'dratsie' is widely used in Scotland and finds its way into a charming Shetland story:

> Along the sea-shore lives the dratsie or otter ... The otter had the reputation of being extremely hard to kill, and it frequently managed to bite its captor before it finally succumbed. An otter's bite was helped to heal by the application of three hairs drawn from the animal's tail. On one occasion a man named Robbie Glen was struggling home with a 'dead' otter on his back when the animal suddenly came to life and sank its teeth into the lower part of Robbie's back. Robbie pulled at the animal's tail but it had a firm grip of his flesh and it was soon clear to both that neither would give way. At last they came to the only sensible solution and agreed to call it quits. 'Lat be for lat be', said Robbie, and as the otter relaxed its grip Robbie let go its tail and both went their respective ways. The expression is still repeated whenever two adversaries agree to call it a draw – 'Lat be for lat be,' as Robbie Glen said, 'tae da otter'.[144]

> Shetlanders have a local story that to descend from a cliff Otters put their tails in their mouths and roll down into the sea like a 'simmond clew' or a ball of straw rope.[145]

The Otter has lived on Shetland for a long time and, together with the Wood Mouse (*Apodemus sylvaticus*, see page 472), is the only mammal to have been there longer than recorded history.

Also from Scotland comes another dubious but endearing belief:

> On the west coast of Scotland, on the wild and rocky shores of the Isles, there are flat rocks here and there worn smooth by the countless feet of hundreds of generations of otters. They are known as 'otters altars'. For centuries they have been used as dining-tables by the otter on migration.[146]

While arguments still ebb and flow on the merits and ethics of Fox-hunting, few country people today will view the idea of hunting Otters with dogs as anything other than an obscenity. It was an archaic pursuit that one imagines would have passed away with cock-fighting and bear-baiting yet, while Otters had been hunted over the centuries, extraordinarily, the first pack of Otterhounds was formed as recently as 1796, and the Otter hunt reached its peak in the 1920s when there were 23 packs. However, for anyone who has

These Victorian gentlemen, shown here with their Otterhounds, were continuing a tradition that persisted into the twentieth century.

seen the animal at close quarters, merely to read an account of Otter-hunting gives a chill.

> Your huntsman early in the morning before he bring foorth your houndes, must goe to the water; and seeke for the new swaging of an Otter, and in the mud or grauell finde out the sealing of his foote, so shall he perceiue perfectly whether hee goe up the water or downe: which done, you must take your houndes to the place where he lodged the night before; and cast your traylors off upon the trayle you thinke best; keeping your whelps stil in the couples: for so must they be entred. Then must there be on either side of the water two men with Otter speares to strike him, if it bee a great water: But if it be a small water you must forbeare to strike him, for the better making of your hounds. The Otter is chiefly to be hunted with slow houndes great mouthed, which to a yong man is a verie earnest sporte, he will vent so oft and put up ouer water at which time the houndes will spend their mouthes verie lustely: Thus may you haue good sport at an Otter two or three houres if you list. An Otter sometimes will be trayled a mile or two before he come to the holt where he lyeth, and the earnestness of the sporte beginneth not till he bee found at which time some must runne up the water, some downe to see where he vents, and so pursue him with great earnestness till he bee kild. But the best hunting of him is in a great water when the banke is full, for then he cannot haue so great succour in his holes, as when it is at an ebbe: And hee maketh the best sporte in a moonshine night, for then he will runne much ouer the land, and not keepe the water as he will in the day.[147]

Although Tarka, and also the exotic Mijbil in Gavin Maxwell's 1960 story *Ring of Bright Water* achieved a sort of immortality,[148] many another Otter has gone down in history less for what it is than what it is not: countless Scottish Otters have been called the Loch Ness Monster.

> One trick of otter behaviour ... is their habit of occasionally swimming in line astern. This is seen mainly around dawn. The bitch and her cubs will do this and on rare occasions several such family parties will follow each other through the water, showing nine to twelve humps with the extended neck and small head of the leading otter showing above water ... Otters have given rise to more stories of mystery animals on land and monsters in water than any other species. Apart from the swimming in line astern a pair of otters will leap at the surface, one behind the other, in pursuit of a large fish. The effect is of a large fast-moving writhing snake. Another trick is to swim with only the head exposed, paddling rapidly with the hind-feet, leaving a line of dark hump-like waves in the almost unseen animal's wake.[149]

Once in serious decline, the Otter *(Lutra lutra)* is now becoming re-established in rivers and waterways all over the British Isles.

An old and widely held belief was that an Otter could be killed only by an attack on a particular spot on its breast.

> There is a spot under his breast, and he can only be killed by wounding this. The rest of his body is protected by enchantment. The smoke of him will fell a man sixty yards away.
>
> There is a little black spot which everyone has on some part of the body. Those who have it 'above the breath,' i.e. above the mouth, can never be drowned. It is called the 'otter spot,' ball dobhrain.[150]

Like a number of other fish-devouring creatures (see for instance Guillemot, page 289), Otters have fallen prey to the religious dictum of not eating anything that is considered flesh on certain days. Isaak Walton touched on it:

> Pisc, I pray, honest huntsman, let me ask you a pleasant question: do you hunt a beast or a fish? Hunt. Sir it is not in my power to resolve you; I leave it to be resolved by the college of Carthusians, who have made their vows never to eat flesh. But I have heard the question hath been debated among many great clerks, and they seem to differ about it; yet most agree that her tail is fish; and if her body be fish too, then I may say that a fish will walk upon land; for an otter does so, sometimes, five or six or ten miles in a night, to catch for her young ones, or to glut herself with fish.[151]

Thomas Pennant in the eighteenth century said that he had seen an Otter being prepared for dinner in the Carthusian monastery near Dijon.

Among various medicinal uses for Otters, quite the most unpleasant was:

> The Liver pulverized ... stops Haemorrhagia's and all manner of Fluxes: The Testicles made into Powder, and drank, help to cure the Epilepsy ... .[152]

Otter skins were once highly prized and, in the reign of the mid-tenth-century Welsh king, Howel Dda, it was said that:

> ... the skin of an ox, a deer, a fox, and an otter are all valued at the same price, eight times as dear as the skin of a sheep or goat.[153]

Their skins were also endowed with magical properties:

> The skin being magic is, of course, considered a charm, also an antidote against fever and smallpox, a safeguard against drowning, and efficacious in child-birth.[154]

There is a River Otter in Devon and Somerset, on which stand Otterford, Ottery St Mary, the charmingly named Upottery and Venn Ottery. Not far away, in Hampshire, is Otterbourne, while, among others, is Otterham in Cornwall ('a ham frequented by Otters') and Ottershaw in Surrey ('Otter wood').

Cats
(Felidae)

Wild Cat *(Felis sylvestris)*

OTHER COMMON NAMES: *catamount; cat-a-mountain.*

Although older natural history books tell you that the 'British Tiger' was not the ancestor of the domestic cat, zoologists have since rather revised their view and, today, the household pussycat is thought to be derived from a North African and Middle Eastern subspecies of the native *Felis sylvestris.* The British Wild Cat is now a rare animal, and even rarer in its pure form, untainted by hybridization with feral animals. There are perhaps about 3,500 British Wild Cats living in the north of Scotland, more than there are Black Rats *(Rattus rattus)* but only a tiny fraction of the number of Foxes, or even of feral cats, which probably exceed three-quarters of a million. They were once much more widespread and Wild Cat bones have been found at many archaeological sites, including the Iron Age Glastonbury lake village in Somerset. Derek Yalden has carefully mapped the dramatic decline:

> It had already gone from southern England before 1800, and disappeared from Lancashire, Westmorland, Herefordshire and Shropshire in the early nineteenth century. The latest county records from England are probably those from Yorkshire about 1840, from Loweswater, Cumberland and Castle Eden, Durham in 1843, and Eslington, Northumberland in 1853.[155]

There were corresponding retreats from Wales and the Scottish Lowlands and the conclusion is that Wild Cats probably have not occurred south of the Great Glen since 1915.

Although superficially similar to the domestic tabby, the Wild Cat is a much larger animal, distinguished especially by its very bushy, black-ringed tail. It is a creature of wild places, the edges of highland forest and, especially in summer, the high moorlands, where it hunts Rabbits, hares, rodents and, to some extent, birds, amphibians, fish and other prey. It has long been an animal of legendary ferocity, and the most celebrated of all Wild Cat tales comes not from Scotland but from Northamptonshire:

> The athletic abilities and ferocity of the wild cat are almost proverbial, but this is based on anecdotes rather than precise descriptions. There was, for example, the ancient and traditional story of the fight between Percival Cresare and a wild cat. The youth was returning from Doncaster through Melton Wood when attacked by a wild cat, he sought refuge in the porch of Barnburgh Church, where the fight continued. In the morning youth and cat were found, the youth dead of severe lacerations, the cat crushed against the wall by his feet.[156]

At Thorpe Gresley, near Rotherham, there is a monument to a human death caused by a Wild Cat.

Wild Cats were once hunted but were probably never numerous enough to be of truly major importance. They were not mentioned in Guillaume Twici's list of beasts of the chase in 1340 (see Pine Marten, page 418), although, in his late sixteenth-century *Treatise on Hunting,* Sir Thomas Cokayne wrote that hunting bucks in summer and hunting hares (with Wild Cats, martens and Foxes as equal second favourites) in winter were then the leading pursuits of the English country gentleman.[157]

The relative paucity of places named after Wild Cats is also a reasonable indication of their low numbers compared with other sporting quarry. Although 'catt' was *cat* in Old English, the possibility that sometimes the Polecat, marten cat or even domestic cat was being referred to (as well as someone called Ceatta, which was an Anglo-Saxon personal name) rather clouds the picture. Nonetheless, among about 50 probable Wild Cat places are Cattishall in Suffolk ('a hill frequented by Wild Cats'), Catford in Kent ('Wild Cat ford') and Catmore in Berkshire ('Wild Cat lake'), as well as the unequivocal Wild Cat Spring in Cumberland and Wildcathishevede in Cheshire.

Although there are under a million feral cats in Britain, there are now believed to be around 9 million non-feral ones; domestic pets like yours and mine. Although not part of Britain's wild fauna, they cannot go unremarked as a recent survey by the Mammal Society pointed to them as the main predators of British wildlife. The survey estimated that 200

Somewhat like a domestic tabby, but larger and with a thick, ringed tail, the Wild Cat *(Felis sylvestris)* is Britain's fiercest predator.

million mammals, 55 million birds, 10 million reptiles and amphibians and 10 million other creatures were killed by cats annually; this is perhaps not surprising given that the domestic cat population is six times greater than that of all wild terrestrial predators combined.[158]

There are numerous reports of 'large black cats', 'panthers', 'pumas' or 'black beasts' that supposedly roam the British countryside but, despite my numerous enquiries, no-one has been able to show me a captured animal, a decent photograph, a clear, unarguable film or any other irrefutable evidence that such creatures exist; there seems no reason for taking them any more seriously than unicorns or Welsh dragons.

Walruses, Seals and Sea Lions

(Pinnepedia)

The Pinnipedia, the fin-footed mammals, are creatures that, like the whales, have returned to the water. Almost all are marine and highly adapted to life in the sea, with subcutaneous stores of blubber for insulation, a streamlined body, oily fur and modifications to enable them to hold their breath for long periods under water. They are beautifully designed for swimming and for catching and feeding on fish and other marine life. However, unlike whales, they must still come ashore to breed, and their breeding grounds, or 'rookeries', may contain tens of thousands of animals. They also come ashore, or 'haul out', to rest. The European species are thought to have two different origins: the Walrus (*Odobenus rosmarus*, Family Odobenidae) has apparently evolved from a bear-like creature while the seals have evolved from an Otter-like animal. A third group, the eared seals, or sea lions, do not occur in Europe.

The Walrus (a word perhaps originally from the Old Norse, meaning a 'horse whale') is a huge, tusked, mollusc-eating animal of the Arctic Ocean that appears occasionally in the far north of Britain and was sometimes brought home alive by whalers. Curiously, a Walrus may have been the earliest British zoological specimen as one was brought to Alfred the Great; another was exhibited in London in the reign of James I.[159]

> There is a story of a Walrus at Longhope [Orkney] which annoyed people crossing the bay on the way to church, by putting its tusks over the gunwale of the boat ... .[160]

There have been at least twelve sightings of Walruses in Shetland within recent history, the most celebrated being predictably christened 'Wally' by the media as it subsequently made a rather prolonged exploration of the east coast of Britain in 1981 before being flown back to the Arctic. A magnificent Walrus was also photographed on the Aberdeenshire coast in 1954. Old items made from Walrus tusks are sometimes seen; there was a historical tradition of using Walrus ivory for the hilts of weapons and also for drinking cups because of beliefs that the ivory would enable its owner to be cured of wounds received in battle and resist the effects of poison. Nonetheless, they have never been familiar enough to enter much of British lore and tradition. Possibly the most charming exception is Lewis Carroll's poem 'The Walrus and the Carpenter' in *Through the Looking Glass and What Alice Found There*:

> 'The time has come,' the Walrus said,
> 'To talk of many things:
> Of shoes – and ships – and sealing-wax –
> Of cabbages – and kings –
> And why the sea is boiling hot –
> And whether pigs have wings.'[161]

My generation, however, is more likely to be familiar with John Lennon and Paul McCartney's curious 1967 song 'I Am the Walrus', from The Beatles' *Magical Mystery Tour* record, which achieved its greatest fame through being banned by the BBC for containing the word 'knickers'.

Despite the rarity of Walruses, Britain is well supplied with seals. One or other of the two native species – the Common Seal *(Phoca vitulina)* and the Grey Seal *(Halichoerus grypus)* – may be found in many places, principally around the northern and western coasts or in the area of The Wash, while at least four Arctic species – the Harp Seal *(Phoca groenlandica)*, the Bearded Seal *(Erignathus barbatus)*, the Hooded Seal *(Cystophora cristata)* and possibly even the smallest of all seals, the tiny Ringed Seal *(Phoca hispida)* – turn up occasionally in the far north of Britain. It is the two native species, however, and especially the Grey Seal, that have entered the traditions and tales of northern people, and in a quite extraordinary way.

The Walrus and the Carpenter in Sir John Tenniel's illustration for Carroll's *Through the Looking Glass and What Alice Found There.*

The name 'seal' is Old English, perhaps with some connection to 'sal', meaning 'salt', or 'sea water', but also linked with the Teutonic word *selhos*, 'fish'. Females are known as 'cows', males as 'bulls', and giving birth is known as 'calving', although the young are usually called 'pups', not 'calves'. Groups of seals are called 'herds', but breeding colonies are 'rookeries'. 'Selchie' and similar names are widely used for seals in Orkney and Shetland. The Gaelic name 'ron' may have come from the Norse *hraun*, meaning 'a rocky, desolate place', and is used especially for the Grey Seal (see below). It is the inspiration for the name of the Hebridean islands of North and South Rona, which now support the greatest concentration of breeding Grey Seals anywhere. North Rona was home to a small colony of people until the early nineteenth century, finally becoming uninhabited in 1844. It was apparently the home of the seventh-century hermit St Ronan, who was said to be have been taken there not by a seal, unfortunately, but by a whale. Humphrey Hewer wondered about Rona and its name:

> Was it once a great seal colony before human habitation? If so it must have been a long time ago for the 16th century author Martin Martin refers to the inhabitants as being of a very ancient race bearing names quite unlike any Scottish or Norse ones.[162]

A number of other northern and western islands are not surprisingly derived from ancient seal names; *Eilean Nan-Ron*, the Isle of Seals, for instance, is off Colonsay.

Seals
(Phocidae)

OTHER COMMON NAMES: *bilder; boca; brineld; brun-swine/ brun-swyne; dog; hran; horeng; jarck: mollewelle; morse; neubling; powart; sael; saelkie; saylch; sea-dog; selch; selchie; seekie; seolbh; silkie; swelchie; willie-powret.*

Common Seal *(Phoca vitulina)*

OTHER COMMON NAMES: *black seal; harbour seal; spotted seal; tang-fish (Shetland); tangie.*

Much smaller than the Grey Seal and rather similarly grey-brown spotted, the Common Seal is readily identified by its concave muzzle and whiskered, dog-like face. Also unlike the Grey Seal, the sexes are of similar size and tend to congregate in smaller numbers, rarely in groups of more than a thousand. They commonly breed on intertidal shores and the pups can swim at the first high tide after birth.

Compared with the Grey, they are much more of an inshore seal, and this is reflected in their local names: 'harbour seal' is particularly appropriate, and their old Shetland name 'tang-fish'

This large Walrus *(Odobenus rosmarus)* was photographed, far from its usual range, on the Aberdeenshire coast in 1954.

('seaweed fish') contrasts with 'haaf-fish ('deep-sea fish'), which is applied to the Grey.

Common Seal populations are difficult to estimate, but a serious disease caused by Phocine Distemper Virus, which struck the population at the end of the 1980s, certainly reduced their numbers, although they have since recovered; the present total of about 50,000 is around half that for Grey Seals. Their main natural enemies are Killer Whales *(Orcinus orca)* and sharks.

Common Seals eat a wide range of food, which varies from one locality to another but includes fish, cephalopods and crustaceans. Although the Grey Seal is generally regarded as the greater threat to Salmon stocks, it was a proposed Common Seal sanctuary on the island of Islay that resulted in the first successful appeal against a designated wildlife conservation area, in 1999, on the grounds that local Salmon fisherman would suffer in consequence.[163]

Grey Seal *(Halichoerus grypus)*

OTHER COMMON NAMES: *Atlantic seal; haaf-fish (Shetland); ron-mor; selchie.*

The Grey Seal is the larger of the two British species and is readily distinguished by its facial profile: the nose and top of the head form a straight line, giving it a 'Roman' nose. Not only is it more common than the Common Seal, it is also not grey, or at least not uniformly so, both sexes being more or less spotted in shades of grey-brown, although the female is strikingly very much smaller than the male.

In British waters, Grey Seals feed mainly on fish, especially Salmon, and fish trapped in nets are particularly vulnerable to seal predation, with the result that the animals are extremely unpopular with fisherman.

Like Common Seals, Grey Seals are still numerous and increasing; the British breeding population is probably about 100,000, around half of the world total for the species. Indeed, they are so prolific that shooting contraceptive drugs into females by means of air-guns has been suggested.[164]

Grey Seals tend to breed on remote, offshore rocks and islands, although they will come ashore to rest on intertidal shores. They sometimes turn up in expected places. Among the most famous individuals was 'Billy the Seal', which arrived many years ago in Cardiff Docks and lived in the pond in Victoria Park, Cardiff, for a very long time, becoming familiar to locals and visitors alike. On occasion, when the city was flooded, he could be seen swimming in the streets, but attempts to find him a spouse proved fruitless, not least because, after his death, Billy the Seal proved to be female.

Unlike Common Seal pups, young Grey Seals have white fur for the first two weeks of life:

> Once the bull had taken up his territory one of the cows joined him and here she gave birth to a creamy-yellow pup, covered with a soft silky fur. At two days

A Grey Seal *(Halichoerus grypus)* in a harbour near John o'Groats receives a reward from this Scottish fisherman for its curiosity.

old the pup would call to its mother with a long, anxious note, and she would answer from the sea with a mournful, wavering call. If I went too near this pup, as he lay on his back, he would roll over and hiss savagely, while great tears welled up and ran down his face.[165]

The apparent pure-white innocence of seal-pup fur, so conspicuously stained red with blood when they are clubbed to death during the culls, has worked as a powerful propaganda tool for those opposed to the activity.

No animal in Britain has a deeper, richer and more extraordinary folklore than Ron-Mor, the great seal of the northern islanders. Countless stories have been woven around it but, most significantly, there is an ancient and tangled belief in the selkie folk, or 'seal people'. These are one form of what in Shetland are called 'Finns', a vague term for any feared sea monster. In one version of the belief, they possessed a skin or covering like that of a seal skin and, when they wore it, they could take to the sea and behave like a seal. Should they lose the garment, they were doomed to a life on land. There is a corollary, in which seals turn into women when their skins are stolen, and marry and bear children, being able to return to the sea only if they recover their skins. Even into the twentieth century, there were Shetlanders who claimed to be descended from Finn women:

> The early ancestors of the Conelly family in Ireland were said to have been seal, so no Conelly would ever kill one, since this would have brought bad luck. The fact that children are occasionally born with a membrane between the fingers or toes would have encouraged the growth of such ideas.[166]

The author and photographer Seton Gordon (1886–1977) told a comparable story of the Clan MacCodrum of North Uist, which had a special affinity with the seals; when the annual autumn seal-killing, or 'battue', took place:

> ... an old woman of the clan was always seized by violent pains out of sympathy with her kinsfolk of the sea that were being murdered in their surf-drenched home.[167]

There are also legends of women having intercourse with either real seals or seal men, and of offspring with flippers being produced:

> Unsatisfied mortal women sometimes sought a seal man lover. If a woman wished to contact a selchie, she had to shed seven tears into the sea at high tide. A young woman gathering shellfish at Brecken, Yell, Shetland, swooned when she met a seal man. Nine months later, she bore a son with a seal's face ... Descendants of a Stronsay (Orkney) woman, who sprang from her union with a seal man she loved, had thick, horny skin on the palms of their hands and soles of their feet.[168]

There are many comparable tales, all no doubt elaborated in the telling around smoky fires on wild, northern winter nights, with the wind howling about the stone-built crofts and the sea noisily shifting the shingle on the shore.

David Thomson cleverly wove the legends into a beautifully written narrative in his extraordinary 1954 book *The People of the Sea*, which told of Sean Sweeney the seal-killer, Mrs Carnoustie with her peculiar legs 'kind of all in one with two flat feet sticking out sideways', and Osie Fea, who said that:

> ... it was aye a joke wi' the women ... that if their men neglected them, they'd away to the selchie folk for comfort, or if a husband was unfaithful they'd do the same.[169]

It comes as no surprise that people in general were very suspicious of seals and that fishermen tended to feel uneasy when they came too close.

> They looked so wise, so knowing with their big round eyes and bewhiskered faces ... some people even went so far to suggest that seals were once angels in heaven before the mutiny described in the book of Revelation which culminated in the expulsion of Satan. The angels which had sided with Satan against Christ were cast out with their leader, and those that fell to the ground became trows or fairies, whilst those that fell into the sea became seals.[170]

The most logical explanation for the seal-people legend must surely lie in a number of well-documented historical accounts of people (real people, presumably Inuits) from northern lands who were seen off the coast or sometimes washed ashore on island beaches in seal-skin-covered kayaks. In 1682 one person was found at Eday in Orkney but, when locals launched boats to investigate, paddled away, never to be seen again. Another was seen two years later at Westray. The Reverend John Brand, in *A Brief Description of Orkney* (1701), said, 'There are frequently Finnmen seen here upon the coasts', and then reported recent occurrences on Westray and Stronsay.[171]

An account published in 1782 described a kayak in the museum of Marischal College, Aberdeen University (where it still remains) as:

> ... a canoe, taken at sea, with an Indian man in it, about the beginning of this century. He was brought alive to Aberdeen but died soon after his arrival, and could give no account of himself. He is supposed to have come from the Labrador coast, and to have lost his way at sea. The canoe is covered with fish skins ... .[172]

It has been pointed out that:

> ... Irish monks made long voyages to Faeroe and Iceland in their coracles, and there is no reason why men from northern countries should not have carried out the journey in reverse. With their bodies concealed inside their seal-skin vessels and with their short paddles dipping into the water, they would certainly have appeared more like seals swimming than men rowing.[173]

Even so, Margaret Fea, in David Thomson's book, did say 'The Finn men were real men ...'.[174]

Despite the islanders' legendary affinity with seals, the animals were nonetheless hunted and killed out of sheer necessity, for their meat, their skins and their invaluable oil, which fuelled lamps and fires (the northern islands are lands without trees and wood). Humphrey Hewer reminded us that:

> Until paraffin became a cheap source of light and fuel, seal oil was used in most peasant communities of North-West Europe, particularly the Hebrides.[175]

The Grey Seal *(Halichoerus grypus)* tends to breed on remote, offshore rocks, and the pups have white fur for the first few weeks.

The Common Seal *(Phoca vitulina)* usually breeds on intertidal shores and is generally much more of an inshore seal.

It had other uses too:

> The oil is collected from boiled seal meat as it drips from the suspended sack in a warm room, and is kept in a bottle, to be used as a cure for rheumatic disorders. Near Brandon creek in Kerry I was offered a drop or two to cure a sprained knee.[176]

The Scottish antiquary and historian, John Pinkerton (1798–1826), a man with the dubiously flattering epitaph that 'his powers of research were greater than his literary talent', described how the natives of Harris would salt seal flesh in the ashes of burned seaweed and then eat it in the hungry days of spring, using a long, pointed stick to avoid the smell clinging to their hands.[177] The seals were killed with a club, in much the same way as the highly publicized Canadian seal culls of modern times that have been filmed and shown around the world in the hope of shaming the Canadian government. Thomas Southwell in his *Seals and Whales of the British Seas* (1881) wrote that from Orkney 'a ship commonly goes once a year to Soliskerry [Sule Skerry, about 50 kilometres/30 miles off the north coast of Scotland] and seldom returns without 200 or 300 seals'.[178] It was on such dangerous journeys in the wild seas of autumn that much of the folklore and superstition concerning seals must have arisen and the voyage also gave birth to the famous, soulfully beautiful ballad 'The Grey Selchie of Sule Skerrie':

> 'I pray thee tell to me thy name,
> Oh tell me where thy dwellin' be?'
> 'Ah, I've come far, frae west of Hoy;
> I earn my livin' in the sea.'
>
> 'I am a man upo' the land,
> I am a selchie in the sea,
> An' when I'm far frae every strand
> My dwellin' is in Sule Skerrie.'
>
> 'Alas, alas, this woeful fate,
> This weary fate's been laid for me –
> A selchie frae the west of Hoy
> To Norway land to mate wi' me!'[179]

Some seal products made their way south and were put to use on the mainland. Seal skin was once used for the manufacture of high-quality shoes, bags and other articles and seal flesh was eaten in many a respectable Catholic dining room on the premise that, like Otter and Puffin, it was more fish than flesh and therefore acceptable on Fridays or in Lent. Rather unexpectedly, 'Seales' along with 'Porpisses' featured on the menu at the enthronement of the Archbishop of York in 1465 (see Bittern, page 212).

Dolphins, Whales and Porpoises

(Cetacea)

Many people find it hard to believe that dolphins, porpoises and whales are British animals. They seem to have their affinities on the one hand with the icy waters of the Poles and on the other with the warm tropical seas where Moby Dick, the fictional Sperm Whale, finally took its revenge. However, whales occur in all the oceans of the world, many of them migrating enormous distances, and while, with one exception, you are unlikely to see a large one from a British shore-line, some of the smaller members of the group, the dolphins and porpoises, regularly come close to land.

Nonetheless, cetaceans have entered British lore and history in many ways. It is often forgotten that whaling was once partly a British-based occupation, although perhaps not on the same scale as some nations, and whales were certainly caught by boats from British ports and whaling men came back with fantastic tales. Closer to home, there were occasionally closer encounters with whales, which are strangely prone to becoming stranded on the coast, either driven there by storms or, more usually but still largely inexplicably, by deliberately beaching themselves. Whenever this occurred, the arrival on some small beach of a dying or decomposing object the size of a decent boat not surprisingly gave rise to all manner of myth and speculation about its nature and origin. Moreover, if the body was barely or newly dead, it would have provided an unlooked-for but very welcome supply of fresh meat; whale bones are unearthed from time to time in coastal prehistoric sites. As recently as the late 1950s, Alister Hardy, writing for an audience who had endured war-time privations, said:

> Before the war it was not unusual to find, among a group of otherwise well-educated people, a number who thought that a whale was just a particular kind of exceptionally large fish. Today, however, there can be few who have not, at any rate once, tried a whale steak

Judging by its thrashing tail, this whale, being hauled ashore with such difficulty at Broadstairs in Kent, is still alive.

This archway, fashioned from the jawbones of a whale, is in Bragor, on the Hebridean island of Lewis.

> for dinner and, finding it not unlike beef, have learnt the fact (at first most astonishing) that whales are just as much warm-blooded red-fleshed mammals as are oxen and sheep.[180]

All cetaceans were long considered 'Fishes Royal' under a statute of Edward II passed around 1324 and, in 1971, when moving the second reading of the Wild Creatures and Forest Laws Bill, which removed certain royal rights to wild creatures, the then Lord Chancellor, Lord Hailsham, commented in the House of Lords that:

> ... whales are royal fish and if any peer says they may be royal but are not fish, I shall refer him to the Gospel according to St Matthew and leave him to the tender mercies of the bench of bishops.[181]

In the event, following objections from the Director of the Natural History Museum, the Royal Prerogative to Royal Fish was retained, largely because it enabled the Museum to examine and record whales stranded on the coast.

Although the fact that whales are mammals, not fish, is now generally recognized, it is still extraordinary to learn that they have evolved from land-living predators with hooves and, on at least one occasion, a modern whale has been found with four atavistic internal legs. Like seals, whales are mammals that have returned to the water but they have adapted to it even more completely; they are also both highly intelligent and biologically sophisticated. Although they are unable to extract oxygen from the water, and consequently need to surface from time to time, they are as functionally aquatic as fish, with the added advantage of being able to echolocate. By means of a complex and still imperfectly understood communication system, animals can maintain contact with each other despite being huge distances, probably thousands of kilometres, apart. The sounds they produce are often both beautiful and haunting. Whales are propelled by their tail flukes, which are horizontal in relation to the body, not vertical like the tail fin of a fish. They are typically very large compared with land mammals, and one species that has occurred in British waters, the Blue Whale *(Balaenoptera musculus)*, is probably the largest animal that has ever existed.

It is has very recently been proved that whales are also the longest-lived of mammals and possibly the longest lived of all animals. Living Bowhead Whales *(Balaena mysticetus)*, also known as Greenland Right Whales (because they were the 'right' whales for the whalers), have been found with the heads of eighteenth-century harpoons embedded in their hides. Also, whales' eyes contain certain chemicals that change in composition at a known rate and these can be analysed and measured. Tests on the eyes of Bowhead Whales killed by Inuits have demonstrated that many may be well over 100 years old; one individual was 211 years old.

The name 'whale' is Old English, with links to Old Norse and Old High German languages. Whales have also attracted a special vocabulary, especially when they occur in a group: while some whales are generally solitary, many species consistently congregate in large numbers, sometimes with very close social bonding between the individuals. Such an assembly of whales is called a 'school', a sixteenth-century word etymologically similar to 'shoal', or a 'pod', which is the nineteenth-century American name.

Whales are divided into two groups: the toothed whales and the baleen whales. The toothed whales (Odontoceti) include over 85 per cent of living species of whale and embrace animals ranging from the relatively small dolphins and porpoises to the very large Sperm Whale *(Physeter macrocephalus)*. They feed on a wide variety of aquatic life. The second group (the baleen whales, or Mysticeti) comprises the ten species of baleen whale (a tautology as the word 'baleen' is derived from the Latin *balaena*, meaning 'whale') and includes all the largest species, sometimes called the 'great whales'. They feed on the smallest food: plankton and, typically, the minute shrimps and other animal life known collectively as 'krill', which they filter from the water by sieve-like structures called 'baleen plates'. All whales, except a very few species of dolphin, are marine. Although hunting is now controlled by international treaty (the International Whaling Commission was established for this purpose in 1946), there are escape clauses in the legislation (most notably the dubious permission to kill whales for 'scientific study') and some nations have failed to sign the treaty. This, combined with pollution and other factors, means that most species, certainly most large ones, are threatened.

Although during the eighteenth and nineteenth centuries, whalers set sail for distant seas from several British ports, particularly Dundee, Peterhead, Hull and Whitby, they returned only with casks of blubber and other whale products, the catch being cut up on site. The principal species taken on these voyages were Arctic Right Whales and tropical Sperm Whales, both of which have the advantage of floating when dead. They were harpooned by hand from small boats, a feat requiring remarkable skill and no little bravery. Although stocks were undoubtedly depleted, the well-chronicled demise of so many other species of great whale has been more recent and can be attributed to the Norwegian Svend Foyn, who invented the harpoon gun and explosive harpoon head in the mid-nineteenth century. There were once seven whaling stations in the British Isles, which functioned at various times from the early part of the nineteenth century until the late 1920s and where whales caught closer to home were landed. There was one at Blacksod Bay, Co. Mayo in Ireland, one on Lewis and at Bunaveneader on Harris in the Hebrides, and most importantly, four established by Norwegian companies in 1903–4 in Shetland: at Collafirth, Olnafirth and two at Ronas Voe.

Most stories and beliefs about whales do not distinguish between the various species, although dolphins have tended to merit their own special lore. Probably about 24 species of whale have been recorded in British waters and/or beached on British shores but several are either vagrants or rare, if regular, visitors. One very rare vagrant, which deserves a mention, is one of the white whales, a group related to the dolphins. The Narwhal *(Monodon monocerus)* is an Arctic whale, although, extraordinarily, one was stranded in the Thames Estuary in 1949. Nonetheless, parts of Narwhals occasionally turn up at antiques sales in auction rooms because the male has one tooth elongated into a forwardly pointing spiral tusk up to 3 metres (3¼ yards) in length, which has fascinated collectors for a long time. The animals seem to use it in sexual contests, rather as a deer uses its antlers, but in centuries past this astonishing structure gave rise to the legend of the unicorn, a creature that is immortalized opposite the lion on the royal coat of arms. Then, as now, these objects commanded extremely high prices, partly due to an old, utterly fallacious but almost inevitable belief that ground-up Narwhal tusk was an aphrodisiac.

The 14 whales described below can be considered both regular visitors and at least reasonably common in British waters. However, they are not necessarily easy to identify because usually all that can be seen of them when they break the surface of the sea is an amorphous hump. Sometimes whales 'breach', leaping from the water in spectacular fashion, and the way in which they re-enter the water tends to differ from one species to another: some re-enter in a clean dive while others produce an enormous splash, called 'crash-breaching'. Sometimes too they may be seen 'blowing', producing a spout, apparently of water, which has a characteristic pattern according to species. In reality, the spout is simply the condensed breath as the whale exhales through its upwardly facing nostril, or blow-hole.

True dolphins
(Delphinidae)

Dolphins are small whales, characterized by an elongated dorsal fin and usually having a pronounced horny beak. Several species are still relatively common but, in parts of the world, dolphins are caught in large numbers for food. Many thousands also die as a result of being trapped in fishing nets, especially Tuna nets, and, in consequence, wildlife conservation groups have at different times sought to boycott Tuna products and the countries that produce them. One species or another may be seen fairly regularly in Britain from boats offshore, especially during the summer months. Their intelligence and adaptability has led to dolphins, particularly the Bottle-nosed Dolphin (named from its bottle-like beak), being displayed in dolphinaria. Some species, most notably the Common Dolphin, have beautiful coloration with yellow flashes on the flanks and there has long been a belief that these colours change as it dies:

> Parting day
> Dies like the dolphin, whom each pang imbues
> With a new colour as it gasps away,
> The last still loveliest, till 'tis gone;
> And all is gray.
> (Lord Byron, *Childe Harold's Pilgrimage*)

Long before it was known that dolphins do indeed make the beautiful sounds, they were thought to be lovers of music. There was also an ancient, mainly classical Mediterranean pastime of boys (and others) riding on their backs, the most celebrated representation of this distinctly curious practice being a wall painting in Pompeii.

Dolphins are, however, seen quite commonly on walls, or at least on hanging signs, in Britain because The Dolphin is a favoured name for seaside inns. The dolphin has also long been a popular heraldic symbol, although the heraldic dolphin is a weird-looking, half-mammal and half-fish creation, often with

The Bottle-nosed Dolphin *(Tursiops truncatus)*, seen here playing near an oil rig, is the species most frequently seen in dolphinaria.

its tail curled beneath itself. It appears in this form on the arms of the Fishmongers' Company and also of the heir to the French throne, who was always known as the Dauphin in consequence. It is possible that some Dolphin inns were so named in tribute to the French Ambassador to Britain.

Dolphin flesh has been highly esteemed for centuries 'roasted and dressed with kindred porpess sauce, crumbs of fine white bread, vinegar, and sugar' and, like the porpoise, it has acquired a 'pig' name: the 'mere-swine', or 'sea pig'. It is not unusual to see 'dolphin' on the menu today, even in respectable fish restaurants, but you should not be put off: the item in question is not the mammal but the Dolphin Fish or Dorado *(Coryphaena hippuris)*, which also changes colour on death.

Common Dolphin *(Delphinus delphis)*
OTHER COMMON NAMES: *Fraser's dolphin; mere; meer-swim; meer-swine.*
One of the commonest marine mammals, Common Dolphins are widely distributed, especially in warm and temperate seas and are frequently seen in British waters. They swim in schools, or 'pods', numbering up to a thousand animals, and are often seen by sailors, leaping clear of the water or riding the bow-waves of boats. Dark above, they have a pale grey belly, alternating light and dark bands on the flanks and a distinctive dark circle around the eye.

Bottle-nosed Dolphin *(Tursiops truncatus)*
Friendly and easily tamed, this is the dolphin most frequently seen in dolphinaria; wild specimens have also been known to associate with bathers and yachtsmen. Small schools may be seen off all British coasts. It is slate-grey in colour, darker above, with a white throat and belly, and a shorter beak than most dolphins.

White-beaked Dolphin *(Lagenorhynchus albirostris)*
This fairly gregarious dolphin is moderately common in the North Sea and North Atlantic. It is generally dark in colour, with a black tail fluke and flippers, longitudinal white areas on the flanks, a white throat and belly and, usually, a white beak, although darker beaked animals occur in some areas.

Atlantic White-sided Dolphin *(Lagenorhynchus acutus)*
Found in the northern part of the Atlantic and around all British coasts, this dolphin is generally black, with a white belly and a light brown band on the flanks. Like the White-beaked Dolphin, it has a short, distinct beak, usually black above and grey below.

False Killer Whale *(Pseudorca crassidens)*
The third largest dolphin, this is similar in outline to the Killer Whale (see below) but smaller and lacking the striking white flanks and underside. It is widely distributed throughout the world but seldom seen.

Killer Whale *(Orcinus orca)*

OTHER COMMON NAMES: *orca.*

The largest and most dramatic looking of the dolphins is the black-and-white Killer Whale, a fast-swimming and ferocious predator, although one that curiously has also adapted to dolphinarium life and even featured as the hero of a popular film, *Free Willy*. Killer Whales eat squid, seals, large fish and, in other parts of the world, penguins; a celebrated recent wildlife film dramatically showed Killer Whales plucking penguins from the surf of a South American shore. Gavin Maxwell said with some accuracy:

> Anyone writing of Killer Whales finds it necessary to quote the discovered contents of one Killer's stomach ...

True to form, he went on to describe one that he had found:

> ... to contain no fewer than thirteen porpoises and fourteen seals.[182]

With its large, triangular dorsal fin, the Killer Whale is often mistaken for a shark when it swims close to the surface, although many a swimmer must have been relieved to discover that the fin was not followed by a corresponding and similar vertical tail. The dorsal fin conveniently bears markings that differ slightly from one animal to another and so enables scientists to identify individuals at a distance.

The Latin name *orca* is an old one for a whale, although, contrary to common belief, 'Orkney' is not 'whale island' but probably the Celtic *Inse Orc*, the Isles of Boars.

Risso's Dolphin *(Grampus griseus)*

OTHER COMMON NAMES: *dunter; lowper; Risso's grampus.*

Risso's Dolphin is named after the Italian naturalist (Giovanni Antonio Risso, 1777–1845). Unusually among dolphins, it lacks a beak and most closely resembles a Pilot Whale, with a grey body that is often marked with long scratches, possibly caused by squid suckers.

Perhaps the most celebrated British specimen of Risso's Dolphin was one caught in the mouth of the Thames in 1758 and brought on a barge to Westminster, where it was examined and dissected by the celebrated surgeon and naturalist John Hunter, founder of the Hunterian Museum at the Royal College of Surgeons.

This photograph, taken in *c.*1905, shows that, even in death, the Killer Whale *(Orcinus orca)* is still a formidable creature.

Long-finned Pilot Whale *(Globicephala melaena)*

OTHER COMMON NAMES: *blackfish; caa'ing whale.*

The Long-finned Pilot Whale is the species that is most often stranded in considerable numbers; there seems to be a very close social bonding between the members of a school, which results in them all suffering the same fate. Their habit of entering bays and other shallow waters in large numbers gave rise to the bloodthirsty practice of the 'caa' ('caa', or 'ca', is an old Scots word for 'drive'; hence 'caa'ing whale', 'the whale that is driven'), which once took place in Orkney and Shetland, and still does in the Faeroes. Boats were launched to prevent the whales escaping back to the open sea, and they were driven ashore and slaughtered with almost anything on which the local people could lay their hands. The sea turned red. Sam Berry reported that, at Sourin Bay, Rousay, Orkney, 60 whales were killed on one occasion in the 1860s and sold for £260.

> In the 1870s 300 were caught in Linga Strand, Stronsay. A dyke at Grainbank Farm, Wideford Hill, was built of the skulls of Caa'ing Whales driven into Kirkwall Bay.[183]

Probably the largest 'caa' ever to take place in the British Isles was at Quendale on the South Mainland of Shetland in 1845, when 1,540 whales were slaughtered; the last was of 83 animals at Weidale in 1903. Perhaps not surprisingly, these spectacular blood-lettings seem to have been recorded far more in Gaelic than in English literature, although the Scottish author John Buchan included a graphic description of one, a 'grind', that took place in the Faeroes, or 'Norland', in the climax to his 1936 adventure *The Island of Sheep*:

> The shores of the voe were dense with people, and its surface and that of the lesser voes black with a multitude of boats. But at the heads of each inlet was a spouting and quivering morass in which uncouth men laboured with bloody spears. It was a scene as macabre as any nightmare ... .[184]

In the Faeroes, the meat is eaten and considered a choice delicacy; in Orkney and Shetland, the whales were simply butchered for their blubber.

Porpoises (Phocaenidae)

Harbour Porpoise *(Phocoena phocoena)*

OTHER COMMON NAMES: *common porpoise; Herring hog.*

This is the smallest and also the commonest British whale, differing from the dolphins in lacking a beak. It can regularly be seen in many coastal areas, especially the North Sea and the English Channel, although its numbers have declined in recent years, probably in part because of the decline in the numbers of Herring, one of the principal fish on which it feeds. The name 'Herring hog' reflects this diet and the word 'porpoise' itself alludes to a pig-like character, coming through Middle English from the Latin *porcus piscis*, meaning 'hog-fish'.

> Tourists on river cruises and people queuing to ride on the London Eye yesterday were entertained by a porpoise in the Thames ... Steve Colclough, Fisheries Officer for the Environment Agency, said ... this is an indication of how clean the Thames is these days. For the last eight or nine years there have been many sea mammals spotted in the river, with a dolphin swimming as far as Hammersmith last autumn.[185]

When Porpoises were seen close to ships at sea, it was said to be time to seek shelter because a storm was threatening. If they were seen swimming to windward, this was a particularly serious sign and bad weather would follow within 12 hours – the 'Porpoise that doth betoken raine or stormes of weather' (Robert Chester, 'Love's Martyr').

> Come, porpoise, where's Haterius?
> His gout keeps him most miserably constant;
> Your dancing shows a tempest.
> (Ben Jonson, *Sejanus*)

Porpoise flesh was once considered a delicacy and, although stranded animals were the usual source, the fact that quantities of porpoise were available when required as far inland as York, for the enthroning feast of the Archbishop in 1465 ('Porpisses and Seales, twelve'; see Bittern, page 212), is evidence that they were also caught deliberately. How they arrived fresh is a matter for speculation but it is worth noting that:

> ... although few sailors in English ships know it, porpoise beef improves vastly by keeping, getting tenderer every day the longer it hangs, until at last it becomes as tasty a viand as one could wish to dine upon.[186]

There is a commonly quoted belief that a porpoise will come to the aid of a drowning man:

> It is a well-known fact that the porpoise is an extremely inquisitive creature and loves to sport and play. Anything floating on or near the surface of the sea will attract his attention. His first action on approaching the object of his curiosity is to roll under it. In doing so, something partly submerged, like the body of a drowning person, is nudged to the surface of the water. The sea does its part and automatically drives floating objects towards the beach. It is therefore within the realm of possibility that on some remote occasion a porpoise did happen to assist unintentionally in the rescue of a drowning person. An observer watching the procedure from a distance might well have assumed that the action of the porpoise was intentional. No doubt it was from such an incident as this that the story originated.[187]

***The Spermaceti Whale*, an engraving by W.H. Lizars (1799–1859), shows early whalers hunting from small boats.**

Sperm Whales
(Physeteridae)

Sperm Whale *(Physeter macrocephalus)*
OTHER COMMON NAMES: *cachalot.*
Arguably, the Sperm Whale is the most famous of all whales because Moby Dick, the eponymous hero of Herman Melville's novel, was a white Sperm Whale that eventually gained the better of its obsessed and vengeful pursuer Captain Ahab in his ship the *Pequod.*[188] The Sperm Whale has certainly made a special contribution to literature. Melville had served aboard a whaling ship in the 1840s and this, his most celebrated novel, was first published in three volumes in England in 1851 as *The Whale* and then, three months later in his native USA. It was not well received at first but gradually, and certainly after the author's death in 1891, it came to be recognized as 'the greatest work of American fiction' and 'the central masterpiece of American literature'. The first edition consisted of only 3,000 copies and, as the unsold volumes were destroyed in a warehouse fire in 1853, it is now especially sought after: a copy was advertised recently for £28,000. The major part of the story was very successfully and convincingly filmed by John Huston, although, as an anonymous critic once said, 'I can no longer see a Sperm Whale without thinking of Gregory Peck'. Whatever one's feelings about the morality of whaling, and despite Britain's maritime heritage, there is nothing in British literary fiction to compare with this wonderful and evocative tale of the sea.

> Through the serene tranquilities of the tropical seas,
> among waves whose hand-clappings were suspended
> by exceeding rapture, Moby Dick moved on, still
> withholding from sight the full terrors of his
> submerged trunk, entirely hiding the wretched
> hideousness of his jaw.[189]

A compelling, home-grown, true-life whaling tale is *The Cruise of the 'Cachalot': Round the World after Sperm Whales* by Frank Bullen (1857–1915), sometime 'errand boy, nomad etc.', who claimed to have received no education after the age of 11. It was first published in 1898 and Rudyard Kipling endorsed the work: 'It is immense, there is no other word. I've never read anything that equals it in its deep-sea wonder and mystery.' *Cachalot*, the name of Bullen's ship, is an alternative name for the Sperm Whale itself and comes from the French, probably meaning 'toothed'.

The Sperm Whale was a prime target of the whalers because of the range of products it yielded: meat, blubber, ambergris and spermaceti. Ambergris is a foul-smelling, greasy black substance of unknown function produced in the whale's intestine. On exposure to the air, it becomes paler, harder and better smelling and is used in the cosmetics industry to fix and intensify perfume (I have long puzzled how anyone discovered this). Spermaceti and sperm-oil are liquid esters of special quality produced in the body of the Sperm Whale, the more valuable spermaceti being concentrated in the relatively enormous head. 'Spermaceti' was so-called because it was once thought to be the whale's sperm and it is from this substance that the species

The Sperm Whale *(Physeter macrocephalus)* is the largest of the toothed whales. British strandings are almost always of male animals.

takes its name. It was widely used for a variety of purposes:

> And telling me the sovreign'st thing on earth
> Was parmaceti for an inward bruise ...
> (*Henry IV Part 1*, Act 1, Scene 3)

The legends and reputation of the Sperm Whale have been enhanced by the knowledge that its preferred prey is another of the ocean's great mysteries and terrors: the giant squid. Frank Bullen vividly described watching the two entangled at the surface one moonlit night off Sumatra, although, oddly, it is often and widely said that no giant squid has ever been seen alive.

Most Sperm Whales in the North Atlantic, and those that reach the British coast, are males; the females tend to stay in warmer waters closer to the equator. Many a British seaside town has a whale relic, often a large jaw-bone or vertebra, as a trophy of some ancient corpse.

> Until it was destroyed by bombing in the 1939–45 war, a curious 'seat' formed of the base of the skull and the first vertebrae of a sperm whale washed up on Caister beach in 1582, stood in a niche by the west door of St Nicholas's Church in Great Yarmouth. In earlier times it was placed by the church gate, where it acquired the name of 'The Devil's Seat' and was thought to bring disaster to anyone who sat in it. Once inside the church, however, it lost this reputation and the custom arose of each newly-married couple racing to sit in it, in the belief that the one who did so first would rule the household.[190]

Today, the coastal authorities report all strandings to the Natural History Museum and are thus a valuable source of information, although the ownership of stranded whales and, rather more importantly, the responsibility for disposing of several tons of rotting flesh is legally complex.

Although the term 'whalebone' is used erroneously for articles made from baleen (see below), real whale bone, and especially whales' teeth, sometimes called 'whale ivory', has provided the material for ornamental carvings, in much the same way as elephant ivory. A characteristic art form that is associated uniquely with whale bones and teeth, especially Sperm Whale teeth, is known as 'scrimshaw'. The name is mid-nineteenth century and may originally have been someone's surname, but the art originated among the whalers, who spent their free time at sea engraving pictures on pieces of bone and teeth, later selling them as souvenirs. Old scrimshaw can be still be purchased at antique shops but it is easily imitated and it is almost impossible to distinguish genuine old scrimshaw from the copies. Cheap souvenir shops offer even cruder versions made of plastic.

Beaked Whales

Northern Bottle-nosed Whale *(Hyperodon ampullatus)*
Rather like the dolphins, the beaked whales have an elongated beak and this is especially pronounced in the appropriately named Northern Bottle-nosed Whale. Any large whale with a beak in British waters is most likely to be this species, although it has become much less frequent in recent years.

Rorquals
(Balaenopteridae)

'Rorqual' is a name derived from Norwegian and Old Icelandic *röhrval*, meaning 'red whale'. The rorquals are strikingly streamlined whales and have characteristic deep grooves in the skin, stretching back from just under the jaw. The family includes some very large species, most notably the Blue Whale itself.

The baleen plates, the mistakenly called 'whalebone', proved a useful by-product of the whaling industry before the invention of plastics and contributed to the decorum and appearance of many a society lady. From the early seventeenth century, whalebone 'stays' were, literally, the foundations of women's corsetry, thin strips of the pliable material being sewn into the cloth to mould and shape the body. In a hilarious debate during the discussion of the Wild Creatures and Forest Laws Bill in the House of Lords in 1971, it emerged that, according to an old tradition, a whale taken as a 'Royal Fish' by the Crown should be divided between the King and Queen, the head being the King's property and the tail the Queen's. The reason for this 'whimsical division' was that the Queen would require the whalebone for the royal corsets. Unfortunately, with whalebone originating in the whale's mouth, it would in reality have fallen to the King instead.(191)

'Whalehide' is another word that has survived when the original substance has passed from use. Gardeners can still buy flexible whalehide plant pots but fortunately today no whale has contributed to their manufacture as they are produced from paper impregnated with bitumen.

Fin Whale *(Balaenoptera physalus)*
OTHER COMMON NAMES: *common rorqual; razorback.*
The Fin Whale, or 'razorback', with its pronounced dorsal ridge between the fin and flukes, has a coloration that is extremely unusual among large animals because of its asymmetry: the right side of the lower jaw is white and the left side dark.

Sei Whale *(Balaenoptera borealis)*
OTHER COMMON NAMES: *Rudolphi's rorqual.*
The Sei Whale takes its unusual name from a small fish called *seje*, which commonly forms part of its diet. Its alternative name, 'Rudolphi's rorqual', honours a German Professor of Anatomy K.A. Rudolphi (1771–1832).

Minke Whale *(Balaenoptera acutorostrata)*
OTHER COMMON NAMES: *lesser rorqual; piked whale.*
The Minke Whale, at only about 8 metres ($8\frac{3}{4}$ yards) in length, is the largest whale that you are likely to see from a British shoreline; there is a chance of spotting one from almost any headland on the west coast of Scotland, Ireland, Orkney or Shetland. 'Minke' is derived from a Norwegian word meaning 'lesser' and is not connected with Mink (see page 416).

Even-toed Ungulates
(Artiodactyla)

Pigs
(Suidae)

Generally, domesticated pigs are called 'pigs' and their undomesticated relatives are called 'boars', which can cause confusion because male pigs, wild or not, are also called 'boars' while all females are called 'sows'. They are considered by zoologists to be the most primitive of their group, the even-toed ungulates (or Artiodactyla). 'Ungulate' is from Latin and means 'hoofed'. There are also odd-toed ungulates (or Perissodactyla), but it is a long time since any horses and rhinoceroses were wild in Britain, and the third members of this group, the tapirs, have never occurred in this country.

Wild Boar *(Sus scrofa)*
OTHER COMMON NAMES: *wild pig; wild swine.*
The European Wild Boar is unlikely to be mistaken for any domestic breed of pig because it has a long, bristly brown coat with thick brown underfur, a very strong smell and, most impressively, two upwardly pointing canine tusks in the male, which give it a very threatening appearance. While I have never come face to face with one in the wild in Britain, I have done so when alone in remote parts of forest in other countries and can reliably report the experience to be pretty unnerving.

The Wild Boar was once one of the prime beasts of the chase and still is in many parts of Europe; in Germany, for instance, 200,000 animals are killed each year. In Britain, they were once treated in much the same way as deer and protected for hunting; there is considerable evidence that Wild Boar were similarly enclosed in parks and, because they are unable to leap over walls and fences like deer, presumably they stayed put more readily. So valuable were they considered that, in the reign of William I, the illegal killing of deer and Wild Boar was punishable by blinding. Today, when the term 'venison' tends

The Wild Boar *(Sus scrofa)*, one of the supreme beasts of the chase, became extinct in Britain around the thirteenth century.

to be restricted to the meat of the deer, it is forgotten that it was originally applied to the meat of Wild Boar and other beasts of the chase as well. The word comes from the same source as 'venery', or 'venerie', meaning 'hunting' (see Fox, page 404), from the Latin *venari*, 'to hunt'.

Wild Boar were important members of the woodland ecosystem, feeding on acorns, beech mast and other fruits and nuts, as well as bulbs, rhizomes and other vegetation, which they unearthed in a crude way, greatly disturbing the soil in the process. This ability of pigs to unearth subterranean items has found a well-known outlet in France and elsewhere, where trained pigs are used to search for truffles, although it has been exploited very little in England.

Oliver Rackham believes that the Wild Boar became extinct in England in the thirteenth century and that the many later references to boars being hunted and eaten is simply evidence of their having been reintroduced. In favour of this argument, he points out that, while records exist of Wild Boar being supplied from the Forest of Dean to Henry III's court until 1260, the accounts then stop.[192] There seems no argument that James I brought boar from France in 1608 and from Germany three years later, and Charles I certainly attempted to reintroduce them to the New Forest, although they were destroyed during the Civil War. It is possible, nonetheless, that some native British stock was also kept in parks, at least as status symbols, after they had become extinct in the wild.

The boar has featured extensively in English heraldry: it was the emblem of Edward III and, even more famously, of Richard III. Today, it is used as a rallying symbol by the Richard III Society, which seeks to place the king in an accurate historical context and restore his name, which was so blackened by the Shakespearean play. Largely in consequence of its heraldic use, there are a good many Boar inns, and none more famous than the Boar's Head in Eastcheap, where Sir John Falstaff and his chums made merry.

The Wild Boar *(Sus scrofa)* is featured on the coat of arms of Richard III (reigned 1483–85) and his wife, Anne Neville.

There is a frequently repeated story of the Earl of Oxford being killed while hunting boar in 1395. This would have been particularly unfortunate as a blue boar was the symbol of the Vere family, the Earls of Oxford, but the event seems unlikely as no Earl of Oxford appears to have died from any cause in that year. Oxford is, nonetheless, the home of one of the most splendid of Wild Boar traditions, the Boar's Head Gaudy, which is still held annually at Christmas at Queen's College. The event is believed by the College to date back to 'sometime during the Middle Ages' and the earliest version of 'The Boar's Head Carol' is in a printing by Jan van Wynkyn (Wynkyn de Worde) dated 1521. Until the nineteenth century, the event took place on Christmas Day but as Dr Robert Taylor, Physics Professor at Queen's, explains, 'modern family pressures dictate that the gaudy now takes place in the run up to Christmas instead'. A former Provost of Queen's gave a full account of the ceremony in 1923:

> Before dinner on Christmas Day the Boar's Head is brought in procession into the College Hall. At the hour appointed the Provost and Fellows in residence, with any guests who may have been invited, enter the hall and arrange themselves on the east side of the high table facing the door. Grace before meat is said and the trumpet sounded in each quadrangle as summons to dinner. The procession then enters the hall. The head, borne on the silver basin, presented by Sir Joseph Williamson in 1668, is carried by four servants, conducted by the chief singer, generally a member of the College, and followed by the choristers under the direction of the College organist. As the procession begins to move from the cloister into the hall the choir sings the refrain:
>
> *Caput apri defero* [I bring the boar's head],
> *Redens laudes Domino* [Sing thanks to the Lord].
>
> The Boar's Head in hand bear I,
> Bedeck'd with bays and rosemary.
> And I pray you, masters, be merry,
> *Quot estis in convivio?* [How many are at the feast?]
>
> The Boar's Head as I understand,
> Is the bravest dish in all the land,
> When thus bedeck'd with a gay garland.
> *Let us service cantico* [Let us serve it while singing].

> Our Steward hath provided this,
> In honour of the King of Bliss,
> Which on this day to be served is,
> *In Reginensi Atrio* [Within the Queen's Hall].

> Between each verse the procession moves forward, the choir singing the refrain, and the fourth repetition of it brings the head up to the dais, where it is placed upon the table in front of the Provost. The chief singer is presented by the Provost with the orange which has till then been between the front teeth of the boar; and the bays, rosemary, and holly, of which some of the sprigs are gilt, are distributed among the spectators.[193]

I was recently told, in hushed tones, a professional secret: that the Boar's head is boiled upside down in a large saucepan and that weights are attached to its ears to prevent them from flopping when cooked.

And why do they do it? According to tradition:

> ... a student of that college, who, while walking in the neighbouring forest of Shotover, and reading Aristotle, was suddenly attacked by a wild boar. The furious beast came open-mouthed upon the youth who, however, very courageously and with a happy presence of mind, is said to have rammed in the volume and cried '*Graecum est*', fairly choking the savage with the sage.[194]

There is speculation that the student may have been Barnard Gilpin (1517–83), who was certainly a member of the College and whose family had a boar's head on its coat of arms. Or perhaps it was a man called Copcot, whose name can be found on a small stained-glass lancet window facing the door in the church of St Giles at Horspath, near Oxford, below a scene depicting an elderly man in the garb of an apostle holding a spear on which a boar's head is impaled.[195] The date of the church window seems wrong, however, and I prefer the Gilpin version, even though he would have been only four years old at the time when the Boar's Head Carol was printed. Perhaps his exploit was simply grafted onto an earlier College association with boars.

In recent times, the Wild Boar has loomed large again in England, and not only at Queen's College. I can recall recipes for pork suggesting that cooking it in cider would help to re-create the taste of Wild Boar by imparting the flavour of the wild fruits on which the boar would have fed. Now we have the real thing because Wild Boar are being farmed in England for their meat. The British Wild Boar Association was formed in 1989 and about 40 farmers are registered, some rearing Wild Boar/Domestic Pig hybrids, but others pure Wild Boar stock, originating mainly from eastern Europe. Almost inevitably, some Wild Boar have escaped and it is estimated that between 100 and 250 are living free in several areas of southeast England, where they are known to damage crops, kill lambs and cause other hazards.

The stained-glass window at St Giles, Horspath, depicts a boar's head impaled on a spear.

> There have been reports of several accidents involving cars ... on the A268 in East Sussex. The local garage repair shop manager confirmed that he had repaired at least three Wild Boar damaged vehicles in November and December 1996 ... an Austin Metro was considerably damaged.[196]

Identifying places historically associated with boars is not easy because an apparently obvious clue, the Old English word *swin*, meant 'domestic pig', and so rules out such places as Swinton in Lancashire ('pig farm') and Swineshead in Bedfordshire ('pig hill'). *Eofor* meant 'Wild Boar'; so Evershaw in Berkshire was probably 'Wild Boar copse' and Everdon in Northamptonshire 'Wild Boar hill', while there seems little doubt about Wildboarclough in Derbyshire.

Groups of Chinese Water Deer *(Hydropotes inermis)* have become established in and around the Norfolk Broads.

Deer
(Cervidae)

You cannot get away from deer in Britain. Once in a while you will see real wild or semi-wild animals (up to seven species if you look in the right places), but almost everywhere you can find inns called The White Hart or The Stag's Head, mounted heads in the halls of country houses, parts of deer on the top of walking sticks in tourist shops, venison sausages in selected butchers, handsome beasts in full Highland splendour in every municipal art gallery and many a distinguished head on many a municipal crest. All of this reflects an age-old fascination with a group of animals that have been admired, kept, hunted, shot and eaten for centuries.

The deer family includes the only truly wild British hoofed mammals and also the largest British land animals. They share a number of features with many related hoofed creatures, including an ability to run very fast and to ruminate (chew the cud). Most deer (the introduced Chinese Water Deer, *Hydropotes inermis*, being one of the few exceptions) have one unusual and special feature: they produce antlers. These are extraordinary structures, superficially like horns but shed each year, the succeeding year's crop usually being larger than the last, at least until the animal is in its prime. Initially, antlers are covered with a skin and fine hairs, known as 'velvet', which dies and falls away when they finish growing. The anatomy of antlers and how they are formed and shed are both well known but why deer should produce something that drops off annually, whereas their close relatives, cows, keep their horns permanently, is still a matter for speculation. Antlers are used, just as horns tend to be, for trials of strength between males; the Reindeer *(Rangifer tarandus)* is the only species of deer in which both sexes possess them.

Deer are also unusual in being referred to in lore and tradition by such a wide range of names. They are called 'stags', 'harts', 'hinds', 'does', 'bucks', 'calves' and 'fawns', almost anything except 'deer', although that word itself has a curious origin in Old English. Originally it simply meant a 'beast' and was not applied specifically to the animals we call deer until later. Nonetheless, it surely demonstrates that the Red Deer, *the* beast, was an animal of enormous importance at that time. Among the intriguing and wide vocabulary that has become attached to deer, mainly through their importance in hunting, is the word 'slot' applied to their hoof prints. This is a sixteenth-century word, from the Old French *esclot*, 'a horse's hoof-print', and was perhaps originally Scandinavian.

Chinese Water Deer *(Hydropotes inermis)*

This is one of only two species of deer with no antlers but the two slightly downward-pointing canine tusks in the male are very distinctive. They are shy, dog-sized deer from China, reddish-brown in summer but pale grey-brown in winter. A few escaped from a herd at Whipsnade during the Second World War and others subsequently absconded from various private collections. Groups are now established in the Norfolk Broads and some adjoining counties, but it seems unlikely that they will ever spread very far. Indeed, they may already be declining because they are vulnerable to cold winters and to being taken by Foxes.

Reeves' Muntjac *(Muntiacus reevesi)*

OTHER COMMON NAMES: *barking deer; Chinese muntjac.*

Surprisingly, this little deer from the subtropical forests of South-East Asia has not only survived but also become established in Britain. It is a truly tiny beast, rich red-brown in summer and darker brown in winter (and presumably also changes colour with the seasons in its native land). The antlers are appropriately tiny and unbranched.

Like many another exotic mammal now living wild in Britain, it is an escape, initially from the Duke of Bedford's Park at Woburn, where it was brought first in 1894. (There is evidence that the 'escape' may have been supplemented by intentional releases; Derek Yalden pointedly suggests there was deliberate obfuscation of the affair).[197] They were first seen in the wild in the 1920s and have since extended their range, probably supplemented by other escapes and human assistance until they are now present across a wide swathe of the English Midlands, East Anglia and the south of England, with isolated pockets elsewhere.

Reeves' Muntjac *(Muntiacus reevesi)* is well established in most southern counties of Britain and is spreading northwards.

> It is quite an achievement to see a muntjac but one can often hear their raucous, eerie bark which is loud and repetitive.[198]

Although they live in pairs or as isolated individuals rather than in herds, they can be a great nuisance if they enter rural areas or gardens and have also caused significant damage to vegetation in nature reserves. In February 2001, one that had lived for some time roaming free in the centre of Coventry was finally cornered in the men's lavatory of a multistorey car park.[199]

> In my view, the muntjac, now living wild even in London, is the ungulate equivalent of the Brown Rat.[200]

There has been more than a little confusion because a number of Indian Muntjac *(Muntiacus muntjak)* were thought to have escaped from Woburn and hybridized with their Chinese cousins in the wild but this now seems improbable.

The name 'muntjac' originates in Indonesia, but exactly where is uncertain. Options are the Javanese word *mindjangan*, meaning 'a deer', or 'a Sunda Straits dialect word' perhaps meaning 'springing', or 'graceful'. The specific name *reevesi* refers to John Reeves (1774–1856), one-time Inspector of Tea for the East India Company, who found, described and gave his name to a wide range of South-East Asian fauna and flora, such as Reeves' Pheasant (see Partridges, Quails and Pheasants (Phasianidae), page 261) and the plant genus *Reevesia.*

Red Deer *(Cervus elaphus)*

Perhaps it is poetic justice that this magnificent animal, the largest British land mammal and the second largest European deer, should have been blessed with such a tautological scientific name: literally, it translates as 'deer deer' because both the Latin *cervus* and the Greek *elephus* mean 'deer'. Although large by comparison with other native mammals, Red Deer in Britain are not nearly as large overall, nor in the spread of their antlers, as some Continental races, especially those from Germany; moreover, the animals that so characteristically strut the Scottish Highlands are smaller than English animals.

The Red Deer is a true British native but its range has been very much reduced and today it is found mainly in the Scottish Highlands, on Exmoor and in the Quantocks. Nonetheless, its numbers are increasing and there are now probably around 300,000 animals in Britain. However, except in some Scottish populations, the purity of the stock has been diluted by hybridization with the introduced Sika Deer *(Cervus nippon)*, and it has been suggested that, in the future, pure-bred Red Deer may survive only on the Scottish islands.

A Red Deer (*Cervus elephus*) stag against a backdrop of mountains and a stormy sky is an impressive sight.

The food tends to vary with habitat. In woodland, Red Deer browse on the shoots of trees and shrubs (causing much damage in the process); on moorlands, they graze on heather, grass and other vegetation.

The Red Deer is a rich red-brown animal, almost invariably without spots except when young. The male is known as a 'stag' and the female a 'hind. 'Stag' is Middle English, and perhaps Old English for any male animal in its prime, although it has always been applied especially to the Red Deer. A 'hart' (Middle English, perhaps associated with 'horned') is a male deer, especially a Red Deer after its fifth year, but it is now rarely used and 'stag' is the more general word. In old place names, however, 'hart' is much more common; Hartanger in Kent ('hart slope'), Hartfield in Sussex ('an open area frequented by harts') and Harthill in Cheshire ('a hill frequented by harts') are among numerous examples. Ironically, one of the few 'stag' place names, Stagenhoe ('a spur of land frequented by stags') is in Hertfordshire ('hart's ford'). 'Hind' is Old English and may perhaps be Greek in origin. Hindhead in Surrey is probably 'a hill frequented by hinds'. Young Red Deer are called 'calves' and a male in its second year with its first antlers is a 'brocket', although other terms are also used.

There is an extensive lore relating to the Red Deer and also a large, complex vocabulary, principally because it has long been an important beast of the chase. The Normans kept Red Deer with other animals in their hunting parks and forests.

> Peasants were, by law, forbidden to hunt larger animals. No layman with an income of less than 40s. per year, or priest receiving less than 10s. per year, was allowed to keep hunting dogs, nor were they entitled to use 'fyrets, heys, nets, harepipes nor cords nor other engines for to take or destroy deer, hares nor conies nor other gentlemens game.' If they had been caught, the penalty was a year's imprisonment.[201]

At Cranham in Gloucestershire, an annual three-day feast and roast, ending on the second Monday in August, is thought to commemorate an event on some date long forgotten, when the villagers asserted their right to common land by roasting a deer on it in the presence of the lord of the manor. An ancient right of peers to take a deer on their way to Parliament after sounding a horn was considered as an option by Lord Hailsham when Lord Chancellor. He reported that he once told his father:

> 'If ever I succeed to your peerage, I shall shoot a deer in Richmond Park, blowing a motor horn first'. My father went into the matter and told me the privilege was wholly obsolete and, so far as he could ascertain, had never been exercised.[202]

Over the centuries, other interests, such as the requirement for oak timber to supply the Navy, grew to compete with the needs of the deer.

> I mean the red-deer, which toward the beginning of this century amounted to about five hundred head, and made a stately appearance. There is an old keeper, now alive, named Adams, whose great-grandfather (mentioned in a perambulation taken in 1635), grandfather, father and self, enjoyed the head keepership of Wolmer Forest in succession for more than an hundred years. This person assures me, that his father has often told him, that Queen Anne, as she was journeying on the Portsmouth road, did not think the forest of Wolmer beneath her royal regard. For she came out of the great road at Lippock, which is just by, and, reposing herself on a bank smoothed for that purpose, lying about half a mile to the east of Wolmer Pond, and still called Queen's Bank, saw with great complacency and satisfaction the whole herd of red deer brought by the keepers along the vale before her, consisting then of about five hundred head. A sight this, worthy the attention of the greatest sovereign! But he

> farther adds that, by means of the Waltham blacks, or, to use his own expression, as soon as they began blacking, they were reduced to about fifty head, and so continued decreasing till the time of the late Duke of Cumberland. It is now more than thirty years ago that his Highness sent down an huntsman, and six yeoman-prickers, in scarlet jackets laced with gold, attended by the stag-hounds; ordering them to take every deer in this forest alive, and to convey them in carts to Windsor.[203]

From the Middle Ages, therefore, Red Deer began to disappear from much of England and were, in any event, being replaced as a sporting favourite by the introduced Fallow Deer *(Dama dama)*. The popularity and importance of deer meat, however, should not be underestimated. Consider, for instance, the marriage of Margaret, the daughter of Henry III, to Alexander III of Scotland at York on 26 December 1251:

> Planning began at least by the summer, and by the end of July, beasts were being bought in York and other fairs, with orders for them to be pastured until needed. They were not slaughtered until just before the wedding. At about the same time orders for the catching, slaughtering and salting of 300 red and fallow deer were issued. In November nearly 1,000 more roe, fallow and red deer were ordered ... .[204]

The methods used to hunt and kill deer, of all species, also changed over the centuries.

> Before the advent of the express rifle the deer were hunted with deerhounds, and this amounted in many cases to a military operation. While the gentry slept, several hundred of their retainers were out in the darkness taking up positions over a wide area of country to drive as many of the deer as possible to a central point in the bottom of the glen. The drive would commence at first light and by the time the gentry had breakfasted and taken up their positions in hiding in the glen bottom the herdsmen moving the deer downwind were closing in. With most of the deer on the flat the ambush was loosed and they were run to ground and killed by the powerful hounds. Guns, arrows and daggers were used by those who could get close enough, and in the high spirited free-for-all the carnage was wrought. Probably ten times more escaped than were killed, but there was no refinement in the choice of those which were killed. The herd was savagely mauled with stags, hinds and calves being killed indiscriminately ... During the second world war large numbers of deer were slaughtered indiscriminately to supply meat for the lawful and 'black' markets. The

Edward, Prince of Wales, and Lord Cork, with the Royal Stag Hounds in 1885, in a painting by George Bouverie Goddard (1832–86).

means of killing by spotlighting the herds by the roadside in the winter night and shooting them down with all manner of shotguns and rifles was a repetition of the cruelty of the deerhound ambushes.[205]

Today, the Red Deer is again a prime sporting animal, although much less as a beast of the chase than as a target for the precision stalker. The traditional stag hunt still survives in England but has declined greatly in recent years, not least because the National Trust has banned hunting on its land. This was motivated partly by scientific evidence presented by Professor Patrick Bateson in a study commissioned by the National Trust, which showed that, by the end of a hunt, the deer are suffering considerably. Although there remain arguments about some of the findings, it is now generally agreed that:

> ... towards, or at the end of a chase ... there is clear evidence of very low levels of carbohydrate (glycogen) in the deer's muscles and ... this largely explains why the deer ceases running.[206]

The feats that a hunted stag will perform to escape its pursuers are legendary and in many parts of the country there are places called Stag's Leap or Deer Leap, which claim to commemorate some such gigantic bound across a gorge or other obstacle. There are still three registered packs of stag hounds in the West Country but they account for only about 160 Red Deer each year, estimated by Lord Burns Committee of Inquiry into Hunting with Dogs (see Fox, page 404) to be about 15 per cent of the numbers that need to be culled in the area to maintain a stable population. The Committee also believed that:

> Stalking, if carried out to a high standard and with the availability of a dog or dogs to help find any wounded deer that escape, is in principle the better method of culling deer from an animal welfare perspective. In particular, it obviates the need to chase the deer in the way which occurs in hunting.[207]

This selection of trophies, displayed on one of the walls at Glaslough Castle in Ireland, includes a magnificent set of antlers.

Today, the management of Red Deer populations is a careful balancing act between maintaining a biologically sustainable population and satisfying the demand for sport and venison while at the same time protecting forests and woodlands from damage.

Although historically 'venison' had a wider meaning (see Wild Boar, page 443), the term is now used almost exclusively for the flesh of deer. It is moderately popular with the meat-buying public, although many potential customers are dissuaded from trying it because of a belief that it is strongly flavoured and 'gamey'.

Venison roasts feature in a number of rural diversions and fairs but none is more exclusive than the Hanley Venison Festival in Stoke-on-Trent, which originated in 1783 under the aegis of the Ancient Corporation of Hanley.

> The Duke of Sutherland supplied the venison for the first feast from his estate at Trentham and it is said to have been supplied from Trentham for every festival since ... There is a waiting list of potential members which is by invitation only ... not on a first come first served basis ... A traditional local venison song goes as follows:
>
> Some ancient men did once agree
> To have a day for mirth and glee
> Which they agreed upon should be
> A day of feasting.
>
> Chorus:
> Then let us sing
> Long live the Queen
> The Mayor, long live he,
> Next year may we meet again
> In peace and harmony.
> And thus these good men and true
> A corporation formed for you
> To cheer the men of Hanley through
> Centuries of feasting.[208]

Rutting Red Deer stags *(Cervus elephus)* put their antlers to good use during this battle to establish supremacy and mating rights.

But do the venison-roasters of Stoke-on-Trent adhere to the tradional advice?

> Rosted Veneson must have vinegre sugre and sinamon/and butter boyled upon a chafyng dishe with coles/but the sauce maie not be to tarte and then laie the veneson upon the sauce.[209]

While the goal of the deer-stalker is partly the meat, part is the trophy afforded by the head with its antlers – and the more branched the antlers, the better. The terminology of antler branching is complex and the different numbers and patterns of terminal spikes are all named. Many British Red Deer stags have more than 10 points; a 'Royal' has antlers bearing 12 points, and this tends to be the maximum attained in Britain, although the age at which it occurs varies; English Royals are generally younger than Scottish. Elsewhere in Europe, up to 24 points may be found. Occasionally stags have no antlers and these too have special names: 'notts' in the west of England and 'hummels' in Scotland. Contrary to popular belief, they can still sire young.

Antlers are used by the stags in fearsome trials of strength during the mating ritual, or 'rut' (a Middle English word from the Latin *rugire* meaning 'to roar'). A large rutting Red Deer stag is an awesome sight as it bellows defiance at other stags and protects its harem of hinds.

> As the night fell I crawled out to the no-man's land between the two territories. After only a short wait the larger of the two beasts roared once, ran twenty yards, roared again, and ran on to stop only twenty feet from me. The stag, standing so close that I could hear his breathing, then moved away to challenge his rival from a few yards before sweeping back like a gust of wind to re-shepherd his wandering wives.[210]

An enraged or injured stag can be a dangerous animal: Richard, Duke of Bernay, the elder brother of King William II (who was himself later killed, perhaps not accidentally, by a hunter's arrow in the New Forest), was gored to death by a New Forest stag. Rutting is generally perceived as one of the more blatantly sexual mammalian performances and the word has, unsurprisingly, been applied to young men of comparable demeanour; and, of course, the human ritual of the stag night is hardly lacking in sexual overtones. I do wonder however why the female equivalent is not a 'hind party' but a 'hen party'; perhaps it is a case of the word becoming corrupted.

Close inspection shows that Sir Edwin Landseer's *Monarch of the Glen* is a 'Royal' stag, with twelve points to its antlers.

The hornless hart carries off the harem.
Magnificent antlers are nothing in love.
Great tines are only a drawback and danger
To the noble stag that must bear them.

Crowned as with an oak tree he goes,
A sacrifice for the ruck of his race,
Knowing full well that his towering pointsy
Single him out, a mark for his foes.

Yet no polled head's triumphs since the world began
In love or war have made a high heart thrill
Like the sight of a Royal with its Rights and Crockets,
Its Pearls, and Beam, and Span.
(Hugh MacDiarmid, 'The Royal Stag')

The Red Deer, especially the stag, has featured in numerous artistic creations and deer-stalking has been the background setting for many a stirring tale, and never more famously than in John Buchan's 1925 novel *John Macnab*, in which the stalking of a stag in secret is the central thrust of the plot.[211] However, no deer can be more celebrated in British culture than the one in Sir Edwin Landseer's archetypal Victorian painting of a Royal stag, *Monarch of the Glen*, copies of which hang on countless drawing room walls. The word 'deer-stalker', as applied to the close-fitting tweed hat that people stalking deer are supposed to wear, is as recent as 1881. It had been in use for only six years when its most famous wearer, Sherlock Holmes, first appeared in the *Strand Magazine*.

In the Scottish Highlands, deer were called 'fairy cattle' and were believed to be milked by fairies on the mountain tops. Deer were also believed to be among the animals that shed tears (actually an oily secretion from the eyes) and to be adept at healing their injuries. As Christopher Marlowe wrote in *Edward II*, 'The forest deer being struck runs to an herb that closeth up the wounds.'

Parts of Red Deer, such as heads with antlers, feature in some Morris dancing displays, although the famous Abbots Bromley Horn Dance, which is far older than Morris dancing, involves a quite different species (see Reindeer, page 456). There is still a popular children's playground game in which one child chases the others until he manages to touch one; the two then join hands, continue the chase and round up the remaining children. I played this game, which I think we called 'tig', as a child in Derbyshire, but was quite unaware that it was once called Stag Warning and that the first chaser was known as 'the stag'. The last one left free at the end would become the new stag and all would then sing:

Stag warning, stag warning
Come out tomorrow morning.

Or:

Stig, Stag, a rumping, rumping stag,
The first one as I catch I'll put in a bag.[212]

The 'White Hart' (as in the name of many English inns) probably dates from the time of Richard II, whose emblems were a swan and a collared antelope, although a white antelope with a golden collar is a depiction of an animal that is variously said to have been caught by Alexander the Great or by Charlemagne. The antelope then became the White Hart, which was used as an emblem by Henry V and Edward IV, among others. This appears in a number of medieval illustrations, most notably on the panel painting known as the 'Wilton Diptych' (English or French, *c.*1395–99), which depicts the accession of Richard II. In Hampshire, a more specific tradition links the White Hart with Henry VII:

According to tradition, while the King and his court were visiting Ringwood a hunt in the New Forest was arranged. Henry and his companions quickly brought down several of the deer driven before them, but a renowned white hart named Albert gave them a magnificent run before it turned to bay in the meadows by the Avon. The ladies of the party requested that the hounds be whipped off and the noble stag decorated with a collar of gold. To commemorate the event it is said that the inn where Henry lodged hung out a portrait of the gallant beast resplendent in its golden collar and chain. Moreover, the White Hart inn of Ringwood claimed to be the first to bear this now famous sign of the collared white hart.[213]

There are innumerable instances of Red Deer, generally stags, being used as heraldic emblems and they have also found their way onto several municipal crests:

The deer's head in the arms of Windsor, based on a fifteenth century seal, and the stag and oak-tree forming the crest of Berkshire, represent the forest land where the Norman kings were wont to hunt ... The arms of Rawtenstall, commemorating the Forest of Rossendale, contain a squirrel, a wolf, and a red hand cut off at the wrist – an allusion to the penalty for killing the king's deer.[214]

One of my early childhood memories is of being taken to Derby and travelling on a trolley bus. The buses were painted green and, at the bus stop, the old Derby Corporation crest emblazoned on their sides was at my eye-level. This crest featured a seated Red Deer, known as a 'Stag at Lodge', or more colloquially in Derby as 'The Buck in the Park'. In *c.*1760, the animal was the inspiration for one of the early figures produced by William Duesbury at the old Derby porcelain factory, which was very appropriate as the name 'Derby' is from the Old Norse 'a place where deer were seen'.

***The Vision of St Eustace* shows the patron saint of hunters converted to Christianity by seeing a stag with a crucifix between its antlers.**

The parish church at Tavistock has a Red Deer stag's head with a crucifix on its head. This is known as the vision of St Eustace (a vision also attributed to St Hubert and St Ignatius) who, when hunting in the forest, saw a vision of a Red Deer stag with the crucifixion between its antlers, making him turn to Christianity. This symbol is also on a heraldic shield at the Dean's House, Exeter Cathedral.[215] [St Eustace and St Hubert (d.727) are both patron saints of huntsmen.]

Sika Deer *(Cervus nippon)*

This close relative of the Red Deer originates from eastern Asia and can be distinguished by its smaller size and the possession of some whitish spots on its red-brown coat. Three different geographical races have been kept in collections in Britain but only the Japanese form ('Sika' is a Japanese name) seems to have become established in the wild. London Zoo obtained a pair in 1860 and, at the same time, four animals were acquired by Viscount Powerscourt for his estate in County Wicklow. From this collection, animals were then sent to many other estates throughout Britain and inevitably some escaped. In some places, animals were also deliberately released to provide sporting quarry. The most dramatic burst for freedom came from a herd that was kept on Brownsea Island in Poole Harbour in 1896 and that promptly swam to the mainland (many deer are excellent swimmers); these were largely the source of the present population in the conifer plantations of the area.

The most extensive populations of Sika Deer, however, are in the coniferous forests of Scotland, where they not only cause considerable damage to the trees but have also extensively hybridized with native Red Deer (see page 447), seriously diluting the latter's genetic purity. The situation almost certainly cannot be halted, not least because many of the hybrid animals are externally indistinguishable from pure Red Deer. Interestingly, the Sika might have been expected to compete with another introduced deer species, the very similarly sized Fallow Deer, but this does not seem to have happened because, as Derek Yalding has pointed out, the two just do not meet:

> ... in the New Forest, Fallow Deer live north-west of the London to Bournemouth railway, and Sika to the south-east.[216]

Autumn is the mating season for Sika Deer:

> A visitor to the New Forest in October may hear a strange rising and falling whistle followed by a grunting sound. This is the Sika stag's rutting or mating call. It is quite unlike the call of any other British mammal and is mostly heard in the early morning and at dusk. The stag may call three or four times each quavering whistle lasting about five seconds. There follows a long period of silence from ten minutes up to an hour ... It was certainly one of the most difficult wildlife sounds that I have ever recorded, since if you miss the first call the next one may be too far and anywhere in relation to the proper angle of the reflector microphone ... Coming as it does suddenly from a concealed animal I found the whistle quite startling and it always made the hairs on my neck stand up.[217]

Fallow Deer *(Dama dama)*

The Fallow Deer is probably the most familiar deer, especially in England. It is the predominant species in most British parks and great estates and is the large deer that is most likely to be seen in 'the wild'. A beautiful animal, it is smaller than the Red Deer, mainly chestnut brown in summer, more grey in winter, and usually with an abundance of white spots, although the colour ranges from black through various browns to white. The black form is said to be descended from some animals presented to King James I by the King of Norway. Their most characteristic feature is the palmate (partially flattened) antlers in the mature male.

The Fallow Deer is native to southern Europe and once occurred naturally in Britain, disappearing after the last Ice Age. The date of its reintroduction to Britain has long been debated: the Romans were once considered to have been responsible but many zoologists now believe that all contemporary archaeological remains are of other deer, especially Roe, and that the Normans are the most likely source. As Red Deer are called 'stags' and 'hinds', so Fallow Deer are known as 'bucks' and 'does', and their young as 'fawns'. Both 'buck' and 'doe' are Old English words, but their use during the Saxon period is certainly no proof that the terms were referring to Fallow Deer because they could equally have been alluding to the native Roe. Certainly, however, by the late fifteenth century the introduced species was in need of a

Sika Deer *(Cervus nippon)* have become established in Britain, particularly Scotland, where they hybridize with Red Deer *(C. elephas)*.

distinctive name and 'fallow', meaning 'pale-coloured' was adopted; the word has nothing to do with fields lying fallow.

Fallow Deer very soon outnumbered both native British deer. Derek Yalden quotes a typical medieval excavation at Launceston which yielded 439 Fallow Deer bones, compared with 53 Roe Deer and 49 Red Deer, while Oliver Rackham calculated that Henry III's annual requirements for his feasts between 1231 and 1272 were 607 Fallow, 159 Red and 45 Roe.[218]

One of the most famous herds of Fallow Deer in England is that at Magdalen College, Oxford, which probably dates from the very beginning of the eighteenth century; the earliest record in the College is of charges made for slaughtering deer in the deer park, which later, in 1705–7, became known as The Grove. According to custom, the herd should number about forty, the number of college scholars under the founder's constitution, although in recent years there has been something of a population explosion. Inevitably in an academic institution, traditions and stories have accumulated around and about the deer. Some tutors have found the noise of rutting 'not conducive to academic thought', while T.E. Lawrence, who was educated at Magdalen but later became a Fellow of All Souls, was said once to have plotted to rustle some deer and shut them in All Souls to enliven matters there. More recently, during a performance of Ben Jonson's *Volpone* in The Grove, some deer escaped through the open gates of New Buildings and had to be pursued down the High Street by actors in costume.

The name 'fawn', unlike 'fallow', has nothing to do with colour and comes from Old French and then the Latin *fetus*, 'an offspring'. The enchanting image of a young deer has become inseparable in many people's perception from 'Bambi', in the 1942 Walt Disney cartoon film. In reality, Bambi was a European deer, almost certainly a Fallow fawn, because Disney's film was based on the 1926 novel *Bambi: A Life in the Woods* by the Austrian author Felix Salten (born Siegmund Salzmann; 1869–1945), who became entranced by the wildlife he saw on a holiday in the Alps.[219] The emotional strings that are pulled in the public conscience whenever the idea of deer culling is discussed are known in shooting circles as 'the Bambi factor'.

Reindeer *(Rangifer tarandus)*

Outside the month of December, most people in Britain never give the Reindeer a second thought, which is a pity because it a remarkable beast, perfectly adapted to living in a cold climate. For people in cold climates, such as the Lapps, it has provided most of the basic essentials of life for centuries: meat, milk, hide and transport (the Reindeer is 'the deer that can be reined').

It ceased to be a British native sometime in the Mesolithic period and, apart from Father Christmas's Rudolph, it has not entered the country's lore and tradition. Nonetheless, it is back again: about 80 animals are wandering around the Cairngorms having been reintroduced over the past 50 or so years by the Reindeer Council of the UK.

The famous herd of Fallow Deer *(Dama dama)* at Richmond Park in London.

Participants in the Abbots Bromley Horn Dance, early in the twentieth century.

In one sense, it has never really been away, at least not in Abbots Bromley. On the Monday following the Sunday after 4 September, Wakes Monday, this Staffordshire village stages one of those truly wonderful old British customs that are the envy of other nations. The Abbots Bromley Horn Dance may originally have had some association with the nearby old Forest of Needwood, and the custom must surely have a long-forgotten connection with deer-hunting. What is undeniably strange, however, is that the 'horns' that take so pivotal role in the ceremony are not, as might be expected, the antlers of the native Red Deer but those of Reindeer. Most of them are mounted on wooden heads and are thought to be sixteenth century in origin but at least one set of antlers is much older and has been radiocarbon dated to 1065, plus or minus 80 years. No-one knows where the 'horns' originated or how they came to Abbots Bromley but, nonetheless, at 7.30 a.m. on the appointed day, the six sets of antlers are borne on short poles from the village church. Half of the antlers are painted white with brown tips and half are brown tipped with gold. The performance involves six dancers, a hobby horse, a Fool, Maid Marian, a boy with a bow and arrow (Robin Hood), a triangle-player and another musician. One local family, the Fowells, is said to have been involved with the dance for 400 years; it was first mentioned in 1532 and is thought to be the oldest traditional dance in Europe.

> A series of interwoven steps with dancers in a line is followed by the formation of a circle, and then the teams of dancers skip off in opposite directions. The stag dance follows when the dancers with antlers mime the fighting clash of rutting stags, advancing and retreating three times. The other characters snap their sticks or bows or bang a ladle in time with the music, which now is a variety of folk tunes, because the original 'genuine' Horn Dance tune was lost at the end of the 1880s. The dancers meet and cross over three times and 'fight' three times, then form a line and continue on their way, repeating the pattern countless times during a long day of dancing.[220]

Roe Deer *(Capreolus capreolus)*

Hanging on the wall of my study is a small, old wooden shield, which I rescued from a junk shop; the words 'Thetford Chase' are engraved on the bottom and attached to it is the upper

The most numerous of British deer, the Roe Deer *(Capreolus capreolus)* has a characteristic white rump patch and a large black nose.

part of a deer skull with a pair of three-pointed antlers about 16 centimetres (6 inches) long. They once belonged to a Roe Deer, a delightful small animal, itself only about 65 centimetres (26 inches) high at the shoulder and known rather unflatteringly as *Capreolus*, 'the little goat'.

Roe Deer are mid-brown with a characteristic white rump patch, a big black nose and a white chin. They are the only native British species apart from the Red Deer and are typically found in ones and twos or small groups at the edge of woodland, generally close to fields where they can cause damage among crops, especially cereals. The habitat needs of the three most important British deer are therefore significantly different: 'young thicket stage favours Roe Deer, more mature woodland is better for Red Deer, and open woodland with glades favours Fallow.'[221]

Roe are the most numerous of British deer with a population of around half a million animals, although in other parts of Europe, where they are targeted for venison, populations are much higher – about 700,000 are shot annually in Germany alone. The Roe:

> ... will make delicate meate, if your Cooke season it, lard it and bake it well.
>
> When your hounds have kild a Roe, the best man in the companie is to take the assay, which he must doo crosse ouer the tewell. Then must the hounds be taken away out of sight and small space distant for troubling the Huntsman, who must first slit the legges and cut them off at the first ioynt: then must he slit the throte downe the brisket to the nether end, and take the skinne cleane of: which done, he must slit his little bellie, taking out the panache with all bloud in the bodie, and lay it uppon the skinne with the foure feete. If any towne be neere hand you must send for bread, for the better reliefe of your hounds to be broken in the bloud.[222]

In Britain, Thetford Chase (the home of my set of antlers) and the surrounding area have been one of their smaller strongholds although the bulk of the population is in Scotland, the north of England and the south and west. In all areas, however, Roe, like almost all other deer, are expanding in numbers and there is genuine concern about the danger they pose to traffic.

'Roe' is from an old Teutonic word of unknown meaning but it is probably safe to assume that Rogate in Sussex and Reigate in Surrey are both 'a gate for Roe Deer'. Similarly, some old place names (ones that pre-date the arrival of the Fallow Deer with the Normans) containing 'buck' also refer to Roe: Buckfast in Devon is 'the stronghold of the buck' and Buckden in Yorkshire and Buckden in Huntingdonshire are 'the valley of the bucks'. Care is needed, however, because while *bucc* is Old English for 'buck', *bōcen* is Old English for beech.

Sheep, Goats and Cattle
(Bovidae)

Soay Sheep *(Ovis aries)*
Sheep have never been native British animals. Domestic sheep originated from a Eurasian animal called *Ovis orientalis* and they were taken by early man wherever he settled; the earliest remains of sheep in Britain have been dated to just over 5,000 years ago. Numerous breeds have been developed with characteristics suitable for particular climates and regions. Many have since died out or been superseded, but the Soay Sheep of Britain are widely thought to represent the survivors of perhaps the oldest domesticated breed in Europe; according to one view they were originally of Viking origin. Soay is a Hebridean outlier, an island in the St Kilda group, far out in the Atlantic off the west coast of Scotland, and the small, long-haired, horned sheep were described by Hector Boerce, who visited the island in the early sixteenth century, as being:

> ... So wild that they cannot be taken with a snare; their hair is long ... neither like the wool of a sheep or a goat ... with horns longer and thicker than those of an ox ... .[223]

After the evacuation of St Kilda in 1930 (see Fulmar, page 205), the Soay Sheep were left behind, and there they remain. Additional introductions have been made to some of the Welsh islands, to Lundy and to Ailsa Craig, as well as to collections on the mainland, although the sheep are now much less common than formerly in zoological gardens.

Feral Goat *(Capra hircus)*
It is unlikely that there have ever been wild goats in Britain and the earliest signs of them in archaeological remains seem to be those at Windmill Hill in Wiltshire, which are around 4,500 years old. In truth, goats have never featured greatly in this country, even as domestic animals, and they are marginal to any consideration of what constitutes the livestock of a typical British farm. As a result, their part in the folklore and traditions of Britain is small compared with that in the Mediterranean lands, where the domestic goat originated. They loom large in the Bible and references do turn up in British historical records to people petitioning for grazing rights for their goats; conversely, there are accounts of landowners, especially forest owners, not permitting goats on their property because of the damage they cause.

The Soay Sheep *(Ovis aries)*, which lives on the St Kildan island of Soay in the Hebrides, may be of Viking origin.

The Great Orme goat, the regimental mascot of the Royal Welch Fusiliers, 1889.

The omnivorous diet of goats is legendary and there is a common belief that they will eat anything. They are undeniably valuable domestic animals, providing most importantly meat, milk, hides, horn, wool and tallow for candles.

> ... The hill people believed that Goats were better able to tell them of coming bad weather by descending to lower levels, by killing adders and preventing abortion in cattle.[224]

That goats will trample and even eat Adders is well authenticated; the name of one species of Himalayan goat, the Markhor *(Capra falconeri)*, even means 'snake-eater' in Persian.

Eventually, however, goats declined further in importance in Britain as more productive and hardier breeds of sheep and cattle became available, and it is largely the descendants of discarded stock that survive on upland areas, especially in Wales and Scotland and some of the offshore islands. It is estimated that there are over 3,500 Feral Goats in Britain today, their wide range of colours, which include all possible combinations of brown, grey, black and white, reflecting the enormous number of domestic strains from which they have originated; pure white goats have always been especially prized.

Among the most celebrated Feral Goats in Britain now are the Great Orme goats that roam on the peninsula above Llandudno in north Wales.

> The Great Orme goat arrived in this country during the reign of King George IV from the Himalayas via Persia and into Paris ... it has become famous as the regimental Mascot of the Royal Welch Fusiliers, and was the original Royal herd of cashmere goats maintained by Queen Victoria in Windsor Great Park. The British history of this breed began in 1823 when a gentleman named Christopher Tower imported two pairs from Paris to his estate at Weald Hall in Essex. They provided enough cashmere to make a shawl and this was greatly admired by King George IV who was presented with a pair which were put into Windsor Great Park. Queen Victoria took a particular interest in the herd and set up an industry to produce yet more shawls. In 1889 Colonel Mostyn acquired a breeding group and took them to North Wales with the intention of providing regimental mascots for the Royal Welch Fusiliers. When the herd had increased sufficiently, he released them onto the Great Orme ... .[225]

Domestic Cattle *(Bos taurus)*

OTHER COMMON NAMES: *wild white cattle.*

The extinct prehistoric cow of Britain and elsewhere was the Aurochs *(Bos primigenius)*, which was hunted and eaten by early man in Britain but had died out by Roman times. However, other ancient cows, mysterious animals whose origins even now are unclear, are still in existence. These are the so-called 'wild white cattle' that lived, and live, in enclosed parks. Those at Chillingham Park near Alnwick in Northumberland are thought to be the genetically purest of the few surviving herds; the others have probably been crossed with Highland cattle and Longhorns.

Chillingham cattle are white, long-horned, unfriendly and certainly highly inbred; at one point after the Second World War, the herd was reduced to 13 animals. Although the earliest specific reference to them is dated 1646, there is good reason for thinking that they are considerably older, perhaps thirteenth century. They are always referred to as 'Wild' and always by the thirteenth-century Norman French word 'cattle', rather than the Old English 'cow'. 'Wild cattle' are referred to constantly in medieval and later accounts, often as hunting quarry in enclosed parks, and differentiated from other types of cow. For instance, at Archbishop George Neville's feast in 1465, the bill of fare included 'Oxen, one hundred and foure' as distinct from 'Wilde Bull, six' (see Bittern, page 212).

There is a notion that the Chillingham herd might be survivors from Roman times and, having seen herds of the rather similar, fearsome-looking white, long-horned Arezzo cattle of the Roman Campagna, I can understand the reasoning behind this. However, there is no real evidence that these modern Italian beasts originated where they now occur; indeed, they may have come from the area today called Hungary. The Roman link for Chillingham cattle therefore seems specious. Speculation on their origins continues but at least modern genetic studies have clarified some issues: it is now perfectly clear that they have nothing in common with the Aurochs and are probably the survivors of an early domestic breed of unknown origin. They owe their survival, according to Oliver Rackham, to 'being a Mediaeval status symbol', and they certainly afford a better glimpse than other cattle in Britain today of how some of the ancient beasts of these islands may have appeared.

Rodents

(Rodentia)

One in every four mammals is a rodent; rodents constitute the largest mammalian order and there are around 1,700 species in the world. This pattern is reflected in Britain where the 15 or so species outnumber any other type of mammal except bats. The group includes some of the most unpopular and yet most common and familiar of mammals, united by the possession of jaws with self-sharpening teeth that are designed to gnaw. The principal groups of rodents are the beavers, the squirrels and their allies, the voles, mice and rats, the dormice, the porcupines, the Coypu, and the birch and jumping mice, most of which have, or have had, at least temporarily, representatives in Britain. Because so many of them either live closely with human beings or have considerable impact on human lives, they have acquired a lore and literature that enchants and appals almost equally. The Rabbits and hares (see page 482) were once classed with the Rodents but they have been found to have very different teeth and may not even be closely related.

Squirrels

(Sciuridae)

Grey Squirrel *(Sciurus carolinensis)*

> In some places in the London district a light grey Squirrel may be seen, and thought to be a colour variation of our native species. It is really an American visitor, distinct in colour and without tufts to the ears ... British naturalists of a not-distant future will probably have to include two species of Squirrels in their lists.[226]

This was written by Edward Step in 1921 and how prophetic he proved. The Grey Squirrel is now so widespread and has proved so destructive to vegetation that it is commonly

Two-week-old Grey Squirrels huddle together in their nest.

called the 'tree rat', a name that no-one would ever dream of applying to the native Red Squirrel. The Grey eats shoots of many kinds, roots, acorns, beech mast, flowers, bark, fruits and other vegetation and, like the Red, will cache surplus nuts and beech mast below ground level or in hollow trees. This is why people who habitually collect and hoard things are commonly referred to as 'squirrels'.

The natural home of the Grey Squirrel is the entire eastern half of the USA and part of southern Canada, where it is known as the Eastern Gray Squirrel (to distinguish it from the very similar Western Gray Squirrel, *Sciurus griseus*); I have found that the North American animals seem to have a leaner, hungrier look than their British counterparts. Most Grey Squirrels are genuinely grey and while melanistic and albino animals occur, they are usually thought to be rare, although they seem to be present in Kent:

> We have an albino squirrel in Meopham ... it spends a fair amount of time in next door's walnut tree ... it has been seen to the north of us, approximately 2 miles away in an area known as New Barn and 2 miles south of us in Trosley country park, which provides an indication of the range of the family in question.[227]

The Grey Squirrel *(Sciurus carolinensis)* has taken over many habitats once occupied by the Red Squirrel *(S. vulgaris)*.

> Note on my road atlas advises 'white squirrel seen 12.15 p.m. 5-2-97 Bidborough Ridge, ran across road into Waghorn's Wood'. This was published in our local paper 'Kent and Sussex Courier' and produced a photograph taken on Christmas Day 1996 in a garden on Frant Hill. Also advised by R.D. Hall of Hadlow (near Tonbridge) of a well known 'white', often seen in the car park to the rear of Tesco in West Malling.[228]

Unlike many naturalized mammals, most of the British Grey Squirrel population, which is now in excess of quarter of a million, originates from deliberate introductions rather than escapes. Between 1876 and 1929, Grey Squirrels were liberated on over twenty occasions. They spread with enormous rapidity (even individual animals will move considerable distances) and although the importing or keeping of Grey Squirrels was made illegal in 1938, this was truly a classic instance of locking the stable door after the horse had bolted. Grey Squirrels were here to stay and are now to be found over much of England and Wales, although they are much less common in Scotland and, in general, less numerous in areas of extensive conifer forest or plantations.

> The grey squirrel is not a native of Ireland but was introduced into the country at Castle Forbes, Co. Longford in 1911 ... a wicker hamper containing eight or 12 squirrels, a wedding present from the Duke of Buckingham to one of the daughters of the house, was opened on the lawn after the wedding breakfast, whereupon the bushy-tailed creatures quickly leapt out and scampered off into the woods where they went forth and multiplied.[229]

The Grey Squirrel is almost twice the size and weight of the Red and characteristically builds its nest, or 'drey', rather differently: well away from the main trunk of a tree rather than close to it. There is a close correlation between the spread of the Grey Squirrel and the retreat of the Red, and the question of a causal relationship has long been argued.

On balance, it seems that there has been no serious fighting or killing and that the Grey has simply moved into areas, especially of deciduous woodland, at the same time as the Red vacated them. Why the Red should have vacated them remains to be solved and, in some districts, 'Red Squirrels disappeared up to 18 years before Greys appeared, in others they coexisted for up to 16 years'.[230] The Red is also declining in other areas at the extremes of its range and where the Grey does not occur.

Numerous attempts have been made to limit the spread of Grey Squirrels. An anti-Grey Squirrel campaign was launched in 1931 and, for many years until the practice ceased in 1958, the Government paid a bounty of 1s or 2s for every Grey Squirrel tail. Free shotgun cartridges were supplied to squirrel clubs. Ron Murton reported that 'there were those enlightened gentlemen who trapped the animals, cut off their tails and released the creatures in hope that the progeny would provide further remuneration'.[231]

Red Squirrel *(Sciurus vulgaris)*

OTHER COMMON NAMES: *puggy (Suffolk); scorel; scropel; scrug; scruggy (Wiltshire); skarale; squaggy; squirrell (Surrey); squerylle; swirrel (northern England).*

The Red Squirrel is an altogether smaller, daintier and prettier animal than the Grey and is particularly endearing because of its beautiful colour and long ear-tufts. The colour is usually a rich red-brown but varies widely, with distinctly red, distinctly brown and black phases existing, although melanistic (all-black) animals are less common in Britain than elsewhere in Europe. Although they were once widespread, today the best chances of seeing a Red Squirrel are in Scotland, central Wales, parts of East Anglia, such as Thetford Forest where conifers occur, and, intriguingly, the Isle of Wight, which the Grey Squirrel has not yet reached.

> Conservationists, who believe that the best future for the red [squirrel] is on islands such as Anglesey and the Isle of Wight, have culled 3,500 grey squirrels so far on Anglesey after trapping them ... Andrew Tyler, the director of the charity Animal Aid said: 'This is an obscene and pointless cull. The grey squirrel is being killed for no other reason than that it is grey and foreign'.[232]

The Red Squirrel has not always been considered such a delightful creature, and its 'temper', which it demonstrates by a characteristic harsh chattering, has long been known.

> He sees me, and at once, swift as a bird
> Ascends the neighb'ring beech; there whisks his brush,
> And perks his ears, and stamps and scolds aloud,
> With all the prettiness of feigned alarm.
> (William Cowper, 'The Task')

Although they are now protected, for many years Red Squirrels were deliberately persecuted, partly for the perceived damage that they cause to trees, partly for purely fanciful notions ('They are very harmful and will eat all manner of woollen garments,' was Edward Topsell's unhelpful contribution), and partly for no reason whatsoever.

> Groups of men and youths would stone it from tree to tree until they had forced it into one that stood alone. There it was stoned until, in an effort to escape it dropped to the ground, usually to succumb under a shower of stones. Some years ago the squirrels that added to the attractions of Richmond Park were shot by the keepers to prevent them being killed in this way by gangs of youths coming from London.[233]

> Then, as a nimble Squirrel from the wood,
> Ranging the hedges for his filbert-food,
> Sits partly on a bough his browne nuts cracking,
> And from the shell the sweet white kernell taking,
> Till (with their crookes and bags) a sort of boyes,
> (To share with him) come with so great a noyse,
> That he is forc'd to leave a nut night broke,
> And for his life leape to a neighbour oake;
> Thence to a beeche, thence to a row of ashes;
> Whilst through the quagmires, and red water plashes,
> The boyes runne dabling through thicke and thin,
> One tears his hose, another breakes his shin:
> This, torn and tatter'd, hath with much adoe
> Got by the bryers; and that hath lost his shoe;
> This drops his hand; that headlong falls for haste;
> Another cryes behinde for being last;
> With stickes and stones, and many a sounding hollow,
> The little folle, with no small sport, they follow,
> Whilst he, from tree to tree, from spray to spray,
> Gets to the wood, and hides him in his dray.
> (William Browne, 'The Squirrel-hunt')

The Squirrel Hunt was part of rural Boxing Day or St Stephen's Day celebrations, although in some districts, Good Friday was preferred. Roy Palmer tells of Good Friday Squirrel Hunts by the 'Dursley congregation, often numbering a hundred' on Stinchcombe Hill in Gloucestershire. Elsewhere, it was held on St Andrew's Day, 30 November. In Sussex:

> ... the bricklayers used to take their annual holiday on this day, and would go in gangs into the woods to hunt squirrels and other small animals by the primitive but effective method of stunning them with short stout sticks; this they called 'going St Andring'. Afterwards, they went to an inn for a celebratory supper and drinking session. The dead squirrels were taken home to be eaten.[234]

On the same day, in the Parish of Easling in Kent, the Squirrel Hunt was clearly little more than an excuse for rural hooliganism:

> ... the labourers and lower kind of people, assembling together formed a lawless rabble, and provided with guns, poles, clubs, and other such weapons, spent the greatest part of the day in parading through the woods and grounds, with loud shoutings. Under the pretence of demolishing the squirrels, some few of which they killed, they destroyed numbers of hares, pheasants, partridges, and in short whatever came in their way, breaking down the hedges, and doing much other mischief, and in the evening betaking themselves to the alehouses, finished their career there, as is usual with such sort of gentry.[235]

In Suffolk, they seemed to prefer Christmas morning when:

> ... half the idle fellows and boys in a Parish assemble in any wood or plantation ... .[236]

The history of squirrel-eating is probably as long as that of the squirrel itself and it is evident that the Grey has easily replaced the Red among the nation's culinary delights. Many rural people still eat Grey Squirrels, although strangely few people readily admit it, and the general agreement seems to be that the taste is similar to Rabbit.

The name 'squirrel' is Late Middle English but has a pedigree extending back through French and Latin to a Greek word meaning 'shade tail', in essence conjuring up a delightful image of an animal that sits in the shade of its own tail. The squirrels' nest has been called a 'drey' (or 'dray') since the early seventeenth century but the origin of the word is not known. A sixteenth-century name for the squirrel was 'bun', which was subsequently conferred on the Rabbit (see page 490). In Surrey, where squirrels were called 'squaggies', the nest was a 'jug', or 'squaggy-jug'.

> One day, when all the woods where bare and blea,
> I wandered out to take a pleasant walk
> And saw a strange-formed nest on stoven tree
> Where startled pigeon buzzed from bouncing hawk.
> I wondered strangely what the nest could be
> And thought besure it was some foreign bird,
> So up I scrambled in the highest glee,
> And my heart jumpt at every thing that stirred.
> 'Twas oval shaped; strange wonder filled my breast;
> I hoped to catch the old one on the nest
> When something bolted out – I turned to see –
> And a brown squirrel pattered up the tree.
> 'Twas lined with moss and leaves, compact
> and strong;
> I sluthered down and wondering went along.
> (John Clare, 'The Squirrel's Nest')

Squirrels are good swimmers and no doubt climb onto floating logs or other debris when the opportunity presents itself; this can be the only reason for the widespread but utterly fallacious belief that they use their tail as a sail.

> The admirable wit of the beast appeareth in her swimming or passing over the waters; for when hunger or some convenient prey of meat constraineth her to pass over a river, she seeketh out some rinde or small bark of a tree, which she setteth upon the water and goeth into it, and holding her tail like a sail letteth the wind drive her to the other side.[237]

> This was later interpreted as Christians crossing the troubled seas of this life on the Cross of Christ.[238]

In modern times, the responsibility for perpetuating this myth can be laid squarely at the door of Beatrix Potter, who illustrated *The Tale of Squirrel Nutkin* (1903) with an image of a whole tribe of squirrels setting sail. In truth, the whole of the book is:

> ... a Tale about a tail ... they made little rafts out of twigs ... Each squirrel had a little sack and a large oar, and spread out his tail for a sail.[239]

This is no great contribution to natural history but, after all, it is a story about a Tawny Owl that eats not only mice and Moles, but also minnows, eggs and honey – and children love it.

Many parts of the country have their versions of an old rhyme well known in the Cotswolds:

> A squirrel can hop from Swell to Stow
> Without resting his foot or wetting his toe.

In neighbouring Warwickshire, it was once said that a squirrel could jump from tree to tree from one end of the county to the other; they would have to make some remarkably long leaps these days, especially since elm disease took its toll. In Charnwood Forest in Leicestershire, a squirrel might hop 10 kilometres (6 miles) from tree to tree without once touching the ground; but then there were so many trees that 'a traveller might journey from Beaumanor to Bardon on a summer day without once seeing the sun'.[240] In Rossendale Forest in Lancashire, the saying was that a squirrel could leap from tree to tree all the way from Rawtenstall to Sharneyford and, also in Lancashire:

> From Birchen have to Hilbre
> A squirrel might hop from tree to tree.

Most curiously, as far as I can discover, nowhere in Britain has been named after these so familiar and once so very common creatures.

The native Red Squirrel *(Sciurus vulgaris)* is both smaller and daintier than the Grey Squirrel (*S. carolinensis*), with prominent ear-tufts.

Beavers
(Castoridae)

Beavers are very distinctive and always likely to attract attention because of their large, flattened, scaly tails, which they slap onto the water surface as a warning, their aquatic habit and their general destructiveness to waterside vegetation. They fell trees to eat the bark, to build homes or lodges and to construct extensive dams and canal systems in order to maintain the water level and provide them with reliable routes to feeding areas.

European Beaver *(Castor fiber)*

British readers' knowledge of beavers is probably confined to the Canadian Beaver *(Castor canadensis)*, as portrayed in American cartoon films that make great play of their ability to fell trees and build dams through sheer hard work (the phrase 'eager beaver' has passed into everyday currency). The European Beaver, the largest native European rodent, is more of an unknown quantity. In Britain today, you will have to visit a zoo to see one but it was not always so and beavers were British animals into recorded history.

Archaeological studies suggest that the beaver was once fairly widespread but the relative paucity of beaver place names (fewer than twenty) indicate that it was already becoming rare by the Anglo-Saxon period. Beverley in Yorkshire is probably 'beaver stream' (originally 'Beverlic', from the Old English *lecc*, meaning 'a stream', rather than from *leah*, 'an open area'), Bevercotes in Nottinghamshire is probably 'beavers' huts', and Beverstone in Gloucestershire is probably 'beaver's stone'. Nonetheless, as so often happens, there are confounding factors because 'to beaver' meant to ferment woad in a pit called a beaver pit. 'Beaver' itself, however, is a name much older than Old English and has been traced to an Old Aryan word, perhaps Sanskrit, meaning 'brown'.

According to a widely quoted statement attributed to the Welsh King Howel Dda and dating from the mid-tenth century: 'the King is to have the worth of Beavers, Martens and Ermines in whatsoever spot they are killed, because from them the borders of the King's garments are made'.[241] This is probably reasonable evidence that beavers were still inhabiting Wales at that time, although the last reliable written account of British beavers seems to have been that by Giraldus

The Beaver *(Castor fiber)* is extinct in the British Isles, although it was once fairly widespread. Small reintroductions are underway.

Cambrensis, who toured Wales in 1188 and described the creatures with reasonable accuracy, referring to the Teifi as the only river in Wales or England where they still occurred.[242] They may have persisted for slightly longer in Scotland but, by the twelfth century, their time was clearly all but past. The demand for beaver fur to make fashion garments (a 'beaver' came to mean a type of hat as long ago as the sixteenth century) had to be satisfied by pelts imported from Europe and then, much later, from North America.

The possibility of reintroducing beavers to Britain has been discussed for some time.

> The first breeding colony of beavers in Scotland for more than 400 years was established this weekend at a secret location ... Derek Gow, project leader, explained 'We have released them in a secure wetland area in Scotland but kept the exact location secret ... The matter is in ministerial hands and when the green light is given we shall step up the reintroduction. At present we have gone as far as we can without jumping ahead of the law.'[243]

Voles, Muskrats, Mice and Rats
(Muridae)

VOLES AND MUSKRATS

Although there were about 100 million voles in Britain at the last count, they feature very little in the literature and folklore of the country. Yet, collectively, the three principal species (the Bank Vole, Water Vole and Field Vole) outnumber every other type of mammal by a long way and they are essential prey for a large number of predators. It may be simply because most of them are small, live in fields and woods and do not usually come into houses, although gardeners will be familiar enough with their activities.

Their origin was neatly and succinctly explained by David Macdonald:

> A niche for diminutive grazers was made possible by the evolution of grasses in the Miocene and was occupied by voles which arose from the hamster lineage in the late Pliocene.[244]

And they are still here, this group of diminutive, grass-eating creatures, rather like mice but with short, stubby faces and rather short, rather stubby tails, although they have been adaptable enough to turn their attention to other things besides grasses; seeds, fruits, bulbs, corms and tubers are all very much to their liking.

'Vole' is one of the relatively few words to have come into the English language from Norwegian. It was originally *vollmus*, meaning 'field mouse', and for a long time was 'vole-mouse' in English, being shortened to its present form only in 1805. (In modern English, therefore, the little creature is distinguished from a mouse by being called a 'field'.) There is no obvious reason why a Norwegian word should have been borrowed for what has long been a very common British animal, although Norway is certainly a traditional home for rodents, being the territory of a close relative of the voles, the Norway Lemming *(Lemmus lemmus)*. This is the animal that is subject to enormous population explosions (even more enormous than those of voles) and large-scale migrations that, in legend at least, involve the creatures hurling themselves over cliffs. In prehistoric times, lemmings lived in Britain but they retreated with the ice and never returned.

The Bank Vole *(Clethrionomys glareolus)* is primarily a woodland species, nesting in a burrow among the roots of deciduous trees.

Apart from the native species of vole, one large alien, the Muskrat *(Ondatra zibethicus)* has been introduced and exterminated (see Muskrat, page 469).

Bank Vole *(Clethrionomys glareolus)*
This rather beautifully coloured little vole, distinctly reddish with grey on its sides, is principally a woodland animal and typically lives in earth banks beneath deciduous trees, where it burrows among the roots. It constructs a pretty little spherical nest of grass and moss.

Bank Voles are eaten by several species of owl, Weasels and other predators, and they are commonly troublesome in gardens, not least because they are generally less timid than their relative, the Field Vole. (It is the only vole bold enough to come into my house, despite the presence of a cat.)

It occurs throughout Britain and, although absent historically from Ireland, it is now spreading outwards from a presumed introduction to the southwest about 50 years ago. On some of the offshore islands, such as Mull and Skomer, a distinct, larger subspecies occurs. The Bank Vole is outnumbered by the Field Vole and there are estimated to be a mere 23 million in Britain.

Water Voles *(Arvicola terrestris)* were once common along many rivers, but the numbers have now dropped markedly.

Water Vole *(Arvicola terrestris)*

OTHER COMMON NAMES: *northern water vole; water rat.*

As a boy, I lived beside a small river and Water Voles were by far the commonest mammals on the river bank, their holes visible everywhere in the waterside mud. While fishing, I often heard one loudly gnawing vegetation long before I spotted the animal itself, sitting endearingly on its haunches, holding its lunch in its forefeet. From time to time, when another approached too closely, an unholy fracas would ensue.

Since then, Water Voles have declined dramatically everywhere and they currently number only about 1 million. The reasons for their disappearance are complex but loss of suitable river bank habitat is a major factor, no doubt exacerbated by the recent climate change that has resulted in high rainfall and flooding, and consequently the assiduous clearing of river-drainage channels by water companies. A survey of Water Vole populations in the late 1980s revealed that only around 30 per cent of sites known to have been occupied between 1900 and 1940 still supported voles. Predation by American Mink has been a further disaster for them.

The Water Vole was once, apparently logically for an aquatic creature, called *Arvicola amphibius* (*amphibius* meaning 'leading two lives', i.e. on land and water), a name that has since been superseded by *terrestris* ('of the ground'). The reason for this change in specific name is perhaps more understandable if it is known that, elsewhere in Europe, the Water Vole is not closely associated with watercourses and it tunnels extensively into pasture land and similar sites. I have been told that, on Read's Island in the Humber estuary, they behave in a similar fashion; a sensible precaution against being swept out to sea.[245] The nest is constructed within a burrow, above the water-line and is lined with grass, chopped reed stalks and leaves.

> I kept, bred and even tamed Water Voles in large cages for something between twelve and twenty years. One would even climb onto my lap to get treats and many fed from my hands.[246]

Water Voles, whether or not they live close to water, are wonderful swimmers and it is always a delight to see them crossing a river or pond with a mouthful of nesting material, which they somehow manage to keep dry. They are much bigger than Bank and Field Voles, with rather long, dark brown fur; almost black in the northern part of the British Isles (they are absent from Ireland). In truth, they are about the size of a rat, and rats do sometimes live on river banks, although there is no justification for confusing them because rats have pointed faces and much longer tails. In the sixteenth century, however, things were evidently different. The Water Vole was first called the water rat in 1552, long before the Brown Rat arrived here. The Black Rat was the only rat in Britain and any rodent approximately the same size, stubby nose or not, was likely to be regarded in the same ill light.

Not that this excuses the confusion that has arisen from Kenneth Grahame's *The Wind in the Willows*, which was not written until 1908 and in which one of the prinicpal characters, variously called Water Rat, Rat and Ratty, is in fact a Water Vole.[247] (Grahame's biographer, Peter Green, cryptically believes that the Rat represents 'the repressed Bohemian Ulysses'.[248]) Grahame's two principal illustrators knew their natural history well enough: Arthur Rackham depicts Ratty quite unmistakably as a stubby-nosed Water Vole and so does Ernest Shepard, whose drawing of the encounter between the Water Rat and the Sea Rat intriguingly shows a stubby-nosed vole face to face with a dark, long-nosed beast – the Black Rat *(Rattus rattus)* perhaps? However, the principal illustrator of the modern editions, John Burningham, shows Ratty very clearly as the Brown Rat *(Rattus norvegicus)*. Thus a generation of children may have grown up unable to distinguish between them all.

Field Vole *(Microtus agrestis)*

OTHER COMMON NAMES: *dog mouse (Surrey); field mouse; grass mouse; marsh mouse (East Anglia); ranny; short-tailed field mouse; short-tailed vole.*

A darker, more brown-and-grey-coloured animal than the Bank Vole, and with a shorter tail, this was for many years called the 'short-tailed vole'. Field Voles occur throughout the British mainland but are absent from Ireland and many of the offshore islands, although they are present on most of the Inner Hebrides. Unexpectedly, their place on Orkney, and also on Guernsey, is taken by a paler-coloured subspecies of the continental Common Vole *(Microtus arvalis)*, which does not exist anywhere else in Britain; these were once called 'Orkney Voles' and 'Guernsey Voles'.

There are about 75 million Field Voles *(Microtus agrestis)* in Britain, especially in grassy habitats.

Although Field Voles are typically creatures of grassy places, in practice they occur almost everywhere – all 75 million of them. Like most voles, they are almost exclusively herbivorous, preferring to eat grass but also taking other vegetation, including tree bark. They tend to be noisier than Bank Voles and chirp loudly and aggressively. Field Voles are an important food for many birds of prey, especially owls, and also many carnivorous mammals, but none of this predation seems to have any effect on their overall numbers. They burrow in soil but generally nest above it, in a small spherical nest at the base of a clump of grass; unlike Bank Voles, they seem reluctant to climb shrubs and trees. Field Vole populations peak on a four-year cycle and, in years of 'vole plagues', the damage they do to vegetation can be considerable. However, the owls and other predators react accordingly and the populations crash again.

Muskrat *(Ondatra zibethicus)*

Musquash was one of the new furs of the twentieth century and proved especially popular because the animal that provided it was amenable to being farmed. The name 'musquash' originated in one of the languages of the Algonquin people from the region of the Ottawa river in North America and promised a rather exotic and glamorous image to ladies of fashion, who might have taken exception to wrapping themselves in the skin of an animal whose name meant little more than 'smelly rat'. Muskrat farms opened throughout Europe, and the first escapes in Britain occurred in the late 1920s.

Muskrats are rather like large Water Voles and live in similar places but the escaped animals caused extensive damage to river banks and drainage systems in the Severn valley, central Scotland and southeast England, as well as in Ireland. A concerted trapping campaign killed around 5,000 animals and succeeded in eradicating them by 1937. Thus native Scotsmen were denied the opportunity to wear Muskrat sporrans, a practice adopted by some of their transatlantic cousins.

Muskrats have become naturalized in other parts of Europe, although it remains to be seen whether Europeans will develop the obsessive interest in them that exists in North America, where they are hunted, trapped, eaten and doted over, which has resulted in around 100 websites being devoted to the beast.

MICE

The mouse must be a strong contender for being the best known of all animals. Everyone knows instantly what a mouse looks like and everyone has seen one; they have featured in countless books and stories (generally in sympathetic and rather enchanting portrayals); many people (not just children) keep white mice as pets; women traditionally stand on chairs when they see one (presumably because they mistakenly believe that it will run up their legs); and they are universally synonymous with smallness.

In Britain, there are three mainly 'outdoor' mice: the relatively uncommon and local Harvest Mouse *(Micromys minutus)*, the Yellow-necked Mouse *(Apodemus flavicollis)* and the phenomenally common and abundant Wood Mouse *(Apodemus sylvaticus)*. There is also one mainly 'indoor' mouse, the House Mouse *(Mus musculus)*, the species with which urban people will be most familiar. The outdoor mice are long-established natives but the House Mouse, like both of the British rats (see Black Rat and Brown Rat, pages 479 and 477) was introduced, albeit many centuries ago and this species is now, after man, the most widely distributed mammal in the world.

Mice differ most obviously from their relatives, the voles, by having a pointed face and a relatively long tail. In practice, they are more likely to be confused with shrews, to which they are quite unrelated, but shrews, like voles, are short-tailed and very unlikely to come into houses.

'Mouse' is an ancient word. It reached modern English from Old English but has a pedigree extending back through Latin and Greek to Sanskrit. It is said to have originally meant 'thief', its Sanskrit roots having to do with stealing. As so often with plurals, it is not obvious when 'mouses' became 'mice' but regrettably the excellent Middle English verb 'to mouse' ('to hunt or catch mice') has largely passed from use, finding expression now only when someone describes their cat as 'a good mouser'.

Unlike voles, which are mainly vegetarian and largely grazing animals, mice as a group are fairly omnivorous, although I have always been mystified why, when and by whom it was decreed that the best bait for a mousetrap is cheese: bread is far more effective.

Harvest Mouse *(Micromys minutus)*

OTHER COMMON NAMES: *harvest-row (southern England).*

This charming beast is the smallest European rodent and has been seen by very few people in present-day Britain. It is superficially like a tiny, long-tailed vole, with a russet-red back and striking white underparts. It is appropriately named because its home is among tall grasses and cereal crops, where it builds its grassy, more or less spherical breeding nest conspicuously attached to the stems. The nest is especially obvious in winter, when stems have been left standing, although it will by then have been vacated because, with the onset of cold weather, the Harvest Mouse moves closer to ground level, or sometimes below it, and constructs a different, larger, home. Now that wheat varieties have shorter stems, it has begun to return to some of its old habitats in reed beds and other damp places. It is self-evidently vulnerable to modern farming practices and also falls prey to a large number of different predators, including owls, Foxes, Stoats, Weasels and cats. Nonetheless, it is still present in considerable numbers, mainly in England south of central Yorkshire.

The name 'Harvest Mouse' is recent (early nineteenth century) for the simple but extraordinary reason that the animal was not recognized as a distinct species until Gilbert White sent specimens to Thomas Pennant:

> Sent two field-mice, a species very common in these
> parts (tho' unknown to the zoologists) to Thomas
> Pennant Esq of Downing in Flintshire. They resemble
> much in colour ye Mus domesticus medius of Ray; but
> are smaller than the Mus domesticus vula: seu minor
> of the same great Naturalist. They never enter houses;
> are carried into ricks, & barns with ye sheaves, abound
> in harvest; & build three nests, composed of blades of
> corn, up from the ground among the standing wheat:
> & sometimes in thistles. They breed as many as eight
> Young at one time.[249]

It must have been the Harvest Mouse that John Clare described (such a distinctive nest could hardly have gone unnoticed by one of the greatest literary nest-observers):

> I found a ball of grass among the hay
> And progged it as I passed and went away;
> And when I looked I fancied something stirred
> And turned agen and hoped to catch the bird –
> When out an old mouse bolted in the wheats
> With all her young ones hanging at her teats;
> She looked so odd and so grotesque to me,
> I ran and wondered what the thing could be,
> And pushed the knapweed bunches where I stood;
> When the mouse hurried from the craking brood.
> The young ones squeaked, and when I went away
> She found her nest again among the hay.
> The water o'er the pebbles scarce could run
> And broad old cesspools glittered in the sun.
> (John Clare, 'Mouse's Nest')

Its early history is vague because there is no archaeological evidence of Harvest Mice in Britain until Roman times. Zoologists seem undecided whether this is because both the mouse and its habitat were very rare before then (although Neolithic man certainly grew crops); or because the mouse was introduced.

The Harvest Mouse *(Micromys minutus)* builds its almost spherical nest among tall grasses and cereal crops.

The Wood Mouse *(Apodemus sylvaticus)* is often seen in gardens or among stores of fruit or seeds in garden sheds.

Wood Mouse *(Apodemus sylvaticus)*
OTHER COMMON NAMES: *long-tailed field mouse.*
See Yellow-necked Mouse below.

Yellow-necked Mouse *(Apodemus flavicollis)*
For a long time, there was widely believed to be only one British species of outdoor mouse. This was not only because the Wood Mouse had been confused with the Harvest Mouse *(Micromys minutus)* but also because it was not appreciated that the Wood Mouse was in reality two distinct species: *Apodemus sylvaticus* and *Apodemus flavicollis* (the Yellow-necked Mouse). The name *sylvaticus*, meaning 'of the woods', was assigned to it by Linnaeus and, at least as far as most British Wood Mice are concerned, was an unfortunate choice because these are mice of grassland, gardens and hedgerows rather than woods. Nonetheless, it is a long-established native species with a reliable archaeological record.

In 1894, careful study revealed that some Wood Mice were generally larger and more boldly coloured, with a complete and rather striking band of yellow fur around the neck, as opposed to the simple patch of yellow found on most others. These proved to be a distinct but closely related species, the Yellow-necked Mouse, and studies of fossil remains revealed that it too had been in Britain for a long time. It has always been present, but in much smaller numbers than the Wood Mouse, and its present British population is only about 2 per cent of that of the Wood Mouse. Unusually, its present distribution is very heavily concentrated in the southeast of England and in the counties around the Severn Valley, in areas of ancient woodland, so 'Wood Mouse' would, in truth, be a far more appropriate name for it.

When your cat catches a mouse in the garden, it will almost certainly be a Wood Mouse. Likewise, when you find a mouse in your house as a casual visitor, as tends to happen in the autumn, it will also almost certainly be a Wood Mouse (or, in a few areas, just possibly a Yellow-necked Mouse).

> Whoever wrote 'Quiet as a mouse' could not have been thinking of yellow-necks. Each autumn they invade my house. They scramble up through the cavity walls and the noise they make is quite considerable. They race behind the skirting and under the floorboards, and when one of them is heard, in the quiet of night, bounding between the joists the sound is like that of heavy human footsteps. It is highly probable that yellow-necks may be responsible for some reports of ghostly footsteps.[250]

House Mice have a characteristic smell and the absence of this is another good indicator of an *Apodemus* mouse. If you are unwise enough to leave your garden apple-store or boxes of seeds unprotected, it is Wood Mice that will eat them. If you are unwise enough to try to control them by trapping, you will catch yet another mouse every time you set a trap. And it can only have been a wandering brown Wood Mouse, not a grey House Mouse, that was the inspiration for 'Appley Dapply', an old rhyme that Beatrix Potter adopted:

> Appley Dapply, a little
> brown mouse,
> Goes to the cupboard in
> somebody's house.
> In somebody's cupboard
> There's everything nice,
> Cake, cheese, jam, biscuits,
> – All charming for mice!
> Appley Dapply has little
> sharp eyes,
> And Apply Dapply is so fond
> of pies!

Wood Mice are extremely common, extremely numerous and occur throughout the British Isles. They are a very important food source for a wide range of predators, including owls, Foxes and other carnivores; 'mouse-hunt' is an obsolete dialect word for a Weasel. Indeed, there are so many mice in the fields, gardens and hedgerows that it was once thought that normal processes of reproduction could not be entirely responsible:

> The generation and procreation of Mice is not only by copulation, but also nature worketh wonderfully in engendering them from earth and small showers of rain ... A female Mouse having free liberty to litter, within less than half a year she hath brought forth a hundred and twenty young ones.[251]

While I am sure that John Clare found his Harvest Mouse nest on the grass stalks (see Harvest Mouse, page 470), I am equally certain that Robert Burns found his Wood Mouse's home below ground:

> Wee, sleekit, cowrin', tim'rous beastie,
> O, what a panic's in thy breastie!
> Thou need na start awa sae hasty,
> Wi' bickering brattle!
> I wad be laith to rin an' chase thee,
> Wi' murd'ring pattle!
>
> I'm truly sorry man's dominion
> Has broken Nature's social union,
> An' justifies that ill opinion
> Which makes thee startle
> At me, thy poor earth-born companion
> An' fellow-mortal!
>
> I doubt na, whyles, but thou may thieve;
> What then? poor beastie, thou maun live!
> A daimen-icker in a thrave
> 'S a sma' request:
> I'll get a blessin wi' the lave,
> An' never miss't!
>
> Thy wee-bit housie, too, in ruin!
> Its silly wa's the win's are strewin!
> An' naething, now, to big a new ane,
> O' foggage green!
> An' bleak December's winds ensuin,
> Baith snell an' keen!
>
> Thou saw the fields laid bare an' wast,
> An' weary winter comin fast,
> An' cozie here, beneath the blast,
> Thou thought to dwell,
> Till crash! the cruel coulter past
> Out thro' thy cell.
>
> That wee bit heap o' leaves an' stibble,
> Has cost thee mony a weary nibble!
> Now thou's turned out, for a' thy trouble,
> But house or hald,
> To thole the winter's sleety dribble,
> An' cranreuch cauld!
>
> But, Mousie, thou art no thy lane,
> In provin foresight may be vain:
> The best laid schemes o' Mice an' Men
> Gang aft agley,
> An' lea'e us nought but grief an' pain
> For promis'd joy!
>
> Still thou art blest, compared wi' me!
> The present only toucheth thee:
> But Och! I backward cast my e'e
> On prospects drear!
> An' forward, tho' I canna see,
> I guess an' fear!
> (Robert Burns, 'To a Mouse, On Turning Her up in Her Nest with the Plough, November, 1785')

Burns described his mouse as 'cowering' and 'timid', and timidity is a trait often associated with mice. They are extremely difficult to approach closely because of their very good eyesight, acute hearing and sensitivity to vibration. The painter Charles Leslie, writing of John Constable, said:

> I remember to have heard him say, 'When I sit down to make a sketch from nature, the first thing I try to

do is to forget that I have ever seen a picture;' and it is related as a curious proof of the stillness with which he used to sit whilst painting in the open air, that one day a field mouse was found in his pocket.[252]

House Mouse *(Mus musculus)*

Where did House Mice live before there were houses? Probably somewhere in the steppes of central Asia, where a wild variant still exists, although they have long been present in the Middle East as well. It is currently believed that they switched their attention from wild seeds to cereal grains, moved indoors with the harvested crop, liked what they found and then stayed close to man as he moved along the trade routes of Europe, Asia and ultimately the world. When did it happen? Perhaps initially in the Early Palaeolithic period, although the oldest House Mouse remains in Britain date from the Iron Age.

The House Mouse is more uniformly grey-brown than the Wood Mouse and the tail, which is the same length as the body (not longer), is distinctly ringed and scaly. It is accompanied by a characteristic, musty smell that once experienced will be stored in the olfactory memory for ever. However, it should not be forgotten that the House Mouse still also lives outdoors, away from human habitation.

> Frank Fraser Darling found House Mice on one of the Treshnish Isles although no-one had lived on it for eighty years ... the populations in Britain can be divided conveniently into three groups. There is a large urban group found in houses, factories and warehouses, a second rural one associated with farms and a smaller population found in the fields.[253]

It was inevitable that this closest of all wild mammalian companions of man should have slipped effortlessly into folklore. There is a very widespread belief, not only in Britain, that it is a sign of death if a house is suddenly over-run with mice; there is no obvious logic to this, no deeply buried historic knowledge in the human psyche, because, unlike rats, mice do not carry the plague or very much else in the way of harmful disease. There is even less logic in such fears as:

> If a mouse runs over anyone, he or she will not live long, and if one squeaks behind an invalid's bed, recovery is unlikely.

There has also long been a related but inexplicable notion that a mouse is the embodiment of a human soul:

> It was a good thing for a milk tooth to be put in a mouse's hole so as to ensure that the new tooth would be as small as a mouse's.[254]

Stewed, boiled, fried or roast mouse (the precise recipe varying from district to district) is a widely quoted traditional remedy for the more common childhood complaints, such as diarrhoea, bed-wetting and whooping cough.

The longest-running play in theatrical history is Agatha Christie's *The Mousetrap*.

> In 1936, an Oxfordshire woman stated that whooping-cough could be cured by skinning a mouse, removing all its bones, and giving the fried meat to the sick child. She said the taste was not at all unpleasant, and that if the dish was properly prepared, the patient need never know what he was eating.[255]

> A supposed cure for bed wetting was to boil a mouse on top of the washing boiler and serve it in a bread sandwich to the unlucky incontinent. This cure came from my maternal grandmother who was born in the 1880s.[256]

The 'best known round in the world' was how Iona and Peter Opie described the following rhyme:

> Three blind mice, three blind mice, see how they run!
> They all ran after the farmer's wife,
> And she cut off their tails with a carving knife,
> Did you ever see such a thing in your life as three blind mice![257]

It exists in various versions, the oldest apparently dating from 1609, and 'Did you ever see such fools in your life' occurs sometimes in the fourth line. But why *three* mice and why they were blind seems utterly elusive.

There are many comparable versions of another well-known mouse rhyme, one of which I have in a nursery song book of 1825 and runs:

Hiccory, diccory dock,
The mouse ran up the clock;
The clock struck one,
And the mouse came down,
Hiccory, diccory, dock.[258]

There is no special reason why mice should be involved but I am sure that they did run up and down old long-case clocks, and the rhyme itself was certainly used as an aid to counting. And while the three mice that were blind might just have been Wood Mice (it was a farmer's wife in hot pursuit after all), surely any mouse going up a clock must have been a House Mouse *(Mus musculus)*.

One of the harvest of treasures that came to light during the restoration of my old house was an early nineteenth-century mousetrap, which had slipped down a wall cavity; a fairly grim device, it would almost certainly have decapitated any unfortunate victim. People have trapped mice for as long as they have had houses: the word 'mousetrap' dates from 1475; and has the very pleasing but rarely heard alternative plural 'mice-traps'. 'If a man ... make a better mousetrap than his neighbour ... the world will make a beaten path to his door' has often been attributed, erroneously, to Ralph Waldo Emerson (1803–82), but it was shortly after Emerson's death that someone finally managed to produce one. John Mast of Lititz, Pennsylvania, filed for a patent in 1889 on the now-familiar modern spring-loaded snap-trap, once made of wood and wire but today very often all-metal. It has become inseparable from the House Mouse in cartoons and other stories, although dozens of people the world over still file patents for new mousetraps every year.

There is an immediate although not widely appreciated link between blind mice and mousetraps. The longest-running play in theatrical history is Agatha Christie's thriller *The Mousetrap*, which opened in London in 1952. It was originally written in 1947 as a radio play called *Three Blind Mice*, after being commissioned by the BBC for the occasion of the eightieth birthday of Queen Mary.

A 'mousehole' is a hole in which a mouse lives or, more generally, if children's stories and cartoons are to be believed, a hole in a skirting board through which a mouse flees to safety from the family cat. Mousehole, pronounced locally as 'mowzle', is also a village in Cornwall and was once the home of Dolly Pentreath, reputedly the last person to speak Cornish as a native tongue; it is said to have been named after a mousehole-shaped smugglers' cave to the south of the village. There are also less obvious places named after mice.

Mousehole, Cornwall, 1934. This village is said to be named after a mousehole-shaped smugglers' cave.

Muscoates in Yorkshire is 'mouse-infested huts', Musbury in Durham is probably 'mouse burrow', Musgrave in Westmorland is 'a grove infested by mice' and Muston in Leicestershire may be 'mouse-infested tun', although 'mus' can also have a connection with the old word for moss.

It is a telling commentary on our times that when the word 'mouse' is mentioned today, people immediately think of the little plastic object that scurries across the mouse-mat of their computer rather than of *Mus musculus* scampering across the kitchen floor.

RATS

How now! A rat? Dead, for a ducat, dead!
(*Hamlet*, Act 3, Scene 4)

With some exceptions, such as spiders and snakes, the overall title of the most loathed of animals would probably go to the rat every time. This deep hatred and fear of rats is quite unlike any other human emotion, and even the most experienced and knowledgeable of zoologists, put on the spot, would generally place rats top of their distaste list.

Biologically, there is little difference between rats and mice, except that rats are bigger. Also, as far as most of the world, including Britain, is concerned, they are relatively newcomers. It comes as a huge surprise for most people to realize that the common rat of Britain today, the Brown Rat *(Rattus norvegicus)*, has been in this country for under 300 years. Its forerunner, the Black Rat *(Rattus rattus)*, has a longer history but it is not a native animal either.

'Rat' is an Old English word of unknown origin and was originally confined to the Black Rat. As long ago as the sixteenth century, it was being used as 'an opprobrious epithet', a use that has gathered momentum over the centuries and more recently has spawned the verb 'to rat', meaning 'to tell tales'. In 1999, the Metropolitan Police began a poster campaign urging people to identify suspected drug-dealers by using the slogan 'Rat on a rat'. In truth, calling anyone a 'rat' could not be more opprobrious. Almost anything unpleasant is likely to have the word 'rat' attached

This seventeenth-century woodcut illustrates people fleeing London to escape the plague, thought to be carried by rats.

to it – the 'rat race', the 'rat run', the 'rat pack' – although perhaps the most famous rat saying of modern times was the American 'You dirty rat', attributed erroneously to the actor James Cagney. Nonetheless, rats have filled the silver screen often enough: they have featured in many a horror film and they achieved modern literary immortality through the rat-based torture described, and later filmed, in George Orwell's *Nineteen Eighty-four*. So what has given these particular rodents such an unenviable reputation?

Rats are omnivorous and will eat human foodstuffs and human possessions, causing considerable loss and spoilage, although the same is true of mice and other creatures. However, rats live in holes and skulk, they inhabit unpleasant places such as drains and sewers, they run quickly and elusively (and might dash up your leg or trousers), they bite when cornered and they indulge in cannibalism; all of which are fairly unsavoury attributes. They can also have a serious impact on human health and I believe that it is this knowledge, implanted in our collective psyche, that underpins our collective 'ratophobia'. The Black Rat, through its lice and fleas, transmits typhus. Also, accordingly to popular belief, it was rat fleas that brought the bacterium *Yersinia pestis*, the cause of bubonic plague, to Europe, the form known as the Black Death reducing the human population by about a half between the fourteenth and sixteenth centuries. However, this popular belief, and my theory, may not stand up to closer scrutiny.

Classic, more or less contemporary accounts of the plague, such as Daniel Defoe's *A Journal of the Plague Year*, which admittedly was written some time later in 1722, do not mention rats.[259] This is with good reason because the Black Rat was not firmly identified as a carrier of plague until 1897. The only reference to rats in historical accounts seems to be that by Sir Theodore Mayerne, physician to Charles I (who published the first English book on entomology), who wrote in 1631 that rats and other vermin 'running from house to house and creeping over stuff may receive and carry the infection'. No two historians seem to agree on the matter, but it is clear that other rodents, not just Black Rats, carry plague: marmots in central Asia and squirrels in North America, for instance. Even if the Black Rat initially brought the disease to Europe, and even if it was an important plague vector in towns, it is now questioned whether simple rat transmission could account for the vast death rates in the countryside; spread by human fleas moving from person to person seems more likely. Also, where were all the dead rats that one might expect to find lying around in the streets of plague-affected towns; there are just no contemporary illustrations. It is not even agreed that the Black Death was bubonic plague; the variant disease, pneumonic plague, which can be spread directly from human to human without any vector, may have been at least partially the cause of the vast mortality and suffering.[260]

In addition, it was conventionally thought that Brown Rats were relatively unimportant as carriers of plague, although they are now known to have been significant reservoirs of the disease in modern Vietnam. Even in Europe, the Brown Rat has a modern human health impact of its own, partly as a carrier of the protozoon *Toxoplasma gondii*, the cause of toxoplasmosis, as well as *Salmonella* food-poisoning bacteria, but most importantly as a vector of leptospirosis. The form known as Weil's disease is transmitted by *Leptospira* bacteria present in rat's urine and well over half of all Brown Rats in Britain are thought to be infected.

Despite the fear and loathing that rats engender, people have eaten them, in some parts of the world from choice although in Britain only in times of extreme hardship. Nonetheless, there is a Rat Inn at Hexham in Northumberland where rat appears on the menu. Fortunately it appears on the back, where the inn name is explained as:

> ... shrouded in mystery ... one popular story is that a former landlord acted as a traitor during the Jacobite uprising and was regarded locally as 'a rat'. Another is that local rat-catchers always met and drank in the bar and a third tale is that a giant rodent was caught and put on display at the inn.[261]

Nonetheless, closer examination of the front of the menu reveals that they do indeed offer ratburger and chips.

Perhaps not surprisingly, nowhere of any significance in Britain appears to have been named after rats, although there are some likely candidates. For instance, the residents of Ratby, in Leicestershire, probably spend much time explaining that there was once a great local landowner called Rota, but who would believe the people from Rattery, in Devon, who will tell you that their village is simply named after a 'red tree'.

Brown Rat *(Rattus norvegicus)*

OTHER COMMON NAMES: *common rat; Hanoverian rat; Norway rat; Russian rat.*

Although generally greyish brown above and paler beneath, some Brown Rats are very dark and may be mistaken for Black Rats by the inexperienced. Brown Rats, however, have shorter ears and a thicker, shorter tail and, in any event, the chances of seeing a Black Rat in Britain today are very remote.

The Brown Rat is now the ubiquitous British rat, living in almost every type of habitat. Although it generally stays close to human habitation and is very common around farms, there are genuinely remote rural populations, including some on uninhabited Scottish islands. They frequently occur close to water, but the river-bank 'rat' is more correctly, and once more commonly, the Water Vole (see page 468). Rats are good swimmers and, although they must presumably at some time have crossed the real Rubicon, the Brown Rat certainly crossed a metaphorical one when large numbers swam the River Volga in 1727 on their expansion westwards.

The name 'Norway rat' stems from a mistaken belief that the animal originated in Norway and 'Hanoverian rat' from a similarly mistaken belief that it came to England by ship with George I – a convenient myth for his political opponents to perpetuate. The truth is that the Brown Rat originated in central Asia, perhaps in China, and was first recorded in England shortly before 1730, after the Hanoverians and well before it reached Norway. It very swiftly replaced the Black Rat everywhere and, unlike the modern replacement of the Red Squirrel by the Grey, it appears to have resulted from direct competition for food and habitat, perhaps with some fighting and killing too. At the end of 1777, Gilbert White wrote, 'The Norway rats destroy all the indigenous [sic] ones.'(262)

Although they have probably had no serious affect on the populations of any other mainland mammal, it is undeniable that Brown Rats account for a great many ground-nesting birds and have had a major impact on the life of small islands. A commonly recounted episode, dating from 1816, tells of Brown Rats swimming ashore from a wrecked ship to the tiny island of Priestholm in the Menai Strait, where they all but removed the resident Puffin population. In turn, Brown Rats now form an important food for a number of predators, including Foxes, domestic cats and, above all, Barn Owls.

> With accounts of rats migrating in columns, stories are sometimes linked of one of the rats carrying a straw in its mouth, with another rat holding on; or it may be that the story is of just two rats, one hanging on to the straw. In all such instances it is questioned whether this a case of a blind rat being led by another.(263)

> One evening I was walking round a Suffolk farm with a farmer friend who was carrying a .410 shotgun. We opened a barn door and there we saw two Brown Rats walking slowly across the floor. The one behind had the tail of the one in front in its mouth and seemed to be led by the front animal. My friend fired and killed both rats. I examined both and found that the one behind was quite blind with a dense cataract over both eyes. The assumption was that one rat had been leading its handicapped companion. There have been occasional stories of a sighted rat leading a blind one by means of a straw in their mouths.(264)

> I was once solemnly assured that rats perform enemas on each other by spitting through a straw but there are rare, incredible but authenticated stories of 'rat kings' where numbers of Black Rats were found together with their tails entwined, unable to get away.(265)

Brown Rats have inevitably become very important pests and, although chemical poisons are still the preferred method of control, the rats are becoming increasingly resistant to anticoagulants like warfarin. Short of some natural viral epidemic, comparable with myxomatosis, nothing seems likely to limit their continued existence and spread. On a domestic level, a larger version of the old spring-loaded mousetrap was developed to kill them, although, for many years, individual communities all had their own rat-catcher, a role now taken over by the Local Authority Rodent Control Operative. It is too often claimed that there are more rats than people in Britain today, but this is a considerable exaggeration; around 7 million is probably more accurate and means they are only the fourth commonest rodent. Nonetheless, it is estimated that about 45 per cent of agricultural premises and 3.5 per cent of urban domestic ones are rat infested.(266)

> Another 'great untrue' about rats is the story that if a rat is cornered it will leap for one's throat ... When provoked, rats can be very aggressive and savage – though they take considerable provocation before they will retaliate ... Alas, formidable as they are, they lack the biological equipment to enable them to leap as high as a man's throat. When I did a great deal of live rat catching, I kept my captives in barrels only three

At one time, rat-catchers were employed on London's underground railway to rid the system of vermin.

feet high, and I found that no rat could escape ... So, unless you are a small child, or an adult crouching on your knees, or a pygmy over in Britain for an exchange ratting scheme, your throat is really quite safe.[267]

At one time, ratting was a popular rural pastime, a dog, generally some form of terrier, was placed in a pit with a number of rats and bets placed on how long it would take to kill them.

> A rat pit at The Ship in Soar Lane [Leicester] existed until after the First World War. An open-topped cage was filled with rats and a dog was put in for one minute. The dead rats were counted and replaced by live ones. Then a second dog was put in. The process was repeated as many times as desired. The winner was the dog which killed the largest number of rats. Rats were taken clandestinely to the pit by women who carried them next to the skin under their blouses, through which they could be seen moving about. They relied on the knowledge that a rat kept in the dark and allowed relative freedom of movement will not bite. In the same public house one man used to nail a live rat to the counter and tear it to pieces with his teeth. Another man won bets at Braunston by biting live rats to death within living memory.[268]

> Although ratting was primarily a rural sport, Richard Callaby, a Cambridge dog fancier of the late nineteenth and early twentieth centuries, used to arrange rat hunts in Girton, mainly for the amusement of undergraduates, but also for any interested townspeople. According to the late Mr Noel Teulon Porter, there was, early in this century, a cellar under a house in Grantchester Meadows, Cambridge, in which undergraduates and others held rat hunts with terriers hired from Callaby. In the event of a visit from the Proctors the students could escape over the nearby fields.[269]

Black Rat *(Rattus rattus)*

OTHER COMMON NAMES: *roof rat; ship rat.*

All references to rats in Britain before the early eighteenth century are to the Black Rat *(Rattus rattus)*. Like the Brown Rat, it was originally an Asian animal, perhaps from China or South-East Asia, although it was established in India at an early date. It spread westwards along prehistoric trade routes and so the recent television re-creation of Neanderthal people living in rat-infested caves in southern France is probably quite accurate. It was long thought that Black Rats did not reach Britain until the Middle Ages, but unequivocal fifth-century remains in a Roman well in York have proved that they arrived earlier. (Rats, like Rabbits, may burrow down into early deposits where they die, much to the confusion of archaeologists.) Black Rats have travelled the world by sea and the name 'ship rat' is entirely appropriate; this is also the rat that is popularly believed to leave the sinking ship. It was also Black Rats that brought about the end of the 30 people who lived on the remote island of North Rona in 1685:

Rats leave a sinking ship in this illustration by I.N. Taber in *Century* magazine, 1894.

> The rats came ashore from a wrecked ship, and ate up the barley meal stored in sheep skins and the people eventually starved to death, the steward of St Kilda finding the last woman lying dead on the rocks with her child on her breast. The rats also starved thereafter, probably because, as ship rats, they were ill-fitted for life on North Rona and because the immense swell on the island prevents any hunting of the rock surfaces of the intertidal zone.[270]

North Rona was repopulated, at least until the nineteenth century, and gradually the Black Rat became confined to ports and warehouses. Today, the British Black Rat, especially on the mainland, is a very rare creature; there may be scarcely more than a thousand animals left. This is not least because it is very susceptible to rodenticides and, unlike the Brown Rat, has developed no resistance to them; were it not so destructive and tainted a beast it would presumably be protected. Small populations still live on Lundy, where three colour variants, one truly black and the others grey-brown, exist side by side.

Dormice
(Myoxidae)

Neither of the two species of dormouse, one native to Britain and the other introduced, is seen very often, partly because they have a limited distribution, partly because they are nocturnal and sleep in the daytime but also because they hibernate and sleep from October to April.

'Dormouse' is a wholly appropriate name but its origin is puzzling. It is Middle English and is presumed to be related to the French verb *dormir*, meaning 'to sleep', although the *Oxford English Dictionary* points out that the most obvious correlation is with the French *dormeuse*, 'a sleeper', which was unknown before the seventeenth century.

Dormice are delightful creatures, intermediate in many respects between true mice and squirrels and, like squirrels, endowed with beautiful bushy tails. They feed principally on nuts, fruit and fungi and build spherical summer breeding nests, usually fairly high in a tree, although their winter hibernation nests are generally below ground.

Edible Dormouse *(Myoxus glis)*

OTHER COMMON NAMES: *fat dormouse; glis; glis glis.*

The Romans ate dormice and this was the species that they 'farmed'. It is a grey-brown animal, the largest of all dormice and rather similar to a young Grey Squirrel. It occurs widely in Europe, although it is not native to Britain. The Romans fed their dormice on chestnuts and currants and then, finally and grotesquely, fattened them in earthenware jars called *gliaria.* They were stuffed for cooking and were very highly regarded. (They still are: I was recently served dormouse with spaghetti in Rome.) It is perfectly possible that the Romans brought Edible Dormice to Britain but no traces or records of them have been found. Strangely, remains of another European species, the Garden Dormouse *(Eliomys quercinus)*, have been discovered on Roman sites and so presumably the Romans also ate this.

The Edible Dormice in Britain today are descendants of a group that was imported from central Europe by the Hon. Walter Rothschild and released at Tring Park in Hertfordshire in 1902. They multiplied rapidly and spread further into Hertfordshire and Buckinghamshire, where there may now be

The Edible Dormouse *(Myoxus glis)* is larger and more bushy-tailed than the Common Dormouse *(Muscardinus avellanarius)*.

The strictly nocturnal Common Dormouse *(Muscardinus avellanarius)* has a deserved reputation for sleeping.

around 10,000 and where they have become familiar in attics, thatched roofs and similar places, generally unseen but revealing their presence by their noisy activities. Why they have never extended their range beyond the Chilterns, when there is nothing to stop them, is a complete mystery.

> When I lived in Little Gaddesden, Berkhamsted, Hertfordshire (from 1953–1989), we had charming little animals called Glis-Glis living in our attic. They were like small squirrels with bushy tails but had a white ring round their eyes which made their eyes look enormous. They did not eat anything in the house but used to go out into the woods to feed. Occasionally we used to see them in the kitchen, clinging to the curtains.[271]

Common Dormouse *(Muscardinus avellanarius)*
OTHER COMMON NAMES: *chestle-crum; derry-mouse; derrymouse; dorymouse; dozing-mouse; hazel dormouse; sleeper; sleepmouse.*
This is the native dormouse, smaller and less bushy tailed than the Edible Dormouse and largely restricted to deciduous woodland in southern England and Wales, infrequently coming down to ground level. It is typically found among Hazel trees (hence *avellanarius*, from *Corylus avellana*, the Hazel), feeding on the nuts, and has suffered from the decline in routine coppicing of Hazel plantations as well as the overall loss of suitable habitats.

Dormice construct a pretty little summer nest using the stripped bark of Honeysuckle but, although they are said to be 'strictly nocturnal' and generally hibernate from October onwards, clearly not all individuals have read the textbooks:

> In November for five days, a dormouse fed at our bird feeder. We had never seen one before.[272]

Dormice are beautiful and endearing, and Shakespeare used them as a symbol of sleep.

> She did show favour to the youth in your sight only to exasperate you, to awake your dormouse valour, to put fire in your heart, and brimstone in your liver
> (*Twelfth Night*, Act 3, Scene 2)

Lewis Carroll, however, was their undoubted champion:

> The Dormouse slowly opened its eyes. 'I wasn't asleep,' it said in a hoarse, feeble voice, 'I heard every word you fellows were saying.'[273]

Coypu
(Myocastoridae)

Coypu *(Myocaster coypus)*
OTHER COMMON NAMES: *nutria.*
After the Muskrat (see page 469), the Coypu is the second of the large exotic rodents that have become established in the British Isles and then eliminated (apart from the two species of porcupine, *Hystrix cristata* and *Hystrix indica*, which once escaped from collections but lived wild and free for only a very few years). Coypu were in a rather special league, however, in terms of the threat that they posed to large areas of waterside habitat. Indeed, 40 years ago, there seemed every likelihood that they would become not only permanent members of the British fauna but also an extremely serious problem.

They are impressive South American animals, among the largest of rodents and superficially like very big rats with sumptuous bright orange incisor teeth. (More than one has graced a fairground side-show as 'The largest rat in the world' or 'Giant rat caught in the sewers of London'.) Like Muskrats, they were introduced to Britain and other parts of Europe through the medium of fur farms, their fur being known as 'nutria'.

'Coypu' is a native South American name that came into English at the end of the eighteenth century, although 'nutria' is the Spanish word for Otter (from *Lutra*, the Latin name of the Otter) and became applied to Coypu fur in the nineteenth century. When the Spaniards arrived in South America and first saw Coypu, they thought they were Otters. The first animals reached Britain in 1929 and, within three years, some were at large, at first mainly in Hampshire, Sussex, Surrey and the East Anglian counties of Norfolk and Suffolk. The East Anglian populations were the most successful and self-sustaining, and they found the habitat of the fens and the Broads very much to their liking.

The impact of these large animals was not at first apparent, partly because they were believed to be slow breeders and vulnerable to cold winters. Nonetheless, by the 1950s it was evident that the Coypu had been misjudged. They did suffer in the very cold winter of 1946–7 but recovered with renewed vigour. They

Wild Coypu *(Myocastor coypus)* became established soon after its introduction in 1929, wreaking havoc along waterways.

undermined river banks, damaged reed beds and other waterside plant life (thus bringing about further instability and enhancing the likelihood of flooding), disturbed the breeding sites of several aquatic bird species, caused damage to commercial crops such as potatoes and sugar beet, and also raided gardens.

> I was anxious to obtain the first sound recordings of this strange animal and so I went to Horsey Mere ... on a dark December night in 1953. Just after ten-thirty we heard a sound like the distant moan of a destroyer's siren ... then we heard a splashing as the newcomer reached the shore and climbed the bank. Once more came the long moan; then, suddenly, came a strange cacophony – a series of deep, weird pumping screams that grew in intensity and pitch, broken only by brief pauses in which anxious breaths were taken. Slowly the anguished screams died away; then once more the moaning scream broke out and reached a new peak of sound and passion.[274]

By the time that the Ministry of Agriculture, Fisheries and Food began its anti-Coypu campaign in 1962, there were around 200,000 animals across East Anglia. The exceptionally hard winter of 1962–3, however, accounted for vast numbers (probably up to 90 per cent) and this gave a huge boost to what became an eradication programme. Whereas Muskrats had been killed directly by the type of trap used against them, and much other wildlife suffered in consequence (vast numbers of Moorhens and Water Voles were killed), the Coypu campaign used live traps that enabled the trappers to shoot the Coypu and release other creatures. By 1987, the Coypu was extinct in Britain. 'It required 24 trappers, setting 48 traps per day, or 216,000 trap-nights per year, and cost £2.5 million.'[275] More than any other episode in Britain in recent times, the Coypu affair demonstrated the folly of liberating or allowing the escape of exotic animals into an alien environment.

Hares and Related Mammals
(Lagomorpha)

Once classed as rodents, although it is now thought possible that they are not even closely related to them, lagomorphs are among Britain's most numerous mammals by virtue of the fact that, while they are few in species, one of those species is the Rabbit. Apart from a small group of burrowing creatures called pikas, which do not occur in Europe, all lagomorphs are either rabbits or hares, herbivorous animals with long back legs, long ears and an ability to run very fast. They all share an unusual habit called 'refection', a delicate way of saying that they eat a special form of their own faeces because their food must pass twice through the digestive system and they will die if this is denied them for any length of time.

Hares and Rabbits
(Leporidae)

Both species of British hare may be native and, although never particularly common or abundant, they have accumulated what is probably the greatest fund of lore, myth and mystery of any British mammal. However, no other animal has suffered a greater decline in public recognition; largely, I am sure, because that relative newcomer, the Rabbit, has caught up and overtaken it. From a folklore standpoint, the hare is an animal of the past; sadly almost no-one in Britain now adheres to, or even remembers, the numerous old hare-based customs and traditions.

Brown Hare *(Lepus europaeus)*

OTHER COMMON NAMES: *aunt Sarah (East Anglia); bandy (Suffolk); bautie/ bawtie/ bawd/ bawtie/ bawty/ betty (Scotland, northern England, East Anglia); bowen (Suffolk); bun (Scotland, northern England); capron; caproun; common hare; cuttie; donie; European hare; farmer; fennel (female animals); fuddie; great-hare (three-year old animals); hallan-chacker; jack-hare (male) (Devon); Katie (Cumberland); lagos; lavrock/ laverock; leveret (young animals); malkin/ maudkin/ maukin/ mawken (and numerous variants) (Scotland, northern England); old aunt (East Anglia); old Sarah/ Sarah/ old Sally/ Sally (East Anglia); poor wat (northern England, Norfolk, Cornwall); puss (northern England); scavernick (Cornwall); skyper; whiddie (Aberdeenshire, Banffshire); wintail (East Yorkshire).*

Most people are familiar with the appearance of a Rabbit but, when they first see a hare at close quarters, they are invariably surprised by its much greater size. A fully grown Brown Hare is a large animal and it runs extremely fast; the expression 'to kiss the hare's foot', meaning 'to be late', alludes to the notion that, if you hesitate, the hare will have gone and its footprint will be all that is left. As large, fleet-footed animals, it is small wonder that hares have been beasts of the chase for centuries, sometimes rated even higher than deer, and 'to run with the hare and hunt with the hounds' has long been a synonym for playing a double game. In earlier centuries, they were hunted with dogs and bows and arrows. The advent of firearms meant that many were inevitably shot, although in some areas, especially Ireland, they were killed with a 'throwing stick' (a primitive boomerang), a stone or even caught in pit traps.

Today, hares, after Foxes, are the most numerous prey of hunting activities in Britain. In 2000, Lord Burns' Inquiry (see Fox, page 411) estimated that about 1,650 were killed annually by around 100 registered packs of Beagles, Basset Hounds and harriers, the Beagles being the most numerous; this was a very small percentage of the number killed by shooting, which the inquiry estimated at around 300,000. ('Harrier' is a sixteenth-century name for a dog that pursues hares; its origin seems different from that of the harrier birds, see Marsh Harrier, page 246). While hunting with harriers is on horseback, hunting with Beagles and Basset Hounds is on foot. The hares are hunted between late August and the end of March, although there is no statutory close season, and the figures cited show that hunting is of no significance in the control of hare numbers; nor is hare-coursing.

Hare-coursing is an ancient sport, probably introduced to Britain by the Romans, who also set lions against Christians, and this does not make it any more justifiable. There are still about 24 hare-coursing clubs in Britain, as well as much unregulated activity. Typically a hare is driven by beaters onto the coursing field, where dogs are released to pursue it. The average chase takes 40 seconds. Most of the dogs in regulated coursing are Greyhounds, but Whippets, Salukis and others are also used; Lurchers are generally used in unregulated coursing. The objective is to test the dogs' skill and, if a hare is caught, it is expected that an attendant will quickly despatch it. For the hare, being pursued by a Greyhound is generally the best option because Greyhounds hunt by sight alone and so, if the hare is able to run into long grass where the dog cannot see it, the chase is over. An even more acceptable pursuit, if

Hares are traditionally hunted on foot and these Old Berkeley Beagles at Penn, Buckinghamshire are straining to pick up the scent.

you happen to be a hare, takes place in Greyhound track-racing, where the fast-moving 'hare' that provides the stimulus for the dogs to race is simply a small, wire-borne electronic device. However, most acceptable of all, for both dogs and hares, is the nineteenth-century public-school game Hare and Hounds, which is also the commonest of the hare-inspired inn names. The game is sometimes called a paper-chase, all parties are human and one participant, the hare, lays a trail of pieces of paper for the pursuers, or hounds, to follow. Understandably, in this environmentally conscious age, throwing waste-paper around the countryside is no longer as tolerable as it was to Tom Brown and his contemporaries in Thomas Hughes' *Tom Brown's Schooldays* (1857):

> 'Please sir, we've been out Big-side Hare-and-hounds
> and lost our way.'[276]

The origin of the Brown Hare as a British mammal is something of a mystery. There is plenty of archaeological evidence of its presence from the Iron Age onwards but not much before. Julius Caesar's remark about the Ancient Britons is often quoted: 'Hares, fowl and geese they think it unlawful to eat, but rear them for amusement and pleasure'.[277] This clearly confirms that the hare was here in pre-Roman times and, if it was introduced before then, begs the question by whom. If it was native then, why are there no more archaeological remains; or is the problem simply the difficulty of distinguishing the bones of the Brown Hare from those of the Mountain Hare, which is certainly native?

As befits an animal with a deep, long-standing folklore, the hare has accumulated more regional names than any other British mammal. 'Hare' itself is from Old English; the female is a 'doe' and the male is a 'buck', often called a 'jack-hare'. 'Jack' is a rather common way of denoting a male animal (witness 'jack rabbit' and many bird names) and sometimes the female hare may be called a 'gill' in consequence. 'Leveret' for the young is Late Middle English, from the Old French *lièvre*, meaning 'a hare', from which 'laverock' presumably comes. The name 'puss' betrays an old but mysterious linking of hares with cats; many hare names in Wales include *cath*, the Welsh word for cat, while 'mawken', 'Sally' and comparable names are said to be derived from the names frequently given to pet cats. Cats, like hares, were once considered the familiars of witches, and cats were also linked with hares in being the alternative choices of sacrificial animal to incorporate into a new building. 'Bandy' is a reference to the curved hind legs. I wonder, however, about 'bautie', 'bawd', 'bawty' and similar words. It is sometimes said that these too are cat names, but most words with comparable spellings have an association with sexuality or lewdness, and I cannot help but link them

In the old public-school game of Hare and Hounds, one boy (the 'hare') sets off first, laying a paper trail for the others (the 'hounds').

In spring, Brown Hares *(Lepus europaeus)* may sometimes be seen 'boxing', behaviour that led to the expression 'Mad March hare'.

with the once very widely held but obscure belief that hares change sex from one year to the next.

Apart from their much greater size, hares differ most obviously from Rabbits in their behaviour. They are mainly solitary, not colonial, and do not burrow. Instead they use shallow resting hollows, or depressions in fields or long grass, which have been called 'forms' (because they take the form of the animal that rests in it) since 1575. (The expression 'to start a hare' once meant to disturb it from its form, although it has since passed into the language as meaning the introduction of an irrelevant point in an argument.) Like Rabbits, hares are very much animals of the open countryside, where they may cause significant damage on farmland by feeding on cereals and root crops. Hares sometimes feed in groups but usually, when more than one animal is seen, they are indulging in courtship or mating behaviour: several males pursuing one female, older or dominant males chasing away potential rivals or, most famously, the celebrated 'Mad March hare' behaviour in spring, in which two animals rear up on their hind legs and appear to be boxing. Almost invariably, one of the two boxers is a male and the other an unreceptive female spurning his advances. It is this rearing-up on their ungainly long hind legs that has resulted in the grotesque and slightly disturbing semi-human form in which hares are often depicted. The apparent stupidity of the hare's behaviour was most famously commented on by Lewis Carroll in *Alice's Adventures in Wonderland*:

> The Hatter shook his head mournfully. 'Not I!' he replied. 'We quarrelled last March – just before he went mad, you know – ' (pointing with his teaspoon at the March Hare).[278]

However, the imagined madness of hares goes back much further, and led to the expression 'hare-brained' (1548):

> A hare-brain'd Hotspur, govern'd by a spleen ...
> (*Henry IV Part 1*, Act 5, Scene 2)

Brown Hares occur naturally throughout Britain except the northern part of Scotland, where they are replaced by the Mountain Hare. However, they have a patchy distribution; for instance, they are particularly numerous in East Anglia but sparser in the West Country and much of Wales. The Mountain Hare is the hare of Ireland, although Brown Hares were introduced there on several occasions, mainly in the nineteenth century for coursing, and have become established in the north-west. Brown Hares are still fairly numerous in Britain (the national population is in excess of 800,000) but, judging from the evidence of game bags, have clearly declined over the past century; the reasons seem mostly related to changes in farming practices, including the removal of hedgerows that give them valuable cover. An increase in the number of predators (principally Foxes) may also be significant.

The Brown Hare really is a beautiful brown in colour but has black tips to its ears. Before I had even seen a live hare at close quarters, I, like many another boy, was familiar with its fur because I used it to make that classic, ubiquitous and extremely good Brown Trout fly, the March Brown. Brown Hares do not change colour in winter but albino hares are not rare, and neither are black, melanistic animals. In folklore, a white hare had special significance, sometimes simply bringing bad luck (whereas a black one was good), but were rather commonly thought to represent the returning soul of someone, such as a crossed lover, who had died in tragic circumstances. Although many people (including some of my own family) still say 'Rabbits' or 'White Rabbits' on the first day of the month (if they remember), rather fewer still say, as folk once did, 'Hares, hares' out loud as they go to bed on the evening before.

Although eating hares, like eating geese, has traditionally been taboo in Celtic parts of Britain, elsewhere they have generally been appreciated as offering a substantial and very tasty meal, although there have always been caveats:

> ... some old country people refuse to eat hare, even where it still thrives, for it seems to have a greater volume of blood in its body than other animals of its size, which makes the meat dark. As a result, it is believed that if a pregnant woman eats hare she will have a miscarriage.[279]

On a related theme, there was a widespread belief that, if a pregnant woman saw a hare, her child would have the congenital facial abnormality of a fissure in the mid-line of the upper lip, known by analogy with the face of the hare as a 'hare lip'. A hare's paw, if carried in the right-hand pocket, was widely believed to ward off cramp and rheumatism. Samuel Pepys swore by it, and not only for rheumatism:

> Now I am at a losse to know whether it be my hare's foot which is my preservative against wind, for I never had a fit of the collique since I wore it.[280]

One of the most widely quoted culinary aphorisms is 'First catch your hare ...' but, as far as I can determine, no-one actually said it. It is most often attributed to Hannah Glasse, author of *The Art of Cookery Made Plain and Easy* (1747), but her advice was different: 'Take your hare when it is cased,' in which sense 'case' does not mean 'to catch' but 'to skin'.[281] In any event, once the hare is to hand, it may be hung, roasted and rather grotesquely served in the traditional manner, as if sitting on the plate and complete with its head and ears. A better, tastier, more succulent and certainly more famous way of preparing hare is jugged, which Mrs Beeton considered 'Very Good':

> Ingredients – 1 hare, 1½ lb gravy beef, ½ lb butter, 1 onion, 1 lemon, 6 cloves, pepper, cayenne, and salt to taste, ½ pint port wine.
> Mode – Skin, paunch, and wash the hare, cut it into pieces, dredge them with flour, and fry in boiling butter. Have ready 1½ pints of gravy made from the above proportions of beef, and thickened with a little flour. Put this into a jar; add the fried pieces of hare, an onion stuck with 6 cloves, a lemon peeled and cut in half, and a good seasoning of pepper, cayenne and salt; cover the jar down tightly, put it up to the neck into a stewpan of boiling water, and let it stew until the hare is quite tender. When nearly done, pour in the wine, and add a few forcemeat balls, fried or baked in the oven for a few minutes before they are put to the gravy. Serve with red currant jelly.
> Time – 3½ to 4 hours. If the hare is old, allow 4½ hours. Average cost – 7s 6d. Sufficient for 7 or 8 persons. Seasonable from September to the end of February.[282]

There is certainly general agreement that the hare is winter fare. 'A hare is ever good but best from October to Lent,' according to *A Propre New Booke of Cokery* of 1545.[283]

I can find no special or obvious reason why hares should have accumulated a lore and legend so much out of proportion with their familiarity or behaviour. They have never been common enough for many people to see one every day and therefore feel they had to comment, yet in Britain and further afield, there are whole books about them. George Evans and David Thomson devoted 262 pages of *The Leaping Hare* (1972) to a survey of just about everything ever written or said about hares but still could not offer an explanation.[284] Admittedly, hares do a few strange things, most notably indulging in the 'Mad March' ritual, which is readily explained; they also sometimes appear in considerable numbers sitting or running in rings, which is not; and for parts of the year, they seem to disappear from view and presumably live a secretive existence. However, none of this persuades me that they are deserving of such remarkable attention.

Like a number of other animals, the hare has long held a special significance for people with normally hazardous occupations. Fishermen would never utter the word 'hare' while at sea, and might not even set off if they saw a hare on the way to their boats. Miners, similarly, would stay away from the pit. The commonest and most widespread of hare beliefs, however, at least in Britain, relate to their association with witches. Witches were thought capable of turning themselves into hares and the resulting so-called 'witch-hare' could be killed only with a silver bullet 'or by placing Rowan or Vervain behind the gunstock'.[285] If a witch-hare was wounded and comparable wounds appeared at around the same time on an old woman of the

The characteristic black tips to the ears are clearly shown in this Brown Hare *(Lepus europaeus)*, sitting in a stubble field in Norfolk.

neighbourhood, this was considered evidence enough that she was a witch. A typical witch-hare story from Somerset concerns the oddly named Black Smock Inn in the village of Stathe. It is said that an old woman who once lived in the house cast spells on the local cattle and generally made a nuisance of herself to local farmers. Finally, they decided to have their revenge but she spotted them approaching. She was unable to escape through the back of the property, however, and instead escaped up the chimney and ran off over the moors in the shape of a hare. While making good her escape, her smock was burned and so the building was named in memory of the event. There is a common association between hares and fire, the sight of a hare being thought to presage a fire. Enid Porter cited a real-life example in the *Cambridge Independent Press and Chronicle* in 1956:

> A curious point arises in connection with a fire at Fordham. It is part of the old Cambridgeshire folk lore that a fire always follows if a hare runs down the main street of a village. The week before last a hare did run down the main street of Fordham. It was pursued by Mr Richard Nicholls, a septuagenarian, and it was killed in a shed within three yards of the place where this fire broke out.[286]

If you buy an Easter greetings card today, there is a good chance that it will bear an illustration of a Rabbit, an Easter Bunny. This connection is relatively recent, however, and the traditional Easter animal was once the hare, an alliance that pre-dated Christianity and probably originated with pagan rites of fertility and the return of spring. The modern word 'Easter' is derived, through Old English, from Eastre, a pagan festival held at the vernal equinox; according to the Venerable Bede, this in turn took its name from Eostre, the Teutonic goddess of the dawn, whose sacred animal was the hare. At one time, hare hunts were held at Easter in many parts of Britain but today they survive only in the still fairly widespread children's game of finding Easter eggs hidden in a supposed hare's nest. At Coleshill in Warwickshire an old challenge was for the young men of the parish to catch a hare and bring it to the parson before 10 o'clock on Easter Monday. In return, they would be rewarded with a calf's head, a hundred eggs for their breakfast and a groat. I am assured that Vicars of Coleshill do not regret the passing of the custom.

Leicestershire seems to have been a stronghold of Easter hare activities. Roy Palmer describes one in Leicester itself:

> Part of the area, near to the present Convent of St Catherine, was known as Black Annis' Bower-Close. On Easter Monday a fair was held there, attended by the mayor in his scarlet gown and other officials, accompanied by the waits. In the morning there were various entertainments and sporting events, followed by a hare-hunt at noon. Originally a real hare was hunted, but later the hounds and huntsmen followed the trail of a dead cat soaked in aniseed. Traditionally, the trail ended at the mayor's door, so the hunt went in full cry through the streets of the town. The custom was already ancient when it was first mentioned in the town records, in 1668. It seems to have fallen into disuse about a century later, though the fair continued.[287]

In pre-Christian times, the hare was the traditional spring animal but, on Easter cards now, the Rabbit has taken its place.

The beautiful village of Hallaton in southeast Leicestershire, however, still stages its Hare Pie Scramble and Bottle-Kicking every Easter Monday, The excuse is that, 'hundreds of years ago' (Roy Palmer thinks it was probably medieval), a local woman was saved from a charging bull by a hare running across the bull's path and distracting it. As a token of her appreciation, she bequeathed some land to the rector, the sole condition being that a hare pie should be distributed annually to the parishioners (so much for the hare's good turn), together with a quantity of ale. The pie is cooked on Easter Monday in an 8-centimetre (20-inch) square tin at the Bewick

Arms or at the premises of the Torch Trust for the Blind. The ingredients include two hares, 4 lb flour, 2 lb lard, 3 lb onions, 7 lb potatoes and seasoning (which sounds pretty solid fare). The pie is paraded through the village, from the Fox Inn to the gate of the magnificent St Michael's Church. Borne behind the pie in the procession are the three 'bottles' that will be used in the bottle-kicking match, which appears to be a more recent addition to the ritual. In reality, these are small casks rather than bottles; two are brown and filled with about a gallon of ale while the other is painted red and white and empty. Sometimes the effigy of a hare is borne aloft on a pole.

The pie is then cut up and distributed by the rector to the crowd. The remainder is put into sacks and carried away in a further procession that ends at the Hare Pie Bank, a site that is believed to have once been a place of pagan worship. There the remains of the pies are scattered and scrambled for. At one time, penny loaves were distributed as well, 'a common form of survival in sacrificial customs'. Meanwhile the Bottle-Kicking match takes place between Hallaton and the neighbouring village of Medbourne. Mayhem ensues as the two teams of unlimited numbers of competitors endeavour to kick one of the casks over their respective touchlines, which are about 1.2 kilometres (¾ mile) apart. Inevitably, and in the best traditions of British village life, it all ends in beers; the losers have to watch the winners drink the casks dry.

However, hare mystery is not all ancient and in recent times there has been considerable attention and publicity given to the extraordinary accumulation of hares on airfields. It was noticed almost as soon as there were aeroplanes. A First World War pilot was quoted by George Evans and David Thomson:

> Invariably two or three hares would get up and race parallel to our line of flight. Partridges did the same thing. When we had a bag of two, three or four hares Captain Vincent used to land; and I got out and collected the game.[288]

This has continued, unabated and unexplained, as a report in *The Times* in 1966 shows:

> The hares of Aldergrove Airport, Belfast, are by now well known in many parts of the world. Daily they put on performances for passengers who arrive and fly out in jet airliners. Eyebrows go up and fingers point as aircraft sweep to the tarmac, and green turf rushes past the windows. Up to seventy or eighty hares often course alongside.[289]

Rather surprisingly, hares crop up rarely in heraldry, although, for reasons best known to themselves and now surely long forgotten, an ancient Derbyshire family improbably named FitzErcald had a coat of arms officially described as 'Argent, three hares playing bagpipes gules'. George Evans and David Thomson referred to a curious emblem of three creatures in a circular pattern, which appears on a number of church roofs in the vicinity of Dartmoor. This was apparently a symbol or craft sign of the old Dartmoor tin-miners and is known as Tinner's Rabbits. However, because of the long ears (and their own interests), the authors believed the representations were hares, although they offered no explanation of why they should be especially asociated with tin-mining. It might, I suppose, be yet another manifestation of their link with hazardous occupations but I have a different explanation, which depends on the creatures really being Rabbits (see Rabbit, page 490). The hare was famously the symbol of Corinium, Roman Cirencester, and a beautiful fourth-century floor mosaic of a hare from there is today displayed in the town's Corinium Museum.

The many people with the surname 'Hare' claim, probably justifiably, that their ancestors were fast runners. Likewise, we presumably need look no further than the obvious to find the inspiration behind the epithet of the only English monarch to have been named after a native animal: Harold I, known as Harefoot, the son of the Danish King Cnut and his mistress Aelgifu of Northampton, who reigned unspectacularly from 1035 to 1040. Among the many place names, Harehope in Northumberland is 'hares' valley', Hargrave in Cheshire is 'hares' grove' and Harley in Shropshire may be 'hares' wood'. There are several pitfalls, however; Harefield in Middlesex is nothing to do with hares and is probably the 'feld of the army', while Harewood in Hampshire is probably 'grey wood' because *hār* means 'grey' in Old English.

A fourth-century mosaic of a hare, now displayed in the Corinium Museum, Cirencester.

The Mountain Hare *(Lepus timidus)* has shorter ears than the Brown Hare and its grey summer coat turns white in winter.

Mountain Hare *(Lepus timidus)*

OTHER COMMON NAMES: *alpine hare; blue hare; Irish hare; tundra hare; variable hare; white hare.*

Older books will tell you that there are three British species of hare because the large hares of Ireland were long thought to be distinct from those of Scotland. Now they are both considered to be subspecies of the Mountain Hare, a creature with the best claim to be the native British hare. It has been hunted and eaten for millennia; Stone Age cave remains of Mountain Hares commonly show the cut marks of stone tools on the bones. It is overall a smaller animal than the Brown Hare, with shorter ears and a greyish summer coat that turns white in winter, although only partially so in Irish animals. The name 'blue hare' refers to the beautiful bluish tint to the undercoat, which shows through on the animal's sides. It is a hare of the north and the uplands and its range extends as far as the snow-line. Its general behaviour is similar to that of the Brown Hare, although it seems to lead a less solitary existence.

> On the mountains Blue Hares lope away from you in their curious dog-like fashion or sit quite still in small groups. They seem vulnerable on Highland roads; I once counted 39 corpses on 10 km of mountain road.[290]

Like Brown Hares, they live in an above-ground form, sometimes called a 'couch', although they may dig shallow shelters for the leverets. Mountain Hares occur on the Scottish islands and, unlike Rabbits, are good and enthusiastic swimmers and have been known to venture across really rough sea. Mountain Hares were introduced to the Derbyshire Peak District in the 1920s.

Rabbit *(Oryctolagus cuniculus)*

OTHER COMMON NAMES: *batty; bun; bunk; bunny; capron; caproun; clargyman/ clergyman (Cheshire: black rabbit); coney; conig; cony; cuning; cunning; cunyng; jack-sharp; kinnen; kinning; kinnon; kiunin; kjunen; kyoneen; map; mappy; parson (Somerset: black rabbit); rabbert; riote; rump (young animals); scurel; sharpling; sharpnails.*

> The primroses were over. Towards the edge of the wood, where the ground became open and sloped down to an old fence and a brambly ditch beyond, only a few fading patches of pale yellow still showed among the dog's mercury and oak-tree roots. On the other side of the fence, the upper part of the field was full of rabbit-holes.[291]

No animal has more swiftly become fully integrated into British life and literature than the Rabbit, which occurs in greater or lesser numbers throughout the country. It is now impossible to imagine Britain without these endearing furry creatures, adored by children through their stories and their pets, detested by farmers for their voracious appetites, and with a legendary fecundity and resilience to the worst that can be thrown at them. Yet, historically, they are aliens. There are no archaeological remains of Rabbits in Britain before the twelfth century, they have no Anglo-Saxon name, and they were not mentioned in *Domesday Book*. The firm conclusion today is that they were introduced by the Normans for food and fur.

Rabbits are probably native to southern France and the Iberian peninsula, from where they were taken to Italy by the Romans, who kept them in small, walled enclosures known as *leporaria*. One supposition is that the Normans first obtained Rabbits from their kingdom in Sicily and later brought them to Britain, where they kept them in similar more or less enclosed areas, called 'warrens', from which they were caught and killed. The name 'warren' is Late Middle English, with an origin in Old French meaning 'somewhere guarded'; it later came to mean anywhere densely populated, by humans as well as animals. Houses with numerous residents are still commonly called warrens, and it was at one time a familiar term for a brothel.

Rabbits are smaller and plumper than hares, with shorter back legs, and are overall fawn above, paler below, and with brown, not black, tips to the shorter ears. The short tail is white beneath and readily visible when the animal runs.

> Nature never did a crueller thing than when she gave rabbits white tails: it makes it possible to shoot them long after it is too dark to see any other quarry.[292]

There is a range of colour variants that have been used to produce the multiplicity of domestic forms, reared for pets, meat and fur. Both white and black Rabbits are seen in the wild and melanistic (black) animals are surprisingly common in some areas; the fact that they have acquired their own names is testimony to this and they have, of course, attracted their own lore.

> In Northumberland, in the 60s and 70s, it was considered by farmers etc. very unlucky and 'not done' to shoot a black rabbit.[293]

Rabbits are generally vegetarian and voracious feeders, which is why they have not endeared themselves to farmers. Unlike hares, they are also highly social animals and, in the summer, with large numbers of young in the colonies, several hundred Rabbits can commonly be seen in a single field. As every gardener knows, they are particularly adept at selecting the most succulent green crops to eat. Although Rabbits are important prey for a number of predators, including Foxes, domestic cats and the larger birds of prey, their very effective reproduction, early sexual maturity and year-round breeding mean that their numbers are seldom seriously depleted.

The name 'Rabbit' is Late Middle English, perhaps from the Norman French (*rabet* was a French term of endearment for the young, like 'bunny'); alternatively, it could be a rare, even unique, example of a British animal name with an older origin in the Walloon language because there is a Walloon word *robett* and a Flemish one *robbe*. It has often been used as a term of contempt or cowardice:

> Away, you whoreson upright rabbit, away!
> (*Henry IV Part 2*, Act 2, Scene 2)

Until the eighteenth century, the name 'Rabbit' was used only for the young animals, the adult being known as a 'coney' (plural 'coneys') or 'cony' (plural 'conies') and it was as 'conyes' that they featured in large numbers on medieval menus (see Bittern, page 212). 'Cony' is a word with echoes in other European languages and came, through Old French, from the Latin *cuniculus*, which survives in the scientific name; but note that the biblical cony was a Hyrax, an unrelated animal. The heraldic Rabbit is still called a cony and the word is sometimes used today for the fur; as with other fashion furs, like nutria and musquash, its wearing seems less objectionable if it has a more appealing name. When Rabbits were called 'conies', warrens were often called 'coneygarths'. The name 'bunny', which survives in children's language and as a local name, dates from the early seventeenth century, when the word 'bun', of unknown origin, was used for the Red Squirrel; it was transferred to Rabbits through dialect usage. Today, it is sometimes used as a term of endearment and sometimes as a female personal name; from it has come the frightful modern 'bunnikins', made familiar by the range of children's pottery manufactured by the Wedgwood company.

The Normans often chose offshore islands for their warrens, probably because the Rabbits were likely to be safer from Foxes and other potential predators, as well as from the local human population. Unlike hares, Rabbits do not like water and are most reluctant to swim. The earliest reference to a Rabbit warren in Britain is to one on the Isles of Scilly in 1176; years later, in the nineteenth century, Augustus Smith, who 'did much to haul the islands, protesting and kicking, into modern times', attempted to

'The Three Conies' inn sign at Thorpe Mandeville in Northamptonshire is a rare usage of the old name for the Rabbit.

populate the different islands of the group with differently coloured Rabbits. Even today, they are said to be mainly white on St Helen's, mainly black on Samson and dun-coloured on St Mary's. There is a reference to 2,000 skins or pelts being taken from Lundy in 1272 and many early accounts of Rabbits being caught with Ferrets (see Polecat, page 417). Warrens were also established in parks and other secure places on the mainland too.

> Initially, the right of warren was a right to hunt, and have the exclusive right to hunt, 'lesser game', an ill-defined term which seems to have included Hares, Foxes and Martens (as opposed to 'greater game', deer and boar).[294]

Among the best documented of the old mainland warrens is that in Guildford Park, which was a royal hunting park associated with Guildford Castle. There are many references to the warren there in the Calendars of Close, Liberate and Patent Rolls, the earliest in 1235:

> The sheriff of Surrey is commanded that he cause Galfrido Dispensatore to have ten live conies [*decem cuninos vivos*] in the king's park of Guildford, of the king's gift.

Then, in 1245:

> Bailiffs of Gudeford to send 60 conies to Westminster by the Friday before Christmas.

And in 1254:

> Of hunting for the use of the king's freemen. The keeper of the park of Guildford is ordered to allow the bearer of these presents to take in the king's warren [*cuningera regis*] of Guildford 40 conies for the use of the king's freemen staying at Windsor.[295]

It does appear that medieval Rabbits were less keen on excavation than their modern counterparts. The old warrens were constructed on light, free-draining soils, and pipes and artificial burrows were installed to encourage them to breed; a necessity that seems laughable today. Sometimes, the soil was mounded to improve drainage and some of these apparently inexplicable features still exist in parts of the country. By the sixteenth century, there was specific legal encouragement for the construction of warrens, along with dovecotes, parks and fish ponds, and comparable incentive for ne'er-do-wells to poach them.

> In the Middle Ages, rising waters made Brean Down [Somerset] almost an island and rabbit warrens were established here and rabbits sold for fur and food ... Artificial warrens were made on the Down and young rabbits introduced to encourage breeding. There is a well documented account of a case taken to Star Chamber. Thomas Bond of London owned the ancient Brean Down Rabbit warren, being 400 acres in size and adjoining the sea side (from the size of this, it must be that area called Warren Road). He employed two men to attend the warren and grounds. They had a dwelling place and storage for nets, ferrets, dogs and weapons. The warreners had guard against poachers who might well be armed and vicious. One Friday night in 1607 the warreners were awakened by noises from the Warren. When they went to investigate they found ten men armed with shafts, daggers and pistols, who had forced their way into the enclosure and had hunted and killed 40 rabbits. When they confronted the thieves, they were insulted and attacked physically. Nicholas Battersby of Uphill was taken before the Star Chamber and accused of the crime, unfortunately the findings of the Court are not recorded.[296]

Some old warren names survive, even though their original use is generally forgotten; Dawlish Warren in Devon is one of the best known. On light heathland soils, adjacent warrens merged together over large areas; among the largest of these was Lakenheath Warren in Suffolk (now partly beneath the modern airfield), whose history, from the twelfth century to the Second World War, was outlined by Oliver Rackham.[297] In the neighbouring Thetford Warren, the ruins of the warren lodge still exist: 'almost a castle, to remind us of the importance of Rabbits in the ecclesiastical economy and the security risk of storing quantities of valuable Rabbit-skins and salt rabbits.' Some lodges have since been put to other use; a famous Dartmoor hostelry is Warren House Inn, around which Grouse have been introduced for sport.

> Portland Bill in Dorset has its own peculiar Rabbit lore. If you want to know about Portland and all the superstitions and names surrounding rabbits I'm afraid you will have to ask a Portlander, who of course will never mention them by name.[298]
>
> Portland has always been a penal colony and when hard labour was the norm, prisoners were made to work in the famous quarries to extract Portland stone. Unfortunately these were riddled with Rabbit warrens and when an inexperienced prisoner put his pick into the stone it often caused a cave-in, sometimes resulting in injury or even death. The cry 'Rabbits' became the warning shouted to fellow workers and consequently that word is now associated still with death and bad luck. It is a word never said on the island. If you want to buy a Rabbit in the butcher's shop, you point, or ask for a furry thing or underground mutton! If a prankster writes the word on a crane or a fishing boat, they will not be used that day. And not all that long ago, a pub run by a non-Portlander on the island put the Chas and Dave record 'Rabbit' on the juke box and lost all their customers.[299]

As the centuries passed, the socio-economic status of the Rabbit changed. It inevitably escaped from the warrens (an

Large numbers of Rabbits *(Oryctolagus cuniculus)* can cause serious damage to crops and their burrowing can undermine farmland.

event that tends to be referred to as 'leakage'), although it was not until the eighteenth century that it was recognized as the serious agricultural pest that it is today. The spread of feral and ultimately wild animals meant that the Rabbit became much more readily available as a source of meat. No-one any longer needed to buy Rabbits from someone's warren; you simply went out and caught, trapped or shot one. From being a luxury food in the thirteenth century, costing 3d per carcass, by the nineteenth and early twentieth centuries it had become the poor man's supper.

> Rabbits of course were a normal part of our meals [Shropshire, 1920s]. At harvest time most of the time was spent chasing them with a stick, the sheaves lying on the ground slowing them down. Some farmers made us share them out, if we were lucky, a few were sold at six pence or a shilling.[300]

In times of hardship, the people of the British Isles still turn to Rabbits. Shortly after the Second World War, my father used to bring home Rabbits when he was on leave from an RAF base in Pembrokeshire, where they abounded on the cliff tops, and I recall being told of the welcome they received. Even more recently, I am reliably informed that Rabbits all but disappeared from large areas of Yorkshire during the coal-miners' strike of 1984–5, when local mining families had little or no income.

Rather like hare and much game, Rabbit meat is tasty but can be dry unless cooked carefully. Simply roasting it is useless. My grandfather assured me that, by eating one of his favoured delicacies, Rabbit's brains on toast, I would grow up to be intelligent. It would be invidious to comment on how accurate his prediction was but they are an acquired taste and most people wanting to eat Rabbit today will turn to the dish for which countless cooks through the ages have produced countless recipes: Rabbit pie takes some matching.

> Rich Rabbit Pie
> Time to bake, one hour and a quarter
> One rabbit; a quarter of a pound of ham, or bacon; pepper; salt, and a little pounded mace; four hard-boiled eggs; some forcemeat balls; a large cupful of good stock, or gravy; puff paste.
> Cut up and bone a young rabbit; put the bones into a stewpan with a few herbs, pepper, salt, and any trimmings you may have. Just cover with water, and let it simmer to about half a pint. Put a puff paste round the edge of a pie-dish, put the slices of rabbit in, then the bacon, or ham, hard-boiled eggs cut into pieces, and some forcemeat balls. Season the whole with pepper, salt, and pounded mace, pour in a cupful of stock, or water, cover the top with a good puff paste,

Yeovil factory workers in 1939 with their Rabbits' feet, which they hope will bring luck to their football team.

make a hole with a knife in the top, and bake it in a
rather hot oven, taking care it does not too much
colour. When done, pour through the hole at the top
the gravy made from the bones, and serve.[301]

Today, your Rabbit, if wild, will have been shot. Over the years, however, almost any method of catching and killing Rabbits was deemed acceptable, although modern legislation has outlawed the less humane, such as the gin trap, which employed a noose of thin wire, and the snare.

I hear a sudden cry of pain!
There is a rabbit in a snare:
Now I hear the cry again,
But I cannot tell from where ...
Little one! Oh, little one!
I am searching everywhere!
(James Stephens, 'The Snare')

Lying in an iron trap
He cries all through the deafened night –
Until his smiling murderer comes
To kill him in the morning light.
(William Davies, 'The Rabbit')

The killing was generally by a sharp blow to the back of the neck: the Rabbit punch, outlawed for humans in the boxing ring, but legal in the countryside. And the Rabbit, normally a silent creature, does produce an appalling scream when trapped or otherwise in pain.

By the early 1950s, the Rabbit population of Britain was vast: perhaps 100 million animals. Around 40 million were being killed each year for their fur and meat yet the annual cost to agriculture of the damage they caused was still approaching £50 million. However, from this apparently impregnable position as almost certainly the most numerous British mammal, the Rabbit was about to suffer the most catastrophic decline of any warm-blooded animal in the recorded history of these islands.

The South American Myxoma Virus was responsible. It is carried by Rabbit fleas and mosquitoes, is almost specific to *Oryctolagus* Rabbits and has been known since the end of the nineteenth century. It was introduced into Australia in 1950, initially with some difficulty, in an attempt to control the Rabbit populations there. Then Swiss scientists studying the disease liberated some infected rabbits near Paris in 1952 and, either carried by vectors crossing the English Channel or perhaps by human intervention, the disease reached England; the first myxomatosis-affected animals were found in Kent in

October 1953. Legislation was passed to prohibit the deliberate spread of the disease and affected animals but to no avail. It spread through the country like wildfire and killed around 99 per cent of the Rabbit population. Affected animals are a vision of pathos. The first time I encountered one was as a boy, in Derbyshire in the late 1950s, and I recall being very disturbed by its grotesquely swollen head and eyelids.

> An infected Rabbit ... is not a pleasant sight. It may wander about aimlessly and I once saw one trying to enter an hotel in Orford.[302]

> A baby rabbit
> With eyes full of pus
> Is the work of scientific us.
> (Spike Milligan, 'Myxomatosis')

The ecological impact of the loss of most of the Rabbit population was considerable but little that was irreversible seems to have occurred. The Rabbit population soon began to recover, partly through increased immunity to the virus, partly to the virus itself becoming attenuated, or weakened, and partly perhaps through Rabbits tending to emulate hares and nest above ground where the flea vectors are less prolific than in burrows. There are now about 37 million Rabbits in Britain.

As the Rabbit rapidly overtook the hare in both numbers and familiarity, so it took over much of the hare's folklore. It became good luck to see a black Rabbit and bad luck to see a white one. It became important to say 'Rabbits' or 'White Rabbits' on the first day of the month; with regional variants in the number of times it must be repeated. In some places, it must be said before the foot has first touched the floor when getting out of bed. I was brought up to believe that 'White Rabbits' must be said three times before any other words are spoken. There is no obvious reason why a Rabbit's foot should be a good-luck token but countless children have carried one in their pocket when taking an examination, they have been placed in babies' cots or prams to guard against evil, and no doubt many women have worn Rabbit's foot brooches, just like the one my old aunt used to pin to her blouse.

The emblem of the Tinner's Rabbits on West Country church roofs, which George Evans and David Thomson thought represented hares (see Brown Hare, page 482), might really be Rabbits after all.

> The two ears belong to each of three Rabbits as you look at it so it reminds one of the Trinity. The miners built warrens to breed Rabbits for food so they were important to them.[303]

It is easy to see why children adore Rabbits, why they are such popular pets – and why children's authors have been swift to capitalize on this. They are warm, furry, cuddly, do not usually bite, are easy to keep in captivity (albeit often in hutches far too small) and are generally undemanding and relatively long-lived (up to nine years, even in the wild). I was never without at least one Rabbit among my childhood pets. The most famous fictional Rabbit in England is still Beatrix Potter's Peter, who first saw the light of day in 1902 and has been adored ever since. More recently, however, Peter Rabbit has been supplanted, if not in long-term popularity at least in the length of the story and arguably in literary quality, by Richard Adams' hugely successful 1972 novel *Watership Down*, which made his fortune and enabled him to retire from the Civil Service.[304] It told of a colony of anthropomorphic Rabbits living in Hampshire. Unusually for a fantasy story, the places mentioned were real and the site of Watership Down itself was later bought by the composer Andrew Lloyd-Webber. In 1978 the story was filmed with huge success, not least because of the song 'Bright Eyes' sung by Art Garfunkel, and *Watership Down* has since become a minor industry, almost on a par with *Star Trek*.

Because of its relatively late arrival in the British landscape, the Rabbit has not really made its mark on British maps, apart from the places named after relict warrens. There are many towns and villages that appear to commemorate conies but it is easy to be misled: pre-Norman names like Coneysthorpe in Yorkshire or Coney Weston in Suffolk have nothing to do with four-footed beasts, only two-footed ones, and owe their origin to the Old Scandinavian word *konungr* for king.

The story of Peter Rabbit and his exploits in Mr McGregor's garden is one of the most enchanting of Beatrix Potter's tales.

SOURCE NOTES

Invertebrates

1 Philip Henry Gosse, cited in: Barrett, J., *Life on the Sea Shore*, Collins, London, 1974
2 *Pliny's Natural History*, cited in: 'Animal fossils as charms' by Kenneth P. Oakley, in: Porter, J.R. & Russell, W.M.S. [eds], *Animals in Folklore*, D.S. Brewer and Rowman & Littlefield, Cambridge, for the Folklore Society, 1978
3 Porter, Enid, *The Folklore of East Anglia*, Batsford, London, 1974
4 Sue Daly, Jersey
5 Swainson, C., *A Handbook of Weather Folk-lore*, William Blackwood & Sons, Edinburgh and London, 1873
6 Walton, Izaak, *The Complete Angler*, 6th edition, Rivington et al., London, 1797 [First published 1653]
7 Beith, Mary, *Healing Threads: Traditional Medicines of the Highlands and Islands*, Polygon, Edinburgh, 1995
8 Polson, 'Our Highland Folklore Heritage', *Northern Chronicle*, Inverness, 1926
9 Buckland, F.T., *Curiosities of Natural History*, Series I, Macmillan & Co., London, 1857
10 Swainson, 1873, see Note 5
11 Beith, 1995, see Note 7
12 Sue Daly, Jersey
13 Lovell, M.S., *The Edible Mollusks of Great Britain and Ireland*, Reeve & Co., London, 1867
14 Cited in: Lovell, 1867, see Note 13
15 Martin, Martin, *A Description of the Western Isles of Scotland*, Andrew Bell, London, 1703
16 Lovell, 1867, see Note 13
17 Step, Edward, *Shell Life*, Frederick Warne & Son, London, 1945
18 Fox-Davies, A.C., *Complete Guide to Heraldry*, T.C. & E.C. Jack, London, 1925
19 Mrs B.J. Scandrett, Oxfordshire
20 Michael Ratsey, Somerset
21 Tony Vaughan, National Snail Farming Centre, cited at www.home-workers.com
22 *Conchologists' Newsletter*, No. 26, September 1968
23 *Conchologists' Newsletter*, No. 59, December 1976
24 Stuart Bailey, Malacological Society of London
25 Wodehouse, P.G., *Over Seventy – An Autobiography With Digressions*, Herbert Jenkins, London, 1957
26 *Conchologists' Newsletter*, No. 34, September, 1970
27 Thompson, Flora, *Candleford Green*, Oxford University Press, Oxford, 1943
28 Palmer, Roy, *The Folklore of Warwickshire*, B.T. Batsford, London, 1976
29 See Note 26
30 *Conchologists' Newsletter*, No. 43, December 1972
31 See Note 30
32 Udal, John Symonds, *Dorsetshire Folklore*, 2nd edition, Toucan Press, St Peter Port, 1970 [First published by Stephen Austin & Sons, Hertford, 1922]
33 See Note 30
34 Anon., *Songs for the Nursery*, William Darton, London, 1825
35 White, Gilbert, 'Naturalist's Calendar', in: *The Journals of Gilbert White*, Vol. 1, [ed. Francesca Greenoak], Century, London, 1986
36 Stella M. Davies, in: *Conchologists' Newsletter*, No. 44, March 1973
37 Opie, Iona & Peter, *The Lore and Language of Schoolchildren*, Clarendon Press, Oxford, 1959
38 *Sunday Post*, 9 July, 1978
39 *Western Morning News*, 6 December 1999
40 *The Times*, 7 August 1865
41 Environment Agency, Press Release, 9 October 1998
42 See Note 30
43 *Conchologists' Newsletter*, No. 68, March, 1979
44 Cox, Ian [ed.], *The Scallop*, Shell Transport & Trading Co., London, 1957
45 Bolitho, H. [ed.], *The Glorious Oyster*, Sidgwick & Jackson, London, 1960
46 Bolitho, 1960, see Note 45
47 Lucas Phillips, C.E., *Cockleshell Heroes*, William Heinemann, London, 1956
48 Sue Daly, Jersey
49 Wesley-Smith, Peter, *The Ombley-Gombley*, Angus & Robertson, Sydney, 1969
50 Communicated by Stella Turk, Cornwall
51 Chinery, M., *Garden Creepy-crawlies*, Whittet Books, London, 1986
52 Michael Ratsey, Somerset
53 Chinery, 1986, see Note 51
54 Mrs Rita Stone, Hertfordshire
55 Hillyard, Paul, *The Book of the Spider*, Pimlico, London, 1995 [First published 1994]
56 White, Gilbert, *Natural History of Selborne*, new edition, George Routledge, London, 1880 [First published 1789]
57 Robert Hooke, quoted in: Bristowe, W.S., *The World of Spiders*, Collins, London, 1971 [First published 1958]
58 Dr W.G.G. Loyn, Aberystwyth
59 Keith Richard Power, Lancashire
60 Anon., *Ladybird Nursery Rhymes*, Ladybird Books, Loughborough, 1994
61 Marwick, Ernest W., *Folklore of Orkney and Shetland*, Batsford, London, 1975
62 Hillyard, 1994, see Note 55
63 Muffet, Thomas, *Edward Topsell's History of Four-footed Beasts and Serpents and Insects*, Vol. 3. *The Theater of Insects*, Frank Cass & Co., London, 1967 [First published 1658]
64 Opie, Iona & Peter [eds], *The Oxford Dictionary of Nursery Rhymes*, 12th edition, Oxford University Press, Oxford, 1980 [First published 1951]
65 Bristowe, 1958, see Note 57
66 Scott, Walter, *Tales of a Grandfather*, Robert Cadell, Whittaker & Co., London, 1829
67 Martin Hannan, at Silver Scots Place, http://personal.nbnet.ca/
68 Hillyard, 1994, see Note 55
69 Gunther, R.T. [ed.], *The Diary and Will of Elias Ashmole*, [edited and extended from original manuscripts] Oxford University Press, Oxford, 1927
70 A.G. Lounsbach, Somerset
71 Bristowe, 1958, see Note 57
72 Gibson, F., *Superstitions About Animals*, Scott Publishing Co. Ltd, London, 1904
73 Mary Shave, Suffolk
74 Miss E. M. Howes, West Sussex
75 Dr A. P. Radford, Somerset
76 Long, W.H., *The Dialect of the Isle of Wight with Tales and Anecdotes*, Reeves & Turner, London, and G.A. Brannon & Co., Newport, 1886
77 Corbet, Philip S., Longfield, Cynthia & Moore, N.W., *Dragonflies*, Collins, London, 1960
78 Hammond, Cyril O., *The Dragonflies of Great Britain and Ireland*, 2nd edition [revised by Robert Merritt], Harley Books, Colchester, 1983 [First published 1977]
79 Chinery, M., *Insects of Britain and Northern Europe*. Collins, London, 1979
80 Dickens, Charles, *The Cricket on the Hearth. A Fairy Tale of Home*, printed and published for the author by Bradbury & Evans, London, 1846
81 Porter, Enid, *Cambridgeshire Customs and Folklore*, Routledge & Kegan Paul, London, 1969
82 Keogh, John, *Zoologia Medicinalis Hibernica*, S. Powell, Dublin, 1739
83 Burke, Edmund, *A Letter to a Member of the National Assembly; in Answer to his Book on French Affairs*, Dodsley, Owen & Rivington, London, 1791
84 Muffet, 1658, see Note 63
85 Gordon, D.G., *The Compleat Cockroach*, Ten Speed Press, Berkeley, California, 1996
86 Muffet, 1658, see Note 63
87 Captain John Smith, quoted in: Gordon, 1996, see Note 85

88 Hunt, R., *Popular Romances of the West of England*, 3rd edition, Chatto & Windus, London, 1881
89 Opie, Iona and Peter, 1959, see Note 37
90 Keogh, 1739, see Note 82
91 Southall, John, *A Treatise of Buggs*, J. Roberts, London, 1730
92 Opie, I. & P., *The Oxford Nursery Rhyme Book*, 8th edition, Oxford University Press, Oxford 1977 [First published 1955]
93 Harris, Moses, *The Aurelian*, printed for the author, London, 1766
94 Petiver, J., *Papilionum Britanniae Icones*, printed for the author, London, 1717
95 Williams, C.B., *Insect Migration*, Collins, London, 1958
96 Salmon, Michael A., *The Aurelian Legacy*, Harley Books, Colchester, 2000
97 Ford, E.B., *Butterflies*, Collins, London, 1945
98 Susan Major, St Agnes C of E Primary School, Isles of Scilly
99 Ford, 1945, see Note 97
100 Tony Loy, Sorby Natural History Society, Sheffield
101 Clinton Keeling, Surrey
102 Ford, 1945, see Note 97
103 Harry Gleason, Cambridgeshire
104 Susan Major, St Agnes, C of E Primary School, Isles of Scilly
105 Udal, 1970, see Note 32
106 Topsell, Edward, *The History of Four-Footed Beasts and Serpents and Insects*, Da Capo Press, New York, 1967 [Reprint of 1658 edition]
107 Golding, William, *Lord of the Flies*, Faber & Faber, London, 1954
108 Bingley, William, *History of Animated Nature*, Lovell Reeve, London, 1968
109 Reiter, Paul, 'From Shakespeare to Defoe: Malaria in England in the Little Ice Age', *Emerging Infectious Diseases*, Vol. 6, 2000
110 Defoe, Daniel, *A Tour Through the Whole Island of Great Britain*, Folio Society, London, 1983 [First published 1724–26]
111 Muffet, 1658, see Note 63
112 Muffet, 1658, see Note 63
113 White, 1789, see Note 56
114 Erzinclioglu, Zakaria, *Blowflies*, Richmond Publishing Co., Richmond, 1996
115 Erzinclioglu, 1996, see Note 115
116 Muffet, 1658, see Note 63
117 Anon., 1994, see Note 60
118 Muffet, 1658, see Note 63
119 Tony Loy, Sorby Natural History Society, Sheffield
120 *Daily Telegraph*, 14 December 1998
121 Miss Elizabeth St. B. Anderson, Somerset
122 Radford, E. & M.A., *The Encyclopedia of Superstitions*, 6th edition [edited and revised by Christina Hole], Helicon Publishing Ltd, Oxford, 1995 [First published 1948]
123 Mrs K.M. Jackson, Dorset
124 Mrs Rita Stone, Hertfordshire
125 Barry J. Colledge, Newcastle upon Tyne
126 Palmer, 1976, see Note 28
127 Opie, Iona & Peter, 1955, see Note 92
128 Gwynn Jones, T., *Welsh Folklore and Folk Custom*, Methuen & Co., London, 1930
129 *Folklore Journal*, No. 27, 1916
130 *The Times*, 16 June 1999
131 Mrs M.A. Trendall, Berkshire
132 *Guardian*, 16 May 2001
133 Exell, A.W., *The History of the Ladybird*, Phoenix, London, 1999
134 Majerus, M., *Ladybirds*, Collins, London, 1994
135 Majerus, 1994, see Note 136
136 Opie, Iona & Peter, 1955, see Note 92
137 Majerus, 1994, see Note 136
138 Hugh Collins, Dorset
139 Radford, E. & M.A., 1948, see Note 124
140 Majerus, 1994, see Note 136
141 Mrs Edmund Craster, d.1864

Fish

1 Williamson, Henry, *Salar the Salmon*, Faber & Faber, London 1935
2 Waters, Brian, *The Severn Tide*, J.M. Dent, London, 1947
3 Waters, 1947, see Note 2
4 Brewer, Clifford, *The Death of Kings*, Abson Books, London, 2000
5 Waters, 1947, see Note 2
6 Royal Archives, RA PP/VIC/1900/7704
7 Mrs J. Kelsey, Royal Archives, Windsor
8 Lt Col Sir Malcolm Ross, Lord Chamberlain's Office, Buckingham Palace, London
9 Jude Tyrrell, *Science*, London
10 Marwick, Ernest W., *Folklore of Orkney and Shetland*, Batsford, London, 1975
11 Maxwell, Gavin, *Harpoon at a Venture*, Hart-Davis, London, 1952
12 Geddes, Tex, *Hebridean Sharker*, Routledge, London, 1954
13 *Daily Telegraph*, 20 September 2000
14 Buchan, David [ed.], *Folk Tradition and Folk Medicine in Scotland – The Writings of David Rorie*, Canongate Academic, Edinburgh, 1994
15 *De Prerogativa Regis* 17 Edw.II st.1, *c*.11, *c*.1324
16 Sir Malcolm Ross, see Note 8
17 Moore, A.W., *Folk-Lore of the Isle of Man*, S.R. Publishers Ltd, Yorkshire, 1971 [First published 1891 in Isle of Man and London]
18 Hardy, Alister, *The Open Sea*. Vol. II. *Fish and Fisheries*, Collins, London, 1959
19 Darwin, Bernard, *Receipts and Relishes Being a Vade Mecum for the Epicure in the British Isles*, Whitbread, London, 1950
20 Keogh, John, *Zoologia Medicinalis Hibernica*, S. Powell, Dublin, 1739
21 Hull, Eleanor, *Folklore of the British Isles*, Methuen & Co., London, 1928
22 Fox, Lilla M., *Costumes and Customs of the British Isles*, Chatto & Windus, London, 1974
23 Courtney, M.A., *Cornish Feasts and Folklore*, Beare & Son, Penzance, 1890 [Revised and reprinted from the Folklore Society Journals, 1886–7]
24 *Daily Telegraph*, 18 December 1996
25 Baker, Jenny, *Simply Fish*, Faber & Faber, London, 1988
26 Walton, Izaak, *The Complete Angler*, 6th edition, Rivington et al., London, 1797 [First published 1653]
27 Jones, J.W., *The Salmon*, Collins, London, 1959
28 Walton, 1653, see Note 26
29 Williamson, 1935, see Note 1
30 Graham Paskett, Staffordshire
31 Christian, Roy, *The Country Life Book of Old English Customs*, Country Life, London, 1966
32 Waters, 1947, see Note 2
33 Walton, 1653, see Note 26
34 James, H.F. & Bartholomew, A.T. [eds], *The Complete Works of Samuel Butler*, Bartholomew, London, 1923–26
35 Simpson, Jacqueline, *The Folklore of the Welsh Border*, Batsford, London, 1976
36 Exell, A.W., *Blockley As It Was*, Hendon Publishing, Hendon, 1990
37 Maitland, Peter S. & Campbell, R. Miall, *Freshwater Fishes*, Collins, London, 1992
38 Walton, 1653, see Note 26
39 Walton, 1653, see Note 26
40 Walton, 1653, see Note 26
41 Berners, Juliana, *The Boke of St Albans*, cited in: Maitland & Campbell, 1992, see Note 37
42 John Parsons, Norfolk
43 Pepys, Samuel, *The Diary*, George Bell & Sons, London, 1893
44 Clinton Keeling, Surrey
45 Saki (H.H. Munro), *Reginald*, Methuen, London, 1904
46 Keogh, 1739, see Note 20
47 Walton, 1653, see Note 26
48 Walton, 1653, see Note 26
49 Anon., *Ladybird Nursery Rhymes*, Ladybird Books Ltd, Loughborough, 1994
50 *Mistress King's Receipt Book*, quoted in: Kightly, Charles, *The Perpetual Almanack of Folklore*, Thames & Hudson, London, 1987 [Reprinted 1994]

51 Walton, 1653, see Note 26
52 Porter, Enid, *Cambridgeshire Customs and Folklore*, Routledge & Kegan Paul, London, 1969
53 Barrett, W.H., *Tales from the Fens*, Routledge & Kegan Paul, London, 1966
54 Walton, 1653, see Note 26
55 Marwick, 1975, see Note 10
56 Goodrich-Freer, A. 'More folklore from the Hebrides', *Folklore Journal*, Vol. 13, 1902
57 Hammond, P.W., *Food and Feast in Medieval England*, Wrens Park Publishing, Gloucestershire, 1993
58 Sue Daly, Jersey
59 Kurlansky, Mark, *Cod*, Jonathan Cape, London, 1998
60 *Daily Telegraph*, 25 January 2001
61 World Wide Fund for Nature, Press Release, 24 January 2001
62 Kurlansky, 1998, see Note 59
63 Wilde [Lady], *Ancient Cures, Charms and Usages of Ireland*, Ward & Downey, London, 1890
64 Baker, 1988, see Note 25
65 Walton, 1653, see Note 26
66 Buckland, F.T., *Curiosities of Natural History*, Series I, Macmillan & Co., London, 1857
67 Walton, 1653, see Note 26
68 Nigel Melville, Dorset
69 Simpson, Jacqueline, *The Folklore of Sussex*, Batsford, London, 1973
70 Darwin, 1950, see Note 19

Amphibians

1 Clinton Keeling, Surrey
2 Palmer, Roy, *The Folklore of Warwickshire*, B.T. Batsford, London, 1976
3 Ann Peters, St John's College School, Cambridge
4 Aflalo, F.G., *Country Life*, 22 October 1904
5 *Daily Telegraph*, 14 January 1999
6 Lever, Christopher, *The Naturalized Animals of the British Isles*, Hutchinson, London, 1977
7 Pennant, Thomas, *British Zoology*, Benjamin White, London, 1776
8 Elizabeth Derges, Devon
9 Thorndike, Lynn, *A History of Magic and Experimental Science*, Columbia University Press, New York, 1934, quoted from *Philosophical Transactions of the Royal Society*
10 White, Gilbert, *Natural History of Selborne*, new edition, George Routledge, London, 1880 [First published 1789]
11 Buckland, F.T., *Curiosities of Natural History*, Series I, Macmillan & Co., London, 1857
12 Prynne, William, *The History of King John*, Thomas Ratcliff, London, 1670
13 Brewer, Clifford, *The Death of Kings*, Abson Books, London, 2000
14 *Daily Telegraph*, 5 January 1999
15 Black, Beaver H. [ed.], *Gloucestershire Notes and Queries*, William Kent & Co., London, 1884
16 M. Trennel, Norfolk
17 Radford, E. & M.A. *The Encyclopedia of Superstitions*, 6th edition [edited and revised by Christina Hole], Helicon Publishing Ltd, Oxford, 1995 [First published 1948]
18 Radford, E. & M.A., 1995, see Note 17
19 Porter, Enid, *Cambridgeshire Customs and Folklore*, Routledge & Kegan Paul, London, 1969
20 Mrs Rita Stone, Hertfordshire
21 Florence Brojer, Berkshire
22 Morgan, Adrian, *Toads and Toadstools: The Natural History, Folklore and Cultural Oddities of a Strange Association*, Celestial Arts, Berkeley, California, *c.*1995
23 Ainsworth, G.C., *Introduction to the History of Mycology*, Cambridge University Press, Cambridge, 1976
24 Findlay, W.P.K., *Fungi, Folklore, Fiction and Fact*, Richmond Publishing, Richmond, 1982
25 Beebee, Trevor, *Frogs and Toads*, Whittet Books, London, 1985
26 Grahame, Kenneth, *The Wind in the Willows*, Methuen & Co., London, 1908
27 Medawar, P.B. & Medawar, J.D., *Aristotle to Zoos: A Philosophical Dictionary of Biology*, Harvard University Press, Cambridge, 1983
28 John Parsons, Norfolk

Reptiles

1 Mrs Jo Baker, Worcestershire
2 Michael Ratsey, Somerset
3 Bob Howard, Keynsham
4 Gurdon, Lady E.C. [collector & editor] *Country Folk Lore: Gloucestershire, Suffolk, Leicestershire and Rutland. Printed extracts. No.2. Suffolk.* Published for the Folklore Society by D. Nutt & J. Loder, Woodbridge. Pawsey & Hayes, Ipswich, 1893
5 Wagner, Leopold, *Manners, Customs and Observances*, Heinemann, London, 1895
6 Aubrey, John, *Remaines of Gentilisme and Judaisme, by John Aubrey R. S. S.*, W. Satchell, Peyton & Co., for the Folklore Society, 1881 [First published 1686–7]
7 Anon., *Choice Notes from 'Notes and Queries'*, Bell & Daldy, London, 1859
8 Reverend John Samways, Keynsham
9 Topsell, Edward, *The History of Four-Footed Beasts and Serpents and Insects*, Da Capo Press, New York, 1967 [Reprint of 1658 edition]
10 John Wood, Kent
11 Smith, Malcolm, *The British Amphibians and Reptiles*, Collins, London, 1951
12 Bob Howard, Keynsham
13 Harrison, William, *The Description of England, London* [ed. Frederick J. Furnival], N. Trubner for the New York Shakespeare Society, 1877–8 [First published 1577]
14 S. Murton, North Yorkshire
15 John Wood, Kent
16 White, Gilbert. 'Naturalist's Journal', in: *The Journals of Gilbert White*, Vol. 1 [ed. Francesca Greenoak], Century, London, 1986
17 Beebee, Trevor & Griffiths, Richard, *Amphibians and Reptiles*, HarperCollins, London, 2000
18 Glasse, Hannah, *The Art of Cookery Made Plain and Easy*, W. Strahan, London, 1774 edition [First published 1747]
19 Carroll, Lewis, *Alice's Adventures in Wonderland*, Macmillan & Co., London, 1865
20 Pennant, Thomas, *British Zoology*, Benjamin White, London, 1776

Birds

1 Swainson, C., *The Folklore and Provincial Names of British Birds*, Eliot Stock, London, 1886
2 Stevenson, R.L., *Kidnapped*, Cassell & Co. Ltd, London, 1886
3 Whitlock, Ralph, *The Folklore of Wiltshire*, B.T. Batsford Ltd, London, 1976
4 Fisher, James, *The Fulmar*, Collins, London, 1952
5 Martin, Martin, *A Late Voyage to St Kilda*, D. Brown & T. Goodwin, London, 1698
6 MacKenzie, Neil, 'Notes on the birds of St Kilda', *Annals of Scottish Natural History*, 1905
7 Clinton Keeling, Surrey
8 Penhallurick, R.D., *Birds of the Cornish Coast*, D. Bradford Barton, Truro, 1965
9 Lockwood, William, B., *The Oxford Book of British Bird Names*, Oxford University Press, Oxford, 1984
10 Reverend Eddie McKenna, North Berwick
11 Hamish Brown, Fife
12 Clinton Keeling, Surrey
13 Penhallurick, 1965, see Note 8
14 Taylor, John, *The Pennyless Pilgrimage*, London, 1618, cited in: Ferrier, Walter, *The North Berwick Story*, North Berwick Community Council, Edinburgh, 1980
15 Penhallurick, 1965, see Note 8
16 Swainson, 1886, see Note 1
17 Murton, R.K., *Man and Birds*, Collins, London, 1971

18 'Nigel', Isles of Scilly
19 Clinton Keeling, Surrey
20 Mead, Chris, *The State of the Nation's Birds*, Whittett, Stowmarket, 2000
21 Rackham, Oliver, *The History of the Countryside*, Dent, London, 1986
22 Anon., *The Great Feast At the Inthronization of the Reverend Father in God, George Neavill Archbishop of Yorke, Chancellor of England, in the Sixth Yeere of Edward the Fourth*, Edward Husbands, London, 1645
23 Anon., *A Propre New Booke of Cokery*, London, 1545
24 Lowe, Frank A., *The Heron*, Collins, London, 1954
25 Ekwall, Eilert, *The Concise Oxford Dictionary of English Place Names*, 4th edition, Oxford University Press, Oxford, 1960
26 Boisseau, S. & Yalden, D.W., 'The former status of the Crane *Grus grus* in Britain', *Ibis*, Vol. 140, 1998
27 Walton, Izaak, *The Complete Angler*, 6th edition, Rivington et al., London, 1797
28 Lowe, 1954, see Note 24
29 Williamson, John, *The British Angler*, John Hodges, London, 1740
30 Boisseau, S. & Yalden, D.W., see Note 26
31 Lowe, 1954, see Note 24
32 Lowe, 1954, see Note 24
33 Swan, John, *Speculum Mundi*, London, 1635
34 *Daily Telegraph*, 6 June 2000
35 *The Times*, 25 January 1971
36 Walpole, Horace, *Letters To the Countess of Upper Ossory*, 1 December 1786
37 Andersen, Hans Christian, *The Ugly Duckling*, Macmillan, New York, 1927 [1843]
38 Radford, E. & M.A., *The Encyclopedia of Superstitions*, 6th edition [edited and revised by Christina Hole], Helicon Publishing Ltd, Oxford, 1995
39 Earl of Northampton, *Defensative Against The Pyson of Supposed Superstition*, quoted in: Radford, E. & M.A., 1995, see Note 38
40 Lloyd, A.H., 'The College game of swans', in: Rackham, H., *Christ's College in Former Days*, Cambridge University Press, Cambridge, 1939
41 Lloyd, 1939, see Note 40
42 Jonathan Harrison, St John's College, Cambridge
43 Greenoak, Francesca, *All the Birds of the Air – The Names, Lore and Literature of British Birds*, André Deutsch, London, 1979
44 Jonathan Harrison, St John's College, Cambridge
45 Tom Sharpe, *Porterhouse Blue*, Martin Secker & Walburg, London, 1974
46 Darwin, Bernard, *Receipts and Relishes Being a Vade Mecum for the Epicure in the British Isles*, Whitbread, London, 1950
47 www.magd.ox.ac.uk
48 Darwin, 1950, see Note 46
49 Jeremy Furber, Somerset
50 Nicholson, James R., *Shetland Folklore*, Robert Hale, London, 1981
51 Armstrong, Edward A., *The Folklore of Birds*, Collins, London, 1958
52 Bewick, Thomas, *History of British Birds*, F. Graham, London, 1972
53 Lockwood, 1984, see Note 9
54 Pennant, Thomas, *British Zoology*, Benjamin White, London,1776
55 Sample, Geoff, *Bird Songs and Calls of Britain and Northern Europe*, HarperCollins, London, 1996
56 Opie, Iona & Peter, 'The Fox's Foray', in *The Lore and Language of Schoolchildren*, Clarendon Press, Oxford, 1959
57 Swainson, 1886, see Note 1
58 Dickens, Charles, *A Christmas Carol*, Chapman & Hall, London, 1843
59 Wagner, Leopold, *Manners, Customs and Observances*, Heinemann, London, 1895
60 Blount, Thomas, *Fragmenta Antiquitatis* ..., Blackhead & Co., York, 1784.
61 Armstrong, 1958, see Note 51
62 Gomme, George Laurence [ed.], *The Gentleman's Magazine Library: Popular Superstitions, Collection from 1731–1868*, Eliot Stock, London, 1884
63 Gomme, 1884, see Note 62
64 Beeton, Isabella, *Mrs Beeton's Cookery Book*, new and improved edition, Ward, Lock & Co., London, n.d. [*c.*1890]
65 Sillitoe, A., *Saturday Night and Sunday Morning*, W.H. Allen, London, 1958
66 Gomme, 1884, see Note 62
67 Hont, R., *Popular Romances of the West of England*, 3rd edition, Chatto & Windus, London, 1881 [1865]
68 Opie, I. & P., *The Oxford Nursery Rhyme Book*, 8th edition, Oxford University Press, Oxford, 1977
69 Opie, I. & P., 1977, see Note 68
70 Perrault d' Armancour, *Contes de Ma Mere L'Oye*, Paris, 1697
71 Clinton Keeling, Surrey
72 Opie, I. & P., 1977, see Note 68
73 Courtney, M.A., *Cornish Feasts and Folklore*, Beare & Son, Penzance, 1890
74 Opie, Iona & Peter, *The Lore and Language of Schoolchildren*, Clarendon Press, Oxford, 1959
75 Topsell, E., *The History of Four-footed Beasts and Serpents*, Vol. 2 *The History of Serpents*, Frank Cass, London, 1967 [Reprint of edition published in London, 1658]
76 Keogh, John, *Zoologia Medicinalis Hibernica*, S. Powell, Dublin, 1739
77 *Daily Telegraph*, 7 September 2000
78 *The Times*, 18 December 2000
79 Turner, William, *On Birds*, London, 1544
80 Gerard, John, *Herball*, London, 1597
81 Cambrensis, Giraldus, *Topographia Hibernica*, 1185, cited in: Greenoak, 1979, see Note 43
82 Greenoak, 1979, see Note 43
83 Lockwood, 1984, see Note 9; Eckwall, 1960, see Note 25
84 Lockwood, 1984, see Note 9; Clinton Keeling, Surrey
85 Matthew Engel, Herefordshire [*Daily News*, 1868]
86 Eddowes, John, *The Language of Cricket*, Carcanet, London, 1997
87 Channel 4 Television, interview with Michael Atherton, 3 September 2000
88 Opie, I. & P., 1977, see Note 68
89 V. Joan Jackman, Somerset
90 Clinton Keeling, Surrey
91 Beeton, *c.*1890, see Note 64
92 Lockwood, 1984, see Note 9
93 Murton, 1971, see Note 17
94 Morris, James, *Oxford*, Faber & Faber, London, 1965
95 Greenoak, 1979, see Note 43
96 Greenoak, 1979, see Note 43
97 Norma Aubertin-Potter, All Soul's College, Oxford
98 www.all-souls.ox.ac.uk/events/mallard.html
99 See Note 98
100 Anonymous Fellow, All Soul's College, Oxford
101 Anon., 1545, see Note 23
102 Markham, Gervase, *The English House-wife*, C.T. Milford, London, 1942 [First published in 1660]
103 Darwin, 1950, see Note 46
104 Greenoak 1979, see Note 43
105 Ibbott, Selena, *Folklore, Legends and Spells*, Ashgrove Press Ltd, Bath, 1994
106 Simpson, Jacqueline, *The Folklore of Sussex*, B.T. Batsford Ltd, London, 1973
107 Swainson, C., *A Handbook of Weather Folk-lore*, William Blackwood & Sons, Edinburgh and London, 1873
108 *The Folklore Journal*
109 Swainson, 1873, see Note 107
110 Black, W.G., *Folk Medicine; A Chapter in the History of Culture*, Eliot Stock for the Folklore Society, London, 1883
111 Mouffet, Thomas, *The History of Four-footed Beasts and Serpents and Insects*, Vol. 3, *The Theater of Insects*, London, 1658
112 Ingersoll, Ernest, *Birds in Legend Fable and Folklore*, Longmans, Green & Co., New York, 1923
113 Murton, 1971, see Note 17

114 *Daily Telegraph*, 14 December 2000
115 Salvin, Francis Henry & Brodrick, William, *Falconry in the British Isles*, Beech Publishing, Midhurst, 1997 [Reprint of 2nd edition of 1873]
116 Airfield Wildlife Management, Training Manual, cited by Tom Dewick, Moray
117 Ray, John, *The Ornithology of Francis Willughby* [translated into English, and enlarged by ... John Ray], John Martyn, London, 1678
118 Philip Rowell, Powys
119 Lockwood, 1984, see Note 9
120 Greenoak, 1979, see Note 43
121 Keogh, 1739, see Note 76
122 Brown, Leslie, *British Birds of Prey*, Collins, London, 1976
123 *Daily Telegraph*, 4 February 2000
124 Montagu, G., *Ornithological Dictionary*, London, 1802
125 Mead, 2000, see Note 20
126 Murton, 1971, see Note 17
127 Aubrey, John, *Miscellanies upon Various Subjects*, London, 1696
128 Wilson, John Marius, *A Memoir of Field Marshal- The Duke of Wellington*, Vol. II, A. Fullarton, London, Edinburgh, Dublin, n.d. [*c*.1855]
129 Tony Loy, Sheffield
130 Wilde [Lady], *Ancient Cures, Charms and Usages of Ireland*, Ward & Downey, London, 1890
131 Mead, 2000, see Note 20; Brown, 1976, see Note 122
132 Kathryn H. Baker, British Golf Museum, St Andrews, Fife
133 Barry Woodward, Ormskirk, Lancashire
134 Robin, P. Ansell, *Animal Lore in English Literature*, John Murray, London, 1932
135 Brand, John, *Observations on Popular Antiquities*, Chatto & Windus, London, 1900
136 Fraser-Darling, F. & Morton Boyd, J., *The Highlands and Islands*, Collins, London, 1964
137 Wolley, John, *Ootheca Wolleyana*, London, 1864
138 Potter, S. & Sargent, L., *Pedigree Words from Nature*, Collins, London, 1973
139 Drayton, Michael, *Poly-Olbion*, London, 1622
140 Peel, George, 'play' [source not traced], cited in: Samstag, Tony, *For Love of Birds – The Story of the Royal Society for the Protection of Birds 1889–1988*, RSPB, London, 1988
141 Samstag, Tony, *For Love of Birds – The Story of the Royal Society for the Protection of Birds 1889–1988*, RSPB, London, 1988
142 Clinton Keeling, Surrey
143 Brown, 1976, see Note 122
144 Lockwood, 1984, see Note 9
145 Anon., *Country Life*, 1904, cited in: Armstrong, 1958, see Note 51
146 Salvin, Francis Henry, & Brodrick, William, *Falconry in the British Isles*, Beech Publishing, Midhurst, 1997 [Reprint of 2nd edition of 1873], see Note 115
147 Salvin & Brodrick, 1873, see Note 115
148 Brown, 1976, see Note 122
149 Baker, J.A., *The Peregrine*, Collins, London, 1967
150 Royal Society for the Protection of Birds, Press Release, June 2000
151 *Daily Telegraph*, 4 February 2000
152 Salvin & Brodrick, 1873, see Note 115
153 Stevenson, R.L., 1886, see Note 2
154 Murton, 1971, see Note 17
155 Vesey-Fitzgerald, Brian, *British Game*, Collins, London, 1946
156 Buchan, John, *The Thirty-Nine Steps*, William Blackwood, Edinburgh & London, 1915; Colegate, Isobel, *The Shooting Party*, Hamish Hamilton, London, 1980
157 Darwin, 1950, see Note 46
158 Beeton, *c*.1890, see Note 64
159 Vesey-Fitzgerald, 1946, see Note 155
160 Lockwood, 1984, see Note 9
161 Mead, 2000, see Note 20
162 Clinton Keeling, Surrey
163 Vesey-Fitzgerald, 1946, see Note 155
164 Hare, C.E., *Bird Lore*, Country Life Ltd, London, 1952
165 Darwin, 1950, see Note 46
166 Hammond, P.W., *Food and Feast in Medieval England*, Wrens Park Publishing, Gloucestershire, 1993
167 Keogh, 1739, see Note 76
168 George Walton, Hereford
169 *The Times*, 4 January 2000
170 Ransome, Arthur, *Coot Club*, Jonathan Cape, London, 1930
171 Vesey-Fitzgerald, 1946, see Note 155
172 Beeton, *c*.1890, see Note 64
173 Jesse, Edward, *Scenes and Occupations of Country Life*, John Murray, London, 1853 [1844]
174 Arthur R. Willett, Cambridge
175 Clinton Keeling, Surrey
176 Armstrong, 1958, see Note 51
177 Lockwood, 1984, see Note 9
178 Ray, 1678, see Note 117
179 Swainson, 1873, see Note 107
180 Moore, A.W., *Folk-Lore of the Isle of Man*, S.R. Publishers Ltd, Yorkshire, 1971 [First published 1891 in Isle of Man and London]
181 Moore, 1891, see Note 180
182 Murton, 1971, see Note 17
183 Weaver, P., *The Lapwing*, Shire Publications Ltd, Aylesbury, 1985
184 Lees, Edward, *Pictures of Nature in the Silurian Region Around the Malvern Hills*, 1856
185 Hole, Christina, *English Folklore*, Batsford, London, 1940
186 Anon., *Country Life*, Vol. XV, 1904
187 Gibson, F., *Superstitions About Animals*, Scott Publishing Co. Ltd, London, 1904
188 Ingersoll, 1923, see Note 112
189 Robert Ford, *Children's Rhymes, Children's Games, Children's Songs, Children's Stories, A Book for Bairns & Big Folk*, Alexander Gardner, Paisley, 1904.
190 Beeton, Isabella, *c*.1890, see Note 64
191 Vesey-Fitzgerald, 1946, see Note 155
192 'Nigel', Isles of Scilly
193 Prynne, Michael, *Egg-shells*, Barrie & Rockliff, London,1963
194 Lockwood, 1984, see Note 9
195 Parker, Eric, *Shooting by Moor, Field and Shore*, Seeley, Service & Co., London, n.d. [*c*.1910]
196 Beeton, *c*.1890, see Note 64
197 Godfrey Blunt, Bridgnorth, Shropshire
198 Sample, 1996, see Note 55
199 John Wood, Aldington, Kent
200 Francesca Quint, Surrey
201 Lockwood, 1984, see Note 9
202 Browne, Thomas, *Account of Birds ... Found in Norfolk*, 1682
203 Ray, 1678, see Note 117
204 Forbes, A.R., *Gaelic Names of Beasts, Birds, Fishes, Insects, and Reptiles*, Oliver & Boyd, Edinburgh, 1905
205 Newall, Venetia, *Discovering the Folklore of Birds and Beasts*, Shire Publications, Tring, 1971
206 Greenoak, 1979, see Note 43
207 Gwynn Jones, T., *Welsh Folklore and Folk Custom*, Methuen & Co. Ltd, London, 1930
208 Armstrong, 1958, see Note 51
209 Nethersole-Thompson, Desmond, *The Greenshank*, Collins, London, 1951
210 Nethersole-Thompson, 1951, see Note 209
211 Latham, John, *A General Synopsis of Birds*, Vol. 3, London, 1785
212 Colin Winters, Kent
213 Greenoak, 1979, see Note 43
214 de Garis, M., *Folklore of Guernsey*, Channel Islands: Mrs Marie de Garis, Guernsey, Channel Islands, 1975
215 Radford, E. & M.A., 1995, see Note 38
216 'Nigel', Isles of Scilly
217 Keogh, 1739, see Note 76
218 Clinton Keeling, Surrey
219 Ingersoll, 1923, see Note 112
220 Murton, 1971, see Note 17
221 'Nigel', Isles of Scilly
222 Grieve, Symington, *The Great Auk or Garefowl*, Chapman, London, 1885; Fuller, Errol, *The Great Auk*, Southborough, Kent, 1999; Eckert, Allan, *The Last Great Auk*, Collins, London, 1964

223 Lockwood, 1984, see Note 9
224 Greenoak, 1979, see Note 43
225 Boyd, J.M. & Boyd, I.L., *The Hebrides*, Collins, London, 1990
226 MacKenzie, 1905, see Note 6
227 Thompson, Francis, *St Kilda and Other Hebridean Outliers*, David & Charles, Newton Abbot, 1970
228 Montagu, George, *Ornithological Dictionary*, London, 1802
229 Kath Trelissick, Cornwall
230 Greenoak, 1979, see Note 43
231 Gesner, Konrad von, *De Avibus*, *c.*1555
232 John Caius, cited in: Gesner, *c.*1555, see Note 231
233 Mouffet, Thomas, *Health's Improvement*, Thomas Newcomb for Samuel Thomson, London, 1655
234 Owen, George, *Description of Pembrokeshire*, London, 1603
235 Penhallurick, 1965, see Note 8
236 Lockwood, 1984, see Note 9
237 *Daily Telegraph*, 8 February 2001
238 *Daily Telegraph*, 1 August 2000
239 *The Falconer*, Vol. 2, 1948
240 Andrew Mackay, Worthing
241 Anon., *Folklore, Myths and Legends of Britain*, The Reader's Digest Association, London, 1973
242 Murton, R.K., *The Wood Pigeon*, Collins, London, 1965 [Martial, Richard cited in],
243 Darwin, 1950, see Note 46
244 White, Gilbert, *Natural History of Selborne*, new edition, George Routledge, London, 1880
245 Pepys, Samuel, *The Diary*, London [1893 edition]
246 Keogh, John, *Zoologia Medicinalis Hibernica*, S. Powell, Dublin, 1739, see Note 76
247 BTO, quoted in: Mead, 2000, see Note 20
248 Walton, Izaak, *The Complete Angler*, 6th edition, Rivington et al., London, 1797 [First published 1653], see Note 27
249 Kim Spickett and family, Twickenham, Surrey
250 Gregory, Kenneth, *The First Cuckoo: Letters to The Times* 1900–1975, Times Books, London, 1976
251 Opie, Iona & Peter, 1959, see Note 74
252 Liz Trundle, Northamptonshire
253 Jean Hayes, Hertfordshire
254 Swainson, Charles, *The Folklore and Provincial Names of British Birds*, Llanerch Publishers, Felinfach, 1998 [Facsimile reprint; first published 1886]
255 Swainson, 1886, see Note 254
256 Opie, I. & P., 1977, see Note 68
257 Mrs Harvey, quoted in: Palmer, Roy, *The Folklore of Leicestershire and Rutland*, Sycamore Press Ltd, Wymondham, 1985
258 Bett, Henry, *English Myths and Traditions*, B.T. Batsford Ltd, London, 1952
259 White, Gilbert, 1880, see Note 244
260 Miss Trump, quoted in: Baker, Margaret, *The Gardener's Folklore*, David & Charles, Newton Abbot, 1977
261 Armstrong, 1958, see Note 51
262 Armstrong, 1958, see Note 51
263 Armstrong, 1958, see Note 51
264 Armstrong, 1958, see Note 51
265 Armstrong, 1958, see Note 51
266 Swainson, 1886, see Note 254
267 Swainson, 1886, see Note 254
268 Swainson, 1886, see Note 254
269 Field, John Edward, *The Myth of the Pent Cuckoo, A Study in Folklore*, Eliot Stock, London, 1913
270 National Trust Press Release, Spring 1999
271 Brand, John, *Observations on Popular Antiquities*, Chatto & Windus, London, 1900
272 Anon., *Songs for the Nursery*, William Darton, London, 1825
273 Pilgrim, Jane, *Blackberry Farm*, Brockhampton Press, London, 1964; Potter, Beatrix *The Tale of Squirrel Nutkin*, Frederick Warne, London, 1903; Milne, A.A. *Winnie-the-Pooh*, Methuen, London, 1926
274 *Guardian*, 6 September 2000
275 Armstrong, 1958, see Note 51
276 Philip Rowell, Powys
277 G.E. Horton, Cheshire
278 Penelope Fletcher, Berkshire
279 Shirley, James, *Triumph of Peace*, London, 1633
280 Swan, John, *Speculum Mundi*, 1635, see Note 33
281 Udal, John Symonds, *Dorsetshire Folklore*, 2nd edition, Toucan Press, 1970
282 Lilford, [Lord], *Coloured Figures of the Birds of the British Islands*, Vols 1–7 [illustrated by Arthur Thorburn et al.], R.H. Porter, London, 1885–97
283 Liz Trundle, Northamptonshire
284 Margaret Courtney, Baden-Powell House, London
285 Ewing, Juliana Horatia, *The Brownies and Other Tales*, SPCK, London, 1871
286 White, 1880, see Note 244
287 Gibson, 1904, see Note 187
288 Liz Trundle, Northamptonshire
289 Giraldus Cambrensis, quoted in: Swainson, 1886, see Note 254
290 Keogh, 1739, see Note 76
291 Mead, 2000, see Note 20
292 Swainson, 1886, see Note 254
293 Armstrong, 1958, see Note 51
294 P. Trennel, North Lopham
295 Simms, Eric, Lincolnshire
296 Ray, 1678, see Note 117
297 Arabella Clark, Derbyshire
298 Evans, I.H., *Brewer's Dictionary of Phrase and Fable*, Cassell, London, 1959
299 Audrey Pitt, Manchester
300 Kightly, Charles, *The Perpetual Almanack of Folklore*, Thames & Hudson, London, 1987 [Reprinted 1994]
301 White, 1880, see Note 244
302 Dr Johnson, quoted in: Gibson, 1904, see Note 187
303 Anon., *Mistress Jane Hussey's Still-Room Book*, 1692, quoted in: Greenoak, 1979, see Note 43
304 Nicholson, E.R., *Birds and Men*, Collins, London, 1951
305 C.I. Paton, 1938, quoted in: Greenoak, 1979, see Note 43
306 *Bulletin of the British Museum*, cited in Simms, E., see Note 308
307 Viscount Grey of Falloden, cited in Simms, E., see Note 308
308 Simms, E., *British Larks, Pipits and Wagtails*, Collins, London, 1922
309 Nicholson, 1951, see Note 304
310 Kim Spickett and family, Twickenham, Surrey
311 Peter P. Anderson, Cumbria
312 Doreen Blott, Bedfordshire
313 Armstrong, Edward A., *The Wren*, Collins, London, 1955
314 Roger M.C. Sims, Isle of Man
315 Denyce Biggs, Co. Kerry
316 Opie, I. & P., 1959, see Note 68
317 O'Brian, Conor, *Countryman*, September, 1999
318 Armstrong, 1955, see Note 313
319 Nicholson, 1951, see Note 304
320 Clinton Keeling, Surrey
321 A.P. Radford, Somerset
322 Anon., 1825 see Note 272
323 Anon., 1825 see Note 272
324 Lack, David, *The Life of the Robin*, H.F. & G. Witherby, London, 1943
325 Lack, 1943, see Note 312
326 Swainson, 1886, see Note 1
327 Courtney, M.A., *Cornish Feast and Folklore*, Beare & Son, Penzance, 1890 [Revised and reprinted from the Folklore Society Journals, 1886–7]
328 Udal, John Symonds, see Note 281
329 Anon., 1825, see Note 272
330 Anon., 1825, see Note 272
331 Lupton, Thomas, *A Thousand Notable Things of Sundrie Sortes*, London, 1579
332 Keith Richard Power, Lancashire
333 Rita Stone, Hertfordshire
334 Catherine Nicholls, North Yorkshire
335 Florence Brojer, Berkshire
336 Jean Alderton-Menz, Devon
337 Mary Shave, Suffolk
338 Gordon Cawthorne, Suffolk
339 Maureen Jarman, London
340 Jean Hayes, Hertfordshire
341 Penelope Fletcher, Berkshire
342 Gilly Brown, Norfolk
343 Martin Williams, Hurstpierpoint College, West Sussex

344 See Note 343
345 Palmer, Kingsley, *The Folklore of Somerset*, B.T. Batsford Ltd, London, 1976
346 Swainson, 1886, see Note 1
347 Swainson, 1886, see Note 1
348 Armstrong, 1958, see Note 51
349 W.H. Hudson, quoted in: Harrison, Thomas P., 'Meredith as poet of birds', *Bird Notes and News*, Vol. 27, 1956
350 Lodge, Thomas, *Scillaes Metemorphoses*, London 1589
351 Armstrong, 1958, see Note 51
352 Keogh, 1739, see Note 76
353 Walton, 1653, see Note 27
354 Greenoak, 1979, see Note 43
355 Richard Power, Lancashire
356 Swainson, 1886, see Note 1
357 Beeton, *c*.1890, see Note 64
358 Trail, cited in: Greenoak, 1979, see Note 43
359 White, 1880, see Note 244
360 Elizabeth Derges, Devon
361 A.G. Blunt, Barnsley
362 Swainson, 1886, see Note 1
363 Opie, I. & P., 1977, see Note 68
364 Opie, I. & P., 1977, see Note 68
365 Keogh, 1739, see Note 76
366 Andrew Boorde, *Dietary of Health*, London, 1547
367 Arnold Fetters, Nottinghamshire
368 Swainson, 1886, see Note 1
369 Buckland, Frank, *Curiosities of Natural History*, Series II, Richard Bentley, London, 1875
370 Simms, Eric, *British Warblers*, Collins, London, 1985
371 Yeates, George, *Bird Life in Two Deltas*, Faber & Faber, London, 1946
372 White, Gilbert, 1880, see Note 244
373 Edward Thomas, quoted in: Simms, 1985, see Note 370
374 Pennant, Thomas, quoted in: Simms, 1985, see Note 370
375 White, 1880, see Note 244
376 Edward Thomas, quoted in: Simms, 1985, see Note 370
377 Simms, 1985, see Note 370
378 Swainson, 1886, see Note 1
379 Simms, 1985, see Note 370
380 A.P. Radford, Somerset
381 Kightly, 1987, see Note 300
382 Kightly, 1987, see Note 300
383 Florence Brojer, Berkshire
384 K.B. Stratton, Kent
385 Williamson, Henry, *Tarka the Otter*, G.P. Putnam's Sons, London, 1927
386 Perrins, Christopher, *British Tits*, Collins, London, 1979
387 Father Jean Imberdis [1693; translated by Laughton, Eric], quoted in: Perrins, 1979, see Note 386
389 Opie, I. & P., 1977, see Note 68
389 Mary Shave, Suffolk
390 Opie, I. & P., 1977, see Note 68
391 Anon., *Ladybird Nursery Rhymes*, Ladybird Books Ltd, Loughborough, 1994
392 Rita Stone, Hertfordshire
393 Anon., 1994 see Note 391
394 Cedric B. Shipley, Stockton-on-Tees
395 Keith Richard Power, Lancashire
396 Jean Alderton-Menz, Devon
397 Gilly Brown, Norfolk
398 Stevenson, 1886, see Note 2
399 Opie, I. & P., 1977, see Note 68
400 Anne M. Oakley, Canterbury Cathedral Archives
401 Jesse, Edward, *Scenes and Occupations of Country Life*, new edition, John Murray, London, 1853
402 Jean Alderton-Menz, Devon
403 Nicholson, 1951, see Note 304
404 Marshall, William, *Rural Economy of Norfolk*, T. Cadell, London, 1787
405 Darwin, 1950, see Note 46
406 Clive Barton, Norfolk
407 A.G. Blunt, Barnsley
408 Murton, 1971, see Note 17
409 A.G. Blunt, Barnsley
410 Pennant, Thomas, *A Tour in Scotland 1769*, B. White, London, 2nd edition, 1772 [First published in 1772]
411 Andrea Dalton, Cheshire
412 Opie, I. & P., 1977, see Note 68
413 Opie, I. & P. 1977, see Note 68
414 Edwards, George, *From Crow-Scaring to Westminster*, National Union of Agricultural Workers, London, 1922
415 *The History of Little King Pippin*, cited in Opie, I. & P., 1977, see Note 68
416 Palmer, Roy, *The Folklore of Warwickshire*, B.T. Batsford, London,1976
417 Palmer, 1976, see Note 416
418 Palmer, Roy, *The Folklore of Gloucestershire*, West Country Books, Tiverton, 1994
419 Baker, 1977, see Note 260
420 Opie, I. & P., 1977, see Note 68
421 Swainson, 1886, see Note 1
422 Palmer, Roy, 1994, see Note 418
423 E.M. Leather, cited in: Simpson, Jacqueline, *The Folklore of the Welsh Border*, B.T. Batsford Ltd, London, 1976
424 Colin Titcombe, Monmouthshire
425 Newall, Venetia, 1971, see Note 205
426 A.G. Blunt, Barnsley
427 Dickens, Charles, *Barnaby Rudge*, Chapman & Hall, London, 1840
428 Cervantes, Miguel de, *Don Quixote de la Mancha*, Antonio Sancha, Madrid, 1777 [1605]
429 Greenoak, 1979, see Note 43
430 Jane Wight, Derbyshire
431 C.T. Cluness, cited in: Nicholson, 1981, see Note 50
432 Smith, cited in: Murton, 1971, see Note 17
433 Keogh, 1739, see Note 76
434 Florence Brojer, Berkshire
435 Clinton Keeling, Surrey
436 John A. Thickett, Sheffield
437 Heywood, Thomas, quoted in: Anon., *Bird Notes and News*, Vol. 9, 1920
438 Murton, 1971, see Note 17
439 Jack Shea, Dublin
440 Janie Andrew, Herefordshire
441 Radford, E. & M.A., 1995, see Note 38
442 Murton, 1971, see Note 17
443 Boorde, 1547, see Note 336
444 Anon., 'Old nocturnal field sports', *Country Life*, Vol. 16, 1904
445 Kipps, Clare, *Sold for a Farthing*, Muller, London, 1953
446 Anon., *Wisden Cricketers' Almanac 1937*, John Wisden, London, 1937
447 Newton, I., *Finches*, Collins, London, 1972
448 Peter Condor, cited in: Newton, 1972, see Note 447
449 Opie, I. & P., 1977, see Note 68
450 Keogh, 1739, see Note 76
451 Nicholson, 1951, see Note 304
452 Lawson, William, *A New Orchard and Garden*, London, 1618
453 Beeton, *c*.1890, see Note 64
454 Cam Stirling, Perthshire
455 Greenoak, 1979, see Note 43
456 Radford, E. & M.A., 1995, see Note 38
457 Mead, 2000, see Note 20

Mammals

1 Andrew Warner, Sheffield
2 Clinton Keeling, Surrey
3 Jane Carter, Stoke on Trent
4 Diana Stenson, Cheshire
5 King Smith, Dick, *The Hodgeheg*, Hamish Hamilton, London, 1987
6 Eric Simms, Lincolnshire
7 *Chamber's Encyclopaedia*, J.P. Lippincott & Co., Philadelphia, 1923
8 Eric Roberts, *Express and Echo*, Exeter, 4 December 1981
9 *Daily Telegraph*, 11 October 2000
10 Topsell, Edward, *The History of Four-footed Beasts and Serpents and Insects*, Da Capo Press, New York, 1967 [Reprint of 1658 edition]
11 Price, Lawrence, *The Shepherd's Prognostication*, London, 1652
12 Kinahan, G.H., 'Notes on Irish folklore', *Folklore Journal*, Vol. 4, 1881
13 Gillian Brown, Luton
14 Hendrickson, Robert, *The Ocean Almanac*, Hutchinson, London, 1992
15 Godfrey Blunt, Shropshire
16 Harrison-Mathews, Leo, *British Mammals*, Collins, London, 1952
17 Palmer, Roy, *The Folklore of Leicestershire and Rutland*, Sycamore Press, Wymondham, 1985

18 Ellis, William, *Country House-wife's Family Companion*, London, 1750
19 Browne, Thomas, cited in: Black, W.G., *Folk Medicine; A Chapter in the History of Culture*, Elliot Stock for the Folklore Society, London, 1883
20 *Fairfax Household Book, seventeenth to eighteenth centuries*, cited in: Kightly, Charles, *The Perpetual Almanack of Folklore*, Thames & Hudson, London, 1987 [Reprinted 1994]
21 Keogh, John, *Zoologia Medicinalis Hibernica*, S. Powell, Dublin, 1739
22 Palmer, Roy, *The Folklore of Gloucestershire*, West Country Books, Tiverton, 1994
23 Topsell, 1658, see Note 10
24 White, Gilbert, *Natural History of Selborne*, new edition, George Routledge, London, 1880
25 Burne, Charlotte Sophia [ed.], *Shropshire Folklore* [from the collections of Georgina F. Jackson], Trubner & Co., London, 1883
26 Susan Major, St Agnes, C of E Primary School, Isles of Scilly
27 Grahame, Kenneth, *The Wind in the Willows*, Methuen & Co., London, 1908s
28 Anon, *Ladybird Nursery Rhymes*, Ladybird Books Ltd, Loughborough, 1994
29 Brewer, Clifford, *The Death of Kings*, Abson Books, London, 2000
30 Richard Holworthy, *Cours de Monsegur*, France
31 Harrison-Mathews, 1952, see Note 16
32 Browne, Thomas, *Enquiries into Vulgar Errors*, in: *The Works of the Learned Sr. Thomas Brown, Kt., Doctor of Physick, late of Norwich*, Thos. Basset, Ric. Chiswell, Tho. Sawbridge, Charles Mearn & Charles Brome, London, 1686
33 Hawker, R.S., 'The first Cornish mole', in: *Footprints of Former Men in Far Cornwall*, James G. Commin, Walter Weighell & Joseph Pollard, London, Exeter, Launceston & Truro, 1908 [Reissue of original 1870 edition]
34 Roper, William, *A Man of Singular Virtue, Being a Life of Sir Thomas More by his Son-in-law William* Roper, Folio Society, London, 1980 [Reprint of Paris edition of 1626]
35 Swan, John, *Speculum Mundi*, London, 1635
36 Browne, 1686, see Note 32
37 Worlidge, John, *Systema Agriculturae: The Mystery of Husbandry Discovered*, 2nd edition, Thomas Dring, London, 1675
38 Coles, William, *The Art of Simpling*, Nathaniel Brook, London, 1656
39 N.A. Moyle, Shrewsbury
40 David Elwood, Truro
41 Mitford, Mary, *Our Village*, Sidgwick & Jackson, London, 1986
42 Mellanby, Kenneth, *The Mole*, Collins, London, 1971
43 Swan, 1635, see Note 35
44 Murray, James *et al.* [eds], *The Oxford English Dictionary*, Clarendon Press, Oxford, 1961
45 Liz Trundle, Northamptonshire
46 Carroll, Lewis, *Alice's Adventures in Wonderland*, Macmillan & Co., London, 1865
47 White, 1880, see Note 24
48 Bob Howard, Keynsham
49 Graham Loveluck, Anglesey, in: *Daily Telegraph*, 9 August 2000
50 Nicholson, John, *Folklore of East Yorkshire*, Thomas Holderness, Driffield, 1890
51 Nicholson, 1890, see Note 50
52 Radford, E. & M.A., *The Encyclopedia of Superstitions*, 6th edition [edited and revised by Christina Hole], Helicon Publishing Ltd, Oxford, 1995
53 Opie, I. & P., *The Oxford Nursery Rhyme Book*, 8th edition, Oxford University Press, Oxford, 1977
54 Rita Stone, Hertfordshire
55 Mary Shave, Suffolk
56 Dr A.P. Radford, Somerset
57 Radford, E. & M.A., 1995, see Note 52
58 Michael Clarke, Hertfordshire
59 Martin, Laura C., *Wildlife Folklore*, The Globe Pequot Press, Connecticut, 1994
60 Keogh, 1739, see Note 21
61 Eric Simms, Lincolnshire
62 Holland, Philemon (1552–1637), *Pliny's Natural History: A Selection from Philemon Holland's Translation*, [ed. J. Newsome], Clarendon Press, Oxford, 1964
63 Eric Simms, Lincolnshire
64 White, 1880, see Note 24
65 Kim Spickett and family, Twickenham, Surrey
66 White, Gilbert, 'Naturalist's Journal, 22 April 1769', in: *The Journals of Gilbert White*, Vol. 2 [ed. Francesca Greenoak], Century, London, 1988
67 Howard, Robert W., *Auritus, A Natural History of the Brown Long-eared Bat*, William Sessions, York, 1995
68 Fraser Darling, F. & Boyd Morton, J. *The Highlands and Islands*, Collins, London, 1964
69 Smurthwaite, David, *Battlefields of Britain*, Michael Joseph, London, 1993
70 Whitelock, Dorothy [ed.], *The Anglo-Saxon Chronicle: A Revised Translation*, Eyre & Spottiswoode; London, 1961
71 Harting, James, *Extinct British Animals*, Trubner & Co., London, 1880
72 Hunt, R., *Popular Romances of the West of England*, 3rd edition, Chatto & Windus, London, 1881
73 N. Corbett, University of Stirling
74 Scrope, William, *The Art of Deer Stalking Illustrated by a Narrative of a Few Days Sport in the Forest of Atholl*, John Murray, London, 1838
75 Yalden, Derek, *The History of British Mammals*, T. & A.D. Poyser, London, 1999
76 White, Gilbert, 'Naturalist's Journal, 13 February 1778', in: *The Journals of Gilbert White*, Vol. 2 [ed. Francesca Greenoak], Century, London, 1988
77 James Preston, *Daily Telegraph*, 22 August 1983
78 Vesey-Fitzgerald, Brian, *Town Fox, Country Fox*, Eyre & Spottiswoode, London, 1965
79 E.A. Timson, Leicester
80 John Wood, Aldington, Kent
81 Judith Cawthorne, Suffolk
82 Udal, John Symonds, *Dorsetshire Folk-Lore*, 2nd edition, Toucan Press, London, 1970
83 Palmer, 1985, see Note 17
84 *Le Roman de Renart* (The Romance of Renart)
85 Potter, Beatrix, *The Tale of Mr Tod*, Frederick Warne, London, 1912
86 Newall, Venetia, *Discovering the Folklore of Birds and Beasts*, Shire Publications, Tring, 1971
87 De Crespigny, R.C. & Hutchinson, New Forest, John Murray, London, 1903 [Re-issue]
88 Turberville, George, *The Noble Arte of Venerie or Hunting*, 1576
89 Palmer, 1985, see Note 17
90 Palmer, Roy, *Britain's Living Folklore*, David & Charles, Newton Abbot and London, 1991
91 Austen, Jane, *Pride and Prejudice*, T. Egerton, London, 1813
92 Surtees, R.S., *Handley Cross; Or, The Spa Hunt. A Sporting Tale*, A.K. Newman & Co., London, 1846
93 Sassoon, Siegfried, *Memoirs of a Fox-hunting Man*, Faber & Faber, London, 1928
94 Rook, David, *The Ballad of the Belston Fox*, Hodder & Stoughton, London, 1970
95 Dahl, Roald, *Fantastic Mr Fox*, Allen & Unwin, London, 1970
96 Burns, Lord [Chairman], *Final Report of the Committee of Inquiry into Hunting with Dogs*, HMSO, London, 2000
97 Dickens, Charles, *Oliver Twist*, Richard Bentley, London, 1838
98 Wilkinson, J., in *Barnsley Chronicle*, 1871, cited by: A.B. Loy, Sheffield
99 Yalden, 1999, see Note 75
100 Yalden, 1999, see Note 75
101 Grahame, 1908, see Note 27
102 Jefferies, Richard, *The Gamekeeper at Home. Sketches of Natural History and Rural Life*, Smith, Elder, London, 1878

103 Nicholson, James R., *Shetland Folklore*, Robert Hale, London, 1981
104 Jean Alderton-Menz, Devon
105 Christopher Allan, Ede & Ravenscroft, London
106 Jill Lucas, by e-mail
107 Strong, Roy, *Gloriana*, Thames & Hudson, London, 1987
108 Petrarch, Francesco, *The Triumphs of Francesco Petrarch: Florentine Poet Laureate* [trans. Henry Boyd], Little Brown, Boston, 1906
109 Strong, 1987, see Note 107
110 Eric Simms, Lincolnshire
111 Topsell, 1658, see Note 10
112 Grahame, 1908, see Note 27
113 Drabble, Phil, *A Weasel in my Meatsafe*, Collins, London, 1957
114 Partridge, Eric, *Dictionary of Historical Slang*, [abridged by Jacqueline Simpson], Penguin, London, 1986
115 *Guardian*, 1 November 1999
116 *Luttrell Psalter*, British Library, London [Add. MS 42130, f. 202V]
117 Burton, M., *Wild Animals of the British Isles*, 2nd edition, Frederick Warne & Co., London, 1973
118 Eric Simms, Lincolnshire
119 Twici, Guillaume, *Treatise on the Craft of Hunting*, *c.*1340 cited in Watson, J.W.P., *The Book of Foxhunting*, B.T. Batsford, London, 1977
120 Eric Simms, Lincolnshire
121 Grahame, 1908, see Note 27
122 Neal, Ernest, *The Badger*, Collins, London, 1948
123 Wentworth Day, J., *Sporting Adventure*, George Harrap & Co., London, 1937
124 Harrison-Mathews, 1952, see Note 16
125 Neal, 1948, see Note 122
126 Graham, Ritson, 'The Badger in Cumberland', *Transactions of the Carlisle Natural History Society*, Vol. 7, 1946
127 Eric Simms, Lincolnshire
128 Rutty, John, *Essay Towards a Natural History of the County of Dublin*, 1724
129 Baker, Jenny, *Kettle Broth to Gooseberry Fool*, Faber & Faber, London, 1996
130 Jocelyn Morris, Oxford
131 Neal, 1948, see Note 122
132 Forbes, A.R., *Gaelic Names of Beasts, Birds, Fishes, Insects, and Reptiles*, Oliver & Boyd, Edinburgh, 1905
133 Topsell, 1658, see Note 10
134 Neal, 1948, see Note 122
135 Neal, 1948, see Note 122
136 Roger Favell, Lincolnshire
137 E.A.Timson, Leicester
138 Howard, Robert W., *Badgers without Bias*, Abson Books, Bristol, 1981
139 Howard, 1981, see Note 138
140 Howard, 1981, see Note 138
141 Williamson, Henry, *Tarka the Otter*, G.B. Putnam's Sons, London, 1927
142 *The Daily Telegraph*, 21 June 2000
143 Eric Simms, Lincolnshire
144 Nicholson, James R., see Note 103
145 Eric Simms, Lincolnshire
146 Wentworth Day, 1937, see Note 123
147 Cokayne, Thomas, 'A Short Treatise of Hunting' [1591], in: *Journal of the Derbyshire Archaeological and Natural History Society*, Vol. 3, January 1881, Bemrose & Sons, London and Derby
148 Maxwell, Gavin, *Ring of Bright Water*, Longman, London, 1960
149 Burton, 1973, see Note 117
150 Goodrich-Freer, A., 'More folklore from the Hebrides', *Folklore Journal*, Vol. 13, 1902
151 Walton, Izaak, *The Complete Angler*, 6th edition, Rivington et al., London, 1797 [1653]
152 Keogh, 1739, see Note 21
153 Forbes, 1905, see Note 132
154 Beith, Mary, *Healing Threads: Traditional Medicines of the Highlands and Islands*, Polygon, Edinburgh, 1995
155 Yalden, 1999, see Note 75
156 Burton, 1973, see Note 117
157 Cokayne, 1591, see Note 147
158 *Daily Telegraph*, 31 January 2001
159 Jennison, George, *Natural History: Animals, An Illustrated Who's Who of the Animal Kingdom*, A. & C. Black, London 1927
160 Berry, R.J., *The Natural History of Orkney*, Collins, London, 1985
161 Carroll, Lewis, *Through the Looking Glass and What Alice Found There*, Macmillan, London, 1872
162 Hewer, H.R., *British Seals*, Collins, London, 1974
163 *Daily Telegraph*, 12 April 1999
164 *Daily Telegraph*, 10 July 1995
165 Eric Simms, Lincolnshire
166 Newall, Venetia, *Discovering the Folklore of Birds and Beasts*, Shire Publications, Tring, 1971
167 Lockley, R.M., *Grey Seal, Common Seal*, White Lion edition, London, 1977 [Originally published by André Deutsch Ltd, 1966]
168 Marwick, Ernest W, *Folklore of Orkney and Shetland*, Batsford, London, 1975
169 Thomson, David, *The People of the Sea*, Turnstile Press, London, 1954
170 Nicholson, James R, see Note 103
171 Brand, John, *A Brief Description of Orkney*, Zetland, Pightland Firth & Caithess, Brown, Edinburgh, 1701
172 Douglas, Francis, *A General Description of the East Coast of Scotland from Edinburgh to Cullen*, F. Douglas, Paisley, 1782
173 Nicholson, James R., see Note 103
174 Thomson, 1954, se Note 169
175 Hewer, 1974, see Note 162
176 Lockley, 1966, see Note 167
177 Pinkerton, John, *A General Collection of the Best and Most Interesting Voyages and Travels*, Longman, Hurst, Rees and Orme, London, 1808–14
178 Southwell, Thomas, *Seals and Whales of the British Seas*, Jarrold & Son, London, 1881
179 Lockley, 1966, see Note 167
180 Hardy, Alister, *The Open Sea II Fish and Fisheries*, Collins, London, 1959
181 *The Times*, 25 January 1971
182 Maxwell, Gavin, *Ring of Bright Water*, Longman, London, 1960
183 Berry, 1985, see Note 160
184 Buchan, John, *The Island of Sheep*, Hodder & Stoughton, London, 1936
185 *Guardian*, 14 March 2000
186 Bullen, Frank, *The Cruise of the 'Cachalot' – Round the World after Sperm Whales*, Smith, Elder, London, 1898
187 Goodwin, George, 'Porpoise – friend of Man?', *Natural History*, Vol. 106, Pt 8, October 1947
188 Melville, Herman, *Moby Dick or The Whale*, Longman, London, 1907
189 Melville, 1907, see Note 188
190 Porter, Enid, *The Folklore of East Anglia*, Batsford, London, 1974
191 *The Times*, 25 January 1971
192 Rackham, Oliver, *The History of the Countryside*, J.M. Dent, London, 1986
193 Jonathan Bengston, Queen's College, Oxford
194 Harting, 1880, see Note 71
195 Jonathan Bengston, Queen's College, Oxford
196 Martin Goulding, *Current Status and Potential Impact of Wild Boar* (Sus scrofa) *in the English Countryside: A Risk Assessment*, MAFF Central Science Laboratory, York, 1998
197 Yalden, 1999, see Note 75
198 Eric Simms, Lincolnshire
199 *Daily Telegraph*, 21 February 2001
200 Clinton Keeling, Surrey
201 Hammond, P.W., *Food and Feast in Medieval England*, Wrens Park Publishing, Gloucestershire, 1993
202 *The Times*, 25 January 1971
203 White, 1880, see Note 24
204 Hammond, 1993, see Note 201
205 Fraser Darling & Boyd Morton, 1964, see Note 68
206 Burns, 2000, see Note 96
207 Burns, 2000, see Note 96
208 Lawton, Jeanette, 'Hanley Venison Festival', *Deer Journal*, Vol. 7, 1987
209 Anon., *A Propre New Booke of Cokery*, London, 1545
210 Eric Simms, Lincolnshire

211 Buchan, John, *John Macnab*, Hodder & Stoughton, London, 1925
212 Raven, Jon, *The Folklore of Staffordshire*, Batsford Ltd, London, 1978
213 Boase, Wendy, *The Folklore of Hampshire and the Isle of Wight*, Batsford Ltd, London, 1976
214 Scott-Giles, C. Wilfred, *Civic Heraldry of England and Wales*, J.M. Dent & Sons Ltd, London, [1st edition 1933, revised 1953]
215 Christopher Borthen, Editor, *Stalking Magazine*
216 Yalden, 1999, see Note 75
217 Eric Simms, Lincolnshire
218 Yalden, 1999, see Note 75; Rackham, 1986, see Note 192
219 Salten, Felix, *Bambi: A Life in the Woods*, Jonathan Cape, London, 1928
220 Green, Marian, *A Calendar of Festivals*, Element, Dorset, 1991
221 Yalden, Derek, 1999, see Note 75
222 Cokayne, 1591, see Note 147
223 Boece, Hector, *Scotorum Historiae*, Paris, 1526
224 Clinton Keeling, Surrey
225 Annette Cleaner, Great Orme Goat Breeders' Association, Henley-on-Thames
226 Step, Edward, *Animal Life of the British Isles*, Frederick Warne & Co., London, 1921
227 Dr Martin Brimble, Meopham, Kent
228 Howard Dodsworth, Tunbridge Wells, Kent
229 Cabot, David, *Ireland*, Collins, London, 1999
230 Yalden, 1999, see Note 75
231 Murton, Ron
232 *Daily Telegraph*, 30 January 2001
233 Topsell, 1658, see Note 10
234 Simpson, Jacqueline, *The Folklore of Sussex*, Batsford, London, 1973
235 Brand, John, *Observations on Popular Antiquities*, Chatto & Windus, London, 1900
236 Forby, Robert, *The Vocabulary of East Anglia; An Attempt to Record the Vulgar Tongue of the Twin Sister Counties, Norfolk and Suffolk, as it Existed in the Last Twenty Years of the Eighteenth Century, and Still Exists: with Proof of its Antiquity from Etymology and Authority*, J.B Nicols & Son, London, 1830
237 Shorten, Monica, *Squirrels*, Collins, London, 1954
238 Jill Lucas, by e-mail
239 Potter, Beatrix, *The Tale of Squirrel Nutkin*, Frederick Warne, London, 1903
240 Bett, Henry, *English Myths and Traditions*, Batsford, London, 1952
241 Yalden, 1999, see Note 75
242 Cambrensis, Giraldus, *The Itinerary Through Wales and the Description of Wales* [trans. Thomas Roscoe], Everyman, London, 1912
243 *Daily Telegraph*, 14 August 2000.
244 Macdonald, David & Barrett, Priscilla, *Mammals of Britain and Europe*, HarperCollins, London, 1993
245 Clinton Keeling, Surrey
246 Stephanie Ryder, Bristol
247 Grahame, 1908, see Note 27
248 Green, Peter, *Beyond the Wild Wood*, Webb & Bower, London, 1982
249 White, Gilbert, 'Garden Kalendar 4 December 1767', in: *The Journals of Gilbert White*, Vol. 1 [ed. Francesca Greenoak], Century, London, 1986
250 Burton, 1973, see Note 117
251 Topsell, 1658, see Note 10
252 Leslie, Charles, *Memoirs of The Life of John Constable, Esq. R.A. Composed Chiefly of His Letters*, Longman, Brown, Green & Longmans, London, 1845
253 Eric Simms, Lincolnshire
254 Hole, Christina, *English Folklore*, Batsford, London, 1940
255 Radford, E. & M.A., 1995, see Note 52
256 Keith Richard Power, Preston
257 Opie, Iona & Peter [eds], *The Oxford Dictionary of Nursery Rhymes*, Oxford University Press, Oxford [First published 1951, 12th edition 1980]
258 Anon., *Songs for the Nursery*, William Darton, London, 1825
259 Defoe, Daniel, *Journal of the Plague Year*, London, 1722
260 Paul Slack, Linacre College, Oxford
261 Shelagh M. Carter, Hexham, Northumberland
262 White, Gilbert, 'Naturalist's Journal, 16 December 1777', in: *The Journals of Gilbert White*, Vol. 2 [ed. Francesca Greenoak], Century, London, 1988
263 Burton, 1973, see Note 117
264 Eric Simms, Lincolnshire
265 Clinton Keeling, Surrey
266 Yalden, 1999, see Note 75
267 Plummer, Brian D., *Tales of a Rat-hunting Man*, Boydell Press, Ipswich, 1978
268 Palmer, 1985, see Note 17
269 Porter, 1974, see Note 190
270 Fraser Darling & Boyd Morton, 1964, see Note 68
271 Phillipa Foord-Kelsey, Banbury, Oxfordshire
272 Jean Alderton-Menz, Devon
273 Carroll, 1865, see Note 46
274 Eric Simms, Lincolnshire
275 Yalden, 1999, see Note 75
276 Hughes, Thomas, *Tom Brown's Schooldays*, Macmillan, London, 1857
277 Caesar, Julius, *De Bello Gallico*, Book 5, Chapter 12
278 Carroll, 1865, see Note 46
279 Page, Robin, *The Fox and the Orchid – Country Sports and the Countryside*, Quiller Press Ltd, London, 1987
280 Pepys, Samuel, *The Diary*, George Bell & Sons, London, 1893
281 Glasse, Hannah, *The Art of Cookery Made Plain and Easy*, W. Strahan, London, 1774 edition [1747]
282 Beeton, Isabella, *Mrs Beeton's Cookery Book*, New and improved edition, Ward Lock & Co., London, n.d. [*c.*1890]
283 Anon., 1545, see Note 209
284 Evans, George & Thomson, David, *The Leaping Hare*, Faber & Faber, London, 1972
285 Kightly, Charles, *The Perpetual Almanack of Folklore*, Thames & Hudson, London, 1987 [Reprinted 1994]
286 Porter, 1974, see Note 190
287 Palmer, 1985, see Note 17
288 Evans & Thomson, 1972, see Note 284
289 *The Times*, 26 November 1966
290 Eric Simms, Lincolnshire
291 Adams, Richard, *Watership Down*, Rex Collings, London, 1972
292 Lucas, Edward, *Harvest Home*, London, 1913
293 Catherine Nicholls, Yorkshire
294 Yalden, 1999, see Note 75
295 Mary Alexander, Guildford Borough
296 Joan Jackman, Somerset
297 Rackham, 1986, see Note 192
298 Hugh Collins, Wool, Dorset
299 Barbara Howe, Weymouth, Dorset
300 A. Moyle, Shrewsbury
301 Arabella Clarke, Derbyshire
302 Eric Simms, Lincolnshire
303 Margaret Inglis, London
304 Adams, 1972, see Note 291

SELECT BIBLIOGRAPHY

The dates cited are those of the editions consulted; the publication date of the first or original edition, where different and known, is given in brackets.

Anon., *Words of Old Worcestershire*, Kidderminster, Kenneth Tomkinson, Kidderminster [n.d.]

Anon., *Choice Notes from Notes and Queries*, Bell & Daldy, London, 1859

Anon., *Folklore, Myths and Legends of Britain*, Reader's Digest Association Ltd, London, 1973

Anon., *The Oxford Dictionary of Quotations*, Oxford University Press, Oxford, 1987

Anon., *Brewer's Dictionary of 20th-Century Phrase and Fable*, Cassell, London, 1991

Anon., *Ladybird Nursery Rhymes*, Ladybird Books Ltd, Loughbrough, 1994

Anon., *Who's Who 1897–1998* [CD-ROM], A. & C. Black, London, & Oxford University Press, Oxford, 1998

Alford, Violet & Gallop, Rodney, *The Traditional Dance*, Methuen & Co., London, 1935

Andrews, William, *Bygone Derbyshire*, S.R. Publishers, Wakefield, 1971 [1892]

Armstrong, Edward A., *The Wren*, Collins, London, 1955

Armstrong, Edward A., *The Folklore of Birds*, Collins, London, 1958

Asher, J., Warren, M., Fox, R., Harding, P., Jeffcoate, G. & Jeffcoate, S., *The Millennium Atlas of Butterflies in Britain and Ireland*, Oxford University Press, Oxford, 2001

Baer, Florence E., *Folklore and Literature of the British Isles*, Garland Publishing, New York, 1986

Baker, Margaret, *The Gardener's Folklore*, David & Charles, Newton Abbot, 1977

Barrett, J., *Life on the Sea Shore*, Collins, London, 1974

Beebee, Trevor, *Frogs and Toads*, Whittet Books, London, 1985

Beebee, Trevor & Griffiths, Richard, *Amphibians and Reptiles*, HarperCollins, London, 2000

Beith, Mary, *Healing Threads: Traditional Medicines of the Highlands and Islands*, Polygon, Edinburgh, 1995

Berry, R.J., *The Natural History of Orkney*, Collins, London, 1985

Bett, Henry, *English Legends*, Batsford, London, 1950

Bett, Henry, *English Myths and Traditions*, Batsford, London, 1952

Billson, C.J. [collector and ed.], *County Folklore: Gloucestershire, Suffolk, Leicestershire and Rutland. Printed Extracts, No. 3. Leicestershire and Rutland*, published for the Folklore Society by D. Nutt, 1895

Black, G.F. [collector], *County Folklore: Vol. 3. Orkney and Shetland Islands*, Folklore Society. Reprinted by Kraus Reprint, Nendlen/Liechtenstein, 1967

Black, W.G., *Folk Medicine; A Chapter in the History of Culture*, published for the Folklore Society by Elliot Stock, London, 1883

Bland, T. Clifford et al., *Wildfowling*, Seeley, Service & Co., London [n.d.]

Boase, Wendy, *The Folklore of Hampshire and the Isle of Wight*, Batsford, London, 1976

Bolitho, H. [ed.], *The Glorious Oyster*, Sidgwick & Jackson, London, 1960

Borradaile, L.A., Eastham, L.E.S., Potts, F.A., & Saunders, J.T., *The Invertebrata*, 4th edition [revised by G.A. Kerket], Cambridge University Press, Cambridge, 1961 [1958]

Bosworth Smith, R., *Bird Life and Bird Lore*, John Murray, London, 1905

Boulenger, E.G., *British Angler's Natural History*, Collins, London, 1946

Boyd, J.M. & I.L., *The Hebrides*, Collins, London, 1990

Brand, John, *Observations on Popular Antiquities*, Chatto & Windus, London, 1900

Brewer, Clifford, *The Death of Kings*, Abson Books, London, 2000

Brewer, Jo, *Butterflies*, Harry N. Abrams, New York, 1978

Brian, M.V., *Ants*, Collins, London, 1977

Briggs, Katherine M., *The Folklore of the Cotswolds*, Batsford, London, 1974

Bristowe, W.S., *The World of Spiders*, Collins, London, 1971 [1958]

British Ornithologists' Union, *The Official List of Birds of Great Britain with Lists for Northern Ireland and the Isle of Man*, BOU, Tring , 1999

Brown, Leslie, *British Birds of Prey*, Collins, London, 1976

Brown, M.E. & Frost, W.E., *The Trout*, Collins, London, 1967

Buchan, David [ed.], *Folk Tradition and Folk Medicine in Scotland – The Writings of David Rorie*, Canongate Academic, Edinburgh, 1994

Buckland, F.T., *Curiosities of Natural History*, Macmillan, London, 1857– [Various series and dates]

Buczacki, S.T. & Harris, K.M., *Pests, Diseases and Disorders of Garden Plants*, 2nd edition, HarperCollins, London, 1998 [1981]

Burne, Charlotte Sophia [ed.], *Shropshire Folklore* [from the collections of Georgina F. Jackson], Trübner & Co., London, 1883

Burton, M., *Wild Animals of the British Isles*, 2nd edition, Frederick Warne & Co., London, 1973 [1968]

Butler, Colin G., *The World of the Honeybee*, Collins, London, 1954

Buxton, John, *The Redstart*, Collins, London, 1950

Cabot, David, *Ireland*, Collins, London, 1999

Campbell, Gordon [ed.], *John Milton – The Complete Poems*, J.M. Dent, London, 1980 [Everyman's Library, 1909]

Carr, Samuel [ed.], *The Batsford Book of Country Verse*, Batsford, London, 1979

Chinery, M., *Insects of Britain and Northern Europe*, Collins, London, 1979

Chinery, M., *Garden Creepy-crawlies*, Whittet Books, London, 1986

Christian, Roy, *The Country Life Book of Old English Customs*, Country Life, London, 1966

Christian, Roy, *Old English Customs*, David & Charles, Newton Abbot, 1972

Clausen, Lucy, *Insect Fact and Folklore*, Collier Books, New York, 1962

Cokayne, Thomas, 'A Short Treatise of Hunting' [1591], in: *Journal of the Derbyshire Archaeological and Natural History Society*, Vol.3, January 1881, Bemrose & Sons, London and Derby

Condry, William M., *The Natural History of Wales*, Collins, London, 1981

Cooke, M.C., *Our Reptiles and Batrachians*, W.H. Allen & Co., London, 1893

Corbet, Philip S., Longfield, Cynthia & Moore, N.W., *Dragonflies*, Collins, London, 1960

Couch, T.Q., 'A list of obsolete words, still in use among the folk of East Cornwall', *Journal of the Royal Institution of Cornwall*, No. 1, 1864

Courtney, M.A., *Cornish Feasts and Folklore*, Beare & Son, Penzance, 1890 [Revised and reprinted from the Folklore Society Journals, 1886–7]

Cowan, Frank, *Curious Facts in the History of Insects*, J.B. Lippincott & Co., Philadelphia, 1865

Cowden-Clarke, Mary, *The Complete Concordance to Shakespeare*, Bickers & Son, London, 1894

Craig, W.J. [ed.], *The Complete Works of William Shakespeare*, Clarendon Press, Oxford [*c.*1890]

Crossland, J.R. & Parrish, J.M. [eds], *Britain's Wonderland of Nature*, Collins, London [n.d.]

Crowcroft, Peter, *The Life of the Shrew*, Stellar Press, Barnet, 1957

Culpeper, *Complete Herbal*, Wordsworth Editions, Hertfordshire, 1995

Darwin, Bernard, *Receipts and Relishes Being a Vade Mecum for the Epicure in the British Isles*, Whitbread, London 1950

De Crespigny, R.C., & Hutchinson, H., *New Forest*, John Murray, London, 1905 [Re-issue]

de Garis, M., *Folklore of Guernsey*, Mrs Marie de Garis, Guernsey, Channel Islands, 1975

Drake-Carnell, F.J., *Old English Customs and Ceremonies*, Batsford, London, 1938

Durrell, Gerald & Lee, *The Amateur Naturalist*, Dorling Kindersley, London, 1983 [1982]

Eadie, John [ed.], *A New and Complete Concordance to the Holy Scriptures*, John J. Griffin & Co., London, 1853

Easterbrook, Michael, *Butterflies of the British Isles: The Lycaenidae*, Shire Publications, Aylesbury, 1988
Easterbrook, Michael, *Butterflies of the British Isles: The Pieridae*, Shire Publications, Aylesbury, 1989
Ekwall, Eilert, *The Concise Oxford Dictionary of English Place-names*, 4th edition, Clarendon Press, Oxford, 1987
Erzinclioglu, Zakaria, *Blowflies*, Richmond Publishing Co., Richmond, 1996
Evans, George & Thomson, David, *The Leaping Hare*, Faber & Faber, London, 1972
Evans, Glyn, *The Life of Beetles*, George Allen & Unwin, London, 1975
Evans, I.H. [ed.], *Brewer's Dictionary of Phrase and Fable*, 14th edition, Cassell, London, 1989 [1870]
Fairley, J.S., *An Irish Beast Book*, Blackstaff Press, Belfast,1975
Fisher, Arthur T., *Outdoor Life in England*, Richard Bentley & Son, London, 1896
Fisher, James, *The Fulmar*, Collins, London, 1952
Fitter, R. & Manuel, R., *Freshwater Life*, HarperCollins, London, 1986
Flavell, L. & R., *Dictionary of Proverbs and Their Origins*, Kyle Cathie, London, 1993
Forbes, A.R., *Gaelic Names of Beasts, Birds, Fishes, Insects, and Reptiles*, Oliver & Boyd, Edinburgh, 1905
Ford, E.B., *Butterflies*, Collins, London, 1945
Ford, E.B, *Moths*, Collins, London, 1955
Fox, Lilla M., *Costumes and Customs of the British Isles*, Chatto & Windus, London, 1974
Fox-Davies, A.C., *Complete Guide to Heraldry*, T.C. & E.C. Jack, London, 1925
Franklyn, Julian, *A Dictionary of Rhyming Slang*, Routledge & Kegan Paul, London, 1961 [1960]
Fraser Darling, Frank & Boyd, J.M., *The Highlands and Islands*, Collins, London, 1964
Frazer, Deryk, *Reptiles and Amphibians in Britain*, Collins, London, 1983
Frederick, Sir Charles, *Foxhunting*, Seeley, Service & Co., London [n.d.]
Garstang, Walter, *Larval Forms and Other Zoological Verses*, Basil Blackwell Ltd, Oxford, 1951
Gascoigne, M., *Discovering English Customs and Traditions*, 3rd edition [revised and updated by George Monger], Shire Publications, Princes Risborough, 1998 [1969]
Gibson, F., *Superstitions About Animals*, Scott Publishing Co. Ltd, London, 1904
Gibson, James [ed.], *The Complete Poems of Thomas Hardy*, Macmillan, London, 1976
Glasse, Hannah, *The Art of Cookery Made Plain and Easy*, W. Strahan, London, 1774 [1747]
Glyde, John, *The New Suffolk Garland*, published by the author, Ipswich, 1866
Gomme, George Laurence [ed.], *The Gentleman's Magazine Library: Popular Superstitions, Collection from 1731–1868*, Elliot Stock, London, 1884
Gordon, D.G., *The Compleat Cockroach*, Ten Speed Press, Berkeley, California, 1996
Green, Marian, *A Calendar of Festivals*, Element, Dorset 1991
Greenoak, Francesca, *All the Birds of the Air – The Names, Lore and Literature of British Birds*, André Deutsch, London, 1979
Gurdon, Lady E.C. [collector and ed.], *County Folk Lore: Gloucestershire, Suffolk, Leicestershire and Rutland. Printed Extracts, No. 2. Suffolk*, published for the Folklore Society by D. Nutt & J. Loder, Ipswich, and Pawsey & Hayes, Woodbridge, 1893
Gwynn Jones, T., *Welsh Folklore and Folk Custom*, Methuen & Co., London, 1930
Hale, W.G., *Waders*, Collins, London, 1980
Hammond, Cyril O., *The Dragonflies of Great Britain and Ireland*, 2nd edition [revised by Robert Merritt], Harley Books, Colchester, 1983 [1977]
Hammond, P.W., *Food and Feast in Medieval England*, Wrens Park Publishing, Gloucestershire, 1993
Hardy, Alister, *The Open Sea. II Fish and Fisheries*, Collins, London, 1959
Hare, C.E., *Bird Lore*, Country Life Ltd, London, 1952
Harris, J.R., *An Angler's Entomology*, Collins, London, 1952
Harrison-Matthews, Leo, *British Mammals*, Collins, London, 1952
Harrowren, Jean, *Origins of Festivals and Feasts*, Kaye & Ward, London, 1980
Harting, James, *Extinct British Animals*, Trübner & Co., London, 1880
Harting, James Edmund, *The Birds of Shakespeare*, John Van Voorst, London, 1965 [An updated version of *The Ornithology of Shakespeare Critically Examined, Explained and Illustrated*, 1871]
Hartland, F.S.A. [ed.], *County Folk Lore: Gloucestershire, Suffolk, Leicestershire and Rutland. Printed Extracts, No. 1. Gloucestershire.* Reprinted by Kraus Reprint, Nendeln/Liechtenstein, 1967
Harvey, L.A. & Leger-Gordon, D. St, *Dartmoor*, Collins, London, 1953
Hayward, P., Nelson-Smith, T. & Shields, C., *Sea Shore of Britain and Europe*, HarperCollins, London, 1996
Heaney, Seamus, *Death of a Naturalist*, Faber & Faber, London, 1966
Hendrickson, Robert, *The Ocean Almanac*, Hutchinson, 1992
Hewer, H.R., *British Seals*, Collins, London, 1974
Higgins, L.G. & Riley, N.D., *Butterflies of Britain and Europe*, Collins, London, 1984
Hillyard, Paul, *The Book of the Spider*, Pimlico, London, 1995 [1994]
Hinde, Thomas [ed.], *The Domesday Book*, Hutchinson, London, 1985
Hole, Christina, *English Custom and Usage*, Batsford, London, 1941–2
Hole, Christina, *English Folklore*, Batsford, London, 1940
Horsley, Reverend Canon, *Our British Snails*, SPCK, London, 1915
Hull, Eleanor, *Folklore of the British Isles*, Methuen & Co., London, 1928
Hull, Robin, *Scottish Birds – Culture and Tradition*, Mercat Press, Edinburgh, 2001
Hulme, F.E., *Natural History, Lore and Legend*, Bernard Quaritch, London, 1895
Hunt, R., *Popular Romances of the West of England*, 3rd edition, Chatto & Windus, London, 1881
Hutchinson, Thomas [ed.], *Wordsworth's Poetical Works* [revised by Ernest de Selincourt], Oxford University Press, Oxford, 1984 [1904]
Ibbott, Selena, *Folklore, Legends and Spells*, Ashgrove Press Ltd, Bath, 1994
Imms, A.D., *Insect Natural History*, Collins, London, 1947
Ingersoll, Ernest, *Birds in Legend Fable and Folklore*, Longmans, Green & Co., New York, 1923 [Reissued by Singing Tree Press, 1968]
Ingle, R.W., *British Crabs*, British Museum (Natural History), London, and Oxford University Press, Oxford, 1980
Jones, J.W., *The Salmon*, Collins, London, 1959
Jones-Baker, Doris, *The Folklore of Hertfordshire*, Batsford, London, 1977
Kearton, R., *Birds' Nests, Eggs, and Egg-collecting*, Cassell, London, 1903 [1890]
Keogh, John, *Zoologia Medicinalis Hibernica*, S. Powell, Dublin, 1739
Kerney, M.P. & Cameron, R.A.D., *Land Snails of Britain and Europe*, Collins, London, 1979
Kightly, Charles, *The Customs and Ceremonies of Britain*, Thames & Hudson, London, 1987 [Reprinted 1994]
Kightly, Charles, *The Perpetual Almanack of Folklore*, Thames & Hudson, London, 1987 [Reprinted 1994]
Kirby, W. Egmont, *Insects: Foes and Friends*, S.W. Partridge & Co., London, 1898
Kirby, William Forsell, *Butterflies and Moths in Romance and Reality*, Sheldon Press, London, 1913
Klingender, Francis, *Animals in Art and Thought to the End of the Middle Ages*, Routledge & Kegan Paul, London, 1971
Lamb, G.F., *Animal Quotations*, Longman, London, 1985
Langton, T., *Snakes and Lizards*, Whittet Books, London, 1989
Lankaster, E., *The Uses of Animals in Relation to the Industry of Man*, Robert Hardwicke, London, 1860
Larwood, Jacob & Hotten, John Camden, *English Inn Signs*, Chatto & Windus, London, 1951

Lemprière, Raoul, *Customs, Ceremonies and Traditions of the Channel Islands*, Robert Hale, London, 1976

Lever, Christopher, *The Naturalized Animals of the British Isles*, Hutchinson, London, 1977

Lever, Christopher, *They Dined on Eland: The Story of the Acclimatisation Societies*, Quiller Press, London, 1992

Lockley, R.M., *Grey Seal, Common Seal*, White Lion, London, 1977 [André Deutsch, 1966]

Lockwood, William B., *The Oxford Book of British Bird Names*, Oxford University Press, Oxford, 1984

Long, W.H., *The Dialect of the Isle of Wight with Tales and Anecdotes*, Reeves & Turner, London, and G.A. Brannon & Co., Newport, 1886

Lovell, M.S., *The Edible Mollusks of Great Britain and Ireland*, Reeve & Co., London 1867

Lowe, Frank A., *The Heron*, Collins, London, 1954

Lucie-Smith, Edward [ed.], *British Poetry Since 1945*, Penguin Books, London, 1985 [1970]

Lythgoe, J. & G., *Fishes of the Sea*, Blandford Press, London, 1971

Macdonald, David & Barrett, Priscilla, *Mammals of Britain and Europe*, HarperCollins, London, 1993

Macleod, *British Calendar Customs – Scotland*, Vol. 1, Folklore Society, London, 1937

Maitland, Peter S. & Campbell, R. Miall, *Freshwater Fishes*, Collins, London, 1992

Majerus, M., *Ladybirds*, Collins, London, 1994

Martin, Laura C., *Wildlife Folklore*, The Globe Pequot Press, Connecticut, 1994

Marwick, Ernest W., *Folklore of Orkney and Shetland*, Batsford, London, 1975

McDonald, James, *A Dictionary of Obscenity, Taboo and Euphemism*, Sphere Books, London, 1988

McMillan, N., *British Shells*, Frederick Warne & Co. London, 1968

Mead, Chris, *The State of the Nation's Birds*, Whittett Books, Stowmarket, 2000

Mellanby, Kenneth, *The Mole*, Collins, London, 1971

Mitford, Mary Russell, *Our Village*, Sidgwick & Jackson, London, 1986

Moore, A.W., *Folk-Lore of the Isle of Man*, S.R. Publishers, Yorkshire, 1971 [1891]

Moore, Clifford B., *Ways of Mammals – In Fact and Fancy*, The Ronald Press Company, New York, 1953

More, Daphne, *The Bee Book*, Universe Books, New York, 1976

Morgan, Adrian, *Toads and Toadstools: The Natural History, Folklore and Cultural Oddities of a Strange Association*, Celestial Arts, Berkeley, California, *c.*1995

Morris, P.A., *The Hedgehog*, 3rd edition, Shire Publications, Princes Risborough, 1997 [1988]

Morris, Reverend M.C.F., *Yorkshire Folk-Talk*, Henry Frowde, London, and John Sampson, York, 1892

Morrison, Blake & Motion, Andrew [eds], *The Penguin Book of Contemporary British Poetry*, Penguin, London, 1982

Moule, Thomas, *The Heraldry of Fish*, John Van Voorst, London, 1842

Mountford, G., *The Hawfinch*, Collins, London, 1957

Mourier, H., Winding, O. & Sunesen, E., *Wild Life in House and Home*, Collins, London, 1975

Muffet, Thomas, *Edward Topsell's History of Four-footed Beasts and Serpents and Insects, Vol. 3, The Theater of Insects*, Frank Cass & Co., London, 1967 [1658]

Mullarney, K., Svensson, L., Zetterström, D., & Grant, P.J., *Collins Bird Guide*, HarperCollins, London, 2000

Murray, James et al. [eds], *The Oxford English Dictionary*, Clarendon Press, Oxford, 1961

Murton, R.K., *The Wood Pigeon*, Collins, London, 1965

Murton, R.K., *Man and Birds*, Collins, London, 1971

Nance, R. Morton, *A Glossary of Cornish Sea Words*, Federation of Old Cornwall Societies, Marazion, 1963

Napier, James, *Folk Lore of Superstitious Beliefs in the West of Scotland within this Century*, Alexander Gardner, Paisley, 1879

Neal, Ernest, *The Badger*, Collins, London, 1948

Nethersole-Thompson, Desmond, *The Greenshank*, Collins, London, 1951

Newall, Venetia, *Discovering the Folklore of Birds and Beasts*, Shire Publications, Tring, 1971

Newton, I., *Finches*, Collins, London, 1972

Nichols, D. & Cooke, J., *The Oxford Book of Invertebrates*, Oxford University Press, Oxford, 1979 [1971]

Nicholson, E.R., *Birds and Men*, Collins, London, 1951

Nicholson, James R., *Shetland Folklore*, Robert Hale, London, 1981

Nicholson, John, *Folklore of East Yorkshire*, Thomas Holderness, Driffield, 1890

Northcote, W. Thomas [ed.], *County Folk-Lore: Vol. 4, Northumberland*, David Nutt, London, 1904

Oldroyd, H., *The Natural History of Flies*, World Naturalist series, Weidenfeld & Nicolson, London, 1964

Opie, Iona & Peter [eds], *The Oxford Dictionary of Nursery Rhymes*, 12th edition, Oxford University Press, Oxford, 1980 [1951]

Opie, Iona & Peter, *The Oxford Nursery Rhyme Book*, 8th edition, Oxford University Press, Oxford, 1977 [1955]

Opie, Iona & Peter, *The Lore and Language of Schoolchildren*, Clarendon Press, Oxford, 1959

Page, Robin, *The Fox and the Orchid – Country Sports and the Countryside*, Quiller Press, London, 1987

Paisley, Robert Ford, *Children's Rhymes, Games, Songs, and Stories*, Alexander Gardner, Paisley, 1904

Palmer, Kingsley, *The Folklore of Somerset*, Batsford, 1976

Palmer, Roy, *The Folklore of Warwickshire*, Batsford, London, 1976

Palmer, Roy, *The Folklore of Leicestershire and Rutland*, Sycamore Press, Wymondham, 1985

Palmer, Roy, *Britain's Living Folklore*, David & Charles, Newton Abbot, 1991

Palmer, Roy, *The Folklore of Gloucestershire*, West Country Books, Tiverton, 1994

Parker, Eric, *Shooting by Moor, Field and Shore*, Seeley, Service & Co., London [n.d.]

Parker, Eric, *Ethics of Egg-Collecting*, The Field, London, 1935

Partridge, Eric, *The Penguin Dictionary of Historical Slang*, [abridged by Jacqueline Simpson], Penguin, London, 1986

Paterson, Wilma, *Salmon and Women*, H.F. & G. Witherby, London, 1990

Patrides, C.A. [ed.], *The Complete English Poems of John Donne*, 4th edition, Everyman, J.M. Dent, London, 1990 [1985]

Pennant, Thomas, *British Zoology*, Benjamin White, London, 1776

Perrins, Christopher, *British Tits*, Collins, London, 1979

Plummer, D. Brian, *Tales of a Rat-hunting Man*, Boydell Press, Ipswich, 1988 [1978]

Polson, A., 'Our Highland folklore heritage', *Northern Chronicle*, Inverness, 1926

Porter, Enid, *Cambridgeshire Customs and Folklore*, Routledge & Kegan Paul, London, 1969

Porter, Enid, *The Folklore of East Anglia*, Batsford, London, 1974

Porter, J.R. & Russell, W.M.S. [eds], *Animals in Folklore*, published for the Folklore Society by D.S. Brewer and Rowman & Littlefield, Cambridge, 1978

Potter, Beatrix, *Beatrix Potter's Nursery Rhyme Book*, F. Warne & Co./Penguin Books, London, 1975

Potter, S. & Sargent, L., *Pedigree Words from Nature*, Collins, London, 1973

Powell, D. & Robinson, E. [eds], *John Clare*, [Oxford Authors series], Oxford University Press, Oxford, 1984

Price, Martin, *The Restoration and the Eighteenth Century*, Oxford University Press, New York, 1973

Prynne, Michael, *Egg-shells*, Barrie & Rockliff, London, 1963

Rackham, Oliver, *The History of the Countryside*, J.M. Dent, London, 1986

Radford, E. & M.A., *The Encyclopedia of Superstitions*, 6th edition [edited and revised by Christina Hole], Helicon Publishing, Oxford, 1995 [1948]

Ransome, Arthur, *Coot Club*, Jonathan Cape, London, 1930

Ransome, H.M., *The Sacred Bee in Ancient Times and Folklore*, George Allen & Unwin, London, 1937

Raven, Jon, *The Folklore of Staffordshire*, Batsford, London, 1978
Renfrew, Jane, *Food and Cooking in Roman Britain*, English Heritage, Birmingham, 1985
Richardson, A.E., *The Old Inns of England*, 4th edition, Batsford, London, 1942 [1934]
Ridout, R. & Witting, C, *The Macmillan Dictionary of English Proverbs Explained*, Macmillan, London, 1995 [Heinemann, 1967]
Roberts, M.J., *Spiders of Britain and Northern Europe*, HarperCollins, London, 1995
Roberts, Michael [ed.], [revised by Porter, Peter], *The Faber Book of Modern Verse*, 4th edition, Faber & Faber, London, 1982 [1936]
Ansell, Robin P., *Animal Lore in English Literature*, John Murray, London, 1932
Robinson, F.N. [ed.] & Benson, L. [general ed.], *The Riverside Chaucer*, 3rd edition, Oxford University Press, Oxford, 1992 [1987]
Rodd, E.H., *The Birds of Cornwall and the Isles of Scilly*, Trübner & Co., London, 1880
Room, Adrian, *The Naming of Animals*, McFarland & Co., Jefferson, 1993
Ross, Anne, *The Folklore of the Scottish Highlands*, Batsford, London, 1976
Rowland, Beryl, *Blind Beasts: Chaucer's Animal World*, Kent State University Press, Kent, Ohio, 1971
Rowling, Marjorie, *The Folklore of the Lake District*, Batsford, London, 1976
Rudkin, Ethel H., *Lincolnshire Folklore*, E.P. Publishing, Wakefield, 1973 [Beltons, 1936]
Salmon, Michael A., *The Aurelian Legacy*, Harley Books, Colchester, 2000
Salvin, Francis Henry & Brodrick, William, *Falconry in the British Isles*, Beech Publishing, Midhurst, 1997 [Reprint of 2nd edition of 1873]
Samstag, Tony, *For Love of Birds – The Story of the Royal Society for the Protection of Birds 1889–1988*, RSPB, Bedfordshire, 1988
Sarot, Eden Emanuel, *Folklore of the Dragonfly, A Linguistic Approach*, Edizioni di Storia e Letteratura, Rome, 1958
Schmitt, W.L., *Crustaceans*, University of Michigan, 1965 and 1968 [UK edition, David & Charles, Newton Abbot, 1973]
Scott-Giles, C. Wilfred, *Civic Heraldry of England and Wales*, J.M. Dent, London, 1953 [1933]
Shorten, Monica, *Squirrels*, Collins, London, 1954
Simms, Eric, *Woodland Birds*, Collins, London, 1971
Simms, Eric, *British Thrushes*, Collins, London, 1978
Simms, Eric, *British Warblers*, Collins, London, 1985
Simms, Eric, *British Larks, Pipits and Wagtails*, Collins, London, 1992
Simpson, Jacqueline, *The Folklore of Sussex*, Batsford, London, 1973
Simpson, Jacqueline, *The Folklore of the Welsh Border*, Batsford, London, 1976
Simpson, Jacqueline, *The Concise Oxford Dictionary of Proverbs*, 2nd edition, Oxford University Press, Oxford, 1992 [1982]
Singer, Charles, *Early English Magic and Medicine*, Oxford University Press, for the British Academy, London, 1920
Smith, Malcolm, *The British Amphibians and Reptiles*, Collins, London, 1951
Smith, Stuart, *The Yellow Wagtail*, Collins, London, 1950
St Clair, Sheila, *Folklore of the Ulster People*, Mercier Press, Cork, 1971
Stafford, P., *The Adder*, Shire Publications, Princes Risborough, 1987
Staveley, E.F., *British Spiders*, Reeve & Co., London, 1866
Staveley, E.F., *British Insects*, Reeve & Co., London, 1871
Stebbing, Reverend Thomas R.R., *A History of Crustacea*, Kegan Paul, Trench, Trübner & Co., London, 1893
Step, Edward, *British Insect Life*, T. Werner Laurie, London, 1929 [Folklore Society, 1921]
Step, Edward, *Nature Rambles – Winter to Spring*, Frederick Warne & Son, London, 1930
Step, Edward, *Shell Life*, Frederick Warne & Son, London, 1945
Stephen, Sir Leslie & Lee, Sir Sidney [eds], *Dictionary of National Biography*, Oxford University Press, Oxford, 1973 [Smith, Elder & Co., 1885]
Sternberg, Thomas, *The Dialect and Folklore of Northamptonshire*, John Russell Smith, London, 1851
Summers-Smith, J.D., *The House Sparrow*, Collins, London
Sutton, Stephen, *Invertebrate Types – Woodlice*, Ginn & Co., London, 1972
Swainson, C., *A Handbook of Weather Folk-lore*, William Blackwood & Sons, Edinburgh and London, 1873
Swainson, C. *The Folklore and Provincial Names of British Birds*, Llanerch Publishers, Felinfach, 1998 [Facsimile reprint; first published 1886]
Swire, Otta F., *The Outer Hebrides and Their Legends*, Oliver & Boyd, Edinburgh,1966
Thomas, Jeremy, *The Butterflies of Britain and Ireland*, Dorling Kindersley, London, 1991
Thompson, Harry V. & Worden, Alastair, N., *The Rabbit*, Collins, London, 1956
Thompson, W.H., *The Speech of Holderness and East Yorkshire*, A. Brown & Sons, Hull, 1890
Thorndike, Lynn, *A History of Magic and Experimental Science*, Colombia University Press, New York, 1934
Tinbergen, Niko, *The Herring Gull's World*, Collins, London, 1953
Topsell, Edward, *The History of Four-Footed Beasts and Serpents and Insects*, Da Capo Press, New York, 1967 [Reprint of 1658 edition]
Tubbs, Colin R., *The New Forest*, Collins, London, 1986
Udal, John Symonds, *Dorsetshire Folk-Lore*, 2nd edition, Toucan Press, London, 1970 [Stephen Austin & Sons 1922]
Wagner, Leopold, *Manners, Customs and Observances*, Heinemann, London, 1895
Walton, Izaak, *The Complete Angler*, 6th edition, Rivington et al., London, 1797 [1653]
Waters, Brian, *Severn Tide*, J.M. Dent, London, 1947
Watson, J.N.P., *The Book of Foxhunting*, Batsford, London, 1977
Weaver, P., *The Lapwing*, Shire Publications, Aylesbury, 1985
Wentworth Day, J., *Sporting Adventure*, George Harrap & Co., London, 1937
White, Gilbert, *Natural History of Selborne*, new edition, George Routledge, London, 1880 [1789]
White, Gilbert, *The Journals of Gilbert White*, Vol. 1 [ed. Francesca Greenoak], Century, London, 1986
White, Gilbert, *The Journals of Gilbert White*, Vol. 2 [ed. Francesca Greenoak], Century, London, 1988
Whitlock, Ralph, *The Folklore of Wiltshire*, Batsford, London, 1976
Wilde [Lady], *Ancient Cures, Charms and Usages of Ireland*, Ward & Downey, London, 1890
Wilde, Sir William, *Irish Popular Superstitions*, Irish University Press, Ireland. 1972 [1872]
Yalden, Derek, *The History of British Mammals*, T. & A.D. Poyser, London, 1999
Yonge, C.M., *The Sea Shore*, Collins, London, 1949
Yonge, C.M., *Oysters*, Collins, London, 1960
Young, Mark, *The Natural History of Moths*, T. & A.D. Poyser, London, 1997

Principal serial publications referred to are:
British Wildlife
Country Life
Countryman
Daily Telegraph
Guardian
The Times
and the publications of:
British Herpetological Society
British Ornithologists' Union
British Trust for Ornithology
Folklore Society
Linnean Society
Mammal Society
Royal Entomological Society
Royal Society
Royal Society for the Protection of Birds
Woolhope Naturalists' Field Club
Zoological Society of London

Over 400 Websites were also consulted.

INDEX

Page numbers in *italic* refer to illustrations

AUTHOR'S ACKNOWLEDGEMENTS

I am deeply grateful to HRH The Prince of Wales for so kindly writing the Foreword and taking such a close interest in the book.

This book is the result of many individual contributions. *Fauna Britannica* was the idea of Laura Bamford of Octopus Publishing, to whom I am greatly indebted for the opportunity to explore and collect a wealth of remarkable and intriguing information about the wild animal life of Britain that has been my fascination for so many years. Alison Goff at Hamlyn has been most supportive and has shown great understanding both of the way the material should be presented and the always tricky subject of author sensitivities. Jane Birch skilfully saw the text through to publication in a time scale that could so easily have caused many of us sleepless nights, while Maggie O'Hanlon has edited my script with great thoroughness, care and understanding. Other staff at Hamlyn have been unstintingly helpful in many different ways and it is a pleasure to acknowledge the contributions of Kristy Richardson, Cathy Lowne, Lydia Darbyshire, Zoë Holtermann and Christine Junemann, and Geoff Fennell and Mark Stevens, who spent countless hours over the design.

I cannot apportion sufficient thanks to my most diligent and thorough researchers, Sam Ward-Dutton and Claire Musters, who, with Rhoda Sweeting, uncovered and handled a mountain of material in the most carefully organized and logical fashion.

I am most grateful to Susannah Charlton of the *Daily Telegraph* for publicizing the project and thereby putting me in touch with many fascinating and remarkable contributions and contributors.

Janice Light gave up much time to supply information on conchological matters, while Stella Turk made her own extensive Cornish library available and was greatly assisted by the Cornwall Wildlife Trust. Among many librarians and archivists, Caroline Oates of the Folklore Society was especially helpful and took a great interest in the study. Other libraries that were consulted extensively, and for whose assistance and interest I am very grateful, are those at the Natural History Museum in London, the Royal Society for the Protection of Birds and the Birmingham University Shakespeare Institute in Stratford-upon-Avon, as well as Richmond Public Library, Wimbledon Public Library and the library of the Linnean Society.

Many people have read and commented on parts of the script. One person has read it all. To my old friend and zoological mentor, Clinton Keeling, I owe a particular debt for his wise and typically maverick counsel. Nonetheless, I dissociate him and everyone else from any errors that may have slipped through simply because I could not match their standards of exactitude.

My wife Beverley has lived with *Fauna Britannica* for a very long time and has been an ever-present and encouraging counsellor when the going became sticky.

Above all, however, my debt is to the people of these islands, both ancient and modern, who have created and chronicled our animal folklore and traditions. Those of our own time who responded to my numerous queries, or who volunteered information and experience, are listed below, in alphabetical order, and it is with the most sincere pleasure that I thank them all. Although it is perhaps unfair to pick out individuals, I must give special thanks to one of the great observers and recorders of British animal life, Eric Simms of Lincolnshire, who supplied me with copious invaluable facts and reminiscences. I have listed contributors' names as they have signed themselves and apologise if I have misread anyone's handwriting.

Jean Alderton-Menz, Lapford, Devon
Mary Alexander, Leisure Services, Guildford Borough, Surrey
Christopher Allan, Ede & Ravenscroft, London
Miss Elizabeth St. B. Anderson, Yeovil, Somerset
Peter P. Anderson, Cumbria
Janie Andrew, Herefordshire
Mary Ankrett, Sutton Coldfield, West Midlands
Anonymous Fellow, All Soul's College, Oxford
Norma Aubertin-Potter, All Soul's College, Oxford
Grace Ayling, St Ives, Cornwall
Stuart Bailey, Malacological Society of London
Mrs Jo Baker, Bromsgrove, Worcestershire
Kathryn H. Baker, British Golf Museum, St Andrews, Fife
Mrs M.A. Baker, Caernarvon, Gwynedd
K. Barker, Ilford, Essex
Valerie Barkham, Kew, Surrey
Joan Bartholomew, Alicante, Spain
Clive Barton, Norfolk
Rosemary Beel, Somerset
Mrs C.J. Beesley, Cheltenham, Gloucestershire
Patricia Belford, Leeds, West Yorkshire
Jonathan Bengston, Queen's College, Oxford
Mrs Patricia Bennett, Eastbourne, East Sussex
David Berkley, Bridgwater, Somerset
Denyce Biggs, Co. Kerry
Jocelyn C. Blakey, Brandon, Suffolk
Joan Bliss, Bath, Somerset
Doreen Blott, Bedfordshire
A.G. Blunt, Barnsley, South Yorkshire
Godfrey Blunt, Bridgnorth, Shropshire
Nina Bollen, Stourbridge, West Midlands
Les Borg, Lee Valley, Essex
Christopher Borthen, *Stalking Magazine*, Exeter, Devon
Katie Boyle, London
J.A. Boynton, Crowborough, East Sussex
A. Brace, York Minster Archives, York, North Yorkshire
Christine Bridson, Twickenham, Greater London
Dr Martin Brimble, Meopham, Kent
Tony Britten, London
Florence Brojer, Cookham, Berkshire
Phyllis May Brooks, Epsom Downs, Surrey
Gillian Brown, Luton, Bedfordshire
Gilly Brown, Norfolk
Hamish Brown, Burntisland, Fife
Norah Buckley, Salisbury, Wiltshire
Mary Campion, Bromley, Kent
Patrick Carden, Fordingbridge, Hampshire
Jane Carter, Stoke-on-Trent, Staffordshire
Shelagh M. Carter, Hexham, Northumberland
Mrs Sylvia Carvell, Capel, Surrey
Sheila H. Casey, Bexleyheath, Kent
Gordon & Judith Cawthorne, Stanton, Suffolk
Don Chaney, Twickenham, Greater London
Church of Scotland: Press Office
Arabella Clark, Derbyshire
Elizabeth A. Clark, Kempsey, Worcestershire
Michael Clarke, Welwyn, Hertfordshire
Charles Clayton, Westmorland

Annette Cleaner, Great Orme Goat Breeders' Association, Henley-on-Thames, Oxfordshire
C. Clifford, Coventry, West Midlands
Val Clinging, Sheffield, South Yorkshire
Sybil Coady, Loughborough, Leicestershire
Mrs Joy Cobb, Twickenham, Greater London
Mrs B.B. Cole, Sheffield, South Yorkshire
Harry Coles, Essex
Barry J. Colledge, Newcastle upon Tyne, Tyne and Wear
Hugh Collins, Wool, Dorset
Patrica Colson, Gwent
Lillian Cook, Brig o'Turk, Perthshire
Lindsay Corbett, University of Stirling
R.S. Cornish, Cheshire
Margaret Courtney, Guides Association, Baden-Powell House, London
Margaret Coutu, Farnham, Surrey
Mr & Mrs M.J. Crampton, Whitney-on-Wye, Herefordshire
Andrea Dalton, Cheshire
Sue Daly, Jersey
Priscilla & Tony Dart, Welwyn, Hertfordshire
Mrs A.R.F. Dastor, Chipping Norton, Oxfordshire
Clive de Boer, Essex
Miss R. de Falbe, London
Jeffrey Dench, Stratford-upon-Avon, Warwickshire
Elizabeth Derges, Paignton, Devon
Norma Dewhurst, Stratford-upon-Avon, Warwickshire
Tom Dewick, Elgin, Moray
Norma & Ray Dicker, Newport-on-Tay, Fifeshire
Andrea M. Dobbin, Cheadle Hulme, Cheshire
Howard Dodsworth, Tunbridge Wells, Kent
Evelyn Doherty, Hawarden, Flintshire
Emily & Oliver Ducker, Priors Marston, Warwickshire
Mary Dunmall, Taunton, Somerset
Michael Ecob, Stourport-on-Severn, Worcestershire
Liz Elias, Chobham, Surrey
David Elwood, Truro, Cornwall
Matthew Engel, Herefordshire
Roger Favell, Deeping St James, Lincolnshire
Arnold Fetters, Nottinghamshire
Dr Daphne C. Fielding, Billingshurst, West Sussex
Penelope Fletcher, Hungerford, Berkshire
Flt Lt Simon Flynn, Ministry of Defence, London
Phillipa Foord-Kelsey, Banbury, Oxfordshire
Catherine Fox, Bishop's Stortford, Herfordshire
H.D. Fox, Heswall, Merseyside
Eileen Frees, Somerset
Jeremy Furber, Queen Charlton, Somerset
Mrs Jean Gill, Frome, Somerset
Harry Gleason, Cambridgeshire
G. Godman, Bath, Somerset
John M. Goodier, Lymm, Cheshire
Mrs P.M. Gore, Weston-Super-Mare, Somerset
Loris Goring, Brixham, Devon
Martin Goulding, Rye, East Sussex
Jennifer Griffin, Bradford, West Yorkshire
Mrs Susie Groom, Chichester, West Sussex
Judith L. Hall, Warwick, Warwickshire
Louise Hampson, Archivist, York Minster Library, York, North Yorkshire
Martin Hannan, at Silver Scots Place, http://personal.nbnet.ca/
Robin Harcourt-Williams, Archivist, Hatfield House, Hertfordshire
Jean Harris, Sheffield, South Yorkshire
Jonathan Harrison, St John's College, Cambridge
Julian Harrison, Belper, Derbyshire
Mrs B. Hart, Bromley, Kent
Mollie Harten, Plymouth, Devon
Susan Hatfield, Sevenoaks, Kent
Jean Hayes, Buntingford, Hertfordshire
S.J. Hayhow, Museum of Lancashire, Fleetwood
John M. Hazledine, Weston-Super-Mare, Somerset
Kate Hebditch, Dorset Natural History and Archaeological Society, Dorchester, Dorset
Mrs Jeanne Hewitt, Epsom, Surrey
Sonia Higgins, King's Stanley, Gloucestershire
Denis E.W. Hillman, Bexhill-on-Sea, Essex
Mrs S. Hirschfield, Hull, East Riding of Yorkshire
Professor Nigel Hitchin, New College, Oxford
Dr Keith Hollinshead, Leighton Buzzard, Bedfordshire
Richard Holworthy, Cours de Monegur, France
Mrs Sue Hood, Witham, Essex
Vera J. Hopley, Lichfield, Staffordshire
G.E. Horton, Chester, Cheshire
Bob Howard, Keynsham, Somerset
Barbara Howe, Weymouth, Dorset
Miss E.M. Howes, Chichester, West Sussex
Peter Howlett, National Museum of Wales, Cardiff
Jennifer Huber, Westcott, Surrey
Mrs Marjorie Hulett, Thame, Oxfordshire
Margaret Inglis, London
V. Joan Jackman, Brean, Somerset
Mrs K.M. Jackson, Blandford Forum, Dorset
David James, Swansea
Maureen Jarman, London
T.W. Jenkins, London
Mrs Linda Jones, Ceredigion
Clinton Keeling, Surrey
Mrs J. Kelsey, Royal Archives, Windsor, Berkshire
E.C. King, Repton, Derbyshire
Dick King-Smith, Somerset
Alan Knight, British Marine Life Study Society
George Lambert, Burscough, Lancashire
Ginny Lapage, Bodmin, Cornwall
Jeanette Lawton, The British Deer Society, Stoke-on-Trent, Staffordshire
Mrs O. Stirling Lee, Tewkesbury, Gloucestershire
Mrs S. A. Lewis, Cleethorpes, Lincolnshire
Tom Linger, South Nutfield, Surrey
Tony Lintern, Leicester
Mrs Doreen Lomas, Cheadle, Cheshire
A.G. Lounsbach, Portishead, Somerset
Tony Loy, Sheffield, South Yorkshire
Tony Loy, Sorby Natural History Society, Sheffield South Yorkshire
Dr W. G. G. Loyn, Aberystwyth, Ceredigion
Jill Lucas
M. Lunt, Surbiton, Surrey
Suzanne Lyonnet, Redhill, Surrey
Andrew Mackay, Worthing
Dr Michael Majerus, University of Cambridge
Susan Major, St Agnes C. of E. Primary School, Isles of Scilly
P.G.N. Mason, Newport, Shropshire
David Massa, London
Mrs C. McCartan, Kidderminster, Worcestershire
Mrs Ann McCutcheon, Wokingham, Berkshire
Johnnie McHoy, Belfast
Reverend Eddie McKenna, North Berwick
Nigel Melville, Abbotsbury, Dorset
Norah Meyer, Manchester
Mrs. E.A. Moneypenny, Ashford, Kent
Marjorie Morgan, Tunbridge Wells, Kent
Mrs R. Morley, Nottingham
Jocelyn Morris, Oxford
Imogen Mottram, Oakham, Rutland
A. Moyle, Shrewsbury, Shropshire
S. Murton, Thirsk, North Yorkshire
Catherine Nicholls, North Yorkshire
'Nigel', Isles of Scilly
Mrs T.F.F. Nixon, Pampisford, Cambridge
J. & J.M. Norman, Wisbech, Cambridgeshire
Helen O'Brien, Tranent, East Lothian
Anne M. Oakley, Canterbury Cathedral Archives
Sybil M. Ogden, Woking, Surrey
Catherine Oldham, Library, Forest Research Station, Farnham, Surrey
Tom Ormond, London
C. Orton, Burnham-on-Sea, Somerset
Mrs Pat Owers, Ilford, Essex
David E. Oxford, Hampton, Middlesex
Elizabeth Paget, Falmouth, Cornwall
Jean Palmer, Southampton, Hampshire
Katy Pannell, Southam, Warwickshire
Mrs G.M. Parker, Esher, Surrey
Patricia Parkyn, Bridgnorth, Shropshire
John Parsons, Diss, Norfolk
Graham Paskett, Uttoxeter, Staffordshire
Val Perris, Hemel Hempstead, Hertfordshire
Denis E. Perry, Croydon, Surrey
Lin Perry, Frodsham, Cheshire
Ann Peters, St John's College School, Cambridge
Alan R. Phillips, Ryde, Isle of Wight
Audrey Pitt, Manchester
John S. Powell, York Minster Library, York, North Yorkshire
Keith Richard Power, Preston, Lancashire

Emeritus Professor Thomas A. Preston, Christchurch, Dorset
Shirley Price, Quorn, Leicestershire
Cyril R. Prouse, Leighton Buzzard, Bedfordshire
Mrs Judith Prust, The Shrine Church of St Melangell, Llangynog, Carmarthenshire
Carole M. Pyett, Burton Pidsea, Yorkshire
Francesca Quint, Surrey
Dr A.P. Radford, Taunton, Somerset
Brian Radford, Southampton, Hampshire
Valerie Rambaut, Beverley, East Riding of Yorkshire
Michael Ratsey, Martock, Somerset
Patricia Renwick, Warrington, Cheshire
Barry Rice, Shoreham-by-Sea, West Sussex
Dr Tim Rich, National Museum of Wales, Cardiff
Mr & Mrs L. Roberts, Potters Bar, Hertfordshire
Lynn Robins, Enfield, Greater London
Angela Rogers, London
Lt Col Sir Malcolm Ross, Lord Chamberlain's Office, Buckingham Palace, London
Philip Rowell, Newtown, Powys
D. Rudkin, Hertford
Stephanie Ryder, Bristol
Reverend John Samways, Keynsham, Somerset
Diana Sandes, Ballyduff, Co. Waterford
Mrs B. J. Scandrett, Banbury, Oxfordshire
Ian Scott, Blairgowrie, Perthshire
Mary Shave, Suffolk
Jack Shea, Dublin
Cedric B. Shipley, Stockton-on-Tees, Durham
Jean Simmonds, Chalfont St Peter, Buckinghamshire
Catherine Simmons, Virginia Water, Surrey
Eric Simms, Lincolnshire
Roger M.C. Sims, Manx National Heritage, Douglas, Isle of Man
Patricia Skelton, Somerset
Alan Skene, Stromness, Orkney
Paul Slack, Linacre College, Oxford
Mrs N. Small, Bromley, Kent
Rosemary Sorfleet, Canadian High Commission, London
David Spencer, Oakham, Rutland
Kim Spickett & family, Twickenham, Surrey
James H. Stamps, Pebworth, Warwickshire
Diana Stenson, Wilmslow, Cheshire
Cam Stirling, Perthshire
Mrs J.F. Stokes, Emsworth, Hampshire
Mrs Rita Stone, Buntingford, Hertfordshire
K.B. Stratton, Wrotham, Kent
Sir Roy Strong, Hertfordshire
Benita Tapster, Morpeth, Northumberland
Audrey Taylor, Ewell, Surrey
Barbara J. Taylor, Houghton on the Hill, Leicestershire
John A. Thickett, Sheffield, South Yorkshire
Mrs José Thomas, Kidderminster, Worcestershire
Edward Thorpe, Bath, Somerset
Jean Tibbitts, Stratford-upon-Avon, Warwickshire
Pat Tichener, Hinckley, Leicestershire
E.A. Timson, Leicester
Colin Titcombe, Llandogo, Monmouthshire
Angela Townsend, House of Lords, London
Pauline E. Treadwell, Lichfield, Staffordshire
Kath Trelissick, Cornwall
Mrs M.A. Trendall, Crowthorne, Berkshire
M. Trennel, North Lopham, Norfolk
Jennifer M. Trodd, Havant, Hampshire
Liz Trundle, Northamptonshire
Stella Turk, Cornwall
Brian Turner, Warrington, Cheshire
Mrs Jill J. Turner, Birmingham, West Midlands
Jude Tyrrell, Science, London
Jean Usoro, Freshwater East, Pembrokeshire
Mrs E. Vallance, Matlock, Derbyshire
Tony Vaughan, National Snail Farming Centre, cited at www.home-workers.com
Kathleen Walsh, Knutsford, Cheshire
Mrs M.M. Walter, Somerset
George Walton, Hereford
Mr & Mrs B. Ward, Boston, Lincolnshire
Andrew Warner, Sheffield, South Yorkshire
Hilary Waterhouse, Sennybridge, Powys
Malcolm J. Watkins, Heritage Manager, City of Gloucester
Dr Sarah Watkinson, St Hilda's College, Oxford
Mrs J.E. Watson, Blackpool, Lancashire
Rosemary Watson, Weymouth, Dorset
Christine Weightman, Ascot, Berkshire
Mrs S.M. White, Carmarthen
Maureen Whitworth, Stourbridge, West Midlands
Bernard Widdowson, Carlisle, Cumbria
Jane Wight, Derbyshire
Arthur R. Willett, Cambridge
Martin Williams, Hurstpierpoint College, West Sussex
Mrs D.R. Willoughby, Bridport, Dorset
Colin Winters, Kent
Ida Wolstenholme, Mayfield, Sussex
John Wood, Aldington, Kent
Sarah Woodman, Buckfastleigh Town Council, Devon
Barry Woodward, Ormskirk, Lancashire
Joan C. Woodward, Rustington, West Sussex
Marjorie Young, West Kirby, Wirral

PUBLISHER'S ACKNOWLEDGEMENTS

The publisher is grateful for permission to reproduce the following copyright material:

p.18, Barrett, J., *Life on the Sea Shore*, Collins, London, © 1974, J. Barrett; p.20, Porter, J.R. and Russell, W.M.S. [eds], *Animals in Folklore*, D.S. Brewer and Rowman and Littlefield, Cambridge, for the Folklore Society, 1978; p.20, 422, 479, 488, Porter, Enid, *The Folklore of East Anglia*, Batsford, London, 1974; p.21, 25, 238, 271, Swainson, C., *A Handbook of Weather Folk-lore*, William Blackwood and Sons, Edinburgh and London, 1873; p.23, 130, 132, 135, 140, 141, 145, 146, 152, 162, 164, 215, 299, 338, Walton, Izaac, *The Complete Angler*, 6th edition, Rivington et al., London, 1797 [First published 1653]; p.24, 27, 319, 331, 388, Hutchinson, Thomas [Editor] *Wordsworth's Poetical Works*, Oxford, Oxford University Press, 1984 [1904]; p.24, 25, 428, Beith, Mary, *Healing Threads: Traditional Medicines of the Highlands and Islands*, Polygon, Edinburgh, 1995; p.25, Spenser, Edmund. *The Works of Edmund Spenser*, London, Routledge, Warne, and Routledge 1863; p.25, 163, 179, 345, Buckland, F.T., *Curiosities of Natural History*, Series I, Macmillan and Co., London, 1857; p.25, Wodehouse, P.G., *Over Seventy – An Autobiography With Digressions*, Herbert Jenkins, London, 1957; p.27, 28, 30, Lovell, M.S., *The Edible Mollusks of Great Britain and Ireland*, Reeve and Co., London, 1867; p.28, 206, Martin, Martin, *A Late Voyage to St Kilda*, D. Brown and T. Goodwin, London, 1698; p.31, 38, Step, Edward, *Shell Life*, Frederick Warne and Son, London, 1945; p.31, Fox-Davies, A.C., *Complete Guide to Heraldry*, T.C. and E.C. Jack, London, 1925; p.35, Thompson, Flora, *Candleford Green*, Oxford University Press, Oxford, 1943. Reprinted by permission of Oxford University Press; p.35, 103, 175, 366, Palmer, Roy, *The Folklore of Warwickshire*, B.T. Batsford, London, 1976; p.35, 87, 307, 332, 406, Udal, John Symonds, *Dorsetshire Folk-lore*, 2nd edition, Toucan Press, st Peter Port,1970 [First published by Stephen Austin and Sons, Hertford, 1922]; p.35, 304, 330, 333, 475, Anon., *Songs for the Nursery*, William Darton, London, 1825; p.35, 198, 401, 404, 470, 478, White, Gilbert, *The Journals of Gilbert White,* Vol. 1, [ed. Francesca Greenoak], Century, London, 1986; p.36, 71, 225, 229, 300, Opie, Iona and Peter, *The Lore and Language of Schoolchildren*, Clarendon Press, Oxford, 1959; p.36, 87, 278, Gay, John. *Poems On Several Occasions*, London, H. Lintot, J. and R. Tonson and S. Draper, 1745; p.37, Baring-Gould, Sabine, *A Book of Nursery Songs and Rhymes*, London, Methuen, 1895; p.41, Cox, Ian [ed.], *The Scallop*, Shell Transport and Trading Co., London, 1957; p.41, 42, Bolitho, H. [ed.], *The Glorious Oyster*, Sidgwick and Jackson, London, 1960; p.45, Lucas Phillips, C.E., *Cockleshell Heroes*, William Heinemann, London, 1956; p.46, Wesley-Smith, Peter, *The Ombley-Gombley*, Angus and Robertson, Sydney, 1969; p.49, Chinery, M., *Garden Creepy-crawlies*, Whittet Books,

London, 1986; p.54, 59, from *The Book of the Spider* by Paul Hillyard, published by Hutchinson/Pimlico. Reproduced by permission of the Random House Group Limited; p.55, 94, 179, 298, 302, 309, 322, 341, 347, 391, 397, 400, 449, White, Gilbert, *Natural History of Selborne*, new edition, George Routledge, London, 1880 [First published 1789]; p.55, 58, 59, Bristowe, W.S., *The World of Spiders*, Collins, London, © 1958, W.S. Bristowe; p.56, 102, 440, Jonson, Ben, *Plays and Poems*, London, George Routledge and Sons, 1885; p.57, 98, 146, 358, 393, Anon., *Ladybird Nursery Rhymes*, Ladybird Books, Loughborough, 1994; p.57, 121, 153, 432, Marwick, Ernest W., *Folklore of Orkney and Shetland*, Batsford, London, 1975; p.57, 68, 69, 88, 92, 93, 95, 100, 197, 230, 238, 386, 390, 415, 423, 463, 473, Topsell, Edward, *The History of Four-footed Beasts and Serpents and Insects*, Da Capo Press, New York, 1967 [Reprint of 1658 edition]; p.57, Flanders, Michael and Swan, Donald, *The Songs of Michael Flanders and Donald Swan*, London, Elm Tree Books, 1986; p.58, 190, 193, Scott, Sir Walter, *The Poetical Works of Sir Walter Scott*, Bart. Edinburgh, Adam and Charles Black, 1852; p.58, 332, 474, Opie, Iona and Peter [eds], *The Oxford Dictionary of Nursery Rhymes*, 12th edition, Oxford University Press, Oxford, 1980 [First published 1951]; p.58, Scott, Walter, *Tales of a Grandfather*, Robert Cadell, Whittaker and Co., London, 1829; p.59, Gunther, R.T. [ed.], *The Diary and Will of Elias Ashmole*, [edited and extended from original manuscripts] Oxford University Press, Oxford, 1927; p.59, 274, 312, 322, Gibson, F., *Superstitions About Animals*, Scott Publishing Co. Limited, London, 1904; p.60, Howitt, Mary, *Mary Howitt's Complete Poetical Works*, Boston, Wentworth, Hewes and Co. 1858; p.65, Rossetti, Dante Gabriel, *Ballads and Sonnets*, London, Ellis and White, 1881; p.65, 219, 268, Tennyson, Alfred, *Poems of Tennyson*, London: Oxford University Press,1918; p.65, Moore, Thomas, *Paradise And The Peri*, London, Day and Son, 1860; p.65, Long, W.H., *The Dialect of the Isle of Wight with Tales and Anecdotes*, Reeves and Turner, London, and G.A. Brannon and Co., Newport, 1886; p.65, Corbet, Philip S., Longfield, Cynthia and Moore, N.W., *Dragonflies*, Collins, London, 1960; p.65, Hammond, Cyril O., *The Dragonflies of Great Britain and Ireland*, 2nd edition [revised by Robert Merritt], Harley Books, Colchester, 1983 [First published 1977]; p.65, Chinery, M., *Insects of Britain and Northern Europe*. Collins, London, © 1979, M. Chinery; p.66, Dickens, Charles, *The Cricket on the Hearth. A Fairy Tale of Home*, printed and published for the author by Bradbury and Evans, London, 1846; p.66, 312, Milton, John, *The Poems of John Milton* [Edited by Helen Darbishire], Oxford, Oxford University Press,1961; p.66, 152, 181, Porter, Enid, *Cambridgeshire Customs and Folklore*, Routledge and Kegan Paul, London, 1969; p.67, 71, 127, 145, 230, 246, 265, 286, 298, 312, 338, 343, 370, 376, 389, 399, 407, 428, Keogh, John, *Zoologia Medicinalis Hibernica*, S. Powell, Dublin, 1739; p.67, Burke, Edmund, *A Letter to a Member of the National Assembly; in Answer to his Book on French Affairs*, Dodsley, Owen and Rivington, London, 1791; p.69, Gordon, D.G., *The Compleat Cockroach*, Ten Speed Press, Berkeley, California, 1996; p.70, 227, 403, Hunt, R., *Popular Romances of the West of England*, 3rd edition, Chatto and Windus, London, 1881; p.71, Southall, John, *A Treatise of Buggs*, J. Roberts, London, 1730; p.74, 93, Nash, Ogden, *Selected Poetry of Ogden Nash*, New York, Black Dog and Leventhal, 1995; p.74, Hall, Donald [Editor] *The Oxford Book of Children's Verse in America*, Oxford, Oxford University Press, 1985; p.74, 105, 110, 228, 229, 233, 301, 327, 343, 358, 360, 356, 366, 376, 398, Opie, I. and P., *The Oxford Nursery Rhyme Book*, 8th edition, Oxford University Press, Oxford, © Iona and Peter Opie 1955. Reprinted from *The Oxford Nursery Rhyme Book* assembled by Iona and Peter Opie (1963) by permission of Oxford University Press; p.76, Harris, Moses, *The Aurelian*, printed for the author, London, 1766; p.76, Petiver, J., *Papilionum Britanniae Icones*, printed for the author, London, 1717; p.78, Williams, C.B., *Insect Migration*, Collins, London, 1958; p.78, Salmon, Michael A., *The Aurelian Legacy*, Harley Books, Colchester, 2000; p.78, p.80, 83, Ford, E.B., *Butterflies*, Collins, London, 1945; p.82, Patmore, Coventry, *Poems by Coventry Patmore*, London, George Bell and Sons, 1886; p.90, Golding, William, *Lord of the Flies*, Faber and Faber, London, 1954; p.91, Bingley, William, *History of Animated Nature*, Lovell Reeve, London, 1968; p.91, Defoe, Daniel, *A Tour Through the Whole Island of Great Britain*, Folio Society, London, 1983 [First published 1724-26; p.95, Erzinclioglu, Zakaria, *Blowflies*, Richmond Publishing Co., Richmond, 1996; p.95, Swift, Jonathan, *On Poetry: A Rhapsody*, London, J. Huggonson, 1733; p.102, Quiller-Couch, Arthur [Editor], *The Oxford Book of English Verse: 1250–1900*, Oxford, Oxford University Press, 1919; p.103, 141, 180, 181, 219, 286, 373, 380, 398, 474, Radford, E. and M.A., *The Encyclopedia of Superstitions*, 6th edition [edited and revised by Christina Hole], Helicon Publishing Limited, Oxford, 1995 [First published 1948]; p.104, 179, 190, Herrick, Robert, *The Poems of Herrick*, London, Gibbings and Co, 1897; p.106, 334, Webster, John, *The Duchess of Malfi and The White Devil*, London, Bodley Head 1930; p.106, 281, Gwynn Jones, T., *Welsh Folklore and Folk Custom*, Methuen and Co., London, 1930; p.110, Exell, A.W., *The History of the Ladybird*, Phoenix, London, 1999; p.110, p.142, Majerus, M., *Ladybirds*, Collins, London, © 1994, M. Majerus; p.118, Williamson, Henry, *Salar the Salmon*, Faber and Faber, London 1935; p.118, 119, 134, Waters, Brian, *The Severn Tide*, J.M. Dent, London, 1947; p.118, 179, 393, Brewer, Clifford, *The Death of Kings*, Abson Books, London, 2000; p.120, Jude Tyrrell, *Science*, London; p.122, Maxwell, Gavin, *Harpoon at a Venture*, Hart-Davis, London, 1952; p.122, Geddes, Tex, *Hebridean Sharker*, Routledge, London, 1954; p.122, Buchan, David [ed.], *Folk Tradition and Folk Medicine in Scotland – The Writings of David Rorie*, Canongate Academic, Edinburgh, 1994; p.125, 272, 273, Moore, A.W., *Folk-Lore of the Isle of Man*, S.R. Publishers Limited, Yorkshire, 1971 [First published 1891 in Isle of Man and London]; p.125, Hardy, Alister, *The Open Sea*. Vol. II. *Fish and Fisheries*, Collins, London, © 1959, Alister Hardy; p.126, Yeats, W. B., *Collected Poems*, London, Macmillan,1985; p.127, 167, Darwin, Bernard, *Receipts and Relishes Being a Vade Mecum for the Epicure in the British Isles*, Whitbread, London, 1950; p.127, Hull, Eleanor, *Folklore of the British Isles*, Methuen and Co., London, 1928; p.128, from *Costumes and Customs of the British Isles* by Lilla M Fox, published by Chatto and Windus. Reprinted by permission of the Random House Group Limited; p.129, 229, 332, Courtney, M.A., *Cornish Feasts and Folklore*, Beare and Son, Penzance, 1890 [Revised and reprinted from the Folklore Society Journals, 1886–7]; p.129, 161, Baker, Jenny, *Simply Fish*, Faber and Faber, London, 1988; p.132, Jones, J.W., *The Salmon*, Collins, London, © 1959, J.W. Jones; p.134, Christian, Roy, *The Country Life Book of Old English Customs*, Country Life, London, 1966; p.134, 307, 334, Drayton, Michael, *Works*, [Edited J. William Hebel], Stratford upon Avon, Shakespeare Head Press, 1931-41; p.136, James, H.F. and Bartholomew, A.T. [eds], *The Complete Works of Samuel Butler*, Bartholomew, London, 1923–26; p.136, 367, Simpson, Jacqueline, *The Folklore of the Welsh Border*, Batsford, London, 1976; p.136, Exell, A.W., *Blockley As It Was*, Hendon Publishing, Hendon, 1990; p.139, 142, Maitland, Peter S. and Campbell, R. Miall, *Freshwater Fishes*, Collins, London, 1992; p.141, 345, Roberts, Michael, and Porter, Peter, [Editors], *The Faber Book of Modern Verse*, 4th Edition, London, Faber, 1982; p.144, Gray, Thomas, *Poems*, Eton, College Press, 1894; p.144, 298, 486, Pepys, Samuel, *The Diary*, George Bell and Sons, London, 1893; p.144, Saki (H.H. Munro), *Reginald*, Methuen, London, 1904; p.151, 321, 352, 389, 486, Kightly, Charles, *The Perpetual Almanack of Folklore*, Thames and Hudson, London, 1987 [Reprinted 1994]; p.153, Barrett, W.H., *Tales from the Fens*, Routledge and Kegan Paul, London, 1966; p.154, 265, 448, 449, Hammond, P.W., *Food and Feast in Medieval England*, Wrens Park Publishing, Gloucestershire, 1993; p.156, 158, Extract from COD by Mark Kurlansky published by Jonathan Cape. Used by permission of the Random House Group Limited; p.159, 250, Wilde [Lady], *Ancient Cures, Charms and Usages of Ireland*, Ward and Downey, London, 1890; p.169, 238, 463, Simpson, Jacqueline, *The Folklore of Sussex*, Batsford, London, 1973; p.169, Coward, Noel, *Cowardy Custard - The World of Noel Coward*, [Edited John Hadfield], London, Heinemann, 1973; p.177, Lever, Christopher, *The Naturalized Animals of the British Isles*, Hutchinson, London, 1977; p.178, 225, Pennant, Thomas, *British Zoology*, Benjamin White, London, 1776; p.178, Thorndike, Lynn, *A History of Magic and Experimental Science*, Columbia University Press, New York, 1934, quoted from *Philosophical Transactions of the Royal Society*; p.179, Browne, Thomas, *The Works of Sir Thomas Browne*, [Edited Charles Sayle], Edinburgh, John Grant 1912; p.179, Prynne, William, *The History of King John*, Thomas Ratcliff, London, 1670; p.181, Bunyan, John, *The Pilgrim's Progress*, Everyman Library, Dent, London 1985; p.182, Morgan, Adrian, *Toads and Toadstools: The Natural History, Folklore and Cultural Oddities of a Strange Association*, Celestial Arts, Berkeley, California, *c.*1995; p.182, Ainsworth, G.C., *Introduction to the History of Mycology*, Cambridge University Press, Cambridge, 1976; p.182, Findlay, W.P.K., *Fungi, Folklore, Fiction and Fact*, Richmond Publishing, Richmond, 1982; p.182, FROGS AND TOADS by Trevor Beebee (Whittet Books, Stowmarket); p.182, 393, 412, 416, 419, 469, Grahame, Kenneth, *The Wind in the Willows*, Methuen and Co., London, 1908; p.184, Medawar, P.B. and Medawar, J.D., *Aristotle to Zoos: A Philosophical Dictionary of Biology*, Harvard University Press, Cambridge, 1983; p.190, Gurdon, Lady E.C. [Collector and Editor] *Country Folk Lore: Gloucestershire, Suffolk, Leicestershire and Rutland.* Printed extracts, No.2. *Suffolk*. Published for the Folklore Society, D. Nutt, J.Loder, Woodbridge, Pawsey and Hayes, Ipswitch, 1893; p.192, 226, Wagner, Leopold, *Manners, Customs and Observances*, William Heinmann, London, 1895; p.192, Aubery, John, *Remaines of Gentilisme and Judaisme by John Aubery* R.S.S. W. Satchell, Peyton and Co. For the Folklore Society, London, 1881 [1686–7]; p.193, Anon. *Choice Notes from Quotes and Queries*, Bell and Daldy, London, 1859; p.195, Smith, Malcom, *The British Amphibians and Reptiles*, Collins, London, 1951; p.196, Harrison, William, *The Description of England*, [Edited by Frederick J. Furnivall], N. Trubner for the New Shakespeare Society, London, 1877–8 [1577]; p.199, Beebee, Trevor, Griffiths, Richard, and Halliday, Tim, *Amphibians and Reptiles*, HaperCollins, London, © 2000 Trevor Beebee, Richard Griffiths and Tim Halliday; p.199, 486, Glasse, Hannah, *The Art of Cookery made Plain and Easy*, W. Strahan, London, 1774 edition [1747]; p.199, 397, 481, 485, Carroll, Lewis, *Alice's Adventures in Wonderland*, Macmillan and Co., London, 1865; p.202, 211, 225, 301, 303, 312, 316, 332, 336, 340, 342, 345, 350, 366, Swainson, C., *The Folklore and Provincial Names of British Birds*, Eliot Stock, London, 1886; p.203, 260, 359, Stevenson, R.L., *Kidnapped*, Cassell and Co. Limited, London, 1886; p.204, Whitlock, Ralph, *The Folklore of Wiltshire*, B.T. Batsford Limited, London, 1976; p.206, Fisher, James, *The Fulmar*, Collins, London, 1952; p.207, 292, MacKenzie, Neil, 'Notes on the birds of St Kilda', *Annals of Scottish Natural History*, 1905; p.209, 210, 294, Penhallurick, R.D., *Birds of the Cornish Coast*, D. Bradford Barton, Truro, 1965; p.209, 224, 232, 235, 246, 256, 263, 270, 277, 280, 292, 295, Lockwood, William, B., *The Oxford Book of British Bird Names*, Oxford University Press, Oxford, 1984 by permission of Oxford University Press; p.210, Ferrier, Walter, *The North Berwick Story*, North

Berwick Community Council, Edinburgh, 1980; p.211, 236, 241, 248, 260, 273, 288, 363, 370, 372, 373, Murton, R.K., *Man and Birds*, Collins, London, 1971; p.212, 248, 251, 263, 299, 312, 381, Mead, Chris, *The State of the Nation's Birds*, Whittett, Stowmarket, 2000; p.212, 445, 456, 492, Rackham, Oliver, *The History of the Countryside*, Dent, London, 1986; p.213, Anon., *The Great Feast At the Inthronization of the Reverend Father in God, George Neavill Archbishop of Yorke, Chancellor of England, in the Sixth Yeere of Edward the Fourth*, Edward Husbands, London, 1645; p.213, 238, 451, 486, Anon., *A Propre New Booke of Cokery*, London, 1545; p.214, 215, Lowe, Frank A., *The Heron*, Collins, London, 1954; p.214, 232, Eckwall, Eilert, *The Concise Oxford Dictionary of English Place Names*, 4th edition, Oxford University Press, Oxford, 1960; p.215, Williamson, John, *The British Angler*, John Hodges, London, 1740; p.216, 307, 394, 396, Swan, John, *Speculum Mundi*, London, 1635; p.219, Coleridge, S.T., *The Poetical and Dramatic Works of S.T. Coleridge*, London, William Pickering, 1844; p.219, Walpole, Horace, *Letters To the Countess of Upper Ossory*, 1 December 1786; p.219, Andersen, Hans Christian, *The Ugly Duckling*, Macmillan, New York, 1927 [1843]; p.220, Rackham, H., *Christ's College in Former Days*, Cambridge University Press, Cambridge, 1939; p.220, 231, 236, 237, 238, 246, 281, 285, 292, 294, 322, 323, 338, 341, 370, 380, Greenoak, Francesca, *All the Birds of the Air – The Names, Lore and Literature of British Birds*, André Deutsch, London, 1979; p.220, Tom Sharpe, *Porterhouse Blue*, Martin Secker and Walburg, London, 1974; p.221, 370, 413, 426, 433, Nicholson, James R., *Shetland Folklore*, Robert Hale, London, 1981; p.221, 226, 257, 269, 282, 302, 303, 305, 316, 337, 338, Armstrong, Edward A., *The Folklore of Birds*, Collins, London, 1958; p.222, Bewick, Thomas, *History of British Birds*, F. Graham, London, 1972; p.225, 279, Sample, Geoff, *Bird Songs and Calls of Britain and Northern Europe*, HarperCollins, London, 1996; p.225, Dickens, C., *A Christmas Carol*, Chapman and Hall, London, 1843; p.226, Blount, Thomas, *Fragmenta Antiquitatis* ..., Blackhead and Co., York, 1784; p.227, 227, Gomme, George Laurence [ed.], *The Gentleman's Magazine Library: Popular Superstitions, Collection from 1731–1868*, Eliot Stock, London, 1884; p.227, 235, 262, 268, 275, 278, 347, 379, 486, Beeton, Isabella, *Mrs Beeton's Cookery Book*, new and improved edition, Ward, Lock and Co., London, n.d. [c.1890]; p.227, Sillitoe, A., *Saturday Night and Sunday Morning*, W.H. Allen, London, 1958; p.228, Perrault d'Armancour, *Contes de Ma Mere L'Oye*, Paris, 1697; p.231, Turner, William, *On Birds*, London, 1544; p.231, Gerard, John, *Herball*, London, 1597; p.232, Eddowes, John, *The Language of Cricket*, Carcanet, London, 1997; p.236, Morris, James, *Oxford*, Faber and Faber, London, 1965; p.238, Markham, Gervase, *The English House-wife*, C.T. Milford, London, 1942 [First published in 1660]; p.238, Ibbott, Selena, *Folklore, Legends and Spells*, Ashgrove Press Limited, Bath, 1994; p.238, 389, Black, W.G., *Folk Medicine; A Chapter in the History of Culture*, Eliot Stock for the Folklore Society, London, 1883; p.241, 274, 287, Ingersoll, Ernest, *Birds in Legend Fable and Folklore*, Longmans, Green and Co., New York, 1923; p.244, 257, 258, 259, Salvin, Francis Henry and Brodrick, William, *Falconry in the British Isles*, Beech Publishing, Midhurst, 1997 [Reprint of 2nd edition of 1873]; p.244, 271, 280, 318, Ray, John, *The Ornithology of Francis Willughby* [translated into English, and enlarged by John Ray], John Martyn, London, 1678; p.247, 251, 256, 258, Brown, Leslie, *British Birds of Prey*, Collins, London, 1976; p.248, 293, Montagu, G., *Ornithological Dictionary*, London, 1802; p.249, Aubrey, John, *Miscellanies upon Various Subjects*, London, 1696; p.249, Wilson, John Marius, *A Memoir of Field Marshal The Duke of Wellington*, Vol. II, A. Fullarton, London, Edinburgh, Dublin, n.d. [c.1855]; p.253, Robin, P. Ansell, *Animal Lore in English Literature*, John Murray, London, 19; p.253, 304, 464, Brand, John, *Observations on Popular Antiquities*, Chatto and Windus, London, 1900; p.254, 402, 450, 479, Fraser-Darling, F. and Morton Boyd, J., *The Highlands and Islands*, Collins, London, 1964; p.254, Wolley, John, *Ootheca Wolleyana*, London, 1864; p.255, Potter, S. and Sargent, L., *Pedigree Words from Nature*, Collins, London, 1973; p.255, Drayton, Michael, *Poly-Olbion*, London, 1622; p.255, Samstag, N.A., *For Love of Birds*, RSPB, London, 1988; p.258, Baker, J.A., *The Peregrine*, Collins, London, 1967; p.261, 262, 264, 268, 276, Vesey-Fitzgerald, Brian, *British Game*, Collins, London, 1946; p.261, Buchan, John, *The Thirty-Nine Steps*, William Blackwood, Edinburgh and London, 1915; p.261, Colegate, Isabel, *The Shooting Party*, Hamish Hamilton, London, 1980; p.264, Hare, C.E., *Bird Lore*, Country Life Limited, London, 1952; p.268, Ransome, Arthur, *Swallows and Amazons*, Jonathan Cape, London, 1930; p.269, 362, Jesse, Edward, *Scenes and Occupations of Country Life*, John Murray, London, 1853 [1844]; p.273, Weaver, P., *The Lapwing*, Shire Publications Limited, Tring, 1987; p.274, Leyden, John, The Poetical Works of Dr John Leyden, London, Nimmo,1987; p.274, Lees, Edward, *Pictures of Nature in the Silurian Region Around the Malvern Hills*, 1856; p.274, 474, Hole, Christina, *English Folklore*, Batsford, London, 1940; p.274, Robert Ford, *Children's Rhymes, Children's Games, Children's Songs, Children's Stories, A Book for Bairns and Big Folk*, Alexander Gardner, Paisley, 1904; p.276, Prynne, Michael, *Egg-shells*, Barrie and Rockliff, London,1963; p.278, Parker, Eric, *Shooting by Moor, Field and Shore*, Seeley, Service and Co., London, n.d. [*c.*1955]; p.280, Browne, Thomas, *Account of Birds ... Found in Norfolk*, 1682; p.281, 423, 428, Forbes, A.R., *Gaelic Names of Beasts, Birds, Fishes, Insects, and Reptiles*, Oliver and Boyd, Edinburgh, 1905; p.281, 367, 432, Newall, Venetia, *Discovering the Folklore of Birds and Beasts*, Shire Publications, Tring, 1971; p.282, 283, Nethersole-Thompson, Desmond, *The Greenshank*, Collins, London, 1951; p.283, Latham, John, *A General Synopsis of Birds*, Vol. 3, London, 1785; p.286, de Garis, M., *Folklore of Guernsey*, Channel Islands: Mrs Marie de Garis, Guernsey, Channel Islands, 1975; p.291, Grieve, Symington, *The Great Auk or Garefowl*, Chapman, London, 1885; p.291, Fuller, Errol, *The Great Auk*, Southborough, Kent, 1999; p.291, Eckert, Allan, *The Last Great Auk*, Collins, London, 1964; p.292, Boyd, J.M. and Boyd, I.L., *The Hebrides*, Collins, London, 1990; p.292, Thompson, Francis, *St Kilda and Other Hebridean Outliers*, David and Charles, Newton Abbot, 1970; p.294, Gesner, Konrad von, *De Avibus*, *c.*1555; p.294, Mouffet, Thomas, *Health's Improvement*, Thomas Newcomb for Samuel Thomson, London, 1655; p.294, Owen, George, *Description of Pembrokeshire*, London, 1603; p.297, Anon., *Folklore, Myths and Legends of Britain*, The Reader's Digest Association, London, 1973; p.297, Murton, R.K., *The Wood Pigeon*, Collins, London, 1965; p.300, Gregory, Kenneth, *The First Cuckoo: Letters to The Times 1900–1975*, Times Books, London, 1976; p.301, 309, 406, 408, 479, 488, Palmer, Roy, *The Folklore of Leicestershire and Rutland*, Sycamore Press Limited, Wymondham, 1985; p.301, 464, Bett, Henry, *English Myths and Traditions*, B.T. Batsford Limited, London, 1952; p.302, 366, Baker, Margaret, *The Gardener's Folklore*, David and Charles, Newton Abbot, 1977; p.303, Field, John Edward, *The Myth of the Pent Cuckoo, A Study in Folklore*, Eliot Stock, London, 1913; p.305, Pilgrim, Jane, *Blackberry Farm* Brockhampton Press, London, 1964; p.305, 464, THE TALE OF SQUIRREL NUTKIN by Beatrix Potter. Copyright © Frederick Warne and Co., 1903, 2002; p.305, Milne, A.A., *Winnie-the-Pooh*, Methuen, London, 1926; p.307, 473, Kinsley, James [Editor], *Burns, Poems and Songs*, Oxford University Press, Oxford, 1969; p.307, Shirley, James, *Triumph of Peace*, London, 1633; p.308, Lilford, [Lord], *Coloured Figures of the Birds of the British Islands*, Vols 1–7 [illustrated by Arthur Thorburn et al.], 1885–97 R.H. Porter, London; p.309, Ewing, Juliana Horatia, *The Brownies and Other Tales*, SPCK, London, 1871; p.316, Marvell, Andrew, The Poetical Works of Andrew Marvell, M.P. for Hull, 1658, London, Alexander Murray, 1870; p.318, Hopkins Gerard Manley, The Works of Gerard Manley Hopkins, Ware, Wordsworth Editions, 1994; p.321, Evans, Ivor H., *Brewer's Dictionary of Phrase and Fable*, Cassell, London, 1959; p.322, 343, 345, 369, 380, Hardy, Thomas, The Complete Poems, London, Macmillan, 1976; p.322, Carew, Thomas, The Works of Thomas Carew, Edinburgh, W. and C. Tate,1824; p.322, 325, 328, 362, 377, Nicholson, E.R., *Birds and Men*, Collins, London, 1951; p.323, 324, Simms, E., *British Larks, Pipits and Wagtails*, Collins, London, 1992; p.328, Armstrong, Edward A., *The Wren*, Collins, London, 1955; p.329, 357, 376, Chaucer, Geoffrey, The Riverside Chaucer, Oxford, Oxford University Press, 1988; p.331, Rossetti, Christina Georgina, The Poetical Works, London, Macmillan, 1904; p.331, Lack, David, *The Life of the Robin*, H.F. and G. Witherby, London, 1943; p.334, Lupton, Thomas, *A Thousand Notable Things of Sundrie Sortes*, London, 1579; p.336, Palmer, Kingsley, *The Folklore of Somerset*, B.T. Batsford Limited, London, 1976; p.337, Harrison, Thomas P., 'Meredith as poet of birds', *Bird Notes and News*, Vol. 27, 1956; p.338, Lodge, Thomas, *Scillaes Metamorphosis*, London 1589; p.343, Andrew Boorde, *Dietary of Health*, London, 1547; p.346, 347, 349, 350, Simms, Eric, *British Warblers*, Collins, London, 1985; p.346, Yeates, George, *Bird Life in Two Deltas*, Faber and Faber, London, 1946; p.353, 473, Potter, Beatrix, *Beatrix Potter's Nursery Rhyme Book*, Frederick Warne and Co. and Penguin Books, London, 1975; p.354, 424, Williamson, Henry, *Tarka the Otter*, G.P. Putnam's Sons, London, 1927; p.354, Perrins, Christopher, *British Tits*, Collins, London, 1979; p.363, Marshall, William, *Rural Economy of Norfolk*, T. Cadell, London, 1787; p.364, Pennant, Thomas, *A Tour in Scotland 1769*, B. White, London, 2nd edition, 1772 [First published in 1772]; p.365, Edwards, George, *From Crow-Scaring to Westminster*, National Union of Agricultural Workers, London, 1922; p.366, 367, 389, Palmer, Roy, *The Folklore of Gloucestershire*, West Country Books, Tiverton, 1994, reprinted 2001; p.368, Dickens, Charles, *Barnaby Rudge*, Chapman and Hall, London, 1840; p.369, Cervantes, Miguel de, *Don Quixote de la Mancha*, Antonio Sancha, Madrid, 1777, [1605; p.374, Skelton, John, *Poems*, London, Heinemann, 1924; p.374, Kipps, Clare, *Sold for a Farthing*, Muller, London, 1953; p.374, Anon., *Wisden Cricketers' Almanac 1937*, John Wisden, London, 1937; p.375, 376, Newton, I., *Finches*, Collins, London, 1972; p.379, Lawson, William, *A New Orchard and Garden*, London, 1618; p.385, King Smith, Dick, *The Hodgehog*, Hamish Hamilton, London, 1987; p.385, *Chamber's Encyclopaedia*, J.P. Lippincott and Co., Philadelphia, 1923; p.387, Price, Lawrence, *The Shepherd's Prognostication*, London, 1652; p.387, Kinahan, G.H., 'Notes on Irish folklore', *Folklore Journal*, Vol. 4, 1881; p.388, Hendrickson, Robert, *The Ocean Almanac*, Hutchinson, London, 1992; p.388, 393, 420, Harrison-Mathews, Leo, *British Mammals*, Collins, London, 1952; p.389, 395, 464, 470, Tibble, J.W. [Editor], *The Poems of John Clare, 2 Vol.s*, J.M. Dent and Sons, London, 1935; p.389, Ellis, William, *Country House-wife's Family Companion*, London, 1750; p.391, Burne, Charlotte Sophia [ed.], *Shropshire Folklore*, [from the collections of Georgina F. Jackson], Trubner and Co., London, 1883; p.394, Browne, Thomas, *Enquiries into Vulgar Errors*, in: *The Works of the Learned Sr. Thomas Brown, Kt., Doctor of Physick, late of Norwich*, Thos. Basset, Ric. Chiswell, Tho. Sawbridge, Charles Mearn and Charles Brome, London, 1686; p.394, Hawker, R.S., 'The first Cornish mole', in: *Footprints of Former Men in Far Cornwall*, James G. Commin, Walter Weighell and Joseph Pollard, London, Exeter, Launceston and Truro, 1908 [Reissue of original 1870 edition]; p.394, Roper, William, *A Man of Singular Virtue, Being a Life of Sir Thomas More by his Son-*

in-law William, Roper, Folio Society, London, 1980 [Reprint of Paris edition of 1626]; p.394, Worlidge, John, *Systema Agriculturae: The Mystery of Husbandry Discovered*, 2nd edition, Thomas Dring, London, 1675; p.394, Coles, William, *The Art of Simpling*, Nathaniel Brook, London, 1656; p.395, Mitford, Mary, *Our Village*, Macmillan and Co., London, 1893; p.395, Mellanby, Kenneth, *The Mole*, Collins, London, © 1971, Kenneth Mellanby; p.396, Murray, James et al. [eds], *The Oxford English Dictionary*, Clarendon Press, Oxford, 1961 © Oxford University Press 1989. Definition of 'moleskin' reprinted from The Oxford English Dictionary (2nd edition, 1989) by permission of Oxford University Press; p.397, 402, 419, 463, 494, 495, Lamb, G.F., *Animal Quotations*, Longman, London, 1985; p.398, Nicholson, John, *Folklore of East Yorkshire*, Thomas Holderness, Driffield, 1890; p.399, Martin, Laura C., *Wildlife Folklore*, The Globe Pequot Press, Connecticut, 1994; p.400, Holland, Philemon (1552–1637), *Pliny's Natural History: A Selection from Philemon Holland's Translation*, [ed. J. Newsome], Clarendon Press, Oxford, 1964. Reprinted by permission of Oxford University Press; p.402, Howard, Robert W., *Auritus, A Natural History of the Brown Long-eared Bat*, William Sessions, York, 1995; p.402, 450, 479, Fraser Darling, F. and Boyd Morton, J. *The Highlands and Islands*, Collins, London, © 1964, F. Fraser Darling and J. Boyd Morton; p.403, Smurthwaite, David, *Battlefields of Britain*, Michael Joseph, London, 1993; p.403, Whitelock, Dorothy [ed.], *The Anglo-Saxon Chronicle: A Revised Translation*, Eyre and Spottiswoode; London, 1961; p.403, 445, Harting, James, *Extinct British Animals*, Trubner and Co., London, 1880; p.403, Scrope, William, *The Art of Deer Stalking Illustrated by a Narrative of a Few Days Sport in the Forest of Atholl*, John Murray, London, 1838; p.403, 411, 428, 447, 455, 456, 458, 463, 466, 478, 482, 492, Yalden, Derek, *The History of British Mammals*, T. and A.D. Poyser, London, 1999; p.404, Vesey-Fitzgerald, Brian, *Town Fox, Country Fox*, Eyre and Spottiswoode, London,1965; p.406, *Le Roman de Reynart* (The Romance of Renart); p.406, THE TALE OF MR TOD by Beatrix Potter. Copyright © Frederick Warne and Co., 1912, 2002; p.407, De Crespigny, R.C. and Hutchinson, *New Forest*, John Murray, London, 1903 [Re-issue]; p.407, Turberville, George, *The Noble Arte of Venerie or Hunting*, 1576; p.408, Palmer, Roy, *Britain's Living Folklore*, David and Charles, Newton Abbot and London, 1991; p.408, Austen, Jane, *Pride and Prejudice*, T.Egerton, London, 1813; p.408, Surtees, R.S., *Handley Cross; Or, The Spa Hunt. A Sporting Tale*, A.K. Newman and Co., London, 1846; p.409, Masefield, John, *Poems*, William Heinemann, London, 1946; p.409, Sassoon, Siegfried, *Memoirs of a Fox-hunting Man*, Faber and Faber, London, 1928; p.409, Rook, David, *The Ballad of the Belston Fox*, Hodder and Stoughton, London, 1970; p.409, Dahl, Roald, *Fantastic Mr Fox*, Allen and Unwin, London, 1970; p.411, 450, Burns, Lord [Chairman], *Final Report of the Committee of Inquiry into Hunting with Dogs*, HMSO, London, 2000. Crown copyright material is reproduced with the permission of the Controller of HMSO and the Queen's Printer for Scotland; p.411, Dickens, Charles, *Oliver Twist*, Richard Bentley, London, 1838; p.411, Wilkinson, J., in *Barnsley Chronicle*, 1871, cited by: A.B. Loy, Sheffield; p.412, Jefferies, Richard, *The Gamekeeper at Home. Sketches of Natural History and Rural Life*, Smith, Elder, London, 1878; p.414, p.415, Strong, Roy, *Gloriana*, Thames and Hudson, London, 1987; p.415, Petrarch, Francesco, *The Triumphs of Francesco Petrarch: Florentine Poet Laureate*, [trans. Henry Boyd], Little Brown, Boston, 1906; p.416, Drabble, Phil, *A Weasel in my Meatsafe*, Collins, London, 1957; p.416, Partridge, Eric, *Dictionary of Historical Slang*, abridged edition, Penguin, London, 1972; p.418, 427, 428, 472, 478, Burton, M., *Wild Animals of the British Isles*, 2nd edition, Frederick Warne and Co., London, 1973; p.419, Watson, J.W.P., *The Book of Foxhunting*, B.T. Batsford, London, 1977; p.420, 421, 422, 423, Neal, Ernest, *The Badger*, Collins, London, 1948; p.420, 426, Wentworth Day, J., *Sporting Adventure*, George Harrap and Co., London, 1937; p.421, Graham, Ritson, 'The Badger in Cumberland', *Transactions of the Carlisle Natural History Society*, Vol. 7, 1946; p.422, Rutty, John, *Essay Towards a Natural History of the County of Dublin*, 1724; p.422, Baker, Jenny, *Kettle Broth to Gooseberry Fool*, Faber and Faber, London, 1996; p.423, 424, Howard, Robert W., *Badgers without Bias*, Abson Books, Bristol, 1981; p.427, 428, 458, Cokayne, Thomas, 'A Short Treatise of Hunting' [1591], in: *Journal of the Derbyshire Archaeological and Natural History Society*, Vol. 3, January 1881, Bemrose and Sons, London and Derby; p.427, Maxwell, Gavin, *Ring of Bright Water*, Longman, London, 1960; p.427, Walton, Isaak, *The Compleat Angler*, George G. Harrap and Co. Limited, London, 1931 [Text from 5th edition, 1676, with spelling modernized]; p.429, Jennison, George, *Natural History: Animals, An Illustrated Who's Who of the Animal Kingdom*, A. and C. Black, London 1927; p.429, 440, Berry, R.J., *The Natural History of Orkney*, Collins, London, 1985; p.430, Carroll, Lewis, *Through the Looking Glass and What Alice Found There*, Macmillan, London, 1872; p.430, 433, Hewer, H.R., *British Seals*, Collins, London, © 1974, H.R. Hewer; p.432, 434, Lockley, R.M., *Grey Seal, Common Seal*, White Lion edition, London, 1977 [Originally published by André Deutsch Limited, 1966]; p.432, 433, From *The People of the Sea* by David Thomson, 1954, extract taken from the Canongate Classic edition, Canongate Books, Edinburgh; p.434, Pinkerton, John, *A General Collection of the Best and Most Interesting Voyages and Travels*, Longman, Hurst, Rees and Orme, London, 1808–14; p.434, Southwell, Thomas, *Seals and Whales of the British Seas*, Jarrold and Son, London, 1881; p.436, Hardy, Alister, *The Open Sea II Fish and Fisheries*, Collins, London, © 1959, Alister Hardy; p.439, Maxwell, Gavin, *Ring of Bright Water*, Longman Green and Co., London, 1960; p.440, Buchan, John, *The Island of Sheep*, Hodder and Stoughton, London, 1936; p.440, Bullen, Frank, *The Cruise of the 'Cachalot' – Round the World after Sperm Whales*, Smith, Elder, London, 1898; p.440, Goodwin, George, 'Porpoise – friend of Man?', *Natural History*, Vol. 106, Pt 8, October 1947; p.441, Melville, Herman, *Moby Dick or The Whale*, Longman, London, 1907; p.453, Roberts, Michael [Editor], Porter, Peter, *The Faber Book of Modern Verse*, Fourth Edition, Faber and Faber Limited, London, 1982; p.453, Buchan, John, *John Macnab*, Hodder and Stoughton, London, 1925; p.453, Raven, Jon, *The Folklore of Staffordshire*, Batsford Limited, London, 1978; p.453, Boase, Wendy, *The Folklore of Hampshire and the Isle of Wight*, Batsford Limited, London, 1976; p.453, Scott-Giles, C. Wilfred, *Civic Heraldry of England and Wales*, J.M. Dent and Sons Limited, London, [1st edition 1933, revised 1953]; p.456, Salten, Felix, *Bambi: A Life in the Woods*, Jonathan Cape, London, 1928; p.457, Green, Marian, *A Calendar of Festivals*, Element, Dorset, 1991; p.459, Boece, Hector, *Scotorum Historia*, Paris, 1526; p.461, Step, Edward, *Animal Life of the British Isles*, Frederick Warne and Co., London, 1921; p.462, Cabot, David, *Ireland*, Collins, London, © 1999, David Cabot; p.463, Car, Samuel, [Editor], *The Batsford Book of Country Verse*, Batsford, London, 1979; p.464, Forby, Robert, *The Vocabulary of East Anglia; An Attempt to Record the Vulgar Tongue of the Twin Sister Counties, Norfolk and Suffolk, as it Existed in the Last Twenty Years of the Eighteenth Century, and Still Exists: with Proof of its Antiquity from Etymology and Authority*, J.B Nicols and Son, London, 1830; p.464, Shorten, Monica, *Squirrels*, Collins, London, © 1954, Monica Shorten; p.467, Cambrensis, Giraldus, *The Itinerary Through Wales and the Description of Wales*, [trans. Thomas Roscoe], Everyman, London, 1912; p.467, Macdonald, David and Barrett, Priscilla, *Mammals of Britain and Europe*, HarperCollins, London, © 1993, David Macdonald and Priscilla Barrett; p.469, Green, Peter, *Beyond the Wild Wood*, Webb and Bower, London, 1982; p.474, Leslie, Charles, *Memoirs of The Life of John Constable, Esq. R.A. Composed Chiefly of His Letters*, Longman, Brown, Green and Longmans, London, 1845; p.477, Defoe, Daniel, *Journal of the Plague Year*, London, 1722; p.479, Plummer, Brian, *Tales of a Rat-hunting Man*, Boydell Press, Ipswich, 1978; p.484, Plummer, Brian, *Tales of a Rat-hunting Man*, Boydell Press, Ipswich, 1978; p.486, Page, Robin, *The Fox and the Orchid – Country Sports and the Countryside*, Quiller Press Limited, London, 1987; p.486, 489, Evans, George and Thomson, David, *The Leaping Hare*, Faber and Faber, London, 1972; p.490, 495, Adams, Richard, *Watership Down*, Rex Collings, London, 1972; p.491, Lucas, Edward, *Harvest Home*, Methuen, London, 1913

Despite every effort to trace and contact copyright holders prior to publication, this has not always been possible. If notified, the publisher will be pleased to rectify any errors or omissions at the earliest opportunity.

PHOTOGRAPH AND ILLUSTRATION ACKNOWLEDGEMENTS

The publisher wishes to thank the organizations listed below for their kind permission to reproduce the photographs and illustrations. Every effort has been made to acknowledge the pictures properly. We apologise for any unintentional omissions.

AA World Travel Library 236.
Ardea /Jack. A. Bailey 421, 468, /Brian Bevan 148, /Liz Bomford 461, /J.B & S. Bottomley 27, /Francois Gohier 442, /Johan De Meester 472, /S. Meyers 9, /P. Morris 137, 159 top, /M. Watson 480.
The Art Archive 50.
Bridgeman Art Library, London /New York /British Library, London, UK 103, /Christie's Images, London, UK 132, 408, /Corinium Museum, Cirencester, UK 489, /Crawford Municipal Art Gallery, Cork, Ireland 143, /Drewcatt Neate Fine Art Auctioneers, Newbury, Berkshire, UK 135, /Fitzwilliam Museum, University of Cambridge, UK 386, 403, /The Maas Gallery, London, UK 332, /National Gallery, London, UK 454, /Natural History Museum, London, UK 441, /John Noot Galleries, Broadway, Worcestershire 263, /Philips Fine Art Auctioneers, Scotland 38, /S.J.Phillips, London, UK 409, /Private Collection 141, 327, 444, 476, /Russell-Cotes Art Gallery and Museum, Bournemouth, UK 359, /United Distillers and Vintners 452.
B&B Photographs /Professor Stefan Buczacki 420, 491.
Bruce Coleman Ltd /Sarah Cook 245, /Janos Jurka 268, /Gordon Langsbury 366, /Wayne Lankinen 243, /Robert Maier 226, /Allan G Potts 2, /Andrew Purcell 140, 162, /Hans Reinhard 147, /Kim Taylor 89, /Colin Varndell 218.
Collections /Ashley Cooper 109, /Mike England 42, /Geoff Howard 364, 379, /Robert Pilgrim 445, /Brain Shuel 266, /Paul Watts 44.
Collins /Bruce Forman 431.
Corbis UK Ltd /Academy of Natural Sciences of Philadelphia 204, /Bettmann 90, 488, /Frank Lane Picture Agency /Ronald Thompson 354, /Historical Picture Archive /Philip de Bay 246, /Hulton Collection 119, 124, 206, /The Mariners Museum 285, /Philadelphia Museum of Art 390, /Roger Tidman 228, /Lawson Wood 116.
Cornish Picture Library /John Credland 166.
Curtis Brown /Line illustration copyright E.H. Shepard under the Berne Convention, colouring copyright 1970, 1973 by E.H. Shepard and Egmont Books Ltd, reproduced by permission of the Curtis Brown Group Ltd. 305.
East Sussex County Council 311.
Ecoscene /Beames 223, /Frank Blackburn 249, 376.
Effingham Wilson, Royal Exchange 1835 406.
Mary Evans Picture Library 41, 58, 59, 69, 153, 180, 192, 199, 222, 229, 244, 260, 265, 299, 300, 302, 310, 326, 328, 331, 334, 343, 368, 398, 426, 430, 443, 449, 457, 460, 479, /Arthur Rackham Collection 393.
The Edrington Group 259.
Frank Lane Picture Agency /Charlie Brown 352, /Richard Brooks 306, 349, 373, /P Busling /Foto Natura 388, /Michael Callan 261, /Robin Chittenden 269, 315, /Hugh Clark 309, /D. Dugan 387, /A R Hamblin 358, /John Hawkins 200, 355, /David Hosking 10, 456, /M. Jones 304, /Derek Middleton 233, 329, /Ronald Thompson 131, /R Tidman 318, 381, /John Watkins 213, 248, 353, /A Wharton 380, /D.P. Wilson 161, /Roger Wilmshurst 344, /Martin B Withers 257, 361.
John Frost Newspapers 156.
Getty Images /Hulton Archive /Fox Photos 227, /Hulton Archive 29 Top, 30, 34, 43, 47, 75, 127, 151, 155, 157, 158, 232, 256, 284, 295, 296, 369, 395, 407, 414, 435, 475, 478, 483, /Hulton Archive /Alan Webb 494 /Hulton Archive /Haywood Magee 450, /Hulton Archive /Paul Martin 412.
Grombridge & Sons 1885 319.
Reproduced with kind permission of Imperial Tobacco Ltd 145.
Landmark Trust /courtesy of Lundy Island 293.
Longmans, Green & Co. /engraving by Stanley Berkley, 1887. 484.
Billie Love Historical Collection 66, 101, 152, 193, 422.
Mander & Mitchenson Theatre Collection 182.
Maple Fine Arts, London 57.
Nature Picture Library /Niall Benvie 55, 98, 240, 308, 341, /Dan Burton 39, 154, /John Cancalosi 14, /David Cottridge 273, /Sue Daly 18 bottom, 122, /Adrian Davies 288, /Martin Dohrn 108, /Geoff Dore 56, 73, 111, /Bjorn Forsberg 347, /Alan James 121, /Kevin J Keatley 87, /David Kjaer 335, /Steve Knell 348, /Brian Lightfoot 385, 396, /Dietmar Nill 83, /Chris Packham 183, 194, /Colin Preston 467/Jeff Rotman 160, /Colin Seddom 425, /Geoff Simpson 186, 189, /David Tipling 214, 235, 384, 459, /Colin Varndell 372, /Tom Walmsley 184, /Mike Wilkes 205, 316, /Mark Yates 64, 297.
Nature Photographers Ltd /Colin Carver 195, /Hugh Clark 96, /Andrew Cleave 21, /Geoff Du Feu 427, /Michael J. Hammett 169, /Barry Hughes 271, /Hugh Miles 146, /David Osborn 287, /Nicholas Phelps 60, /Paul Sterry 71, 107, 208, 231, 239.
N.H.P.A. /Jim Bain 31, 52, /Matt Bain 40, /A.P. Barnes 86, /G.I. Bernard 136, /Mark Bowler 105, /N.A. Callow 112, /Laurie Campbell 72, 97, 224, 241, 251, 429, 436, /Bill Coster 45, /Stephen Dalton 23, 37, 63, 68, 79, 85, 179, 392, 469, /Manfred Danegger 234, 281, /Melvin Grey 338, 400, 401, /E A Janes 346, 365, /Mike Lane 250, /Lutra 138, 139, 142, /Trevor McDonald 25, /Jean-Louis Le Moigne 290, /Linda & Brian Pitkin 29 bottom, /Eric Soder 410, /Roger Tidman 120, /Roy Waller 19, 51, 113, 171, /Alan Williams 267, 325, 447.
Octopus Publishing Group Limited 217.
Robert Opie Collection 102.
Oxford Scientific Films /David Boag 313.
Papilio /Dennis Johnson 275, 280, /Peter Tatton 159 bottom.
Royal Geographical Society /Rev. Edward Patterson endpapers.
The Ronald Grant Archive 474.
RSPB 270, /Painting of Marsh Harrier flying over the Minsmere reed beds by John Reaney. Reproduced with kind permission of the RSPB 247.
RSPCA Photolibrary /Dave Bevan 32, /John Bracegirdle /Wild Images 36, 62, /Peter Cairns 448, /Colin Carver 177, /Martin Dohrn /Wild Images 104, /Geoff du Feu 61, /Robert Glover 176, /Mark Hamblin 13, 164, 197, 258, 292, /Tony Hamblin 337, 350, 367, /E.A. Janes 99, /Mark Layton 221, /John P. Lee /Wild Images 170, /George McCarthy /Wild Images 33, /Jonathan Plant 230, /S Thompson 462.
The Marquess of Salisbury /Hatfield House 1, 399, 415
Science Photo Library /Jeremy Burgess 100, /Jean-Loup Charmet 24, /Eye of Science 94.
Scotland in Focus /A. Barnes 88, 481, /L. Campbell 18 top, 67, 74, 84, 133, 175, 212, 264, 276, 282, 289, 323, 324, 340, 362, 377, 434, 493, /P Davies 432, /A.G. Firth 134, /John MacPherson 254, /P Smith 378.
Sheffield Wednesday Football Club Limited 307.
Still Pictures /Dino Simeonidis 129.
Sir Benjamin Stone's Pictures 165.
Topham Picturepoint 28.
Travel Ink /Julian Parton 322.
Uglybug Images 92, 174.
Charles Walker Photographic /'Killer', photograph of circa 1905 by A.H. Cocks. Photogravure plate from J.G. Millais, The Mammals of Great Britain, 1906. 439.
Copyright © Frederick Warne & Co. 1902, 2002 495.
Guy Wentworth Memorabilia 203.
Windrush Photos 70, /Peter Cairns 252, /Colin Carver 371, /David Cottridge 301, /Pentti Johansson 357, /Tim Loseby 314, /George McCarthy 178, 190, 417, /Malcolm Rains 356, /Richard Revels FRPS 77, 81, 82, 149, 185, 191, /David Tipling 172, 255, 262, 277, 294, 465, /Steve Young 211.
Woodfall Wild Images /Steve Austin 272, 455, 490, /Maurizio Biancarelli 418, 482, /Guy Edwardes 76, /Bob Glover 320, /Mark Hamblin 242, 317, 330, 342, 360, 382, 451, /Andy Harmer 339, /Paul Hicks 207, /E. A. Janes 446, /C. Johnson 458, /Paul Kay 20, 46, 53, 167, 168, /Mike Lane, FRPS 202, 209, 375, /Hilary Mackay 413, /D. Mason 487, /Tapani Rasanen 283, 278, 286, 466, /John Robinson 279, 351, 471, /Andy Rouse 405, /Sue Scott 128, /Gary Smith 485, /Ben Wilson 438, /David Woodfall 363, 433.

Editorial Manager: Jane Birch
Executive Art Editor: Geoff Fennell
Designer: Mark Stevens
Jacket design: Geoff Fennell
Copy editor: Maggie O'Hanlon
Proofreader: Lydia Darbyshire

Additional editorial assistance: Anne Crane, Cathy Lowne, Kristy Richardson
Picture Researchers: Charlotte Deane, Claire Gouldstone, Zoe Holtermann, Christine Junemann
Index: Richard Bird
Production Manager: Louise Hall